# Preface

Mathematical finance and financial engineering have been rapidly expanding fields of science over the past three decades. The main reason behind this phenomenon has been the success of sophisticated quantitative methodologies in helping professionals manage financial risks. It is expected that the newly developed credit derivatives industry will also benefit from the use of advanced mathematics. This industry has grown around the need to handle credit risk, which is one of the fundamental factors of financial risk. In recent years, we have witnessed a tremendous acceleration in research efforts aimed at better comprehending, modeling and hedging this kind of risk.

Although in the first chapter we provide a brief overview of issues related to credit risk, our goal was to introduce the basic concepts and related notation, rather than to describe the financial and economical aspects of this important sector of financial market. The interested reader may consult, for instance, Francis et al. (1999) or Nelken (1999) for a much more exhaustive description of the credit derivatives industry.

The main objective of this monograph is to present a comprehensive survey of the past developments in the area of credit risk research, as well as to put forth the most recent advancements in this field. An important aspect of this text is that it attempts to bridge the gap between the mathematical theory of credit risk and financial practice, which serves as the motivation for the mathematical modeling studied in this book. Mathematical developments are presented in a thorough manner and cover the structural (value-of-the-firm) and the reduced-form (intensity-based) approaches to credit risk modeling, applied both to single and to multiple defaults. In particular, this book offers a detailed study of various arbitrage-free models of defaultable term structures of interest rates with several rating grades.

This book is divided into three parts. Part I, consisting of Chapters 1-3, is mainly devoted to the classic *value-of-the-firm approach* to the valuation and hedging of corporate debt. The starting point is the modeling of the dynamics of the total value of the firm's assets (combined value of the firm's debt and equity) and the specification of the capital structure of the assets of the firm. For this reason, the name *structural approach* is frequently attributed to this approach. For the sake of brevity, we have chosen to follow the latter convention throughout this text.

Modern financial contracts, which are either traded between financial institutions or offered over-the-counter to investors, are typically rather complex and they involve risks of several kinds. One of them, commonly referred to as a *market risk* (such as, for instance, the interest rate risk) is relatively well understood nowadays. Both theoretical and practical methods dealing with this kind of risk are presented in detail, and at various levels of mathematical sophistication, in several textbooks and monographs. For this reason, we shall pay relatively little attention to the market risk involved in a given contract, and instead we shall focus on the credit risk component.

As mentioned already, Chapter 1 provides an introduction to the basic concepts that underlie the area of credit risk valuation and management. We introduce the terminology and notation related to defaultable claims, and we give an overview of basic market instruments associated with credit risk. We provide an introductory description of the three types of credit-risk sensitive instruments that are subsequently analyzed using mathematical tools presented later in the text. These instruments are: corporate bonds, vulnerable claims and credit derivatives. So far, most analyses of credit risk have been conducted with direct reference to corporate debt. In this context, the contract-selling party is typically referred to as the borrower or the obligor, and the purchasing party is usually termed the creditor or the lender. However, methodologies developed in order to value corporate debt are also applicable to vulnerable claims and credit derivatives.

To value and to hedge credit risk in a consistent way, one needs to develop a quantitative model. Existing academic models of credit risk fall into two broad categories: the *structural models* and the *reduced-form models*, also known as the *intensity-based models*. Our main purpose is to give a thorough analysis of both approaches and to provide a sound mathematical basis for credit risk modeling. It is essential to make a clear distinction between stochastic models of credit risk and the less sophisticated models developed by commercial companies for the purpose of measuring and managing the credit risk. The latter approaches are not covered in detail in this text.

The subsequent two chapters are devoted to the so-called *structural approach*. In Chapter 2, we offer a detailed study of the classic Merton (1974) approach and its variants due to, among others, Geske (1977), Mason and Bhattacharya (1981), Shimko et al. (1993), Zhou (1996), and Buffet (2000). This method is sometimes referred to as the *option-theoretic approach*, since it was directly inspired by the Black-Scholes-Merton methodology for valuation of financial options. Subsequently, in Chapter 3, a detailed study of the Black and Cox (1976) ideas is presented. We also discuss some generalizations of their approach that are due to, among others, Brennan and Schwartz (1977, 1980), Kim et al. (1993a), Nielsen et al. (1993), Longstaff and Schwartz (1995), Briys and de Varenne (1997), and Cathcart and El-Jahel (1998). Due to the way in which the default time is specified, the models worked out in the references quoted above are referred to as the *first-passage-time models*.

Springer Finance

Springer
Berlin
Heidelberg
New York
Hong Kong
London
Milan
Paris
Tokyo

# Springer Finance

*Springer Finance* is a programme of books aimed at students, academics and practitioners working on increasingly technical approaches to the analysis of financial markets. It aims to cover a variety of topics, not only mathematical finance but foreign exchanges, term structure, risk management, portfolio theory, equity derivatives, and financial economics.

*M. Ammann*, Credit Risk Valuation: Methods, Models, and Application (2001)

*E. Barucci*, Financial Markets Theory. Equilibrium, Efficiency and Information (2003)

*T.R. Bielecki and M. Rutkowski*, Credit Risk: Modeling, Valuation and Hedging (2002)

*N.H. Bingham and R. Kiesel*, Risk-Neutral Valuation: Pricing and Hedging of Financial Derivatives (1998, 2nd ed. 2004)

*D. Brigo and F. Mercurio*, Interest Rate Models: Theory and Practice (2001)

*R. Buff*, Uncertain Volatility Models-Theory and Application (2002)

*R.A. Dana and M. Jeanblanc*, Financial Markets in Continuous Time (2002)

*G. Deboeck and T. Kohonen (Editors)*, Visual Explorations in Finance with Self-Organizing Maps (1998)

*R.J. Elliott and P.E. Kopp*, Mathematics of Financial Markets (1999)

*H. Geman, D. Madan, S.R. Pliska and T. Vorst (Editors)*, Mathematical Finance-Bachelier Congress 2000 (2001)

*Y.-K. Kwok*, Mathematical Models of Financial Derivatives (1998)

*A. Pelsser*, Efficient Methods for Valuing Interest Rate Derivatives (2000)

*J.-L. Prigent*, Weak Convergence of Financial Markets (2003)

*M. Yor*, Exponential Functionals of Brownian Motion and Related Processes (2001)

*R. Zagst*, Interest-Rate Management (2002)

*A. Ziegler*, Incomple Information and Heterogeneous Beliefs in Continous-time Finance (2003)

Tomasz R. Bielecki
Marek Rutkowski

# Credit Risk:
# Modeling, Valuation
# and Hedging

 Springer

*Tomasz R. Bielecki*

Applied Mathematics Department
Illinois Institute of Technology
Engineering 1 Building
10 West 32nd Street
Chicago, IL 60616
USA
*e-mail: bielecki@iit.edu*

*Marek Rutkowski*

Faculty of Mathematics
and Information Science
Politechnika Warszawska
pl. Politechniki 1
00-661 Warszawa
Poland
*e-mail: markrut@alpha.mini.pw.edu.pl*

Mathematics Subject Classification (2000): 91B28, 91B70, 60G44, 60H05, 60J27, 35Q80
JEL Classification: C00, G12, G13, G32, G33

Cataloging-in-Publication Data applied for
A catalog record for this book is available from the Library of Congress.
Bibliographic information published by Die Deutsche Bibliothek
Die Deutsche Bibliothek lists this publication in the Deutsche Nationalbibliografie;
detailed bibliographic data is available in the Internet at http://dnb.ddb.de

1st edition 2002. Corrected 2nd printing

ISBN 978-3-642-08707-3

1007400738

Springer-Verlag Berlin Heidelberg New York
Springer-Verlag is a part of Springer Science+Business Media

springeronline.com

© Springer-Verlag Berlin Heidelberg 2010
Printed in Germany

The use of general descriptive names, registered names, trademarks, etc. in this publication does not imply, even in the absence of a specific statement, that such names are exempt from the relevant protective laws and regulations and therefore free for general use.

Cover design: *design & production*, Heidelberg

Printed on acid-free paper        41/3111db - 5 4 3 2 1

Within the framework of the structural approach, the default time is defined as the first crossing time of the value process through a default triggering barrier. Both the value process and the default triggering barrier are the model's primitives. Consequently, the main issue is the joint modeling of the firm's value and the barrier process that is usually specified in relation to the value of the firm's debt. Since the default time is defined in terms of the model's primitives, it is common to state that it is given *endogenously* within the model. Another important ingredient in both structural and reduced-form models is the amount of the promised cash flows recovered in case of default, typically specified in terms of the so-called *recovery rate* at default or, equivalently, in terms of the *loss-given-default*. Formally, it is thus possible to single out the *recovery risk* as a specific part of the credit risk; needless to say, the spread, the default and the recovery risks are intertwined both in practice and in most existing models of credit risk. Let us finally mention that econometric studies of recovery rates of corporate bond are rather scarce; the interested reader may consult, for instance, the studies by Altman and Kishore (1996) or Carty and Lieberman (1996).

The original Merton model focuses on the case of defaultable debt instruments with finite maturity, and it postulates that the default may occur only at the debt's maturity date. By contrast, the first-passage-time technique not only allows valuation of debt instruments with both a finite and an infinite maturity, but, more importantly, it allows for the default to arrive during the entire life-time of the reference debt instrument or entity.

The structural approach is attractive from the economic point of view as it directly links default events to the evolution of the firm's capital structure, and thus it refers to market fundamentals. Another appealing feature of this set-up is that the derivation of hedging strategies for defaultable claims is straightforward. An important aspect of this method is that it allows for a study of the optimal capital structure of the firm. In particular, one can study the most favorable timing for the decision to declare bankruptcy as a dynamic optimization problem. This line of research was originated by Black and Cox (1976), and it was subsequently continued by Leland (1994), Anderson and Sundaresan (1996), Anderson, Sundaresan and Tychon (1996), Leland and Toft (1996), Fan and Sundaresan, (1997), Mella-Barral and Perraudin (1997), Mella-Barral and Tychon (1999), Ericsson (2000), Anderson, Pan and Sundaresan (2000), Anderson and Sundaresan (2000).

Some authors use this methodology to forecast default events; however, this issue is not discussed in much detail in this text. Let us notice that the structural approach leads to modeling of default times in a way which does not provide any elements of surprise – in the sense that the resulting random times are predictable with respect to the underlying filtrations. This feature is the source of the observed discrepancy between the credit spreads for short maturities predicted by structural models and the market data.

In Part II, we provide a systematic exposition of technical tools that are needed for an alternative approach to credit risk modeling – the reduced-form approach that allows for modeling of unpredictable random times of defaults or other credit events. The main objective of Part II is to work out various mathematical results underlying the reduced-form approach. Much attention is paid to characterization of random times in terms of hazard functions, hazard processes, and martingale hazard processes, as well as to evaluating relevant (conditional) probabilities and (conditional) expectations in terms of these functions and processes. In this part, the reader will find various pertinent versions of Girsanov's theorem and the martingale representation theorem. Finally, we present a comprehensive study of the problems related to the modeling of several random times within the framework of the intensity-based approach.

The majority of results presented in this part were already known; however, it is not possible to quote all relevant references here. The following works deserve a special mention: Dellacherie (1970, 1972), Chou and Meyer (1975), Dellacherie and Meyer (1978a, 1978b), Davis (1976), Elliott (1977), Jeulin and Yor (1978), Mazziotto and Szpirglas (1979), Jeulin (1980), Brémaud (1981), Artzner and Delbaen (1995), Duffie et al. (1996), Duffie (1998b), Lando (1998), Kusuoka (1999), Elliott et al. (2000), Bélanger et al. (2001), and Israel et al. (2001). Let us emphasize that the exposition in Part II is adapted from papers by Jeanblanc and Rutkowski (2000a, 2000b, 2002).

Part III is dedicated to an investigation of diverse aspects of the *reduced-form approach*, also commonly referred to as the *intensity-based approach*. To the best of our knowledge, this approach was initiated by Pye (1974) and Litterman and Iben (1991), and then formalized independently by Lando (1994), Jarrow and Turnbull (1995), and Madan and Unal (1998). Further developments of this approach can be found in papers by, among others, Hull and White (1995), Das and Tufano (1996), Duffie et al. (1996), Schönbucher (1996), Lando (1997, 1998), Monkkonen (1997), Lotz (1998, 1999), and Collin-Dufresne and Solnik (2001).

In many respects, Part III, where we illustrate the developed theory through examples of real-life credit derivatives and we describe market methods related to risk management, is the most practical part of the book. In Chapter 8, we discuss the most fundamental issues regarding the intensity-based valuation and hedging of defaultable claims in case of single reference credit. From the mathematical perspective, the intensity-based modeling of random times hinges on the techniques of modeling random times developed in the reliability theory. The key concept in this methodology is the survival probability of a reference instrument or entity, or, more specifically, the hazard rate that represents the intensity of default. In the most simple version of the intensity-based approach, nothing is assumed about the factors generating this hazard rate. More sophisticated versions additionally include factor processes that possibly impact the dynamics of the credit spreads.

Important modeling aspects include: the choice of the underlying probability measure (real-world or risk-neutral – depending on the particular application), the goal of modeling (risk management or valuation of derivatives), and the source of intensities. In a typical case, the value of the firm is not included in the model; the specification of intensities is based either on the model's calibration to market data or on the estimation based on historical observations. In this sense, the default time is *exogenously* specified. It is worth noting that in the reduced-form approach the default time is not a predictable stopping time with respect to the underlying information flow. In contrast to the structural approach, the reduced-form methodology thus allows for an element of surprise, which is in this context a practically appealing feature. Also, there is no need to specify the priority structure of the firm's liabilities, as it is often the case within the structural approach. However, in the so-called hybrid approach, the value of the firm process, or some other processes representing the economic fundamentals, are used to model the hazard rate of default, and thus they are used indirectly to define the default time.

Chapters 9 and 10 deal with the case of several reference credit entities. The main goal is to value basket derivatives and to study default correlations. In case of conditionally independent random times, the closed-form solutions for typical basket derivatives are derived. We also give some formulae related to default correlations and conditional expectations. In a more general situation of mutually dependent intensities of default, we show that the problem of quasi-explicit valuation of defaultable bonds is solvable. This should be contrasted with the previous results obtained, in particular, by Kusuoka (1999) and Jarrow and Yu (2001), who seemed to suggest that the valuation problem is intractable through the standard approach, without certain additional restrictions.

In view of the important role played in the modeling of credit migrations by the methodologies based on the theory of Markov chains, in Chapter 11 we offer a presentation of the relevant aspects of this theory.

In Chapter 12, we examine various aspects of credit risk models with multiple ratings. Both in case of credit risk management and in case of valuation of credit derivatives, the possibility of migrations of underlying credit name between different rating grades may need to be accounted for. This reflects the basic feature of the real-life market of credit risk sensitive instruments (corporate bonds and loans). In practice, credit ratings are the natural attributes of credit names. Most authors were approaching the issue of modeling of the credit migrations from the Markovian perspective. Chapter 12 is mainly devoted to a methodical survey of Markov models developed by, among others, Das and Tufano (1996), Jarrow et al. (1997), Nakazato (1997), Duffie and Singleton (1998a), Arvanitis et al. (1998), Kijima (1998), Kijima and Komoribayashi (1998), Thomas et al. (1998), Lando (2000a), Wei (2000), and Lando and Skødeberg (2002).

The topics touched upon in Chapter 12 are continued and further developed in Chapter 13. Following, in particular, Bielecki and Rutkowski (1999, 2000a, 2000b, 2001a) and Schönbucher (2000), we present the most recent developments, which combine the HJM methodology of modeling of instantaneous forward rates with a conditionally Markov model of credit migrations. Probabilistic interpretation of the market price of interest rate risk and the market price of the credit risk is highlighted. The latter is used as the motivation for our mathematical developments, based on martingale methods combined with the analysis of random times and the theory of time-inhomogeneous conditionally Markov chains and jump processes.

As is well known, there are several alternative approaches to the modeling of the default-free term structure of interest rates, based on the short-term rate, instantaneous forward rates, or the so-called market rates (such as, LIBOR rates or swap rates). As we have mentioned above, a model of defaultable term structure based on the instantaneous forward rates is presented in Chapter 13. In Chapters 14 and 15, which in a sense complement the content of Chapter 13, various typical examples of defaultable forward contracts and the associated types of defaultable market rates are introduced. We conclude by presenting the BGM model of forward LIBOR rates, Jamshidian's model of forward swap rates, as well as some ideas related to the modeling of defaultable LIBOR and swap rates.

We hope that this book may serve as a valuable reference for the financial analysts and traders involved with credit derivatives. Some aspects of the text may also be useful for market practitioners involved with managing credit-risk sensitive portfolios. Graduate students and researchers in areas such as finance theory, mathematical finance, financial engineering and probability theory will also benefit from this book. Although it provides a comprehensive treatment of most issues relevant to the theory and practice of credit risk, some aspects are not examined at all or are treated only very succinctly; these include: liquidity risk, credit portfolio management and econometric studies.

Let us once more stress that the main purpose of models presented in this text is the valuation of credit-risk-sensitive financial derivatives. For this reason, we focus on the arbitrage-free (or *martingale*) approach to the modeling of credit risk. Although *hedging* appears in the title of this monograph, we were able to provide only a brief account of the theoretical results related to the problem of hedging against the credit risk. A complete and thorough treatment of this aspect would deserve a separate text.

On the technical side, readers are assumed to be familiar with graduate level probability theory, theory of stochastic processes, elements of stochastic analysis and PDEs. As already mentioned, a systematic exposition of mathematical techniques underlying the intensity-based approach is provided in Part II of the text.

For the mathematical background, including the most fundamental definitions and concepts from the theory of stochastic process and the stochastic analysis based on the Itô integral, the reader may consult, for instance, Dellacherie (1972), Elliott (1977), Dellacherie and Meyer (1978a), Brémaud (1981), Jacod and Shiryaev (1987), Ikeda and Watanabe (1989), Protter (1990), Karatzas and Shreve (1991), Revuz and Yor (1991), Williams (1991), He et al. (1992), Davis (1993), Krylov (1995), Neftci (1996), Øksendal (1998), Rolski et al. (1998), Rogers and Williams (2000) or Steele (2000). In particular, for the definition and properties of the standard Brownian motion, we refer to Chap. 1 in Itô and McKean (1965), Chap. 2 in Karatzas and Shreve (1991) or Chap. II in Krylov (1995).

Some acquaintance with arbitrage pricing theory and fundamentals on financial derivatives is also expected. For an exhaustive treatment of arbitrage pricing theory, modeling of the term structure of interest rates and other relevant aspects of financial engineering, we refer to the numerous monographs available; to mention a few: Baxter and Rennie (1996), Duffie (1996), Lamberton and Lapeyre (1996), Neftci (1996), Musiela and Rutkowski (1997a), Pliska (1997), Bingham and Kiesel (1998), Björk (1998), Karatzas and Shreve (1998), Shiryaev (1998), Elliott and Kopp (1999), Mel'nikov (1999), Hunt and Kennedy (2000), James and Webber (2000), Jarrow and Turnbull (2000a), Pelsser (2000), Brigo and Mercurio (2001), and Martellini and Priaulet (2001). More specific issues related to credit risk derivatives and management of credit risk are discussed in Duffee and Zhou (1996), Das (1998a, 1998b), Caouette et al. (1998), Tavakoli (1998), Cossin and Pirotte (2000), Ammann (1999, 2001), and Duffie and Singleton (2003).

It is essential to stress that we make, without further mention, the common standard technical assumptions:

- all reference probability spaces are assumed to be complete (with respect to the reference probability measure),
- all filtrations satisfy the *usual conditions* of right-continuity and completeness (see Page 20 in Karatzas and Shreve (1991)),
- the sample paths of all stochastic processes are right-continuous functions, with finite left-limits, with probability one; in other words, all stochastic processes are assumed to be RCLL (i.e., càdlàg),
- all random variables and stochastic processes satisfy suitable integrability conditions, which ensure the existence of considered conditional expectations, deterministic or stochastic integrals, etc. For the sake of expositional simplicity, we frequently postulate the boundedness of relevant random variables and stochastic processes.

As a rule, we adopt the notation and terminology from the monograph by Musiela and Rutkowski (1997a). For the sake of the reader's convenience, an index of the most frequently used symbols is also provided. Although we have made an effort to use uniform notation throughout the text, in some places an ad hoc notation was also used.

We are very grateful to Monique Jeanblanc for her numerous helpful comments on the previous versions of our manuscript, which have led to several essential improvements in the text. The second-named author is happy to have opportunity to thank Monique Jeanblanc for fruitful and enjoyable collaboration.

During the process of writing, we have also profited from valuable remarks by, among others, John Fuqua, Marek Musiela, Ben Goldys, Ashay Kadam, Atsushi Kawai, Volker Läger, and Jochen Georg Sutor. They also discovered numerous typographical errors in the previous drafts; all remaining errors are, of course, ours. A large portion of the manuscript was completed during the one-year stay of Marek Rutkowski in Australia. He would like to thank the colleagues from the School of Mathematics at the University of New South Wales in Sydney for their hospitality.

Tomasz Bielecki gratefully acknowledges partial support obtained from the National Science Foundation under grant DMS-9971307 and from the State Committee for Scientific Research (Komitet Badań Naukowych) under grant PBZ-KBN-016/P03/1999.

Marek Rutkowski gratefully acknowledges partial support obtained from the State Committee for Scientific Research (Komitet Badań Naukowych) under grant PBZ-KBN-016/P03/1999.

We would like to express our gratitude to the staff of Springer-Verlag. We thank Catriona Byrne for her encouragement and relentless editorial supervision since our first conversation in the Banach Centre in Warsaw in June 1998, as well as Daniela Brandt and Susanne Denskus for their invaluable technical assistance. We are also grateful to Katarzyna Rutkowska who helped with the editing of the text.

Last, but not least, we would like to thank our wives. Tomasz Bielecki is grateful to Małgosia for her patience, and Marek Rutkowski thanks Ola for her support.

| | |
|---|---|
| Chicago | Tomasz R. Bielecki |
| Warszawa | Marek Rutkowski |

# Table of Contents

# Part I

# Structural Approach

# 1. Introduction to Credit Risk

A *default risk* is a possibility that a counterparty in a financial contract will not fulfill a contractual commitment to meet her/his obligations stated in the contract. If this actually happens, we say that the party defaults, or that the default event occurs. More generally, by a *credit risk* we mean the risk associated with any kind of credit-linked events, such as: changes in the credit quality (including downgrades or upgrades in credit ratings), variations of credit spreads, and the default event. The *spread risk* is thus another components of credit risk. To facilitate the analysis of complex agreements, it is important to make a clear distinction between the *reference (credit) risk* and the *counterparty (credit) risk*. The first generic term refers to the situation when both parties of a contract are assumed to be default-free, but due to specific features of the contract the credit risk of some reference entity appears to play an essential role in the contract's settlement. In other words, the reference risk is that part of the contract's risk, which is associated with the third party; i.e., with the entity, which is not a party in a given agreement. In the present context, the third party is referred to as the reference entity of a given contract. *Credit derivatives* are recently developed financial instruments that allow market participants to isolate and trade the reference credit risk. The main goal of a credit derivative is to t ransfer the reference risk, either completely or partially, between the counterparties. In most cases, one of the parties can be seen as a buyer of an insurance against the reference risk. Such a party is called the seller of the reference risk; consequently, the party that bears the reference risk is referred to as its buyer.

Let us now focus on the counterparty risk. An important feature of all over-the-counter derivatives is that, unlike the exchange-traded contracts, they are not backed by the guarantee of a clearinghouse or an exchange, so that each counterparty is exposed to the default risk of the other party. In practice, parties are sometimes required to post collateral or mark to market periodically, though. The counterparty risk emerges in a clear way in such contracts as *vulnerable claims* and *defaultable swaps*. In both these cases, one needs to quantify the default risk of both parties in order to correctly assess the contract's value. Depending on whether the default risk of one or both parties is taken into account, we say that a contract involves the unilateral (one-sided) or the bilateral (two-sided) default risk.

## 1.1 Corporate Bonds

*Corporate bonds* are debt instruments issued by corporations. They are a part of the capital structure of the firm (just like the equity). By issuing bonds a corporation commits itself to make specified payments to the bondholders at some future dates, and the corporation charges a fee for this commitment. However, the corporation (or firm) may default on its commitment in which case the bondholders will not receive the promised payment in full, and thus they will suffer a financial loss. Of course, the occurrence of default, possibly caused by the firm's bankruptcy, is meaningful only during the lifetime of a particular bond – that is, during the time period between the bond's inception and its maturity.

A corporate bond is an example of a *defaultable claim* (a formal definition of a defaultable claim is provided in Sect. 2.1). We set the notional amount, or the face value, of the bond equal to $L$ units of cash (e.g., U.S. dollars). For the moment, we shall concentrate on a discount bond – that is, we assume that the bond pays no coupons. We also fix the maturity date of the bond, denoted by $T$. Usually, the arbitrage price at time $t$ of a $T$-maturity defaultable bond will be denoted by $D(t, T)$, in particular, $D(T, T) = L$ provided that default has not occurred prior to, or at the maturity date $T$. By contrast, the notation $B(t, T)$ is used to denote the arbitrage price at time $t$ of a $T$-maturity default-free bond with the face value 1; hence necessarily $B(T, T) = 1$.

The *defaultable term structure* is the term structure of interest rates implied by the yields on the default prone *corporate bonds* or on the default prone *sovereign bonds*. A large portion of the credit risk literature is devoted to the modeling of a defaultable term structure as well as to pricing related credit derivatives. Much of the theory presented later in this book applies to both types of bonds. We shall use a generic term *defaultable bond* for any kind of bond with the possibility of default. By contrast, a *default-free bond* pays surely both the coupons and the face value to the at bondholders predetermined dates. Defaultable bonds are also known as *risky bonds,* and the default-free bonds are commonly referred to as *risk-free bonds* or *Treasury bonds.* Holders of all bonds are, of course, exposed to the market (interest rate) risk. The adjective risk-free refers to the presumed absence (or at least negligibility) of the credit risk in bonds of the highest credit quality.

The mathematical techniques presented in this text are applicable to the valuation of general corporate liabilities (or corporate debt). Corporate bonds are one example of such liabilities; corporate loans are another one. Most of the discussion that follows applies to corporate debt in general, although we choose to specify it to corporate bonds. Corporate bonds are characterized by various attributes that we shall now briefly describe. Let us observe that the features such as: recovery rules, safety covenants, credit ratings and default correlations are related not only to corporate bonds but indeed to general defaultable claims, as defined in Sect. 2.1.

### 1.1.1 Recovery Rules

We first provide a simplified description of recovery rules, which is suitable for formulating a mathematical model. In practice, the specific recovery rules will typically include clauses such as, for instance, priority payments upon default based on the debt's seniority (*seniority rules* or the *priority structure*). Generally speaking, *recovery schemes* (or *recovery covenants*, or *recovery rules*) determine the timing and the amount of *recovery payment* that is paid to creditors if the default occurs before the bond's maturity.

The recovery payment is frequently specified by the *recovery rate* $\delta$; i.e., the fraction of the bond's face amount paid to the bondholders in case of default. The timing of the recovery payoff is, of course, another essential ingredient. If a fixed fraction of bond's face value is paid to the bondholders at time of default, usually denoted by $\tau$ in what follows, then the recovery scheme is referred to as the *fractional recovery of par value*. Assuming that the bond's face value equals 1 (that is, $L = 1$), we may represent it as the contingent claim $\tilde{D}^{\delta}(T, T)$, which settles at time $T$ and equals

$$\tilde{D}^{\delta}(T, T) = \mathbb{1}_{\{\tau > T\}} + \delta B^{-1}(\tau, T)\mathbb{1}_{\{\tau \leq T\}}.$$

If, in case of default, the fixed fraction of bond's face value is repaid at maturity date $T$, the recovery scheme is termed the *fractional recovery of Treasury value*. Under this rule, the bond is formally equivalent to the payoff

$$D^{\delta}(T, T) = \mathbb{1}_{\{\tau > T\}} + \delta\mathbb{1}_{\{\tau \leq T\}},$$

and its value at time $t$ is denoted as $D^{\delta}(t, T)$. It is clear that $\tilde{D}^{0}(t, T) = D^{0}(t, T)$. A still another convention – the *fractional recovery of market value* – postulates that at time of default the bondholders receive a fraction of the pre-default market value of a corporate bond. The equivalent contingent claim now takes the following form:

$$D(T, T) = \mathbb{1}_{\{\tau > T\}} + \delta D(\tau-, T)B^{-1}(\tau, T)\mathbb{1}_{\{\tau \leq T\}},$$

where $D(\tau-, T)$ stands for the value of the bond just before the default time. In financial literature, it is not uncommon to use the generic term *loss given default* (LGD, for short) to describe the likely loss of value in case of default (in principle, LGD thus equals 1 minus the recovery rate).

Let us now focus on a more abstract formulation of a recovery rule. In most works on credit risk, it is assumed that if a bond defaults during its lifetime then the recovery payment is made either at the default time $\tau$, or at the maturity $T$ of the bond. In the former case, the recovery payment is determined by the value $Z_{\tau}$ at default time of the *recovery process Z*. In the latter, the recovery payment is determined by the realization of the *recovery claim $\tilde{X}$*. Formally, the two cases described above correspond to the claims $DCT^2$ and $DCT^1$ of Sect. 2.1, where they are termed defaultable claims with *recovery at default* and *recovery at maturity*, respectively.

It should be stressed that the recovery process and/or the recovery value may be specified either exogenously or endogenously with respect to the current market value of the bond. In fact, the specification of the recovery rules may become quite intricate from the mathematical standpoint. As an example of an endogenous recovery rule, let us consider the situation when the recovery payment is paid at maturity $T$ in the amount of $V_T/L_T$ per unit of the bond's face value, provided that $V_T < L_T$ (otherwise, the firm's debt is paid back to the lenders in full). Here, $V_T$ stands for the total value of the firm's assets at the bond's maturity, and $L_T$ represents the total value of the firm's liabilities at this date. The (random) ratio $V_T/L_T$, commonly referred to as the *recovery ratio,* plays an essential role in the bond's valuation in most structural models of defaultable term structure of interest rates such as, for instance, Merton's (1974) model or the Black and Cox (1976) model.

### 1.1.2 Safety Covenants

There are numerous ways in which default (or bankruptcy) occurs in market practice. Typically, bankruptcy implies that the firm's bondholders take control over the firm, and the firm undergoes a reorganization. In practice, an important part of the bankruptcy procedure is the bargaining process. For the sake of (relative) simplicity, this particular aspect is not taken into account in what follows.

*Exogenous bankruptcy* refers to the case when bankruptcy is specified in form of some protective covenants, such as positive net-worth covenant, or when bankruptcy is triggered at an exogenously specified asset value, for instance, the principal value of the debt.

The notion of an *endogenous bankruptcy* covers the situations when bankruptcy is declared by the firm's stockholders if the firm's value falls below certain pre-specified level. Within the framework of the optimal capital structure approach, this level is selected so that the firm's equity value is maximized, so that the value of firm's debt is minimized. We shall discuss some results related to the optimal capital structure in Sect. 3.3. For a more exhaustive analysis of the strategic debt service, we refer to the original papers by Leland (1994, 1998), Anderson and Sundaresan (1996, 2000), Leland and Toft (1996), Mella-Barral and Perraudin (1997), Ericsson and Reneby (1998), Mella-Barral (1999) or Ericsson (2000).

The mathematical concept of *safety covenants* associated with a corporate debt was introduced in literature dealing with the structural approach to credit risk in order to specify the default event. Generally speaking, a safety covenant is modeled as a *barrier process* (also called a *threshold process*), usually denoted as $v$ in what follows. In most cases, the default event is triggered when the firm value process $V$ falls below the barrier process $v$ either prior, or at the maturity date $T$. For the purpose of this text, we choose to use the term *safety covenants* in order to describe any mechanism, which triggers default event before the maturity of the debt.

### 1.1.3 Credit Spreads

A *credit spread* measures the excess return on a corporate bond over the return on an equivalent Treasury bond, i.e., a bond, which is assumed to be free of the credit risk. Depending on the situation, a credit spread may be expressed, e.g., as the difference between respective yields to maturity, or as the difference between respective instantaneous forward rates. The generic term *term structure of credit spreads* will refer to the term structure of such differences. The determination of the credit spread is in fact the ultimate goal of most credit risk models. It is also the topic of several econometric studies. Some authors concentrate on direct modeling of the credit spread, rather than on its derivation from other fundamentals. Such an approach appears to be very convenient when one deals with these credit derivatives that have the credit spread as the underlying instrument.

A large credit spread of a corporate security over the comparable risk-free security is a widely accepted practical measure of the firm's financial distress. *Distressed securities* can thus be defined by directly referring to the high level of credit spreads yielded by some corporate securities, should they not default. A more narrow definition of a distressed security encompasses publicly held and traded debt or equity securities of firms that have defaulted or have filed for protection under the bankruptcy code. For a detailed analysis of the concept of a distressed security, we refer to Altman (1998).

### 1.1.4 Credit Ratings

A firm's *credit rating* is a measure of the firm's propensity to default. Credit ratings are typically identified with elements of a finite set, also referred to as the set of *credit classes* or *credit grades*. In some cases, the credit classes may correspond to credit ratings attributed by a commercial rating agency, such as Moody's Investors Service, Standard & Poor's Corporation, or Fitch IBCA, Duff & Phelps.[1] This does not mean, however, that in the theoretical approach credit ratings should necessarily be understood as being attributed by a commercial rating agency. First, many major financial institutions maintain their own credit rating systems, based on internally developed methodologies, and therefore known as the *internal ratings*. Second, the official credit ratings primarily reflect the likelihood of default, and thus do not necessarily provide the most adequate assessment of the debt's credit quality. Finally, the improvement (deterioration, resp.) of the firm's credit quality typically does not result in an immediate upgrade (downgrade, resp.) of its rating. For more information on existing rating systems, we refer to Altman (1997), Carty (1997), Crouhy at al. (2001) and Krahnen and Weber (2001). In this text, the generic term *credit rating* (or *credit quality*) is used to describe any classification of corporate debt that can be justified for specific purposes.

---

[1] The interested reader may consult, e.g., the Moody's Investors Service at *www.moodys.com* or Standard & Poor's Corporation at *www.standardpoors.com*.

### 1.1.5 Corporate Coupon Bonds

A clear-cut distinction needs to be made between the corporate coupon bond, which pays a coupon rate continuously in time, and a similar bond, which pays coupons at discrete time instants. The former one should be seen as a theoretical construct that is meant to facilitate the analysis of the latter. The corporate coupon bond with continuous coupon rate is widely used in financial literature, particularly in relation with the structural approach, in order to study quantitative and qualitative behavior of corporate debt.

Let us briefly describe a corporate coupon bond with coupon payments occurring at discrete time intervals. It should be stressed here that the coupon payments are only made prior to the default time. A coupon bond may thus be considered as a portfolio composed of the following securities:

- *defaultable coupons*, which sometimes are equivalent to defaultable zero-coupon bonds with zero recovery,[2]
- *defaultable face value*, which can be seen as a defaultable zero-coupon bond with, generally speaking, non-zero recovery.

Consider a corporate coupon bond with face value $L$, which matures at time $T = T_n$ and promises to pay (fixed or variable) coupons $c_i$ at times $T_1 < T_2 < \cdots < T_n$. Assume, for instance, that the recovery payment is proportional to the face value and that it is made at maturity $T$, in case the default event occurs before or at the maturity date. Under this convention, the bond's cash flows are:

$$\sum_{i=1}^{n} c_i \mathbb{1}_{\{\tau > T_i\}} \mathbb{1}_{T_i}(t) + \left(L\mathbb{1}_{\{\tau > T\}} + \delta L \mathbb{1}_{\{\tau \leq T\}}\right)\mathbb{1}_T(t), \qquad (1.1)$$

where $\tau$ stands for the bond's default time and the variable $t$ represents the running time. Notice that only the last term, which corresponds to the recovery payment at bond's maturity, depends on the choice of a particular recovery scheme (recall that the coupon payments are subject to the zero recovery). A corporate coupon bond described by (1.1) can also be formally represented as a single cash flow $D_c(T, T)$, which settles at the bond's maturity date $T$ and is given by the following formula:

$$D_c(T, T) = \sum_{i=1}^{n} c_i B^{-1}(T_i, T)\mathbb{1}_{\{\tau > T_i\}} + L\mathbb{1}_{\{\tau > T\}} + \delta L \mathbb{1}_{\{\tau \leq T\}}.$$

The arbitrage price at time $t < \tau$ of this contingent claim – that is, the value of a corporate coupon bond prior to default – is denoted by $D_c(t, T)$. We find it convenient to refer to $D_c(t, T)$ on the set $\{\tau > t\}$ as the bond's *pre-default value*, the random variable $D_c(\tau, T)$ is called the *post-default value* of a corporate coupon bond. A similar terminological convention applies to defaultable zero-coupon bonds and, indeed, to all kinds of defaultable claims.

---

[2] Formally, this equivalence holds if all the defaultable zero-coupon bonds, including the defaultable face value, have the same default time.

### 1.1.6 Fixed and Floating Rate Notes

If a debt contract stipulates that the coupon payments are fixed, we deal with a *fixed-coupon bond* (or, a *fixed-rate note*). Consider the two fixed-coupon bonds, a risk-free bond and a defaultable one, with otherwise identical covenants. If both bonds trade at par (i.e., their prices equal the face values), it is natural to expect that in order to compensate an investor for the default risk, the coupon rate of a corporate bond would be greater than that of a risk-free bond. This is indeed observed in the market practice, and the corresponding discrepancy is referred to as the *fixed-rate credit spread* over Treasury for a given corporate bond. As already noticed, the credit spread reflects the credit quality of the issuer, as perceived by the market – the financial market requires a higher risk premium for lower quality debt, so that the cost of capital for a debtor of lower credit quality is higher. Let us mention that the terms *credit risk* and *spread risk* are used interchangeably by market practitioners.

To quantify the fixed-rate credit spread, for $t = 0$, assume that $B(0, T_i)$ $(D(0, T_i)$, resp.) are known market prices of zero-coupon Treasury (corporate, resp.) bonds with unit face value. Then the spread equals $S := c' - c$, where the coupon rates $c$ and $c'$ can be easily found from the equalities

$$\sum_{i=1}^{n-1} cB(0, T_i) + (1 + c)B(0, T_n) = 1$$

and

$$\sum_{i=1}^{n-1} c'D(0, T_i) + (1 + c')D(0, T_n) = 1.$$

The last equality is based on an implicit assumption that the price of a corporate coupon bond equals the sum of its zero-coupon components, and this holds if all coupons default simultaneously and the recovery rate $\delta$ for the coupons equals zero. The credit spread varies with both the time $t$ and the maturity $T = T_n$, thereby giving rise to a particular term structure of credit spreads. A corporate *floating rate note* (FRNs, for short) is another important example of a defaultable debt. In contrast to a fixed rate note, each coupon payment of an FRN is made according to the floating interest rate prevailing on this coupon's date (or more precisely, on the *reset date*). Consider an FRN specified as follows: the face value is $L$, the maturity is $T = T_n$, and the coupon payments are made at the dates $T_0 = 0 < T_1 < \cdots < T_n$. Let us denote by $L(T_i)$ the floating interest rate for the risk-free borrowing and lending over the accrual period $[T_i, T_{i+1}]$. On each coupon date $T_i$, the coupon payment of an FRN is made according to the credit-risk adjusted floating rate $\hat{L}(T_i) = L(T_i) + s$, where $L(T_i)$ is the risk-free floating rate, and the non-negative constant $s$ represents the bond-specific *floating-rate credit spread*. For otherwise comparable notes, the higher level of the credit spread $s$ usually corresponds to the lower credit quality of the issuer.

Let us examine the credit spread over the risk-free floating rate of an FRN that trades at par. In view of our current convention concerning the recovery scheme (cf. expression (1.1)), the cash flows of such a corporate FRN can be formally represented as follows:

$$\sum_{i=0}^{n} \hat{L}(T_i) \mathbb{1}_{\{\tau > T_i\}} \mathbb{1}_{T_i}(t) + \left( \mathbb{1}_{\{\tau > T\}} + \delta \mathbb{1}_{\{\tau \leq T\}} \right) \mathbb{1}_{T}(t),$$

where, without loss of generality, we have set $L = 1$. Alternatively, an FRN can be treated as a single cash flow, which settles at $T$, and equals

$$\text{FRN}(T, T) = \sum_{i=0}^{n} \hat{L}(T_i) B^{-1}(T_i, T) \mathbb{1}_{\{\tau > T_i\}} + \mathbb{1}_{\{\tau > T\}} + \delta \mathbb{1}_{\{\tau \leq T\}}.$$

Recall that $\tau$ stands for the default time, and $\delta$ is the constant recovery rate. The arbitrage price at time $t < \tau$ of such a contingent claim will be denoted as $\text{FRN}(t, T)$. Using simple no-arbitrage arguments, it is not difficult to show that if the floating rate $L(T_i)$ satisfies

$$L(T_i) = \frac{1}{(T_{i+1} - T_i)} \left( \frac{1}{B(T_i, T_{i+1})} - 1 \right)$$

then a risk-free floating-rate note necessarily trades at par at time 0. Assume that $\hat{L}(T_i) = L(T_i) + s$ and the FRN trades at par at time 0. Then the credit spread $s$ at time 0 can be found from the equality

$$\sum_{i=0}^{n} \tilde{D}(0, T_i) + s \sum_{i=0}^{n} D^0(0, T_i) + D^\delta(0, T_n) = 1,$$

where $\tilde{D}(0, T_i)$ denotes the value at time 0 of the random payoff $L(T_i) \mathbb{1}_{\{\tau > T_i\}}$, which settles at $T_i$, and $D^0(0, T_i)$, $i = 1, \ldots, n$ and $D^\delta(0, T_n)$ are the prices of zero-coupon bonds issued by the same entity as the FRN in question (by virtue of the *cross-default* covenant all these bonds default simultaneously). Notice that the level of the spread $s$ depends, in particular, on the correlation between the risk-free rate $L(T_i)$ and the default event $\{\tau > T_i\}$. For the sake of computational simplicity, it is frequently assumed in financial modeling that these two random factors are mutually independent.

In the market practice, both *callable* and *putable* FRNs are common. The issuer of a callable FRN has the right to redeem the note before its maturity. The holder of a putable FRN has the right to force an early redemption. The changes in credit quality determine whether option exercise is advantageous. This means that typically a floating-rate note has also an embedded credit derivative, specifically, a call or put option on the value of a note.

Let us finally observe that the default risk of the bond's buyer has no relevance whatsoever to the value of a corporate bond. Therefore, a corporate (fixed- or floating-rate) bond may serve as a natural example of a credit-risk sensitive contract with unilateral default risk.

### 1.1.7 Bank Loans and Sovereign Debt

Apart from the market of corporate bonds, there exist two other important sections of the defaultable debt market, namely, the market of *syndicated bank loans* and the *sovereign debt* market.

**Syndicated bank loans.** Syndicated bank loans (SBLs) are primarily large, high grade commercial loans. In recent years, a considerable growth has been observed in the secondary trading on the market of syndicated bank loans. This has been paralleled by the emergence of bank loan ratings. In many respects the syndicated bank loans are similar to corporate bonds, and thus investors are now considering SBLs as substitutes or complements to corporate bonds. We refer to a recent article by Altman and Suggitt (2000), who present a thorough empirical analysis of the default rates on the market of syndicated bank loans.

**Sovereign debt.** As an important sector of the sovereign debt market, let us mention the so-called *Brady bonds*. Brady bonds were issued by several less developed countries. They are primarily denominated in U.S. dollars and traded in the global bond markets. Typically, they contain various forms of credit guarantees and protections, so that it is rather hard to isolate the country-specific credit spread that is embedded in yields on Brady bonds.

### 1.1.8 Cross Default

Due to the complexity of debt indentures, the definitions of the *cross-default event* existing in various accessible sources are rather cumbersome, and thus are subject to differing interpretations. The definition that we want to adopt reflects the fact that the cross-default covenant basically corresponds to provisions in loan agreements or bond indentures, which trigger an event of default if the counterparty (borrower or issuer) defaults on another obligation. Such a description agrees with the definition provided by the International Finance and Commodities Institute, which states that the cross default is: "A provision of a loan or swap agreement stating that any default on another loan or swap will be considered a default on the issue with the cross-default provision. The purpose of this provision is to protect a creditor or counterparty from actions favoring another creditor."

### 1.1.9 Default Correlations

Consider two different defaultable claims whose lifetimes intersect, formally defined as the two random variables on a common probability space. For instance, abstract corporate bonds with the same maturity date are considered to be different defaultable claims if they have different initial credit ratings, or if their recovery covenants differ. Consider also a period of time contained within the lifetimes of the two claims. Let $X$ stand for the random variable, which takes value 1, if the first claim defaults during the specified period of time, and takes value 0 otherwise; an analogous random variable associated with the second claim is denoted by $Y$. By convention, the *default correlation* between the two defaultable claims is defined as the correlation coefficient between the random variables $X$ and $Y$. Default correlations are an important building block of credit risk measurement and management methodologies for credit-risk sensitive portfolios, mentioned in Sect. 1.4. From the theoretical perspective, the issue of modeling correlated defaults was addressed by, among others, Duffie and Singleton (1999), Davis and Lo (1999, 2001), Jarrow et al. (1999), Kijima and Muromachi (2000), Frey and McNeal (2000), Kijima (2000), Embrechts et al. (2002, 2003), Jarrow and Yu (2001), and Zhou (2001). We refer also to Sect. 3.6 and 12.3 for an exhaustive discussion of this important topic.

## 1.2 Vulnerable Claims

*Vulnerable claims* are contingent agreements that are traded over-the-counter between default-prone parties; each side of the contract is thus exposed to the counterparty risk of the other party. The default risk of a counterparty (or of both parties) is thus an important component of financial risk embedded in a vulnerable claim; it should necessarily be taken it into account in valuation and hedging procedures for vulnerable claims. On the other hand, the underlying (reference) assets are assumed to be insensitive to credit risk. *Credit derivatives* are recently developed financial instruments, which allow for a secluded trading in the reference credit risk. In contrast to vulnerable claims, in which the counterparty risk appears as a nuisance or a side effect, credit derivatives are tailored as highly specialized and effective devices to handle or transfer the reference credit risk. Since credit derivatives are offered over-the-counter, a credit derivative typically represents also a vulnerable claim, though, unless the counterparty risk is negligible.

### 1.2.1 Vulnerable Claims with Unilateral Default Risk

The classic example of a vulnerable contingent claim with unilateral default risk is a European *vulnerable option* – that is, an option contract in which the option writer may default on his obligations . In other words, this is an option whose payoff at maturity depends on whether a default event, associated with the option's writer, has occurred before or on the maturity date, or not. The default risk of the holder of the option is manifestly not relevant.

Consider a vulnerable European call option on a default-free $U$-maturity zero-coupon bond – that is, a vulnerable claim with no reference risk. Let $T < U$ be the option expiration date, $B(T, U)$ be the price of the underlying bond, and $K$ be the option's strike price. The payoff $C_T$ at exercise date

$T$ of a European call option written on a default-free zero-coupon bond of maturity $U$ is equal to $C_T = (B(T,U) - K)^+$. Let $\mathcal{D} = \{\tilde{\tau} \leq T\}$ denote the event that the call writer defaults either before or on the exercise date $T$, where $\tilde{\tau}$ is the default time of the call writer. If default occurs, only a fraction $\tilde{\delta}$ of the call's intrinsic value is redeemed by the call owner. Thus, the payoff at the settlement date of this option may be written as

$$C_T^d = C_T 1\!\!1_{\{\tilde{\tau}>T\}} + \tilde{\delta} C_T 1\!\!1_{\{\tilde{\tau}\leq T\}}.$$

It is essential to distinguish between the above vulnerable option on a default-free bond, and a standard (non-vulnerable) option on a defaultable bond, with the payoff at expiry $D_T = \big(D(T,U) - K\big)^+$, where $D(T,U)$ represents the price of a $U$-maturity corporate bond at time $T$ (cf. Sect. 1.1). Finally, we may also consider a vulnerable option on a defaultable bond, with the payoff at maturity given by the formula

$$\tilde{C}_T^d = D_T 1\!\!1_{\{\tilde{\tau}>T\}} + \tilde{\delta} D_T 1\!\!1_{\{\tilde{\tau}\leq T\}}.$$

The last option may serve as a simple example of a *hybrid derivative*; its valuation will involve both the reference and the counterparty risks.

### 1.2.2 Vulnerable Claims with Bilateral Default Risk

Vulnerable claims with *bilateral* (or *two-sided*) default risk are these contracts in which both counterparties are susceptible to default risk. The prime example are here swap agreements between two default-prone entities, known as *defaultable swaps*. Despite the similarity of names, a defaultable swap should not be confused with a default swap, which is in fact a form of insurance against the reference risk (the latter kind of contracts is explained in Sect. 1.3.1 below). In contrast to default-free swaps, alternative settlement rules in case of default may largely influence the valuation of defaultable swaps. In addition, we also need to specify the debt's seniority. Typically, it is assumed that swaps are subordinate to debt in bankruptcy. We shall follow this convention here. Thus, it is natural to assume that if the party that is in default on its original debt is due to make a swap payment, it will default also on the swap contract. If, on the other hand, the party in default is due to receive a swap payment, two alternative settlement rules can be examined: (i) the swap payment is received, or (ii) the swap payment is withheld. If the latter rule is adopted a swap becomes valueless in case of default. In the former case, the swap payment at default has option-like features, and thus the total value of a swap contract depends, in particular, on the value of the embedded option. Various aspects of defaultable swaps were analyzed by several authors, to mention a few: Cooper and Mello (1991), Rendleman (1992), Abken (1993), Duffie and Huang (1996), Huge and Lando (1999), Li (1998), Laurent (2000), Lotz and Schlögl (2000), and Schönbucher (2000b). We shall now introduce some basic notions related to default-prone interest rate contracts; for more details, see Chap. 14.

### 1.2.3 Defaultable Interest Rate Contracts

Let us compare some basic types of spot default-free interest rate agreements and swaps with the corresponding contracts that are susceptible to the default risk of a counterparty. A more detailed study of selected types of single- and multi-period defaultable interest rate contracts is postponed to Chap. 14.

We start by describing the basic type of a (default-free) spot *interest rate agreement* (or a *credit agreement*) with the notional amount $L$ and a nominal interest rate $\kappa$, for the accrual period $[T, U]$. We shall refer to $T$ as the *reset date* and to $U$ as the *settlement date*. An interest rate agreement can be described as a financial contract between two parties, a *receiver* and a *payer*, which is subject to the following covenants:

– at time $T$ the receiver passes the notional amount $L$ to the payer,
– he receives from the payer the accrued amount $L(1 + \kappa(U - T))$ at time $U$.

It is apparent that the covenants of this agreement assume that the payer (of the fixed rate $\kappa$) is certain to deliver the promised payment to the receiver at time $U$. The following features of default-free interest rate agreements are worth stressing. First, the actual timing of the payments is not essential. For instance, the covenants of the agreement introduced above might be equally well restated as follows:

– at the settlement date $U$, the receiver passes to the payer the notional amount $LB^{-1}(T, U)$ discounted from time $T$ to time $U$,
– he receives at time $U$ from the payer the accrued amount $L(1 + \kappa(U - T))$

or, equivalently,

– the receiver passes at time $T$ to the payer the notional amount $L$,
– at the reset date $T$, the receiver collects from the payer the accrued amount $L(1 + \kappa(U - T))B(T, U)$ discounted from time $U$ to time $T$.

The above equivalences are, of course, valid from the perspective of the inception time $T$ only. Second, the covenants of the interest rate agreement described above invoke exchange of principal payments. Thereby, the agreement is in fact equivalent to a loan subject to a fixed interest rate $\kappa$, where the receiver is the lending party, and the payer is the borrowing party. Such an agreement gives rise to the concept of an abstract (default-free) *spot LIBOR rate* $L(T)$ – that is, the level of the fixed rate $\kappa$, which makes the contract have a value of zero at the inception date $T$. One easily checks that

$$L(T) = \frac{1}{(U - T)} \left( \frac{1}{B(T, U)} - 1 \right).$$

More generally, if the contract's inception date $t$ precedes the reset date $T$, the corresponding *forward LIBOR rate* $L(t, T)$ equals

$$L(t, T) = \frac{1}{(U - T)} \left( \frac{B(t, T)}{B(t, U)} - 1 \right).$$

Of course, both $L(T)$ and $L(t, T)$ depend on $U$; for the sake of brevity, though, this is not reflected in our notation.

We turn our attention to the corresponding *defaultable interest rate agreement* (or, equivalently, a *defaultable credit agreement*) in which only the payer party is prone to default. The contract is subject to the following covenants (we assume that none of the parties has gone bankrupt before the date $T$):
- at time $T$, the receiver passes to the payer the notional amount $L$,
- if the payer does not default in the time interval $(T, U]$, then at the settlement date $U$ he pays to the receiver the accrued amount $L(1 + \kappa(U - T))$,
- if the payer defaults in $(T, U]$, then he pays to the receiver at time $U$ the reduced amount $\delta L(1 + \kappa(U - T))$, where $\delta$ is the recovery rate.

Essentially, we deal here with a loan in which the debtor (the borrower, or the payer in the present context) may default on his obligation to repay the debt. An abstract defaultable spot LIBOR rate is the interest rate associated with such a loan (see Chap. 14 for details. In Chap. 14, we study also defaultable credit agreements in which the receiver is the only default-prone party, as well as contracts with bilateral default risk – that is, agreements in which both parties are prone to default).

The basic type of a spot *default-free interest rate swap* is the spot fixed-for-floating swap for the accrual period $[T, U]$, settled in arrears, with the spot default-free LIBOR rate $L(T)$ being the reference floating rate. The covenants of such an agreement, entered into at the reset date $T$, may be summarized as follows: there are two parties to the agreement, where one party is the payer of the fixed rate $\kappa$ and the other is the payer of the floating rate $L(T)$; the parties agree to exchange at the settlement date $U$ the nominal interest payments based on the notional amount $L$; thus, if $L = 1$, the net cash flow at the contract's settlement date $U$ to one of the parties, to the payer of the fixed rate $\kappa$ say, is equal to $(L(T) - \kappa)(U - T)$ (of course, the other party receives the negative of this cash flow).

The value of the fixed rate $\kappa$, which makes this cash flow have a value zero at the inception date $T$, is called the (default-free) *spot swap rate*. It is evident that in the default-free environment the spot swap rate and the spot LIBOR rate coincide. Essentially, in the default-free environment, the loan agreements and the interest rate swaps are equivalent.

As an example of a defaultable counterpart of the default-free spot swap introduced above, consider the following agreement (again, we set $L = 1$):
- the receiver passes at the settlement date $U$ to the payer the full floating amount due: $L(T)(U - T)$,
- if the payer does not default in the time period $(T, U]$, the receiver collects at the settlement date $U$ the full fixed amount due: $\kappa(U - T)$,
- if the payer defaults in $(T, U]$, the receiver gets at time $U$ the reduced amount: $\delta\kappa(U - T)$.

The fixed rate $\kappa$ that makes the above contract valueless at the inception time $T$, is termed the *defaultable spot swap rate*. As shown in Chap. 14, it typically differs from the defaultable spot LIBOR rate. In other words, in the presence of a counterparty risk, the loan and the swap contract are not equivalent to each other.

## 1.3 Credit Derivatives

*Credit derivatives* are privately negotiated derivative securities that are linked to a credit-sensitive asset (index) as the underlying asset (index). More specifically, the reference security of a credit derivative can be any financial instrument that is subject to risk of default (or, more generally, to the credit risk). For example, an actively-traded corporate or sovereign bond, or a portfolio of these bonds, may serve as an underlying asset or index for such a derivative. A credit derivative can also have a loan (or a portfolio of loans) as the underlying reference credit. It is clear that a credit derivative derives its value from the price – and thus from the credit quality – of the underlying default prone credit instrument.

The first agreements for a secluded transfer of credit risk were only signed in the early 1990s. It is thus worthwhile to mention that financial arrangements with features similar to credit derivatives – such as the *letter of credit* or the *bond insurance* – were largely used by commercial banks much earlier. Under a letter of credit, an issuer pays a bank an annual fee in exchange for the bank's promise to make debt payments on behalf of the issuer, should the issuer fail to do so. Under a bond insurance, an issuer pays an insurer to guarantee the performance on a bond (for more details, see, e.g., Fabozzi (2000)). However, in contrast to credit derivatives, these more traditional credit risk protections are not tradeable separately from the underlying obligation.

Credit derivatives can be structured in a large variety of ways; they are typically complex agreements, customized to the specific needs of an investor. Due to the rapidly growing demand, in the past few years the credit derivatives were the worldwide fastest-growing derivative products. As estimated by J.P. Morgan, in April 2000 the total nominal value of outstanding credit derivatives has passed the 1,000 billion U.S. dollar line. The common feature of all credit derivatives is the fact that they allow for the transfer of the credit risk from one counterparty to another; they thus constitute a natural and convenient tool to control the credit risk exposure. The overall risk an investor is concerned with involves two components: market risk and asset-specific credit risk. In contrast to 'standard' interest-rate sensitive derivatives, credit derivatives allow to isolate the firm-specific credit risk from the overall market risk. They also provide a way to synthesize assets that are otherwise not available to a particular investor (in this case, an investor 'buys' – rather than 'sells' – a specific credit risk). For an extensive analysis of economical reasons that support the use of these products, as well as expositions of various aspects of credit derivatives markets, we refer to Duffee and Zhou (1996), Das (1998a, 1998b), Tavakoli (1998), Francis et al. (1999) and Nelken (1999).

We shall now focus for a moment on credit derivatives associated with the defaultable term structure. Similarly as in the case of derivative securities associated with the risk-free term structure, we may formally distinguish three main types of agreements: forward contracts, swaps, and options.

A *forward contract* commits the buyer to purchasing a specified instrument at a specified future date at a price predetermined at contract inception. In a forward contract, the default risk is normally borne by the long party. If a credit event occurs, the transaction is marked to market and unwound. Forward contracts can also be transacted in spread form – that is, the agreement can be based on the specified bond's spread over a benchmark asset. In market practice, the most popular credit-sensitive swap contracts are: *total rate of return swap, asset swap* and *default swap* (explained in some detail in Sect. 1.3.1–1.3.2). *Credit options* are typically embedded in complex credit-sensitive agreements, though the over-the-counter traded credit options, such as: *options on an asset swap* or *credit spread options,* are also available.

Credit derivatives can also be classified into three groups. The first class consists of these credit derivatives that are linked exclusively to the default event; that is, it includes contracts with the payoff determined by the default event, as opposed to the changes in the credit quality of the underlying instrument. *Default swaps, default options,* and *first-to-default swaps* may serve as typical examples of such *default products.* The second group includes the so-called *spread products* – that is, credit derivatives whose payoff is primarily determined by the changes in the credit quality of the underlying security, e.g., *credit spread swaps, credit spread options* and *credit linked notes.* One may distinguish here the subclass of credit derivatives that are linked directly to rating upgrades or downgrades. Finally, the last class encompasses derivatives that allow to transfer the total risk of an asset between two parties. As an example of a *synthetic securitization,* we may cite the *total rate of return swap.* Let us observe that, due to the complexity of traded credit derivatives, the above classification is not definitive. Unfortunately, the terminological conventions relative to these contracts are not yet fully uniform, though a continuous standardization of OTC credit derivatives should be acknowledged. In 1999, the International Swaps and Derivatives Association (ISDA) published the ISDA 1999 *Credit Derivatives Definitions* providing a basic framework for documenting privately negotiated credit derivative transactions, and developed a standard documentation for credit derivatives – the ISDA 1999 *Master Agreement.* Through the ISDA standard documentation, the comparability and objectivity of credit derivatives has increased considerably.

In 2000, *Risk* conducted a survey aiming at determining the scope and the size of credit derivatives market. The following numbers represent percentage of the credit derivatives business generated by particular credit derivative products (in terms of notional outstanding): credit default swaps – 45%, synthetic securitizations – 26%, asset swaps – 12%, credit linked notes and asset repackaging – 9%, basket default swaps – 5%, credit spread options – 3%. The total notional amount of outstanding contracts of the participants in the *Risk* survey was around $810 billion. For more details, see Patel (2001).

### 1.3.1 Default Swaps and Options

*Default swaps* and *default options* may be considered as some sort of debt insurance contracts; that is why they are also known as *default insurance* or *default protection* (they are also sometimes called *credit default swaps/options*). In these agreements, periodic fixed payments (in the case of a default swap) or an upfront fee (in the case of a default option) from the protection buyer is exchanged for the promise of some specified payment from the protection seller to be made only if a particular, pre-specified *credit event* occurs. If a credit event occurs during the lifetime of the default swap/option, the seller pays the buyer an amount to cover the loss, and the contract then terminates. If no credit event has occurred prior to maturity of the contract, both sides end their obligations to each other. Let us stress that we deal here with a protection against the reference credit risk; the default risk of counterparties is thus neglected; in most practical cases it has only a minor impact on the valuation and hedging of a default swap (Hull and White (2001) and Lando (2000b) examine default swaps in the presence of the counterparty risk; see Sect. 12.3.1). The most important covenants of a default swap/option are:
- the specification of the credit event that is formally referred to as the 'default' (in practice, it may include: bankruptcy, insolvency, or payment default, a stipulated price decline for the reference asset, a rating downgrade of the reference entity, etc.),
- the contingent default payment, which may be structured in a number of ways; for instance, it may be linked to the price movement of the reference asset, or it can be set at a predetermined level (e.g., a fixed percentage of the notional amount of the transaction),

and, in the case of the default swap:
- the specification of periodic payments, which largely depend on the credit quality of the reference asset.

Default swaps/options are usually settled in cash; the agreement may also provide for physical delivery, though. For example, it may involve payment at par by the seller of the contract in exchange for the delivery of the defaulted reference asset.

**Standard default swaps/options.** To describe the cash flows in these contracts, let us consider a default swap/option with the maturity date $T$. As the reference asset, we first take a defaultable zero-coupon bond with the face value $L$ and maturity date $U \geq T$. The contingent payment is triggered at the default time $\tau$ of the reference entity. In case of the default option, the protection buyer pays a lump sum (commonly referred to as the *premium*) at the contract's inception, and the payoff at default time is given as

$$(L - D(\tau, U)) \, \mathbb{1}_{\{\tau \leq T\}} = (L - D(\tau, U))^+ \mathbb{1}_{\{\tau \leq T\}}. \tag{1.2}$$

It is not surprising that such a contract is also commonly known as the *default put option*. Since the long party – i.e., the protection buyer – pays an upfront fee, the buyer's credit quality has no relevance in this kind of contract.

In case of the default swap, the protection buyer pays an *annuity*, also called *credit swap premium,* at times $T_i$, $i = 1, \ldots, m$ prior to default or maturity date, whichever comes first, and the payoff at default is as given above. From the buyer's perspective, the cash flows of the default swap can be represented as follows:

$$(L - D(\tau, U))\, \mathbb{1}_{\{\tau \leq T\}}\, \mathbb{1}_{\{\tau\}}(t) - \sum_{i=1}^{m} \kappa \mathbb{1}_{\{\min(\tau, T) > T_i\}}\, \mathbb{1}_{\{T_i\}}(t), \qquad (1.3)$$

where $\kappa$ denotes the annuity amount. Since we are dealing here with a contract with bilateral default risk and a reference risk, the credit quality of both counterparties should be taken into account.

Let us emphasize once more that descriptions of cash flows of a default swap/option given above are valid only under an implicit assumption that the counterparty default risk can be disregarded.

*Remarks.* Alternative covenants for the recovery payment may be considered, for instance, a contract may stipulate that the recovery payment equals

$$(LB(\tau, U) - D(\tau, U))\, \mathbb{1}_{\{\tau \leq T\}},$$

or

$$(D(\tau-, U) - D(\tau, U))\, \mathbb{1}_{\{\tau \leq T\}},$$

where $D(\tau, U)$ represents the value of the defaultable bond immediately after the bond has defaulted – that is, its post-default value. Thus, in the former case, the protection buyer is compensated for the loss of value of the defaultable bond relative to the value of the Treasury bond at time $\tau$. In the latter case, he is compensated for the loss in the value of the defaultable bond immediately after the default, relative to the value of the bond immediately before the default.

The generic expressions (1.2)–(1.3) can be given a more explicit form, as soon as a particular recovery scheme for the underlying bond is adopted. For instance, in the case of fractional recovery of Treasury value, with the fixed recovery rate $\delta$, formulae (1.2)–(1.3) become: $L(1 - \delta B(\tau, U))\mathbb{1}_{\{\tau \leq T\}}$ and

$$L\big(1 - \delta B(\tau, U)\big)\, \mathbb{1}_{\{\tau \leq T\}}\, \mathbb{1}_{\{\tau\}}(t) - \sum_{i=1}^{m} \kappa \mathbb{1}_{\{\min(\tau, T) > t_i\}}\, \mathbb{1}_{\{t_i\}}(t),$$

respectively. If we postulate the fractional recovery of par value instead, then we obtain the following expressions: $L\big(1 - \delta\big)\mathbb{1}_{\{\tau \leq T\}}$ and

$$L\big(1 - \delta\big)\, \mathbb{1}_{\{\tau \leq T\}}\, \mathbb{1}_{\{\tau\}}(t) - \sum_{i=1}^{m} \kappa \mathbb{1}_{\{\min(\tau, T) > t_i\}}\, \mathbb{1}_{\{t_i\}}(t),$$

respectively. Finally, when the fractional recovery of market value covenants are in place, then formulae (1.2)–(1.3) become $(L - \delta D(\tau-, U))\mathbb{1}_{\{\tau \leq T\}}$ and

$$\left(L - \delta D(\tau-, U)\right)\mathbb{1}_{\{\tau \leq T\}}\mathbb{1}_{\{\tau\}}(t) - \sum_{i=1}^{m}\kappa\mathbb{1}_{\{\min(\tau,T)>t_i\}}\mathbb{1}_{\{t_i\}}(t),$$

respectively. In practice, the typical default swap also requires the buyer to pay an accrued premium, based on the time since the previous annuity payment, if there is a default (expression (1.3) can be easily modified to account for this covenant). However, the accrued coupon payment is not covered by this protection. It is in fact more common for a default swap/option to have a corporate coupon bond (e.g., a floating-rate note), rather than a corporate zero-coupon bond, as the underlying asset. In such a case, formulae (1.2)–(1.3) become: $(L - \text{FRN}(\tau, U))\mathbb{1}_{\{\tau \leq T\}}$ and

$$\left(L - \text{FRN}(\tau, U)\right)\mathbb{1}_{\{\tau \leq T\}}\mathbb{1}_{\{\tau\}}(t) - \sum_{i=1}^{m}\kappa\mathbb{1}_{\{\tau>t_i\}}\mathbb{1}_{\{t_i\}}(t),$$

respectively, where $\text{FRN}(\tau, U)$ denotes the market value of a corporate floating-rate note (FRN) just after the issuer's default. Since an FRN typically trades at a price close to par, the last two payoffs reflect more adequately than (1.2)–(1.3) the bond-specific default risk (as opposed to the market interest rate risk). If the default swap has an FRN as the underlying asset, the annuity $\kappa$ is usually close to the credit spread $s$, between the floating rate on the note and the risk-free floating rate (or LIBOR rate).

*Remarks.* Let us note that FRNs may also serve as the underlying instruments of a *credit option.* If the investor takes a certain view regarding the credit quality at time $T < U$ of the firm issuing the underlying FRN, then he may purchase a call option with the payoff $(\text{FRN}(T, U) - K)^+$, where $K$ is the strike price and $T$ the option's expiration date, or a put option with the payoff $(K - \text{FRN}(T, U))^+$. The valuation of the credit call option described above requires modeling of the underlying corporate FRN.

**Exotic default swaps/options.** There are other variations of the standard default swap/option, referred to as *exotic default swaps/options.* These variations may regard the covenants determining the triggering of credit event and/or covenants determining the amount of insurance payoff. For instance, in a *digital default swap/option,* the payment to the long party at default is a pre-specified fixed amount. The *basket default swap/option* is a form of the first-to-default contract: the insurance payment takes place if the first one in a group of specified reference entities defaults (see Sect. 1.3.5 below). The *contingent default swap/option* is a contract in which the insurance payoff requires both the underlying credit event to occur, as well as an additional trigger, such as a credit event with respect to some other reference defaultable claim. Finally, the *dynamic default swap/option* is a variant of a default swap/option in which the notional amount, determining the amount of the insurance payoff, is the marked to market value of a designated portfolio of default swaps/options.

## 1.3.2 Total Rate of Return Swaps

The *total rate of return swap* – also known as the *total return swap* ( *TROR* or *TRS,* for short) – is an agreement in which the total return on some reference entity (a basket of assets, an index, etc.) is exchanged for some other cash flows. One party, referred to as the *payer,* agrees to pay the total return of the reference entity (coupons plus or minus any change in the capital value) on a notional principal amount to another party, referred to as the *receiver.* In return, the *receiver* agrees to make periodic payments according to an agreed (fixed or floating) interest rate on the same notional amount. From the receiver's perspective, a total rate of return swap is thus similar to a synthetic purchase of the underlying entity.

If default of the reference entity occurs during the lifetime of a TROR, the contract terminates immediately; no further coupon or interest rate payment change hands. The receiver has the obligation to cover the change in value of the underlying asset by paying the payer the difference since the start of the swap, though. This means that the receiver accepts the price risk, including the credit risk, of the underlying reference security. Put another way, a TROR has an embedded default swap in which the payer is the protection buyer. The most relevant features of a TROR can be summarized as follows:

– no principal amounts are exchanged and no physical change of ownership occurs,
– the nominal principal of the swap may differ from the nominal value of the reference entity (a bond, a loan, etc.),
– the maturity of the total return swap agreement need not match that of the underlying reference entity,
– at the swap termination (i.e., either at its maturity, or upon default of the reference entity) a price settlement, based on the change in the value of the reference asset, is made.

We shall now give an example of a total return swap. As a reference asset, we take a corporate coupon bond with the promised coupons $c_i$ at times $T_i$. We assume that the notional principal equals 1, and the maturity date of the swap is $U \leq T_n$, so that it expires before the bond's maturity date. In addition, suppose that the fixed *annuity payments* (*reference rate payments*) $\kappa$ are made by the receiver at some scheduled dates $U_1 < U_2 < \cdots < U_m \leq U$. The receiver is entitled to all coupon payments during the lifetime of the contract (prior to default), as well as to the change in value of the underlying bond. From the receiver's perspective, the cash flows are:

$$\sum_{i=1}^{n} c_i \, \mathbb{1}_{\{\tilde{\tau}>T_i\}} \, \mathbb{1}_{\{T_i\}}(t) + \big(D_c(\tilde{\tau},T) - D_c(0,T)\big) \, \mathbb{1}_{\{\tilde{\tau}\}}(t) - \kappa \sum_{i=1}^{m} \mathbb{1}_{\{\tilde{\tau}>U_i\}} \, \mathbb{1}_{\{U_i\}}(t),$$

where $\tilde{\tau} = \min(\tau, U) =: \tau \wedge U$. As before, $\tau$ stands for the default time of the bond, and $D_c(t,T)$ denotes the price of a defaultable coupon bond with unit face value at time $t$.

Let us observe that cash flows described above correspond to an implicit assumption that the counterparty risk can be neglected. In other words, the last expression embeds only the credit risk of the reference entity and the market risk, reflected here in the variations of the price $D_c(t,T)$ of a corporate coupon bond. More complex total return swaps can also incorporate put and call options (to establish caps and floors on the returns of the reference asset), as well as caps and floors on the floating reference rate.

### 1.3.3 Credit Linked Notes

A *credit linked note* (CLN) is a note paying an enhanced coupon to investors for bearing the credit risk of a reference entity. The buyer of the note funds the credit protection that the issuer of the note may thus sell to a third party; in exchange the note pays a higher-than-normal yield. Therefore, the buyer of a CLN can also be seen as a protection seller; the issuer of the CLN is thus the protection buyer. The basic credit linked note can be described by the following covenants:

– at the contract inception, an investor purchases a CLN by paying the issue price of the CLN in cash,
– during the life of the CLN, prior to default, she or he receives regular coupon payments,
– if there is no default by the reference entity during the lifetime of the CLN, the investor receives the full nominal value of the CLN,
– in case of default by the reference entity prior to maturity, the CLN terminates immediately, and the issuer of the CLN redeems the nominal value of the CLN in form of physical delivery of bonds issued by the reference entity (or the contract is settled in cash). No further coupon payments are distributed.

Since the investor makes an upfront fee, the default risk of the counterparty (the issuer of the CLN) is also essential – the enhanced coupon should compensate for both the reference credit risk and the unilateral default risk. Formally, a CLN may be seen as a combination of the default swap with the bond seen as a collateral. It protects the buyer of the default swap from the default risk of the protection seller – i.e., from the default risk of the buyer of a CLN. To summarize, these instruments involve a specific combination of a fixed-income instrument with an embedded credit derivative.

In the market practice, the CLNs are rather complex credit instruments. A credit linked note covenant may stipulate that the principal repayment is reduced to a certain level below par if the external corporate or sovereign debt defaults before the maturity of the note. The first-to-default CLN is linked to credit events from more than one reference entity. If there is a credit event, the note redeems early and there is no more exposure to other reference credit risks. As expected, the investor is compensated for bearing an additional credit risk by higher returns of the first-to-default CLNs.

CLNs are primarily issued by the special purpose corporations, also known as special purpose vehicles (SPVs for short). We shall henceforth refer to the issuer of the CLN as the investment bank. As an example of a credit linked note, let us examine a specific CLN, tied to a default swap associated with a reference corporate zero-coupon bond. We assume that the investment bank holds a high-rated collateral – a coupon bond with face value $L$, coupon rate $c$ and maturity date $T$. At time $t' > 0$, the investment bank enters into a (reference) default swap with a third party. The default swap is relative to some reference credit, for instance, a low-rated corporate bond maturing at time $T$. In this default swap, the investment bank sells default protection in return for an annuity premium $\kappa$. Recall that the cash flows corresponding to a default swap were given in formula (1.3).

Here, the annuity payment dates $t_i$ satisfy $t' < t_i$. At the same time $t'$, the investment bank issues a CLN with the expiration date $T' < T$. An investor purchases the note at time $t'$, for the par value $L$. If the reference credit does not experience a pre-specified credit event before the expiration date $T'$ then the note pays coupon at the rate $c'$ up until its expiration date, and the investor receives the par value back at the expiration date. Otherwise, the note pays the coupon at the rate $c'$ only until the time when the reference credit experiences the specified credit event, and at this time the investor collects only some recovery portion of the par value.

The recovery portion of the par value is determined as follows: at the time when the reference credit experiences the specified credit event, the collateral coupon bond is liquidated, and the investor receives the proceeds only after the third party (i.e., the default swap counterparty) is paid the contingent payment. The investor thus bears credit risks of both the reference and collateral credits, and the enhanced coupon rate $c'$ must compensate the investor for the two credit risks involved.

To describe the cash flows corresponding to the above structure, as seen from the perspective of the investment bank, we shall assume that the collateral bond promises to pay coupons on dates $T_1 < \cdots < T_k < T$, and that the CLN promises to pay coupons on dates $T_1' < \cdots < T_m' < T'$, where $T_1' > t'$. We denote the price process of the collateral bond as $D_c(t, T)$. As usual, $\tau$ stands for the random time of the underlying credit event, which in this case may be declared as the default event of the reference corporate bond (by assumption $\tau > t'$). From the perspective of the bank, the cash flows are

$$
L\mathbb{1}_{\{t'\}}(t) - \left(L - D(\tau, T)\right)\mathbb{1}_{\{\tau \leq T\}}\mathbb{1}_{\{\tau\}}(t) + \sum_{i=1}^{n} \kappa\mathbb{1}_{\{\tau > t_i\}}\mathbb{1}_{\{t_i\}}(t)
$$

$$
+ \sum_{j=1}^{k} cL\mathbb{1}_{\{\tau > T_j\}}\mathbb{1}_{\{T_j\}}(t) - \sum_{l=1}^{m} c'L\mathbb{1}_{\{\tau > T_l'\}}\mathbb{1}_{\{T_l'\}}(t) -
$$

$$
- \left(D_c(\tau, T) - \left(L - D(\tau, T)\right)\right)^{+}\mathbb{1}_{\{\tau \leq T\}}\mathbb{1}_{\{\tau\}}(t) - L\mathbb{1}_{\{\tau > T'\}}\mathbb{1}_{\{T'\}}(t),
$$

where $D(\tau, T)$ represents the post-default price of the reference bond.

### 1.3.4 Asset Swaps

Consider an investor that holds a corporate bond paying fixed rate coupon. Such an investor may be interested in entering into a swap, in which he will pay the fixed coupon and will receive a floating-rate coupon. The combination of the bond and the swap represents the so-called *asset swap*. It is apparent that a position in an asset swap is similar to holding a floating-rate note issued by the corporation of reference (in most cases, the floating rate is given as the LIBOR rate plus a spread, referred to as the *asset swap spread*). Typically, if there is default on the underlying bond, the swap obligation remains in force.

Asset swaps may serve as the underlying securities of *credit options*. A *call on an asset swap* is an option to buy the underlying bond for the predetermined strike price, and to enter at the same time the associated asset swap. The strike price and the strike spread are specified in option's contract (usually, the strike price equals the par value of the underlying bond). A holder of a *put on an asset swap* has the right to sell a corporate bond of reference for the strike price, and to simultaneously enter an asset swap as a receiver of the fixed coupon.

### 1.3.5 First-to-Default Contracts

In a *first-to-default contract,* the protection seller is exposed to the first entity within a portfolio (a basket) of credit risk sensitive instruments, which defaults (or, more generally, which experiences a pre-specified credit event). Such a contract is typically unwound immediately after the first credit event.

Consider a portfolio of $n$ credit instruments. Let us denote by $\tau_i$ the time of the credit event of interest (e.g., default) associated with the $i^{\text{th}}$ instrument. We denote by $\tau$ the time when the first of these credit events occurs – that is, $\tau = \min(\tau_1, \tau_2, \ldots, \tau_n)$. We assume that $\tau_i \neq \tau_j$ for any $i \neq j$, and we consider a contingent claim with the settlement date $T$. If the first credit event happens no later than at time $T$, i.e., when $\tau \leq T$, a contingent payment is made. The amount of the contingent payment depends on the name (the instrument) that defaulted first. Denoting by $A_i$ the contingent payment associated with the $i^{\text{th}}$ instrument, we have that the amount received at time $\tau$ is $A_i$ if $\tau = \tau_i$. If none of the names experiences the credit event prior to or at time $T$ then a payment $A$ is made at the maturity date $T$. The payoff associated with the abstract first-to-default contract can be formally represented as a single cash flow, which settles at time $T$, and equals

$$\mathbb{1}_{\{\tau \leq T\}} B^{-1}(\tau, T) \sum_{i=1}^{n} A_i \mathbb{1}_{\{\tau = \tau_i\}} + A \mathbb{1}_{\{\tau > T\}}.$$

*First-to-default swaps* – also known as *basket default swaps* – are default swaps linked to a portfolio of credit-sensitive securities. The first-to-default contracts are a special case of the $i^{\text{th}}$-to-default contracts. We refer to Chap. 9 and 10 for a detailed study of the latter ones.

### 1.3.6 Credit Spread Swaps and Options

Spread derivatives are credit derivatives that are constructed from market observable credit spread levels in such a way, that they allow for an easy transfer of credit price/spread risk between parties. The *credit spread swap,* also known as the *relative performance total return swap,* is a credit-risk sensitive agreement under which one party makes payments based on the yield-to-maturity of a specific issuer's debt, and the other party makes payments based on comparable Treasury yields (or some other benchmark rate). As expected, in practice the payments' specifications are done in a large variety of ways. For instance, under a *credit spread forward*[3] the payoff at the contract exercise date depends on the difference between the credit spread prevailing on the exercise date and some pre-specified strike level (or some benchmark rate). The payment may be executed in either direction, depending on whether it has a positive or a negative value.

To be more specific, let us assume that for a fee paid periodically, the investor receives (or makes) a payment at the contract exercise date depending on the difference between the credit spread on an underlying corporate discount bond and some specified strike. Here, the credit spread is measured as the difference between the yield of the underlying corporate discount bond and the yield of an equivalent Treasury bond. Formally, at the exercise time $T < U$ the investor receives the difference

$$D(T,U) - LB(T,U) - K,$$

where $K$ is a predetermined strike. Recall that $D(t,U)$ is the price at time $t$ of a $U$-maturity corporate discount bond with face value $L$, and $B(t,U)$ is the price of the $U$-maturity Treasury discount bond with face value 1. As a fee for this contract, the investor is required to pay annuities at times $t_1 < t_2 < \cdots < t_n < T$. Thus, the associated cash flows are:

$$\big(D(T,U) - LB(T,U) - K\big)\mathbb{1}_{\{T\}}(t) - \kappa \sum_{i=1}^{n} \mathbb{1}_{\{t_i\}}(t).$$

If the fee payment is made up front, the above contract is termed the credit spread option (such a contract is similar to the credit default option).

*Credit spread options* are option-like agreements whose payoff is associated with the yield difference of two credit-sensitive assets. For instance, the reference rate of the contract can be a spread of a corporate bond over a benchmark asset of comparable maturity. The option can be settled either in cash or through physical delivery of the underlying bond, at a price whose yield spread over the benchmark asset equals the strike spread. Options on credit spreads allow to isolate the firm-specific credit risk from the market risk. When preferable, the credit spread options can also incorporate interest rate risk by using a fixed strike or yield.

---

[3] Sect. 6.2.1.1 in Ammann (1999) discusses *credit forward contracts.* These contracts are similar to the credit spread forwards.

## 1.4 Quantitative Models of Credit Risk

Formally, by *credit event* we mean any random event whose occurrence affects the ability of the counterparty in a financial contract to fulfill a contractual commitment to meet his or her obligations stated in the contract. The default event is of course a credit event; other examples of credit events include, e.g., changes in credit quality of a corporate bond. Note that credit events may not be directly observed by the parties in a financial contract. We want to stress here that the ability of the counterparty in a financial contract to fulfill a contractual commitment to meet its obligations stated in the contract is not necessarily adversely affected by the occurrence of a credit event. For example, if the credit event occurs due to the increase in the credit quality of a corporate bond then, manifestly, the above mentioned ability is not adversely affected.

A vast majority of mathematical research devoted to the credit risk is concerned with the modeling of the random time when the default event occurs, i.e., the *default time*. Some approaches to the defaultable term structure allow for a possibility of intermediate credit events that are associated with changes in the credit quality of a corporate bond, which migrates between various rating classes. In this case, the modeling of random times of credit migration also becomes an important issue.

Another important problem arising in the quantitative modeling of the credit risk is the issue of mathematical modeling of the so-called *recovery rates*. As already mentioned, the recovery rate specifies the payment to the contract holder in case of default. Recovery payments together with the notional amount of the contract determine the potential cash flows associated with the contract. The main objective of the quantitative models of the credit risk is to provide ways to price and to hedge financial contracts that are sensitive to credit risk. Needless to say that any approach to pricing credit risk should aim at producing an internally consistent (that is, an arbitrage-free) financial model. As already noted, two competing methodologies have emerged in order to model the default/migration times and the recovery rates: the structural approach and the reduced-form approach.

### 1.4.1 Structural Models

*Structural models* are concerned with modeling and pricing credit risk that is specific to a particular corporate obligor (a firm). Credit events are triggered by movements of the firm's value relative to some (random or non-random) credit-event-triggering threshold (or barrier). Consequently, a major issue within this framework is the modeling of the evolution of the firm's value and of the firm's capital structure. For this reason, the structural approach is frequently referred to as the *firm value approach*. Through the modeling of credit events in terms of the value of the firm, the structural methodology links the credit events to the firm's economic fundamentals.

Most structural models are concerned with only one type of credit events, namely, the firm's default. The time of default is typically specified as the first moment when the value of the firm reaches a certain lower threshold, so that it is defined *endogenously* within the model. Such a default triggering mechanism has a natural interpretation as the *safety covenant,* which aims to protect the interests of bondholders against those of stockholders.

An alternative approach within the structural framework postulates that the bankruptcy decision is at the discretion of the stockholders; such an approach leads to the important problem of the specification of the optimal capital structure and the strategic debt service. Let us mention that the recovery rates are also frequently given endogenously within the model, as some function of the value of the firm. From the long list of works devoted to the structural approach, let us mention here: Merton (1974), Black and Cox (1976), Galai and Masulis (1976), Geske (1977), Brennan and Schwartz (1977, 1978, 1980), Pitts and Selby (1983), Cooper and Mello (1991), Rendleman (1992), Kim et al. (1993a), Nielsen et al. (1993), Leland (1994), Longstaff and Schwartz (1995), Anderson and Sundaresan (1996, 2000), Leland and Toft (1996), Briys and de Varenne (1997), Ericsson and Reneby (1998), Mella-Barral and Tychon (1999), and Ericsson (2000).

### 1.4.2 Reduced-Form Models

In this approach, the value of the firm's assets and its capital structure are not modeled at all, and the credit events are specified in terms of some exogenously specified jump process (as a rule, the recovery rates at default are also given exogenously). We can distinguish between the reduced-form models that are only concerned with the modeling of the default time, and that are henceforth referred to as the *intensity-based models,* and the reduced-form models with migrations between credit rating classes, called the *credit migration models.*

**Intensity-based approach.** The main emphasis in the intensity-based approach is put on the modeling of the random time of default, as well as evaluating conditional expectations under a risk-neutral probability of functionals of the default time and corresponding cash flows. Typically, the random default time is defined as the jump time of some one-jump process. As we shall see, a pivotal role in evaluating respective conditional expectations is played by the so-called *default intensity process.* Modeling of the intensity process, which is also known as the hazard rate process, is the starting point in the intensity approach. We need to emphasize here the crucial role of the conditioning information in the modeling of the intensity process. The intensity-based approach is examined in detail in the second part of the present text. The interested reader is also referred to the original papers by, among others, Pye (1974), Ramaswamy and Sundaresan (1986), Litterman and Iben (1991), Jarrow and Turnbull (1995), Duffie et al. (1996), Schönbucher (1996, 1998a, 1998b), Lando (1997, 1998), Monkkonen (1997), and Madan and Unal (1998).

**Credit migrations.** If migrations between credit ratings are not allowed, the model is referred to as the *single credit rating model*, though formally we always have another rating grade in the model, namely, the default state. Otherwise we deal with the *multiple credit ratings model*. The traditional intensity-based approach focuses on the pre-default value of a corporate bond in a single rating class model. More recent studies have extended the intensity-based approach to the case of multiple credit rating classes.

Assume that the credit quality of corporate debt is quantified and categorized into a finite number of disjoint *credit rating classes* (or *credit grades*). Each credit class is represented by an element of a finite set, denoted by $\mathcal{K}$. It is natural to distinguish a particular element $K$ of the set $\mathcal{K} = \{1, \ldots, K\}$, which formally corresponds to the default event. As observed in practice, the credit quality of a given corporate debt changes over time. We shall refer to this feature by saying that the credit quality *migrates* between various credit classes. This migration is frequently modeled in terms of a (conditional) Markov chain, denoted by $C$, with finite state space $\mathcal{K}$ and either discrete or continuous time parameter. The process $C$ is referred to as the *credit migration process*. In most cases, the multiple defaults are excluded, so that the default class represents the absorbing state for the chain $C$. The main issue in this approach is thus the modeling of the transition intensities matrix for the migration process, both under the risk-neutral and the real-world probabilities. The next step is the evaluation of conditional expectations under a risk neutral probability of certain functionals, typically related to the default time. Let us stress here the special role of the so-called *factor models* of credit migration. In these models, based on the theory of Cox processes, the intensities of default and/or migrations are specified as functions of both macro- and micro-economic factors. References dealing with the stochastic modeling of credit migrations include: Das and Tufano (1996), Jarrow et al. (1997), Duffie and Singleton (1998a), Arvanitis et al. (1998), Kijima (1998), Kijima and Komoribayashi (1998), Thomas et al. (1998), Huge and Lando (1999), Bielecki and Rutkowski (2000), Lando (2000a), and Schönbucher (2000).

**Defaultable term structure.** A direct modeling of the defaultable term structure of interest rates is in fact not much different from the modeling of the default-free term structure. It is customary to start with some model of the term structure of interest rates for each credit rating class to be included in the overall model of the defaultable term structure. For example, one may prefer to do the modeling in terms of the instantaneous short-term rate (Vasicek (1977), Cox et al. (1985b)), or in terms of instantaneous forward rates (Heath et al. (1992)). Alternatively, one may adopt the framework of market rates such as: the forward LIBOR rates (Brace et al. (1997), Miltersen et al. (1997)) or the forward swap rates (Jamshidian (1997)). If migrations between rating classes are also modeled, one needs to describe the mechanism that governs the observed variations in credit quality. Otherwise, it is enough to specify the mechanism that triggers the default event.

### 1.4.3 Credit Risk Management

**Hedging of defaultable claims.** In most structural models of credit risk, the default time is predictable with respect to the information carried by traded primary (non-defaultable) securities. These models are typically complete with regard to both the market risk and the credit risk, and thus perfect replication of both default-free and defaultable claims is feasible. On the contrary, in reduced-form models, the default time tends to be an unpredictable stopping time, and perfect hedging of defaultable claims with the use of non-defaultable securities is impossible. To overcome this difficulty, one may either use some defaultable securities as hedging instruments (see Wong (1998), Bélanger et al. (2001), Blanchet-Scalliet and Jeanblanc (2001)), or one may apply other hedging and pricing principles, such as: the local risk minimization (see Lotz (1998)) or the optimal shortfall hedging (see Lotz (1999)), or the utility-based pricing (see Collin-Dufresne and Hugonnier (1999)).

**Integration of risks.** The market risk associated with a financial instrument is the risk resulting from adverse movements in the level or volatility of the value of this instrument. If a given instrument is sensitive to both market and credit risks, the two types of risk are intertwined, and they can not be easily disentangled. Most of the quantitative models of credit risk account for this feature by an appropriate integration of market and credit risks. In the structural approach, credit events are contingent on the movements of the firm value process and, in case of some models, on the dynamics of the value process of all firm's liabilities. This property provides thus an obvious link between the two risks involved. In the reduced-form approach, the intensities of default (or credit migrations) are sometimes postulated to depend on various financial factors such as, e.g., credit spreads or convenience yields. These factors are generally tied to the market risk associated with a given defaultable claim. Jarrow and Turnbull (2000b) provide an interesting discussion on the intersection of the market risk and the credit risk.

**Portfolio management.** Recently developed practical approaches to the active management of credit risk are typically concerned with calculating the probability distribution of losses in case of default for portfolios of credit-risk sensitive instruments. This probability distribution is assessed under the real-world (statistical) probability, so that these methods do not yield directly pricing models for credit derivatives. The credit risk measurement techniques that have recently gained a considerable prominence include:
- the KMV[4] Credit Monitor and Portfolio Manager (see Crosbie (1997)),
- J.P. Morgan's methodology CreditMetrics (see Gupton et al. (1997)),
- the CSFP[5] CreditRisk+ (1997),
- Moody's methodologies Creditscore and RiskCalc,
- McKinsey's CreditPortfolioView (see Wilson (1997a, 1997b)).

---

[4] KMV Corporation was founded by S. Kealhofer, J. McQuown and O. Vasicek.
[5] CSFP is an acronym of Credit Suisse Financial Products [www.csfb.com].

We refer to Saunders (1999), Crouhy et al. (2000), Gordy (2000), Nyfeler (2000), and Cossin and Pirotte (2000) for up-to-date surveys and comparative studies of the credit risk measurement and management methodologies.

### 1.4.4 Liquidity Risk

It is not uncommon to argue that the credit spread inherent in many financial instruments could be more aptly explained as being caused by the presence of liquidity risk, rather than the credit risk. Such a view is supported by rather convincing financial arguments and/or related econometric studies, reported, for instance, in Amihud and Mendelson (1991), Boudoukh and Whitelaw (1991), Longstaff (1995) or Bangia et al. (1999). It seems reasonable to expect that in practice each of these risks may dominate the other. Unfortunately, both these kinds of financial risks are rather difficult to separate. It is also noteworthy that from the purely mathematical viewpoint, the modeling of liquidity risk is not much different from the modeling of credit risk, and, to the best of our knowledge, no original mathematical techniques were developed to deal with the former kind of risk. In Merton's framework, models that allow for a separate treatment of credit and liquidity risks were developed by Longstaff (1995) and Ericsson and Renault (2000). In the reduced-form approach, the liquidity effect is already included as a component of the total spread, and thus the credit risk premium and liquidity premium can not be easily separated. In this text, we have chosen to use the generic term credit risk to cover both kinds of risk, especially when dealing with the reduced-form approach. For an analysis of liquidity risk and further references, the reader may consult the recent paper by Longstaff (2001).

### 1.4.5 Econometric Studies

An essential step in the practical implementation of any mathematical model of credit risk relies on the model's calibration against the real-life data. A detailed presentation of empirical studies related to the credit/liquidity risk is beyond the scope of the present text, though. The interested reader is thus referred to original papers by Sarig and Warga (1989), Sun et al. (1993), Altman and Bencivenga (1995), Altman and Kishore (1996), Duffee (1996, 1998), Carty and Lieberman (1997), Duffie and Singleton (1997), Monkkonen (1997), Altman and Saunders (1998), Kiesel et al. (1999, 2002), Taurén (1999), Altman and Suggit (2000), Liu et al. (2000), Rachev et al. (2000), Shumway (2001), Bakshi et al. (2001), Collin-Dufresne et al. (2001), Carey and Hrycay (2001), Houweling et al. (2001) or Christiansen (2002). To the best of our knowledge, Jonkhart (1979) and Iben and Litterman (1991) were the first researchers who proposed to impute implied probabilities of default from the term structure of yield spreads between default-free and defaultable corporate securities (also see the recent paper by Delianidis and Geske (2001) in this regard).

# 2. Corporate Debt

The structural approach is primarily directed at pricing the firm's liabilities. In this methodology, the firm's liabilities are seen as contingent claims issued against the total value of firm's assets; for this reason this approach is also referred to as the *firm value approach* or the *option-theoretic approach*.

Default event is specified in terms of the evolution of the total value of the firm's assets, henceforth denoted by $V$, as well as in terms of some default triggering barrier. This means that the firm's ability to meet its contractual liabilities, known as the firm's *solvency*, is assumed to be completely determined by the current level of value process $V_t$ in conjunction with appropriately specified bankruptcy covenants.

The terms 'total value of the firm's assets' and 'value of the firm' are assumed to have the same meaning in the present text; they both refer to the combined value of the firm's debt and equity. In the presence of tax benefits and/or bankruptcy costs, it is convenient to introduce also the concept of the *total value of the firm $G(V_t)$*, which makes account for these additional factors.

One of the major shortcomings of the structural approach, raised by several authors, lies in the presumption that the value of the firm can be directly observed. The recent study by Duffie and Lando (1997) addresses this issue by making a less stringent assumption that the issuer's assets cannot be directly observed by investors; instead they receive periodic and imperfect accounting reports. In such a framework, the study of the corporate debt can be done through techniques of the intensity-based methodology. Another flaw of the structural approach is that, in order to make valid the standard no-arbitrage argument, it requires an explicit or implicit assumption that the firm's assets represent a tradeable security (or at least that the firm's value process can be replicated by means of some traded securities, such as: the firm's shares, corporate bonds and default-free bonds).

Various existing structural models may be classified with regard to the specification of the following components:
- the dynamics of the total value of the assets of the firm,
- the structure of the firm's liabilities,
- the default event (in particular, the default triggering barrier),
- the recovery rule in case of default,
- other relevant economic quantities (like the short-term interest rate).

In all classic structural models, the default is triggered when the firm's value process falls below some default triggering barrier. This boundary is specified either exogenously or endogenously with respect to the total value of the firm. In the latter case, the default triggering barrier is typically derived from considerations regarding the optimal capital structure of the firm. Due to the assumed presence of market frictions, like bankruptcy costs and/or corporate taxes, the Modigliani-Miller theorem (see, e.g., Bodie and Merton (1998)) may not hold, and thus the issue of the optimal capital structure arises here in a natural way. In this variant of the structural approach, the default event is designed in an optimal way by the firm's stockholders.

In some structural models, the distinction is made between the total value of the firm's assets and the total value of the firm: the total value of the firm is equal to the value of the assets of the firm, plus the tax deduction (of coupon payments), less the value of the bankruptcy costs. Thus, if the two latter terms are neglected, the value of the firm and the value of the firm's assets coincide. It is essential to distinguish between the total value of the firm, which is difficult to observe, and the total market value of firm's shares. The latter can be observed, at least in the case of publicly traded firms.

The structural approach is attractive from the standpoint of financial economics, as it attempts to link the valuation of corporate debt to economic fundamentals. However, it is aimed at the valuation of the liabilities of a specific firm, rather than on the modeling of defaultable term structures for the various classes of corporate debt. The structural approach, initiated by Black and Scholes (1973), Merton (1974), Galai and Masulis (1976), Black and Cox (1976), and Geske (1977), was subsequently developed in various directions. The non-exhaustive list of references includes: Brennan and Schwartz (1977, 1980), Pitts and Selby (1983), Rendleman (1992), Kim et al. (1993b), Nielsen et al. (1993), Leland (1994), Longstaff and Schwartz (1995), Leland and Toft (1996), Mella-Barral and Tychon (1999), Briys and de Varenne (1997), and Crouhy et al. (1998).

This chapter is organized as follows. We first describe an abstract structural model of credit risk that may be used for valuation of credit risk sensitive instruments. Subsequently, we present the basic mathematical results that underpin the structural approach, and we discuss the classic Merton's model and its variants. The study of the first passage time structural models is postponed to the next chapter.

Let us observe that mathematical methodologies developed within the structural approach can also be applied to valuing callable corporate debt, as well as to convertible bonds. A call event is triggered when the process of the value of the assets of the firm is greater or equal to some call triggering barrier, typically referred to as the *upper threshold* (as opposed to the default triggering barrier, which is also known as the *lower threshold*). We do not deal with the callable debt and convertible bonds in the present text, though.

## 2.1 Defaultable Claims

We fix a finite horizon date $T^* > 0$, and we suppose that the underlying probability space $(\Omega, \mathcal{F}, \mathbb{P})$, endowed with some filtration $\mathbb{F} = (\mathcal{F}_t)_{0 \le t \le T^*}$, is sufficiently rich to support the following objects:

- the *short-term interest rate* process $r$,
- the *firm's value process* $V$, which models the total value of the firm's assets,
- the *barrier process* $v$, which will serve to specify the default time,
- the *promised contingent claim* $X$ representing the firm's liabilities to be redeemed at time $T \le T^*$,
- the process $A$, which models the *promised dividends*, i.e., the firm's liabilities stream that is redeemed continuously or discretely over time to the holder of a defaultable claim,
- the *recovery claim* $\tilde{X}$, which represents the recovery payoff received at time $T$, if default occurs prior to or at the claim's maturity date $T$,
- the *recovery process* $Z$, which specifies the recovery payoff at time of default, if it occurs prior to or at the maturity date $T$.

The probability measure $\mathbb{P}$ is assumed to represent the *real-world* (or *statistical*) probability, as opposed to the *spot martingale measure* (or the *risk-neutral probability*). The latter probability is denoted by $\mathbb{P}^*$ in what follows.

**Technical assumptions.** We postulate that the processes $V$, $Z$, $A$, and $v$ are progressively measurable with respect to the filtration $\mathbb{F}$, and that the random variables $X$ and $\tilde{X}$ are $\mathcal{F}_T$-measurable. In addition, $A$ is assumed to be a process of finite variation, with $A_0 = 0$. We assume without mentioning that all random objects introduced above satisfy suitable integrability conditions that are needed for evaluating the functionals defined in the sequel.

**Default time.** Let us denote by $\tau$ the random time of default. At this stage, it is essential to stress that the various approaches to valuing and hedging of defaultable securities differ between themselves with regard to the ways in which the default event – and thus also the default time $\tau$ – are modeled. In the structural approach, the default time $\tau$ will be typically defined in terms of the value process $V$ and the barrier process $v$. Specifically, we shall set

$$\tau := \inf \{ t > 0 : t \in \mathcal{T}, V_t < v_t \} \tag{2.1}$$

with the usual convention that the infimum over the empty set equals $+\infty$. In (2.1), the set $\mathcal{T}$ is assumed to be a Borel measurable subset of the time interval $[0, T]$ (or $[0, \infty)$ in the case of perpetual claims). From the mathematical standpoint, we shall frequently be justified in substituting the strict inequality '<' with the '≤' in (2.1), and in analogous definitions, without altering the probabilistic content of the definition. Furthermore, $\tau$ will be an $\mathbb{F}$-stopping time, and since the underlying filtration $\mathbb{F}$ in most structural models is generated by a standard Brownian motion, $\tau$ will be an $\mathbb{F}$-predictable stopping time (as any stopping time with respect to a Brownian filtration).

The latter property means that within the framework of the structural approach there exists a sequence of increasing stopping times announcing the default time; in this sense, the default time can be forecasted with some degree of certainty. By contrast, in the intensity-based approach, the default time will not be a predictable stopping time with respect to the 'enlarged' filtration, denoted by $\mathbb{G}$ in Part III of the text. In typical examples, the filtration $\mathbb{G}$ will encompass some Brownian filtration $\mathbb{F}$, but $\mathbb{G}$ will be strictly larger than $\mathbb{F}$. At the intuitive level, in the intensity-based approach the occurrence of the default event comes as a total surprise. For any date $t$, the present value of the default intensity yields the conditional probability of the occurrence of default over an infinitesimally small time interval $[t, t + dt]$.

**Recovery rules.** If default occurs after time $T$, the promised claim $X$ is paid in full at time $T$. Otherwise, depending on the adopted model, either the amount $Z_\tau$ is paid at time $\tau$, or the amount $\tilde{X}$ is paid at the maturity date $T$. In a general setting, we consider simultaneously both kinds of recovery payoff, and thus a defaultable claim is formally defined as a quintuple $DCT = (X, A, \tilde{X}, Z, \tau)$. In most practical situations, however, we shall deal with only one type of recovery payoff – that is, we shall set either $\tilde{X} = 0$ or $Z \equiv 0$. Thus, a typical defaultable claim can be seen either the quadruplet $DCT^1 = (X, A, \tilde{X}, \tau)$ or as $DCT^2 = (X, A, Z, \tau)$, depending on the recovery scheme. The former is called a defaultable claim with *recovery at maturity* ($DCT$ of the *first type*), and the latter a defaultable claim with *recovery at default* ($DCT$ of the *second type*). The absence of the superscript $i$ suggests that a particular expression is valid for a generic defaultable claim. Notice that the date $T$, the information structure $\mathbb{F}$ and the real-world probability $\mathbb{P}$ are also intrinsic components of the definition of a defaultable claim.

### 2.1.1 Risk-Neutral Valuation Formula

Suppose now that our underlying financial market model is arbitrage-free, in the sense that there exists a *spot martingale measure* $\mathbb{P}^*$ (also referred to as a *risk-neutral probability*), meaning that price process of any tradeable security, which pays no coupons or dividends, follows an $\mathbb{F}$-martingale under $\mathbb{P}^*$, when discounted by the *savings account* $B$, given as

$$B_t := \exp\left(\int_0^t r_u\, du\right).$$

We introduce the process $H_t = \mathbb{1}_{\{\tau \leq t\}}$, and we denote by $D$ the process that models all the cash flows received by the owner of a defaultable claim. Let us set $X^d(T) = X \mathbb{1}_{\{\tau > T\}} + \tilde{X} \mathbb{1}_{\{\tau \leq T\}}$.

**Definition 2.1.1.** The *dividend process* $D$ of a defaultable contingent claim $DCT = (X, A, \tilde{X}, Z, \tau)$, which settles at time $T$, equals

$$D_t = X^d(T) \mathbb{1}_{[T,\infty[}(t) + \int_{]0,t]} (1 - H_u)\, dA_u + \int_{]0,t]} Z_u\, dH_u.$$

It is clear that $D$ is a process of finite variation over $[0, T]$. Since

$$\int_{]0,t]} (1 - H_u)\, dA_u = \int_{]0,t]} \mathbb{1}_{\{\tau > u\}}\, dA_u = A_{\tau -}\mathbb{1}_{\{\tau \le t\}} + A_t\mathbb{1}_{\{\tau > t\}},$$

it is apparent that in case the default occurs at some date $t$, the promised dividend $A_t - A_{t-}$, that is due to be paid at this date, is not actually passed over to the holder of a defaultable claim. Furthermore, we have

$$\int_{]0,t]} Z_u\, dH_u = Z_{\tau \wedge t}\mathbb{1}_{\{\tau \le t\}} = Z_\tau \mathbb{1}_{\{\tau \le t\}},$$

where $\tau \wedge t = \min(\tau, t)$. At the formal level, the promised payoff $X$ could be considered as a part of the promised dividends process $A$. However, such a convention would be inconvenient, since in practice the recovery rules concerning the promised dividends $A$ and the promised claim $X$ are generally different. For instance, in the case of a defaultable coupon bond, it is frequently postulated that in case of default the future coupons are lost (formally, they are subject to the zero recovery scheme), but a strictly positive fraction of the bond's face value is usually received by the bondholder. We adopt the following definition of the ex-dividend price $X^d(t, T)$ of a defaultable claim. At any time $t < T$, the random variable $X^d(t, T)$ is meant to represent the current value of all future cash flows associated with a given defaultable claim $DCT$. (By convention, we also set $X^d(T, T) = X^d(T)$.) A formal justification for expression (2.2) is postponed to Sect. 2.1.3.

**Definition 2.1.2.** The (ex-dividend) *price process* $X^d(\cdot, T)$ of the defaultable claim $DCT = (X, A, \tilde{X}, Z, \tau)$, which settles at time $T$, is given as

$$X^d(t, T) = B_t\, \mathbb{E}_{\mathbb{P}^*}\left( \int_{]t,T]} B_u^{-1}\, dD_u \,\Big|\, \mathcal{F}_t \right), \quad \forall t \in [0, T). \qquad (2.2)$$

One easily recognizes (2.2) as a variant of the *risk-neutral valuation formula* that is known to give the arbitrage price of attainable contingent claims (see, for instance, Harrison and Pliska (1981), Duffie (1996), Musiela and Rutkowski (1997a) or Elliott and Kopp (1999)). Attainability of a defaultable claim $DCT$ is not obvious, though. Structural models typically assume that assets of the firm represent a tradeable security (in practice, the total market value of firm's shares is usually taken as the proxy for $V$). Consequently, the issue of existence of replicating strategies for defaultable claims can be analyzed in a similar way as in standard default-free financial models. In particular, it is essential to assume that the reference filtration $\mathbb{F}$ is generated by the price processes of tradeable assets. Otherwise, for instance, when the default time $\tau$ is the first passage time of $V$ to a lower threshold, which does not represent the price of a tradeable asset (so that $\tau$ is not a stopping time with respect to the filtration generated by some tradeable assets), the issue of attainability of defaultable contingent claims becomes more delicate. To summarize, the validity of the valuation formula (2.2) is not obvious a priori, so that it needs to be examined on a case by case basis.

For the ease of future reference, we shall now examine in some detail the two special cases of expression (2.2). It follows immediately from (2.2) that the price process $X^{d,i}(\cdot, T)$ of a defaultable claim $DCT^i$ equals, for $i = 1, 2$:

$$X^{d,i}(t, T) := B_t \, \mathbb{E}_{\mathbb{P}^*} \left( \int_{]t,T]} B_u^{-1} \, dD_u^i \, \Big| \, \mathcal{F}_t \right), \tag{2.3}$$

where

$$D_t^1 = \left( X \mathbb{1}_{\{\tau > T\}} + \tilde{X} \mathbb{1}_{\{\tau \leq T\}} \right) \mathbb{1}_{\{t \geq T\}} + \int_{]0,t]} (1 - H_u) \, dA_u,$$

and

$$D_t^2 = X \mathbb{1}_{\{\tau > T\}} \mathbb{1}_{\{t \geq T\}} + \int_{]0,t]} (1 - H_u) \, dA_u + \int_{]0,t]} Z_u \, dH_u.$$

Consider first a defaultable claim with recovery at maturity – that is, $DCT^1$. In the absence of the promised dividends (i.e., when $A \equiv 0$), the valuation formula (2.3) reduces to, for $t < T$,

$$X^{d,1}(t, T) := B_t \, \mathbb{E}_{\mathbb{P}^*} \left( B_T^{-1} X^{d,1}(T) \, \big| \, \mathcal{F}_t \right), \tag{2.4}$$

where the terminal payoff $X^{d,1}(T)$, which equals

$$X^{d,1}(T) = X \mathbb{1}_{\{\tau > T\}} + \tilde{X} \mathbb{1}_{\{\tau \leq T\}}, \tag{2.5}$$

represents the cash flow at time $T$ of a given defaultable claim with recovery at maturity. It is thus clear that, in the absence of promised dividends, the discounted price process $X^{d,1}(t, T)/B_t$, $t < T$, follows an $\mathbb{F}$-martingale under $\mathbb{P}^*$, provided, of course, that a usual integrability condition is imposed on $X^{d,1}(T)$. Under a set of technical assumptions, a suitable version of the martingale representation theorem with respect to the Brownian filtration will formally ensure the attainability of the terminal payoff $X^{d,1}(T)$.

We turn our attention to a defaultable claim with recovery at default (once more we assume that $A \equiv 0$). In this case, expression (2.3) defines only the pre-default value of a defaultable claim. Indeed, it is apparent that the value process $X^{d,2}(t, T)$ vanishes identically on the random interval $[\tau, T[$. Consequently, it is natural to expect that the discounted price process $X^{d,2}(t, T)/B_t$, $t < T$, will follow an $\mathbb{F}$-martingale under $\mathbb{P}^*$, only when we consider this process prior to the default time $\tau$. In the PDE approach, described in Sect. 2.2 below, this feature of the process $X^{d,2}(t, T)$ is dealt with through a judicious specification of boundary and terminal conditions.

Another possible solution would be to extend the process $X^{d,2}(t, T)$ on $[\tau, T[$ by assuming, for instance, that the recovery payoff $Z_\tau$ is invested in default-free zero-coupon bonds of maturity $T$. Under this convention, one finishes with a defaultable claim of the first type with $\tilde{X} = Z_\tau B^{-1}(\tau, T)$. It is clear that such a convention does not affect the valuation problem for $DCT^2$ when we are interested in valuing defaultable claim with recovery at default only strictly prior to the default time – that is, when we search for the *pre-default value* of a defaultable claim.

## 2.1.2 Self-Financing Trading Strategies

We are now going to provide a formal justification of Definition 2.1.2, based on the no-arbitrage arguments. We write $S^i$, $i = 1, \ldots, k$ to denote the price processes of $k$ primary securities in an arbitrage-free financial model. We make the standard assumption that the processes $S^i$, $i = 1, \ldots, k - 1$ follow semimartingales. In addition, we set $S_t^k = B_t$ so that $S^k$ represents the value process of the savings account. For the sake of convenience, we assume that $S^i$, $i = 1, \ldots, k - 1$ are non-dividend-paying assets, and we introduce the discounted price processes $\tilde{S}^i$ by setting $\tilde{S}_t^i = S_t^i / B_t$.

Let us now also assume that we have an additional security that pays dividends during its lifespan – assumed to be the time interval $[0, T]$ – according to a process of finite variation $D$, with $D_0 = 0$. Let $S^0$ denote the yet unspecified price process of this security. In particular, we refrain from postulating that $S^0$ follows a semimartingale. Of course, we do not necessarily need to interpret $S^0$ as the value process of a defaultable claim, though we have here this particular interpretation in mind.

Let an $\mathbb{F}$-predictable process $\phi = (\phi^0, \ldots, \phi^k)$ stand for a trading strategy. At this stage, it will be enough to examine a simple trading strategy involving a defaultable claim. In fact, since we do not assume a priori that $S^0$ follows a semimartingale, we are not yet in a position to consider general trading strategies involving the defaultable claim anyway.

Suppose that we purchase at time 0 one unit of the $0^{\text{th}}$ asset at the initial price $S_0^0$, we hold it until time $T$, and we invest all the proceeds from dividends in a savings account. More specifically, we consider a buy-and-hold strategy $\psi = (1, 0, \ldots, 0, \psi^k)$. The associated *wealth process* $U(\psi)$ equals

$$U_t(\psi) = S_t^0 + \psi_t^k B_t, \quad \forall t \in [0, T], \tag{2.6}$$

with some initial value $U_0(\psi) = S_0^0 + \psi_0^k$. We assume that the strategy $\psi$ introduced above is *self-financing*; i.e., we postulate that for every $t \in [0, T]$

$$U_t(\psi) - U_0(\psi) = S_t^0 - S_0^0 + D_t + \int_{]0,t]} \psi_u^k \, dB_u. \tag{2.7}$$

**Lemma 2.1.1.** *The discounted wealth $\tilde{U}_t(\psi) = B_t^{-1} U_t(\psi)$ of a self-financing trading strategy $\psi$ satisfies, for every $t \in [0, T]$,*

$$\tilde{U}_t(\psi) = \tilde{U}_0(\psi) + \tilde{S}_t^0 - \tilde{S}_0^0 + \int_{]0,t]} B_u^{-1} dD_u. \tag{2.8}$$

*Proof.* We define an auxiliary process $\hat{U}(\psi)$ by setting $\hat{U}_t(\psi) := U_t(\psi) - S_t^0 = \psi_t^k B_t$. In view of (2.7), we have

$$\hat{U}_t(\psi) = \hat{U}_0(\psi) + D_t + \int_{]0,t]} \psi_u^k \, dB_u,$$

and so the process $\hat{U}(\psi)$ follows a semimartingale.

An application of Itô's product rule yields

$$
\begin{aligned}
d\big(B_t^{-1}\hat{U}_t(\psi)\big) &= B_t^{-1}d\hat{U}_t(\psi) + \hat{U}_t(\psi)\, dB_t^{-1} \\
&= B_t^{-1}dD_t + \psi_t^k B_t^{-1}dB_t + \psi_t^k B_t\, dB_t^{-1} \\
&= B_t^{-1}dD_t,
\end{aligned}
$$

where we have used the obvious equality $B_t^{-1}dB_t + B_t\, dB_t^{-1} = 0$. Integrating the last equality, we obtain

$$
B_t^{-1}\big(U_t(\psi) - S_t^0\big) = B_0^{-1}\big(U_0(\psi) - S_0^0\big) + \int_{]0,t]} B_u^{-1}dD_u,
$$

and this immediately yields (2.8). $\qquad\square$

In view of Lemma 2.1.1, for every $t \in [0,T]$ we also have:

$$
\tilde{U}_T(\psi) - \tilde{U}_t(\psi) = \tilde{S}_T^0 - \tilde{S}_t^0 + \int_{]t,T]} B_u^{-1}\, dD_u. \tag{2.9}
$$

### 2.1.3 Martingale Measures

We are ready to derive the risk-neutral valuation formula for the ex-dividend price $S_t^0$. To this end, we assume that our model is arbitrage-free, meaning here that it admits a (not necessarily unique) spot martingale measure $\mathbb{P}^*$ equivalent to $\mathbb{P}$. In particular, this implies that the discounted price $\tilde{S}^i$ of any non-dividend paying primary security, as well as the discounted wealth process $\tilde{U}(\phi)$ of any *admissible* self-financing trading strategy $\phi = (0, \phi^1, \dots, \phi^k)$, follow martingales under $\mathbb{P}^*$. In addition, we postulate that the trading strategy $\psi$ introduced in Sect. 2.1.2 is also *admissible*, so that the discounted wealth process $\tilde{U}(\psi)$ follows a $\mathbb{P}^*$-martingale with respect to the filtration $\mathbb{F}$.

We make an assumption that the market value at time $t$ of the $0^{\text{th}}$ security comes exclusively from the future dividends stream; this amounts to postulate that $S_T^0 = \tilde{S}_T^0 = 0$. In view of this convention, we shall refer to $S^0$ as the *ex-dividend price* of the $0^{\text{th}}$ asset, e.g., a defaultable claim.

**Proposition 2.1.1.** *The ex-dividend price process $S^0$ satisfies, for $t \in [0,T]$,*

$$
S_t^0 = B_t\, \mathbb{E}_{\mathbb{P}^*}\left( \int_{]t,T]} B_u^{-1}\, dD_u \,\Big|\, \mathcal{F}_t \right). \tag{2.10}
$$

*Proof.* In view of the martingale property of the discounted wealth process $\tilde{U}(\psi)$, for any $t \in [0,T]$ we have

$$
\mathbb{E}_{\mathbb{P}^*}\big(\tilde{U}_T(\psi) - \tilde{U}_t(\psi)\,\big|\, \mathcal{F}_t\big) = 0.
$$

Taking into account (2.9), we thus obtain

$$
\tilde{S}_t^0 = \mathbb{E}_{\mathbb{P}^*}\left( \tilde{S}_T^0 + \int_{]t,T]} B_u^{-1}\, dD_u \,\Big|\, \mathcal{F}_t \right).
$$

Since by assumption $S_T^0 = \tilde{S}_T^0 = 0$, the last formula yields (2.10). $\qquad\square$

Let us now examine a general trading strategy $\phi = (\phi^0, \ldots, \phi^k)$. The associated *wealth process* $U(\phi)$ equals $U_t(\phi) = \sum_{i=0}^{k} \phi_t^i S_t^i$. A strategy $\phi$ is said to be *self-financing* if $U_t(\phi) = U_0(\phi) + G_t(\phi)$ for every $t \in [0, T]$, where the *gains process* $G(\phi)$ is defined as follows:

$$G_t(\phi) := \int_{]0,t]} \phi_u^0 \, dD_u + \sum_{i=0}^{k} \int_{]0,t]} \phi_u^i \, dS_u^i.$$

**Corollary 2.1.1.** *For any self-financing trading strategy $\phi$, the discounted wealth process $\tilde{U}(\phi) := B_t^{-1} U_t(\phi)$ follows a local martingale under $\mathbb{P}^*$.*

*Proof.* Since $B$ is a continuous process of finite variation, Itô's product rule gives

$$d\tilde{S}_t^i = S_t^i \, dB_t^{-1} + B_t^{-1} dS_t^i$$

for $i = 0, \ldots, k$, and so

$$
\begin{aligned}
d\tilde{U}_t(\phi) &= U_t(\phi) \, dB_t^{-1} + B_t^{-1} dU_t(\phi) \\
&= U_t(\phi) \, dB_t^{-1} + B_t^{-1} \left( \sum_{i=0}^{k} \phi_t^i \, dS_t^i + \phi_t^0 \, dD_t \right) \\
&= \sum_{i=0}^{k} \phi_t^i \left( S_t^i \, dB_t^{-1} + B_t^{-1} dS_t^i \right) + \phi_t^0 B_t^{-1} dD_t \\
&= \sum_{i=1}^{k-1} \phi_t^i \, d\tilde{S}_t^i + \phi_t^0 \left( d\tilde{S}_t^0 + B_t^{-1} dD_t \right) = \sum_{i=1}^{k-1} \phi_t^i \, d\tilde{S}_t^i + \phi_t^0 \, d\hat{S}_t^0,
\end{aligned}
$$

where the process $\hat{S}^0$ is given by the formula

$$\hat{S}_t^0 := \tilde{S}_t^0 + \int_{]0,t]} B_u^{-1} \, dD_u.$$

To conclude, it suffices to observe that in view of (2.10) the process $\hat{S}^0$ satisfies

$$\hat{S}_t^0 = \mathbb{E}_{\mathbb{P}^*} \left( \int_{]0,T]} B_u^{-1} \, dD_u \, \Big| \, \mathcal{F}_t \right),$$

and thus it follows a martingale under $\mathbb{P}^*$.    □

*Remarks.* (i) It is worth noticing that $\hat{S}_t^0$ represents the discounted cum-dividend price at time $t$ of the $0^{\text{th}}$ asset.

(ii) Under the assumption of uniqueness of a spot martingale measure $\mathbb{P}^*$, any $\mathbb{P}^*$-integrable contingent claim is attainable, and the valuation formula can be justified by means of replication. Otherwise – that is, when a martingale probability measure is not unique – the right-hand side of (2.10) may depend on the choice of a particular martingale probability. In this case, a process defined by (2.10) for an arbitrarily chosen spot martingale measure $\mathbb{P}^*$ can be taken as the no-arbitrage price process of a defaultable claim.

## 2.2 PDE Approach

The aim of this section is to provide mathematical tools for valuing and hedging defaultable claims according to the various structural theories. We now make the following, rather stringent, assumption that will allow for the no-arbitrage arguments to be used. The discussion of the practical validity of Assumption (A.1) is postponed to Sect. 2.4.3. Let us only mention here that the assumption that the firm's asset represent a tradeable security is almost unanimously seen as a major drawback of the value-of-the-firm approach.

**Assumption (A.1)** The firm's assets, default-free zero-coupon bonds, as well as the defaultable claims are traded securities. The trading is frictionless (no transaction costs, no taxes, infinite divisibility of assets, etc.) and it takes place continuously in time.

The next two assumptions set up a strong Markov diffusion model of the combined dynamics for the short-term interest process and the value process of the firm's assets under the spot martingale measure, denoted by $\mathbb{P}^*$ in what follows. Versions of this model underly most structural models discussed hereafter. For the sake of convenience, we fix the horizon date $T > 0$. Such a convention is not restrictive, unless someone considers perpetual contingent claims.

**Assumption (A.2)** The risk-neutral dynamics of the short-term interest rate process $r_t$, $t \geq 0$, are given as:

$$dr_t = \mu_r(r_t, t)\, dt + \sigma_r(r_t, t)\, d\tilde{W}_t, \quad r_0 > 0, \tag{2.11}$$

where $\tilde{W}$ is a standard Brownian motion under the probability measure $\mathbb{P}^*$, with respect to the reference filtration $\mathbb{F}$. We shall assume that the coefficients $\mu_r : \mathbb{R} \times [0, T] \to \mathbb{R}$ and $\sigma_r : \mathbb{R} \times [0, T] \to \mathbb{R}$ are sufficiently regular deterministic functions, so that the SDE (2.11) admits a unique, global, strong solution $r_t$, $t \in [0, T]$, for any initial condition $r_0 \in \mathbb{R}_+$.

*Remarks.* Note that the market price of interest rate risk is already embedded in the drift coefficient $\mu_r$. To implement the valuation results provided below, one would need to estimate the function $\mu_r$ and thus, perhaps, to estimate the market price of interest rate risk. Kim et al. (1993a) set the market price of interest rate risk to zero in their model, and report that their results are not sensitive to this specification. In several other structural models, both coefficients $\mu_r$ and $\sigma_r$ are set to zero, so that the short-term rate $r$ is constant.

The next assumption postulates a diffusion-type dynamics for the process $V$ that models the total value of the firm's assets. Notice that in formula (2.12) below, the coefficient $\kappa$ represents a *payout ratio*; it is aimed to represent the net total payout made (or inflow received) by the firm. The appearance of $r_t$ in the drift term in (2.12) is thus an immediate consequence of Assumption (A.1) combined with standard no-arbitrage arguments.

**Assumption (A.3)** The risk-neutral dynamics of the process $V$ are given by

$$\frac{dV_t}{V_t} = \left(r_t - \kappa(V_t, r_t, t)\right) dt + \sigma_V(V_t, t) \, dW_t^* \tag{2.12}$$

with $V_0 > 0$, where $W^*$ follows a standard Brownian motion under $\mathbb{P}^*$ with respect to $\mathbb{F}$. We assume that the functions $\kappa : \mathbb{R}_+ \times \mathbb{R} \times [0, T] \to \mathbb{R}$ and $\sigma_V : \mathbb{R}_+ \times [0, T] \to \mathbb{R}$ are sufficiently regular (e.g., globally Lipschitz) to guarantee the existence of a unique, strong, global solution to the stochastic differential equation (2.12).

For the sake of convenience, we shall postulate that the process $V$ is non-negative. In addition, the instantaneous correlation coefficient between the Brownian motions $\tilde{W}$ and $W^*$ is assumed to be constant; we henceforth denote it by $\rho_{Vr}$.

Process $V$, following dynamics like (2.12), should rather be considered as a 'default-free' process that would be followed by the total value of the firm's assets if there were no default covenants in place, and should be more appropriately referred to as the process of the 'pre-default total value of the firm's assets.' Because of various default covenants, which will be later described in detail, the stochastic process of the total value of the firm's assets follows the dynamics (2.12) only up to the default time, at which time the characteristics of the process change. After discounting, an appropriately modified gains process corresponding to $V$, typically would not follow a $\mathbb{P}^*$-martingale with respect to the filtration $\mathbb{F}$. Nevertheless, most structural models studied in literature consider the process $V$ with unrestricted dynamics like (2.12), or the jump-diffusion versions of these, as a process modeling the total value of the firm's assets. The corresponding discounted gains process obviously satisfies the martingale property under the spot martingale measure $\mathbb{P}^*$ (cf. Sect. 6.2 in Musiela and Rutkowski (1997a)) and is considered an underlying traded security in the majority of papers devoted to the structural approach. We shall follow this convention in this chapter.

In order to apply the PDE approach, we need to impose some technical assumptions regarding the recovery rules, the promised dividends process $A$, and the default time $\tau$. Let us emphasize that Assumptions (A.4)–(A.6) are also postulated in most of literature dealing with structural models.

**Assumption (A.4)** The promised contingent claim $X$, the recovery payoff $\tilde{X}$ and the recovery process $Z$ satisfy

$$X = g(V_T, r_T), \quad \tilde{X} = h(V_T, r_T), \quad Z_t = z(V_t, r_t, t), \quad \forall t \in [0, T],$$

for some measurable functions $g, h : \mathbb{R}_+ \times \mathbb{R} \to \mathbb{R}$ and $z : \mathbb{R}_+ \times \mathbb{R} \times [0, T] \to \mathbb{R}$.

**Assumption (A.5)** The promised dividends process $A$ is given as:

$$A_t = \int_0^t c(V_u, r_u, u) \, du, \quad \forall t \in [0, T],$$

for some integrable coupon-rate function $c : \mathbb{R}_+ \times \mathbb{R} \times [0, T] \to \mathbb{R}$.

**Assumption (A.6)** The default triggering barrier process $v$ equals

$$v_t = \bar{v}(V_t, r_t, t), \quad \forall t \in [0, T],$$

for some measurable function $\bar{v} : \mathbb{R}_+ \times \mathbb{R} \times [0, T] \to \mathbb{R}$.

The next lemma is a consequence of the strong Markov property of the two-dimensional diffusion process $(r, V)$.

**Lemma 2.2.1.** *Under Assumptions* (A.1)–(A.6) *we have*

$$X^d(t, T) = u(V_t, r_t, t), \quad \forall t \in [0, T], \tag{2.13}$$

*for some measurable function* $u : \mathbb{R}_+ \times \mathbb{R} \times [0, T] \to \mathbb{R}$.

*Proof.* In view of the risk-neutral valuation formula (2.3) and Assumptions (A.4)–(A.5), it is clear that the price process $X^{d,1}(\cdot, T)$ may be represented as follows (recall that $\tau$ is an $\mathbb{F}$-stopping time):

$$X^{d,1}(t, T) = B_t \, \mathbb{E}_{\mathbb{P}^*} \left( \mathbb{1}_{\{\tau > t\}} \int_{]t, T]} B_u^{-1} dD_u^1 \,\Big|\, \mathcal{F}_t \right)$$

$$+ B_t \, \mathbb{E}_{\mathbb{P}^*} \left( \mathbb{1}_{\{\tau \le t\}} \int_{]t, T]} B_u^{-1} dD_u^1 \,\Big|\, \mathcal{F}_t \right)$$

$$= B_t \, \mathbb{E}_{\mathbb{P}^*} \left( B_T^{-1} \big( X \mathbb{1}_{\{\tau > T\}} + \tilde{X} \mathbb{1}_{\{\tau \le T\}} \big) \,\Big|\, \mathcal{F}_t \right) \mathbb{1}_{\{\tau > t\}}$$

$$+ B_t \, \mathbb{E}_{\mathbb{P}^*} \left( B_T^{-1} \tilde{X} \mathbb{1}_{\{\tau \le T\}} \,\Big|\, \mathcal{F}_t \right) \mathbb{1}_{\{\tau \le t\}}$$

$$+ B_t \, \mathbb{E}_{\mathbb{P}^*} \left( \int_t^{T \wedge \tau} B_u^{-1} c(V_u, r_u, u) \, du \,\Big|\, \mathcal{F}_t \right) \mathbb{1}_{\{\tau > t\}}.$$

Consequently, on the set $\{\tau > t\}$, i.e., prior to default, we have:

$$X^{d,1}(t, T) = \mathbb{E}_{\mathbb{P}^*} \left( e^{-\int_t^T r_u du} g(V_T, r_T) \mathbb{1}_{\{\tau > T\}} \,\Big|\, \mathcal{F}_t \right)$$

$$+ \mathbb{E}_{\mathbb{P}^*} \left( e^{-\int_t^T r_u du} h(V_T, r_T) \mathbb{1}_{\{\tau \le T\}} \,\Big|\, \mathcal{F}_t \right)$$

$$+ \mathbb{E}_{\mathbb{P}^*} \left( \int_t^{T \wedge \tau} e^{-\int_t^u r_s ds} c(V_u, r_u, u) \, du \,\Big|\, \mathcal{F}_t \right).$$

After default, i.e., on the set $\{\tau \le t\}$, the price of a defaultable claim represents the current value of the recovery payment $\tilde{X}$, namely

$$X^{d,1}(t, T) = \mathbb{E}_{\mathbb{P}^*} \left( e^{-\int_t^T r_u du} h(V_T, r_T) \mathbb{1}_{\{\tau \le T\}} \,\Big|\, \mathcal{F}_t \right).$$

Using (2.1), Assumption (A.6), and the strong Markov property of the process $(V, r)$, we conclude that

$$X^{d,1}(t, T) = u^{(1)}(V_t, r_t, t), \quad \forall t \in [0, T], \tag{2.14}$$

for some measurable function $u^{(1)}$ defined on $\mathbb{R}_+ \times \mathbb{R} \times [0, T]$.

Likewise, the price process $X^{d,2}(\cdot, T)$ satisfies

$$
X^{d,2}(t,T) = B_t \, \mathbb{E}_{\mathbb{P}^*} \left( \mathbb{1}_{\{\tau > t\}} \int_{]t,T]} B_u^{-1} dD_u^2 \,\Big|\, \mathcal{F}_t \right)
$$

$$
+ B_t \, \mathbb{E}_{\mathbb{P}^*} \left( \mathbb{1}_{\{\tau \le t\}} \int_{]t,T]} B_u^{-1} dD_u^2 \,\Big|\, \mathcal{F}_t \right)
$$

$$
= B_t \, \mathbb{E}_{\mathbb{P}^*} \left( B_T^{-1} X \mathbb{1}_{\{\tau > T\}} + B_\tau^{-1} Z_\tau \mathbb{1}_{\{\tau \le T\}} \,\Big|\, \mathcal{F}_t \right) \mathbb{1}_{\{\tau > t\}}
$$

$$
+ B_t \, \mathbb{E}_{\mathbb{P}^*} \left( \int_t^{T \wedge \tau} B_u^{-1} c(V_u, r_u, u) \, du \,\Big|\, \mathcal{F}_t \right) \mathbb{1}_{\{\tau > t\}}.
$$

Thus, $X^{d,2}(t,T) = 0$ on the set $\{\tau \le t\}$, and

$$
X^{d,2}(t,T) = \mathbb{E}_{\mathbb{P}^*} \left( e^{-\int_t^T r_u \, du} g(V_T, r_T) \mathbb{1}_{\{\tau > T\}} \,\Big|\, \mathcal{F}_t \right)
$$

$$
+ \mathbb{E}_{\mathbb{P}^*} \left( e^{-\int_t^\tau r_u \, du} z(V_\tau, r_\tau, \tau) \mathbb{1}_{\{\tau \le T\}} \,\Big|\, \mathcal{F}_t \right)
$$

$$
+ \mathbb{E}_{\mathbb{P}^*} \left( \int_t^{T \wedge \tau} e^{-\int_t^u r_s \, ds} c(V_u, r_u, u) \, du \,\Big|\, \mathcal{F}_t \right)
$$

on the set $\{\tau > t\}$. Invoking again (2.1), Assumption (A.6) and the strong Markov property of $(V, r)$, we obtain

$$
X^{d,2}(t,T) = u^{(2)}(V_t, r_t, t), \quad t \in [0,T], \tag{2.15}
$$

for some function $u^{(2)}$ defined on $\mathbb{R}_+ \times \mathbb{R} \times [0,T]$. The general case can be treated along similar lines. $\qquad\square$

Our next goal is to derive a partial differential equation (PDE, for short) that would allow us to find explicitly, at least in some cases, the 'pricing function' $u$. As expected, the basic tool we shall use here is the Itô formula. Therefore, we shall henceforth assume that the function $u$ is smooth enough; in particular, we shall write $u_t$, $u_V$, $u_r$, $u_{VV}$, $u_{rr}$ and $u_{Vr}$ to denote first and second order partial derivatives of the function $u = u(V, r, t)$. Applying Itô's rule to the process $u(V_t, r_t, t)$ and using assumptions (2.11)–(2.12), we obtain the following expression for the Itô differential of the process $X^d(t, T)$:

$$
dX^d(t,T) = du(V_t, r_t, t) = \mu_X(t) \, dt + \sigma_{X,V}(t) \, dW_t^* + \sigma_{X,r}(t) \, d\tilde{W}_t, \tag{2.16}
$$

where

$$
\mu_X(t) = u_V(V_t, r_t, t)(r_t - \kappa(V_t, r_t, t))V_t + u_r(V_t, r_t, t)\mu_r(r_t, t)
$$
$$
+ \tfrac{1}{2} u_{VV}(V_t, r_t, t)\sigma_V^2(V_t, t)V_t^2 + \tfrac{1}{2} u_{rr}(V_t, r_t, t)\sigma_r^2(r_t, t) \tag{2.17}
$$
$$
+ u_{Vr}(V_t, r_t, t)\sigma_V(V_t, t)\sigma_r(r_t, t)V_t\rho_{Vr} + u_t(V_t, r_t, t)
$$

and

$$
\begin{cases} \sigma_{X,V}(t) = u_V(V_t, r_t, t)\sigma_V(V_t, t)V_t, \\ \sigma_{X,r}(t) = u_r(V_t, r_t, t)\sigma_r(r_t, t). \end{cases} \tag{2.18}
$$

It should be stressed that the right-hand side of (2.16) is valid for every $t \in [0, T]$ in the case of a claim with recovery at maturity (i.e., when $i = 1$). On the other hand, it holds for $t \in [0, \tau \wedge T]$ only, when we consider a claim with recovery at default (that is, for $i = 2$).

Consider now a unit default-free zero-coupon bond maturing at $T$. The price at time $t$ of such a bond is given as

$$B(t,T) = \mathbb{E}_{\mathbb{P}^*}\left(e^{-\int_t^T r_u \, du} \,\Big|\, \mathcal{F}_t\right) = B_t \, \mathbb{E}_{\mathbb{P}^*}\left(B_T^{-1} \,\big|\, \mathcal{F}_t\right).$$

Given our Markovian assumptions, we have $B(t,T) = p(r_t, t, T)$ for some function $p(r, t, T)$, where $r \in \mathbb{R}$ and $t \in [0, T]$. Assuming that the function $p$ is sufficiently smooth in the first two arguments, we conclude that the price process $B(t,T)$ has the following dynamics under $\mathbb{P}^*$ (see, e.g., Musiela and Rutkowski (1997a))

$$\frac{dB(t,T)}{B(t,T)} = r_t \, dt + \sigma_B(r_t, t, T) \, d\tilde{W}_t,$$

where

$$\sigma_B(r_t, t, T) = p_r(r_t, t, T)\sigma_r(r_t, t). \tag{2.19}$$

*Remarks.* Under Markovian assumptions, the price at time $t$ of a default-free discount bond of a given maturity date $T$ is well known to be a function of the two state variables: $r_t$ and $t$. Equation (2.13) shows that, within the framework of structural models satisfying Assumptions (A.4)–(A.6), the price of a corporate bond is a function of the following three state variables: the firm's value $V_t$, the short-term rate $r_t$ and the running time $t$.

### 2.2.1 PDE for the Value Function

Our next aim is to derive the pricing PDE that need to be obeyed by the function $u$ in (2.13). To this end, we shall proceed with a formal analysis involving the Itô formula and the arbitrage argument (our derivation parallels the one in Merton (1974) or Brennan and Schwartz (1980)). We shall examine a self-financing trading strategy $\phi_t = (\phi_t^0, \phi_t^1, \phi_t^2, \phi_t^3)$ with the value process $U(\phi)$ given by the equality:

$$U_t(\phi) = \phi_t^0 X^d(t, T) + \phi_t^1 V_t + \phi_t^2 B(t, T) + \phi_t^3 B_t \tag{2.20}$$

for every $t \in [0, T]$. Since a strategy $\phi$ is self-financing, we also have[1]

$$dU_t(\phi) = \phi_t^0 \left(dX^d(t, T) + c(V_t, r_t, t) \, dt\right) \\ + \phi_t^1 \left(dV_t + \kappa(V_t, r_t, t)V_t \, dt\right) + \phi_t^2 \, dB(t, T) + \phi_t^3 \, dB_t. \tag{2.21}$$

---

[1] Recall that the defaultable claim pays continuously a coupon rate $c(V_t, r_t, t)$, and the firm is assumed to pay continuously cash flows at a rate $\kappa(V_t, r_t, t)$; the latter cash flows are assumed to be proportional to the current value of the firm.

The following result gives a PDE that is obeyed by the pricing function, as well as a generic expression for the replicating strategy. By solving this PDE, subject to appropriate terminal and boundary conditions, one can verify that a given defaultable claim is attainable and find its arbitrage price. We shall rarely rely on the PDE approach in the present text, though.

**Proposition 2.2.1.** *Suppose that the function $u$ in (2.13) belongs to the class $C^{2,2,1}(\mathbb{R}_+ \times \mathbb{R} \times [0,T])$. Then $u$ obeys the fundamental PDE*[2]

$$
\begin{aligned}
&u_t(V,r,t) + (r - \kappa(V,r,t))Vu_V(V,r,t) + \mu_r(r,t)u_r(V,r,t) \\
&+ \tfrac{1}{2}\sigma_V^2(V,t)V^2 u_{VV}(V,r,t) + \tfrac{1}{2}\sigma_r^2(r,t)u_{rr}(V,r,t) \\
&+ \sigma_V(V,t)\sigma_r(r,t)V\rho_{Vr}u_{Vr}(V,r,t) + c(V,r,t) - ru(V,r,t) = 0.
\end{aligned}
\tag{2.22}
$$

*The replicating self-financing strategy for the defaultable claim satisfies*

$$
\phi_t^1 = u_V(V_t, r_t, t), \qquad \phi_t^2 = \frac{u_r(V_t, r_t, t)}{B(t,T)p_r(r_t, t, T)},
$$

*and*

$$
\phi_t^3 = B_t^{-1}\left(u(V_t, r_t, t) - u_V(V_t, r_t, t)V_t - \frac{u_r(V_t, r_t, t)}{p_r(r_t, t, T)}\right).
$$

*Proof.* For arbitrary processes $\phi_t^0, \phi_t^1, \phi_t^2$, we may choose $\phi_t^3$ in such a way that $\phi$ is self-financing and $U(\phi) \equiv 0$. In particular, we necessarily have

$$
\phi_t^3 = -B_t^{-1}\big(\phi_t^0 X^d(t,T) + \phi_t^1 V_t + \phi_t^2 B(t,T)\big).
$$

Combining the last formula with (2.21) and using the equality $dB_t = r_t B_t\, dt$, we obtain an equivalent form of the self-financing condition (2.21)

$$
\begin{aligned}
&\phi_t^0\big(dX^d(t,T) + c(V_t, r_t, t)\, dt\big) + \phi_t^1\big(dV_t + \kappa(V_t, r_t, t)V_t\, dt\big) \\
&+ \phi_t^2\, dB(t,T) - r_t\big(\phi_t^0 X^d(t,T) + \phi_t^1 V_t + \phi_t^2 B(t,T)\big)\, dt = 0,
\end{aligned}
$$

where, in view of (2.12), we have

$$
dV_t + \kappa(V_t, r_t, t)V_t\, dt = V_t\big(r_t\, dt + \sigma_V(V_t, t)\, dW_t^*\big).
$$

Since we wish to replicate a defaultable claim, it is convenient to take $\phi^0 \equiv -1$. Then, using (2.16), we obtain

$$
\begin{aligned}
&-\mu_X(t)\, dt - \sigma_{X,V}(t)\, dW_t^* - \sigma_{X,r}(t)\, d\tilde{W}_t - c(V_t, r_t, t)\, dt \\
&+ \phi_t^1\, V_t\big(r_t dt + \sigma_V(V_t, t)\, dW_t^*\big) + \phi_t^2 B(t,T)\big(r_t\, dt + \sigma_B(r_t, t, T)\, d\tilde{W}_t\big) \\
&- r_t\big(-X^d(t,T) + \phi_t^1 V_t + \phi_t^2 B(t,T)\big)\, dt = 0,
\end{aligned}
$$

which in turn yields

---

[2] Let us stress that the domain on which the fundamental PDE is satisfied depends on whether a claim is settled at maturity or at default time.

$$-\mu_X(t)\,dt - \sigma_{X,V}(t)\,dW_t^* - \sigma_{X,r}(t)\,d\tilde{W}_t - c(V_t, r_t, t)\,dt$$
$$+ \phi_t^1 V_t \sigma_V(V_t, t)\,dW_t^* + \phi_t^2 B(t, T)\sigma_B(r_t, t, T)\,d\tilde{W}_t + r_t X^d(t, T)\,dt = 0.$$

The next step is to further specify processes $\phi^1$ and $\phi^2$ in such a way that the martingale components in the last formula vanish. To this end, we set (cf. (2.18))

$$\phi_t^1 V_t \sigma_V(V_t, t) = \sigma_{X,V}(t) = u_V(V_t, r_t, t)V_t \sigma_V(V_t, t), \tag{2.23}$$

and

$$\phi_t^2 B(t, T)\sigma_B(r_t, t, T) = \sigma_{X,r}(t) = u_r(V_t, r_t, t)\sigma_r(r_t, t),$$

where, by virtue of (2.19),

$$\sigma_B(r_t, t, T) = p_r(r_t, t, T)\sigma_r(r_t, t).$$

Put more explicitly, we have $\phi_t^1 = u_V(V_t, r_t, t)$ and

$$\phi_t^2 = \frac{u_r(V_t, r_t, t)}{B(t, T)p_r(r_t, t, T)}.$$

We end up with the following equality, in the integrated form,

$$\int_0^t \left(\mu_X(s) + c(V_s, r_s, s) - r_s u(V_s, r_s, s)\right)ds = 0. \tag{2.24}$$

Combining (2.24) with (2.17), we obtain

$$u_t(V_t, r_t, t) + (r - \kappa(V_t, r_t, t))V_t u_V(V_t, r_t, t) + \mu_r(r_t, t)u_r(V_t, r_t, t)$$
$$+ \tfrac{1}{2}\sigma_V^2(V_t, t)V_t^2 u_{VV}(V_t, r_t, t) + \tfrac{1}{2}\sigma_r^2(r_t, t)u_{rr}(V_t, r_t, t)$$
$$+ \sigma_V(V_t, t)\sigma_r(r_t, t)V_t \rho_{Vr} u_{Vr}(V_t, r_t, t) + c(V_t, r_t, t) - r_t u(V_t, r_t, t) = 0.$$

The last equality holds when the function $u$ obeys the fundamental PDE (2.22). The form of the replicating strategy is also clear. $\qquad\square$

If the running time $t$ is changed to $\mathbf{t} := T - t$, which represents time to the claim's maturity, and the model is time-homogeneous, the PDE (2.22) takes the following form

$$-u_{\mathbf{t}}(V, r, \mathbf{t}) + (r - \kappa(V, r))V u_V(V, r, \mathbf{t}) + \mu_r(r)u_r(V, r, \mathbf{t})$$
$$+ \tfrac{1}{2}\sigma_V^2(V)V^2 u_{VV}(V, r, \mathbf{t}) + \tfrac{1}{2}\sigma_r^2(r)u_{rr}(V, r, \mathbf{t})$$
$$+ \sigma_V(V)\sigma_r(r)V \rho_{Vr} u_{Vr}(V, r, \mathbf{t}) + c(V, r) - ru(V, r, \mathbf{t}) = 0.$$

Notice that the fundamental PDE reduces the classic Black-Scholes PDE

$$u_t(V, t) + (r - \kappa(V, t))V u_V(V, t) + \tfrac{1}{2}\sigma_V^2(V, t)V^2 u_{VV}(V, t) - ru(V, t) = 0$$

when we set $\mu_r(r, t) = \sigma_r(r, t) = c(V, r, t) = 0$ and suppress in the notation the dependence on the (constant) interest rate $r$. It also gives the PDE for the arbitrage price of a default-free interest-rate sensitive security

$$u_t(r, t) + \mu_r(r, t)u_r(r, t) + \tfrac{1}{2}\sigma_r^2(r, t)u_{rr}(r, t) + c(r, t) - ru(r, t) = 0,$$

provided that the security in question does not depend on the process $V$.

### 2.2.2 Corporate Zero-Coupon Bonds

Assume that $A \equiv 0$ and $X = L$ for some constant $L > 0$. Then the process given by (2.4) may be seen as the arbitrage price of a defaultable (corporate) zero-coupon bond with the face value $L$. The price $D(t, T)$ of such a bond equals

$$D(t, T) = B_t \, \mathbb{E}_{\mathbb{P}^*} \big( B_T^{-1}(L\mathbb{1}_{\{\tau > T\}} + \tilde{X}\mathbb{1}_{\{\tau \leq T\}}) \,|\, \mathcal{F}_t \big).$$

It is convenient to rewrite the last formula as follows:

$$D(t, T) = LB_t \, \mathbb{E}_{\mathbb{P}^*} \big( B_T^{-1}(\mathbb{1}_{\{\tau > T\}} + \delta(T)\mathbb{1}_{\{\tau \leq T\}}) \,|\, \mathcal{F}_t \big), \qquad (2.25)$$

where the random variable $\delta(T) = \tilde{X}/L$ represents the *recovery rate upon default*. It is natural to assume that $0 \leq \tilde{X} \leq L$ so that $0 \leq \delta(T) \leq 1$; this assumption is not essential, though. Alternatively, we may re-express the bond price as follows:

$$D(t, T) = L\big(B(t, T) - B_t \, \mathbb{E}_{\mathbb{P}^*}(B_T^{-1} w(T)\mathbb{1}_{\{\tau \leq T\}} \,|\, \mathcal{F}_t)\big), \qquad (2.26)$$

where $B(t, T) := B_t \, \mathbb{E}_{\mathbb{P}^*}(B_T^{-1} \,|\, \mathcal{F}_t)$ denotes the price of a unit default-free zero-coupon bond, and $w(T) := 1 - \delta(T)$ is the so-called *writedown rate upon default*. As apparent from (2.25)–(2.26), the value of a corporate bond depends on the joint probability distribution under $\mathbb{P}^*$ of the three-dimensional random variable $(B_T, \delta(T), \tau)$ or, equivalently, $(B_T, w(T), \tau)$.

*Example 2.2.1.* Merton's (1974) model (see Sect. 2.3) postulates that the recovery payoff upon default equals $\tilde{X} = V_T$, where the random variable $V_T$ represents the value of the firm at time $T$. Consequently, the random recovery rate equals $\delta(T) = V_T/L$, and the writedown rate equals $w(T) = 1 - V_T/L$.

**Case of non-random interest rates.** Assume that the savings account $B$ is non-random. Then the price of a default-free zero-coupon bond equals $B(t, T) = B_t B_T^{-1}$, and thus $D(t, T) = L_t(1 - w^*(t, T))$, where $L_t = LB(t, T)$ is the present value of future liabilities, and $w^*(t, T)$ is the *conditional expected writedown rate* under $\mathbb{P}^*$, specifically,

$$w^*(t, T) := \mathbb{E}_{\mathbb{P}^*} \big( w(T)\mathbb{1}_{\{\tau \leq T\}} \,|\, \mathcal{F}_t \big).$$

The *conditional expected writedown rate upon default* equals, under $\mathbb{P}^*$,

$$w_t^* := \frac{\mathbb{E}_{\mathbb{P}^*} \big( w(T)\mathbb{1}_{\{\tau \leq T\}} \,|\, \mathcal{F}_t \big)}{\mathbb{P}^*\{\tau \leq T \,|\, \mathcal{F}_t\}} = \frac{w^*(t, T)}{p_t^*},$$

where $p_t^* := \mathbb{P}^*\{\tau \leq T \,|\, \mathcal{F}_t\}$ is the *conditional risk-neutral probability of default*. Finally, let $\delta_t^* := 1 - w_t^*$ be the *conditional expected recovery rate upon default* under $\mathbb{P}^*$. In terms of $p_t^*, \delta_t^*$ and $p_t^*$, we obtain

$$D(t, T) = L_t(1 - p_t^*) + L_t p_t^* \delta_t^* = L_t(1 - p_t^* w_t^*).$$

If the random variables $w(T)$ and $\tau$ are conditionally independent with respect to the $\sigma$-field $\mathcal{F}_t$ under $\mathbb{P}^*$, then we have $w_t^* = \mathbb{E}_{\mathbb{P}^*}(w(T) \,|\, \mathcal{F}_t)$.

*Example 2.2.2.* Let the recovery rate $\delta(T)$ be constant; $\delta(T) = \delta$ for some real number $\delta$ (so that the writedown rate $w(T) = w := 1 - \delta$ is non-random as well). Then $w^*(t, T) = wp_t^*$ and $w_t^* = w$ for every $0 \le t \le T$. Furthermore,

$$D(t, T) = L_t(1 - p_t^*) + \delta L_t p_t^* = L_t(1 - wp_t^*).$$

**Case of random interest rates.** We return to the general case of random interest rates. We denote by $\mathbb{P}_T$ the *forward martingale measure* for the date $T$, associated with the spot martingale measure $\mathbb{P}^*$. Recall that, for a fixed $T > 0$, the probability measure $\mathbb{P}_T$, equivalent to $\mathbb{P}^*$ on $(\Omega, \mathcal{F}_T)$, is specified by its Radon-Nikodým density:[3]

$$\frac{d\mathbb{P}_T}{d\mathbb{P}^*} = \frac{1}{B(0, T)B_T}, \quad \mathbb{P}^*\text{-a.s.}$$

It is known that the price process of any tradeable asset, as well as the wealth process of any self-financing trading strategy, follow (local) martingales under $\mathbb{P}_T$, when discounted by the bond price $B(t, T)$. Using, for instance, Lemma 13.2.3 in Musiela and Rutkowski (1997a), we obtain the following representation for the price $D(t, T)$ of a $T$-maturity defaultable bond:

$$D(t, T) = L_t \, \mathbb{P}_T\{\tau > T \,|\, \mathcal{F}_t\} + L_t \, \mathbb{E}_{\mathbb{P}_T}(\delta(T)\mathbb{1}_{\{\tau \le T\}} \,|\, \mathcal{F}_t).$$

Put another way, $D(t, T) = L_t(1 - w^T(t, T))$, where $w^T(t, T)$ is the conditional expected writedown rate under $\mathbb{P}_T$, that is,

$$w^T(t, T) := \mathbb{E}_{\mathbb{P}_T}\big(w(T)\mathbb{1}_{\{\tau \le T\}} \,|\, \mathcal{F}_t\big).$$

It is clear that $D(t, T)$ depends on the joint probability distribution under $\mathbb{P}_T$ of the two-dimensional random variable $(\delta(T), \tau)$ or, equivalently, $(w(T), \tau)$. The conditional expected writedown rate upon default equals, under $\mathbb{P}_T$,

$$w_t^T := \frac{\mathbb{E}_{\mathbb{P}_T}(w(T)\mathbb{1}_{\{\tau \le T\}} \,|\, \mathcal{F}_t)}{\mathbb{P}_T\{\tau \le T \,|\, \mathcal{F}_t\}} = \frac{w^T(t, T)}{p_t^T},$$

where $p_t^T = \mathbb{P}_T\{\tau \le T \,|\, \mathcal{F}_t\}$ is the conditional, forward risk-adjusted probability of default. It is easily seen that $D(t, T) = L_t(1 - p_t^T w_t^T)$. Finally, when the random variables $w(T)$ and $\tau$ are conditionally independent with respect to the $\sigma$-field $\mathcal{F}_t$ under $\mathbb{P}_T$ we have $w_t^T = \mathbb{E}_{\mathbb{P}_T}(w(T) \,|\, \mathcal{F}_t)$.

*Example 2.2.3.* Assume the recovery rate $\delta(T) = \delta$ is constant. Then $w_t^T = w$ and

$$D(t, T) = L_t(1 - p_t^T) + \delta L_t p_t^T = L_t(1 - wp_t^T).$$

When $\tau$ is an $\mathbb{F}$-stopping time, the last formula yields

$$D(t, T) = \begin{cases} (1 - w)LB(t, T), & \text{on } \{\tau \le t\}, \\ (1 - wp_t^T)LB(t, T), & \text{on } \{\tau > t\}. \end{cases}$$

---

[3] For the properties of the forward martingale measure $\mathbb{P}_T$, also known as the *forward risk-adjusted probability*, see Sect. 13.2.2 in Musiela and Rutkowski (1997a). Let us only observe that $\mathbb{P}_T = \mathbb{P}^*$ when $B$ is non-random.

**Credit spreads.** We assume, without loss of generality, that $L = 1$, and, for the sake of convenience, we consider the case of a constant recovery rate $\delta$. By definition, the *default-free yield-to-maturity* $Y(t, T)$ and the *defaultable yield-to-maturity* $Y^d(t, T)$ satisfy

$$Y(t, T) = -\frac{\ln B(t, T)}{T - t}, \quad Y^d(t, T) = -\frac{\ln D(t, T)}{T - t}, \qquad (2.27)$$

where the formula for $Y^d(t, T)$ is valid only prior to default. After default, the yield $Y^d(t, T)$ is defined through the equality $D(t, T) = (1-w)e^{-Y^d(t,T)(T-t)}$, and thus $Y(t, T) = Y^d(t, T)$ on the set $\{\tau \le t\}$. Prior to default – that is, on the set $\{\tau > t\}$ – the *credit spread* $S(t, T)$ equals

$$S(t, T) := Y^d(t, T) - Y(t, T) = -\frac{\ln(1 - w \, \mathbb{P}_T\{\tau \le T \mid \mathcal{F}_t\})}{T - t}.$$

The *instantaneous forward rate* $f(t, T)$ and its defaultable counterpart $g(t, T)$, are defined through the following equations

$$B(t, T) = \exp\left(-\int_t^T f(t, u) \, du\right), \quad D(t, T) = \exp\left(-\int_t^T g(t, u) \, du\right),$$

that is,

$$f(t, T) = -\frac{\partial \ln B(t, T)}{\partial T}, \quad g(t, T) = -\frac{\partial \ln D(t, T)}{\partial T}.$$

Obviously $f(t, T) = g(t, T)$ after default, whereas on the set $\{\tau > t\}$ we have

$$g(t, T) = f(t, T) - \frac{\partial \ln(1 - w \, \mathbb{P}_T\{\tau \le T \mid \mathcal{F}_t\})}{\partial T}.$$

Consequently, the *instantaneous forward credit spread* $s(t, T)$, defined as $s(t, T) := g(t, T) - f(t, T)$, satisfies

$$s(t, T) = \frac{w}{1 - w \, \mathbb{P}_T\{\tau \le T \mid \mathcal{F}_t\}} \frac{\partial \mathbb{P}_T\{\tau \le T \mid \mathcal{F}_t\}}{\partial T}.$$

Assume that $\mathbb{P}_T\{\tau \le T \mid \mathcal{F}_t\} < 1$. Then the credit spread equals

$$s(t, T) = g(t, T) - f(t, T) = l(t, T)\gamma(t, T),$$

where the *default loss rate* $l(t, T)$ is given by the expression

$$l(t, T) = \frac{w \, \mathbb{P}_T\{\tau > T \mid \mathcal{F}_t\}}{1 - w \, \mathbb{P}_T\{\tau \le T \mid \mathcal{F}_t\}},$$

and $\gamma(t, T)$ equals

$$\gamma(t, T) = \frac{1}{\mathbb{P}_T\{\tau > T \mid \mathcal{F}_t\}} \frac{\partial \mathbb{P}_T\{\tau \le T \mid \mathcal{F}_t\}}{\partial T}.$$

The interpretation of $\gamma(t, T)$ is clear if, for instance, the default-free interest rates are deterministic. In this case, we have $\mathbb{P}_T = \mathbb{P}^*$ for every $T > 0$, and thus $\gamma(t, T)$ can be interpreted as the *forward hazard rate* of the default time $\tau$ under $\mathbb{P}^*$, as of time $t$, for the future date $T$.

### 2.2.3 Corporate Coupon Bond

Consider a corporate bond maturing at time $T$, which continuously pays constant coupon at rate $c$, and has a principal $L$. In addition, suppose that the bond defaults at the first time $\tau$, when the firm's asset value process $V$ hits a constant lower threshold $\bar{v}$, where $V_0 > \bar{v}$. Upon default a fraction $\beta_2$ of $\bar{v}$ is paid to the bondholders at the default time $\tau$. Let us assume that the spot interest rate $r$ is constant. As we shall see later in this chapter, such assumptions are typical for a class of structural models analyzing the optimal capital structure of a firm. Notice that we deal here with recovery scheme of the second type. In terms of our abstract model, we have

$$X = L, \quad Z \equiv \beta_2 \bar{v}, \quad A_t = ct.$$

For $t = 0$, the valuation formula (2.3) yields

$$X^{d,2}(0,T) = e^{-rT} L \, \mathbb{P}^*\{\tau > T\} + \beta_2 \bar{v} \, \mathbb{E}_{\mathbb{P}^*}\left(e^{-r\tau} \mathbb{1}_{\{\tau \leq T\}}\right)$$
$$+ c \int_0^T e^{-rs} \, \mathbb{P}^*\{\tau > s\} \, ds. \tag{2.28}$$

Notice that expression (2) in Leland and Toft (1996) is equivalent to the last formula (for more details, see Sect. 3.3.3 below).

*Example 2.2.4.* Several authors examined the valuation of corporate *consol bonds,* also known as corporate *perpetuities.* These are coupon-bearing bonds that pay a constant coupon rate and have an infinite maturity (i.e., $T = \infty$). The price of such a bond at time $t$ satisfies

$$X^{d,2}(t,\infty) = \mathbb{E}_{\mathbb{P}^*}\left(\beta_2 \bar{v} e^{r(t-\tau)} \mathbb{1}_{\{\tau \geq t\}} + c \mathbb{1}_{\{\tau \geq t\}} \int_t^\tau e^{r(t-s)} \, ds \,\Big|\, \mathcal{F}_t\right), \tag{2.29}$$

where $\bar{v}$ is a positive constant. In the special case when $t = 0$, equality (2.29) becomes

$$X^{d,2}(0,\infty) = \beta_2 \bar{v} \, \mathbb{E}_{\mathbb{P}^*}\left(e^{-r\tau}\right) + c \int_0^\infty e^{-rs} \, \mathbb{P}^*\{\tau > s\} \, ds. \tag{2.30}$$

For instance, the valuation formula (7) in Leland (1994) corresponds to expression (2.30) (for more information on consol bonds, see Sect. 3.3.2 below). Let us also assume that the functions $\kappa$ and $\sigma_V$ in Assumption (A.3) do not depend on the time variable $t$ – that is, the value process $V$ is time-homogeneous (recall that the spot interest process $r$ is assumed to be constant here). Then the price $X^{d,2}(t,\infty)$ depends on $V_t$ only:

$$X^{d,2}(t,\infty) = u^\infty(V_t)$$

for some deterministic function $u^\infty : \mathbb{R}_+ \to \mathbb{R}$. In this case, the fundamental pricing PDE takes a simpler form (in fact, it becomes an ODE):

$$\tfrac{1}{2}\sigma_V^2(V)V^2 u_{VV}^\infty + (r - \kappa(V))V u_V^\infty + c - r u^\infty = 0.$$

Appropriate boundary conditions need, of course, to be specified.

## 2.3 Merton's Approach to Corporate Debt

In his pathbreaking paper, Merton (1974) considers a firm with a single liability carrying a promised (deterministic) terminal payoff $L$. Several standard conditions are imposed on the continuous-time Black-Scholes-type *frictionless market*. Let us recall the most important assumptions:

- trading takes place continuously in time,
- all traded assets are infinitely divisible,
- an unrestricted borrowing and lending of funds is possible at the same interest rate,
- no restrictions on the short-selling of traded securities are present,
- the transaction costs and taxes (or tax benefits) are disregarded,
- the bankruptcy and/or reorganization costs in case of default are negligible.

### 2.3.1 Merton's Model with Deterministic Interest Rates

One of the simplifying assumptions in the original Merton's model is that the short-term interest rate is constant and equals $r$. Therefore, the price at time $t$ of the unit default-free zero-coupon bond with maturity $T$ is easily seen to be $B(t, T) = e^{-r(T-t)}$. The latter formula can be extended to the case of a deterministic continuously compounded interest rate $r : \mathbb{R}_+ \to \mathbb{R}$. In this case, the price of a $T$-maturity zero-coupon bond equals:

$$B(t, T) = \exp\left( -\int_t^T r(u)\, du \right), \quad \forall\, t \in [0, T].$$

In the sequel, we denote by $E(V_t)$ ($D(V_t)$, resp.) the value of the firm's equity (debt, resp.) at time $t$; hence, the total value of firm's assets satisfies $V_t = E(V_t) + D(V_t)$. We postulate that the firm's value process $V$ follows a geometric Brownian motion under the spot martingale measure $\mathbb{P}^*$, specifically,

$$dV_t = V_t\big((r - \kappa)\, dt + \sigma_V\, dW_t^*\big), \tag{2.31}$$

where $\sigma_V$ is the constant volatility coefficient of the value process $V$ and the constant $\kappa$ represents the payout ratio, provided that it is non-negative. Otherwise, $\kappa$ reflects an inflow of capital to the firm. The process $W^*$ is the one-dimensional standard Brownian motion under $\mathbb{P}^*$, with respect to some reference filtration $\mathbb{F}$ (it is common to take $\mathbb{F} = \mathbb{F}^{W^*}$; this is not essential, though). Notice that dynamics (2.31) is justified only under the assumption that the total value of the firm's assets represents a traded security.

We postulate that the default event may only occur at the debt's maturity date $T$. Specifically, if at the maturity $T$ the total value $V_T$ of the firm's assets is less than the notional value $L$ of the firm's debt, the firm defaults and the bondholders receive the amount $V_T$. Otherwise, the firm does not default, and its liability is repaid in full. We are thus dealing here with a rather elementary example of a defaultable claim with recovery at maturity.

In terms of the generic model introduced in Sect. 2.1, we have

$$X = L, \ A \equiv 0, \ \tilde{X} = V_T, \ \tau = T\mathbb{1}_{\{V_T < L\}} + \infty\mathbb{1}_{\{V_T \geq L\}},$$

where, as usual, $\infty \times 0 = 0$. Put another way (cf. (2.5)),

$$X^{d,1}(T) = L\mathbb{1}_{\{\tau > T\}} + V_T\mathbb{1}_{\{\tau \leq T\}} = L\mathbb{1}_{\{V_T \geq L\}} + V_T\mathbb{1}_{\{V_T < L\}}$$

or, equivalently,

$$X^{d,1}(T) = \min\left(V_T, L\right)\mathbb{1}_{\{V_T \geq L\}} + \min\left(V_T, L\right)\mathbb{1}_{\{V_T < L\}} = \min\left(V_T, L\right).$$

The fixed amount $L$ may be interpreted as the face value (or par value) of a corporate zero-coupon bond maturing at time $T$. Since

$$X^{d,1}(T) = \min\left(V_T, L\right) = L - (L - V_T)^+,$$

where $x^+ = \max(x, 0)$ for every $x \in \mathbb{R}$, the price process $X^{d,1}(t, T)$ of a defaultable zero-coupon bond is manifestly equal to the difference of the value of a default-free zero-coupon bond with the face value $L$ and the value of a European put option written on the firm's assets, with the strike price $L$ and the exercise date $T$. This put option, with the terminal payoff $(L - V_T)^+$, is commonly referred in the present context as the *put-to-default*. Formally, the value of the firm's debt at time $t$ thus equals

$$D(V_t) = D(t, T) = LB(t, T) - P_t, \tag{2.32}$$

where $P_t$ is the price of the put-to-default, and where, for the sake of notational convenience, we write $D(t, T)$ to denote the price of a defaultable bond:

$$D(t, T) := X^{d,1}(t, T) = B_t \, \mathbb{E}_{\mathbb{P}^*}(B_T^{-1} X^{d,1}(T) \,|\, \mathcal{F}_t).$$

It is apparent from (2.32) that the value at time $t$ of the firm's equity satisfies

$$E(V_t) = V_t - D(V_t) = V_t - LB(t, T) + P_t = C_t, \tag{2.33}$$

where $C_t$ stands in turn for the price at time $t$ of a call option written on the firm's assets, with the strike price $L$ and the exercise date $T$. To justify the last equality in (2.33), we may observe that at time $T$ we have

$$E(V_T) = V_T - D(V_T) = V_T - \min\left(V_T, L\right) = (V_T - L)^+,$$

and thus the firm's equity can be seen as a call option on the firm's assets. Alternatively, we may directly use the so-called *put-call parity* relationship for European-style options:

$$C_t - P_t = V_t - LB(t, T).$$

Combining (2.32) with the classic Black-Scholes formula for the arbitrage price of a European put option, Merton (1974) derived a closed-form expression for the arbitrage price of a corporate bond. In what follows, $N$ denotes the standard Gaussian cumulative distribution function:

$$N(x) = \frac{1}{\sqrt{2\pi}} \int_{-\infty}^{x} e^{-u^2/2} \, du, \quad \forall\, x \in \mathbb{R}.$$

**Proposition 2.3.1.** *We have*

$$D(t,T) = V_t e^{-\kappa(T-t)} N\big(-d_1(V_t, T-t)\big) + LB(t,T) N\big(d_2(V_t, T-t)\big), \quad (2.34)$$

*where for every $t \in [0, T]$*

$$d_{1,2}(V_t, T-t) = \frac{\ln(V_t/L) + \big(r - \kappa \pm \frac{1}{2}\sigma_V^2\big)(T-t)}{\sigma_V \sqrt{T-t}}. \quad (2.35)$$

*Proof.* Suppose first that we take the classic Black-Scholes options valuation formula for granted. Recall that the Black-Scholes price of a European put option with the strike price $L$, written on a dividend-paying stock equals (see, for instance, Proposition 6.2.1 in Musiela and Rutkowski (1997a))

$$P_t = LB(t,T) N\big(-d_2(V_t, T-t)\big) - V_t e^{-\kappa(T-t)} N\big(-d_1(V_t, T-t)\big),$$

so that

$$D(t,T) = V_t e^{-\kappa(T-t)} N\big(-d_1(V_t, T-t)\big) + LB(t,T)\big(1 - N\big(-d_2(V_t, T-t)\big)\big).$$

Since obviously $N(-x) = 1 - N(x)$, the last expression is easily seen to be equivalent to Merton's formula (2.34).

For the reader's convenience, we provide below the direct derivation of expression (2.34), based on the risk-neutral valuation formula (2.10). For the sake of notational convenience, we shall write $\sigma$ rather than $\sigma_V$, and we denote $\tilde{r} = r - \kappa$. When applied to a defaultable bond, (2.10) yields

$$D(t,T) = B(t,T) \, \mathbb{E}_{\mathbb{P}^*}\big(L\mathbb{1}_{\{V_T \geq L\}} + V_T \mathbb{1}_{\{V_T < L\}} \,\big|\, \mathcal{F}_t\big),$$

so that

$$D(t,T) = LB(t,T) \, \mathbb{P}^*\{V_T \geq L \,|\, \mathcal{F}_t\} + B(t,T) \, \mathbb{E}_{\mathbb{P}^*}(V_T \mathbb{1}_{\{V_T < L\}} \,|\, \mathcal{F}_t). \quad (2.36)$$

Put another way, $D(t,T) = LB(t,T)J_1 + B(t,T)J_2$ with

$$J_1 = \mathbb{P}^*\{V_T \geq L \,|\, \mathcal{F}_t\}, \quad J_2 = \mathbb{E}_{\mathbb{P}^*}(V_T \mathbb{1}_{\{V_T < L\}} \,|\, \mathcal{F}_t).$$

Solving SDE (2.31), for every $t \in [0, T]$ we obtain

$$V_T = V_t \exp\big(\sigma(W_T^* - W_t^*) + (\tilde{r} - \tfrac{1}{2}\sigma^2)(T-t)\big). \quad (2.37)$$

For $J_1$, we have (recall that $L > 0$)

$$J_1 = \mathbb{P}^*\Big\{ V_t \exp\big(\sigma(W_T^* - W_t^*) + (\tilde{r} - \tfrac{1}{2}\sigma^2)(T-t)\big) \geq L \,\Big|\, \mathcal{F}_t\Big\}$$

$$= \mathbb{P}^*\Big\{-\sigma(W_T^* - W_t^*) \leq \ln(V_t/L) + (\tilde{r} - \tfrac{1}{2}\sigma^2)(T-t) \,\Big|\, \mathcal{F}_t\Big\}$$

$$= \mathbb{P}^*\Big\{\xi \leq \frac{\ln(x/L) + (\tilde{r} - \tfrac{1}{2}\sigma^2)(T-t)}{\sigma\sqrt{T-t}}\Big\}_{x=V_t}$$

$$= N\big(d_2(V_t, T-t)\big),$$

since the random variable $\xi := -(W_T^* - W_t^*)/\sqrt{T-t}$ is independent of the $\sigma$-field $\mathcal{F}_t$, and has the standard Gaussian law $N(0,1)$ under $\mathbb{P}^*$.

To evaluate $J_2$, it is convenient to introduce an auxiliary probability measure $\bar{\mathbb{P}}$ on $(\Omega, \mathcal{F}_T)$ by setting

$$\frac{d\bar{\mathbb{P}}}{d\mathbb{P}^*} = \exp\left(\sigma W_T^* - \tfrac{1}{2}\sigma^2 T\right) =: \eta_T, \quad \mathbb{P}^*\text{-a.s.}$$

It is well known that for every $t \in [0, T]$ we have

$$\frac{d\bar{\mathbb{P}}}{d\mathbb{P}^*}\bigg|_{\mathcal{F}_t} = \exp\left(\sigma W_t^* - \tfrac{1}{2}\sigma^2 t\right) = \eta_t, \quad \mathbb{P}^*\text{-a.s.}$$

Let us denote $A = \{V_T < L\}$. It is clear that

$$J_2 = \mathbb{E}_{\mathbb{P}^*}\left(V_T \mathbb{1}_A \mid \mathcal{F}_t\right) = V_0 \, e^{\tilde{r}T} \, \mathbb{E}_{\mathbb{P}^*}\left(\eta_T \mathbb{1}_A \mid \mathcal{F}_t\right).$$

Consequently, using the abstract Bayes rule, we obtain

$$J_2 = V_0 \, e^{\tilde{r}T} \, \eta_t \, \bar{\mathbb{P}}\{A \mid \mathcal{F}_t\} = B^{-1}(t, T) V_t e^{-\kappa(T-t)} \, \bar{\mathbb{P}}\{A \mid \mathcal{F}_t\},$$

where the last equality is a consequence of the following chain of equalities:

$$V_0 \, e^{\tilde{r}T} \, \eta_t = V_0 \exp\left(\sigma W_t^* - \tfrac{1}{2}\sigma^2 t + (r - \kappa)T\right) = V_t e^{(r-\kappa)(T-t)}.$$

By virtue of Girsanov's theorem, the process $\bar{W}_t = W_t^* - \sigma t$ follows a standard Brownian motion on the space $(\Omega, \mathbb{F}, \bar{\mathbb{P}})$. The dynamics of $V$ under $\bar{\mathbb{P}}$ are

$$dV_t = V_t\left((\tilde{r} + \sigma^2)\, dt + \sigma \, d\bar{W}_t\right),$$

and thus for every $t \in [0, T]$ we have

$$V_T = V_t \exp\left(\sigma(\bar{W}_T - \bar{W}_t) + (\tilde{r} + \tfrac{1}{2}\sigma^2)(T - t)\right).$$

Consequently,

$$\begin{aligned}
\bar{\mathbb{P}}\{A \mid \mathcal{F}_t\} &= \bar{\mathbb{P}}\left\{V_t \exp\left(\sigma(\bar{W}_T - \bar{W}_t) + (\tilde{r} + \tfrac{1}{2}\sigma^2)(T - t)\right) < L \,\Big|\, \mathcal{F}_t\right\} \\
&= \bar{\mathbb{P}}\left\{\sigma(\bar{W}_T - \bar{W}_t) < -\ln(V_t/L) - (\tilde{r} + \tfrac{1}{2}\sigma^2)\,(T - t) \,\Big|\, \mathcal{F}_t\right\} \\
&= \bar{\mathbb{P}}\left\{\bar{\xi} < \frac{-\ln(x/L) - (\tilde{r} + \tfrac{1}{2}\sigma^2)(T - t)}{\sigma\sqrt{T - t}}\right\}_{x=V_t} \\
&= N\left(-d_1(V_t, T - t)\right),
\end{aligned}$$

since $\bar{\xi} := (\bar{W}_T - \bar{W}_t)/\sqrt{T - t}$ is independent of the $\sigma$-field $\mathcal{F}_t$, and has the standard Gaussian law $N(0, 1)$ under $\bar{\mathbb{P}}$. We conclude that

$$J_2 = B^{-1}(t, T) V_t e^{-\kappa(T-t)} N\left(-d_1(V_t, T - t)\right).$$

This completes the derivation of formula (2.34).     $\square$

From the proof of Proposition 2.3.1, we deduce also that

$$D(t, T) = LB(t, T)\, \mathbb{P}^*\{V_T \geq L \mid \mathcal{F}_t\} + V_t e^{-\kappa(T-t)} \, \bar{\mathbb{P}}\{V_T < L \mid \mathcal{F}_t\}.$$

When $\kappa = 0$, it is not difficult to verify that $\bar{\mathbb{P}}$ is a martingale measure corresponding to the choice of $V$ as a discount factor. In other words, $\bar{\mathbb{P}}$ is equivalent to $\mathbb{P}^*$ and the process $B_t/V_t$ follows a martingale under $\bar{\mathbb{P}}$.

We preserve here the notation and terminology introduced in Sect. 2.2.2. Notice that the conditional probabilities of default are:

$$p_t^* = \mathbb{P}^*\{V_T < L \,|\, \mathcal{F}_t\} = N\big(-d_2(V_t, T - t)\big),$$

and

$$\bar{p}_t = \bar{\mathbb{P}}\{V_T < L \,|\, \mathcal{F}_t\} = N\big(-d_1(V_t, T - t)\big).$$

It is customary to refer to $p_t^*$ as the conditional risk-neutral probability of default. When $\kappa = 0$, $\bar{p}_t$ can also be seen as the 'risk-neutral probability of default' (associated with a different choice of the discount factor, however). Merton's valuation formula can be re-expressed as follows:

$$D(t, T) = L_t(1 - p_t^*) + L_t p_t^* \delta_t^* = L_t(1 - p_t^* w_t^*),$$

where $L_t = LB(t, T)$ is the present value of the promised claim (as well as the present value of the exposure at default), and $\delta_t^*$ is the conditional risk-neutral expected recovery rate upon default. Specifically,

$$\delta_t^* := \frac{\mathbb{E}_{\mathbb{P}^*}\{V_T \mathbb{1}_{\{V_T < L\}} \,|\, \mathcal{F}_t\}}{L\,\mathbb{P}^*\{V_T < L \,|\, \mathcal{F}_t\}} = \frac{V_t e^{-\kappa(T-t)} N\big(-d_1(V_t, T - t)\big)}{LB(t, T) N\big(-d_2(V_t, T - t)\big)}.$$

Recall also that $w_t^* = 1 - \delta_t^*$ is called the conditional risk-neutral expected writedown rate upon default. Let $l_t := L_t/V_t$ stand for the firm's *leverage ratio*. In terms of the process $l_t$, formula (2.34) becomes

$$\frac{D(t, T)}{L_t} = l_t^{-1} e^{-\kappa(T-t)} N\big(-h_1(l_t, T - t)\big) + N\big(h_2(l_t, T - t)\big), \qquad (2.38)$$

where

$$h_{1,2}(l_t, T - t) = \frac{-\ln l_t - \kappa(T - t) \pm \frac{1}{2}\sigma_V^2(T - t)}{\sigma_V \sqrt{T - t}}. \qquad (2.39)$$

Notice that the quantity $l_t$ gives the 'nominal' value of the firm's leverage ratio. Indeed, $L_t$ represents the default-free value of the firm's debt, as opposed to the actual market value $D(t, T)$ of the firm's debt.

**Hedging of a corporate bond.** Since Merton's formula can be seen as a variant of the Black-Scholes valuation result, the form of the replicating (self-financing) trading strategy for a defaultable bond can be easily deduced from the well-known expressions for the Black-Scholes hedging strategy for a European put option. For the sake of completeness, we state the following corollary to Proposition 2.3.1, in which we write $D(t, T) = u(V_t, t)$.

**Corollary 2.3.1.** *The unique replicating strategy for a defaultable bond involves holding at any time $t \le T$ the $\phi_t^1 V_t$ units of cash invested in the firm's value and $\phi_t^2 B(t, T)$ units of cash invested in default-free bonds, where for every $t \in [0, T]$*

$$\phi_t^1 = u_V(V_t, t) = e^{-\kappa(T-t)} N\big(-d_1(V_t, T - t)\big)$$

*and*

$$\phi_t^2 = \frac{D(t, T) - \phi_t^1 V_t}{B(t, T)} = LN\big(d_2(V_t, T - t)\big).$$

**Credit spreads.** An important characteristic of a defaultable bond is the difference between its yield and the yield of an equivalent default-free bond, i.e., the *credit spread*. Recall that the credit spread $S(t,T)$ is defined through the formula $S(t,T) = Y^d(t,T) - Y(t,T)$, where $Y^d(t,T)$ and $Y(t,T)$ are given by (2.27). In Merton's model the yield on a default-free bond is equal to the short-term interest rate; i.e., $Y(t,T) = r$. Using (2.38) with $L = 1$, we arrive at the following representation for the credit spread in Merton's model

$$S(t,T) = -\frac{\ln\left(l_t^{-1} e^{-\kappa(T-t)} N\big(-h_1(l_t,T-t)\big) + N\big(h_2(l_t,T-t)\big)\right)}{T-t}.$$

Let us now analyze the behavior of the credit spread when time converges to the debt's maturity. For this purpose, observe that: $\lim_{t \to T} l_t = L/V_T$,

$$\lim_{t \to T} N\big(-h_1(l_t,T-t)\big) = \begin{cases} 1, & \text{on } \{V_T < L\}, \\ 0, & \text{on } \{V_T > L\}, \end{cases}$$

and

$$\lim_{t \to T} N\big(h_2(l_t,T-t)\big) = \begin{cases} 0, & \text{on } \{V_T < L\}, \\ 1, & \text{on } \{V_T > L\}. \end{cases}$$

The reader can readily verify that

$$\lim_{t \to T} S(t,T) = \begin{cases} +\infty, & \text{on } \{V_T < L\}, \\ 0, & \text{on } \{V_T > L\}. \end{cases} \tag{2.40}$$

An essential feature of Merton's model is that the default time $\tau$ appears to be a predictable stopping time with respect to the filtration $\mathbb{F}^V$ generated by the value process $V$, as it is announced, for instance, by the following sequence of $\mathbb{F}^V$-stopping times:

$$\tau_n = \{\, t \geq T - \tfrac{1}{n} : V_t < L \,\} \tag{2.41}$$

with the usual convention that $\inf \emptyset = \infty$.

*Remarks.* It follows from (2.37) that $\mathbb{F}^V = \mathbb{F}^{W^*}$, where $\mathbb{F}^{W^*}$ is the filtration generated by the Brownian motion $W^*$. It is well known that any stopping time with respect to a Brownian filtration is predictable.

The conditional probability $\mathbb{P}^*\{V_T > L \mid V_{T-\varepsilon} > L\}$ can be made arbitrarily close to 1 by selecting sufficiently small $\varepsilon > 0$. Thus, if a firm is not in financial distress at a date very close to the maturity $T$ (that is, if $V_{T-\varepsilon} > L$), it is very unlikely to default. As it is apparent from formula (2.40), near the maturity, the credit spread for such a firm is close to zero. The feature that the short-term credit spreads are close to zero is contradicted by the empirical evidence (see, e.g., Jones et al. (1984)), and it is frequently quoted as a major shortcoming of Merton's approach. The so-called *first-passage-time models* (discussed at some length in Chap. 3) suffer from the same shortcoming, though. One of possible remedies is to introduce jumps in the dynamics of the firm's value as was postulated, for instance, by Mason and Bhattacharya (1981) and Zhou (1996).

## 2.3.2 Distance-to-Default

In practice, the contractual liabilities of a firm rarely have the same maturity. It is thus important to classify the debt outstanding by identifying, for instance, the short-term debt (i.e., the current liabilities), medium-term debt, and long-term debt. Empirical studies show that, in general, firms do not default when their asset value falls below the book value of their total liabilities. Typically, the *default point* – that is, the asset value at which the firm will actually default – will lie somewhere between short-term liabilities and total liabilities. A version of Merton's model accounting for the existence of several categories of debt was put forward by Vasicek (1984). We do not present details of this approach, but we focus only on the concept of the *distance-to-default*. We shall first check that the knowledge of the current value and the volatility of the firm's equity, allows us to determine the value and the volatility of the firm's assets. For the sake of convenience, we assume as before that the maturity of the total firm's debt is $T$ and we set $\kappa = 0$ in dynamics (2.31). Then the market value of the firm's equity satisfies (cf. (2.33))

$$E(V_t) = C_t = V_t N\big(d_1(V_t, T - t)\big) - LB(t, T) N\big(d_2(V_t, T - t)\big) \qquad (2.42)$$

with the functions $d_1$ and $d_2$ given by (2.35). Using the Itô formula, one can check that the dynamics of the price process $C$ under $\mathbb{P}^*$ are

$$dC_t = rC_t \, dt + V_t N\big(d_1(V_t, T - t)\big)\sigma_V \, dW_t^*,$$

and thus the volatility of the firm's equity admits the following representation

$$\sigma_t^E = \frac{V_t}{E(V_t)} N\big(d_1(V_t, T - t)\big)\sigma_V. \qquad (2.43)$$

Suppose that the current market value $E(V_t)$ and the volatility coefficient $\sigma_t^E$ of the firm's equity are known. Using (2.42)–(2.43), we may find, in principle, the current value $V_t$ of the firm's assets and the volatility coefficient $\sigma_V$. To specify the actual probability of default, we assume that under the real-world probability $\mathbb{P}$ the value process $V$ satisfies

$$dV_t = V_t\big(\mu \, dt + \sigma_V \, dW_t\big)$$

for some constant $\mu \in \mathbb{R}$ and a certain Brownian motion $W$ under $\mathbb{P}$. It results that:

$$\mathbb{P}\{\tau \le T \,|\, \mathcal{F}_t\} = \mathbb{P}\{V_T < L \,|\, \mathcal{F}_t\} = N(-d_t),$$

where the *distance-to-default* at time $t$, denoted by $d_t$, is defined as

$$d_t = \frac{\ln(V_t/L) + \big(\mu - \tfrac{1}{2}\sigma_V^2\big)(T - t)}{\sigma_V \sqrt{T - t}}.$$

It measures, in terms of the 'standard deviation' $\sigma_V \sqrt{T - t}$, the distance of the expected total value of firm's assets from the default point $L$. For a detailed presentation of practical implementations of these ideas, see, for instance, Crouhy et al. (2000) or Part 6 in Cossin and Pirotte (2000).

## 2.4 Extensions of Merton's Approach

In this section, we present an brief survey of papers devoted to various applications of the original Merton approach, or to its extensions.

**Vulnerable claims.** Several authors extend Merton's (1974) approach to the case of more general contingent claims that are subject to the default risk of the issuing party. For instance, Johnson and Stulz (1987) assume that the option is the sole liability of the counterparty, and they model default as occurring when the value of the option is greater than the value of the assets of the counterparty upon exercise. Since typically option positions are not significant with respect to the total value of the firm, and options rarely constitute the sole firm's liability, the original Johnson and Stulz (1987) set-up seems to be unrealistic. Hull and White (1995) extend the Johnson and Stulz approach to the case when the option and a defaultable bond are of equal seniority, with the same recovery ratio. Klein (1996) and Klein and Inglis (2001) consider the two-dimensional process $(S, V)$, where $S$ represents the value of the asset underlying a vulnerable option; they provide valuation formulae for vulnerable options in this setting.

**Corporate debt.** Merton's model makes a simplistic assumption that the firm issues a single zero-coupon bond only. Several authors have modified the original approach to cover the following real-life features of corporate debt:
 - corporate coupon-bonds, Geske (1977, 1979),
 - debt structure (short-term and long-term debt), Vasicek (1984),
 - bond covenants (priority rules, payment schedules, sinking fund, etc.), Ho and Singer (1982, 1984),
 - floating-rate debt, Cox et al. (1980),
 - duration of defaultable zero-coupon bond, Chance (1990).

**Swap contracts.** Cooper and Mello (1991) apply Merton's approach to the case of a stylized single-period swaps in which a fixed payoff is exchanged for a random payoff. They assume, in particular, that swap contracts are subordinate to debt in bankruptcy, and that in the event of default on its own debt of a counterparty that is owed value in a swap, the value of the swap will be paid to the bankrupt firm. They show that if the maturity of the debt coincides with the maturity of a swap, the value of a swap can be found using the valuation formula for an option on a maximum of two underlying assets, specifically, the value of the firm and the random payoff. Papers in a similar vein include, among others, Abken (1993), Baz and Pascutti (1996), Rich and Leipus (1997), Li (1998), Hübner (2001), and Yu and Kwok (2002).

**Random interest rates.** Shirakawa (1999) examines the behavior of credit spreads in Merton's framework. Shimko et al. (1993) extend Merton's model by postulating that short-term interest rate is governed by Vasicek's model, while Wang (1999a) combines Merton's model with the independent CIR model of the short-term rate. Finally, Szatzschneider (2000) extends essentially the latter approach by dispensing with the independence assumption.

### 2.4.1 Models with Stochastic Interest Rates

The short-term interest rate is assumed to be constant in Merton's model, and thus the model disregards the interest rate risk. As already mentioned, Shimko et al. (1993) relax this rather stringent assumption by allowing for a stochastic short-term interest rate that evolves according to Vasicek's (1977) model. Pricing the defaultable bond within Merton's model, but with stochastic term structure of interest rates is essentially equivalent to pricing a European put option on a stock under stochastic interest rates. For Vasicek's model, the latter problem was treated by Jamshidian (1989) and a closed-form expression[4] was derived there for a price of a European call and put options on bonds and stocks. Assume that the interest rate $r$ obeys Vasicek's dynamics:

$$dr_t = (a - br_t)\, dt + \sigma_r\, d\tilde{W}_t \tag{2.44}$$

and the firm's value is governed by the SDE

$$dV_t = V_t\big(r_t\, dt + \sigma_V\, dW_t^*\big), \tag{2.45}$$

where both equations hold under the spot martingale measure $\mathbb{P}^*$, and the Brownian motions $\tilde{W}$ and $W^*$ are correlated, with constant instantaneous correlation coefficient $\rho_{Vr}$. Let us denote, for any $t \le T$,

$$\sigma^2(t, T) = \int_t^T \big(\sigma_V^2 - 2\rho_{Vr}\sigma_V b(u, T) + b^2(u, T)\big)\, du,$$

where $b(t, T)$ stands for the volatility of a default-free zero-coupon bond in Vasicek's model, i.e.,

$$b(t, T) = \sigma_r b^{-1}\big(1 - e^{-b(T-t)}\big), \quad \forall t \in [0, T].$$

Consider a European put option with expiry date $T$ and strike $L$ written on the value process $V$. The valuation formula established in Jamshidian (1989) reads

$$P_t = LB(t, T)N\big(-h_2(V_t, t, T)\big) - V_t N\big(-h_1(V_t, t, T)\big), \tag{2.46}$$

where $B(t, T) = B(t, T, r_t)$, and where, for every $t \le T$,

$$h_{1,2}(V_t, t, T) = \frac{\ln\big(V_t/B(t, T)\big) - \ln L \pm \frac{1}{2}\sigma^2(t, T)}{\sigma(t, T)}. \tag{2.47}$$

The valuation formula for a defaultable bond derived by Shimko et al. (1993) is thus a straightforward consequence of Jamshidian's result (it is enough to combine (2.32) with (2.46)). Of course, Vasicek's model can be substituted here with any term structure model, in which a closed-form expression for the price of a European stock option is available; for instance, the Gaussian HJM set-up, the CIR model, etc. Hedging strategy for a defaultable bond in this framework can thus be seen as a simple consequence of the corresponding result for European stock option.

---

[4] Jamshidian's formula is a special case of a general result for the Gaussian HJM set-up (cf. Heath et al. (1992) or Sect. 13.3 in Musiela and Rutkowski (1997a)).

### 2.4.2 Discontinuous Value Process

Zhou (1996) extends Merton's approach by modeling the firm's value process $V$ as a geometric jump-diffusion process.[5] The main purpose of Zhou's study was to address the issue of predictability of the default time $\tau$, inherent in Merton's model (the time of default remains predictable within the so-called simplified version of Zhou's model presented in this section, though). To state Zhou's equation for the dynamics of the value process $V$, we need to introduce a Poisson process $N$ with the intensity $\lambda$ under the probability measure $\mathbb{P}^*$ and a sequence of independent identically distributed random variables $(U_i)_{i \geq 1}$ with the finite expected value $\nu = \mathbb{E}_{\mathbb{P}^*}(U_i)$. We assume that the $\sigma$-fields generated by the processes $W^*$, $N$ and the sequence $(U_i)_{i \geq 1}$ are mutually independent under $\mathbb{P}^*$. The equation for the dynamics of $V$ under the risk-neutral measure $\mathbb{P}^*$ now takes the following form:

$$dV_t = V_{t-}\big((r - \lambda\nu)\, dt + \sigma_V\, dW_t^* + d\pi_t\big), \tag{2.48}$$

where $\pi$ is a jump process whose jump times are specified by the jump times of the Poisson process $N$, and the size of the $i^{\text{th}}$ jump is $U_i$. In other words, the process $\pi$ is a marked Poisson process:

$$\pi_t = \sum_{i=1}^{N_t} U_i, \quad \forall t \in [0, T].$$

We endow our underlying probability space with the filtration $\mathbb{F}$ generated by processes $W^*$ and $\pi$. It is not difficult to check that the compensated process $\tilde{\pi}_t = \pi_t - \lambda\nu t$ is a $\mathbb{P}^*$-martingale with respect to this filtration. Consequently, the process $V_t^* = e^{-rt}V_t$, which is easily seen to satisfy

$$dV_t^* = V_{t-}^*\big(\sigma_V\, dW_t^* + d\tilde{\pi}_t\big), \tag{2.49}$$

also follows a martingale under $\mathbb{P}^*$ with respect to $\mathbb{F}$. Equation (2.49) can be solved explicitly, yielding

$$V_t^* = V_0^* \exp\big(\tilde{\pi}_t + \sigma_V W_t^* - \tfrac{1}{2}\sigma_V^2 t\big) \prod_{u \leq t} (1 + \Delta\tilde{\pi}_u)\exp(-\Delta\tilde{\pi}_u),$$

where $\Delta\tilde{\pi}_u = \tilde{\pi}_u - \tilde{\pi}_{u-}$ or, equivalently,

$$V_t = V_0 \exp\big(\sigma_V W_t^* + (r - \tfrac{1}{2}\sigma_V^2 - \lambda\nu)t\big) \prod_{i=1}^{N_t} (1 + U_i). \tag{2.50}$$

From now on, we assume, in addition, that $U_i + 1$ has the log-normal distribution under $\mathbb{P}^*$ – that is, $\ln(U_i + 1) \sim N(\mu, \sigma)$. This implies that

$$\nu := \mathbb{E}_{\mathbb{P}^*}(U_i) = \exp\big(\mu + \tfrac{1}{2}\sigma^2\big) - 1.$$

---

[5] Mason and Bhattacharya (1981) use a pure jump process for the firm's value.

The case considered in Sect. 2 of Zhou (1996) corresponds to a defaultable claim with recovery at maturity

$$X = L, \ A \equiv 0, \ \tilde{X} = L(1 - \bar{w}(V_T/L)), \ \tau = T\mathbb{1}_{\{V_T < L\}} + \infty\mathbb{1}_{\{V_T \geq L\}},$$

where $\bar{w} : \mathbb{R}_+ \to \mathbb{R}$, referred to as the *writedown function*, determines the recovery value in case of default. If we choose $\bar{w}(x) = 1 - x$, $\tilde{X}$ reduces to the recovery structure of the original Merton model. In general, we have

$$X^{d,1}(T) = L\mathbb{1}_{\{\tau > T\}} + L(1 - \bar{w}(V_T/L))\mathbb{1}_{\{\tau \leq T\}}$$

or, equivalently,

$$X^{d,1}(T) = L\big(\mathbb{1}_{\{V_T \geq L\}} + \bar{\delta}(V_T/L)\mathbb{1}_{\{V_T < L\}}\big),$$

where $\bar{\delta}(V_T/L) = 1 - \bar{w}(V_T/L)$ is the recovery rate of the defaulted bond. The following auxiliary result establishes the conditional probability law of the default event $\{\tau = T\}$ with respect to the $\sigma$-field $\mathcal{F}_t$.

**Lemma 2.4.1.** *The risk-neutral conditional probability of default satisfies*

$$\mathbb{P}^*\{\tau = T \,|\, \mathcal{F}_t\} = \sum_{i=0}^{\infty} e^{-\lambda(T-t)} \frac{(\lambda(T - t))^i}{i!} N\big(- d_{2,i}(V_t, T - t)\big),$$

*where, for every $i \in \mathbb{N}$ and $t \in \mathbb{R}_+$,*

$$d_{2,i}(V, t) = \frac{\ln(V/L) + \mu_i(t)}{\sigma_i(t)}$$

*with*

$$\mu_i(t) = (r - \tfrac{1}{2}\sigma_V^2 - \lambda\nu)t + i\mu, \quad \sigma_i^2(t) = \sigma_V^2 t + i\sigma^2.$$

*Proof.* Obviously $\mathbb{P}^*\{\tau = T \,|\, \mathcal{F}_t\} = \mathbb{P}^*\{V_T < L \,|\, \mathcal{F}_t\}$. In view of the assumed independence of the Brownian motion $W^*$ and the jump component $\pi$, it is enough to consider the conditional probability with respect to the number of jumps in the interval $[t, T]$ and to use the formula for the total probability. In view of (2.50), on the set $\{N_T - N_t = i\}$ the random variable $V_T$ can be represented as follows

$$V_T = V_t \exp\left(\sigma_V(W_T^* - W_t^*) + (r - \tfrac{1}{2}\sigma_V^2 - \lambda\nu)(T - t) + \sum_{j=1}^{i} \zeta_j\right),$$

where $\zeta_j$, $j = 1, \ldots, i$, are independent identically distributed random variables with the Gaussian law $N(\mu, \sigma)$. In addition, $\zeta_j$s are independent of $W^*$. Put another way, $V_T = V_t e^\zeta$, where $\zeta$ is a Gaussian random variable, independent of $\mathcal{F}_t$, with the expected value:

$$\mathbb{E}_{\mathbb{P}^*}(\zeta) = (r - \tfrac{1}{2}\sigma_V^2 - \lambda\nu)(T - t) + i\mu,$$

and the variance:

$$\text{Var}_{\mathbb{P}^*}(\zeta) = \sigma_V^2(T - t) + i\sigma^2.$$

The above representation for the random variable $V_T$ leads directly to the asserted formula. The details are left to the reader. $\qquad\square$

**Defaultable bond.** We define the price $D(t,T)$ of a defaultable bond by setting

$$D(t,T) = B_t \, \mathbb{E}_{\mathbb{P}^*}(B_T^{-1} X^{d,1}(T) \,|\, \mathcal{F}_t). \tag{2.51}$$

Due to the presence of the jump component in the dynamics of $V$, it is clear that an analytical approach to the valuation of defaultable claims in Zhou's framework requires solving an integro-differential PDE involving the infinitesimal generator of $V$, and this does not seem to be an easy matter. On the other hand, the valuation of a defaultable bond through the probabilistic approach presents no difficulties. Of course, it still remains a problem of validity of formula (2.51), because it is not supported in Zhou's set-up by the existence of a replicating strategy for a defaultable bond. Thus, in contrast to Merton's valuation formula of Proposition 2.3.1, expression (2.51) should be seen as the formal definition of the price process of a defaultable bond.

**Proposition 2.4.1.** *Assume that $\bar{w}(x) = 1 - x$. Then for any $t \in [0,T]$ we have*

$$D(t,T) = LB(t,T)\Big\{1 - \sum_{i=0}^{\infty} e^{-\lambda(T-t)} \frac{\big(\lambda(T-t)\big)^i}{i!} \, N\big(-d_{2,i}(V_t, T-t)\big)$$

$$+ \frac{V_t}{L} \sum_{i=0}^{\infty} e^{\mu_i(T-t)+\sigma_i^2(T-t)/2-\lambda(T-t)} \frac{\big(\lambda(T-t)\big)^i}{i!} \, N\big(-d_{1,i}(V_t, T-t)\big)\Big\},$$

*where, for every $i \in \mathbb{N}$ and $t \in \mathbb{R}_+$, we denote*

$$\mu_i(t) = (r - \tfrac{1}{2}\sigma_V^2 - \lambda\nu)t + i\mu, \quad \sigma_i^2(t) = \sigma_V^2 t + i\sigma^2,$$

*and*

$$d_{2,i}\big(V_t, t\big) = \frac{\ln(V_t/L) + \mu_i(t)}{\sigma_i(t)}, \quad d_{1,i}(V_t, t) = d_{2,i}(V_t, t) + \sigma_i(t).$$

*Proof.* It suffices to apply the valuation formula established in Merton (1973). It extends the Black-Scholes formula to the case of a European put option written on a stock, whose price follows a jump-diffusion process given by (2.48). For a more direct proof, notice that $X^{d,1}(T)$ equals:

$$X^{d,1}(T) = L - L\mathbb{1}_{\{V_T < L\}} + V_T \mathbb{1}_{\{V_T < L\}},$$

so that

$$D(t,T) = LB(t,T) - L\mathbb{P}^*\{V_T < L \,|\, \mathcal{F}_t\} + B(t,T)\,\mathbb{E}_{\mathbb{P}^*}(V_T \mathbb{1}_{\{V_T<L\}} \,|\, \mathcal{F}_t).$$

The second term in the last formula can be found using Lemma 2.4.1. For the last term, it suffices to first condition with respect to the number of jumps of $N$ in the interval $[t,T]$. On the set $\{N_T - N_t = i\}$, we obtain

$$\mathbb{E}_{\mathbb{P}^*}(V_T \mathbb{1}_{\{V_T<L\}} \,|\, \mathcal{F}_t) = V_t \, \mathbb{E}_{\mathbb{P}^*}(e^\zeta \mathbb{1}_{\{xe^\zeta < L\}}) \,|_{x=V_t},$$

where $\zeta$ is an auxiliary Gaussian random variable that was introduced in the proof of Lemma 2.4.1. The valuation formula now follows directly from the elementary Lemma 2.4.2. $\qquad\square$

**Lemma 2.4.2.** *Let $\zeta$ be a Gaussian random variable under $\mathbb{P}$ with the expected value $m$ and the variance $\sigma^2$. Then for any strictly positive $x$ we have*

$$\mathbb{E}_{\mathbb{P}}(e^{\zeta}\mathbb{1}_{\{e^{\zeta}<x\}}) = e^{m+\sigma^2/2} N\left(\frac{\ln x - m - \sigma^2}{\sigma}\right). \tag{2.52}$$

*Proof.* Equality (2.52) can be established by elementary integration. Alternatively, we can make use of Girsanov's theorem. It is clear that

$$I := \mathbb{E}_{\mathbb{P}}(e^{\zeta}\mathbb{1}_{\{e^{\zeta}<x\}}) = \mathbb{E}_{\mathbb{P}}(e^{m+\sigma W_1}\mathbb{1}_{\{e^{m+\sigma W_1}<x\}})$$
$$= e^{m+\sigma^2/2}\,\mathbb{E}_{\mathbb{P}}(e^{\sigma W_1 - \sigma^2/2}\mathbb{1}_{\{e^{m+\sigma W_1}<x\}}),$$

where $W$ follows a standard Brownian motion on some filtered probability space $(\Omega, \mathbb{F}, \mathbb{P})$. Let $\tilde{\mathbb{P}}$ be a probability measure, equivalent to $\mathbb{P}$ on $(\Omega, \mathcal{F}_1)$, with the following Radon-Nikodým density

$$\frac{d\tilde{\mathbb{P}}}{d\mathbb{P}} = \exp\left(\sigma W_1 - \tfrac{1}{2}\sigma^2\right), \quad \mathbb{P}\text{-a.s.}$$

From Girsanov's theorem, the process $\tilde{W}_t = W_t - \sigma t$ follows a standard Brownian motion under $\tilde{\mathbb{P}}$, and thus

$$I = e^{m+\sigma^2/2}\,\tilde{\mathbb{P}}\{e^{m+\sigma W_1} < x\} = e^{m+\sigma^2/2}\,\tilde{\mathbb{P}}\{e^{m+\sigma\tilde{W}_1+\sigma^2} < x\}$$
$$= e^{m+\sigma^2/2}\,\tilde{\mathbb{P}}\{\sigma\tilde{W}_1 < \ln x - m - \sigma^2\}.$$

This immediately yields (2.52). □

Observe that in the case of no jumps – that is, for $\lambda = 0$ – the formula established in Proposition 2.4.1 reduces to Merton's result (2.34). It is noteworthy that the closed-form expressions for the value of a defaultable bond can also be derived for other natural choices of the writedown function, such as: $\bar{w}(x) = w_0 - w_1 x$, $\bar{w}(x) = \min(1, w_0 - w_1 x)$, etc.

*Remarks.* We have discussed only a special Merton-like case of Zhou's approach. The general model examined by Zhou (1996) belongs to the class of first-passage-time models that are studied at some length in the next chapter. He postulates that the default time $\tau$ is the first passage time of the firm's value to a constant barrier. More specifically,

$$\tau = \inf\{t \in [0, T) : V_t \leq \bar{v}\},$$

where $\bar{v} > 0$ is a positive constant. Such an approach was previously adopted by, among others, Kim et al. (1993a) and Longstaff and Schwartz (1995) (see Sect. 3.4 for more details). Furthermore, if default occurs prior to the bond's maturity $T$, the owner receives the payoff $\tilde{X} = L(1 - \bar{w}(V_{\tau}/L))$ at time $T$; equivalently, he gets the amount $Z_{\tau} = B(\tau, T)\tilde{X}$ at default time. An analytical result for the price of a defaultable bond is not available in this set-up; Zhou (1996) provides a tractable way of valuing such a bond, though.

### 2.4.3 Buffet's Approach

A common assumption in classic structural models is that the assets of the firm are tradeable (see Assumption (A.1)), so that the discounted total value of the firm's assets follows a martingale under the risk-neutral probability measure $\mathbb{P}^*$ (see Assumption (A.3)). Although convenient, such an assumption is difficult to sustain from the practical viewpoint, since in practice the value of the firm is not tradeable, nor even observable.

To circumvent, at least on the theoretical level, the issue of non-tradability of the firm's value, Ericsson and Reneby (1999) argue that it is enough to postulate that at least one of the firm's securities (e.g., a common stock) is traded, and the market is complete, in the sense that the firm's value can be replicated by dynamic trading in some securities. In this case, the value process itself can be formally viewed as the price process of a tradeable asset. Although this intuitive argument can be made formal, at least under specific assumptions imposed on the model, it is disputable whether this really closes the case.

In an alternative approach, proposed recently by Buffet (2000), it is also assumed that the firm's shares are traded, but the market completeness, in the sense described above, is not assumed. The main goal is thus to derive the risk-neutral dynamics of the firm's value, starting from economic fundamentals. To derive the dynamics of the profit per unit time, Buffet (2000) specifies first the real-world dynamics for the unit manufacturing cost and for the number of units sold per unit time. Under the assumption that the firm selects the selling price so as to maximize the profit rate, he subsequently finds the formula for profits per unit time and derives the dynamics under the martingale measure of the basic tradeable asset: the discounted value of firm's stock. An interesting feature of his result is that the real-world growth rate is still present in the risk-neutral representation of the tradeable asset. In the last step, he establishes the formula for the value of the firm, under the assumption that it derives from the reinvested profits.

For a corporate zero-coupon bond, he postulates, as in the original Merton's model, that the bond is formally described by the following quantities:

$$X = L, \ A \equiv 0, \ \tilde{X} = V_T, \ \tau = T\mathbb{1}_{\{V_T < L\}} + \infty \mathbb{1}_{\{V_T \geq L\}}.$$

Using the risk-neutral valuation formula

$$D(t, T) = e^{-r(T-t)}\mathbb{E}_{\mathbb{P}^*}\big(\min(V_T, L) \,\big|\, \mathcal{F}_t\big),$$

he derives a closed-form solution for the price of a defaultable bond in terms of two state variables: the profit of the firm and the value of the firm (note that these processes do not represent prices of traded assets). An important contribution of Buffet's paper is raising the issue of the appropriate choice of the basic tradeable assets for the purpose of valuing and hedging of defaultable claims within the framework of the structural approach.

# 3. First-Passage-Time Models

The *first-passage-time approach* extends the original Merton model by accounting for the observed feature that the default may occur not only at the debt's maturity, but also prior to this date. Formally, it associates the default event with the first passage time of some specified random process, most notably the firm's value process, to some pre-specified barrier. The default triggering barrier may be a random process itself, called a barrier process, and may be given either endogenously or exogenously with respect to the model. Consequently, the first-passage-time models allow for a greater flexibility in modeling credit events in comparison with the Merton model of corporate debt. First, they allow for the time of the bankruptcy of the firm to occur before the maturity of the debt instrument issued by the firm. Second, the recovery payoff associated with the default event can be specified in a large variety of ways, in order to reflect more closely the real-life bond covenants and other important factors, such as the bankruptcy costs and/or taxes.

In Sect. 3.1, we present selected probabilistic properties of first passage times that will prove useful in the sequel. Subsequent sections provide a detailed analysis of several specific first-passage-time models that have been developed in financial literature. Although this is done in a unified way, we do not feel that it would be beneficial to elaborate an abstract setting that would cover all existing models as special cases. We prefer to categorize the first-passage-time models into two broad classes. The first class contains models that assume deterministic short-term default-free interest rates; models from the second class assume that interest rates follow stochastic processes. Such a classification roughly reflects the historical developments of the models. In our discussion, we study defaultable bonds with both finite maturity (corporate zero-coupon bonds) and infinite maturity (corporate consol bonds). To derive the values of these instruments we rely on the probabilistic techniques as well as the analytical approach; a much stronger emphasis is put on the former techniques, though. Let us emphasize the important contribution of the first-passage-time model literature to the analysis of the optimal capital structure of a firm; Sect. 3.3 is entirely devoted to this issue. In Sect. 3.4, we present an overview of first-passage-time models with random interest rates. Sect. 3.5 deals very briefly with some recent developments, and Sect. 3.6 examines the case of dependent defaults.

## 3.1 Properties of First Passage Times

We have already briefly discussed the risk-neutral valuation formulae for corporate bonds. In fact, many results in the existing literature rely on the probabilistic approach. For instance, bond valuation formulae in Longstaff and Schwartz (1995) and Saá-Requejo and Santa-Clara (1999) correspond to the following generic expression

$$D(t,T) = B(t,T)\,\mathbb{P}_T\{\tau > T \,|\, \mathcal{F}_t\} + \delta B(t,T)\,\mathbb{P}_T\{\tau \leq T \,|\, \mathcal{F}_t\}$$

in which the default time is defined as the first passage time of the value process to a (constant or variable) barrier. Direct computations based on the above formula require, of course, the knowledge of conditional distribution of the default time $\tau$ with respect to the $\sigma$-field $\mathcal{F}_t$. In this section, we shall provide a few results related to this issue.

Let us first consider two one-dimensional Itô processes $X^1$ and $X^2$ with respective dynamics under the probability measure $\mathbb{P}^*$ given by

$$dX_t^i = X_t^i\big(\mu_i(t)\,dt + \sigma_i(t)\,dW_t^i\big), \quad X_0^i = x^i > 0, \tag{3.1}$$

for $i = 1, 2$, where $W^i$, $i = 1, 2$, are independent $d$-dimensional Brownian motions with respect to the underlying filtration $\mathbb{F}$, and $\mu_i : \mathbb{R}_+ \to \mathbb{R}$, $\sigma_i : \mathbb{R}_+ \to \mathbb{R}^d$ are such that the SDEs (3.1) possess unique, strong, global solutions. Let us also assume that $x^1 > x^2$. Frequently, the default time $\tau$ is modeled as $\tau = \inf\{t \geq 0 : X_t^1 \leq X_t^2\}$. It is convenient to introduce the log-ratio process $Y_t := \ln(X_t^1/X_t^2)$, so that $\tau = \inf\{t \geq 0 : Y_t \leq 0\}$. The dynamics of $Y$ are described in the next lemma. Since the proof of Lemma 3.1.1 relies on a straightforward application of Itô's formula, it is omitted.

**Lemma 3.1.1.** *The process $Y$ satisfies*

$$dY_t = \nu(t)\,dt + \sigma_1(t)\,dW_t^1 - \sigma_2(t)\,dW_t^2 \tag{3.2}$$

*with*

$$\nu(t) = \mu_1(t) - \mu_2(t) + \tfrac{1}{2}|\sigma_2(t)|^2 - \tfrac{1}{2}|\sigma_1(t)|^2, \tag{3.3}$$

*where $|\cdot|$ stands for the Euclidean norm in $\mathbb{R}^d$.*

Suppose now that the coefficients $\mu_i$ are real constants and $\sigma_i$, $i = 1, 2$ are constant vectors in $\mathbb{R}^d$. In this case, the process $Y$ follows a Brownian motion with the standard deviation $\sigma$ and the drift $\nu$, specifically: $dY_t = \nu\,dt + \sigma\,dW_t^*$, $Y_0 = y_0$, where

$$\nu = \mu_1 - \mu_2 + \tfrac{1}{2}|\sigma_2|^2 - \tfrac{1}{2}|\sigma_1|^2, \quad \sigma^2 = |\sigma_1|^2 + |\sigma_2|^2,$$

and $W^*$ is a standard (one-dimensional) Brownian motion under $\mathbb{P}^*$ with respect to $\mathbb{F}$. Put another way:

$$Y_t = y_0 + \nu t + \sigma W_t^*, \quad \forall t \in \mathbb{R}_+, \tag{3.4}$$

for some constants $\nu \in \mathbb{R}$ and $\sigma > 0$. Let us notice that $Y$ inherits from $W^*$ a strong Markov property with respect to $\mathbb{F}$.

### 3.1.1 Probability Law of the First Passage Time

Let $\tau$ stand for the *first passage time to zero* by the process $Y$, that is, $\tau := \inf\{t \geq 0 : Y_t = 0\}$. It is well known that in an arbitrarily small time interval $[0, t]$ the sample path of the Brownian motion started at 0 passes through origin infinitely many times (see, for instance, Page 42 in Krylov (1995)). Using Girsanov's theorem and the strong Markov property of the Brownian motion, it is thus easy to deduce that first passage time by $Y$ to zero coincides with the first crossing time by $Y$ of the level 0, that is, with probability 1:

$$\tau = \inf\{t \geq 0 : Y_t < 0\} = \inf\{t \geq 0 : Y_t \leq 0\}.$$

**Lemma 3.1.2.** *Let $Y$ be given by (3.4), where $\nu \in \mathbb{R}$, $\sigma > 0$, and $W^*$ is a standard Brownian motion under $\mathbb{P}^*$. Then the random variable $\tau$ has an inverse Gaussian probability distribution under $\mathbb{P}^*$. More specifically, for any $0 < s < \infty$,*

$$\mathbb{P}^*\{\tau \leq s\} = \mathbb{P}^*\{\tau < s\} = N(h_1(s)) + e^{-2\nu\sigma^{-2}y_0} N(h_2(s)), \tag{3.5}$$

*where $N$ is the standard Gaussian cumulative distribution function, and*

$$h_1(s) = \frac{-y_0 - \nu s}{\sigma\sqrt{s}}, \quad h_2(s) = \frac{-y_0 + \nu s}{\sigma\sqrt{s}}.$$

*Proof.* Notice first that

$$\mathbb{P}^*\{\tau \geq s\} = \mathbb{P}^*\{\inf_{0 \leq u \leq s} Y_u \geq 0\} = \mathbb{P}^*\{\inf_{0 \leq u \leq s} X_u \geq -y_0\}, \tag{3.6}$$

where $X_u = \nu u + \sigma W_u^*$. Recall that for every $x < 0$ we have (see, e.g., Harrison (1990), Chap. II in Krylov (1995) or Corollary B.3.4 in Musiela and Rutkowski (1997a))

$$\mathbb{P}^*\{\inf_{0 \leq u \leq s} X_u \geq x\} = N\left(\frac{-x + \nu s}{\sigma\sqrt{s}}\right) - e^{2\nu\sigma^{-2}x} N\left(\frac{x + \nu s}{\sigma\sqrt{s}}\right).$$

When combined with (3.6), this yields (3.5). □

The following corollary is a consequence of Lemma 3.1.2 and the strong Markov property of the process $Y$ with respect to the filtration $\mathbb{F}$.

**Corollary 3.1.1.** *Under the assumptions of Lemma 3.1.2 for any $t < s$ we have, on the set $\{\tau > t\}$,*

$$\mathbb{P}^*\{\tau \leq s \,|\, \mathcal{F}_t\} = N\left(\frac{-Y_t - \nu(s-t)}{\sigma\sqrt{s-t}}\right) + e^{-2\nu\sigma^{-2}Y_t} N\left(\frac{-Y_t + \nu(s-t)}{\sigma\sqrt{s-t}}\right).$$

We are in a position to apply the foregoing results to specific examples of default times. In the first example, we examine the case of a constant lower threshold.

*Example 3.1.1.* Let the value process $V$ follow (2.12) with constant coefficients $\kappa$ and $\sigma_V > 0$. In addition, suppose that the short-term interest rate process is constant, i.e., $r_t = r$, $t \geq 0$. We thus have

$$dV_t = V_t\big((r - \kappa)\,dt + \sigma_V\,dW_t^*\big). \tag{3.7}$$

Let us also assume that the barrier process $v$ is constant and equal to $\bar{v}$, where the constant $\bar{v}$ satisfies $\bar{v} < V_0$. We set

$$\tau = \inf\{t \geq 0 : V_t \leq \bar{v}\} = \inf\{t \geq 0 : V_t < \bar{v}\}.$$

Now, letting $X_t^1 = V_t$ and $X_t^2 = \bar{v}$, so that $Y_t = \ln(V_t/\bar{v})$, and identifying the terms in (3.1), we obtain

$$\mu_1 \equiv r - \kappa, \quad \sigma_1 \equiv \sigma_V, \quad x^1 = V_0$$

and

$$\mu_2 \equiv 0, \quad \sigma_2 \equiv 0, \quad x^2 = \bar{v}.$$

Consequently, $\nu = r - \kappa - \frac{1}{2}\sigma_V^2$ and $\sigma = \sigma_V$ in (3.4). Applying Corollary 3.1.1, we obtain for every $s > t$, on the set $\{\tau > t\}$,

$$\mathbb{P}^*\{\tau \leq s \,|\, \mathcal{F}_t\} = N\left(\frac{\ln\frac{\bar{v}}{V_t} - \nu(s - t)}{\sigma_V\sqrt{s - t}}\right) + \left(\frac{\bar{v}}{V_t}\right)^{2a} N\left(\frac{\ln\frac{\bar{v}}{V_t} + \nu(s - t)}{\sigma_V\sqrt{s - t}}\right),$$

where

$$a = \frac{\nu}{\sigma_V^2} = \frac{r - \kappa - \frac{1}{2}\sigma_V^2}{\sigma_V^2}. \tag{3.8}$$

The last result was used in Leland and Toft (1996) (see Sect. 3.3.3 below).

*Example 3.1.2.* Assume that the value process $V$ and the short-term interest rate $r$ are as in Example 3.1.1. For a fixed $\gamma$, let the barrier function be defined as $\bar{v}(t) = Ke^{-\gamma(T-t)}$ for $t \in \mathbb{R}_+$, so that $\bar{v}(t)$ satisfies

$$d\bar{v}(t) = \gamma\bar{v}(t)\,dt, \quad \bar{v}(0) = Ke^{-\gamma T}.$$

Letting $X_t^1 = V_t$, $X_t^2 = \bar{v}(t)$ and identifying the terms in (3.1), we obtain

$$\mu_1 \equiv r - \kappa, \quad \sigma_1 \equiv \sigma_V, \quad x^1 = \bar{v}(0)$$

and

$$\mu_2 \equiv \gamma, \quad \sigma_2 \equiv 0, \quad x^2 = Ke^{-\gamma T},$$

so that the drift and diffusion coefficients in (3.4) are $\tilde{\nu} \equiv r - \kappa - \gamma - \frac{1}{2}\sigma_V^2$ and $\sigma \equiv \sigma_V$. We define the stopping time $\tau$ as $\tau = \inf\{t \geq 0 : V_t \leq \bar{v}(t)\}$. From Corollary 3.1.1, we obtain for every $t < s$, on the set $\{\tau > t\}$,

$$\mathbb{P}^*\{\tau \leq s \,|\, \mathcal{F}_t\} = N\left(\frac{\ln\frac{\bar{v}(t)}{V_t} - \tilde{\nu}(s - t)}{\sigma_V\sqrt{s - t}}\right) + \left(\frac{\bar{v}(t)}{V_t}\right)^{2\tilde{a}} N\left(\frac{\ln\frac{\bar{v}(t)}{V_t} + \tilde{\nu}(s - t)}{\sigma_V\sqrt{s - t}}\right),$$

where

$$\tilde{a} = \frac{\tilde{\nu}}{\sigma_V^2} = \frac{r - \kappa - \gamma - \frac{1}{2}\sigma_V^2}{\sigma_V^2}. \tag{3.9}$$

The last formula was used in Black and Cox (1976) (see Sect. 3.2.1 below).

### 3.1.2 Joint Probability Law of $Y$ and $\tau$

We shall now establish the joint law of $Y$ and $\tau$. To be more specific, we shall find, for every $y \geq 0$,

$$I := \mathbb{P}^*\{Y_s \geq y, \tau \geq s \,|\, \mathcal{F}_t\} = \mathbb{P}^*\{Y_s \geq y, \tau > s \,|\, \mathcal{F}_t\},$$

where $\tau = \inf\{t \geq 0 : Y_t \leq 0\} = \inf\{t \geq 0 : Y_t < 0\}$. Recall that we denote $Y_t = y_0 + X_t$, where $X_t = \nu t + \sigma W_t$. We write

$$m_s^X = \inf_{0 \leq u \leq s} X_u, \quad m_s^Y = \inf_{0 \leq u \leq s} Y_u.$$

Let us quote the following well known result (see, for instance, Corollary B.3.3 in Musiela and Rutkowski (1997a)).

**Lemma 3.1.3.** *For every $s > 0$, the joint distribution of $(X_s, m_s^X)$ satisfies, for every $x, y \in \mathbb{R}$ such that $y \leq 0$ and $x \geq y$,*

$$\mathbb{P}^*\{X_s \geq x, m_s^X \geq y\} = N\left(\frac{-x + \nu s}{\sigma\sqrt{s}}\right) - e^{2\nu\sigma^{-2}y} N\left(\frac{2y - x + \nu s}{\sigma\sqrt{s}}\right).$$

**Corollary 3.1.2.** *For any $s > 0$ and $y \geq 0$ we have*

$$\mathbb{P}^*\{Y_s \geq y, \tau \geq s\} = N\left(\frac{-y + y_0 + \nu s}{\sigma\sqrt{s}}\right) - e^{-2\nu\sigma^{-2}y_0} N\left(\frac{-y - y_0 + \nu s}{\sigma\sqrt{s}}\right).$$

*Proof.* Since

$$\mathbb{P}^*\{Y_s \geq y, \tau \geq s\} = \mathbb{P}^*\{Y_s \geq y, m_s^Y \geq 0\} = \mathbb{P}^*\{X_s \geq y - y_0, m_s^X \geq -y_0\},$$

the formula is obvious. □

More generally, the Markov property of $Y$ justifies the following result.

**Lemma 3.1.4.** *Under the assumptions of Lemma 3.1.2, for any $t < s$ and $y \geq 0$ we have, on the set $\{\tau > t\}$,*

$$\mathbb{P}^*\{Y_s \geq y, \tau \geq s \,|\, \mathcal{F}_t\} = N\left(\frac{-y + Y_t + \nu(s - t)}{\sigma\sqrt{s - t}}\right)$$
$$- e^{-2\nu\sigma^{-2}Y_t} N\left(\frac{-y - Y_t + \nu(s - t)}{\sigma\sqrt{s - t}}\right).$$

*Example 3.1.3.* Assume, as before, that the dynamics of $V$ are

$$dV_t = V_t\big((r - \kappa)\,dt + \sigma_V\,dW_t^*\big) \tag{3.10}$$

and $\tau = \inf\{t \geq 0 : V_t \leq \bar{v}\} = \inf\{t \geq 0 : V_t < \bar{v}\}$, where the constant $\bar{v}$ satisfies $\bar{v} < V_0$. By applying Lemma 3.1.4 to $Y_t = \ln(V_t/\bar{v})$ and $y = \ln(x/\bar{v})$, we obtain the following result, which is valid for $x \geq \bar{v}$, on the set $\{\tau > t\}$,

$$\mathbb{P}^*\{V_s \geq x, \tau \geq s \,|\, \mathcal{F}_t\} = N\left(\frac{\ln(V_t/x) + \nu(s-t)}{\sigma\sqrt{s-t}}\right)$$
$$- \left(\frac{\bar{v}}{V_t}\right)^{2a} N\left(\frac{\ln\bar{v}^2 - \ln(xV_t) + \nu(s-t)}{\sigma\sqrt{s-t}}\right),$$

where $\nu = r - \kappa - \frac{1}{2}\sigma_V^2$ and $a = \nu\sigma_V^{-2}$.

*Example 3.1.4.* Assume that $V$ satisfies (3.10) and that the barrier function equals $\bar{v}(t) = Ke^{-\gamma(T-t)}$ for some positive constant $K$. Using again Lemma 3.1.4, but this time with $Y_t = \ln(V_t/\bar{v}(t))$ and $y = \ln(x/\bar{v}(s))$, we find that for every $t < s \leq T$ and $x \geq \bar{v}(s)$ we have, on the set $\{\tau > t\}$,

$$\mathbb{P}^*\{V_s \geq x, \tau \geq s \,|\, \mathcal{F}_t\} = N\left(\frac{\ln(V_t/\bar{v}(t)) - \ln(x/\bar{v}(s)) + \tilde{\nu}(s-t)}{\sigma_V\sqrt{s-t}}\right)$$
$$- \left(\frac{\bar{v}(t)}{V_t}\right)^{2\tilde{a}} N\left(\frac{-\ln(V_t/\bar{v}(t)) - \ln(x/\bar{v}(s)) + \tilde{\nu}(s-t)}{\sigma_V\sqrt{s-t}}\right),$$

where $\tilde{\nu} = r - \kappa - \gamma - \frac{1}{2}\sigma_V^2$ and $\tilde{a} = \tilde{\nu}\sigma_V^{-2}$. Upon simplification, this yields

$$\mathbb{P}^*\{V_s \geq x, \tau \geq s \,|\, \mathcal{F}_t\} = N\left(\frac{\ln(V_t/x) + \nu(s-t)}{\sigma_V\sqrt{s-t}}\right)$$
$$- \left(\frac{\bar{v}(t)}{V_t}\right)^{2\tilde{a}} N\left(\frac{\ln\bar{v}^2(t) - \ln(xV_t) + \nu(s-t)}{\sigma_V\sqrt{s-t}}\right),$$

where $\nu = r - \kappa - \frac{1}{2}\sigma_V^2$. In particular, by setting $t = 0$ and $s = T$, we obtain for $x \geq \bar{v}(T)$

$$\mathbb{P}^*\{V_T \geq x, \tau \geq T\} = N\left(\frac{\ln(V_0/x) + \nu T}{\sigma_V\sqrt{T}}\right)$$
$$- \left(\frac{\bar{v}(0)}{V_0}\right)^{2\tilde{a}} N\left(\frac{\ln\bar{v}^2(0) - \ln(xV_0) + \nu T}{\sigma_V\sqrt{T}}\right).$$

*Remarks.* Notice that if we take $x = \bar{v}(s) = Ke^{-\gamma(T-s)}$, then clearly

$$1 - \mathbb{P}^*\{V_s \geq \bar{v}(s), \tau \geq s \,|\, \mathcal{F}_t\} = \mathbb{P}^*\{\tau < s \,|\, \mathcal{F}_t\} = \mathbb{P}^*\{\tau \leq s \,|\, \mathcal{F}_t\}.$$

On the other hand, we have

$$1 - N\left(\frac{\ln(V_t/\bar{v}(s)) + \nu(s-t)}{\sigma_V\sqrt{s-t}}\right) = N\left(\frac{\ln(\bar{v}(t)/V_t) - \tilde{\nu}(s-t)}{\sigma_V\sqrt{s-t}}\right)$$

and

$$N\left(\frac{\ln\bar{v}^2(t) - \ln(\bar{v}(s)V_t) + \nu(s-t)}{\sigma_V\sqrt{s-t}}\right) = N\left(\frac{\ln(\bar{v}(t)/V_t) + \tilde{\nu}(s-t)}{\sigma_V\sqrt{s-t}}\right).$$

It is thus easy to see that by setting $x = \bar{v}(s)$ we rediscover the formula previously established in Example 3.1.2.

## 3.2 Black and Cox Model

The original Merton model does not allow for a premature default, in the sense that the default may only occur at the maturity of the claim. Several authors put forward structural-type models in which this restrictive and un-realistic feature is relaxed. In most of these models, the time of default is given as the first passage time of the value process $V$ to a deterministic or random barrier. The default may thus occur at any time before or on the bond's maturity date $T$. The challenge here is to appropriately specify the lower threshold $v$, the recovery process $Z$, and to compute the corresponding functional that appears on the right-hand side of (2.3). As one might eas-ily guess, this is a non-trivial problem, in general. In addition, the practical problem of the lack of direct observations of the value process $V$ largely limits the applicability of the first-passage-time models. In this section, we survey several first passage time structural models corresponding to various specifi-cations of basic components of a credit risk model. In most cases examined below, the default time is denoted by $\tau$; the symbols $\bar{\tau}$, $\hat{\tau}$ and $\tilde{\tau}$ being reserved to some auxiliary random times.

### 3.2.1 Corporate Zero-Coupon Bond

Black and Cox (1976) extend Merton's (1974) research in several directions. In particular, they make account for specific features of debt contracts as: safety covenants, debt subordination, and restrictions on the sale of assets. They assume that the firm's stockholders (or bondholders) receive a con-tinuous dividend payment, proportional to the current value of the firm. Consequently, equation (2.31) takes the following form:

$$dV_t = V_t\big((r - \kappa)\, dt + \sigma_V\, dW_t^*\big), \tag{3.11}$$

where the constant $\kappa \geq 0$ represents the payout ratio, and $\sigma_V > 0$ is the constant volatility coefficient. The short-term interest rate is assumed to be non-random, specifically, $r_t = r$, where $r$ is a constant. This means that the interest rate risk is disregarded in the original Black and Cox (1976) model.

**Safety covenants.** Let us first focus on the safety covenants in the firm's indenture provisions. Generally speaking, safety covenants provide the firm's bondholders with the right to force the firm to bankruptcy or reorganization if the firm is doing poorly according to a set standard. The standard for a poor performance is set in Black and Cox (1976) in terms of a time-dependent deterministic barrier $\bar{v}(t) = Ke^{-\gamma(T-t)}$, $t \in [0,T)$, for some constants $K, \gamma > 0$. They postulate that as soon as the value of firm's assets crosses this lower threshold, the bondholders take over the firm. Otherwise, default takes place at debt's maturity or not depending on whether $V_T < L$ or not. Let us set:

$$v_t = \begin{cases} \bar{v}(t), & \text{for } t < T, \\ L, & \text{for } t = T. \end{cases} \tag{3.12}$$

The default event occurs at the first time $t \in [0, T]$ at which the firm's value $V_t$ falls below the level $v_t$, or the default event does not occur at all. The default time $\tau$ thus equals (as usual, $\inf \emptyset = +\infty$):

$$\tau = \inf \{ t \in [0, T] : V_t < v_t \}.$$

The recovery process $Z$ and the recovery payoff $\tilde{X}$ are proportional to the value process, specifically, $Z \equiv \beta_2 V$ and $\tilde{X} = \beta_1 V_T \mathbb{1}_{\{\bar{\tau} \geq T\}}$ for some constants $\beta_1, \beta_2 \in [0, 1]$. The classic case examined by Black and Cox (1976) corresponds to $\beta_1 = \beta_2 = 1$. To summarize, we consider the following model:

$$X = L, \ A \equiv 0, \ Z \equiv \beta_2 V, \ \tilde{X} = \beta_1 V_T \mathbb{1}_{\{\bar{\tau} \geq T\}}, \ \tau = \bar{\tau} \wedge \hat{\tau},$$

where the *early default time* $\bar{\tau}$ equals

$$\bar{\tau} = \inf \{ t \in [0, T) : V_t < \bar{v}(t) \},$$

and $\hat{\tau}$ stands for Merton's default time: $\hat{\tau} = T \mathbb{1}_{\{V_T < L\}} + \infty \mathbb{1}_{\{V_T \geq L\}}$.

*Remarks.* Assume that $V_0 > \bar{v}(0)$. It is important to notice that since the process $V$ satisfies (3.11) and $\bar{v}$ is a smooth function, $\bar{\tau}$ is also the first passage time of the value process $V$ to the deterministic barrier $\bar{v}$, specifically,

$$\bar{\tau} = \inf \{ t \in [0, T) : V_t \leq \bar{v}(t) \} = \inf \{ t \in [0, T) : V_t = \bar{v}(t) \}.$$

The choice of a strict or large inequality in the definition of the early default time $\bar{\tau}$ is thus a matter of convention. The same observation applies to other examples of first-passage-time structural models considered in the sequel.

In addition, we postulate that $\bar{v}(t) \leq LB(t, T)$ or, more explicitly,

$$K e^{-\gamma(T-t)} \leq L e^{-r(T-t)}, \quad \forall t \in [0, T], \tag{3.13}$$

so that, in particular, $K \leq L$. Condition (3.13) ensures that the payoff to the bondholder at the default time $\tau$ never exceeds the face value of debt, discounted at a risk-free rate. Since the interest rate $r$ is assumed to be constant, the pricing function $u = u(V, t)$ of a defaultable bond solves the following PDE:

$$u_t(V, t) + (r - \kappa) V u_V(V, t) + \tfrac{1}{2} \sigma_V^2 V^2 u_{VV}(V, t) - r u(V, t) = 0$$

with the boundary condition $u(K e^{-\gamma(T-t)}, t) = \beta_2 K e^{-\gamma(T-t)}$ and the terminal condition $u(V, T) = \min(\beta_1 V, L)$. To find an explicit solution to this problem, we prefer to rely on a probabilistic approach, though. To this end, we notice that for any $t < T$ the price $D(t, T) = u(V_t, t)$ of a defaultable bond admits the following probabilistic representation, on the set $\{\tau > t\} = \{\bar{\tau} > t\}$ (see the proof of Lemma 2.2.1):

$$D(t, T) = \mathbb{E}_{\mathbb{P}^*} \left( L e^{-r(T-t)} \mathbb{1}_{\{\bar{\tau} \geq T, V_T \geq L\}} \,\Big|\, \mathcal{F}_t \right)$$

$$+ \mathbb{E}_{\mathbb{P}^*} \left( \beta_1 V_T e^{-r(T-t)} \mathbb{1}_{\{\bar{\tau} \geq T, V_T < L\}} \,\Big|\, \mathcal{F}_t \right)$$

$$+ \mathbb{E}_{\mathbb{P}^*} \left( K \beta_2 e^{-\gamma(T-\bar{\tau})} e^{-r(\bar{\tau}-t)} \mathbb{1}_{\{t < \bar{\tau} < T\}} \,\Big|\, \mathcal{F}_t \right).$$

After default – that is, on the set $\{\tau \le t\} = \{\bar{\tau} \le t\}$, we clearly have

$$D(t,T) = \beta_2 \bar{v}(\tau) B^{-1}(\tau,T) B(t,T) = K\beta_2 e^{-\gamma(T-\tau)} e^{r(t-\tau)}.$$

The first two conditional expectations in the valuation formula for defaultable bond can be computed by using the formula for the conditional probability $\mathbb{P}^*\{V_s \ge x, \tau \ge s \,|\, \mathcal{F}_t\}$, established in Example 3.1.4. To evaluate the third conditional expectation, we shall employ the conditional probability law of the first passage time of the process $V$ to the barrier $\bar{v}(t)$ – this law was already found in Example 3.1.2. We are thus in a position to establish the following valuation result, due to Black and Cox (1976). Recall that we denote:

$$\nu = r - \kappa - \tfrac{1}{2}\sigma_V^2, \quad \tilde{\nu} = \nu - \gamma = r - \kappa - \gamma - \tfrac{1}{2}\sigma_V^2,$$

and $\tilde{a} = \tilde{\nu}\sigma_V^{-2}$. For the sake of brevity, in the statement and the proof of Proposition 3.2.1 we shall write $\sigma$ instead of $\sigma_V$.

**Proposition 3.2.1.** *Assume that $\tilde{\nu}^2 + 2\sigma^2(r-\gamma) > 0$. Then the price process $D(t,T) = u(V_t, t)$ of a defaultable bond equals, on the set $\{\tau > t\}$,*

$$\begin{aligned}
D(t,T) = {} & LB(t,T)\big(N\big(h_1(V_t, T-t)\big) - R_t^{2\tilde{a}} N\big(h_2(V_t, T-t)\big)\big) \\
& + \beta_1 V_t e^{-\kappa(T-t)}\big(N\big(h_3(V_t, T-t)\big) - N\big(h_4(V_t, T-t)\big)\big) \\
& + \beta_1 V_t e^{-\kappa(T-t)} R_t^{2\tilde{a}+2}\big(N\big(h_5(V_t, T-t)\big) - N\big(h_6(V_t, T-t)\big)\big) \\
& + \beta_2 V_t \big(R_t^{\theta+\zeta} N\big(h_7(V_t, T-t)\big) + R_t^{\theta-\zeta} N\big(h_8(V_t, T-t)\big)\big),
\end{aligned}$$

*where $R_t = \bar{v}(t)/V_t$,*

$$\theta = \tilde{a} + 1, \quad \zeta = \sigma^{-2}\sqrt{\tilde{\nu}^2 + 2\sigma^2(r-\gamma)}$$

*and*

$$h_1(V_t, T-t) = \frac{\ln(V_t/L) + \nu(T-t)}{\sigma\sqrt{T-t}},$$

$$h_2(V_t, T-t) = \frac{\ln \bar{v}^2(t) - \ln(LV_t) + \nu(T-t)}{\sigma\sqrt{T-t}},$$

$$h_3(V_t, T-t) = \frac{\ln(L/V_t) - (\nu + \sigma^2)(T-t)}{\sigma\sqrt{T-t}},$$

$$h_4(V_t, T-t) = \frac{\ln(K/V_t) - (\nu + \sigma^2)(T-t)}{\sigma\sqrt{T-t}},$$

$$h_5(V_t, T-t) = \frac{\ln \bar{v}^2(t) - \ln(LV_t) + (\nu + \sigma^2)(T-t)}{\sigma\sqrt{T-t}},$$

$$h_6(V_t, T-t) = \frac{\ln \bar{v}^2(t) - \ln(KV_t) + (\nu + \sigma^2)(T-t)}{\sigma\sqrt{T-t}},$$

$$h_7(V_t, T-t) = \frac{\ln(\bar{v}(t)/V_t) + \zeta\sigma^2(T-t)}{\sigma\sqrt{T-t}},$$

$$h_8(V_t, T-t) = \frac{\ln(\bar{v}(t)/V_t) - \zeta\sigma^2(T-t)}{\sigma\sqrt{T-t}}.$$

Before proceeding to the proof of Proposition 3.2.1, we state an elementary lemma.

**Lemma 3.2.1.** *For any $a \in \mathbb{R}$ and $b > 0$ we have, for every $y > 0$,*

$$\int_0^y x \, dN \left( \frac{\ln x + a}{b} \right) = e^{\frac{1}{2}b^2 - a} N \left( \frac{\ln y + a - b^2}{b} \right) \tag{3.14}$$

*and*

$$\int_0^y x \, dN \left( \frac{-\ln x + a}{b} \right) = e^{\frac{1}{2}b^2 + a} N \left( \frac{-\ln y + a + b^2}{b} \right). \tag{3.15}$$

*Let $a, b, c \in \mathbb{R}$ satisfy $b < 0$ and $c^2 > 2a$. Then for every $y > 0$*

$$\int_0^y e^{ax} \, dN \left( \frac{b - cx}{\sqrt{x}} \right) = \frac{d + c}{2d} g(y) + \frac{d - c}{2d} h(y), \tag{3.16}$$

*where $d = \sqrt{c^2 - 2a}$ and*

$$g(y) = e^{b(c-d)} N \left( \frac{b - dy}{\sqrt{y}} \right), \quad h(y) = e^{b(c+d)} N \left( \frac{b + dy}{\sqrt{y}} \right).$$

*Proof.* The proof of (3.14)–(3.15) is standard. For (3.16), observe that

$$f(y) := \int_0^y e^{ax} \, dN \left( \frac{b - cx}{\sqrt{x}} \right) = \int_0^y e^{ax} n \left( \frac{b - cx}{\sqrt{x}} \right) \left( -\frac{b}{2x^{3/2}} - \frac{c}{2\sqrt{x}} \right) dx,$$

where $n$ is the probability density function of the standard Gaussian law. On the other hand,

$$g'(x) = e^{b(c - \sqrt{c^2 - 2a})} n \left( \frac{b - \sqrt{c^2 - 2a}\, x}{\sqrt{x}} \right) \left( -\frac{b}{2x^{3/2}} - \frac{\sqrt{c^2 - 2a}}{2\sqrt{x}} \right)$$

$$= e^{ax} n \left( \frac{b - cx}{\sqrt{x}} \right) \left( -\frac{b}{2x^{3/2}} - \frac{d}{2\sqrt{x}} \right)$$

and

$$h'(x) = e^{b(c + \sqrt{c^2 - 2a})} n \left( \frac{b + \sqrt{c^2 - 2a}\, x}{\sqrt{x}} \right) \left( -\frac{b}{2x^{3/2}} + \frac{\sqrt{c^2 - 2a}}{2\sqrt{x}} \right)$$

$$= e^{ax} n \left( \frac{b - cx}{\sqrt{x}} \right) \left( -\frac{b}{2x^{3/2}} + \frac{d}{2\sqrt{x}} \right).$$

Consequently,

$$g'(x) + h'(x) = -e^{ax} \frac{b}{x^{3/2}} n \left( \frac{b - cx}{\sqrt{x}} \right)$$

and

$$g'(x) - h'(x) = -e^{ax} \frac{d}{x^{1/2}} n \left( \frac{b - cx}{\sqrt{x}} \right).$$

Thus, $f$ can be represented as follows:

$$f(y) = \frac{1}{2} \int_0^y \left( g'(x) + h'(x) + \frac{c}{d} \left( g'(x) - h'(x) \right) \right) dx.$$

Since $\lim_{y \to 0+} g(y) = \lim_{y \to 0+} h(y) = 0$, we conclude that for every $y > 0$ we have

$$f(y) = \frac{1}{2}(g(y) + h(y)) + \frac{c}{2d} \left( g(y) - h(y) \right).$$

This end the proof of the lemma.                                    □

*Proof of Proposition 3.2.1.* Since the proof relies on calculations that are rather standard, though lengthy, we shall merely sketch the proof. We need to find the following conditional expectations:

$$D_1(t,T) = LB(t,T) \, \mathbb{P}^* \{ V_T \geq L, \, \bar{\tau} \geq T \,|\, \mathcal{F}_t \},$$
$$D_2(t,T) = \beta_1 B(t,T) \, \mathbb{E}_{\mathbb{P}^*} \left( V_T \mathbb{1}_{\{V_T < L, \, \bar{\tau} \geq T\}} \,\big|\, \mathcal{F}_t \right),$$
$$D_3(t,T) = K \beta_2 B_t e^{-\gamma T} \, \mathbb{E}_{\mathbb{P}^*} \left( e^{(\gamma - r)\bar{\tau}} \mathbb{1}_{\{t < \bar{\tau} < T\}} \,\big|\, \mathcal{F}_t \right).$$

For the sake of notational convenience, we set $t = 0$. Let us first evaluate $D_1(0,T)$ – that is, the part of the bond's value corresponding to no-default event. From Example 3.1.4, we know that if $L \geq \bar{v}(T) = K$ then

$$\mathbb{P}^* \{ V_T \geq L, \, \bar{\tau} \geq T \} = N \left( \frac{\ln \frac{V_0}{L} + \nu T}{\sigma \sqrt{T}} \right) - R_0^{2\tilde{a}} N \left( \frac{\ln \frac{\bar{v}^2(0)}{LV_0} + \nu T}{\sigma \sqrt{T}} \right)$$

with $R_0 = \bar{v}(0)/V_0$. It is thus clear that

$$D_1(0,T) = LB(0,T) \left( N \left( h_1(V_0,T) \right) - R_0^{2\tilde{a}} N \left( h_2(V_0,T) \right) \right).$$

Let us now examine $D_2(0,T)$ – that is, the part of the bond's value associated with default at time $T$. It is clear that

$$\frac{D_2(0,T)}{\beta_1 B(0,T)} = \mathbb{E}_{\mathbb{P}^*} \left( V_T \mathbb{1}_{\{V_T < L, \, \bar{\tau} \geq T\}} \right) = \int_K^L x \, d\mathbb{P}^* \{ V_T < x, \, \bar{\tau} \geq T \}.$$

Using again Example 3.1.4 and the fact that $\mathbb{P}^* \{ \bar{\tau} \geq T \}$ does not depend on $x$, we get, for every $x \geq K$,

$$d\mathbb{P}^* \{ V_T < x, \, \bar{\tau} \geq T \} = dN \left( \frac{\ln \frac{x}{V_0} - \nu T}{\sigma \sqrt{T}} \right) + R_0^{2\tilde{a}} dN \left( \frac{\ln \frac{\bar{v}^2(0)}{xV_0} + \nu T}{\sigma \sqrt{T}} \right).$$

Let us denote

$$K_1(0) = \int_K^L x \, dN \left( \frac{\ln x - \ln V_0 - \nu T}{\sigma \sqrt{T}} \right)$$

and

$$K_2(0) = \int_K^L x\, dN \left( \frac{2\ln \bar{v}(0) - \ln x - \ln V_0 + \nu T}{\sigma\sqrt{T}} \right).$$

Using (3.14)–(3.15), we obtain

$$K_1(0) = V_0 e^{(r-\kappa)T} \left( N\left( \frac{\ln\frac{L}{V_0} - \hat{\nu}T}{\sigma\sqrt{T}} \right) - N\left( \frac{\ln\frac{K}{V_0} - \hat{\nu}T}{\sigma\sqrt{T}} \right) \right),$$

where $\hat{\nu} = \nu + \sigma^2 = r - \kappa + \frac{1}{2}\sigma^2$, and

$$K_2(0) = V_0 R_0^2 e^{(r-\kappa)T} \left( N\left( \frac{\ln\frac{\bar{v}^2(0)}{LV_0} + \hat{\nu}T}{\sigma\sqrt{T}} \right) - N\left( \frac{\ln\frac{\bar{v}^2(0)}{KV_0} + \hat{\nu}T}{\sigma\sqrt{T}} \right) \right).$$

Since

$$D_2(0,T) = \beta_1 B(0,T)\left( K_1(0) + R_0^{\tilde{a}} K_2(0) \right),$$

we conclude that

$$\begin{aligned}
D_2(0,T) &= \beta_1 V_0 e^{-\kappa T} \left( N\left( h_3(V_0, T) \right) - N\left( h_4(V_0, T) \right) \right) \\
&\quad + \beta_1 V_0 e^{-\kappa T} R_0^{2\tilde{a}+2} \left( N\left( h_5(V_0, T) \right) - N\left( h_6(V_0, T) \right) \right).
\end{aligned}$$

It remains to find $D_3(0,T)$ – that is, the part of bond's value associated with the possibility of forced bankruptcy before the bond's maturity date $T$. To this end, it is enough to calculate the following expected value

$$\bar{v}(0)\, \mathbb{E}_{\mathbb{P}^*}\left( e^{(\gamma-r)\bar{\tau}} \mathbb{1}_{\{\bar{\tau} < T\}} \right) = \bar{v}(0) \int_0^T e^{(\gamma-r)s}\, d\mathbb{P}^*\{\bar{\tau} \le s\},$$

where (see Example 3.1.2)

$$\mathbb{P}^*\{\bar{\tau} \le s\} = N\left( \frac{\ln(\bar{v}(0)/V_0) - \tilde{\nu}s}{\sigma\sqrt{s}} \right) + \left( \frac{\bar{v}(0)}{V_0} \right)^{2\tilde{a}} N\left( \frac{\ln(\bar{v}(0)/V_0) + \tilde{\nu}s}{\sigma\sqrt{s}} \right).$$

Notice that $\bar{v}(0) < V_0$, and thus $\ln(\bar{v}(0)/V_0) < 0$. Using (3.16), we obtain

$$\begin{aligned}
\bar{v}(0) &\int_0^T e^{(\gamma-r)s}\, dN \left( \frac{\ln(\bar{v}(0)/V_0) - \tilde{\nu}s}{\sigma\sqrt{s}} \right) \\
&= \frac{V_0(\tilde{a}+\zeta)}{2\zeta} R_0^{\theta-\zeta} N\left( h_8(V_0, T) \right) - \frac{V_0(\tilde{a}-\zeta)}{2\zeta} R_0^{\theta+\zeta} N\left( h_7(V_0, T) \right)
\end{aligned}$$

and

$$\begin{aligned}
\frac{\bar{v}(0)^{2\tilde{a}+1}}{V_0^{2\tilde{a}}} &\int_0^T e^{(\gamma-r)s}\, dN \left( \frac{\ln(\bar{v}(0)/V_0) + \tilde{\nu}s}{\sigma\sqrt{s}} \right) \\
&= \frac{V_0(\tilde{a}+\zeta)}{2\zeta} R_0^{\theta+\zeta} N\left( h_7(V_0, T) \right) - \frac{V_0(\tilde{a}-\zeta)}{2\zeta} R_0^{\theta-\zeta} N\left( h_8(V_0, T) \right).
\end{aligned}$$

Consequently,

$$D_3(0,T) = \beta_2 V_0 \left( R_0^{\theta+\zeta} N\left( h_7(V_0, T) \right) + R_0^{\theta-\zeta} N\left( h_8(V_0, T) \right) \right). \tag{3.17}$$

This completes the proof of the proposition. $\qquad\square$

The financial interpretation of the coefficients $\beta_1$ and $\beta_2$ is that they reflect the bankruptcy (or reorganization) costs incurred at the time of default. It is clear that as soon as $\beta_1 < 1$ and/or $\beta_2 < 1$ the value of a defaultable bond is less than in case of zero bankruptcy costs, i.e., when $\beta_1 = \beta_2 = 1$. In some circumstances, the values $\beta_1 < 1$ and/or $\beta_2 < 1$ can be interpreted as reflecting the violation of the strict priority rule.

It should be noted that, similarly as in the case of the Merton model, the Black and Cox model produces credit spreads close to zero for small maturities, a feature that is inconsistent with empirical studies. The reason again is that the default time is predictable with respect to the natural filtration of the value process $V$.

**Strict priority rule.** For the sake of simplicity, we shall assume that $\beta_1 = \beta_2 = 1$, i.e., no bankruptcy/reorganization costs are present. Suppose that the firm's debt can be classified into *senior* bonds and (subordinated) *junior* bonds, with the same maturity date $T$. At debt's maturity, payments can be made to the holders of junior bonds only if the promised payment to the holders of senior bonds has been made. Such a convention is commonly referred to as the *strict* (or *absolute*) *priority rule*. Assume that the total face value $L$ of the firm's liabilities equals $L = L_s + L_j$, where $L_s$ ($L_j$, resp.) is the face value of senior bonds (of junior bonds, resp.) Let $u(V_t, t; L, \bar{v})$ stand for the price $D(t, T)$ – given by Proposition 3.2.1 – of a defaultable bond in the Black and Cox model, where, for the sake of convenience, we have introduced in the notation the face value $L$ and the barrier function $\bar{v}$.

It is clear that the value $D_s(t, T)$ at time $t < T$ of the senior debt equals, on the set $\{\tau > t\}$,

$$D_s(t, T) = u(V_t, t; L_s, \bar{v})$$

and it amounts to $\min(\bar{v}(\tau), L_s B(\tau, T))$ at time of default, provided that default has occurred prior to the maturity date. The total value of firm's debt equals, on the set $\{\tau > t\}$,

$$D(t, T) = u(V_t, t; L, \bar{v})$$

and it equals $\bar{v}(\tau)$ at time of default. Thus, the value of the junior debt is

$$D_j(t, T) = D(t, T) - D_s(t, T) = u(V_t, t; L, \bar{v}) - u(V_t, t; L_s, \bar{v})$$

on the set $\{\tau > t\}$, and it equals $\min(\bar{v}(\tau) - L_s B(\tau, T), L_j B(\tau, T))$ at time of default, provided that the default has occurred prior to the maturity date. For instance, if $\bar{v}(t) = K B(t, T)$ for some constant $K \leq L$ then we have, on the set $\{\tau > t\}$,

$$D_j(t, T) = \begin{cases} L_j B(t, T), & \text{if } K = L, \\ D(t, T) - L_s B(t, T), & \text{if } L_s \leq K < L, \\ D(t, T) - D_s(t, T), & \text{if } K < L_s. \end{cases}$$

As one might easily guess, the above analysis can be extended to cover the case of several classes of subordinated debt.

**Special cases.** Let us now analyze some special cases of the Black-Cox valuation formula. We shall assume that $\beta_1 = \beta_2 = 1$, and the barrier function $\bar{v}$ is chosen in such a way that $K = L$. Then necessarily $\gamma \geq r$ (otherwise, condition (3.13) would be violated). Obviously, if $K = L$, then $K_1(t) = K_2(t) = 0$, and thus $D(t,T) = D_1(t,T) + D_3(t,T)$, where:

$$D_1(t,T) = LB(t,T)\big(N\big(h_1(V_t, T-t)\big) - R_t^{2\tilde{a}}N\big(h_2(V_t, T-t)\big)\big) \qquad (3.18)$$

and

$$D_3(t,T) = V_t\big(R_t^{\theta+\varsigma}N\big(h_7(V_t, T-t)\big) + R_t^{\theta-\varsigma}N\big(h_8(V_t, T-t)\big)\big). \qquad (3.19)$$

**Case $\gamma = r$.** If we also assume that $\gamma = r$, then $\varsigma = -\sigma^{-2}\tilde{\nu}$, and thus

$$V_t R_t^{\theta+\varsigma} = LB(t,T), \quad V_t R_t^{\theta-\varsigma} = V_t R_t^{2\tilde{a}+1} = LB(t,T)R_t^{2\tilde{a}}.$$

Moreover, it is also easy to see that in this case

$$h_1(V_t, T-t) = \frac{\ln(V_t/L) + \nu(T-t)}{\sigma\sqrt{T-t}} = -h_7(V_t, T-t),$$

while

$$h_2(V_t, T-t) = \frac{\ln \bar{v}^2(t) - \ln(LV_t) + \nu(T-t)}{\sigma\sqrt{T-t}} = h_8(V_t, T-t).$$

We conclude that if $\bar{v}(t) = Le^{-r(T-t)} = LB(t,T)$, then $D(t,T) = LB(t,T)$. This result is quite intuitive; a defaultable bond with a safety covenant represented by the barrier function, which equals the discounted value of the bond's face value, is obviously equivalent to a default-free bond with the same face value and maturity. Notice also that when $\gamma = r$ but $K < L$, then we have: $D_3(t,T) = KB(t,T)\mathbb{P}^*\{\tau < T \mid \mathcal{F}_t\}$.

**Case $\gamma > r$.** If $K = L$ but $\gamma > r$ then one would expect that $D(t,T)$ would be smaller than $LB(t,T)$. We shall show that when $\gamma$ tends to infinity (all other parameters being fixed), then the Black and Cox price converges to Merton's price of Proposition 2.3.1:

$$\lim_{\gamma\to\infty} D(t,T) = V_t e^{-\kappa(T-t)}N\big(-d_1(V_t, T-t)\big) + LB(t,T)\big(d_2(V_t, T-t)\big).$$

First, it is clear that $h_1(V_t, T-t) = d_2(V_t, T-t)$. Furthermore, straightforward calculations show that

$$\lim_{\gamma\to\infty} R_t^{2\tilde{a}}N\big(h_2(V_t, T-t)\big) = \lim_{\gamma\to\infty} R_t^{\theta-\varsigma}N\big(h_8(V_t, T-t)\big) = 0$$

and thus the second term on the right-hand side of (3.18), as well as the second term on the right-hand side of (3.19), vanish. Finally,

$$\lim_{\gamma\to\infty} R_t^{\theta+\varsigma}N\big(h_7(V_t, T-t)\big) = e^{-\kappa(T-t)}N\big(-d_1(V_t, T-t)\big),$$

since $\lim_{\gamma\to\infty} R_t^{\theta+\varsigma} = e^{-\kappa(T-t)}$ and $\lim_{\gamma\to\infty} h_7(V_t, T-t) = -d_1(V_t, T-t)$.

### 3.2.2 Corporate Coupon Bond

We shall assume now that $r > 0$ and that a defaultable bond of fixed maturity $T$ and face value $L$ pays continuously coupons at a constant rate $c$, so that $A_t = ct$ for $t \in \mathbb{R}_+$.[1] The coupon payments stop as soon as default occurs. Formally, we consider a defaultable claim specified as follows:

$$X = L, \ A_t = ct, \ Z \equiv \beta_2 V, \ \tilde{X} = \beta_1 V_T, \ \tau = \inf\{t \in [0,T] : V_t < v_t\}$$

with the barrier $v$ given by (3.12). Let us denote by $D_c(t,T)$ the value of such a claim at time $t < T$. It is clear that (cf. (2.28))

$$D_c(t,T) = D(t,T) + \mathbb{E}_{\mathbb{P}^*}\left(\int_t^T ce^{-r(s-t)} \mathbb{1}_{\{\tilde{\tau}>s\}} \, ds \,\Big|\, \mathcal{F}_t\right)$$

$$= D(t,T) + ce^{rt}\int_t^T e^{-rs} \mathbb{P}^*\{\tilde{\tau} > s \,|\, \mathcal{F}_t\} \, ds =: D(t,T) + A(t,T),$$

where $A(t,T)$ stands for the discounted value of future coupon payments. In particular, setting $t = 0$, we obtain

$$D_c(0,T) = D(0,T) + c\int_0^T e^{-rs} \mathbb{P}^*\{\tilde{\tau} > s\} \, ds = D(0,T) + A(0,T),$$

where (we preserve here the convention to write $\sigma$ rather than $\sigma_V$)

$$\mathbb{P}^*\{\tilde{\tau} > s\} = N\left(\frac{\ln(V_0/\bar{v}(0)) + \tilde{\nu}s}{\sigma\sqrt{s}}\right) - \left(\frac{\bar{v}(0)}{V_0}\right)^{2\tilde{a}} N\left(\frac{\ln(\bar{v}(0)/V_0) + \tilde{\nu}s}{\sigma\sqrt{s}}\right).$$

An integration by parts formula yields

$$\int_0^T e^{-rs} \mathbb{P}^*\{\tilde{\tau} > s\} \, ds = \frac{1}{r}\left(1 - e^{-rT}\mathbb{P}^*\{\tilde{\tau} > T\} + \int_0^T e^{-rs} \, d\mathbb{P}^*\{\tilde{\tau} > s\}\right).$$

We assume, as usual, that $V_0 > \bar{v}(0)$, so that $\ln(\bar{v}(0)/V_0) < 0$. Arguing in a similar way as in the last part of the proof of Proposition 3.2.1 (more specifically, using formula (3.16)), we obtain

$$\int_0^T e^{-rs} \, d\mathbb{P}^*\{\tilde{\tau} > s\} = -\left(\frac{\bar{v}(0)}{V_0}\right)^{\tilde{a}+\tilde{\zeta}} N\left(\frac{\ln(\bar{v}(0)/V_0) + \tilde{\zeta}\sigma^2 T}{\sigma\sqrt{T}}\right)$$

$$-\left(\frac{\bar{v}(0)}{V_0}\right)^{\tilde{a}-\tilde{\zeta}} N\left(\frac{\ln(\bar{v}(0)/V_0) - \tilde{\zeta}\sigma^2 T}{\sigma\sqrt{T}}\right),$$

where $\tilde{\nu} = r - \kappa - \gamma - \frac{1}{2}\sigma^2$, $\tilde{a} = \tilde{\nu}\sigma^{-2}$, and $\tilde{\zeta} = \sigma^{-2}\sqrt{\tilde{\nu}^2 + 2\sigma^2 r}$. Although we have focused on the case when $t = 0$, it is apparent that the derivation of the general formula for any $t < T$ hinges on basically the same arguments. Combining the above formulae, we arrive at the following result.

---

[1] Some readers may prefer to write here $cL$, where $L$ is the bond's face value, rather than $c$; this is, of course, a matter of convention only.

**Proposition 3.2.2.** *Consider a defaultable $T$-maturity bond with face value $L$, which pays continuously coupons at a constant rate $c$. The arbitrage price of such a bond equals $D_c(t, T) = D(t, T) + A(t, T)$, where $D(t, T)$ is the value of a defaultable zero-coupon bond given by Proposition 3.2.1, and $A(t, T)$ equals, on the set $\{\tau > t\} = \{\bar{\tau} > t\}$,*

$$A(t, T) = \frac{c}{r}\Big[1 - B(t, T)\Big(N\big(k_1(V_t, T - t)\big) - R_t^{2\tilde{a}}N\big(k_2(V_t, T - t)\big)\Big)$$
$$- R_t^{\tilde{a}+\tilde{\zeta}}N\big(g_1(V_t, T - t)\big) - R_t^{\tilde{a}-\tilde{\zeta}}N\big(g_2(V_t, T - t)\big)\Big],$$

*where $R_t = \bar{v}(t)/V_t$, and*

$$k_1(V_t, T - t) = \frac{\ln(V_t/\bar{v}(t)) + \tilde{\nu}(T - t)}{\sigma\sqrt{T - t}},$$

$$k_2(V_t, T - t) = \frac{\ln(\bar{v}(t)/V_t) + \tilde{\nu}(T - t)}{\sigma\sqrt{T - t}},$$

$$g_1(V_t, T - t) = \frac{\ln(\bar{v}(t)/V_t) + \tilde{\zeta}\sigma^2(T - t)}{\sigma\sqrt{T - t}},$$

$$g_2(V_t, T - t) = \frac{\ln(\bar{v}(t)/V_t) - \tilde{\zeta}\sigma^2(T - t)}{\sigma\sqrt{T - t}}.$$

Some authors apply the general result to the special case when the default triggering barrier is postulated to be a constant. In this special case, the coefficient $\gamma$ equals zero. Consequently, $\tilde{\nu} = \nu$ and

$$\tilde{\zeta} = \sigma^{-2}\sqrt{\nu^2 + 2\sigma^2 r} = \zeta.$$

Assume, in addition, that $\bar{v} \geq L$, so that the firm's insolvency at maturity $T$ is excluded. For the sake of the reader's convenience, we state the following immediate corollary to Propositions 3.2.1 and 3.2.2.

**Corollary 3.2.1.** *Assume that $\gamma = 0$ so that the barrier is constant: $v \equiv \bar{v}$. If $\bar{v} \geq L$ then the arbitrage price of a defaultable coupon bond equals, on the set $\{\tau > t\} = \{\bar{\tau} > t\}$,*

$$D_c(t, T) = \frac{c}{r} + B(t, T)\Big(L - \frac{c}{r}\Big)\Big(N\big(k_1(V_t, T - t)\big) - R_t^{2\tilde{a}}N\big(k_2(V_t, T - t)\big)\Big)$$
$$+ \Big(\beta_2\bar{v} - \frac{c}{r}\Big)\Big(R_t^{\tilde{a}+\tilde{\zeta}}N\big(g_1(V_t, T - t)\big) + R_t^{\tilde{a}-\tilde{\zeta}}N\big(g_2(V_t, T - t)\big)\Big),$$

*where $R_t = \bar{v}/V_t$.*

Let us mention that the valuation formula of Corollary 3.2.1 coincides with expression (3) in Leland and Toft (1996). Letting the bond's maturity $T$ tend to infinity, we obtain the following result for the arbitrage price of a consol bond (see also Corollary 3.2.2 below)

$$D_c(t) = D_c(t, \infty) = \frac{c}{r}\left(1 - \left(\frac{\bar{v}}{V_t}\right)^{\tilde{a}+\tilde{\zeta}}\right) + \beta_2\bar{v}\left(\frac{\bar{v}}{V_t}\right)^{\tilde{a}+\tilde{\zeta}}.$$

### 3.2.3 Corporate Consol Bond

Our next goal is to examine a perpetual coupon bond – that is, a bond with infinite maturity, which pays continuously coupons at a constant rate $c$. Its price $D_c(t)$ at any date $t \in \mathbb{R}_+$ equals

$$
D_c(t) := \lim_{T \to \infty} \mathbb{E}_{\mathbb{P}^*}\left( \int_t^T c e^{-r(s-t)} \, \mathbb{1}_{\{\tilde{\tau} > s\}} \, ds \, \Big| \, \mathcal{F}_t \right)
$$

$$
+ \lim_{T \to \infty} \mathbb{E}_{\mathbb{P}^*}\left( K\beta_2 e^{\gamma(\tilde{\tau}-T)} e^{-r(\tilde{\tau}-t)} \mathbb{1}_{\{t < \tilde{\tau} < T\}} \, \Big| \, \mathcal{F}_t \right)
$$

or, equivalently,

$$
D_c(t) = \lim_{T \to \infty} A(t,T) + \lim_{T \to \infty} D_3(t,T),
$$

where $D_3(t,T)$ is defined as in the proof of Proposition 3.2.1. Using the formula established in this proposition, we obtain

$$
\lim_{T \to \infty} D_3(t,T) = \beta_2 V_t \left( R_t^{\theta+\zeta} N\big(h_7(V_t, T-t)\big) + R_t^{\theta-\zeta} N\big(h_8(V_t, T-t)\big) \right)
$$

$$
= \beta_2 V_t \left( \frac{\bar{v}(t)}{V_t} \right)^{\tilde{a}+1+\zeta} = \beta_2 \bar{v}(t) \left( \frac{\bar{v}(t)}{V_t} \right)^{\tilde{a}+\zeta},
$$

where $\zeta = \sigma^{-2}\sqrt{\bar{\nu}^2 + 2\sigma^2(r-\gamma)}$. On the other hand, the reader can readily verify that

$$
\lim_{T \to \infty} A(t,T) = \frac{c}{r}\left( 1 - \left( \frac{\bar{v}(t)}{V_t} \right)^{\tilde{a}+\tilde{\zeta}} \right)
$$

provided that $V_t > \bar{v}(t)$. We conclude that the following result holds.

**Corollary 3.2.2.** *The price of a defaultable consol bond, which pays continuously coupons at a constant rate $c$, equals, on the set $\{\tau > t\} = \{\tilde{\tau} > t\}$,*

$$
D_c(t) = \frac{c}{r}\left( 1 - \left( \frac{\bar{v}(t)}{V_t} \right)^{\tilde{a}+\tilde{\zeta}} \right) + \beta_2 \bar{v}(t) \left( \frac{\bar{v}(t)}{V_t} \right)^{\tilde{a}+\zeta}. \tag{3.20}
$$

*Assume, in addition, that $\gamma = 0$ so that the barrier is constant: $\bar{v}(t) = \bar{v}$ for every $t \in \mathbb{R}_+$. Let us set $\tilde{\tau} = \inf\{t \geq 0 : V_t \leq \bar{v}\}$. Then $\zeta = \tilde{\zeta}$, and so*

$$
D_c(t) = \frac{c}{r}(1 - \bar{q}_t) + \beta_2 \bar{v}\bar{q}_t, \tag{3.21}
$$

*where*

$$
\bar{q}_t := \left( \frac{\bar{v}}{V_t} \right)^{\tilde{a}+\tilde{\zeta}} = -\int_t^\infty e^{-rs} \, d\mathbb{P}^*\{\tilde{\tau} > s \mid \mathcal{F}_t\}.
$$

*Finally, if $\gamma = \kappa = 0$ then*

$$
D_c(t) = \frac{c}{r}\left( 1 - \left( \frac{\bar{v}}{V_t} \right)^\alpha \right) + \beta_2 \bar{v}\left( \frac{\bar{v}}{V_t} \right)^\alpha, \tag{3.22}
$$

*where $\alpha = 2r/\sigma^2$.*

## 3.3 Optimal Capital Structure

An interesting aspect of the Black and Cox (1976) paper is that it originated studies regarding optimal capital structure of a firm in the context of servicing corporate debt. We shall first give a brief account of some of their considerations. Subsequently, we shall report other classic approaches to the issue of optimality of bankruptcy. The basic idea is that stockholders can choose the bankruptcy policy in such a way that the value of equities will be maximized or, equivalently, that the value of the debt will be minimized.

### 3.3.1 Black and Cox Approach

Following Black and Cox (1976), we consider a firm that has an interest paying bonds outstanding. For simplicity, we assume that it is a consol bond, which pays continuously coupon rate $c$. Assume that $r > 0$ and the payout rate $\kappa$ is equal to zero in (3.11). This condition can be given a financial interpretation as the restriction on the sale of assets, as opposed to issuing of new equity. In other words, we postulate that the firm's liability can only be financed by issuing new equity, as opposed to the sale of existing assets. Equivalently, we may think about a situation in which the stockholders will make payments to the firm to cover the interest payments. However, they have the right to stop making payments at any time and either turn the firm over to the bondholders or pay them a lump payment of $c/r$ per unit of the bond's notional amount. Recall that we denote by $E(V_t)$ ($D(V_t)$, resp.) the value at time $t$ of the firm equity (debt, resp.), hence the total value of the firm's assets satisfies $V_t = E(V_t) + D(V_t)$.

Black and Cox (1976) argue that there is a critical level of the value of the firm, denoted as $v^*$, below which no more equity can be sold. This critical level may be determined by stockholders in the course of a certain optimization procedure that in fact determines the optimal capital structure of the firm. To be more specific, the critical value $v^*$ will be chosen by stockholders, whose aim is to minimize the value of the bonds, and thus to maximize the value of the equity. Notice that $v^*$ is nothing else than a constant default barrier in the problem under consideration; the optimal default time $\tau^*$ thus equals $\tau^* = \inf\{t \geq 0 : V_t < v^*\} = \inf\{t \geq 0 : V_t \leq v^*\}$. To find the value of $v^*$, let us first fix the bankruptcy level $\bar{v}$. Thus, the ODE for the pricing function $u^\infty = u^\infty(V)$ of a consol bond takes the following form[2]

$$\tfrac{1}{2}V^2\sigma^2 u^\infty_{VV} + rV u^\infty_V + c - r u^\infty = 0, \qquad (3.23)$$

subject to the lower boundary condition $u^\infty(\bar{v}) = \min(\bar{v}, c/r)$ and the upper boundary condition

$$\lim_{V \to \infty} u^\infty_V(V) = 0.$$

---

[2] Recall that we have assumed that $\kappa = 0$. Let us also stress that no bankruptcy costs are present; in other words, the coefficients $\beta_1$ and $\beta_2$ are equal to 1.

For the last condition, observe that when the firm's value grows to infinity, the possibility of default becomes meaningless, so that the value of the defaultable consol bond tends to the value $c/r$ of the default-free consol bond. It is well known that the general solution to (3.23) has the following form:

$$u^\infty(V) = \frac{c}{r} + K_1 V + K_2 V^{-\alpha}, \tag{3.24}$$

where $\alpha = 2r/\sigma^2$ and $K_1, K_2$ are some constants, to be determined from boundary conditions. The boundary conditions imply that $K_1 = 0$, and

$$K_2 = \begin{cases} \bar{v}^{\alpha+1} - (c/r)\bar{v}^\alpha, & \text{if } \bar{v} < c/r, \\ 0, & \text{if } \bar{v} \ge c/r, \end{cases}$$

and hence if $\bar{v} < c/r$, then[3]

$$u^\infty(V_t) = \frac{c}{r} + \left(\bar{v}^{\alpha+1} - \frac{c}{r}\bar{v}^\alpha\right)V_t^{-\alpha} = \frac{c}{r}\left(1 - \left(\frac{\bar{v}}{V_t}\right)^\alpha\right) + \bar{v}\left(\frac{\bar{v}}{V_t}\right)^\alpha.$$

We have thus rediscovered, using an analytical approach, formula (3.22) of Corollary 3.2.2, with $\beta_2 = 1$. As was already mentioned, it is in the interest of the stockholders to select the bankruptcy level in such a way that the value of the debt, $D(V_t) = u^\infty(V_t)$, is minimized, and thus the value of firm's equity

$$E(V_t) = V_t - D(V_t) = V_t - \frac{c}{r}(1 - \bar{q}_t) - \bar{v}\bar{q}_t \tag{3.25}$$

is maximized. It is easy to check that the optimal level of the barrier does not depend on the current value of the firm, and is given by an explicit formula

$$v^* = \frac{c}{r}\frac{\alpha}{\alpha+1} = \frac{c}{r+\sigma^2/2}. \tag{3.26}$$

Given the optimal strategy of the stockholders, the price process of the firm's debt (i.e., of a consol bond) takes the form, on the set $\{\tau^* > t\}$,

$$D^*(V_t) = \frac{c}{r} - \frac{1}{V_t^\alpha}\left(\frac{c}{r} - \frac{c}{r+\sigma^2/2}\right)\left(\frac{c}{r+\sigma^2/2}\right)^\alpha = \frac{c}{r} - \frac{1}{\alpha V_t^\alpha}\left(\frac{c}{r+\sigma^2/2}\right)^{\alpha+1}$$

or, equivalently (cf. (3.21)),

$$D^*(V_t) = \frac{c}{r}(1 - q_t^*) + v^* q_t^*, \tag{3.27}$$

where

$$q_t^* = \left(\frac{v^*}{V_t}\right)^\alpha = \frac{1}{V_t^\alpha}\left(\frac{c}{r+\sigma^2/2}\right)^\alpha. \tag{3.28}$$

Let us stress that this result was derived under the assumption that stockholders are not allowed to sell assets to cover outstanding interest payments. It is thus worth noticing that Black and Cox (1976) also examine a situation when the firm's assets can actually be sold to make interest payments. The interested reader may consult the original paper for the corresponding valuation formula for the firm's debt.

---

[3] If $\bar{v} \ge c/r$ then obviously $u^\infty(V_t) = c/r$; this is the maximal value of the firm's debt. It corresponds to the case of a fully protected debt.

A more realistic modeling of the bankruptcy and of the bargaining process, which takes into account other important factors (such as tax benefits and bankruptcy costs) was done by, among others, Leland (1994), Leland and Toft (1996), Anderson and Sundaresan (1996), Mella-Barral and Perraudin (1997), and Mella-Barral (1999). We shall now provide a brief account of some of the results of the first two papers.

### 3.3.2 Leland's Approach

The primary concern of the Leland (1994) paper is the optimal capital structure of a levered firm over an infinite time horizon. More specifically, it is assumed that the firm issues a consol paying a coupon at the rate $c$ while the firm is solvent. The firm becomes insolvent, that is the default occurs, when the process $V$ of the total value of the firm's assets for the first time hits the constant lower reorganization barrier $\bar{v}$. Similarly as in the Black and Cox approach, the optimal bankruptcy level $v^*$ is chosen in such a way that the value of the firm's equity is maximized. As before, the interest rate is constant, and the dynamics of the firm's value under $\mathbb{P}^*$ are given by (3.11) with $r > 0$ and $\kappa = 0$.

**Bankruptcy costs.** We assume that the firm defaults as soon as the value process hits the constant barrier $\bar{v}$, so that the default time $\tau$ equals

$$\tau = \bar{\tau} := \inf \{t \geq 0 : V_t \leq \bar{v}\} = \inf \{t \geq 0 : V_t < \bar{v}\}.$$

We postulate that at the time of default the bondholders receive a recovery payment equal to $\beta_2 \bar{v}$, where $\beta_2 \in [0, 1]$. The fraction $(1 - \beta_2)\bar{v}$, lost upon bankruptcy, represents bankruptcy/reorganization costs in the present set-up. Denoting by $D(V_t)$ the value of the firm's debt at time $t$ and using (3.22), we obtain, on the set $\{\bar{\tau} > t\}$,

$$D(V_t) = \frac{c}{r}\left(1 - \left(\frac{\bar{v}}{V_t}\right)^\alpha\right) + \beta_2\bar{v}\left(\frac{\bar{v}}{V_t}\right)^\alpha = \frac{c}{r}(1 - \bar{q}_t) + \beta_2\bar{v}\bar{q}_t, \qquad (3.29)$$

where $\alpha = 2r/\sigma^2$ and $\bar{q}_t = (\bar{v}/V_t)^\alpha$. Furthermore, it is clear that at any time $t$ the present value of bankruptcy costs, denoted by $B(V_t)$, is given by the following expression: $B(V_t) = (1 - \beta_2)\bar{v}\bar{q}_t$.

**Tax benefits.** The tax benefits associated with debt financing can be interpreted as a security that pays a constant coupon rate equal to the tax-sheltering value of the interest payments, denoted as $\bar{c}$, as long as the firm is solvent. Otherwise, i.e., after default, tax benefits cannot be claimed. Thus, the current value of the tax benefits, denoted by $T(V_t)$, equals

$$T(V_t) = \frac{\bar{c}}{r}(1 - \bar{q}_t),$$

where $\bar{c} < c$ is a given constant. In financial interpretation, it seems natural to represent $\bar{c}$ as the product $\bar{c} = \mu c$, where $\mu$ denotes the corporate tax rate.

Consequently, the *total value of the firm*, denoted by $G(V_t)$, equals

$$G(V_t) := V_t + T(V_t) - B(V_t) = V_t + \frac{\bar{c}}{r}(1 - \bar{q}_t) - (1 - \beta_2)\bar{v}\bar{q}_t. \qquad (3.30)$$

Let us denote by $E(V_t)$ the current value of the firm's equity, hence we now have

$$E(V_t) := G(V_t) - D(V_t).$$

Using (3.29) and (3.30), we obtain the following expression for the value of the firm's equity, on the set $\{\bar{\tau} > t\}$,

$$E(V_t) = V_t - \frac{c - \bar{c}}{r} + \left(\frac{c - \bar{c}}{r} - \bar{v}\right)\bar{q}_t = V_t - \frac{c - \bar{c}}{r}(1 - \bar{q}_t) - \bar{v}\bar{q}_t, \qquad (3.31)$$

where $\bar{q}_t = (\bar{v}/V_t)^\alpha$. It is interesting to notice that, under the convention adopted by Leland (1994), bankruptcy costs do not affect the value of the firm's equity, as opposed to the firm's debt. In other words, in the event of default the bankruptcy costs are paid in full by the bondholders.

**Unprotected debt.** Let us first assume that no lower bound is imposed on the value of the endogenously chosen level of the triggering barrier. However, we still assume that the bankruptcy necessarily occurs whenever the value of the equity hits zero. Also, the value of the equity at the time of bankruptcy is always 0, no matter what is the chosen level $\bar{v}$ of the default triggering barrier.

Comparing (3.31) and (3.25), we conclude that this corresponds to the case examined in Sect. 3.3.1 with one minor modification, namely, the coupon rate $c$ needs to be replaced by the net coupon rate $c - \bar{c}$. Therefore, to find the value of $\bar{v}$ that maximizes the right-hand side of (3.31) we may use the same approach as in the preceding section to obtain (cf. (3.26))

$$v^* = \frac{c - \bar{c}}{r}\frac{\alpha}{\alpha + 1} = \frac{c - \bar{c}}{r + \sigma^2/2}, \qquad (3.32)$$

so that

$$q_t^* = \frac{1}{V_t^\alpha}\left(\frac{c - \bar{c}}{r + \sigma^2/2}\right)^\alpha.$$

Of course, if the tax benefits are neglected, then the optimal value $v^*$ coincides with the value given by (3.26). The maximal value of the firm's equity is

$$E^*(V_t) = V_t - \frac{c - \bar{c}}{r} + \frac{1}{\alpha V_t^\alpha}\left(\frac{c - \bar{c}}{r + \sigma^2/2}\right)^{\alpha+1},$$

and the minimal value of the firm's debt equals

$$D^*(V_t) = \frac{c}{r} - \frac{1}{\alpha V_t^\alpha}\left(\frac{c}{r} - \frac{\beta_2(c - \bar{c})}{r + \sigma^2/2}\right)\left(\frac{c - \bar{c}}{r + \sigma^2/2}\right)^\alpha.$$

At bankruptcy the value of the firm's equity is 0, and the total value of the firm equals $G(V_{\bar{\tau}^*}) = G(v^*) = D^*(v^*) = \beta_2 v^*$. Finally, given the optimal policy of firm's stockholders, the firm's equity is maximized at time 0 if the coupon rate $c$ is chosen to satisfy: $c = \bar{c} + V_0(r + \sigma^2/2)$.

**Protected debt.** Consider now the case when bankruptcy is triggered when the value of the firm's assets falls below the principal value of debt. Assume, in addition, that the principal value of the debt was equal to its value at the inception date 0. The level of the default triggering barrier is thus exogenously given to be $\bar{v} = D(V_0)$. Using (3.29), we obtain the following equation

$$D(V_0) = \frac{c}{r}\left(1 - \left(\frac{D(V_0)}{V_0}\right)^\alpha\right) + \beta_2 D(V_0)\left(\frac{D(V_0)}{V_0}\right)^\alpha, \tag{3.33}$$

which implicitly specifies the value of the protected debt at time 0. Assume first that $\beta_2 = 1$, i.e., there are no bankruptcy costs. Since by assumption the firm is solvent at time 0 (so that $V_0 > D_0$), equation (3.33) yields $D(V_0) = c/r$, as expected. Unfortunately, in the case when $\beta_2 < 1$, the closed-form solution to (3.33) is not available.

### 3.3.3 Leland and Toft Approach

Leland and Toft (1996) are concerned with finite maturity corporate debt, as opposed to corporate perpetuity studied by Leland (1994). They also assume that the short-term interest rate is constant, and that the process of the total value of the firm's assets satisfies, under the spot martingale measure $\mathbb{P}^*$,

$$dV_t = V_t\big((r - \kappa)\,dt + \sigma_V\,dW_t^*\big),$$

where the constant $\kappa \geq 0$ is the payout ratio. As before, we write $\bar{v}$ to denote the constant bankruptcy level. We allow for non-zero bankruptcy costs so that we have $\beta_2 \in [0,1]$. Under these assumptions, the value of a defaultable coupon bond is given by a suitable version of the formula of Corollary 3.2.1.

**Stationary debt structure.** Leland and Toft (1996) assume that the firm has a *stationary debt structure,* in the following sense: at any time $t$, the debt outstanding is composed of coupon bonds of maturities from the interval $[t, t+T]$ with the constant coupon rate $c = C/T$ and the face value uniformly distributed over this interval. This implies that the total coupon paid by all outstanding bonds is $C$ per year. To preserve the debt structure as time elapses, new bonds are issued at a rate $l = L/T$ per year, where $L$ is the total face value of all outstanding bonds. The same amount of principal is retired when the previously issued bonds mature. Therefore, as long as the firm remains solvent, the total face value remains constant at any date $t$, and the outstanding bonds have a uniform distribution of face value in the interval $[t, t+T]$. It is thus clear that at any time $t$ before default the value of the firm's debt equals

$$D(V_t) = \int_t^{t+T} D_c(t, u)\,du,$$

where the price $D_c(t, u)$ is given by the formula of Corollary 3.2.1, with $L$ substituted with $L/T$, $c$ substituted with $C/T$, and $\beta_2$ substituted with $\beta_2/T$.

Let us set $t = 0$. By virtue of Corollary 3.2.1, the price $D_c(0, u)$ equals

$$D_c(0, u) = \frac{C}{rT} + \frac{e^{-ru}}{T}\left(L - \frac{C}{r}\right)g(u) + \frac{1}{T}\left(\beta_2\bar{v} - \frac{C}{r}\right)h(u), \qquad (3.34)$$

where the functions $g, h : [0, T] \to \mathbb{R}$ are given by (as usual, $R_0 = \bar{v}/V_0$)

$$g(u) = N\big(k_1(V_0, u)\big) - R_0^{2\tilde{a}}N\big(k_2(V_0, u)\big)$$

and

$$h(u) = R_0^{\tilde{a}+\tilde{\zeta}}N\big(g_1(V_0, u)\big) + R_0^{\tilde{a}-\tilde{\zeta}}N\big(g_2(V_0, u)\big).$$

Let us establish the following result due to Leland and Toft (1996).

**Proposition 3.3.1.** *The value of the firm's debt at time 0 equals*

$$D(V_0) = \frac{C}{r} + \frac{1}{rT}\left(L - \frac{C}{r}\right)G(T) + \frac{1}{T}\left(\beta_2\bar{v} - \frac{C}{r}\right)H(T),$$

*where*

$$G(T) = \int_0^T e^{-ru}g(u)\,du = r^{-1}\big(1 - h(T) - e^{-rT}g(T)\big), \qquad (3.35)$$

*and* $H(T) := \int_0^T h(u)\,du = \tilde{H}(T)$, *where for every* $T \geq 0$

$$\tilde{H}(T) := \frac{\sqrt{T}}{\zeta\sigma}\left(R_0^{\tilde{a}+\tilde{\zeta}}g_1(V_0, T)N\big(g_1(V_0, T)\big) - R_0^{\tilde{a}-\tilde{\zeta}}g_2(V_0, T)N\big(g_2(V_0, T)\big)\right).$$

*Proof.* We need to evaluate the integral $D(V_0) = \int_0^T D_c(0, u)\,du$, where $D_c(0, u)$ is given by (3.34). Notice that since $g(0) = 1$, we have

$$G(T) = \int_0^T e^{-ru}g(u)\,du = r^{-1}\left(1 - e^{-rT}g(T) - \int_0^T e^{-ru}\,dg(u)\right).$$

Moreover, using Lemma 3.2.1 (see also Sect. 3.2.2), we obtain

$$\int_0^T e^{-ru}\,dg(u) = h(T).$$

This shows that (3.35) holds. It is not hard to verify that $H(0) = \tilde{H}(0) = 0$. Furthermore, the derivative of $H$ obviously equals

$$H'(T) = h(T) = R_0^{\tilde{a}+\tilde{\zeta}}N\big(g_1(V_0, T)\big) + R_0^{\tilde{a}-\tilde{\zeta}}N\big(g_2(V_0, T)\big).$$

Thus, to establish the equality $H(T) = \tilde{H}(T)$, it is enough to check by straightforward differentiation that $\tilde{H}'(T) = h(T)$. □

By making use of Proposition 3.3.1, Leland and Toft (1996) first find an explicit formula for the optimal value $v^*(T)$ of the default triggering barrier. Subsequently, they show that $\lim_{T\to\infty} v^*(T) = v^*$, where $v^*$ is given by (3.32). Finally, they also examine the optimal leverage ratio that maximizes the value of the firm for alternative choices of the debt's maturity.

### 3.3.4 Further Developments

We shall now provide a short overview of some further studies of first-passage-time models with constant interest rate. Various extensions of structural models put forward in recent years are typically much more complex than the classic models. In fact, they are usually based of rather involved economic considerations, which cannot be reported here in detail. Instead, we shall only indicate the most crucial features of these studies.

**State variables.** It is not uncommon in first-passage-time modeling to define the default time not in terms of the firm's value, but rather in terms of some other state variables, which reflect the economic fundamentals (such as: the firm's operating earnings, the price of the firm's product, etc.). Mella-Barral and Tychon (1999) consider a generic first-passage-time model with a state variable that follows a geometric Brownian motion. The default is triggered when the state variable hits a constant threshold level $\bar{x}$ and the value of assets at default is considered an exogenous input, given through some function $V(\bar{x})$. In particular, they derive the valuation formula for a defaultable coupon bond with finite maturity and they examine the term structure of credit spreads under the assumption that the coupon rate is chosen in such a way that the initial bond price is equal to the face value.

**Strategic debt service.** It is widely recognized in financial literature that the presence of bankruptcy/liquidation costs may induce creditors to accept deviations from contractual payments, rather than to force the firm's bankruptcy. Anderson and Sundaresan (1996) and Mella-Barral and Perraudin (1997) were the first to account for this feature by incorporating the possibility of the debt's renegotiation in case of distress; in other terms, they explicitly deal with strategic debt service (recent papers in this vein include Leland (1998), Mella-Barral (1999), and Ericsson (2000)). Mella-Barral and Perraudin (1997) assume that stockholders may persuade bondholders that it is in their interest to renegotiate the debt, and thus to receive less than the originally contracted interest payments. Using the PDE approach, they perform an analysis of the optimality of the debt's renegotiation, in the sense of the maximization of the firm's value. Anderson and Sundaresan (1996) research differs from that of Mella-Barral and Perraudin (1997) in two main aspects. First, they model explicitly the so-called 'bankruptcy game' between the bondholders and the stockholders who act in their own self-interest. A detailed analysis of the bargaining procedure is shown to lead to the discrete-time game-theoretic model of bankruptcy process. Second, they focus on bonds of fixed maturity, as opposed to perpetual bonds studied in Mella-Barral and Perraudin (1997). The continuous-time version of the Anderson and Sundaresan model was subsequently examined by means of the PDE approach in Anderson et al. (1996), who studied perpetual corporate bonds. In all cases, the results of numerical studies seem to support the conclusion that bankruptcy/renegotiation costs may have a significant impact on the level of credit spreads.

**Probabilistic approaches.** Ericsson and Reneby (1998) show that most defaultable claims – even accounting for bankruptcy costs, corporate taxes, or deviations from the strict priority rule – can be decomposed into relatively simple three building blocks: a down-and-out call option, a down-and-out digital option, and a digital claim, which pays one unit of cash in the event of default. Under the assumption of a constant triggering barrier, they obtain the valuation formulae for these 'building blocks', and subsequently for a wide range of defaultable claims. They also show how to apply these results to previously studied problems related to strategic debt service. In particular, they rederive, through a probabilistic approach, some formulae previously established through the PDE approach by Mella-Barral and Perraudin (1997). Barone-Adesi and Colwell (1999) model directly the difference $X_t := V_t - v_t$ by postulating that $X$ is governed by the following SDE:

$$dX_t = \mu X_t \, dt + \sqrt{\sigma^2 X_t^\beta + \kappa^2} \, dW_t$$

for some constants $\mu, \sigma, \beta$ and $\kappa$, and they derive a closed-form solution for the value of a defaultable bond. Sarkar (2001) examines the probability that a callable corporate bond will actually be called within a given period.

**Comparative studies.** Anderson and Sundaresan (2000) consider a corporate perpetuity with a notional amount $L$ and the coupon rate $c$ (as before, we assume that the coupon rate $c$ encompasses the face value $L$). They do not specify any particular model for the dynamics of the value process $V$, but they assume a constant level $r$ of the short-term interest rate. Their valuation equation nests the results for corporate perpetuities derived in Black and Cox (1976), Leland (1994), Anderson et al. (1996), and Mella-Barral and Perraudin (1997). On the set $\{\tau > t\}$, it takes the following generic form:

$$D_c(t) = \frac{c}{r}(1 - \bar{q}_t) + \bar{q}_t \max{(\delta \bar{v} - K, 0)}, \qquad (3.36)$$

where $D_c(t)$ stands for the price of the defaultable perpetual bond at time $t$, $\bar{q}_t$ denotes the 'weighted' probability of default as seen at time $t$, $\bar{v}$ is the bankruptcy level, $\delta$ is the recovery rate, and finally $K$ represents the fixed cost of bankruptcy. Particular specifications of these quantities depend on modeling assumptions that are made regarding activities of the firm and its operating covenants. For example, in Black and Cox (1976), Leland (1994) and Anderson et al. (1996), the coefficient $\bar{q}_t$ is known to satisfy $\bar{q}_t = (\bar{v}/V_t)^\gamma$, where $\gamma > 0$ is a constant, which exact value varies across alternative models. By contrast, in Mella-Barral and Perraudin (1997) the default probability is an explicit function of the price of the firm's product.

Anderson and Sundaresan (2000) provide a comparative study of structural models with and without a strategic debt service, using the time series data for the U.S. corporate bond market. They conclude that the modifications of the classic structural models that allow for the endogenous determination of the default threshold based on economic fundamentals have led to an improvement of structural models.

## 3.4 Models with Stochastic Interest Rates

We shall now examine a natural extension of the Black and Cox approach that covers both the firm-specific credit risk and the market (interest rate) risk. Formally, our goal is to extend the valuation formula of Proposition 3.2.1 to the case of random interest rates and time dependent coefficients $\kappa$ and $\sigma_V$. In order to make such a generalization feasible, we postulate that:

– the random triggering barrier $v$ is chosen in a judicious way, namely, $v_t = KB(t,T)f(t)$ for some constant $K$, and some function $f : [0,T] \to \mathbb{R}_+$,

– the volatility of the forward value of the firm is a deterministic function.

To satisfy the latter requirement, we shall place ourselves in the Gaussian HJM set-up; i.e., we shall assume that the bond price volatility is deterministic. More specifically, we shall now assume that the dynamics under $\mathbb{P}^*$ of the firm's value and of a default-free zero-coupon bond are

$$dV_t = V_t\big((r_t - \kappa(t))\,dt + \sigma_V(t)\,dW_t^*\big),$$

and

$$dB(t,T) = B(t,T)\big(r_t\,dt + b(t,T)\,dW_t^*\big),$$

respectively, where $W^*$ is a $d$-dimensional standard Brownian motion, and $\kappa : [0,T] \to \mathbb{R}$ and $\sigma_V, b : [0,T] \to \mathbb{R}^d$ are (bounded) deterministic functions. In this case, the *forward value* $F_V(t,T) := V_t/B(t,T)$ of the firm satisfies, under the forward martingale measure $\mathbb{P}_T$,

$$dF_V(t,T) = -\kappa(t)F_V(t,T)\,dt + F_V(t,T)\big(\sigma_V(t) - b(t,T)\big)\,dW_t^T,$$

where the process

$$W_t^T = W_t^* - \int_0^t b(u,T)\,du, \quad \forall\, t \in [0,T],$$

is a $d$-dimensional standard Brownian motion under $\mathbb{P}_T$. Consequently, an auxiliary process $F_V^\kappa(t,T)$, given by the formula

$$F_V^\kappa(t,T) = F_V(t,T)e^{-\int_t^T \kappa(u)\,du}, \quad \forall\, t \in [0,T],$$

follows a lognormally distributed martingale under $\mathbb{P}_T$, specifically:

$$dF_V^\kappa(t,T) = F_V^\kappa(t,T)\big(\sigma_V(t) - b(t,T)\big)\,dW_t^T,$$

with the terminal condition $F_V^\kappa(T,T) = F_V(T,T) = V_T$. We consider the following modification of the Black and Cox approach:

$$X = L, \ A \equiv 0, \ Z \equiv \beta_2 V, \ \tilde{X} = \beta_1 V_T, \ \tau = \inf\{t \in [0,T] : V_t < v_t\},$$

where $\beta_2, \beta_1 \in [0,1]$ are constants, and the random barrier $v$ is given by the formula

$$v_t = \begin{cases} KB(t,T)e^{\int_t^T \kappa(u)\,du}, & \text{for } t < T, \\ L, & \text{for } t = T, \end{cases} \tag{3.37}$$

for some constant $0 < K \le L$.

Let us denote, for any $t \leq T$,

$$\kappa(t,T) = \int_t^T \kappa(u)\,du, \quad \sigma^2(t,T) = \int_t^T |\sigma_V(u) - b(u,T)|^2\,du,$$

where $|\cdot|$ stands for the Euclidean norm in $\mathbb{R}^d$. We shall write briefly $F_t$ to denote the forward value of the firm, i.e., $F_t = F_V(t,T)$. Finally, we set

$$\eta_+(t,T) = \kappa(t,T) + \tfrac{1}{2}\sigma^2(t,T), \quad \eta_-(t,T) = \kappa(t,T) - \tfrac{1}{2}\sigma^2(t,T).$$

**Proposition 3.4.1.** *Let the barrier process $v$ be given by (3.37). For any $t < T$, the forward price $F_D(t,T) = D(t,T)/B(t,T)$ of a defaultable bond equals, on the set $\{\tau > t\}$,*

$$\begin{aligned}
F_D(t,T) &= L\big(N\big(\hat{h}_1(F_t,t,T)\big) - (F_t/K)e^{-\kappa(t,T)}N\big(\hat{h}_2(F_t,t,T)\big)\big) \\
&\quad + \beta_1 F_t e^{-\kappa(t,T)}\big(N\big(\hat{h}_3(F_t,t,T)\big) - N\big(\hat{h}_4(F_t,t,T)\big)\big) \\
&\quad + \beta_1 K\big(N\big(\hat{h}_5(F_t,t,T)\big) - N\big(\hat{h}_6(F_t,t,T)\big)\big) \\
&\quad + \beta_2 K J_1(F_t,t,T) + \beta_2 F_t e^{-\kappa(t,T)} J_2(F_t,t,T),
\end{aligned}$$

*where*

$$\hat{h}_1(F_t,t,T) = \frac{\ln(F_t/L) - \eta_+(t,T)}{\sigma(t,T)},$$

$$\hat{h}_2(F_t,T,t) = \frac{2\ln K - \ln(LF_t) + \eta_-(t,T)}{\sigma(t,T)},$$

$$\hat{h}_3(F_t,t,T) = \frac{\ln(L/F_t) + \eta_-(t,T)}{\sigma(t,T)},$$

$$\hat{h}_4(F_t,t,T) = \frac{\ln(K/F_t) + \eta_-(t,T)}{\sigma(t,T)},$$

$$\hat{h}_5(F_t,t,T) = \frac{2\ln K - \ln(LF_t) + \eta_+(t,T)}{\sigma(t,T)},$$

$$\hat{h}_6(F_t,t,T) = \frac{\ln(K/F_t) + \eta_+(t,T)}{\sigma(t,T)},$$

*and for any fixed $0 \leq t < T$ and $F_t > 0$*

$$J_{1,2}(F_t,t,T) = \int_t^T e^{\kappa(u,T)}\,dN\left(\frac{\ln(K/F_t) + \kappa(t,T) \pm \tfrac{1}{2}\sigma^2(t,u)}{\sigma(t,u)}\right).$$

*Remarks.* Let us assume that $\beta_2 = \beta_1 = 1$. It can be checked that when $b \equiv 0$ and the coefficients $\kappa$ and $\sigma_V$ are constant, the valuation formula of Proposition 3.4.1 reduces to the special case of the formula obtained in Proposition 3.2.1, with $\gamma = r - \kappa$. In this case, $J_{1,2}(F_t,t,T)$ can be evaluated explicitly thanks to formula (3.16). Notice also that the choice of the lower threshold as in (3.50), as opposed to expression (3.37), does not lead to a closed-form solution.

*Proof of Proposition 3.4.1.* Under the present assumptions, a defaultable bond with the face value $L$ is equivalent to the payoff $X^d(T)$, which settles at the bond's maturity date $T$, and equals

$$X^d(T) = L\mathbb{1}_{\{V_T \geq L,\, \bar{\tau} \geq T\}} + \beta_1 F_V^\kappa(T,T)\mathbb{1}_{\{V_T < L,\, \bar{\tau} \geq T\}} + \beta_2 v_{\bar{\tau}} B^{-1}(\bar{\tau},T)\mathbb{1}_{\{t < \bar{\tau} < T\}}$$

with

$$\bar{\tau} = \inf\{t < T : F_V^\kappa(t,T) \leq K\} = \inf\{t < T : Y_t \leq 0\},$$

where $Y_t := \ln\left(F_V^\kappa(t,T)/K\right)$. Therefore, the forward price of a defaultable bond admits the following probabilistic representation

$$F_D(t,T) = \mathbb{E}_{\mathbb{P}_T}\left(L\mathbb{1}_{\{V_T \geq L,\, \bar{\tau} \geq T\}}\,\Big|\,\mathcal{F}_t\right) + \beta_1 \mathbb{E}_{\mathbb{P}_T}\left(F_V^\kappa(T,T)\mathbb{1}_{\{V_T < L,\, \bar{\tau} \geq T\}}\,\Big|\,\mathcal{F}_t\right)$$
$$+ \beta_2 \mathbb{E}_{\mathbb{P}_T}\left(v_{\bar{\tau}} B^{-1}(\bar{\tau},T)\mathbb{1}_{\{t < \bar{\tau} < T\}}\,\Big|\,\mathcal{F}_t\right)$$

that is an immediate consequence of the definition of the forward martingale measure $\mathbb{P}_T$. We conclude that, on the set $\{\tau > t\} = \{\bar{\tau} > t\}$,

$$F_D(t,T) = L\,\mathbb{P}_T\{F_V^\kappa(T,T) \geq L,\, \bar{\tau} \geq T \,|\, \mathcal{F}_t\}$$
$$+ \beta_1 \mathbb{E}_{\mathbb{P}_T}\left(F_V^\kappa(T,T)\mathbb{1}_{\{F_V^\kappa(T,T) < L,\, \bar{\tau} \geq T\}}\,\Big|\,\mathcal{F}_t\right)$$
$$+ \beta_2 K \mathbb{E}_{\mathbb{P}_T}\left(e^{\kappa(\bar{\tau},T)}\mathbb{1}_{\{t < \bar{\tau} < T\}}\,\Big|\,\mathcal{F}_t\right) =: I_1(t) + I_2(t) + I_3(t).$$

We note that $Y$ satisfies

$$Y_t = Y_0 + \int_0^t (\sigma_V(u) - b(u,T))\, dW_u^T - \frac{1}{2}\int_0^t |\sigma_V(u) - b(u,T)|^2\, du,$$

and we consider the following deterministic time change $A : [0,T] \to \mathbb{R}_+$ associated with $Y$

$$A_t = \int_0^t |\sigma_V(u) - b(u,T)|^2\, du, \quad \forall t \in [0,T].$$

Let $A^{-1} : [0, A_T] \to [0,T]$ stand for the inverse time change. Then the time-changed process[4] $\tilde{Y}_t := Y_{A_t^{-1}}$ follows under $\mathbb{P}_T$ a one-dimensional Brownian motion with the drift coefficient $\nu = -1/2$, with respect to the time-changed filtration $\tilde{\mathbb{F}}$, where $\tilde{\mathcal{F}}_t = \mathcal{F}_{A_t^{-1}}$ for $t \in [0, A_T]$. Put more explicitly, $\tilde{Y}$ satisfies

$$\tilde{Y}_t = Y_0 + \tilde{W}_t - \tfrac{1}{2}t, \quad \forall t \in [0, A_T],$$

for some $(\mathbb{P}_T, \tilde{\mathbb{F}})$-standard Brownian motion $\tilde{W}$.

Let us examine $I_1(t)$. We set $\tilde{L} = L/K$ and $\tilde{\tau} = \inf\{t < A_T : \tilde{Y}_t \leq 0\}$. Notice that for any fixed $t < T$ we have, on the set $\{\bar{\tau} > t\} = \{\tilde{\tau} > A_t\}$,

$$\mathbb{P}_T\{F_V^\kappa(T,T) \geq L,\, \bar{\tau} \geq T \,|\, \mathcal{F}_t\} = \mathbb{P}_T\{\tilde{Y}_{A_T} \geq \ln \tilde{L},\, \tilde{\tau} \geq A_T \,|\, \tilde{\mathcal{F}}_{A_t}\}.$$

---

[4] See Revuz and Yor (1991).

Using Lemma 3.1.4, with $\nu = -1/2$ and $\sigma = 1$, we obtain

$$\mathbb{P}_T\{\tilde{Y}_{A_T} \geq \ln\tilde{L}, \tilde{\tau} \geq A_T \mid \tilde{\mathcal{F}}_{A_t}\}$$

$$= N\left(\frac{\ln(K/L) + \tilde{Y}_{A_t} - \frac{1}{2}(A_T - A_t)}{\sqrt{A_T - A_t}}\right)$$

$$- e^{\tilde{Y}_{A_t}} N\left(\frac{\ln(K/L) - \tilde{Y}_{A_t} - \frac{1}{2}(A_T - A_t)}{\sqrt{A_T - A_t}}\right).$$

Consequently, we have

$$I_1(t) = L\mathbb{P}_T\{\tilde{Y}_{A_T} \geq \ln\tilde{L}, \tilde{\tau} \geq A_T \mid \tilde{\mathcal{F}}_{A_t}\}$$

$$= LN\left(\frac{\ln(F_t/L) - \kappa(t,T) - \frac{1}{2}\sigma^2(t,T)}{\sigma(t,T)}\right)$$

$$- e^{-\kappa(t,T)}\tilde{L}F_t\, N\left(\frac{2\ln K - \ln(F_tL) + \kappa(t,T) - \frac{1}{2}\sigma^2(t,T)}{\sigma(t,T)}\right).$$

This shows that

$$I_1(t) = L\big(N\big(\hat{h}_1(F_t,t,T)\big) - (F_t/K)e^{-\kappa(t,T)}N\big(\hat{h}_2(F_t,t,T)\big)\big),$$

as expected.

To simplify the notation, we shall evaluate $I_2(t)$ and $I_3(t)$ for $t = 0$ only. The general case follows by the similar arguments as those used in the derivation of the formula for $I_1(t)$, so that it is left to the reader.

We shall now examine $I_2(0)$. Observe that, in view of the definition of processes $\tilde{Y}$ and $A$, we have

$$\mathbb{E}_{\mathbb{P}_T}\left(F_V^\kappa(T,T)\mathbb{1}_{\{F_V^\kappa(T,T)<L,\,\tilde{\tau}\geq T\}}\right) = K\mathbb{E}_{\mathbb{P}_T}\left(e^{\tilde{Y}_{A_T}}\mathbb{1}_{\{\tilde{Y}_{A_T}<\ln\tilde{L},\,\tilde{\tau}\geq A_T\}}\right)$$

so that $I_2(0)$ can also be expressed as follows:

$$I_2(0) = \beta_1 K \int_0^{\ln\tilde{L}} e^x\, d\mathbb{P}_T\{\tilde{Y}_{A_T} < x, \tilde{\tau} \geq A_T\}.$$

Using Lemma 3.1.4 once more, we obtain the following representation

$$d\mathbb{P}_T\{\tilde{Y}_{A_T} < x, \tilde{\tau} \geq A_T\}$$

$$= dN\left(\frac{x - \tilde{Y}_0 + \frac{1}{2}A_T}{\sqrt{A_T}}\right) + e^{\tilde{Y}_0}\, dN\left(\frac{-x - \tilde{Y}_0 - \frac{1}{2}A_T}{\sqrt{A_T}}\right)$$

$$= dN\left(\frac{x - \ln(F_0/K) + \kappa(0,T) + \frac{1}{2}\sigma^2(0,T)}{\sigma(0,T)}\right)$$

$$+ e^{-\kappa(0,T)}\frac{F_0}{K}\, dN\left(\frac{-x - \ln(F_0/K) + \kappa(0,T) - \frac{1}{2}\sigma^2(0,T)}{\sigma(0,T)}\right).$$

Thus, $I_2(0) = I_{21}(0) + I_{22}(0)$, where, by standard calculations,

$$I_{21}(0) = \beta_1 K \int_0^{\ln \tilde{L}} e^x \, dN \left( \frac{x - \ln(F_0/K) + \kappa(0,T) + \frac{1}{2}\sigma^2(0,T)}{\sigma(0,T)} \right)$$

$$= \beta_1 F_0 e^{-\kappa(0,T)} N \left( \frac{\ln(L/F_0) + \kappa(0,T) - \frac{1}{2}\sigma^2(0,T)}{\sigma(0,T)} \right)$$

$$- \beta_1 F_0 e^{-\kappa(0,T)} N \left( \frac{\ln(K/F_0) + \kappa(0,T) - \frac{1}{2}\sigma^2(0,T)}{\sigma(0,T)} \right)$$

$$= \beta_1 F_0 e^{-\kappa(0,T)} \left( N\big(\hat{h}_3(F_0,0,T)\big) - N\big(\hat{h}_4(F_0,0,T)\big) \right)$$

and

$$I_{22}(0) = \beta_1 e^{-\kappa(0,T)} F_0 \int_0^{\ln \tilde{L}} e^x \, dN \left( \frac{-x - \ln(F_0/K) + \kappa(0,T) - \frac{1}{2}\sigma^2(0,T)}{\sigma(0,T)} \right)$$

$$= \beta_1 K N \left( \frac{2\ln K - \ln(LF_0) + \kappa(0,T) + \frac{1}{2}\sigma^2(0,T)}{\sigma(0,T)} \right)$$

$$- \beta_1 K N \left( \frac{\ln(K/F_0) + \kappa(0,T) + \frac{1}{2}\sigma^2(0,T)}{\sigma(0,T)} \right)$$

$$= \beta_1 K \left( N\big(\hat{h}_5(F_0,0,T)\big) - N\big(\hat{h}_6(F_0,0,T)\big) \right).$$

To establish the last two formulae, it was enough to notice that for any $c \neq 0$ and every $a, b, d \in \mathbb{R}$ we have

$$\int_a^b e^x \, dN(cx + d) = e^{\frac{1}{2}(\tilde{d}^2 - d^2)} \left( N(cb + \tilde{d}) - N(ca + \tilde{d}) \right),$$

where $\tilde{d} = d - c^{-1}$. It is clear that $I_{21}(0) > 0$ and $I_{22}(0) < 0$. We always have $I_2(0) > 0$, though.

It remains to evaluate $I_3(0)$, where

$$I_3(0) = \beta_2 K \, \mathbb{E}_{\mathbb{P}_T}\left( e^{\kappa(\tilde{\tau},T)} \mathbb{1}_{\{\tilde{\tau} < T\}} \right) = \beta_2 K \int_0^T e^{\kappa(t,T)} \, d\mathbb{P}_T\{\tilde{\tau} \leq t\}.$$

For this purpose, notice that Lemma 3.1.2 yields

$$\mathbb{P}_T\{\tilde{\tau} \leq s\} = N \left( \frac{-Y_0 + \frac{1}{2}s}{\sqrt{s}} \right) + e^{Y_0} N \left( \frac{-Y_0 - \frac{1}{2}s}{\sqrt{s}} \right),$$

where $\tilde{Y}_0 = Y_0$, and, as before, $\tilde{\tau} = \inf\{t < A_T : \tilde{Y}_t \leq 0\}$. Since clearly $\mathbb{P}_T\{\tilde{\tau} \leq t\} = \mathbb{P}_T\{\tilde{\tau} \leq A_t\}$, we obtain

$$\mathbb{P}_T\{\tilde{\tau} \leq t\} = N \left( \frac{-Y_0 + \frac{1}{2}A_t}{\sqrt{A_t}} \right) + e^{Y_0} N \left( \frac{-Y_0 - \frac{1}{2}A_t}{\sqrt{A_t}} \right)$$

$$= N \left( \frac{\ln \frac{K}{F_0} + \kappa(0,T) + \frac{1}{2}A_t}{\sqrt{A_t}} \right) + e^{-\kappa(0,T)} \frac{F_0}{K} N \left( \frac{\ln \frac{K}{F_0} + \kappa(0,T) - \frac{1}{2}A_t}{\sqrt{A_t}} \right).$$

We conclude that $I_3(0) = I_{31}(0) + I_{32}(0)$, where

$$
\begin{aligned}
I_{31}(0) &= \beta_2 K \int_0^T e^{\kappa(t,T)} \, dN \left( \frac{\ln(K/F_0) + \kappa(0,T) + \frac{1}{2}\sigma^2(0,t)}{\sigma(0,t)} \right) \\
&= \beta_2 K J_1(F_0, 0, T)
\end{aligned}
$$

and

$$
\begin{aligned}
I_{32}(0) &= \beta_2 F_0 e^{-\kappa(0,T)} \int_0^T e^{\kappa(t,T)} \, dN \left( \frac{\ln(K/F_0) + \kappa(0,T) - \frac{1}{2}\sigma^2(0,t)}{\sigma(0,t)} \right) \\
&= \beta_2 F_0 e^{-\kappa(0,T)} J_2(F_0, 0, T).
\end{aligned}
$$

This completes the proof of the proposition. $\qquad\square$

Unfortunately, no explicit formulae for $J_1(F_t, t, T)$ and $J_2(F_t, t, T)$ seem to be available in the general time-dependent set-up. A much more transparent expression for these two terms can be obtained in the case when $\kappa \equiv 0$; i.e., in the absence of dividends. For the ease of further reference, we state the following immediate corollary to Proposition 3.4.1.

**Corollary 3.4.1.** *Under the assumptions of Proposition 3.4.1, if $\kappa \equiv 0$ then*

$$
\begin{aligned}
F_D(t,T) = {}& L\big(N\big(-d_1(F_t, t, T)\big) - (F_t/K) N\big(d_6(F_t, t, T)\big)\big) \\
& + \beta_1 F_t \big(N\big(d_2(F_t, t, T)\big) - N\big(d_4(F_t, t, T)\big)\big) \\
& + \beta_1 K \big(N\big(d_5(F_t, t, T)\big) - N\big(d_3(F_t, t, T)\big)\big) \\
& + \beta_2 K N\big(d_3(F_t, t, T)\big) + \beta_2 F_t N\big(d_4(F_t, t, T)\big),
\end{aligned}
$$

*where*

$$
d_1(F_t, t, T) = \frac{\ln(L/F_t) + \frac{1}{2}\sigma^2(t,T)}{\sigma(t,T)} = d_2(F_t, t, T) + \sigma(t,T),
$$

$$
d_3(F_t, t, T) = \frac{\ln(K/F_t) + \frac{1}{2}\sigma^2(t,T)}{\sigma(t,T)} = d_4(F_t, t, T) + \sigma(t,T),
$$

$$
d_5(F_t, t, T) = \frac{\ln(K^2/F_t L) + \frac{1}{2}\sigma^2(t,T)}{\sigma(t,T)} = d_6(F_t, t, T) + \sigma(t,T).
$$

*Proof.* Since on the set $\{\tau > t\}$ we have $F_t > K$, we now obtain

$$
J_1(F_t, t, T) = \int_t^T dN \left( \frac{\ln(K/F_t) + \frac{1}{2}\sigma^2(t,u)}{\sigma(t,u)} \right) = N \left( \frac{\ln(K/F_t) + \frac{1}{2}\sigma^2(t,T)}{\sigma(t,T)} \right)
$$

and

$$
J_2(F, t, T) = \int_t^T dN \left( \frac{\ln(K/F_t) - \frac{1}{2}\sigma^2(t,u)}{\sigma(t,u)} \right) = N \left( \frac{\ln(K/F_t) - \frac{1}{2}\sigma^2(t,T)}{\sigma(t,T)} \right),
$$

as expected. $\qquad\square$

We shall now give an overview of first-passage-time models with stochastic interest rates, in which the triggering barrier $v$ is assumed to be a strictly positive constant. As one might guess, due to the lognormal property of the firm's value process, such an assumption makes the calculations of a price of a defaultable bond more difficult than in the case of a judiciously chosen random barrier. Typically, no closed-form solution to the bond valuation problem is available for a model with a constant barrier and with random interest rates, and thus some sort of a numerical approach needs to be employed in order to solve this problem.

### 3.4.1 Kim, Ramaswamy and Sundaresan Approach

Kim et al. (1993a) develop a particular model that incorporates both default risk and interest rate risk. The short-term interest rate is driven by the SDE introduced by Cox et al. (1985b)

$$dr_t = (a - br_t)\, dt + \sigma_r \sqrt{r_t}\, d\tilde{W}_t, \tag{3.38}$$

where $\tilde{W}$ is a standard Brownian motion under the spot martingale measure $\mathbb{P}^*$. In financial literature, such a model of the short-term interest rate dynamics is commonly referred to as the CIR term structure model. For the sake of simplicity, we set the risk premium for the interest rate risk as zero (this convention was also adopted by Kim et al. (1993a)). Thus, the short-term rate has identical dynamics (3.38) under the risk-neutral probability measure $\mathbb{P}^*$ and under the real-world probability $\mathbb{P}$. In other words, we postulate that the market price of the interest rate risk vanishes (this condition can be, of course, relaxed).

The value process $V$ is assumed to evolve according to the SDE

$$dV_t = V_t\big((r_t - \kappa)\, dt + \sigma_V\, dW_t^*\big), \tag{3.39}$$

where the two Brownian motions $\tilde{W}$ and $W^*$ are correlated, with the instantaneous correlation coefficient $\rho_{Vr}$. The filtration $\mathbb{F}$ is taken here to be the natural filtration of the pair of Brownian motions $\tilde{W}$ and $W^*$. It can be easily seen that the pair $(V, r)$ follows a strong Markov process with respect to this filtration, under the martingale measure $\mathbb{P}^*$.

Kim et al. (1993a) consider the case when the bond's indenture provisions prohibit the stockholders from selling the firm's assets to pay dividends. The bondholders have priority and must be continuously paid a coupon at the rate of $c$ dollars per unit of time. The firm defaults prior to the debt's maturity $T$ if it is unable to meet its coupon payments, and this happens when the firm's value crosses the lower reorganization barrier $\bar{v} := c/\kappa$. At the intuitive level, the value $\bar{v}$ represents the breakeven point in the following sense: if $V_t = \bar{v}$ the dividends match exactly the coupon payments due over an infinitesimal time interval in the future; if the value process $V$ falls below the threshold $\bar{v}$ ($V$ is above the threshold $\bar{v}$, resp.) the dividends payments are insufficient (largely sufficient, resp.) to cover the coupon payments due.

It is thus natural to assume that the initial value $V_0 > c/\kappa$. If the firm does not default prior to the maturity date then it still may default at the maturity date if at this date the firm's value falls below the bond's notional amount $L$. Let us denote by $B^c(t, T) = B^c(t, T, r_t)$ the price at time $t$ of an equivalent default-free bond with continual interest payments of $c$ per unit time, and the face value $L$ at maturity $T$. Valuation of such a bond is standard within the CIR framework and involves formulae derived therein.

**Defaultable bond.** Kim et al. (1993a) postulate that the payoff to the bondholders upon bankruptcy of the firm is equal to minimum of $V_{\tilde{\tau}}$ and $\delta(T - \tilde{\tau})B^c(\tilde{\tau}, T)$, where a deterministic function $\delta : [0, T] \rightarrow [0, 1]$, with $\delta(0) = 1$, represents the time varying recovery rate. We thus deal here with a model with recovery at default, in which

$$X = L, \; A_t = ct, \; Z_t = \min(V_t, \delta(T-t)B^c(t, T)), \; \tau = \inf\{t \in [0, T] : V_t < v_t\}$$

and the default triggering barrier $v$ is given by the formula

$$v_t = \begin{cases} \bar{v}, & \text{if } t < T, \\ L, & \text{if } t = T. \end{cases} \tag{3.40}$$

In financial interpretation, the difference $V_{\tilde{\tau}} - \delta(T - \tilde{\tau})B^c(\tilde{\tau}, T)$ can be seen as the *bankruptcy cost*.

In view of (3.38)–(3.39), it is clear that Assumptions (A.1)–(A.6) of Sect. 2.2 are satisfied in the present setting. In particular, the short-term rate follows a time-homogeneous diffusion process. It is apparent that the fundamental pricing PDE (2.22) takes the form

$$u_t(V, r, t) + (r - \kappa)V u_V(V, r, t) + (a - br)u_r(V, r, t) + \tfrac{1}{2}\sigma_V^2 V^2 u_{VV}(V, r, t)$$
$$+ \tfrac{1}{2}\sigma_r^2 r u_{rr}(V, r, t) + \sigma_V \sigma_r \sqrt{r} V \rho_{Vr} u_{Vr}(V, r, t) + c - ru(V, r, t) = 0.$$

Analytical valuation of a defaultable bond, requires solving the above PDE subject to the boundary conditions

$$u(\bar{v}, r, t) = \min(\bar{v}, \delta(T - t)B^c(t, T, r)), \quad \lim_{V \to \infty} u(V, r, t) = B^c(t, T, r),$$

that need to be satisfied for $t \in [0, T)$, as well as the terminal condition $u(V, r, T) = \min(V, L)$. The above problem seems to be analytically intractable; also, the probabilistic approach – as developed in Sect. 3.4 – cannot be directly applied to the present setting for two main reasons. First, expression (3.40) for the triggering barrier is not a convenient choice, as opposed to (3.37). Second, it is well known that the volatility of a zero-coupon bond does not follow a deterministic function in the CIR model of the term structure.

It is thus worth mentioning that Kim et al. (1993a) conduct a numerical analysis of the valuation problem. Their analysis indicates, among others, that their model exhibits the undesirable, though common within the structural framework, feature that the credit spreads are close to zero for bonds of short maturities.

### 3.4.2 Longstaff and Schwartz Approach

Longstaff and Schwartz (1995), similarly as Kim et al. (1993a), consider both the default risk and the interest rate risk inherent in a corporate debt. The stochastic short-term interest rate evolves according to the Vasicek (1977) model

$$dr_t = (a - br_t) \, dt + \sigma_r \, d\tilde{W}_t,$$

and the firm's value obeys the SDE

$$dV_t = V_t \left( r_t \, dt + \sigma_V \, dW_t^* \right),$$

where both dynamics are under the martingale measure $\mathbb{P}^*$, and the Brownian motions $\tilde{W}$ and $W^*$ are correlated, with the instantaneous correlation coefficient $\rho_{Vr}$. Following Longstaff and Schwartz (1995), we assume that the default event is triggered if the firm's value process hits a constant threshold $\bar{v}$ during the life of the bond. The recovery payment received by the bondholders is paid at bond's maturity date $T$ and it is proportional to the face value of the bond. We thus have $\tilde{X} = (1 - w)L = \delta L$, where the writedown rate $w$ (and thus also the recovery rate $\delta$) is assumed to be a fixed constant. Equivalently, the owner of a corporate bond receives at time of default $(1 - w)L$ default-free zero-coupon bonds, provided that the default occurs prior to the bond's maturity. It is also implicitly assumed that $\bar{v} \geq L$, where $L$ is the face value of a corporate bond, hence if default does not occur prior to the bond's maturity, the debt is paid off in full. The interpretation of the threshold value $\bar{v}$ is that as long as $V$ is greater than $\bar{v}$, the firm is solvent. If $V$ falls below $\bar{v}$, however, the firm enters financial distress and becomes insolvent, either due to inability to meet the current obligations (the *flow-based insolvency*), or due to the fact that the minimum net worth requirements are violated (the *stock-based insolvency*). In view of cross-default provisions, it is understood that the firm defaults simultaneously on all its obligations. It thus seems natural to formally postulate that $\bar{v} = \sum_{i=1}^k (1 - w_i)L_i$, where for each $i = 1, \ldots, k$ we denote by $L_i$ the total face value of the debt from the $i^{\text{th}}$ class, and by $w_i$ the associated writedown rate. Notice that such a convention assumes that the debt's seniority no longer plays an essential role (it is already reflected in the values of the writedowns). Put another way, the strict absolute priority, as described in the second part of Sect. 3.2.1, is no longer postulated.[5] For each particular class of debt, the corresponding writedown coefficients can be estimated on the basis of historical data. The writedown $w$ can also be made random, provided that it is independent under $\mathbb{P}^*$ (or $\mathbb{P}_T$) of other sources of risk (modeled by $\tilde{W}$ and $W^*$). Since in this case it is enough to replace in the formulae below the constant $w$ by the expected value under $\mathbb{P}^*$ (or $\mathbb{P}_T$) of the random writedown, such an extension is rather trivial from the mathematical viewpoint.

---

[5] This feature is indeed supported by the empirical evidence (see, for instance, Franks and Torous (1989, 1994), Eberhart et al. (1990), or Weiss (1990)).

To summarize, we postulate that a particular defaultable bond corresponds to

$$X = L, \quad A \equiv 0, \quad Z_t = (1-w)LB(t,T), \quad \tau = \inf\{t \in [0,T] : V_t < \bar{v}\},$$

for some constant $\bar{v} > 0$. In particular, the bond's payoff at its maturity date $T$ equals

$$X^{d,1}(T) = (1-w)L\mathbb{1}_{\{\tau \leq T\}} + L\mathbb{1}_{\{\tau > T\}} = L(1 - w\mathbb{1}_{\{\tau \leq T\}}).$$

The analytic valuation of a defaultable zero-coupon bond, with the face value $L$ and maturity date $T$, relies on solving the fundamental PDE

$$u_t(V,r,t) + (r - \kappa)Vu_V(V,r,t) + (a - br)u_r(V,r,t) + \tfrac{1}{2}\sigma_V^2 V^2 u_{VV}(V,r,t)$$
$$+ \tfrac{1}{2}\sigma_r^2 u_{rr}(V,r,t) + \sigma_V\sigma_r V\rho_{Vr}u_{Vr}(V,r,t) + c - ru(V,r,t) = 0$$

subject to the boundary conditions, for $t \in [0,T)$,

$$u(\bar{v},r,t) = (1-w)B(t,T,r)L, \quad \lim_{V \to \infty} u(V,r,t) = LB(t,T,r),$$

as well as the terminal condition $u(V,r,T) = L$. An analytical solution to this problem seems to be rather hard to obtain, and thus we shall focus on the probabilistic representation of the price of a defaultable bond. It is apparent that the price $D(t,T)$ satisfies, on the set $\{\tau > t\}$,

$$D(t,T) = LB(t,T)\big(1 - w\mathbb{P}_T\{\tau \leq T \,|\, \mathcal{F}_t\}\big), \tag{3.41}$$

where $\mathbb{P}_T\{\tau \leq T \,|\, \mathcal{F}_t\}$ is the conditional probability, under the forward martingale measure $\mathbb{P}_T$, that the default occurs prior to the maturity date $T$. As usual, $B(t,T)$ represents the price of the unit default-free zero-coupon bond. For Vasicek's model, the price $B(t,T)$ is known to be given by the closed-form expression[6]

$$B(t,T) = e^{m(t,T)-n(t,T)r_t} =: B(t,T,r_t), \tag{3.42}$$

where

$$n(t,T) = \frac{1}{b}\left(1 - e^{-b(T-t)}\right) \tag{3.43}$$

and

$$m(t,T) = \frac{\sigma_r^2}{2}\int_t^T n^2(u,T)\,du - a\int_t^T n(u,T)\,du.$$

For our further purposes, it is useful to notice that (3.42) yields the following dynamics for the default-free bond price under the martingale measure $\mathbb{P}^*$:

$$dB(t,T) = B(t,T)\big(r_t\,dt + b(t,T)\,d\tilde{W}_t\big).$$

Let us stress that the bond price volatility $b(\cdot,T) : [0,T] \to \mathbb{R}$ follows here a deterministic function of time, specifically, $b(t,T) = \sigma_r n(t,T)$ (although formally $b(t,T) = -\sigma_r n(t,T)$).

---

[6] For the derivation of this formula, see, for instance, Jamshidian (1989) or Sect. 12.3 in Musiela and Rutkowski (1997a).

The most difficult step in the probabilistic derivation of a tractable formula for the price $D(t, T)$ is the study of the conditional probability of default. Although at first glance the form of the default time $\tau$ may seem simple, this problem appears to be rather involved. Notice that, without loss of generality, we may rewrite (2.44)–(2.45) as follows:

$$\begin{cases} dr_t = (a - br_t)\, dt + \sigma_r\, d\tilde{W}_t, \\ dV_t = V_t\big(r_t\, dt + \sigma_V(\rho\, d\tilde{W}_t + \sqrt{1 - \rho^2}\, d\hat{W}_t)\big), \end{cases}$$

where $\rho = \rho_{Vr}$, and $\tilde{W}$ and $\hat{W}$ are mutually independent standard Brownian motions under $\mathbb{P}^*$. Consequently, the dynamics of $r$ and $V$ under the forward martingale measure $\mathbb{P}_T$ are:

$$\begin{cases} dr_t = (a - br_t - \sigma_r^2 n(t, T))\, dt + \sigma_r\, dW_t^T, \\ dV_t = V_t\big((r_t - \sigma_V \sigma_r \rho n(t, T))\, dt + \sigma_V(\rho\, dW_t^T + \sqrt{1 - \rho^2}\, d\hat{W}_t)\big), \end{cases}$$

where

$$W_t^T = \tilde{W}_t - \int_0^t b(u, T)\, du = \tilde{W}_t + \sigma_r \int_0^t n(u, T)\, du.$$

Notice $\hat{W}$ and $W^T$ follow standard Brownian motions under $\mathbb{P}_T$. We conclude that under the forward martingale measure $\mathbb{P}_T$ the value of the firm $V$ follows an Itô process, with the drift coefficient dependent on the short-term interest rate. The two-dimensional process $(V, r)$ follows a two-dimensional Markov diffusion under $\mathbb{P}_T$. To the best of our knowledge, a closed-form solution for the probability distribution of the first passage time to a constant barrier by the process $V$ specified above is not available.

As noted by Longstaff and Schwartz (1995), certain quasi-explicit results can be obtained, though. Notice first that the SDE for the short-term rate can be explicitly solved. Indeed, we have

$$r_t = r_0 + \int_0^t e^{b(s-t)}\big(a\, ds + \sigma_r\, dW_s^T\big).$$

Using Fubini's theorem, we obtain

$$\int_0^t r_u\, du = r_0 t + \int_0^t b^{-1}\big(1 - e^{-b(t-u)}\big)\big(a\, du + \sigma_r\, dW_u^T\big),$$

so that (cf. (3.43))

$$\int_0^t r_u\, du = r_0 t + \int_0^t n(u, t)\big(a\, du + \sigma_r\, dW_u^T\big). \tag{3.44}$$

On the other hand, the discounted value process $V_t^* = V_t / B_t$ solves the SDE

$$\frac{dV_t^*}{V_t^*} = -\sigma_V \sigma_r \rho n(t, T)\, dt + \sigma_V(\rho\, dW_t^T + \sqrt{1 - \rho^2}\, d\hat{W}_t),$$

which also means that

$$V_t^* = V_0^* \exp\left(-\int_0^t \big(\sigma_V \sigma_r \rho n(u, T) + \tfrac{1}{2}\sigma_V^2\big)\, du + \sigma_V\big(\rho\, W_t^T + \sqrt{1 - \rho^2}\, \hat{W}_t\big)\right).$$

Combining the last formula with (3.44), we obtain the following result.

**Lemma 3.4.1.** *The value of the firm* $V$ *equals* $V_t = V_0 \exp(m(t) + \xi(t))$, *where*

$$m(t) = \int_0^t \left( r_0 - \tfrac{1}{2}\sigma_V^2 + an(u,t) - \sigma_V\sigma_r\rho n(u,T) \right) du$$

*and*

$$\xi(t) = \int_0^t \left( \sigma_r n(u,t) + \sigma_V \rho \right) dW_u^T + \sigma_V \sqrt{1 - \rho^2}\, \hat{W}_t.$$

It is thus apparent that the auxiliary process $Y_t := \ln(V_t/\bar{v})$ follows under the forward martingale measure $\mathbb{P}_T$ a Gaussian process with independent increments. Moreover, for any $s \le t$,

$$\mathbb{E}_{\mathbb{P}_T}(Y_t - Y_s) = m(t) - m(s) =: \mu(s,t),$$

and[7]

$$\mathrm{Var}_{\mathbb{P}_T}(Y_t - Y_s) = \int_s^t \left( \sigma_r n(u,t) + \sigma_V \rho \right)^2 du + \sigma_V^2 (1 - \rho^2)(t - s) =: \sigma^2(s,t).$$

It is useful to observe that $Y$ is a one-dimensional time-inhomogeneous continuous Markov process under $\mathbb{P}_T$. Recall that our goal is to find the (conditional) probability law, under the forward measure $\mathbb{P}_T$, of the first passage time of $V$ to $\bar{v}$ or, equivalently, of the first passage of the auxiliary process $Y$ to 0. Due to the Markovian property of $Y$, it suffices to consider the case where $t = 0$. Let $f$ stand for the probability density function of the first passage time to 0, given the initial value $Y_0 > 0$. In fact, we are interested in values of $f$ for $u \in [0, T]$ only. Since $Y$ is a Gaussian process, it is clear that

$$\mathbb{P}_T\{Y_T < 0\} = N\left( \frac{-Y_0 - \mu(0,T)}{\sigma(0,T)} \right).$$

On the other hand, $Y$ is a continuous Markov process, and thus the value of $Y$ is strictly negative at time $T$, if and only if $Y$ reaches 0 at some instant $u < T$ and the increment $Y_T - Y_u$ is strictly negative. We conclude that the probability $\mathbb{P}_T\{Y_T < 0\}$ admits the following representation, which is manifestly a special case of the classic equation:

$$\mathbb{P}_T\{Y_T < 0\} = \int_0^T f(u)N\left( \frac{-\mu(u,T)}{\sigma(u,T)} \right) du.$$

We are in a position to state the following result, used in Longstaff and Schwartz (1995) to approximate the density $f$ of the default time. Since Proposition 3.4.2 follows directly from the considerations above, its proof is omitted.

---

[7] Since the function $n(t,T)$ is given by (3.43), explicit formulae for $n(t,T)$ are available upon simple integration. Note that the expected value $\mu(s,t)$ depends on the maturity $T$, and the variance $\sigma^2(s,t)$ is independent of $T$.

**Proposition 3.4.2.** *Assume that $V_0 > \bar{v}$. The density function $f$ of the first passage time of the value process $V$ to the fixed barrier $\bar{v}$ obeys the following equality*

$$N\left(\frac{\ln(\bar{v}/V_0) - \mu(0, T)}{\sigma(0, T)}\right) = \int_0^T f(u) N\left(\frac{-\mu(u, T)}{\sigma(u, T)}\right) du.$$

*More generally, $f$ satisfies the following integral equation for every $s \in [0, T]$:*

$$N\left(\frac{\ln(\bar{v}/V_0) - \mu(0, s)}{\sigma(0, s)}\right) = \int_0^s f(u) N\left(\frac{-\mu(u, s)}{\sigma(u, s)}\right) du. \qquad (3.45)$$

Using Proposition 3.4.2, Longstaff and Schwartz (1995) provide a series representation for the probability density function $f$ of default time. To this end, we fix a natural number $n$, and we assume that $f$ is constant on each interval $[(i-1)\Delta, i\Delta)$ for $i = 1, \ldots, n$, where $\Delta = T/n$. By discretizing the integral equation (3.45) for $s = i\Delta$ and $i = 1, \ldots, n$, we arrive at the following system of $n$ linear equations:

$$N(\alpha_i) = q_i + \sum_{j=1}^{i-1} q_j N(\beta_{ji}), \quad i = 1, \ldots, n,$$

where $q_j = f(j\Delta)\Delta$ for $j = 1, \ldots, n$ and

$$\alpha_i = \left(\frac{\ln(\bar{v}/V_0) - \mu(0, i\Delta)}{\sigma(0, i\Delta)}\right), \quad \beta_{ji} = N\left(\frac{-\mu(j\Delta, i\Delta)}{\sigma(j\Delta, i\Delta)}\right).$$

Note that the quantity $q_i$ depends not only on $i$, but also on $n$. Furthermore, we have

$$\mathbb{P}_T\{\tau < T\} = \int_0^T f(u)\, du = \lim_{n \to \infty} \sum_{j=1}^n q_j,$$

and thus we may find numerically the value of the bond price $D(0, T)$. Also, it appears that the above approximation converges rapidly. Longstaff and Schwartz (1995) conduct a numerical analysis showing that their approach leads to several desirable features of the defaultable term structure (their model produces credit spreads close to zero for small maturities, though). They also examine the case of a floating-rate corporate debt within the framework of their model, and they derive a quasi-explicit valuation formula that allows them to analyze the qualitative behavior of a floating-rate corporate debt. It appears that the value of a floating-rate corporate debt can be an increasing function of the maturity of the bond (or an increasing function of the level of interest rates) in some situations. For the model's implementation, the interested reader may also consult Lehrbass (1997).

*Remarks.* Collin-Dufresne and Goldstein (2001) develop a numerical procedure for computing the first passage time density of a two-dimensional Gaussian Markov process, and subsequently they apply this procedure to a generalization of the Longstaff and Schwarz model with time-dependent barrier process that depends on the value of the firm.

### 3.4.3 Cathcart and El-Jahel Approach

Following Kim et al. (1993a), Cathcart and El-Jahel (1998) postulate that the short-term interest rate $r$ follows the CIR dynamics, specifically:

$$dr_t = (a - br_t)\, dt + \sigma_r \sqrt{r_t}\, d\tilde{W}_t, \qquad (3.46)$$

where $\tilde{W}$ is a standard Brownian motion under the martingale measure $\mathbb{P}^*$. They postulate, however, that the default time is the first passage time to a constant barrier $\bar{Y}$ by a *signaling process* $Y$, which is assumed to follow under $\mathbb{P}^*$

$$dY_t = Y_t\big(\mu_Y\, dt + \sigma_Y\, dW_t^*\big),$$

where the two Brownian motions $\tilde{W}$ and $W^*$ are assumed to be mutually independent. If default occurs prior to the debt's maturity, the bondholder receives $\delta$ default-free equivalent zero-coupon bonds.

Since the lower threshold level is assumed to be constant, the model proposed by Cathcart and El-Jahel (1998) can also be seen as a modification of the Longstaff and Schwartz (1995) approach. Recall that in the latter model the short-term interest rate follows a Gaussian process, so that the interest rate takes on negative values with positive probability. This flaw disappears if we use the CIR model instead, since the process $r$ defined by (3.46) is non-negative, provided that $r_0 > 0$. On the other hand, though interest rates are modeled through a stochastic process, both coefficients, $\mu_Y$ and $\sigma_Y$, are constant in the dynamics of the signaling process, so that $Y$ follows a geometric Brownian motion under the spot martingale measure.

The authors argue that the relaxation of the usual covenant, which stipulates that the default time is directly associated with the firm's value, should be seen as an advantage of their model. Indeed, the signaling process can be chosen to describe the most relevant economic fundamentals for each particular issuer. Due to this flexibility, the model is suitable to the valuation of defaultable debt issued by entities that do not have identifiable collection of assets (for instance, to the sovereign debt, municipal debt, etc.).

It is apparent that the PDE satisfied by the pricing function of a defaultable bond reads:

$$u_t(Y, r, t) + \mu_Y Y u_Y(Y, r, t) + (a - br)u_r(Y, r, t)$$
$$+ \tfrac{1}{2}\sigma_Y^2 Y^2 u_{YY}(Y, r, t) + \tfrac{1}{2}\sigma_r^2 r u_{rr}(Y, r, t) - ru(Y, r, t) = 0,$$

the boundary conditions are:

$$u(\bar{Y}, r, t) = \delta B(t, T, r), \quad \lim_{Y \to \infty} u(Y, r, t) = B(t, T, r). \qquad (3.47)$$

and the terminal condition is simply $u(\bar{Y}, r, T) = 1$. Since a closed-form solution for the price $B(t, T, r)$ of the default-free zero-coupon bond is available within the CIR framework, the boundary conditions (3.47) involve explicit formulae. Cathcart and El-Jahel (1998) furnish quasi-analytical solutions, in terms of the inverse Laplace transform, to the valuation problem above, as well as to the case of a floating-rate debt.

### 3.4.4 Briys and de Varenne Approach

Briys and de Varenne (1997) put forward a model, which main objective is to correct the apparent deficiencies of some models that were proposed and studied by other researchers. They indicate two such drawbacks:
- first, the pricing equations do not necessarily assure that the payment to bondholders is not greater than the firm's value upon default. This problem appears in the Nielsen et al. (1993) approach, since in their model the payoff upon bankruptcy is independent of the level of the stochastic barrier and of the value of the asset,
- second, some models allow for a possibility that the firm that is in a solvent position at the maturity relative to the threshold may nevertheless have insufficient assets to match the face value of the bond. For instance, in the Longstaff and Schwartz (1995) approach, analyzed in Sect. 3.4.2, the inequality $V_T < (1 - w)L$ is not excluded a priori.

It should be made clear that the model proposed by Briys and de Varenne (1997) is a special case of the Black and Cox model with stochastic interest rates. In particular, their valuation formula for a defaultable bond is an immediate consequence of Proposition 3.4.1. Briys and de Varenne (1997) consider, under the spot martingale measure $\mathbb{P}^*$, the so-called generalized Vasicek's model of the short-term rate

$$dr_t = a(t)(b(t) - r_t) \, dt + \sigma(t) \, d\tilde{W}_t,$$

where $a, b, \sigma : [0, T] \to \mathbb{R}$ are deterministic functions. Thus, the price $B(t, T)$ of a default-free zero-coupon bond satisfies

$$dB(t, T) = B(t, T)\left(r_t \, dt + b(t, T)d\tilde{W}_t\right) \tag{3.48}$$

for some deterministic function $b(\cdot, T) : [0, T] \to \mathbb{R}$. They define the value of the firm process $V$ by setting

$$\frac{dV_t}{V_t} = r_t \, dt + \sigma_V \left(\rho \, d\tilde{W}_t + \sqrt{1 - \rho^2} \, d\hat{W}_t\right), \tag{3.49}$$

where $\sigma_V > 0$ is a constant, $\tilde{W}$ and $\hat{W}$, are mutually independent Brownian motions, and $\rho = \rho_{V,r}$ is the local correlation coefficient between the risk-free interest rate and the firm's value.

**Defaultable bond.** The default barrier is defined as a constant quantity discounted at the risk-free rate up to the bond's maturity so that

$$v_t = \begin{cases} KB(t, T), & \text{if } t < T, \\ L, & \text{if } t = T, \end{cases} \tag{3.50}$$

where $0 < K \le L$. We define the default time $\tau$ in a standard way, specifically,

$$\tau = \inf \{t \in [0, T] : V_t < v_t\}.$$

Let us now focus on the payoff at default. As soon as the threshold is hit (and thus crossed) prior to the maturity date $T$, the bondholders receive an exogenously specified fraction of the remaining assets: $\beta_2 V_\tau = \beta_2 KB(\tau, T)$, where $\beta_2 \in [0, 1]$. If, on the contrary, the bankruptcy barrier is not hit prior to the maturity date $T$, the payoff to the bondholders at time $T$ is equal to

$$L\mathbb{1}_{\{\tau > T\}} + \beta_1 V_T \mathbb{1}_{\{\tau = T\}},$$

where $\beta_1 \in [0, 1]$. Consequently, the bond's cash flow at the maturity date can be represented as follows:

$$X^{d,1}(T) = \beta_2 K \mathbb{1}_{\{\tau < T\}} + \beta_1 V_T \mathbb{1}_{\{\tau = T\}} + L\mathbb{1}_{\{\tau > T\}}.$$

To summarize, we have:

$$X = L, \ A \equiv 0, \ Z \equiv \beta_2 V, \ \tilde{X} = \beta_1 V_T, \ \tau = \inf\{t \in [0, T] : V_t < v_t\},$$

where $v$ is given by (3.50). As already mentioned, an explicit valuation formula for a defaultable bond established by Briys and de Varenne (1997) is a special case of the result provided by Proposition 3.4.1 (since the firm pays no dividends, it is enough to make use of Corollary 3.4.1). We state their result in its original formulation, subject to a suitable change of notation.

**Corollary 3.4.2.** *Suppose that the bond price volatility $b(t, T)$ is a deterministic function. Then the forward price $F_D(t, T) = D(t, T)/B(t, T)$ of a defaultable bond equals*

$$\begin{aligned}
F_D(t, T) = L &- D_1(t, T) + D_2(t, T) \\
&- (1 - \beta_2)\big(F_t N\big(d_4(F_t, t, T)\big) + KN\big(d_3(F_t, t, T)\big)\big) \\
&- (1 - \beta_1)F_t\big(N\big(d_2(F_t, t, T)\big) - N\big(d_4(F_t, t, T)\big)\big) \\
&- (1 - \beta_1)K\big(N\big(d_5(F_t, t, T)\big) - N\big(d_3(F_t, t, T)\big)\big),
\end{aligned}$$

*where $F_t = V_t/B(t, T)$,*

$$\begin{aligned}
D_1(t, T) &= LN\big(d_1(F_t, t, T)\big) - F_t N\big(d_2(F_t, t, T)\big), \\
D_2(t, T) &= KN\big(d_5(F_t, t, T)\big) - (F_t L/K)N\big(d_6(F_t, t, T)\big),
\end{aligned}$$

*and*

$$d_1(F_t, t, T) = \frac{\ln(L/F_t) + \frac{1}{2}\sigma^2(t, T)}{\sigma(t, T)} = d_2(F_t, t, T) + \sigma(t, T),$$

$$d_3(F_t, t, T) = \frac{\ln(K/F_t) + \frac{1}{2}\sigma^2(t, T)}{\sigma(t, T)} = d_4(F_t, t, T) + \sigma(t, T),$$

$$d_5(F_t, t, T) = \frac{\ln(K^2/F_t L) + \frac{1}{2}\sigma^2(t, T)}{\sigma(t, T)} = d_6(F_t, t, T) + \sigma(t, T),$$

*where*

$$\sigma^2(t, T) = \int_t^T \big((\rho\sigma_V - b(u, T))^2 + (1 - \rho^2)\sigma_V^2\big)\,du. \tag{3.51}$$

*Proof.* In view of expressions (3.48)–(3.49), representation (3.51) of the volatility $\sigma(t,T)$ of the forward value of the firm is an immediate consequence of Itô's rule. It is thus enough to observe that the formula we need to prove is in fact a special case of the general valuation result established in Corollary 3.4.1                                                                                            □

Briys and de Varenne (1997) provide intuitive interpretations of each component of their formula. First, $I_1(t) := LB(t,T)$ is obviously the price of a default-free bond with the face value $L$. The negative term $-I_2(t)$, where

$$I_2(t) = B(t,T)D_1(t,T) = LB(t,T)N\big(d_1(F_t,t,T)\big) - V_t N\big(d_2(F_t,t,T)\big) > 0$$

represents the short position in the so-called *put-to-default*; i.e., a European style option with the terminal payoff $(L - V_T)^+$. The next term:

$$I_3(t) = B(t,T)D_2(t,T) = KB(t,T)N\big(d_5(F_t,t,T)\big) - (V_t L/K)N\big(d_6(F_t,t,T)\big)$$

corresponds to the long position in $L/K$ units of European put options with strike level $K^2/L$ and maturity $T$. It is associated with the possibility of an early default triggered by the safety covenant.

Let us first consider the case when a strict priority rule is enforced; that is, when $\beta_1 = \beta_2 = 1$. In such a case, all remaining terms in the bond valuation formula vanish, and we arrive at a simple representation $D(t,T) = I_1(t) - I_2(t) + I_3(t)$, where the difference $I_1(t) - I_2(t)$ represents Merton's price of a defaultable bond, and the term $I_3(t)$ is strictly positive. It is thus clear that the safety covenant is indeed in the interest of the bondholders, in the sense that in all circumstances it increases the value of a defaultable bond. If the strict priority rule is violated, i.e., if either $\beta_1$ or $\beta_2$ (or both) is less than 1, the defaultable bond becomes less valuable than in the previous case. Indeed, it apparent that

$$I_4(t) := (1 - \beta_2)F_t N\big(d_4(F_t,t,T)\big) + KN\big(d_3(F_t,t,T)\big) > 0$$

and

$$I_5(t) := (1 - \beta_1)F_t\big(N\big(d_2(F_t,t,T)\big) - N\big(d_4(F_t,t,T)\big)\big)$$
$$+ (1 - \beta_1)K\big(N\big(d_5(F_t,t,T)\big) - N\big(d_3(F_t,t,T)\big)\big) > 0,$$

where the latter inequality follows from the proof of Proposition 3.4.1. Since $D(t,T) = I_1(t) - I_2(t) + I_3(t) - I_4(t) - I_5(t)$, it is clear that the bond's value declines when $\beta_1$ and/or $\beta_2$ decrease.

**Credit spreads.** Briys and de Varenne (1997) present also a numerical analysis of the credit spread (we set here $L = 1$)

$$S(t,T) = -\frac{1}{(T - t)} \ln \frac{D(t,T)}{B(t,T)} \tag{3.52}$$

as a function of the bond's maturity $T$ and of the other model parameters. The numerical results reported in the paper indicate that the credit spreads produced by their model are generally larger than those obtained using Merton's (1974) approach.

### 3.4.5 Saá-Requejo and Santa-Clara Approach

Nielsen et al. (1993) examine a specific structural model with an endogenously specified stochastic barrier and random interest rates. Their approach was subsequently revised in Saá-Requejo and Santa-Clara (1999). In particular, Nielsen et al. (1993) assume that the short-term interest rate $r$ process follows Vasicek's model. By contrast, no specific choice of the dynamics of $r$ is required for the validity of the approach developed by Saá-Requejo and Santa-Clara (1999). For this reason, we prefer to focus on the latter, rather general, methodology. We make the standard assumption that the firm's value $V$ follows under the spot martingale measure $\mathbb{P}^*$

$$dV_t = V_t\big((r_t - \kappa)\,dt + \sigma_V\,dW_t^*\big). \tag{3.53}$$

The short-term interest rate $r$ is assumed to follow an Itô process driven by a standard Brownian motion $\tilde{W}$

$$dr_t = \mu_r\,dt + \sigma_r\,d\tilde{W}_t,$$

where, as usual, the correlation coefficient between standard Brownian motions $W^*$ and $\tilde{W}$ is assumed to be constant; we denote it by $\rho_{Vr}$. We may and do assume, without loss of generality, that

$$W_t^* = \rho_{Vr}\,\tilde{W}_t + \sqrt{1 - \rho_{Vr}^2}\,\hat{W}_t, \tag{3.54}$$

where $\hat{W}$ is a standard Brownian motion independent of $\tilde{W}$. Let $\mathbb{F}$ be the natural filtration of the Brownian motions driving the dynamics for the processes $V$, $r$ and $v$. The price $B(t,T)$ of a default-free zero-coupon bond is given by the formula

$$B(t,T) = \mathbb{E}_{\mathbb{P}^*}\left(e^{-\int_t^T r_u\,du}\,\big|\,\mathcal{F}_t\right),$$

hence, the bond price dynamics under $\mathbb{P}^*$ are

$$dB(t,T) = B(t,T)\big(r_t\,dt + b(t,T)\,d\tilde{W}_t\big) \tag{3.55}$$

for some $\mathbb{F}$-adapted process $b(t,T)$, $t \le T$. The price $B(t,T)$ depends on the choice of a particular model for the short-term rate. For instance, one may choose Vasicek's model adopted by Longstaff and Schwartz (1995) or the CIR model considered previously in a similar context by Kim et al. (1993a).

The default is triggered at the first instance the value process $V$ reaches critical level, defined by a continuous positive stochastic process $v$. Formally, the barrier process $v$ is modeled as a solution to the SDE

$$\frac{dv_t}{v_t} = (r_t - \zeta)\,dt + \tilde{\sigma}_v\,d\tilde{W}_t + \hat{\sigma}_v\,d\hat{W}_t \tag{3.56}$$

with the initial condition $v_0 < V_0$. The constant coefficients $\kappa$ and $\zeta$ could be replaced by some functions $\kappa(V_t, v_t, B(t,T))$ and $\zeta(V_t, v_t, B(t,T))$.

One possible interpretation of the barrier process, put forward by Saá-Requejo and Santa-Clara (1999), is that it represents the market value of all liabilities of the firm. In this interpretation, the coefficient $\zeta$ should be seen as the constant payout ratio to the debt-holders of the firm. Let us stress that the novelty in this approach is that it is assumed that the value process of the total firm's debt is given exogenously given. Hence, the goal is to develop arbitrage valuation a particular defaultable security that should thus be seen as the derivative security in the financial market model with the following primary tradeable assets: value of the firm process $V$, value of the firm's debt $v$ and default-free bonds.

Other interpretations of the lower threshold $v$ are also possible. For instance, it can be seen as the discounted face value of all of the firm's liabilities, with the price $B(t, T)$ of a default-free bond chosen as a discount factor. It is essential to assume that the barrier process $v$ represents the market value of a tradeable security.

*Remarks.* It should be acknowledged that we adopt here a slightly different notational convention than the original one. Saá-Requejo and Santa-Clara (1999) postulate the following dynamics for the barrier process $v$

$$\frac{dv_t}{v_t} = (r_t - \zeta)\, dt + \sigma_{rv}\, d\tilde{W}_t + \sigma_{Vv}\, dW_t^*. \tag{3.57}$$

A comparison of expressions (3.56) and (3.57) yields

$$\tilde{\sigma}_v = \sigma_{rv} + \sigma_{Vv}\rho_{Vr}, \qquad \hat{\sigma}_v^2 = \sigma_{Vv}^2(1 - \rho_{Vr}^2).$$

It is thus clear that all results given below can also be reformulated in terms of the coefficients $\sigma_{rv}$ and $\sigma_{Vv}$.

**Solvency ratio.** The random character of $v$ makes the task of finding an explicit valuation formula much harder than in the Briys and de Varenne (1997). Notice that we may express the default time $\tau$ as

$$\tau := \inf\{t \in [0, T] : V_t < v_t\} = \inf\{t \in [0, T] : R_t < 0\}, \tag{3.58}$$

where $R_t := \ln(V_t/v_t)$ referred to as the *solvency ratio*, represents the current state of the firm's credit quality. It appears that the solvency ratio has rather simple dynamics under the forward measure; the drift coefficient in the dynamics of $R$ under $\mathbb{P}_T$ is not constant, though. For the sake of convenience, we shall denote $\rho = \rho_{Vr}$.

**Lemma 3.4.2.** *Under the forward martingale measure $\mathbb{P}_T$, the solvency ratio $R$ obeys the SDE:*

$$dR_t = \nu_t\, dt + (\sigma_V\rho - \tilde{\sigma}_v)\, dW_t^T + (\sigma_V\sqrt{1 - \rho^2} - \hat{\sigma}_v)\, d\hat{W}_t, \tag{3.59}$$

*where*
$$\nu_t = \zeta - \kappa + \tfrac{1}{2}(\tilde{\sigma}_v^2 + \hat{\sigma}_v^2 - \sigma_V^2) + (\sigma_V\rho - \tilde{\sigma}_v)b(t, T).$$

*Processes $W^T$ and $\hat{W}$ follow mutually independent standard Brownian motions under $\mathbb{P}_T$.*

*Proof.* It is enough to combine Lemma 3.1.1 with the definition of the forward measure $\mathbb{P}_T$. Indeed, using (3.53)–(3.54), we obtain

$$\frac{dV_t}{V_t} = (r_t - \kappa)\, dt + \sigma_V \rho\, d\tilde{W}_t + \sigma_V \sqrt{1 - \rho^2}\, d\hat{W}_t.$$

On the other hand, we have

$$\frac{dv_t}{v_t} = (r_t - \zeta)\, dt + \tilde{\sigma}_v\, d\tilde{W}_t + \hat{\sigma}_v\, d\hat{W}_t.$$

Therefore, in view of Lemma 3.1.1, the process $R_t = \ln(V_t/v_t)$ satisfies

$$dR_t = \mu_t\, dt + (\sigma_V \rho - \tilde{\sigma}_v)\, d\tilde{W}_t + (\sigma_V \sqrt{1 - \rho^2} - \hat{\sigma}_v)\, d\hat{W}_t$$

with

$$\mu_t = \zeta - \kappa + \tfrac{1}{2}(\tilde{\sigma}_v^2 + \hat{\sigma}_v^2 - \sigma_V^2). \tag{3.60}$$

Since $d\tilde{W}_t = dW_t^T + b(t, T)\, dt$, the asserted formula (3.59) is now obvious. The last statement is also clear, since the Radon-Nikodým density of $\mathbb{P}_T$ with respect to $\mathbb{P}^*$ equals

$$\frac{d\mathbb{P}_T}{d\mathbb{P}^*} = \exp\left( \int_0^T b(u, T)\, d\tilde{W}_u - \frac{1}{2} \int_0^T b^2(u, T)\, du \right),$$

and thus we may apply Girsanov's theorem. $\qquad\square$

**Corollary 3.4.3.** *Suppose that the bond price volatility $b(t, T)$ is a deterministic function. Then the solvency ratio $R$ follows a strong Markov process under $\mathbb{P}_T$. Consequently, the conditional probability of default, $\mathbb{P}_T\{\tau \leq T \mid \mathcal{F}_t\}$, is a function of $R_t$, $t$ and $T$.*

Notice that we may rewrite (3.59) as follows:

$$dR_t = \mu_t\, dt + \sigma_R\, d\bar{W}_t, \tag{3.61}$$

where $\sigma_R$ is a constant diffusion coefficient:

$$\sigma_R^2 := (\sigma_V \rho - \tilde{\sigma}_v)^2 + (\sigma_V \sqrt{1 - \rho^2} - \hat{\sigma}_v)^2, \tag{3.62}$$

and the process $\bar{W}$ follows a one-dimensional standard Brownian motion with respect to $\mathbb{F}$ under $\mathbb{P}_T$. In some circumstances, namely when

$$(\sigma_V \rho - \tilde{\sigma}_v) b(t, T) = 0, \quad \forall t \in [0, T],$$

the solvency ratio $R$ follows a Brownian motion with constant drift under $\mathbb{P}_T$, so that we may directly apply Corollary 3.1.1 in order to compute the probability $\mathbb{P}_T\{\tau \leq T \mid \mathcal{F}_t\}$. The drift coefficient in the dynamics of $R$ is time-dependent, in general. It is worthwhile to notice that it is non-random, provided that the bond price volatility $b(t, T)$ follows a deterministic function as, for instance, in the Gaussian HJM framework.

**Defaultable bond.** We shall now describe the recovery structure of a defaultable zero-coupon bond. Suppose that the bondholders receive at debt's maturity $T$ a fraction $(1 - w)L$ of the nominal value $L$ of the bond upon the firm's default, where $w$ is either constant, or more generally, an $\mathcal{F}_T$-measurable random variable. Hence, we have:

$$X = L, \ A \equiv 0, \ \tilde{X} = (1 - w)L, \ \tau = \inf\{t \in [0, T] : V_t < v_t\},$$

so that the bond's payoff at maturity $T$ equals

$$X^{d,1}(T) = (1 - w)L\mathbb{1}_{\{\tau \leq T\}} + L\mathbb{1}_{\{\tau > T\}} = L(1 - w\mathbb{1}_{\{\tau \leq T\}}).$$

In the case of a constant writedown $w$, this leads to the following representation of the price of a defaultable bond, on the set $\{\tau > t\}$,

$$D(t, T) = LB(t, T)\big(1 - w\mathbb{P}_T\{\tau \leq T \mid \mathcal{F}_t\}\big), \tag{3.63}$$

where $\mathbb{P}_T\{\tau \leq T \mid \mathcal{F}_t\}$ is the conditional probability under the forward martingale measure $\mathbb{P}_T$ that the default occurs during the bond's lifetime.

*Remarks.* Formula (3.63) is also used in the Longstaff and Schwartz (1995) model discussed below. However, the present set-up differs from their approach in one important aspect, namely, the default time $\tau$ in (3.58) is the first passage time of $V$ to a random barrier $v$, rather than the first passage time of $V$ to a constant barrier $\bar{v}$ as in otherwise identical formula (3.41).

An explicit expression for $D(t, T)$ is easily available in the special case when the coefficients in dynamics (3.61) of the solvency process $R$ are constant. The following result is an immediate consequence of formulae (3.58) and (3.63) combined with Corollary 3.1.1.

**Proposition 3.4.3.** *Assume that the equality* $\sigma_V \rho - \tilde{\sigma}_v = 0$ *is valid so that the solvency ratio $R$ follows under $\mathbb{P}_T$ a generalized Brownian motion with drift:* $R_t = R_0 + \mu_R t + \sigma_R \bar{W}_t$, *where in view of (3.60) and (3.62)*

$$\mu_R = \zeta - \kappa + \tfrac{1}{2}(\tilde{\sigma}_v^2 + \hat{\sigma}_v^2 - \sigma_V^2), \quad \sigma_R^2 = (\sigma_V\sqrt{1 - \rho^2} - \hat{\sigma}_v)^2.$$

*Denoting $a = \mu_R \sigma_R^{-2}$ and $u = T - t$, we get, on the set $\{\tau > t\}$,*

$$D(t, T) = LB(t, T)\left(1 - wN\left(\frac{-R_t - \mu_R u}{\sigma_R\sqrt{u}}\right) - we^{-2aR_t}N\left(\frac{-R_t + \mu_R u}{\sigma_R\sqrt{u}}\right)\right).$$

**Credit spreads.** The credit spread equals (we set $L = 1$)

$$S(t, T) := -\frac{1}{(T - t)}\ln\frac{D(t, T)}{B(t, T)} = \frac{\ln(w\mathbb{P}_T\{\tau \leq T \mid \mathcal{F}_t\} - 1)}{T - t}.$$

It is to be expected that if $R_t$ is kept away from zero, for $t$ close to the maturity $T$, so that the firm remains solvent near the bond's maturity date, then the default probability will converge to zero fast enough when $t$ tends to $T$, so that the spread $S(t, T)$ will also tend to zero. This indeed is substantiated by the numerical results reported in Nielsen et al. (1993). As we observed before, this feature of the model is not confirmed by empirical data.

**Fundamental PDE.** As explained above, in the present set-up a closed-form solution for the value of a defaultable bond is not easily obtained through a probabilistic approach. Our next goal is thus to examine the PDE approach to the valuation and hedging of defaultable claims in the present framework. Since the Markovian property of the short-term rate is not imposed here, the PDE (3.69) has different features than the PDE (2.22). We now assume that not only the firm's value $V$, but also the barrier process $v$ represents the price of a tradeable asset. Therefore, we may choose as hedging instruments for a defaultable claim the following assets: default-free zero-coupon bonds, the firm's assets and the firm's liabilities – that is, processes $B(t, T)$, $V_t$ and $v_t$. To derive the associated PDE, we need to postulate in addition that

$$X = g(V_T, v_T), \quad A_t = c(V_t, v_t, B(t, T), t), \quad \tilde{X} = h(V_T, v_T), \quad (3.64)$$

for some functions $g, h : \mathbb{R}_+^2 \to \mathbb{R}$ and $c : \mathbb{R}_+^3 \times [0, T] \to \mathbb{R}$. We shall now sketch the derivation of the fundamental PDE satisfied by the pricing function of a defaultable claim. To this end, let us assume that the pricing function $u : \mathbb{R}_+^3 \times [0, T] \to \mathbb{R}$, where $u = u(V, v, B, t)$ is a smooth function. Using (3.53) and (3.55)–(3.56), we obtain for $u = u(V_t, v_t, B(t, T), t)$

$$\begin{aligned}
du = {} & u_t \, dt + u_V \, dV_t + u_v \, dv_t + u_B \, dB(t, T) \\
& + u_{Vv} \, d\langle V, v \rangle_t + u_{VB} \, d\langle V, B \rangle_t + u_{vB} \, d\langle v, B \rangle_t \\
& + \tfrac{1}{2} u_{VV} \, d\langle V, V \rangle_t + \tfrac{1}{2} u_{vv} \, d\langle v, v \rangle_t + \tfrac{1}{2} u_{BB} \, d\langle B, B \rangle_t,
\end{aligned} \quad (3.65)$$

where the quadratic variations of processes $V, v$ and $B$ equal

$$\begin{aligned}
d\langle V, V \rangle_t &= \sigma_V^2 V_t^2 \, dt, \\
d\langle v, v \rangle_t &= (\tilde{\sigma}_v^2 + \hat{\sigma}_v^2) \, dt, \\
d\langle B, B \rangle_t &= b^2(t, T) B^2(t, T) \, dt,
\end{aligned} \quad (3.66)$$

and the predictable covariations are:

$$\begin{aligned}
d\langle V, v \rangle_t &= \sigma_V (\tilde{\sigma}_v \rho + \hat{\sigma}_v \sqrt{1 - \rho^2}) v_t V_t \, dt, \\
d\langle V, B \rangle_t &= \sigma_V \rho V_t b(t, T) B(t, T) \, dt, \\
d\langle v, B \rangle_t &= \tilde{\sigma}_v v_t b(t, T) B(t, T) \, dt.
\end{aligned} \quad (3.67)$$

Consider a self-financing trading strategy $\phi_t = (\phi_t^0, \phi_t^1, \phi_t^2, \phi_t^3)$ with the value process $U(\phi)$ given by

$$U_t(\phi) = \phi_t^0 u(V_t, v_t, B(t, T), t) + \phi_t^1 V_t + \phi_t^2 v_t + \phi_t^3 B(t, T). \quad (3.68)$$

The self-financing property of $\phi$ reads

$$\begin{aligned}
dU_t(\phi) = {} & \phi_t^0 \big( du(V_t, v_t, B(t, T), t) + c(V_t, v_t, B(t, T)) \big) \, dt \\
& + \phi_t^1 \big( dV_t + \kappa V_t \, dt \big) + \phi_t^2 \big( dv_t + \zeta v_t \, dt \big) + \phi_t^3 \, dB(t, T),
\end{aligned}$$

where the Itô differential $du(V_t, v_t, B(t, T), t)$ is given by (3.65).

Let us take $\phi_t^1 = -1$ for $t \in [0, T]$, and let us postulate that processes $\phi_t^1$, $\phi_t^2$, $\phi_t^3$ satisfy, with suppressed arguments,

$$\phi_t^1 = u_V, \quad \phi_t^2 = u_v, \quad \phi_t^3 = u_B.$$

In addition, we assume that our strategy replicates a defaultable claim so that $U_t(\phi) = 0$ for $t \in [0, T]$. Then (3.68) yields

$$u = u_V V_t + u_v v_t + u_B B(t, T).$$

Combining the last equality with the self-financing condition, we obtain

$$
\begin{aligned}
du &= \phi_t^1\big(dV_t + \kappa V_t\,dt\big) + \phi_t^2\big(dv_t + \zeta v_t\,dt\big) + \phi_t^3\,dB(t,T) - c\,dt \\
&= u_V\big(dV_t + \kappa V_t\,dt\big) + u_v\big(dv_t + \zeta v_t\,dt\big) + u_B\,dB(t,T) - c\,dt.
\end{aligned}
$$

A comparison of the last formula with expression (3.65) provided by Itô's lemma leads to

$$
\begin{aligned}
&u_t\,dt + u_V \kappa V_t\,dt + u_v \zeta v_t\,dt + u_{Vv}\,d\langle V, v\rangle_t + u_{VB}\,d\langle V, B\rangle_t \\
&\quad + u_{vB}\,d\langle v, B\rangle_t + \tfrac{1}{2}u_{VV}\,d\langle V, V\rangle_t + \tfrac{1}{2}u_{vv}\,d\langle v, v\rangle_t + \tfrac{1}{2}u_{BB}\,d\langle B, B\rangle_t.
\end{aligned}
$$

The following result, which is an immediate consequence of the last formula combined with (3.66)–(3.67), gives a fundamental PDE for the pricing function of a defaultable claim, and a generic form of the replicating strategy in the Saá-Requejo and Santa-Clara framework.

**Proposition 3.4.4.** *Suppose that a defaultable claim $(X, A, \tilde{X}, Z, \tau)$ satisfies (3.64) with $\tau$ given by (3.58). In addition, let the bond price volatility $b(t, T)$, $t \in [0, T]$, be deterministic. If the function $u = u(V_t, v_t, B(t, T), t)$ belongs to the class $C^{2,2,2,1}(\mathbb{R}_+^3 \times [0, T])$, then it obeys the following PDE*

$$
\begin{aligned}
&u_t + \kappa V u_V + \zeta v u_v + \sigma_V\big(\tilde{\sigma}_v \rho + \hat{\sigma}_v\sqrt{1-\rho^2}\big)vV u_{Vv} \\
&\quad + \sigma_V \rho b(t,T)V B u_{VB} + \tilde{\sigma}_v b(t,T)v B u_{vB} + \tfrac{1}{2}\sigma_V^2 V^2 u_{VV} \qquad (3.69) \\
&\quad + \tfrac{1}{2}(\tilde{\sigma}_v^2 + \hat{\sigma}_v^2)v^2 u_{vv} + \tfrac{1}{2}b^2(t,T)B^2 u_{BB} = 0.
\end{aligned}
$$

*The replicating self-financing strategy for a defaultable claim satisfies*

$$\phi_t^1 = u_V, \quad \phi_t^2 = u_v, \quad \phi^3 = u_B.$$

Although the valuation PDE of Proposition 3.4.4 may look strange at first glance, it is nothing else than a particular case of the multi-dimensional version of the Black-Scholes-Merton PDE, or more precisely, of its well-known generalization to the Gaussian HJM framework. For instance, when $\rho = \rho_{Vr} = 0$ and $\hat{\sigma}_v = \kappa = \zeta = 0$, equation (3.69) reduces to:

$$
u_t + \tilde{\sigma}_v b(t,T)v B u_{vB} + \tfrac{1}{2}\big(\sigma_V^2 V^2 u_{VV} + \tilde{\sigma}_v^2 v^2 u_{vv} + b^2(t,T)B^2 u_{BB}\big) = 0.
$$

The last equation can be easily recognized as a special case of the PDE for the value function $u$ of a contingent claim $X = g(V_T, v_T)$ in the Gaussian HJM framework.

# 3.5 Further Developments

### 3.5.1 Convertible Bonds

Brennan and Schwartz (1977, 1980) develop models of pricing of convertible bonds. This model can aid the investor in choosing between convertibles and common stock and can help the firm in assessing the feasible trade-offs between various characteristics of an issue. This model allows for the uncertainty inherent in interest rates and accounts for the possibility of senior debt in the firm's capital structure. This is an advantage over the earlier model. However, it also makes the model more complex and substantially increases the computational burden associated with solving the differential equation. Because of this increased complexity, an analysis was performed to determine the error that would result by assuming a known constant interest rate. For a reasonable range of interest rates, the errors from certain interest rate models were found to be slight. Therefore, for practical purposes it may be preferable to use this simpler model for valuing convertible bonds.

Brennan and Schwartz (1978) address the effects of corporate income taxes on the relationship between capital structure and valuation. If the interest tax savings cease once the firm has gone bankrupt, then it is apparent that the issue of additional debt will have two effects on the value of the company. First, it will increase the tax savings to be enjoyed for as long as the firm survives. Second, it will reduce the probability of the firm's survival for any given period. Depending on which one of these conflicting variables prevails, the value of the company may increase or decrease as additional debt is issued. The option pricing framework is applied here to relate the value of a levered firm to the value of an unlevered firm, to the amount of debt, and the time to the debt's maturity.

### 3.5.2 Jump-Diffusion Models

The issue of 'predictability' of a default time in diffusion-type models was addressed by, among others, Schönbucher (1996), Zhou (1996), and Hilberink and Rogers (2000). All these authors introduce unpredictable jumps into the dynamics of the firm's value process. Consequently, the default time is no longer a predictable stopping time in approaches proposed by these authors.

### 3.5.3 Incomplete Accounting Data

We have already pointed out one major drawback of most structural models, namely, that the credit spreads with short maturities are close to zero. There is another important problem – of a more practical nature – with the structural approach. It arises due to the frequent unavailability of observations on the value process $V$. At the theoretical level, this issue was recently addressed by Crouhy et al. (1998) and Duffie and Lando (2001).

## 3.6 Dependent Defaults: Structural Approach

We identify the $n$ credit entities with a collection of $n$ firms; the $i^{\text{th}}$ credit entity is also referred to as the $i^{\text{th}}$ firm. The *credit migration process* $C^i$, to be formally introduced below, describes the evolution of the credit rating of the $i^{\text{th}}$ firm. Suppose that we are given $n \geq 2$ credit entities. Let $\mathcal{K}_i = \{k_1^i, k_2^i, \ldots, k_{m_i}^i\}$, $m_i \geq 2$, denote the set of possible credit ratings for the $i^{\text{th}}$ entity. Each state $k_{m_i}^i$ represents the default state for the $i^{\text{th}}$ firm; it is assumed to be an absorbing state for the corresponding credit migration process $C^i$. The case $m_i = 2$ corresponds to the situation when only two credit categories for the $i^{\text{th}}$ firm are considered – 'no-default' and 'default'.

We fix a filtered probability space $(\Omega, \mathbb{F}, \tilde{\mathbb{P}})$, where $\tilde{\mathbb{P}} = \mathbb{P}$ ($\tilde{\mathbb{P}} = \mathbb{P}^*$, resp.) plays the role of the real-world probability (the risk-neutral probability, resp.), depending on a particular application one has in mind. In accordance with the main paradigm of the structural approach, we assume that the evolution of each rating process $C^i$ is determined by the behavior of the value process $V^i$ of the $i^{\text{th}}$ firm. In particular, the firm specific default for the $i^{\text{th}}$ firm occurs whenever this firm's assets value process falls below the firm specific threshold process, denoted by $v^i$ in the sequel. Formally, the rating process $C^i$ jumps to the absorbing state $k_{m_i}^i$ at the first moment $V^i$ crosses the barrier $v^i$. This means that the $i^{\text{th}}$ default time $\tau_i$ is set to satisfy: $\tau_i = \inf\{t \in \mathbb{R}_+ : V_t^i < v_t^i\} = \inf\{t \in \mathbb{R}_+ : V_t^i \leq v_t^i\}$. Let $D^i(t)$ stand for the event: the $i^{\text{th}}$ firm has defaulted by the time $t$. It is clear that

$$D_i(t) = \{C_t^i = k_{m_i}^i\} = \{\tau_i \leq t\} = \{H_t^i = 1\},$$

where we set $H_t^i = \mathbb{1}_{\{\tau_i \leq t\}}$. By definition, the *default correlation* between the firms $i$ and $j$ over the time interval $[0, t]$ is defined as Pearson's correlation between the random variables $H_t^i$ and $H_t^j$. Denoting this correlation by $\rho_{ij}^D(t)$, it is elementary to show that

$$\rho_{ij}^D(t) = \frac{Q_{ij}(t) - Q_i(t)Q_j(t)}{\sqrt{Q_i(t)(1 - Q_i(t))}\sqrt{Q_j(t)(1 - Q_j(t))}}, \tag{3.70}$$

where we write

$$Q_i(t) = \tilde{\mathbb{P}}\{D_i(t)\} = \tilde{\mathbb{P}}\{\tau_i \leq t\} = \mathbb{E}_{\tilde{\mathbb{P}}}(H_t^i),$$

$$Q_j(t) = \tilde{\mathbb{P}}\{D_j(t)\} = \tilde{\mathbb{P}}\{\tau_j \leq t\} = \mathbb{E}_{\tilde{\mathbb{P}}}(H_t^j),$$

and

$$Q_{ij}(t) = \tilde{\mathbb{P}}\{D_i(t) \cap D_j(t)\} = \tilde{\mathbb{P}}\{\tau_i \leq t, \tau_j \leq t\} = \mathbb{E}_{\tilde{\mathbb{P}}}(H_t^i H_t^j).$$

It is useful to observe that $Q_{ij}(t) = Q_i(t) + Q_j(t) - Q^{ij}(t)$, where

$$Q^{ij}(t) = \tilde{\mathbb{P}}\{D_i(t) \cup D_j(t)\} = \tilde{\mathbb{P}}\{\tau_i \leq t \text{ or } \tau_j \leq t\} = \tilde{\mathbb{P}}\{\tau_{ij} \leq t\},$$

and where $\tau_{ij} = \tau_i \wedge \tau_j = \min(\tau_i, \tau_j)$.

Therefore, we see that if one can compute respective probabilities involving the random times $\tau_i$, $\tau_j$ and $\tau_{ij}$, then one can also find the default correlation $\rho_{ij}^D(t)$. In order to compute these probabilities, one needs to adopt specific model assumptions regarding the evolution of the value processes $V^i$, $i = 1, \ldots, n$ and the barrier processes $v^i$, $i = 1, \ldots, n$.

*Remarks.* One should distinguish between the default correlation coefficient $\rho_{ij}^D(t)$ defined above, and the (linear) correlation coefficient $\rho^{\tau_i, \tau_j}$ between default times, defined as follows:

$$\rho^{\tau_i, \tau_j} = \frac{\mathbb{E}_{\tilde{\mathbb{P}}}(\tau_i \tau_j) - \mathbb{E}_{\tilde{\mathbb{P}}}(\tau_i)\mathbb{E}_{\tilde{\mathbb{P}}}(\tau_j)}{\sqrt{\operatorname{Var}_{\tilde{\mathbb{P}}}(\tau_i)}\sqrt{\operatorname{Var}_{\tilde{\mathbb{P}}}(\tau_j)}},$$

where $\operatorname{Var}_{\tilde{\mathbb{P}}}(\tau_i)$ is the variance of $\tau_i$ under $\tilde{\mathbb{P}}$.

In order to proceed further, we need to formulate a model for the dynamics of the firms' values. We shall follow in the footsteps of, among others, J.P. Morgan's CreditMetrics, KMV's Portfolio Manager, and Zhou (2001), who adopt the Merton (1974) approach by modeling the firms' values as geometric Brownian motions. We prefer to formulate the model in terms of the logarithmic returns: $R_t^i := \ln(V_t^i/V_0^i)$ for $i = 1, \ldots, n$. The dynamics for the $n$-dimensional stochastic process $R = (R^1, \ldots, R^n)'$ (the "prime" denotes a matrix transpose) are:

$$dR_t = \mu\, dt + B\, dW_t, \tag{3.71}$$

where $\mu = (\mu_1, \ldots, \mu_n)'$ is a $n \times 1$ constant vector, $B$ is a $n \times m$ constant matrix, and $W = (W^1, \ldots, W^m)'$ is an $m$-dimensional standard Brownian motion. In view of (3.71), the random variable $R_t$ has a multivariate Gaussian distribution with the mean vector $\mu t$ and the covariance matrix $Ct$, where $C = [c_{ij}]_{n \times n} = BB'$. We postulate that $c_{ij} = \rho_{ij}\sigma_i\sigma_j$, with $-1 \leq \rho_{ij} \leq 1$ and $\rho_{ii} = 1$ for $i, j = 1, \ldots, n$. The coefficients $\sigma_i$ represent the volatilities of returns, whereas the coefficients $\rho_{ij}$ represent the linear local correlations between the returns. The coefficients $\rho_{ij}$ are sometimes (rather informally) referred to as the *asset correlations*. Notice that – according to (3.71) – the dynamics of each asset value process $V^i$ are governed by the SDE

$$dV_t^i = V_t^i\left(\left(\mu_i + \tfrac{1}{2}\sigma_i^2\right) dt + \sigma_i\, dW_t^i\right), \tag{3.72}$$

where for each $i$, the process $W^i$ is a real-valued standard Brownian motion. We make a natural assumption that $V_0^i > v_0^i$ for every $i$, i.e., at time 0 all firms are solvent. In what follows, we shall also make use of the auxiliary processes:

$$\hat{R}_t^i := \frac{R_t^i - \mu_i t}{\sigma_i \sqrt{t}}. \tag{3.73}$$

In what follows, the process $\hat{R}^i$ will be referred to as the *normalized return* for the $i^{\text{th}}$ firm.

We consider here a special case of the default correlation corresponding to the interval $[0, t]$. That's why in (3.70) we use unconditional expectations under the probability $\tilde{\mathbb{P}}$. Formally, these unconditional expectations are indeed conditional expectations with respect to the trivial $\sigma$-field $\mathcal{F}_0 := \{\Omega, \emptyset\}$. In general, one might consider the conditional correlations between default events of the firms $i$ and $j$ on the time interval $[s, t]$ for $0 \leq s < t$, given the $\sigma$-field $\mathcal{F}_s$. Typically, the interval $[s, t]$ spans over a one year period, the date $s$ is the present date. The Markovian nature of dynamics (3.71), combined with the assumptions imposed on the barrier processes for the two special cases considered below, show that there is no loss of generality in considering the time interval $[0, t]$, rather than $[s, t]$.

### 3.6.1 Default Correlations: J.P. Morgan's Approach

The probability $\tilde{\mathbb{P}}$ represents here the empirical (or, the real-world) probability. We assume that the barrier processes are deterministic positive constants: $v_t^i = \hat{v}^i > 0$ for every $t \in \mathbb{R}_+$. KMV's Portfolio Manager and J.P. Morgan's CreditMetrics define the *distance-to-default* $d_i(t)$ at time $t$ for the $i^{\text{th}}$ firm by setting (cf. Sect. 2.3.1)

$$d_i(t) := \frac{\ln(V_0^i/\hat{v}^i) + \mu_i t}{\sigma_i \sqrt{t}}.$$

Recall that in view of (3.71)–(3.72), $\mu_i$ represents the drift coefficient for the logarithmic asset returns, rather than the drift coefficient for the asset returns. If we denote by $\tilde{\mu}_i$ the drift coefficient for the asset returns then, by virtue of (3.72), equality $\tilde{\mu}_i = \mu_i + (1/2)\sigma_i^2$ holds, and thus

$$d_i(t) := \frac{\ln(V_0^i/\hat{v}^i) + (\tilde{\mu}_i - \frac{1}{2}\sigma_i^2)t}{\sigma_i \sqrt{t}}.$$

We already know that the random variable $R_t$ has a multivariate Gaussian distribution. Specifically, the probability distribution of the normalized return $\hat{R}_t^i$, which is given by (3.73), is the univariate standard Gaussian distribution, and the joint probability distribution of $\hat{R}_t^i$ and $\hat{R}_t^j$ is the standard bivariate Gaussian distribution with the correlation coefficient $\rho_{ij}$. More explicitly, the corresponding probability density function equals

$$f(x, y; 1, 1, \rho_{ij}) = \frac{1}{2\pi\sqrt{1 - \rho_{ij}^2}} \exp\left(-\frac{x^2 - 2\rho_{ij}xy + y^2}{2(1 - \rho_{ij}^2)}\right).$$

Let $N$ stand, as usual, for the cumulative probability distribution function of the standard univariate Gaussian probability law. Likewise, we denote by $N_2(x, y; \rho)$ the value at $(x, y) \in \mathbb{R}^2$ of the cumulative probability distribution function of the standard bivariate Gaussian distribution with the correlation coefficient $\rho$.

Given this notation, we see that $\tilde{\mathbb{P}}\{V_t^i \le \hat{v}^i\} = N(-d_i)$, and

$$\tilde{\mathbb{P}}\{V_t^i \le \hat{v}^i, V_t^j \le \hat{v}^j\} = N_2(-d_i, -d_j; \rho_{ij}),$$

where $d_i = d_i(t)$ and $d_j = d_j(t)$. Combining the above observations with (3.70), CreditMetrics computes the value of the default correlation $\rho_{ij}^D(t)$ as (see, e.g., Crouhy et al. (2000))

$$\rho_{ij}^D(t) = \frac{N_2(-d_i, -d_j; \rho_{ij}) - N(-d_i)N(-d_j)}{\sqrt{N(-d_i)(1 - N(-d_i))}\sqrt{N(-d_j)(1 - N(-d_j))}}.$$

This result does not seem to be correct, though; the reason being that, in general, the inclusions $\{V_t^i \le \hat{v}^i\} \subset D_i(t)$ and

$$\{V_t^i \le \hat{v}^i, V_t^j \le \hat{v}^j\} \subset D_i(t) \cap D_i(t)$$

are strict, so that the inclusions may not be replaced with equalities.

The correct application of formula (3.70) may be done in terms of probabilities of hitting the constant barriers $\hat{v}^i$ and $\hat{v}^j$ by one-dimensional diffusion processes; the details of these calculations are omitted. The interested reader is referred to the next section, where analogous calculations are provided.

For a more detailed analysis of the default correlation and interesting simulations of loan portfolios, we refer to Gersbach and Lipponer (1997a, 1997b) and Erlenmaier and Gersbach (2000). Crouhy et al. (2000) show how the asset correlations $\rho_{ij}$'s (called there *asset return correlations*) are calibrated using the factor approach that was developed by, among others, KMV.

## 3.6.2 Default Correlations: Zhou's Approach

We place ourselves in the set-up of Example 3.1.2 with $\kappa = 0$. We assume that the barrier $v^i$ is given as a deterministic function: $v^i(t) = K_i e^{\lambda_i t}$ for some positive constants $\lambda_i$, $i = 1, \dots, n$. By applying formula (3.9), we obtain (recall that processes $V^i$ are given by (3.72))

$$Q_i(t) = \tilde{\mathbb{P}}\{\tau_i \le t\} = 2N\left(-\frac{\ln(V_0^i/K_i)}{\sigma_i \sqrt{t}}\right).$$

Let us fix $i \ne j$. It is apparent from (3.70) that in order to compute the default correlation $\rho_{ij}^D(t)$, one needs to find the following probabilities:

$$Q_i(t) = \tilde{\mathbb{P}}\{\tau_i \le t\}, \quad Q_j(t) = \tilde{\mathbb{P}}\{\tau_j \le t\}, \quad Q_{ij}(t) = \tilde{\mathbb{P}}\{\tau_i \le t, \tau_j \le t\}.$$

Zhou (2001) undertakes the computation of the last probability in terms of $Q_i(t), Q_j(t)$ and the probability $Q^{ij}(t)$, which is defined by the formula:

$$Q^{ij}(t) = \tilde{\mathbb{P}}\{\tau_i \le t \text{ or } \tau_j \le t\} = \tilde{\mathbb{P}}\{\tau_{ij} \le t\}.$$

To proceed further with our calculations, we observe that, under the present assumptions, the default time $\tau_i$ is given by the equality:

$$\tau_i = \inf\{t \geq 0 : V_t^i \leq K_i e^{\lambda_i t}\}.$$

Thus, we also have

$$\tau_i = \inf\{t \geq 0 : X_t^i \geq b_i\},$$

where we set $b_i = -\ln(K_i/V_0^i)$ and the process $X_t^i$ satisfies

$$X_t^i = -\ln\left(e^{-\lambda_i t} V_t^i / V_0^i\right) = -R_t^i + \lambda_i t \tag{3.74}$$

where $R_t^i := \ln(V_t^i/V_0^i)$. Consequently, the random time $\tau_{ij} = \tau_i \wedge \tau_j$ is also equal to the first time $t$ such that either $X_t^i \geq b_i$ or $X_t^j \geq b_j$, so that

$$Q^{ij}(t) = \tilde{\mathbb{P}}\{\tau_i \leq t \text{ or } \tau_j \leq t\} = \tilde{\mathbb{P}}\{M_t^i \geq b_i \text{ or } M_t^j \geq b_j\}, \tag{3.75}$$

where we denote

$$M_t^i = \max_{0 \leq s \leq t} X_s^i, \quad M_t^j = \max_{0 \leq s \leq t} X_s^j.$$

An application of Itô's formula yields the following dynamics for the two-dimensional process $X^{i,j} = (X^i, X^j)'$:

$$dX_t^{i,j} = -B_{i,j}\begin{pmatrix} dW_t^i \\ dW_t^j \end{pmatrix}$$

with $X_0^{i,j} = (0,0)'$, where

$$B_{i,j}B_{i,j}' = \begin{pmatrix} \sigma_i^2 & \rho_{ij}\sigma_i\sigma_j \\ \rho_{ij}\sigma_i\sigma_j & \sigma_j^2 \end{pmatrix}.$$

Since we postulate that $V_0^i > K_i$ for all $i = 1, 2, \ldots, n$, it is clear that $b_i > 0$ for every $i = 1, \ldots, n$. Thus, the random time $\tau_{ij}$ is the first passage time of the planar diffusion process $X^{i,j}$ to a fixed boundary consisting of two intersecting lines: $x_1 = b_1$ and $x_2 = b_2$. This observation was made by Zhou (2001) in order to compute the probability $Q^{ij}(t)$ through the PDE method. Zhou (2001) provides the closed-form expressions for the probability $Q^{ij}(t)$ in terms of modified Bessel functions.

**Special case.** In order to simplify the calculations, Zhou (2001) postulates first that $\lambda_i = \mu_i$ for every $i = 1, \ldots, n$. Let us denote

$$d_i = \frac{\ln(V_0^i/K_i)}{\sigma_i}$$

so that

$$Q_i(t) = \tilde{\mathbb{P}}\{\tau_i \leq t\} = 2N\left(-\frac{\ln(V_0^i/K_i)}{\sigma_i \sqrt{t}}\right) = 2N\left(\frac{-d_i}{\sqrt{t}}\right).$$

In this case, the probability $Q^{ij}(t)$ we are looking for is given by the following result.

**Proposition 3.6.1.** *Assume that $\lambda_i = \mu_i$ and $\lambda_j = \mu_j$. Then*

$$Q^{ij}(t) = 1 - \frac{2r_0}{\sqrt{2\pi t}} e^{-\frac{r_0^2}{4t}} \sum_{n=1,3,\ldots} \frac{\sin(\beta_n \theta_0)}{n} \left[ I_{\frac{1}{2}(\beta_n+1)}\left(\frac{r_0^2}{4t}\right) + I_{\frac{1}{2}(\beta_n-1)}\left(\frac{r_0^2}{4t}\right)\right]$$

*where $r_0 = d_j/\sin\theta_0$, $\beta_n = n\pi/\alpha$ and $I_\nu(z)$ is the modified Bessel of the first kind of order $\nu$. Moreover*

$$\alpha = \begin{cases} \tan^{-1}\left(-\frac{\sqrt{1-\rho_{ij}^2}}{\rho_{ij}}\right), & \text{if } \rho_{ij} < 0, \\ \pi + \tan^{-1}\left(-\frac{\sqrt{1-\rho_{ij}^2}}{\rho_{ij}}\right), & \text{otherwise,} \end{cases}$$

*and*

$$\theta_0 = \begin{cases} \tan^{-1}\left(\frac{d_j\sqrt{1-\rho_{ij}^2}}{d_i-\rho_{ij}d_j}\right), & \text{if } \rho_{ij} < 0, \\ \pi + \tan^{-1}\left(\frac{d_j\sqrt{1-\rho_{ij}^2}}{d_i-\rho_{ij}d_j}\right), & \text{otherwise.} \end{cases}$$

*Proof.* Alternative derivations of the formula are given in Rebholz (1994) and Zhou (2001). □

**General case.** We no longer assume that $\lambda_i = \mu_i$ and $\lambda_j = \mu_j$. Then we have

$$Q_i(t) = N\left(-\frac{d_i}{\sqrt{t}} - \frac{\mu_i - \lambda_i}{\sigma_i}\sqrt{t}\right) + e^{\frac{2(\lambda_i-\mu_i)d_i}{\sigma_i}} N\left(-\frac{d_i}{\sqrt{t}} + \frac{\mu_i - \lambda_i}{\sigma_i}\sqrt{t}\right).$$

For the proof of the following result, the interested reader is referred to Rebholz (1994).

**Proposition 3.6.2.** *Let $\lambda_i, \mu_i, \lambda_j, \mu_j$ be any given constants. Then*

$$Q^{ij}(t) = 1 - \frac{2}{\alpha t} e^{-\frac{r_0^2}{2t}} e^{a_1 x_1 + a_2 x_2 + a_3 t} \sum_{n=1}^{\infty} \sin(\beta_n \theta_0) \int_0^\alpha \sin(\beta_n \theta) g_n(\theta)\, d\theta$$

*where $r_0, \beta_n, \alpha$ and $\theta_0$ are as in Proposition 3.6.1 and*

$$g_n(\theta) = \int_0^\infty r e^{-r^2/2t} e^{d_i r \sin(\theta-\alpha) - d_j r \cos(\theta-\alpha)} I_{\beta_n}\left(\frac{rr_0}{t}\right) dr,$$

$$a_1 = \frac{(\lambda_i - \mu_i)\sigma_j - (\lambda_j - \mu_j)\rho_{ij}\sigma_i}{(1-\rho_{ij}^2)\sigma_i^2\sigma_j},$$

$$a_2 = \frac{(\lambda_j - \mu_j)\sigma_i - (\lambda_i - \mu_i)\rho_{ij}\sigma_j}{(1-\rho_{ij}^2)\sigma_j^2\sigma_i},$$

$$a_3 = \frac{a_1^2\sigma_i^2}{2} + \rho_{ij}a_1 a_2\sigma_i\sigma_j + \frac{a_2^2\sigma_j^2}{2} - a_1(\lambda_i - \mu_i) - a_2(\lambda_j - \mu_j).$$

*Finally, $x_1 = a_1\sigma_i + \rho_{ij}a_2\sigma_j$ and $x_2 = a_2\sigma_j\sqrt{1-\rho_{ij}^2}$.*

*Remarks.* It is perhaps worth noting that the above propositions originate from the following result, which provides a joint density for two correlated drifted Brownian motions and their running minima. Suppose that $W^1$ and $W^2$ are Brownian motions under $\tilde{\mathbb{P}}$, and that the correlation coefficient between $W^1$ and $W^2$ is $\rho$. Let

$$X_t^k = \alpha_k t + \sigma_k W_t^k, \quad k = 1, 2,$$

and let us denote $m_t^k = \min_{0 \le s \le t} X_s^k$. Then for any fixed $t > 0$ the joint law of $(X_t^1, X_t^2, m_t^1, m_t^2)$ is given by

$$\tilde{\mathbb{P}}\big(X_t^k \in dx_k, m_t^k \in dm_k, k = 1, 2\big) = f(x_1, x_2, m_1, m_2; t)\, dx_1 dx_2 dm_1 dm_2,$$

where

$$f(x_1, x_2, m_1, m_2; t) = \frac{2}{\beta t} e^{-\frac{r^2 + r_0^2}{2t}} \frac{e^{A_1 x_1 + A_2 x_2 + A_3 t}}{\sigma_1 \sigma_2 \sqrt{1 - \rho^2}} \times$$

$$\times \sum_{n=1}^{\infty} \sin\left(\frac{n\pi\Theta_0}{\beta}\right) \sin\left(\frac{n\pi\Theta}{\beta}\right) I_{\frac{n\pi}{\beta}}\left(\frac{r r_0}{t}\right)$$

and the constants are specified as follows:

$$A_1 = \frac{\alpha_1 \sigma_2 - \rho \alpha_2 \sigma_1}{(1 - \rho^2)\sigma_1^2 \sigma_2}, \quad A_2 = \frac{\alpha_2 \sigma_1 - \rho \alpha_1 \sigma_2}{(1 - \rho^2)\sigma_1 \sigma_2^2},$$

$$A_3 = -\alpha_1 a_1 - \alpha_2 a_2 + \frac{1}{2}\left(\sigma_1^2 a_1^2 + \sigma_2^2 a_2^2\right) + \rho \sigma_1 \sigma_2 a_1 a_2,$$

$$\beta = \begin{cases} \tan^{-1}\left(-\frac{\sqrt{1-\rho^2}}{\rho}\right), & \text{if } \rho < 0, \\ \pi - \tan^{-1}\left(\frac{\sqrt{1-\rho^2}}{\rho}\right), & \text{if } \rho > 0, \end{cases}$$

$$z_1 = \frac{1}{\sqrt{1-\rho^2}}\left[\left(\frac{x_1 - m_1}{\sigma_1}\right) - \rho\left(\frac{x_2 - m_2}{\sigma_2}\right)\right], \quad z_2 = \frac{x_2 - m_2}{\sigma_2},$$

$$z_{10} = \frac{1}{\sqrt{1-\rho^2}}\left[-\frac{m_1}{\sigma_1} + \rho\frac{m_2}{\sigma_2}\right], \quad z_{20} = -\frac{m_2}{\sigma_2},$$

$$r = \sqrt{z_1^2 + z_2^2}, \quad \tan\Theta = \frac{z_2}{z_1}, \quad \Theta \in [0, \beta],$$

$$r_0 = \sqrt{z_{10}^2 + z_{20}^2}, \quad \tan\Theta_0 = \frac{z_{20}}{z_{10}}, \quad \Theta_0 \in [0, \beta].$$

For details please see, for instance, Iyengar (1985) or He et al. (1998).

Part II

Hazard Processes

# 4. Hazard Function of a Random Time

In this chapter, the problem of a quasi-explicit evaluation of various conditional expectations is studied in the case when the only filtration available in calculations is the natural filtration of a random time. At the intuitive level, we consider here an individual who is able to observe a certain random time $\tau$, but has no access to any other information. A detailed analysis of a more interesting and practically more relevant case – when an additional flow of information is also available – is postponed to the next chapter.

## 4.1 Conditional Expectations w.r.t. Natural Filtrations

Let $\tau : \Omega \to \mathbb{R}_+$ be a non-negative random variable, henceforth referred to as the *random time,* which is defined on a probability space $(\Omega, \mathcal{G}, \mathbb{P})$. For convenience, we assume that $\mathbb{P}\{\tau = 0\} = 0$ and $\mathbb{P}\{\tau > t\} > 0$ for any $t \in \mathbb{R}_+$. The last condition means that $\tau$ is assumed to be unbounded; more precisely, it is not dominated with probability 1 by a constant. A bounded random time can also be studied using techniques presented in what follows, though. Let $F$ stand for the (right-continuous) cumulative distribution function of $\tau$, i.e., $F(t) = \mathbb{P}\{\tau \leq t\}$ for every $t \in \mathbb{R}_+$. The *survival function $G$* of $\tau$ is defined by the formula: $G(t) := 1 - F(t) = \mathbb{P}\{\tau > t\}$ for every $t \in \mathbb{R}_+$.

*Example 4.1.1.* If $\tau$ is exponentially distributed under $\mathbb{P}$ with parameter $\lambda$, then $F(t) = 1 - e^{-\lambda t}$ and thus the survival function equals $G(t) = e^{-\lambda t}$.

We define the *jump process $H$* associated with the random time $\tau$ by setting $H_t = \mathbb{1}_{\{\tau \leq t\}}$ for $t \in \mathbb{R}_+$. It is obvious that the process $H$ has right-continuous sample paths, specifically, each sample paths is equal to 0 before random time $\tau$, and it equals 1 for $t \geq \tau$.

Let $\mathbb{H} = (\mathcal{H}_t)_{t \geq 0}$ stand for the filtration generated by $H$, specifically, for any $t \in \mathbb{R}_+$ we set $\mathcal{H}_t = \sigma(H_u : u \leq t)$. The filtration $\mathbb{H}$ is assumed to be $(\mathbb{P}, \mathcal{G})$-completed. Finally, we set $\mathcal{H}_\infty = \sigma(H_u : u \in \mathbb{R}_+)$. The $\sigma$-field $\mathcal{H}_t$ represents the information generated by the observations of the occurrence of the random time $\tau$ up to time $t$ – that is, on the time interval $[0, t]$.

We use the commonly standard notation $\sigma(\eta)$ for the $\sigma$-field generated by a random variable $\eta$. We also assume that $Y$ is an integrable random variable on the probability space $(\Omega, \mathcal{G}, \mathbb{P})$ – that is, $\mathbb{E}_{\mathbb{P}}|Y| < \infty$.

Let us first enumerate a few basic properties of the filtration $\mathbb{H}$:

(H.1) $\mathcal{H}_t = \sigma(\{\tau \le u\} : u \le t)$,

(H.2) $\mathcal{H}_t = \sigma(\sigma(\tau) \cap \{\tau \le t\})$,

(H.3) $\mathcal{H}_t = \sigma(\tau \wedge t) \vee (\{\tau > t\})$,

(H.4) $\mathcal{H}_t = \mathcal{H}_{t+}$,

(H.5) $\mathcal{H}_\infty = \sigma(\tau)$,

(H.6) for any $A \in \mathcal{H}_\infty$ we have: $A \cap \{\tau \le t\} \in \mathcal{H}_t$.

All properties above are easy to check; let us only mention that in order to establish (H.6), it is enough to consider an arbitrary event $A$ of the following form: $A = \{\tau \le s\}$ for some $s \in \mathbb{R}_+$.

**Lemma 4.1.1.** *Let $Y$ be a $\mathcal{G}$-measurable random variable. Then*

$$\mathbb{1}_{\{\tau \le t\}} \mathbb{E}_{\mathbb{P}}(Y \mid \mathcal{H}_t) = \mathbb{E}_{\mathbb{P}}(\mathbb{1}_{\{\tau \le t\}} Y \mid \mathcal{H}_\infty) = \mathbb{1}_{\{\tau \le t\}} \mathbb{E}_{\mathbb{P}}(Y \mid \tau) \qquad (4.1)$$

*and*

$$\mathbb{1}_{\{\tau > t\}} \mathbb{E}_{\mathbb{P}}(Y \mid \mathcal{H}_t) = \mathbb{1}_{\{\tau > t\}} \frac{\mathbb{E}_{\mathbb{P}}(\mathbb{1}_{\{\tau > t\}} Y)}{\mathbb{P}\{\tau > t\}}.$$

*Proof.* We shall first check that

$$\mathbb{E}_{\mathbb{P}}(\mathbb{1}_{\{\tau \le t\}} Y \mid \mathcal{H}_\infty) = \mathbb{E}_{\mathbb{P}}(\mathbb{1}_{\{\tau \le t\}} Y \mid \mathcal{H}_t).$$

In view of (H.6), we have $A \cap \{\tau \le t\} \in \mathcal{H}_t$ for any $A \in \mathcal{H}_\infty$. Consequently,

$$\int_A \mathbb{E}_{\mathbb{P}}(\mathbb{1}_{\{\tau \le t\}} Y \mid \mathcal{H}_\infty) \, d\mathbb{P} = \int_A \mathbb{1}_{\{\tau \le t\}} Y \, d\mathbb{P} = \int_{A \cap \{\tau \le t\}} Y \, d\mathbb{P}$$

$$= \int_{A \cap \{\tau \le t\}} \mathbb{E}_{\mathbb{P}}(Y \mid \mathcal{H}_t) \, d\mathbb{P} = \int_A \mathbb{1}_{\{\tau \le t\}} \mathbb{E}_{\mathbb{P}}(Y \mid \mathcal{H}_t) \, d\mathbb{P}$$

$$= \int_A \mathbb{E}_{\mathbb{P}}(\mathbb{1}_{\{\tau \le t\}} Y \mid \mathcal{H}_t) \, d\mathbb{P}$$

since the event $\{\tau \le t\}$ belongs to $\mathcal{H}_t$. To establish the second formula, we need to show that

$$\mathbb{E}_{\mathbb{P}}(\mathbb{1}_{\{\tau > t\}} Y \mid \mathcal{H}_t) = c \mathbb{1}_{\{\tau > t\}}, \quad \text{where} \quad c = \frac{\mathbb{E}_{\mathbb{P}}(\mathbb{1}_{\{\tau > t\}} Y)}{\mathbb{P}\{\tau > t\}}.$$

Equivalently, we need to check that for any $A \in \mathcal{H}_t$

$$\int_A \mathbb{E}_{\mathbb{P}}(\mathbb{1}_{\{\tau > t\}} Y \mid \mathcal{H}_t) \, d\mathbb{P} = \int_A c \mathbb{1}_{\{\tau > t\}} \, d\mathbb{P}.$$

In this case, it is enough to consider events of the form: $A = \{\tau \le s\}$ for $s \le t$, as well as the event $A = \{\tau > t\}$. In the former case, both sides of the last equality are equal to 0. Furthermore, since $A = \{\tau > t\} \in \mathcal{H}_t$ we obtain

$$\int_A \mathbb{E}_{\mathbb{P}}(\mathbb{1}_A Y \mid \mathcal{H}_t) \, d\mathbb{P} = \int_A \mathbb{1}_A Y \, d\mathbb{P} = \int_\Omega \mathbb{1}_A Y \, d\mathbb{P} = c \mathbb{P}\{A\} = \int_A c \mathbb{1}_A \, d\mathbb{P}.$$

This completes the proof of the lemma. $\qquad \square$

**Corollary 4.1.1.** *For any $\mathcal{G}$-measurable random variable $Y$ we have*

$$\mathbb{E}_{\mathbb{P}}(Y \,|\, \mathcal{H}_t) = \mathbb{1}_{\{\tau \le t\}} \mathbb{E}_{\mathbb{P}}(Y \,|\, \tau) + \mathbb{1}_{\{\tau > t\}} \frac{\mathbb{E}_{\mathbb{P}}(\mathbb{1}_{\{\tau > t\}} Y)}{\mathbb{P}\{\tau > t\}}. \qquad (4.2)$$

*For any $\mathcal{H}_t$-measurable random variable $Y$ we have*

$$Y = \mathbb{1}_{\{\tau \le t\}} \mathbb{E}_{\mathbb{P}}(Y \,|\, \tau) + \mathbb{1}_{\{\tau > t\}} \frac{\mathbb{E}_{\mathbb{P}}(\mathbb{1}_{\{\tau > t\}} Y)}{\mathbb{P}\{\tau > t\}}, \qquad (4.3)$$

*that is, $Y = h(\tau)$ for a Borel measurable $h : \mathbb{R} \to \mathbb{R}$, which is constant on the open interval $]t, \infty[$.*

The basic formula (4.2), though simple, appears to be quite useful. Let us state some special cases of this result. For any $t < s$ we have

$$\mathbb{P}\{\tau \ge s \,|\, \mathcal{H}_t\} = \mathbb{1}_{\{\tau > t\}} \mathbb{P}\{\tau \ge s \,|\, \tau > t\}$$

and

$$\mathbb{P}\{\tau > s \,|\, \mathcal{H}_t\} = \mathbb{1}_{\{\tau > t\}} \mathbb{P}\{\tau > s \,|\, \tau > t\}. \qquad (4.4)$$

The following result is a straightforward consequence of (4.4).

**Corollary 4.1.2.** *The process $M$ given by the formula*

$$M_t = \frac{1 - H_t}{1 - F(t)}, \qquad \forall t \in \mathbb{R}_+, \qquad (4.5)$$

*follows an $\mathbb{H}$-martingale. Equivalently, for every $0 \le t \le s$,*

$$\mathbb{E}_{\mathbb{P}}(H_s - H_t \,|\, \mathcal{H}_t) = \mathbb{1}_{\{\tau > t\}} \frac{F(s) - F(t)}{1 - F(t)}. \qquad (4.6)$$

*Proof.* Equality (4.4) can be rewritten as follows:

$$\mathbb{E}_{\mathbb{P}}(1 - H_s \,|\, \mathcal{H}_t) = (1 - H_t) \frac{1 - F(s)}{1 - F(t)}.$$

This immediately yields the martingale property of $M$. The second formula is also clear. $\qquad \square$

**Definition 4.1.1.** An increasing function $\varGamma : \mathbb{R}_+ \to \mathbb{R}_+$ given by the formula

$$\varGamma(t) := -\ln G(t) = -\ln(1 - F(t)), \qquad \forall t \in \mathbb{R}_+,$$

is called the *hazard function* of $\tau$. If the cumulative distribution function $F$ is absolutely continuous with respect to the Lebesgue measure – that is, when $F(t) = \int_0^t f(u) \, du$, for a Lebesgue integrable function $f : \mathbb{R}_+ \to \mathbb{R}_+$, then we have

$$F(t) = 1 - e^{-\varGamma(t)} = 1 - e^{-\int_0^t \gamma(u) \, du},$$

where $\gamma(t) = f(t)(1 - F(t))^{-1}$. The function $\gamma$ is called the *intensity function* (or the *hazard rate*) of the random time $\tau$.

Notice that $\Gamma(t)$ is well defined for any $t \in \mathbb{R}_+$, since by assumption $F(t) < 1$ for every $t \in \mathbb{R}_+$. Furthermore, we have

$$\Gamma(\infty) := \lim_{t \to \infty} \Gamma(t) = \infty,$$

since clearly $\lim_{t \to \infty}(1 - F(t)) = 0$. It is also obvious that the intensity function $\gamma : \mathbb{R}_+ \to \mathbb{R}$ (if it exists) is a non-negative function. Finally, $\gamma$ is Lebesgue integrable on any bounded interval $[0, t]$ and $\int_0^\infty \gamma(u)\, du = \infty$.

*Example 4.1.2.* If $\tau$ is exponentially distributed with parameter $\lambda$ under $\mathbb{P}$ the hazard rate of $\tau$ is constant: $\gamma(t) = \lambda$ for every $t \in \mathbb{R}_+$.

Using the hazard function $\Gamma$, we may rewrite (4.2) as follows:

$$\mathbb{E}_{\mathbb{P}}(Y \mid \mathcal{H}_t) = \mathbb{1}_{\{\tau \le t\}} \mathbb{E}_{\mathbb{P}}(Y \mid \tau) + \mathbb{1}_{\{\tau > t\}}\, e^{\Gamma(t)}\, \mathbb{E}_{\mathbb{P}}(\mathbb{1}_{\{\tau > t\}} Y). \tag{4.7}$$

In particular, for any $t \le s$ equality (4.4) takes the following form:

$$\mathbb{P}\{\tau > s \mid \mathcal{H}_t\} = \mathbb{1}_{\{\tau > t\}}\, e^{\Gamma(t) - \Gamma(s)} = \mathbb{1}_{\{\tau > t\}}\, e^{-\int_t^s \gamma(u)\, du},$$

where the second equality holds, provided that $\tau$ admits the hazard rate $\gamma$.

**Corollary 4.1.3.** *Let $Y$ be $\mathcal{H}_\infty$-measurable so that $Y = h(\tau)$ for some Borel measurable function $h : \mathbb{R}_+ \to \mathbb{R}$. Then the following statements are true.*
(i) *If the hazard function $\Gamma$ of $\tau$ is continuous, then we have*

$$\mathbb{E}_{\mathbb{P}}(Y \mid \mathcal{H}_t) = \mathbb{1}_{\{\tau \le t\}} h(\tau) + \mathbb{1}_{\{\tau > t\}} \int_t^\infty h(u) e^{\Gamma(t) - \Gamma(u)}\, d\Gamma(u). \tag{4.8}$$

(ii) *If $\tau$ admits the intensity function $\gamma$, then we have*

$$\mathbb{E}_{\mathbb{P}}(Y \mid \mathcal{H}_t) = \mathbb{1}_{\{\tau \le t\}} h(\tau) + \mathbb{1}_{\{\tau > t\}} \int_t^\infty h(u) \gamma(u) e^{-\int_t^u \gamma(v)\, dv}\, du.$$

*In particular, for any $t \le s$,*

$$\mathbb{P}\{\tau > s \mid \mathcal{H}_t\} = \mathbb{1}_{\{\tau > t\}} e^{-\int_t^s \gamma(v)\, dv}$$

*and*

$$\mathbb{P}\{t < \tau < s \mid \mathcal{H}_t\} = \mathbb{1}_{\{\tau > t\}} \left(1 - e^{-\int_t^s \gamma(v)\, dv}\right).$$

**Lemma 4.1.2.** *The process $L$, given by the formula*

$$L_t := \mathbb{1}_{\{\tau > t\}} e^{\Gamma(t)} = (1 - H_t) e^{\Gamma(t)}, \quad \forall t \in \mathbb{R}_+, \tag{4.9}$$

*follows an $\mathbb{H}$-martingale.*

*Proof.* It suffices to observe that the process $L$ coincides with the process $M$ introduced in Corollary 4.1.2. $\qquad \square$

## 4.2 Martingales Associated with a Continuous Hazard Function

We already know that the $\mathbb{H}$-adapted process of finite variation $L$ given by formula (4.9) is an $\mathbb{H}$-martingale (no matter whether $\Gamma$ is a continuous or a discontinuous function). In this section, we will examine further important examples of martingales associated with the hazard function. We make throughout an additional assumption that the hazard function $\Gamma$ of a random time $\tau$ is continuous.

We shall first assume that the cumulative distribution function $F$ is an absolutely continuous function, so that the random time $\tau$ admits the intensity function $\gamma$. Our goal is to establish a martingale characterization of $\gamma$. More specifically, we shall check directly that the process $\hat{M}$, defined as:

$$\hat{M}_t := H_t - \int_0^t \gamma(u)\mathbb{1}_{\{u \le \tau\}}\,du = H_t - \int_0^{t \wedge \tau} \gamma(u)\,du = H_t - \Gamma(t \wedge \tau),$$

follows an $\mathbb{H}$-martingale. To this end, recall that by virtue of (4.6) we have

$$\mathbb{E}_{\mathbb{P}}(H_s - H_t \mid \mathcal{H}_t) = \mathbb{1}_{\{\tau > t\}}\frac{F(s) - F(t)}{1 - F(t)}.$$

On the other hand, if we denote

$$Y = \int_t^s \gamma(u)\mathbb{1}_{\{u \le \tau\}}\,du = \int_{t \wedge \tau}^{s \wedge \tau} \frac{f(u)}{1 - F(u)}\,du = \ln\frac{1 - F(t \wedge \tau)}{1 - F(s \wedge \tau)},$$

then obviously $Y = \mathbb{1}_{\{\tau > t\}}Y$. Let us set $A = \{\tau > t\}$. Using first (4.2) and then Fubini's theorem, we obtain

$$\mathbb{E}_{\mathbb{P}}(Y \mid \mathcal{H}_t) = \mathbb{E}_{\mathbb{P}}(\mathbb{1}_A Y \mid \mathcal{H}_t) = \mathbb{1}_A \frac{\mathbb{E}_{\mathbb{P}}(Y)}{\mathbb{P}\{A\}} = \mathbb{1}_A \frac{\mathbb{E}_{\mathbb{P}}\left(\int_t^s \gamma(u)\mathbb{1}_{\{u \le \tau\}}\,du\right)}{1 - F(t)}$$

$$= \mathbb{1}_A \frac{\int_t^s \gamma(u)(1 - F(u))\,du}{1 - F(t)} = \mathbb{1}_A \frac{F(s) - F(t)}{1 - F(t)} = \mathbb{E}_{\mathbb{P}}(H_s - H_t \mid \mathcal{H}_t).$$

This shows that the process $\hat{M}$ follows an $\mathbb{H}$-martingale. We have thus established the following simple, but remarkable, result.

**Lemma 4.2.1.** *Assume that*

$$F(t) = 1 - e^{-\int_0^t \gamma(u)\,du}, \quad \forall t \in \mathbb{R}_+,$$

*where $\gamma : \mathbb{R}_+ \to \mathbb{R}_+$ is the hazard rate of $\tau$. Then the process $\hat{M}$*

$$\hat{M}_t = H_t - \int_0^{t \wedge \tau} \gamma(u)\,du, \quad \forall t \in \mathbb{R}_+, \tag{4.10}$$

*follows an $\mathbb{H}$-martingale.*

It appears that a counterpart of Lemma 4.2.1 can be established when $F$ is merely continuous. Before examining this extension, we recall an auxiliary result. For the proof of Lemma 4.2.2, the interested reader is referred, for instance, to Brémaud (1981) or Revuz and Yor (1999).

**Lemma 4.2.2.** *Let $g$ and $h$ be two right-continuous functions with left-hand limits. If $g$ and $h$ are of finite variation on $[0, t]$ then we have*

$$g(t)h(t) = g(0)h(0) + \int_{]0,t]} g(u-) \, dh(u) + \int_{]0,t]} h(u) \, dg(u)$$

$$= g(0)h(0) + \int_{]0,t]} g(u) \, dh(u) + \int_{]0,t]} h(u-) \, dg(u)$$

$$= g(0)h(0) + \int_{]0,t]} g(u-) \, dh(u) + \int_{]0,t]} h(u-) \, dg(u)$$

$$+ \sum_{u \le t} \Delta g(u) \Delta h(u),$$

*where $\Delta g(u) = g(u) - g(u-)$ and $\Delta h(u) = h(u) - h(u-)$.*

Any of the equalities of Lemma 4.2.2 will be referred to as the *integration by parts formula* (or the *product rule*) for functions of finite variation. We shall frequently apply this formula to stochastic processes of finite variation. In such a case, the integrals should be understood as the path-wise integrals, defined with probability 1.

**Proposition 4.2.1.** *Assume that the hazard function $\Gamma$ is continuous. Then the process of finite variation $\hat{M}_t = H_t - \Gamma(t \wedge \tau)$ follows an $\mathbb{H}$-martingale. Furthermore, for every $t \in \mathbb{R}_+$ we have*

$$L_t = 1 - \int_{]0,t]} L_{u-} \, d\hat{M}_u. \tag{4.11}$$

*Proof.* For the sake of brevity, we shall make use of Lemma 4.1.2 (the direct calculations also give, of course, the required result). It is clear that $\hat{M}$ follows an $\mathbb{H}$-adapted integrable process. Using the integration by parts formula for functions of finite variation,[1] we obtain

$$L_t = (1 - H_t)e^{\Gamma(t)} = 1 + \int_{]0,t]} e^{\Gamma(u)} \big( (1 - H_u) \, d\Gamma(u) - dH_u \big) \tag{4.12}$$

since $\Gamma$ is a continuous increasing function. This in turn yields

$$\hat{M}_t = H_t - \Gamma(t \wedge \tau) = \int_{]0,t]} \big( dH_u - (1 - H_u) \, d\Gamma(u) \big) = -\int_{]0,t]} e^{-\Gamma(u)} \, dL_u,$$

---

[1] It can also be seen as a version of Itô's product rule for (discontinuous) semi-martingales (see, e.g., Elliott (1982) or Protter (1990)). Let us mention that since $\hat{M}$ is of finite variation, it is a *purely discontinuous martingale*.

so that $\hat{M}$ is manifestly an $\mathbb{H}$-martingale. Since (4.12) may be rewritten as follows:

$$L_t = 1 + \int_{]0,t]} e^{\Gamma(u)}(1 - H_{u-})(d\Gamma(u \wedge \tau) - dH_u) = 1 - \int_{]0,t]} L_{u-} \, d\hat{M}_u,$$

it is clear that (4.11) is valid. □

**Proposition 4.2.2.** *Assume that the hazard function $\Gamma$ of $\tau$ is continuous. Then for any Borel measurable function $h : \mathbb{R}_+ \to \mathbb{R}$ such that the random variable $h(\tau)$ is integrable, the process $\hat{M}^h$, given by the formula*

$$\hat{M}_t^h = \mathbb{1}_{\{\tau \leq t\}} h(\tau) - \int_0^{t \wedge \tau} h(u) \, d\Gamma(u), \quad \forall t \in \mathbb{R}_+,$$

*is an $\mathbb{H}$-martingale.*

*Proof.* We shall directly verify the martingale property of $\hat{M}^h$. Therefore, the demonstration given below provides also an alternative proof of Proposition 4.2.1. On one hand, formula (4.8) in Corollary 4.1.3 yields

$$I := \mathbb{E}_{\mathbb{P}}\left(h(\tau) \mathbb{1}_{\{t < \tau \leq s\}} \,|\, \mathcal{H}_t\right) = \mathbb{1}_{\{\tau > t\}} e^{\Gamma(t)} \int_t^s h(u) e^{-\Gamma(u)} \, d\Gamma(u).$$

On the other hand, it is clear that

$$J := \mathbb{E}_{\mathbb{P}}\left(\int_{t \wedge \tau}^{s \wedge \tau} h(u) \, d\Gamma(u) \,\Big|\, \mathcal{H}_t\right) = \mathbb{E}_{\mathbb{P}}\left(\tilde{h}(\tau) \mathbb{1}_{\{t < \tau \leq s\}} + \tilde{h}(s) \mathbb{1}_{\{\tau > s\}} \,|\, \mathcal{H}_t\right),$$

where we set $\tilde{h}(s) = \int_t^s h(u) \, d\Gamma(u)$. Consequently, using again formula (4.8), we get

$$J = \mathbb{1}_{\{\tau > t\}} e^{\Gamma(t)} \left(\int_t^s \tilde{h}(u) e^{-\Gamma(u)} \, d\Gamma(u) + e^{-\Gamma(s)} \tilde{h}(s)\right).$$

To conclude the proof, it is enough to observe that the Fubini theorem yields

$$\int_t^s e^{-\Gamma(u)} \int_t^u h(v) \, d\Gamma(v) \, d\Gamma(u) + e^{-\Gamma(s)} \tilde{h}(s)$$

$$= \int_t^s h(u) \int_u^s e^{-\Gamma(v)} \, d\Gamma(v) \, d\Gamma(u) + e^{-\Gamma(s)} \int_t^s h(u) \, d\Gamma(u)$$

$$= \int_t^s h(u) e^{-\Gamma(u)} \, d\Gamma(u),$$

as expected. □

*Remarks.* It is apparent that $\hat{M}^h$ admits the following integral representation

$$\hat{M}_t^h = \int_{]0,t]} h(u) \, d\hat{M}_u.$$

This equality shows that the martingale property of $\hat{M}^h$ is also a straightforward consequence of Proposition 4.2.1.

**Corollary 4.2.1.** *Assume that the hazard function $\Gamma$ of $\tau$ is continuous. Let $h : \mathbb{R}_+ \to \mathbb{R}$ be a Borel measurable function such that the random variable $Y = e^{h(\tau)}$ is integrable. Then the process*

$$\tilde{M}_t^h = \exp\left(\mathbb{1}_{\{\tau \le t\}} h(\tau)\right) - \int_0^{t \wedge \tau} (e^{h(u)} - 1) \, d\Gamma(u)$$

*is an $\mathbb{H}$-martingale.*

*Proof.* Notice that

$$\exp\left(\mathbb{1}_{\{\tau \le t\}} h(\tau)\right) - 1 = \mathbb{1}_{\{\tau \le t\}} e^{h(\tau)} + \mathbb{1}_{\{\tau > t\}} - 1 = \mathbb{1}_{\{\tau \le t\}} e^{h(\tau)} - H_t,$$

so that

$$\tilde{M}_t^h = \mathbb{1}_{\{\tau \le t\}} e^{h(\tau)} - \int_0^{t \wedge \tau} e^{h(u)} \, d\Gamma(u) - \hat{M}_t.$$

To complete the proof of the corollary, it is thus enough to make use of Proposition 4.2.2. □

The next result offers a still another example of an $\mathbb{H}$-martingale associated with a random time $\tau$.

**Corollary 4.2.2.** *Assume that the hazard function $\Gamma$ of $\tau$ is continuous. Let $h : \mathbb{R}_+ \to \mathbb{R}$ be a Borel measurable function such that the random variable $h(\tau)$ is integrable. Then the process*

$$\bar{M}_t^h = (1 + \mathbb{1}_{\{\tau \le t\}} h(\tau)) \exp\left(-\int_0^{t \wedge \tau} h(u) \, d\Gamma(u)\right)$$

*is an $\mathbb{H}$-martingale.*

*Proof.* Let us denote by $U$ the following decreasing continuous process:

$$U_t = \exp\left(-\int_0^{t \wedge \tau} h(u) \, d\Gamma(u)\right).$$

Notice that

$$1 + \mathbb{1}_{\{\tau \le t\}} h(\tau) = 1 + \int_{]0,t]} h(u) \, dH_u =: H_t^h.$$

An application of the product rule yields

$$d\bar{M}_t^h = d(H_t^h U_t) = U_t h(t) \, dH_t - (1 + \mathbb{1}_{\{\tau \le t\}} h(\tau)) U_t h(t) \, d\Gamma(t \wedge \tau).$$

Consequently, we have

$$d\bar{M}_t^h = U_t h(t) \, d(H_t - \Gamma(t \wedge \tau)) = U_t h(t) \, d\hat{M}_t.$$

The last equality makes it clear that the process $\bar{M}^h$ indeed follows an $\mathbb{H}$-martingale. □

## 4.3 Martingale Representation Theorem

The following elementary version of the martingale representation theorem is commonly known (see, for instance, Brémaud (1981)).

**Proposition 4.3.1.** *Assume that $F$ is an absolutely continuous function. Let $M_t^h := \mathbb{E}_{\mathbb{P}}(h(\tau) \mid \mathcal{H}_t)$ for some Borel measurable function $h : \mathbb{R}_+ \to \mathbb{R}$ such that the random variable $h(\tau)$ is integrable. Then*

$$M_t^h = M_0^h + \int_{]0,t]} \hat{h}(u)\, d\hat{M}_u, \tag{4.13}$$

*where $\hat{M}_t = H_t - \int_0^{t \wedge \tau} \gamma_u\, du$ and the function $\hat{h} : \mathbb{R}_+ \to \mathbb{R}$ is given by the formula*

$$\hat{h}(t) = h(t) - e^{\Gamma(t)} \mathbb{E}_{\mathbb{P}}\left(\mathbb{1}_{\{\tau>t\}} h(\tau)\right). \tag{4.14}$$

*Proof.* Observe first that $M_0^h = \mathbb{E}_{\mathbb{P}}(h(\tau))$. Recall also that the random variable $M_t^h$ admits the following representation (cf. (4.8))

$$M_t^h = \mathbb{E}_{\mathbb{P}}(h(\tau) \mid \mathcal{H}_t) = \mathbb{1}_{\{\tau \le t\}} h(\tau) + \mathbb{1}_{\{\tau>t\}} g(t), \tag{4.15}$$

where the function $g : \mathbb{R} \to \mathbb{R}$ equals

$$g(t) := e^{\Gamma(t)} \mathbb{E}_{\mathbb{P}}\left(\mathbb{1}_{\{\tau>t\}} h(\tau)\right) = e^{\Gamma(t)} \int_t^\infty h(u) f(u)\, du. \tag{4.16}$$

If representation (4.13) is valid for some function $\hat{h}$, then we have, on the set $\{\tau > t\}$,

$$M_t^h = \mathbb{E}_{\mathbb{P}}(h(\tau)) - \int_0^t \hat{h}(s)\gamma(s)\, ds = \mathbb{E}_{\mathbb{P}}(h(\tau)) - \int_0^t \hat{h}(s) e^{\Gamma(s)} f(s)\, ds.$$

On the other hand, by virtue of (4.15), equality $M_t^h = g(t)$ holds on this set. Differentiating both sides with respect to $t$, and taking into account the equality $\gamma(t) = e^{\Gamma(t)} f(t)$, we obtain

$$-e^{\Gamma(t)} f(t)\hat{h}(t) = g'(t) = e^{\Gamma(t)} f(t)(g(t) - h(t)).$$

The equality $\hat{h}(t) = h(t) - g(t)$ is thus straightforward on the set $\{t < \tau\}$. Since the process $M^h$ is manifestly continuous on this set, we also have

$$\hat{h}(t) = h(t) - M_t^h = h(t) - M_{t-}^h$$

on the set $\{t < \tau\}$. In view of the last equality, it is clear that, on the event $\{\tau \le t\}$, the right-hand side of (4.13) gives $h(\tau)$, as expected. $\qquad \square$

Proposition 4.3.1 remains valid when the hazard function $\Gamma$ is merely continuous, as the next result shows.

**Proposition 4.3.2.** *Assume that $F$ is a continuous function. Let $M_t^h :=$ $\mathbb{E}_{\mathbb{P}}(h(\tau) \mid \mathcal{H}_t)$ for some Borel measurable function $h : \mathbb{R}_+ \to \mathbb{R}$ such that the random variable $h(\tau)$ is integrable. Then*

$$M_t^h = M_0^h + \int_{]0,t]} \hat{h}(u) \, d\hat{M}_u, \tag{4.17}$$

*where $\hat{M}_t = H_t - \Gamma(t \wedge \tau)$ and $\hat{h}$ satisfies (4.14), i.e., $\hat{h} = h - g$, where $g$ is given by (4.16).*

*Proof.* By virtue of (4.8), the left-hand side of formula (4.17) equals (see also (4.15))

$$I = \mathbb{E}_{\mathbb{P}}(h(\tau) \mid \mathcal{H}_t) = H_t h(\tau) + (1 - H_t)g(t).$$

On the other hand, the right-hand side of (4.17) can be rewritten as follows:

$$
\begin{aligned}
J &= g(0) + \int_{]0,t]} \hat{h}(u) \, d\hat{M}_u \\
&= g(0) + \int_{]0,t]} (h(u) - g(u)) \, d(H_u - \Gamma(u \wedge \tau)) \\
&= g(0) + H_t(h(\tau) - g(\tau)) + \int_0^{t \wedge \tau} (g(u) - h(u)) \, d\Gamma(u) \\
&= g(0) + H_t h(\tau) + (1 - H_t)g(t) - g(t \wedge \tau) + \int_0^{t \wedge \tau} (g(u) - h(u)) \, d\Gamma(u).
\end{aligned}
$$

To check that $I = J$, it suffices to show that

$$g(t \wedge \tau) = g(0) + \int_0^{t \wedge \tau} (g(u) - h(u)) \, d\Gamma(u)$$

or, equivalently, that for any $t \in \mathbb{R}_+$ we have

$$g(t) = g(0) + \int_0^t (g(u) - h(u)) \, d\Gamma(u).$$

Put another way, we need to verify that the following equality holds:

$$e^{\Gamma(t)} \int_t^\infty h(u) \, dF(u) = \int_0^\infty h(u) \, dF(u) + \int_0^t e^{\Gamma(u)} (g(u) - h(u)) \, dF(u).$$

By applying Fubini's theorem, we get (recall that $e^{\Gamma(u)} dF(u) = d\Gamma(u)$)

$$
\begin{aligned}
\int_0^t e^{\Gamma(u)} g(u) \, dF(u) &= \int_0^t e^{2\Gamma(u)} \int_u^\infty h(v) \, dF(v) \, dF(u) \\
&= \int_0^t h(v) \int_0^v e^{\Gamma(u)} \, d\Gamma(u) \, dF(v) + \int_t^\infty h(v) \int_0^t e^{\Gamma(u)} \, d\Gamma(u) \, dF(v) \\
&= \int_0^t h(u) (e^{\Gamma(u)} - 1) \, dF(u) + (e^{\Gamma(t)} - 1) \int_t^\infty h(u) \, dF(u).
\end{aligned}
$$

This completes the proof. $\qquad\qquad\qquad\qquad\qquad\qquad\qquad\qquad\qquad\square$

Notice that representation (4.17) can also be rewritten as follows (cf. formula (5.32)):

$$M_t^h = M_0^h + \int_{]0,t]} (h(u) - M_{u-}^h) \, d\hat{M}_u. \tag{4.18}$$

*Remarks.* Since an arbitrary $\mathcal{H}_\infty$-measurable random variable $X$ has the form $X = h(\tau)$, we may also deduce from Proposition 4.3.2 that any $\mathbb{H}$-martingale admits the representation (4.17). Hence, any $\mathbb{H}$-martingale is a purely discontinuous martingale, as it follows a process of finite variation. Put another way, any continuous $\mathbb{H}$-martingale necessarily follows a constant process.

## 4.4 Change of a Probability Measure

Let $\mathbb{P}^*$ be an arbitrary probability measure on $(\Omega, \mathcal{H}_\infty)$. Assume that $\mathbb{P}^*$ is absolutely continuous with respect to $\mathbb{P}$, i.e., $\mathbb{P}^*\{A\} = 0$ for any event $A \in \mathcal{H}_\infty$ such that $\mathbb{P}\{A\} = 0$. Then there exists a Borel measurable function $h : \mathbb{R}_+ \to \mathbb{R}_+$, which satisfies

$$\mathbb{E}_\mathbb{P}(h(\tau)) = \int_{]0,\infty[} h(u) \, dF(u) = 1,$$

and such that the Radon-Nikodým density of $\mathbb{P}^*$ with respect to $\mathbb{P}$ equals

$$\eta := \frac{d\mathbb{P}^*}{d\mathbb{P}} = h(\tau) \geq 0, \quad \mathbb{P}\text{-a.s.}, \tag{4.19}$$

We shall henceforth write $\mathbb{E}_\mathbb{P}$ ($\mathbb{E}_{\mathbb{P}^*}$, resp.) to denote the expected value with respect to the probability measure $\mathbb{P}$ ($\mathbb{P}^*$, resp.) Probability measure $\mathbb{P}^*$ is equivalent to $\mathbb{P}$ if and only if the inequality in (4.19) is strict, $\mathbb{P}$-a.s.

Furthermore, we shall assume that $\mathbb{P}^*\{\tau = 0\} = 0$ and $\mathbb{P}^*\{\tau > t\} > 0$ for every $t \in \mathbb{R}_+$. The first condition is in fact satisfied for an arbitrary probability measure $\mathbb{P}^*$ absolutely continuous with respect to $\mathbb{P}$. For the second condition to hold, it is sufficient and necessary to postulate that for every $t \in \mathbb{R}_+$

$$\mathbb{P}^*\{\tau > t\} = 1 - F^*(t) = \int_{]t,\infty[} h(u) \, dF(u) > 0, \tag{4.20}$$

where $F^*$ is the cumulative distribution function of $\tau$ under $\mathbb{P}^*$, specifically,

$$F^*(t) := \mathbb{P}^*\{\tau \leq t\} = \int_{]0,t]} h(u) \, dF(u).$$

Condition (4.20) is equivalent to the following one (cf. (4.16))

$$g(t) = e^{\Gamma(t)} \mathbb{E}_\mathbb{P}(\mathbb{1}_{\{\tau > t\}} h(\tau)) = e^{\Gamma(t)} \int_{]t,\infty[} h(u) \, dF(u) = e^{\Gamma(t)} \mathbb{P}^*\{\tau > t\} > 0.$$

From now on, we assume that this is indeed the case, so that the hazard function $\Gamma^*$ of $\tau$ with respect to $\mathbb{P}^*$ is well defined.

Is not difficult to establish the relationship between the hazard functions $\Gamma^*$ and $\Gamma$. Indeed, we have

$$\frac{\Gamma^*(t)}{\Gamma(t)} = \frac{\ln\left(\int_{]t,\infty[} h(u)\,dF(u)\right)}{\ln(1 - F(t))} =: g^*(t),$$

since, by the definition of the hazard function, $\Gamma^*(t) = -\ln(1 - F^*(t))$. Let us now analyze some special cases of the last relationship.

In the first step, we will assume that $F$ is an absolutely continuous function, so that the intensity function $\gamma$ of $\tau$ under $\mathbb{P}$ is well defined. Recall that $\gamma$ is given by the following formula:

$$\gamma(t) = f(t)(1 - F(t))^{-1}, \quad \forall t \in \mathbb{R}_+.$$

Under the present assumptions, the c.d.f. $F^*$ of $\tau$ under $\mathbb{P}^*$ equals

$$F^*(t) := \mathbb{P}^*\{\tau \le t\} = \mathbb{E}_{\mathbb{P}}(\mathbb{1}_{\{\tau \le t\}} h(\tau)) = \int_0^t h(u)f(u)\,du = \int_0^t f^*(u)\,du,$$

where $f^*(u) = h(u)f(u)$, and thus $F^*$ is an absolutely continuous function. Thus, the intensity function $\gamma^*$ of the random time $\tau$ under $\mathbb{P}^*$ exists, and is given by the formula

$$\gamma^*(t) = \frac{f^*(t)}{1 - F^*(t)} = \frac{h(t)f(t)}{1 - \int_0^t h(u)f(u)\,du}.$$

To derive a more explicit relationship between the intensities $\gamma$ and $\gamma^*$, we define an auxiliary function $h^* : \mathbb{R}_+ \to \mathbb{R}$ by setting $h^*(t) = h(t)g^{-1}(t)$. Notice that

$$\gamma^*(t) = \frac{h(t)f(t)}{1 - \int_0^t h(u)f(u)\,du} = \frac{h(t)f(t)}{\int_t^\infty h(u)f(u)\,du}$$
$$= \frac{h(t)f(t)}{e^{-\Gamma(t)}g(t)} = \frac{h^*(t)f(t)}{1 - F(t)} = h^*(t)\gamma(t).$$

This also means that $d\Gamma^*(t) = h^*(t)\,d\Gamma(t)$. It appears that the last equality holds true if $F$ is merely a continuous function. Indeed, if $F$ (and thus $F^*$) is continuous, we get

$$d\Gamma^*(t) = \frac{dF^*(t)}{1 - F^*(t)} = \frac{d(1 - e^{-\Gamma(t)}g(t))}{e^{-\Gamma(t)}g(t)} = \frac{g(t)d\Gamma(t) - dg(t)}{g(t)} = h^*(t)\,d\Gamma(t).$$

We have thus established the following partial result in which, for the sake of convenience, we denote $\kappa(t) = h^*(t) - 1 = h(t)g^{-1}(t) - 1$.

**Proposition 4.4.1.** *Let the two probability measures $\mathbb{P}^*$ and $\mathbb{P}$ be related to each other by means of (4.19). If the hazard function $\Gamma$ of $\tau$ under $\mathbb{P}$ is continuous, then the hazard function $\Gamma^*$ of $\tau$ under $\mathbb{P}^*$ is also continuous and $d\Gamma^*(t) = (1 + \kappa(t))\,d\Gamma(t)$, where $\kappa(t) = h(t)g^{-1}(t) - 1$ and the functions $h$ and $g$ are given by formulae (4.19) and (4.16), respectively.*

Let us now take a closer look at the auxiliary function $\kappa$. To this end, we introduce the following non-negative $\mathbb{P}$-martingale $\eta$:

$$\eta_t := \frac{d\mathbb{P}^*}{d\mathbb{P}}\bigg|_{\mathcal{H}_t} = \mathbb{E}_{\mathbb{P}}(\eta \mid \mathcal{H}_t) = \mathbb{E}_{\mathbb{P}}(h(\tau) \mid \mathcal{H}_t). \tag{4.21}$$

It is clear that $\eta_t = M_t^h$. We shall refer to the process $\eta$ as the *Radon-Nikodým density process* of $\mathbb{P}^*$ with respect to $\mathbb{P}$. In view of (4.7), we have

$$\eta_t = \mathbb{1}_{\{\tau \le t\}} h(\tau) + \mathbb{1}_{\{\tau > t\}} e^{\Gamma(t)} \int_{]t,\infty[} h(u)\, dF(u) = \mathbb{1}_{\{\tau \le t\}} h(\tau) + \mathbb{1}_{\{\tau > t\}} g(t).$$

If, in addition, $F$ is a continuous function then (cf. (4.8))

$$\eta_t = \mathbb{1}_{\{\tau \le t\}} h(\tau) + \mathbb{1}_{\{\tau > t\}} \int_t^\infty h(u) e^{\Gamma(t) - \Gamma(u)}\, d\Gamma(u).$$

On the other hand, using (4.17) and (4.18), we obtain

$$M_t^h = M_0^h + \int_{]0,t]} M_{u-}^h (h^*(u) - 1)\, d\hat{M}_u = M_0^h + \int_{]0,t]} M_{u-}^h \kappa(u)\, d\hat{M}_u,$$

which shows that $\eta$ solves the following SDE:

$$\eta_t = 1 + \int_{]0,t]} \eta_{u-}\kappa(u)\, d\hat{M}_u. \tag{4.22}$$

It is not difficult to find an explicit solution to this equation, namely,

$$\eta_t = \left(1 + \mathbb{1}_{\{\tau \le t\}} \kappa(\tau)\right) \exp\left(-\int_0^{t \wedge \tau} \kappa(u)\, d\Gamma(u)\right). \tag{4.23}$$

In view of the last formula, the martingale property of the process $\eta$ – that is apparent from (4.21) – is thus also a simple consequence of Corollary 4.2.2. The proof of the following classic result is left to the reader.

**Lemma 4.4.1.** *Let $Y$ follow a process of finite variation. Consider the following linear stochastic differential equation*

$$Z_t = 1 + \int_{]0,t]} Z_{u-}\, dY_u. \tag{4.24}$$

*The unique solution $Z_t = \mathcal{E}_t(Y)$ to (4.24), referred to as the Doléans exponential of $Y$, equals*

$$\mathcal{E}_t(Y) = e^{Y_t} \prod_{0 < u \le t} (1 + \Delta Y_u) e^{-\Delta Y_u} = e^{Y_t^c} \prod_{0 < u \le t} (1 + \Delta Y_u), \tag{4.25}$$

*where $Y^c$ is the continuous part of $Y$, i.e., $Y_t^c = Y_t - \sum_{0 < u \le t} \Delta Y_u$.*

Since the process $\eta$ satisfies (4.22), it is clear that it can be represented as follows:

$$\eta_t = \mathcal{E}_t\left(\int_{]0,\cdot]} \kappa(u)\, d\hat{M}_u\right).$$

Expression (4.23) for the random variable $\eta_t$ can thus also be obtained from (4.25), upon setting $dY_u = \kappa(u)\,d\hat{M}_u$. Let us stress that (4.25) is merely a special case of the well known general formula for the Doléans exponential (see Jacod (1979), Elliott (1982), Protter (1990), or Revuz and Yor (1999)). We are in a position to formulate the following result (all statements in Proposition 4.4.2 were already established above).

**Proposition 4.4.2.** *Assume that $F$ is a continuous function. Let $\mathbb{P}^*$ be any probability measure on $(\Omega, \mathcal{H}_\infty)$ absolutely continuous with respect to $\mathbb{P}$, so that (4.19) holds for some function $h$. Assume that $\mathbb{P}^*\{\tau > t\} > 0$ for $t \in \mathbb{R}_+$. Then the Radon-Nikodým density process $\eta$ of $\mathbb{P}^*$ with respect to $\mathbb{P}$ satisfies*

$$\eta_t := \frac{d\mathbb{P}^*}{d\mathbb{P}}\Big|_{\mathcal{H}_t} = \mathcal{E}_t\Big(\int_{]0,\,\cdot\,]} \kappa(u)\,d\hat{M}_u\Big),$$

*where $\kappa(t) = h(t)g^{-1}(t) - 1$ and*

$$g(t) = e^{\Gamma(t)}\int_t^\infty h(u)\,dF(u).$$

*Moreover, the hazard function of $\tau$ under $\mathbb{P}^*$ equals $\Gamma^*(t) = g^*(t)\Gamma(t)$ with*

$$g^*(t) = \frac{\ln\Big(\int_{]t,\infty[} h(u)\,dF(u)\Big)}{\ln(1 - F(t))}.$$

**Corollary 4.4.1.** *If $F$ is continuous then the process $M_t^* = H_t - \Gamma^*(t \wedge \tau)$ is an $\mathbb{H}$-martingale under $\mathbb{P}^*$.*

*Proof.* In view Proposition 4.4.2, the corollary is an immediate consequence of the continuity of $\Gamma^*$, combined with Proposition 4.2.1. Alternatively, we may check directly that the product

$$U_t := \eta_t M_t^* = \eta_t(H_t - \Gamma^*(t \wedge \tau))$$

follows an $\mathbb{H}$-martingale under $\mathbb{P}$. The product rule for processes of finite variation yields (clearly $\Delta M_u^* = \Delta\hat{M}_u$)

$$U_t = \int_{]0,t]} \eta_{t-}\,dM_t^* + \int_{]0,t]} M_{t-}^*\,d\eta_t + \sum_{u \le t} \Delta\hat{M}_u\Delta\eta_u$$

$$= \int_{]0,t]} \eta_{t-}\,dM_t^* + \int_{]0,t]} M_{t-}^*\,d\eta_t + \mathbb{1}_{\{\tau \le t\}}(\eta_\tau - \eta_{\tau-}).$$

Using (4.22), we obtain

$$U_t = \int_{]0,t]} \eta_{t-}\,dM_t^* + \int_{]0,t]} M_{t-}^*\,d\eta_t + \mathbb{1}_{\{\tau \le t\}}\eta_{\tau-}\kappa(\tau)$$

$$= \int_{]0,t]} \eta_{t-}\,d\big(\Gamma(t \wedge \tau) - \Gamma^*(t \wedge \tau) + \mathbb{1}_{\{\tau \le t\}}\kappa(\tau)\big) + N_t,$$

where the process $N$, which equals

$$N_t = \int_{]0,t]} \eta_{t-}\, d\hat{M}_t + \int_{]0,t]} M_{t-}^*\, d\eta_t,$$

is manifestly an $\mathbb{H}$-martingale under $\mathbb{P}$. It remains to show that the process

$$N_t^* := \Gamma(t \wedge \tau) - \Gamma^*(t \wedge \tau) + \mathbb{1}_{\{\tau \le t\}}\kappa(\tau)$$

follows an $\mathbb{H}$-martingale under $\mathbb{P}$. By virtue of Proposition 4.2.2, the process

$$\mathbb{1}_{\{\tau \le t\}}\kappa(\tau) + \Gamma(t \wedge \tau) - \int_0^{t \wedge \tau} (1 + \kappa(u))\, d\Gamma(u)$$

is an $\mathbb{H}$-martingale under $\mathbb{P}$. To finish the proof, it is enough to observe that

$$\int_0^{t \wedge \tau} (1 + \kappa(u))\, d\Gamma(u) - \Gamma^*(t \wedge \tau) = \int_0^{t \wedge \tau} \big((1 + \kappa(u))\, d\Gamma(u) - d\Gamma^*(u)\big) = 0,$$

where the last equality follows from the relationship $d\Gamma^*(t) = (1+\kappa(t))\, d\Gamma(t)$, which was established in Proposition 4.4.1.                                          □

By virtue of Proposition 4.2.1, if the hazard function $\Gamma^*$ is continuous, then the process $M^* = H_t - \Gamma^*(t \wedge \tau)$ follows an $\mathbb{H}$-martingale under $\mathbb{P}^*$. In fact, the martingale property uniquely characterizes the (continuous) hazard function of a random time. In the next section, we shall examine in more detail this important issue.

## 4.5 Martingale Characterization of the Hazard Function

Proposition 4.2.1 raises the natural question whether the martingale property of the process $H_t - \Gamma(t \wedge \tau)$ with respect to the filtration $\mathbb{H}$ uniquely characterizes the hazard function of a random time $\tau$? Our goal is to show that the answer to this question is positive, provided that the hazard function $\Gamma$ is continuous. Notice that for a discontinuous hazard function $\Gamma$, equality (4.12) takes the following form:

$$L_t = L_0 + \int_{]0,t]} (1 - H_u)\, de^{\Gamma(u)} - \int_{]0,t]} e^{\Gamma(u-)}\, dH_u$$

or, equivalently,

$$L_t = 1 + \int_{]0,t]} e^{\Gamma(u-)}\big((1 - H_u)\, d\Gamma(u) - dH_u\big) + \sum_{s \le t,\, s < \tau} \big(\Delta e^{\Gamma(s)} - e^{\Gamma(s-)}\, \Delta\Gamma(s)\big),$$

where

$$\Delta e^{\Gamma(s)} = e^{\Gamma(s)} - e^{\Gamma(s-)}, \quad \Delta\Gamma(s) = \Gamma(s) - \Gamma(s-).$$

The last formula makes in clear that in the case of a discontinuous hazard function $\Gamma$, the process $H_t - \Gamma(t \wedge \tau)$ is not an $\mathbb{H}$-martingale.

Let us recall that $H_t = H_{t \wedge \tau}$; that is, the process $H$ is stopped at time $\tau$. We find it convenient to introduce the notion of a martingale hazard function of a random time.

**Definition 4.5.1.** A function $\Lambda : \mathbb{R}_+ \to \mathbb{R}$ is called a *martingale hazard function* of a random time $\tau$ with respect to its natural filtration $\mathbb{H}$ if and only if the process $H_t - \Lambda(t \wedge \tau)$ follows an $\mathbb{H}$-martingale.

The function $\Lambda$ may also be seen as an $\mathbb{F}^0$-adapted right-continuous stochastic process, where $\mathbb{F}^0$ is the trivial filtration, i.e., $\mathcal{F}_t^0 = \mathcal{F}_0^0 = \{\emptyset, \Omega\}$ for every $t \in \mathbb{R}_+$. We shall sometimes find it useful to refer to the martingale hazard function as the $(\mathbb{F}^0, \mathbb{H})$-*martingale hazard process* of $\tau$. The rationale supporting this convention will become clear in Chap. 6, in which a more general concept of a martingale hazard process with respect to a pair $(\mathbb{F}, \mathbb{G})$ of filtrations is introduced and examined.

**Proposition 4.5.1.** (i) *The unique martingale hazard function of $\tau$ with respect to $\mathbb{H}$ is the right-continuous increasing function $\Lambda$ given by the formula*

$$\Lambda(t) = \int_{]0,t]} \frac{dF(u)}{1 - F(u-)} = \int_{]0,t]} \frac{d\mathbb{P}\{\tau \le u\}}{1 - \mathbb{P}\{\tau < u\}}, \quad \forall t \in \mathbb{R}_+. \tag{4.26}$$

(ii) *The martingale hazard function $\Lambda$ coincides with the hazard function $\Gamma$ if and only if $F$ is a continuous function. In general, for every $t \in \mathbb{R}_+$ we have*

$$e^{-\Gamma(t)} = e^{-\Lambda^c(t)} \prod_{0 < u \le t} (1 - \Delta\Lambda(u)), \tag{4.27}$$

*where $\Lambda^c(t) = \Lambda(t) - \sum_{0 \le u \le t} \Delta\Lambda(u)$ and $\Delta\Lambda(u) = \Lambda(u) - \Lambda(u-)$.*
(iii) *The martingale hazard function $\Lambda$ is continuous if and only if the cumulative distribution function $F$ of $\tau$ is continuous. In this case, $\Lambda$ satisfies $\Lambda(t) = -\ln(1 - F(t)) = \Gamma(t)$ for every $t \in \mathbb{R}_+$.*

*Proof.* Let us first examine the uniqueness. The definition of $\Lambda$ implies that $\mathbb{E}_{\mathbb{P}}(H_t) = \mathbb{E}_{\mathbb{P}}(\Lambda(t \wedge \tau))$. Put more explicitly (recall that $F(0) = 0$),

$$F(t) = \int_{]0,t]} \Lambda(u) \, dF(u) + \Lambda(t)(1 - F(t)), \tag{4.28}$$

so that $\Lambda$ is necessarily a right-continuous function. Furthermore, if $\Lambda_1$ and $\Lambda_2$ are the two right-continuous functions, which satisfy (4.28), then for every $t \in \mathbb{R}_+$ we have

$$\int_{]0,t]} (\Lambda_1(u) - \Lambda_2(u)) \, dF(u) + (\Lambda_1(t) - \Lambda_2(t))(1 - F(t)) = 0.$$

Using the last equality, one can show – by making use of rather standard contraction arguments – that the martingale hazard function $\Lambda$ is unique.

To complete the proof of part (i), we need to establish the martingale property of the process $H_t - \Lambda(t \wedge \tau)$. It is enough to check that for any $t \leq s$ we have

$$\mathbb{E}_{\mathbb{P}}(H_s - H_t \mid \mathcal{H}_t) = \mathbb{1}_{\{\tau > t\}} \frac{F(s) - F(t)}{1 - F(t)} = \mathbb{E}_{\mathbb{P}}(Y \mid \mathcal{H}_t),$$

where the first equality is a consequence of (4.6), and where we have set

$$Y := \Lambda(s \wedge \tau) - \Lambda(t \wedge \tau) = \int_{]t \wedge \tau, s \wedge \tau]} \frac{dF(u)}{1 - F(u-)}.$$

Since $Y = \mathbb{1}_{\{\tau > t\}} Y$, using (4.2), we obtain

$$\mathbb{E}_{\mathbb{P}}(Y \mid \mathcal{H}_t) = \mathbb{E}_{\mathbb{P}}(\mathbb{1}_{\{\tau > t\}} Y \mid \mathcal{H}_t) = \mathbb{1}_{\{\tau > t\}} \frac{\mathbb{E}_{\mathbb{P}}(Y)}{1 - F(t)}.$$

Furthermore, we have

$$\mathbb{E}_{\mathbb{P}}(Y) = \mathbb{P}\{\tau > s\} \int_{]t,s]} \frac{dF(u)}{1 - F(u-)} + \int_{]t,s]} \int_{]t,u]} \frac{dF(v)}{1 - F(v-)} dF(u).$$

Consequently,

$$\mathbb{E}_{\mathbb{P}}(Y) = (\Lambda(s) - \Lambda(t))(1 - F(s)) + \int_{]t,s]} (\Lambda(u) - \Lambda(t)) \, dF(u)$$

$$= (\Lambda(s) - \Lambda(t))(1 - F(s)) - \Lambda(t)(F(s) - F(t)) + \int_{]t,s]} \Lambda(u) \, dF(u).$$

The product rule yields

$$\int_{]t,s]} \Lambda(u) \, dF(u) = \Lambda(s)F(s) - \Lambda(t)F(t) - \int_{]t,s]} F(u-) \, d\Lambda(u). \qquad (4.29)$$

Finally, it is clear from (4.26) that

$$\int_{]t,s]} F(u-) \, d\Lambda(u) = \Lambda(s) - \Lambda(t) - F(s) + F(t).$$

Combining the above equalities, we find that $\mathbb{E}_{\mathbb{P}}(Y) = F(s) - F(t)$ for every $t \leq s$. This completes the proof of (i). To establish (ii), notice that by virtue of (4.26), the survival function $G(t) = 1 - F(t)$ satisfies

$$G(t) = - \int_{]0,t]} G(u-) \, d\Lambda(u).$$

Therefore (cf. (4.24)–(4.25)),

$$e^{-\Gamma(t)} = G(t) = e^{-\Lambda^c(t)} \prod_{0 < u \leq t} (1 - \Delta\Lambda(u)).$$

This completes the proof of (4.27). In particular, the martingale hazard function $\Lambda$ and the hazard function $\Gamma$ are not equal to each other, when the function $F$ is discontinuous. All statements of part (iii) are immediate consequences of part (ii). $\qquad \square$

*Remarks.* Assume that the cumulative distribution function $F$ is absolutely continuous, with the probability density function $f$. Then necessarily

$$\Lambda(t) = \Gamma(t) = \int_0^t f(u)(1 - F(u))^{-1} \, du$$

and thus the martingale hazard function $\Lambda$ is absolutely continuous as well. Specifically, $\Lambda(t) = \int_0^t \lambda(u) \, du$, where $\lambda(u) = \gamma(u) = f(u)(1 - F(u))^{-1}$ for every $u \in \mathbb{R}_+$.

## 4.6 Compensator of a Random Time

By virtue of the properties of the martingale hazard function, the process $C_t := \Lambda(t \wedge \tau)$ satisfies: (i) $C$ is an increasing, right-continuous, $\mathbb{H}$-adapted process, and (ii) the compensated process $H - C$ follows an $\mathbb{H}$-martingale. This shows that the notion of the martingale hazard function is closely related to the concept of the $\mathbb{H}$-compensator of $\tau$ or, more precisely, to the concept of the $\mathbb{H}$-compensator of the associated jump process $H$.

We adopt here the standard convention, which stipulates that $B$ is an *increasing process* if $B$ is an adapted process with non-decreasing, right-continuous sample paths. The process $H$ is, of course, a bounded increasing process, and so also a bounded $\mathbb{H}$-submartingale.

Let us first recall the definition of the compensator of an increasing process (the compensator of an increasing process is also known as its *dual predictable projection*; see, e.g., Dellacherie (1972) or Jacod (1979) in this regard). When specified to our situation, it can be stated as follows.

**Definition 4.6.1.** A process $A$ is called the $\mathbb{H}$-*compensator* of the process $H$ if and only if the following conditions are satisfied: (i) $A$ is an $\mathbb{H}$-predictable increasing process, with $A_0 = 0$, (ii) the compensated process $H - A$ follows an $\mathbb{H}$-martingale.

Existence and uniqueness of the Doob-Meyer decomposition for a bounded submartingale[2] imply that the process $H$ admits a unique $\mathbb{H}$-compensator. We are in a position to prove the following result.

**Lemma 4.6.1.** *Assume that the cumulative distribution function $F$ of $\tau$ is continuous. Then the unique $\mathbb{H}$-compensator $A$ of $\tau$ equals, for every $t \in \mathbb{R}_+$,*

$$A_t = \Lambda(t \wedge \tau) = \Gamma(t \wedge \tau) = -\ln(1 - F(t \wedge \tau)).$$

*Proof.* In view of the definition of the martingale hazard function, part (ii) in Proposition 4.5.1, and Lemma 4.6.1, it is enough to check that the process $A_t = \Lambda(t \wedge \tau)$, $t \in \mathbb{R}_+$, is $\mathbb{H}$-predictable. But this is clear, since the mapping $t \to t \wedge \tau$ defines a continuous, $\mathbb{H}$-adapted process, so that it is an $\mathbb{H}$-predictable process. In view of the continuity of $\Lambda$, we conclude that $A$ is an $\mathbb{H}$-predictable process. □

---

[2] See Theorem 4.10 in Sect. 1.4 of Karatzas and Shreve (1997).

# 5. Hazard Process of a Random Time

The concepts introduced in the previous chapter will now be extended to a more general set-up, when allowance for a larger flow of information – formally represented by some reference filtration $\mathbb{F}$ – is made.

## 5.1 Hazard Process $\Gamma$

We denote by $\tau$ a non-negative random variable on a probability space $(\Omega, \mathcal{G}, \mathbb{P})$, satisfying: $\mathbb{P}\{\tau = 0\} = 0$ and $\mathbb{P}\{\tau > t\} > 0$ for any $t \in \mathbb{R}_+$. We introduce a right-continuous process $H$ by setting $H_t = \mathbb{1}_{\{\tau \le t\}}$ and we denote by $\mathbb{H}$ the associated filtration: $\mathcal{H}_t = \sigma(H_u : u \le t)$. Let $\mathbb{G} = (\mathcal{G}_t)_{t \ge 0}$ be an arbitrary filtration on $(\Omega, \mathcal{G}, \mathbb{P})$. All filtrations are assumed to satisfy the 'usual conditions' of right-continuity and completeness. For each $t \in \mathbb{R}_+$, the information available at time $t$ is captured by the $\sigma$-field $\mathcal{G}_t$. We shall focus on the case described in the following assumption.

**Condition (G.1)** We assume that we are given an auxiliary filtration $\mathbb{F}$ such that $\mathbb{G} = \mathbb{H} \vee \mathbb{F}$; i.e., $\mathcal{G}_t = \mathcal{H}_t \vee \mathcal{F}_t$ for any $t \in \mathbb{R}_+$.

For the sake of simplicity, we assume that the $\sigma$-field $\mathcal{F}_0$ is trivial (so that $\mathcal{G}_0$ is a trivial $\sigma$-field as well). For given filtrations $\mathbb{H} \subseteq \mathbb{G}$, the equality $\mathcal{G}_t = \mathcal{H}_t \vee \mathcal{F}_t$ does not specify uniquely an auxiliary filtration $\mathbb{F}$. For instance, when $\mathcal{G}_t = \mathcal{H}_t$, we may take $\mathbb{F} = \mathbb{F}^0$, but also $\mathbb{F} = \mathbb{H}$ (or indeed any other sub-filtration of $\mathbb{H}$). In most applications, $\mathbb{F}$ will appear in a natural way as the filtration generated by a certain stochastic process.

**Condition (G.1a)** For every $t \in \mathbb{R}_+$, the event $\{\tau \le t\}$ belongs to the $\sigma$-field $\mathcal{F}_t$ (and thus $\tau$ is an $\mathbb{F}$-stopping time).

Under (G.1a), we have $\mathbb{G} = \mathbb{F}$, and thus $\tau$ also is a $\mathbb{G}$-stopping time. In some models, only a partial observation of the random time $\tau$ is postulated. Such a case corresponds to the following condition.

**Condition (G.1b)** For some dates $t \in \mathbb{R}_+$, the event $\{\tau \le t\}$ does not belong to the $\sigma$-field $\mathcal{G}_t$.

Let $\hat{\mathbb{H}} \subset \mathbb{H}$ stand for the filtration associated with the partial observations of $\tau$. Then the enlarged filtration $\mathbb{G}$ equals $\mathbb{G} = \hat{\mathbb{H}} \vee \mathbb{F}$.

Under (G.1), the process $H$ is obviously $\mathbb{G}$-adapted, but it is not necessarily $\mathbb{F}$-adapted. In other words, the random time $\tau$ is a $\mathbb{G}$-stopping time, but it may fail to be an $\mathbb{F}$-stopping time. Under (G.1b), the process $H$ is not $\mathbb{G}$-adapted, i.e., $\tau$ is not a $\mathbb{G}$-stopping time. However, in both cases the following condition is satisfied.

**Condition (G.2)** For every $t \in \mathbb{R}_+$ we have $\mathcal{F}_t \subseteq \mathcal{G}_t \subseteq \mathcal{H}_t \vee \mathcal{F}_t$.

**Lemma 5.1.1.** *Assume that the filtration $\mathbb{G}$ satisfies $\mathbb{G} \subseteq \mathbb{H} \vee \mathbb{F}$, that is, $\mathcal{G}_t \subseteq \mathcal{H}_t \vee \mathcal{F}_t$ for every $t \in \mathbb{R}_+$. Then $\mathbb{G} \subseteq \mathbb{G}^*$, where $\mathbb{G}^* = (\mathcal{G}_t^*)_{t \geq 0}$ with*

$$\mathcal{G}_t^* := \{A \in \mathcal{G} : \exists B \in \mathcal{F}_t, \ A \cap \{\tau > t\} = B \cap \{\tau > t\}\}.$$

*Proof.* It is rather clear that the class $\mathcal{G}_t^*$ is a sub-$\sigma$-field of $\mathcal{G}$. Therefore, it is enough to check that $\mathcal{H}_t \subseteq \mathcal{G}_t^*$ and $\mathcal{F}_t \subseteq \mathcal{G}_t^*$ for every $t \in \mathbb{R}_+$. Put another way, we need to verify that if either $A = \{\tau \leq u\}$ for some $u \leq t$ or $A \in \mathcal{F}_t$, then there exists an event $B \in \mathcal{F}_t$ such that $A \cap \{\tau > t\} = B \cap \{\tau > t\}$. In the former case we may take $B = \emptyset$, and in the latter $B = A$. $\qquad \square$

*Remarks.* By a suitable modification of arguments used in the proof of Lemma 5.1.1, one can show that under (G.2) for any $\mathcal{G}_t$-measurable random variable $Y$ there exists an $\mathcal{F}_t$-measurable random variable $\tilde{Y}$ such that $Y = \tilde{Y}$ on the set $\{\tau > t\}$. Under (G.1), this remarkable property is also a straightforward consequence of part (ii) in Lemma 5.1.2.

For any $t \in \mathbb{R}_+$, we write $F_t = \mathbb{P}\{\tau \leq t \mid \mathcal{F}_t\}$, and we denote by $G$ the $\mathbb{F}$-*survival process* of $\tau$ with respect to the filtration $\mathbb{F}$, given as:

$$G_t := 1 - F_t = \mathbb{P}\{\tau > t \mid \mathcal{F}_t\}, \quad \forall t \in \mathbb{R}_+.$$

Notice that for any $0 \leq t \leq s$ we have $\{\tau \leq t\} \subseteq \{\tau \leq s\}$, and so

$$\mathbb{E}_{\mathbb{P}}(F_s \mid \mathcal{F}_t) = \mathbb{E}_{\mathbb{P}}(\mathbb{P}\{\tau \leq s \mid \mathcal{F}_s\} \mid \mathcal{F}_t) = \mathbb{P}\{\tau \leq s \mid \mathcal{F}_t\} \geq \mathbb{P}\{\tau \leq t \mid \mathcal{F}_t\} = F_t.$$

This shows that the process $F$ ($G$, resp.) follows a bounded, non-negative $\mathbb{F}$-submartingale ($\mathbb{F}$-supermartingale, resp.) under $\mathbb{P}$. We may thus deal with the right-continuous modification of $F$ (of $G$) with finite left-hand limits. The next definition is a straightforward generalization of Definition 4.1.1.

**Definition 5.1.1.** Assume that $F_t < 1$ for $t \in \mathbb{R}_+$. The $\mathbb{F}$-*hazard process* of $\tau$ under $\mathbb{P}$, denoted by $\Gamma$, is defined through the formula $1 - F_t = e^{-\Gamma_t}$. Equivalently, $\Gamma_t = -\ln G_t = -\ln(1 - F_t)$ for every $t \in \mathbb{R}_+$.

Since $G_0 = 1$, it is clear that $\Gamma_0 = 0$. For the sake of conciseness, we shall refer briefly to $\Gamma$ as the $\mathbb{F}$-hazard process, rather than the $\mathbb{F}$-hazard process under $\mathbb{P}$, unless there is a danger of confusion.

In this chapter, we assume that the inequality $F_t < 1$ holds for every $t \in \mathbb{R}_+$, so that the $\mathbb{F}$-hazard process $\Gamma$ is well defined. It should be stressed that the case when $\tau$ is an $\mathbb{F}$-stopping time (i.e., the case when $\mathbb{F} = \mathbb{G}$) is not dealt with in this chapter; an analysis of this case is postponed to Sect. 6.4.2.

### 5.1.1 Conditional Expectations

We shall first focus on the conditional expectation $\mathbb{E}_{\mathbb{P}}(\mathbb{1}_{\{\tau>t\}}Y\,|\,\mathcal{G}_t)$, where $Y$ is a $\mathbb{P}$-integrable random variable. We start by the following result, which is a direct counterpart of Lemma 4.1.1. Unless explicitly stated otherwise, we assume that Condition (G.2) is valid, and thus the filtration $\mathbb{G}$ is the sub-filtration of $\mathbb{G}^*$.

**Lemma 5.1.2.** (i) *Assume that* (G.2) *holds. Then for any $\mathcal{G}$-measurable random variable $Y$ and any $t \in \mathbb{R}_+$ we have*

$$\mathbb{E}_{\mathbb{P}}(\mathbb{1}_{\{\tau>t\}}Y\,|\,\mathcal{G}_t) = \mathbb{P}\{\tau > t\,|\,\mathcal{G}_t\}\frac{\mathbb{E}_{\mathbb{P}}(\mathbb{1}_{\{\tau>t\}}Y\,|\,\mathcal{F}_t)}{\mathbb{P}\{\tau > t\,|\,\mathcal{F}_t\}}. \tag{5.1}$$

(ii) *If, in addition, $\mathcal{H}_t \subseteq \mathcal{G}_t$ (so that* (G.1) *holds) then*

$$\mathbb{E}_{\mathbb{P}}(\mathbb{1}_{\{\tau>t\}}Y\,|\,\mathcal{G}_t) = \mathbb{1}_{\{\tau>t\}}\mathbb{E}_{\mathbb{P}}(Y\,|\,\mathcal{G}_t) = \mathbb{1}_{\{\tau>t\}}\frac{\mathbb{E}_{\mathbb{P}}(\mathbb{1}_{\{\tau>t\}}Y\,|\,\mathcal{F}_t)}{\mathbb{P}\{\tau > t\,|\,\mathcal{F}_t\}}. \tag{5.2}$$

*In particular, for any $t \le s$*

$$\mathbb{P}\{t < \tau \le s\,|\,\mathcal{G}_t\} = \mathbb{1}_{\{\tau>t\}}\frac{\mathbb{P}\{t < \tau \le s\,|\,\mathcal{F}_t\}}{\mathbb{P}\{\tau > t\,|\,\mathcal{F}_t\}}. \tag{5.3}$$

*Proof.* Since (ii) is a straightforward consequence of (i), it is enough to establish the first statement. Let us denote $C = \{\tau > t\}$. To prove (i), we need to verify that (recall that $\mathcal{F}_t \subseteq \mathcal{G}_t$)

$$\mathbb{E}_{\mathbb{P}}\big(\mathbb{1}_C Y \mathbb{P}(C\,|\,\mathcal{F}_t)\,\big|\,\mathcal{G}_t\big) = \mathbb{E}_{\mathbb{P}}\big(\mathbb{1}_C \mathbb{E}_{\mathbb{P}}(\mathbb{1}_C Y\,|\,\mathcal{F}_t)\,\big|\,\mathcal{G}_t\big).$$

Put another way, we need to show that for any $A \in \mathcal{G}_t$ we have

$$\int_A \mathbb{1}_C Y \mathbb{P}(C\,|\,\mathcal{F}_t)\,d\mathbb{P} = \int_A \mathbb{1}_C \mathbb{E}_{\mathbb{P}}(\mathbb{1}_C Y\,|\,\mathcal{F}_t)\,d\mathbb{P}.$$

In view of Lemma 5.1.1, for any $A \in \mathcal{G}_t$ we have $A \cap C = B \cap C$ for some event $B \in \mathcal{F}_t$, and so

$$\int_A \mathbb{1}_C Y \mathbb{P}(C\,|\,\mathcal{F}_t)\,d\mathbb{P} = \int_{A \cap C} Y \mathbb{P}(C\,|\,\mathcal{F}_t)\,d\mathbb{P} = \int_{B \cap C} Y \mathbb{P}(C\,|\,\mathcal{F}_t)\,d\mathbb{P}$$

$$= \int_B \mathbb{1}_C Y \mathbb{P}(C\,|\,\mathcal{F}_t)\,d\mathbb{P} = \int_B \mathbb{E}_{\mathbb{P}}(\mathbb{1}_C Y\,|\,\mathcal{F}_t)\mathbb{P}(C\,|\,\mathcal{F}_t)\,d\mathbb{P}$$

$$= \int_B \mathbb{E}_{\mathbb{P}}(\mathbb{1}_C \mathbb{E}_{\mathbb{P}}(\mathbb{1}_C Y\,|\,\mathcal{F}_t)\,|\,\mathcal{F}_t)\,d\mathbb{P} = \int_{B \cap C} \mathbb{E}_{\mathbb{P}}(\mathbb{1}_C Y\,|\,\mathcal{F}_t)\,d\mathbb{P}$$

$$= \int_{A \cap C} \mathbb{E}_{\mathbb{P}}(\mathbb{1}_C Y\,|\,\mathcal{F}_t)\,d\mathbb{P} = \int_A \mathbb{1}_C \mathbb{E}_{\mathbb{P}}(\mathbb{1}_C Y\,|\,\mathcal{F}_t)\,d\mathbb{P}.$$

This ends the proof.  □

Assume that (G.1) holds. By virtue of part (ii) in Lemma 5.1.2, for any $\mathcal{G}_t$-measurable random variable $Y$, there exists an $\mathcal{F}_t$-measurable random variable $\tilde{Y}$ such that $\mathbb{1}_{\{\tau>t\}}Y = \mathbb{1}_{\{\tau>t\}}\tilde{Y}$. As already mentioned (see remarks after Lemma 5.1.1), this property can also be derived by approximation arguments. If it is taken for granted, the derivation of (5.2) can be substantially simplified. Indeed, suppose that we know that (the first equality below is obvious)

$$\mathbb{1}_{\{\tau>t\}}\mathbb{E}_{\mathbb{P}}(Y\,|\,\mathcal{G}_t) = \mathbb{E}_{\mathbb{P}}(\mathbb{1}_{\{\tau>t\}}Y\,|\,\mathcal{G}_t) = \mathbb{1}_{\{\tau>t\}}\zeta \qquad (5.4)$$

for some integrable $\mathcal{F}_t$-measurable random variable $\zeta$ such that $\zeta = \mathbb{E}_{\mathbb{P}}(Y\,|\,\mathcal{G}_t)$ on $\{\tau > t\}$. By taking the conditional expectation with respect to $\mathcal{F}_t$ of both terms of the second equality in (5.4), we obtain

$$\mathbb{E}_{\mathbb{P}}\big(\mathbb{E}_{\mathbb{P}}(\mathbb{1}_{\{\tau>t\}}Y\,|\,\mathcal{G}_t)\,\big|\,\mathcal{F}_t\big) = \mathbb{E}_{\mathbb{P}}(\mathbb{1}_{\{\tau>t\}}Y\,|\,\mathcal{F}_t) = \zeta\,\mathbb{P}\{\tau > t\,|\,\mathcal{F}_t\},$$

and this immediately yields (5.2). However, it does not seem to be possible to derive (5.1) using this argument. Since (recall that $\mathbb{P}\{\tau > t\,|\,\mathcal{F}_t\} > 0$)

$$\tilde{Y} = \frac{\mathbb{E}_{\mathbb{P}}(\mathbb{1}_{\{\tau>t\}}Y\,|\,\mathcal{F}_t)}{\mathbb{P}\{\tau > t\,|\,\mathcal{F}_t\}}, \qquad (5.5)$$

we have, as expected, $\tilde{Y} = Y$ when $Y$ is an $\mathcal{F}_t$-measurable random variable.

Before we state the next lemma, let us introduce another auxiliary random variable by setting $\hat{Y} = \mathbb{E}_{\mathbb{P}}(Y\,|\,\mathcal{F}_{\tau-})$, where $\mathcal{F}_{\tau-}$ stands for the $\sigma$-field generated by all events that strictly precede the random time $\tau$ (let us stress that $\tau$ is not necessarily an $\mathbb{F}$-stopping time). Since $\mathcal{F}_0$ is trivial, by definition we have (see, e.g., Dellacherie (1972))

$$\mathcal{F}_{\tau-} = \sigma\big(B \cap \{\tau > t\} : B \in \mathcal{F}_t, \, t \in \mathbb{R}_+\big). \qquad (5.6)$$

In particular, the inclusion $\sigma(\tau) \subseteq \mathcal{F}_{\tau-}$ is always valid, and $\sigma(\tau) = \mathcal{F}_{\tau-}$ when $\mathbb{F} = \mathbb{F}^0$ is the trivial filtration. It is also not difficult to check that $\mathcal{G}_{\tau-} = \mathcal{F}_{\tau-}$. Consequently, the equality $\mathbb{E}_{\mathbb{P}}(Y\,|\,\mathcal{F}_{\tau-}) = \mathbb{E}_{\mathbb{P}}(Y\,|\,\mathcal{G}_{\tau-})$ is valid for any $\mathcal{G}$-measurable integrable random variable $Y$.

**Lemma 5.1.3.** *Let $Y$ be an integrable $\mathcal{G}$-measurable random variable and let $\hat{Y} = \mathbb{E}_{\mathbb{P}}(Y\,|\,\mathcal{F}_{\tau-})$. For any $0 \le t \le s$ we have*

$$\mathbb{E}_{\mathbb{P}}(\mathbb{1}_{\{\tau>t\}}Y\,|\,\mathcal{G}_t) = \mathbb{E}_{\mathbb{P}}(\mathbb{1}_{\{\tau>t\}}\hat{Y}\,|\,\mathcal{G}_t), \qquad (5.7)$$

$$\mathbb{E}_{\mathbb{P}}(\mathbb{1}_{\{t<\tau\le s\}}Y\,|\,\mathcal{G}_t) = \mathbb{E}_{\mathbb{P}}(\mathbb{1}_{\{t<\tau\le s\}}\hat{Y}\,|\,\mathcal{G}_t). \qquad (5.8)$$

*Proof.* Consider an arbitrary event $A \in \mathcal{G}_t$. By virtue of Lemma 5.1.1, we may, and do, assume that $A \cap C = B \cap C$, where we write $C = \{\tau > t\}$. Since $B \cap C$ is manifestly in $\mathcal{F}_{\tau-}$, we have

$$\int_A \mathbb{1}_C Y\,d\mathbb{P} = \int_{A\cap C} Y\,d\mathbb{P} = \int_{B\cap C} Y\,d\mathbb{P} = \int_{B\cap C} \mathbb{E}_{\mathbb{P}}(Y\,|\,\mathcal{F}_{\tau-})\,d\mathbb{P}$$

$$= \int_{A\cap C} \mathbb{E}_{\mathbb{P}}(Y\,|\,\mathcal{F}_{\tau-})\,d\mathbb{P} = \int_A \mathbb{1}_C\,\mathbb{E}_{\mathbb{P}}(Y\,|\,\mathcal{F}_{\tau-})\,d\mathbb{P} = \int_A \mathbb{1}_C\hat{Y}\,d\mathbb{P}.$$

This gives (5.7). For (5.8), notice that the event $\{\tau > s\}$ is in $\mathcal{F}_{\tau-}$.     □

It is apparent that formulae (5.1)–(5.3) can be rewritten as follows:

$$\mathbb{E}_{\mathbb{P}}(\mathbb{1}_{\{\tau>t\}}Y \,|\, \mathcal{G}_t) = \mathbb{P}\{\tau > t \,|\, \mathcal{G}_t\}\, \mathbb{E}_{\mathbb{P}}(\mathbb{1}_{\{\tau>t\}}\, e^{\Gamma_t}Y \,|\, \mathcal{F}_t),$$

$$\mathbb{E}_{\mathbb{P}}(\mathbb{1}_{\{\tau>t\}}Y \,|\, \mathcal{G}_t) = \mathbb{1}_{\{\tau>t\}}\, \mathbb{E}_{\mathbb{P}}(\mathbb{1}_{\{\tau>t\}}\, e^{\Gamma_t}Y \,|\, \mathcal{F}_t) \tag{5.9}$$

and

$$\mathbb{P}\{t < \tau \leq s \,|\, \mathcal{G}_t\} = \mathbb{1}_{\{\tau>t\}}\, \mathbb{E}_{\mathbb{P}}\big(1 - e^{\Gamma_t - \Gamma_s} \,|\, \mathcal{F}_t\big).$$

The next corollary deals with some simple, but useful, modifications of these expressions.

**Corollary 5.1.1.** *Let $Y$ be a $\mathcal{G}$-measurable random variable and let $t \leq s$.*
(i) *Assume that* (G.2) *holds. Then*

$$\mathbb{E}_{\mathbb{P}}(\mathbb{1}_{\{\tau>s\}}\, Y \,|\, \mathcal{G}_t) = \mathbb{P}\{\tau > t \,|\, \mathcal{G}_t\}\, \mathbb{E}_{\mathbb{P}}(\mathbb{1}_{\{\tau>s\}}\, e^{\Gamma_t}Y \,|\, \mathcal{F}_t). \tag{5.10}$$

(ii) *If* (G.1) *is valid then*

$$\mathbb{E}_{\mathbb{P}}(\mathbb{1}_{\{\tau>s\}}\, Y \,|\, \mathcal{G}_t) = \mathbb{1}_{\{\tau>t\}}\, \mathbb{E}_{\mathbb{P}}(\mathbb{1}_{\{\tau>s\}}\, e^{\Gamma_t}Y \,|\, \mathcal{F}_t) \tag{5.11}$$

*and*

$$\mathbb{E}_{\mathbb{P}}(\mathbb{1}_{\{t<\tau\leq s\}}\, Y \,|\, \mathcal{G}_t) = \mathbb{1}_{\{\tau>t\}}\, \mathbb{E}_{\mathbb{P}}(\mathbb{1}_{\{t<\tau\leq s\}}\, e^{\Gamma_t}Y \,|\, \mathcal{F}_t). \tag{5.12}$$

*If $Y$ is $\mathcal{F}_s$-measurable, then*

$$\mathbb{E}_{\mathbb{P}}(\mathbb{1}_{\{\tau>s\}}\, Y \,|\, \mathcal{G}_t) = \mathbb{1}_{\{\tau>t\}}\, \mathbb{E}_{\mathbb{P}}\big(e^{\Gamma_t - \Gamma_s}Y \,|\, \mathcal{F}_t\big) \tag{5.13}$$

*and*

$$\mathbb{E}_{\mathbb{P}}(\mathbb{1}_{\{t<\tau\leq s\}}\, Y \,|\, \mathcal{G}_t) = \mathbb{1}_{\{\tau>t\}}\, \mathbb{E}_{\mathbb{P}}\big((\mathbb{1}_{\{\tau>t\}} - e^{-\Gamma_s})e^{\Gamma_t}Y \,|\, \mathcal{F}_t\big).$$

*Proof.* In view of (5.1), to show that (5.10) holds, it is enough to observe that $\mathbb{1}_{\{\tau>t\}}\mathbb{1}_{\{\tau>s\}} = \mathbb{1}_{\{\tau>s\}}$. Equalities (5.11)–(5.12) are immediate consequences of (5.10). For (5.13), notice that, by virtue of (5.11), we obtain

$$\begin{aligned}
\mathbb{E}_{\mathbb{P}}(\mathbb{1}_{\{\tau>s\}}\, Y \,|\, \mathcal{G}_t) &= \mathbb{1}_{\{\tau>t\}}\, \mathbb{E}_{\mathbb{P}}\big(\mathbb{1}_{\{\tau>s\}}\, e^{\Gamma_t}Y \,|\, \mathcal{F}_t\big) \\
&= \mathbb{1}_{\{\tau>t\}}\, \mathbb{E}_{\mathbb{P}}\big(\mathbb{P}\{t > s \,|\, \mathcal{F}_s\}\, e^{\Gamma_t}Y \,|\, \mathcal{F}_t\big) \\
&= \mathbb{1}_{\{\tau>t\}}\, \mathbb{E}_{\mathbb{P}}\big((1 - F_s)\, e^{\Gamma_t}Y \,|\, \mathcal{F}_t\big) \\
&= \mathbb{1}_{\{\tau>t\}}\, \mathbb{E}_{\mathbb{P}}\big(e^{\Gamma_t - \Gamma_s}Y \,|\, \mathcal{F}_t\big).
\end{aligned}$$

To derive the last formula, it suffices to combine (5.9) with (5.13).  □

It is worth noticing that equality (5.13) remains valid when the random variable $Y$ is merely $\mathcal{G}$-measurable, rather than $\mathcal{F}_s$-measurable, provided that, on the right-hand side of (5.13), we substitute $Y$ with the $\mathcal{F}_s$-measurable random variable $\tilde{Y}$ for which $\mathbb{1}_{\{\tau>s\}}Y = \mathbb{1}_{\{\tau>s\}}\tilde{Y}$. More explicitly, we need to replace $Y$ by $\tilde{Y}$ given by the following expression (cf. (5.5)):

$$\tilde{Y} = \frac{\mathbb{E}_{\mathbb{P}}(\mathbb{1}_{\{\tau>s\}}Y \,|\, \mathcal{F}_s)}{\mathbb{P}\{\tau > s \,|\, \mathcal{F}_s\}}.$$

The proof of the next auxiliary result is essentially the same as the proof of part (i) in Lemma 5.1.2.

**Lemma 5.1.4.** *For any $\mathcal{G}$-measurable random variable $Y$ and any sub-$\sigma$-field $\mathcal{F}$ of $\mathcal{G}$ we have*

$$\mathbb{E}_{\mathbb{P}}(\mathbb{1}_{\{\tau>t\}}Y \mid \mathcal{H}_t \vee \mathcal{F}) = \mathbb{1}_{\{\tau>t\}}\frac{\mathbb{E}_{\mathbb{P}}(\mathbb{1}_{\{\tau>t\}}Y \mid \mathcal{F})}{\mathbb{P}\{\tau>t \mid \mathcal{F}\}}. \qquad (5.14)$$

*For any $t \le s$ we have*

$$\mathbb{P}\{\tau>s \mid \mathcal{H}_t \vee \mathcal{F}\} = \mathbb{1}_{\{\tau>t\}}\frac{\mathbb{P}\{\tau>s \mid \mathcal{F}\}}{\mathbb{P}\{\tau>t \mid \mathcal{F}\}}. \qquad (5.15)$$

Our next goal is to examine the conditional expectation $\mathbb{E}_{\mathbb{P}}(\mathbb{1}_{\{\tau \le t\}}Y \mid \mathcal{G}_t)$. Its evaluation under (G.2) is rather difficult, and thus we shall introduce an alternative condition.

**Condition (G.3)** For any $t \in \mathbb{R}_+$ and arbitrary event $A \in \mathcal{H}_\infty \vee \mathcal{F}_t$ we have $A \cap \{\tau \le t\} \in \mathcal{G}_t$.

Under (G.3), for every $t \in \mathbb{R}_+$ we have $\mathcal{H}_t \subseteq \mathcal{G}_t$. It is easy to see that (G.1) is sufficient for (G.3) to hold; however, (G.2) does not imply (G.3). Finally, conditions (G.2) and (G.3), taken together, imply (G.1).

**Lemma 5.1.5.** *Assume that (G.3) holds. For any $\mathcal{G}$-measurable random variable $Y$ we have*

$$\mathbb{E}_{\mathbb{P}}(\mathbb{1}_{\{\tau \le t\}}Y \mid \mathcal{G}_t) = \mathbb{1}_{\{\tau \le t\}}\mathbb{E}_{\mathbb{P}}(Y \mid \mathcal{G}_t) = \mathbb{1}_{\{\tau \le t\}}\mathbb{E}_{\mathbb{P}}(Y \mid \mathcal{H}_\infty \vee \mathcal{F}_t). \qquad (5.16)$$

*Proof.* Let us denote $D = \{\tau \le t\}$. For any $A \in \mathcal{H}_\infty \vee \mathcal{F}_t$ we have (notice that $D \in \mathcal{G}_t$)

$$\int_A \mathbb{E}_{\mathbb{P}}(\mathbb{1}_D Y \mid \mathcal{H}_\infty \vee \mathcal{F}_t)\, d\mathbb{P} = \int_A \mathbb{1}_D Y\, d\mathbb{P} = \int_{A \cap D} Y\, d\mathbb{P}$$

$$= \int_{A \cap D} \mathbb{E}_{\mathbb{P}}(Y \mid \mathcal{G}_t)\, d\mathbb{P} = \int_A \mathbb{1}_D \mathbb{E}_{\mathbb{P}}(Y \mid \mathcal{G}_t)\, d\mathbb{P}.$$

The random variable $\mathbb{1}_D \mathbb{E}_{\mathbb{P}}(Y \mid \mathcal{G}_t)$ is manifestly $\mathcal{H}_t \vee \mathcal{G}_t$-measurable, so that it is also $\mathcal{H}_\infty \vee \mathcal{F}_t$-measurable. We conclude that (5.16) holds.  $\square$

Unless explicitly stated otherwise, we assume from now on that Condition (G.1) holds, i.e., we consider the case when $\mathbb{G} = \mathbb{H} \vee \mathbb{F}$. By combining (5.16) with (5.2), we obtain the following well known result, which is a straightforward generalization of equality (4.2).

**Corollary 5.1.2.** *For any $\mathcal{G}$-measurable random variable $Y$ we have*

$$\mathbb{E}_{\mathbb{P}}(Y \mid \mathcal{G}_t) = \mathbb{1}_{\{\tau \le t\}}\mathbb{E}_{\mathbb{P}}(Y \mid \mathcal{H}_\infty \vee \mathcal{F}_t) + \mathbb{1}_{\{\tau>t\}}\mathbb{E}_{\mathbb{P}}(\mathbb{1}_{\{\tau>t\}}e^{\Gamma_t}Y \mid \mathcal{F}_t).$$

*Any $\mathcal{G}_t$-measurable random variable $Y$ admits the following representation*

$$Y = \mathbb{1}_{\{\tau \le t\}}\mathbb{E}_{\mathbb{P}}(Y \mid \mathcal{H}_\infty \vee \mathcal{F}_t) + \mathbb{1}_{\{\tau>t\}}\mathbb{E}_{\mathbb{P}}(\mathbb{1}_{\{\tau>t\}}e^{\Gamma_t}Y \mid \mathcal{F}_t).$$

**Proposition 5.1.1.** (i) *Let* $h : \mathbb{R}_+ \to \mathbb{R}$ *be a bounded, continuous function. Then for any* $t < s \le \infty$

$$\mathbb{E}_{\mathbb{P}}(\mathbb{1}_{\{t < \tau \le s\}} h(\tau) \mid \mathcal{G}_t) = \mathbb{1}_{\{\tau > t\}} e^{\Gamma_t} \mathbb{E}_{\mathbb{P}}\left(\int_{]t,s]} h(u) \, dF_u \mid \mathcal{F}_t\right). \qquad (5.17)$$

(ii) *Let* $Z$ *be a bounded, $\mathbb{F}$-predictable process. Then for any* $t < s \le \infty$

$$\mathbb{E}_{\mathbb{P}}(\mathbb{1}_{\{t < \tau \le s\}} Z_\tau \mid \mathcal{G}_t) = \mathbb{1}_{\{\tau > t\}} e^{\Gamma_t} \mathbb{E}_{\mathbb{P}}\left(\int_{]t,s]} Z_u \, dF_u \mid \mathcal{F}_t\right). \qquad (5.18)$$

*Proof.* In view of (5.12), to establish (5.17), it is enough to check that

$$\mathbb{E}_{\mathbb{P}}(\mathbb{1}_{\{t < \tau \le s\}} h(\tau) \mid \mathcal{F}_t) = \mathbb{E}_{\mathbb{P}}\left(\int_{]t,s]} h(u) \, dF_u \mid \mathcal{F}_t\right).$$

We first consider a piecewise constant function $h(u) = \sum_{i=0}^n h_i \mathbb{1}_{\{t_i < u \le t_{i+1}\}}$, where, without loss of generality, we take $t_0 = t < \cdots < t_{n+1} = s$. Then

$$\mathbb{E}_{\mathbb{P}}(\mathbb{1}_{\{t < \tau \le s\}} h(\tau) \mid \mathcal{F}_t) = \sum_{i=0}^n \mathbb{E}_{\mathbb{P}}\big(\mathbb{E}_{\mathbb{P}}(h_i \mathbb{1}_{]t_i, t_{i+1}]}(\tau) \mid \mathcal{F}_{t_{i+1}}) \mid \mathcal{F}_t\big)$$

$$= \mathbb{E}_{\mathbb{P}}\left(\sum_{i=0}^n h_i (F_{t_{i+1}} - F_{t_i}) \mid \mathcal{F}_t\right)$$

$$= \mathbb{E}_{\mathbb{P}}\left(\sum_{i=0}^n \int_{]t_i, t_{i+1}]} h(u) \, dF_u \mid \mathcal{F}_t\right)$$

$$= \mathbb{E}_{\mathbb{P}}\left(\int_{]t,s]} h(u) \, dF_u \mid \mathcal{F}_t\right).$$

To complete the proof of part (i), it suffices to approximate an arbitrary continuous function $h$ by a suitable sequence of piecewise constant functions.

The proof of (5.18) relies on similar arguments. We begin by assuming that $Z$ is a stepwise $\mathbb{F}$-predictable process; that is, $Z_u = \sum_{i=0}^n Z_{t_i} \mathbb{1}_{\{t_i < u \le t_{i+1}\}}$ for $t < u \le s$, where $t_0 = t < \cdots < t_{n+1} = s$, and $Z_{t_i}$ is $\mathcal{F}_{t_i}$-measurable random variable for $i = 0, \ldots, n$. We have

$$\mathbb{E}_{\mathbb{P}}\big(\mathbb{1}_{\{t < \tau \le s\}} Z_\tau \mid \mathcal{F}_t\big) = \mathbb{E}_{\mathbb{P}}\left(\sum_{i=0}^n \mathbb{E}_{\mathbb{P}}(\mathbb{1}_{\{t_i < \tau \le t_{i+1}\}} Z_{t_i} \mid \mathcal{F}_{t_{i+1}}) \mid \mathcal{F}_t\right)$$

$$= \mathbb{E}_{\mathbb{P}}\left(\sum_{i=0}^n \mathbb{1}_{\{t_i < \tau \le t_{i+1}\}} Z_{t_i} \mid \mathcal{F}_t\right) = \mathbb{E}_{\mathbb{P}}\left(\sum_{i=0}^n Z_{t_i} (F_{t_{i+1}} - F_{t_i}) \mid \mathcal{F}_t\right).$$

Consequently, for any stepwise, bounded, $\mathbb{F}$-predictable process $Z$ we have

$$\mathbb{E}_{\mathbb{P}}\big(\mathbb{1}_{\{t < \tau \le s\}} Z_\tau \mid \mathcal{F}_t\big) = \mathbb{E}_{\mathbb{P}}\left(\int_{]t,s]} Z_u \, dF_u \mid \mathcal{F}_t\right). \qquad (5.19)$$

In the second step, $Z$ is approximated by a suitable sequence of bounded, stepwise, $\mathbb{F}$-predictable processes. The sum under the sign of the conditional expectation converges to the Itô integral (or to the Lebesque-Stieltjes integral if $F$ is of finite variation). The boundedness of $Z$ and $F$ is a sufficient condition for the convergence of the sequence of conditional expectations.    □

For the validity of (5.17), it suffices to assume that the function $h$ is piecewise continuous. Also, the boundedness of the function $h$ (the process $Z$, resp.) is not a necessary condition for (5.17) ((5.18), resp.) to hold; we have imposed this rather restrictive condition for the sake of convenience. On the other hand, in general the $\mathbb{F}$-predictability of $Z$ cannot be replaced by the weaker condition of the $\mathbb{G}$-predictability of $Z$ in Proposition 5.1.1.

**Corollary 5.1.3.** *Under the assumptions of Proposition 5.1.1, if, in addition, the hazard process $\Gamma$ of $\tau$ is continuous, then*

$$\mathbb{E}_{\mathbb{P}}(\mathbb{1}_{\{t<\tau\leq s\}}h(\tau)\,|\,\mathcal{G}_t) = \mathbb{1}_{\{\tau>t\}}\,\mathbb{E}_{\mathbb{P}}\left(\int_t^s h(u)e^{\Gamma_t-\Gamma_u}\,d\Gamma_u\,\Big|\,\mathcal{F}_t\right) \qquad (5.20)$$

*and*

$$\mathbb{E}_{\mathbb{P}}(\mathbb{1}_{\{t<\tau\leq s\}}Z_\tau\,|\,\mathcal{G}_t) = \mathbb{1}_{\{\tau>t\}}\,\mathbb{E}_{\mathbb{P}}\left(\int_t^s Z_u e^{\Gamma_t-\Gamma_u}\,d\Gamma_u\,\Big|\,\mathcal{F}_t\right). \qquad (5.21)$$

*Proof.* Under the present assumptions, $dF_u = e^{-\Gamma_u}\,d\Gamma_u$, and thus equality (5.20) ((5.21), resp.) is an immediate consequence of (5.17) ((5.18), resp.) □

**Case of a $\mathcal{G}$-measurable random variable.** Let us return to the general case of a $\mathcal{G}$-measurable (bounded) random variable. The following natural and practically important question arises: is it possible to derive an expression similar to (5.21), when $Z_\tau$ is replaced by a $\mathcal{G}$-measurable random variable. We claim that the answer to this question is positive. To show this, we proceed as follows. First, we associate with $Y$ the conditional expectation $\hat{Y} = \mathbb{E}_{\mathbb{P}}(Y\,|\,\mathcal{F}_{\tau-}) = \mathbb{E}_{\mathbb{P}}(Y\,|\,\mathcal{G}_{\tau-})$, where the $\sigma$-field $\mathcal{F}_{\tau-} = \mathcal{G}_{\tau-}$ of all events strictly preceding $\tau$ is formally defined by (5.6). It is known that there exists an $\mathbb{F}$-predictable process $\hat{Z}$ such that $\hat{Z}_\tau = \hat{Y}$ (see p. 126 in Dellacherie and Meyer (1978a)). The following chain of equalities is thus valid (cf. (5.8))

$$\mathbb{E}_{\mathbb{P}}(\mathbb{1}_{\{t<\tau\leq s\}}\,Y\,|\,\mathcal{G}_t) = \mathbb{E}_{\mathbb{P}}(\mathbb{1}_{\{t<\tau\leq s\}}\,\hat{Y}\,|\,\mathcal{G}_t)$$

$$= \mathbb{E}_{\mathbb{P}}(\mathbb{1}_{\{t<\tau\leq s\}}\,\hat{Z}_\tau\,|\,\mathcal{G}_t) = \mathbb{1}_{\{\tau>t\}}e^{\Gamma_t}\,\mathbb{E}_{\mathbb{P}}\left(\int_{]t,s]} \hat{Z}_u\,dF_u\,\Big|\,\mathcal{F}_t\right) \qquad (5.22)$$

$$= \mathbb{1}_{\{\tau>t\}}\,\mathbb{E}_{\mathbb{P}}\left(\int_t^s \hat{Z}_u e^{\Gamma_t-\Gamma_u}\,d\Gamma_u\,\Big|\,\mathcal{F}_t\right),$$

where the last equality holds, provided that the hazard process $\Gamma$ is continuous. It is noteworthy that the uniqueness of the process $\hat{Z}$ is neither claimed, nor required here. In the case when several bounded, $\mathbb{F}$-predictable processes $\hat{Z}$ satisfying the equality $\hat{Z}_\tau = \hat{Y}$ exist, they all yield the same result for the conditional expectation we are interested in.

The next result appears to be useful for the valuation of a defaultable security that promises to pay dividends prior to the default time.

**Proposition 5.1.2.** *Assume that $A$ is a bounded, $\mathbb{F}$-predictable process of finite variation. Then for every $t \leq s$*

$$\mathbb{E}_{\mathbb{P}}\left(\int_{]t,s]} (1 - H_u)\, dA_u \,\Big|\, \mathcal{G}_t\right) = \mathbb{1}_{\{\tau > t\}} e^{\Gamma_t} \mathbb{E}_{\mathbb{P}}\left(\int_{]t,s]} (1 - F_u)\, dA_u \,\Big|\, \mathcal{F}_t\right)$$

*or, equivalently,*

$$\mathbb{E}_{\mathbb{P}}\left(\int_{]t,s]} (1 - H_u)\, dA_u \,\Big|\, \mathcal{G}_t\right) = \mathbb{1}_{\{\tau > t\}} \mathbb{E}_{\mathbb{P}}\left(\int_{]t,s]} e^{\Gamma_t - \Gamma_u}\, dA_u \,\Big|\, \mathcal{F}_t\right).$$

*Proof.* For a fixed, but arbitrary, $t \leq s$, we introduce an auxiliary process $\tilde{A}$ by setting: $\tilde{A}_u = A_u - A_t$ for $u \in [t, s]$. It is clear that $\tilde{A}$ is a bounded, $\mathbb{F}$-predictable process of finite variation; the same remark applies to the process of left-hand limits: $\tilde{A}_{t-}$. Therefore,

$$J_t := \mathbb{E}_{\mathbb{P}}\left(\int_{]t,s]} (1 - H_u)\, dA_u \,\Big|\, \mathcal{G}_t\right)$$

$$= \mathbb{E}_{\mathbb{P}}\left(\int_{]t,s]} \mathbb{1}_{\{\tau > u\}}\, d\tilde{A}_u \,\Big|\, \mathcal{G}_t\right)$$

$$= \mathbb{E}_{\mathbb{P}}\left(\tilde{A}_{\tau-} \mathbb{1}_{\{t < \tau \leq s\}} + \tilde{A}_s \mathbb{1}_{\{\tau > s\}} \,\Big|\, \mathcal{G}_t\right)$$

$$= \mathbb{1}_{\{\tau > t\}} e^{\Gamma_t} \mathbb{E}_{\mathbb{P}}\left(\int_{]t,s]} \tilde{A}_{u-}\, dF_u + \tilde{A}_s(1 - F_s) \,\Big|\, \mathcal{F}_t\right),$$

where the last equality follows from formulae (5.13) and (5.18). Using an obvious equality $G_t = 1 - F_t$, we obtain

$$\mathbb{E}_{\mathbb{P}}\left(\int_{]t,s]} \tilde{A}_{u-}\, dF_u + \tilde{A}_s(1 - F_s) \,\Big|\, \mathcal{F}_t\right) = \mathbb{E}_{\mathbb{P}}\left(-\int_{]t,s]} \tilde{A}_{u-}\, dG_u + \tilde{A}_s G_s \,\Big|\, \mathcal{F}_t\right).$$

Since $\tilde{A}$ follows a process of finite variation (so that its continuous martingale part vanishes), the following version of Itô's product rule is in force

$$\tilde{A}_s G_s = \tilde{A}_t G_t + \int_{]t,s]} \tilde{A}_{u-}\, dG_u + \int_{]t,s]} G_u\, d\tilde{A}_u.$$

But $\tilde{A}_t = 0$, and so

$$\mathbb{E}_{\mathbb{P}}\left(\int_{]t,s]} \tilde{A}_{u-}\, dF_u + \tilde{A}_s(1 - F_s) \,\Big|\, \mathcal{F}_t\right) = \mathbb{E}_{\mathbb{P}}\left(\int_{]t,s]} (1 - F_u)\, dA_u \,\Big|\, \mathcal{F}_t\right).$$

This proves the first asserted formula. The second equality is a simple reformulation of the first. □

### 5.1.2 Semimartingale Representation of the Stopped Process

In the next auxiliary lemma, we assume that an $\mathbb{F}$-predictable process $m$ follows an $\mathbb{F}$-martingale. We are interested in the semimartingale decomposition of the stopped process $\tilde{m}_t := m_{t \wedge \tau}$, $t \in \mathbb{R}_+$, with respect to the enlarged filtration $\mathbb{G}$. We shall occasionally use the commonly standard notation $m^\tau$ for the process $m$ stopped at $\tau$; specifically, we set $m_t^\tau = m_{t \wedge \tau}$ for every $t \in \mathbb{R}_+$.

**Lemma 5.1.6.** *Assume that a process $m$ follows a predictable martingale with respect to the filtration $\mathbb{F}$. Then the following implications are valid.*
*(i) If $F$ is an increasing process, then the stopped process $\tilde{m}_t = m_{t \wedge \tau}$ is a $\mathbb{G}$-martingale.*
*(ii) If $F$ is a continuous submartingale, then the process*

$$\hat{m}_t = \tilde{m}_t + \int_0^{t \wedge \tau} (1 - F_u)^{-1} \, d\langle m, F \rangle_u \tag{5.23}$$

*is a $\mathbb{G}$-martingale.*
*(iii) If $m$ is a continuous process, then the process $\hat{m}$ given by (5.23) follows a $\mathbb{G}$-martingale.*

*Proof.* To establish the first statement, we fix $s > 0$ and we define an $\mathbb{F}$-adapted process $\bar{m}$ by setting $\bar{m}_t = m_{t \wedge s}$ for any $t \in \mathbb{R}_+$. It is clear that for any $t \leq s$ we have $\mathbb{E}_{\mathbb{P}}(\tilde{m}_s \,|\, \mathcal{G}_t) = \mathbb{E}_{\mathbb{P}}(\bar{m}_\tau \,|\, \mathcal{G}_t)$. Furthermore,

$$\mathbb{E}_{\mathbb{P}}(\bar{m}_\tau \,|\, \mathcal{G}_t) = \mathbb{1}_{\{\tau \leq t\}} \bar{m}_\tau + \mathbb{1}_{\{\tau > t\}} e^{\Gamma_t} \mathbb{E}_{\mathbb{P}}\left( \int_{]t,\infty]} \bar{m}_u \, dF_u \,\Big|\, \mathcal{F}_t \right)$$

$$= \mathbb{1}_{\{\tau \leq t\}} m_{\tau \wedge s} + \mathbb{1}_{\{\tau > t\}} e^{\Gamma_t} \mathbb{E}_{\mathbb{P}}\left( \int_{]t,\infty]} m_{u \wedge s} \, dF_u \,\Big|\, \mathcal{F}_t \right)$$

$$= \mathbb{1}_{\{\tau \leq t\}} m_\tau + \mathbb{1}_{\{\tau > t\}} e^{\Gamma_t} J_t,$$

where

$$J_t = \mathbb{E}_{\mathbb{P}}\left( \int_{]t,\infty]} m_{u \wedge s} \, dF_u \,\Big|\, \mathcal{F}_t \right) = -\mathbb{E}_{\mathbb{P}}\left( \int_{]t,s]} m_u \, dG_u + \int_{]s,\infty]} m_s \, dG_u \,\Big|\, \mathcal{F}_t \right),$$

where we write, as usual, $G_t := e^{-\Gamma_t} = 1 - F_t$. Since $G$ is a bounded, $\mathbb{F}$-adapted, decreasing process, by combining the Itô product rule

$$G_s m_s = G_t m_t + \int_{]t,s]} m_u \, dG_u + \int_{]t,s]} G_{u-} \, dm_u$$

with the $\mathbb{F}$-martingale property of $m$, we obtain

$$J_t = \mathbb{E}_{\mathbb{P}}\left( G_t m_t - G_s m_s + \int_{]t,s]} G_{u-} \, dm_u - m_s(G_\infty - G_s) \,\Big|\, \mathcal{F}_t \right) = G_t m_t = e^{-\Gamma_t} m_t,$$

where we have also used the equality $G_\infty = e^{-\Gamma_\infty} = 0$. Consequently, we get

$$\mathbb{E}_{\mathbb{P}}(\tilde{m}_s \,|\, \mathcal{G}_t) = \mathbb{1}_{\{\tau \leq t\}} m_\tau + \mathbb{1}_{\{\tau > t\}} m_t = m_t^\tau = \tilde{m}_t,$$

which is the desired result. This completes the proof of part (i) of the lemma.

For the second statement, we denote, as before, $G_t = e^{-\Gamma_t} = 1 - F_t$, and we note that $G$ now follows a bounded, continuous $\mathbb{F}$-supermartingale. For a fixed $s > 0$, we introduce an auxiliary process $\bar{m}$

$$\bar{m}_t := m_{t \wedge s} + \int_0^{t \wedge s} G_u^{-1} \, d\langle m, F \rangle_u = m_{t \wedge s} + m_{t \wedge s}^*,$$

where we set $m_t^* = \int_0^t G_u^{-1} d\langle m, F \rangle_u$. Obviously, for any $t \leq s$ we have $\mathbb{E}_{\mathbb{P}}(\hat{m}_s \,|\, \mathcal{G}_t) = \mathbb{E}_{\mathbb{P}}(\bar{m}_\tau \,|\, \mathcal{G}_t)$. It suffices thus to show that $\mathbb{E}_{\mathbb{P}}(\bar{m}_\tau \,|\, \mathcal{G}_t) = \hat{m}_t$. To this end, notice that

$$\mathbb{E}_{\mathbb{P}}(\bar{m}_\tau \,|\, \mathcal{G}_t) = \mathbb{1}_{\{\tau \leq t\}} \hat{m}_\tau + e^{\Gamma_t} \mathbb{1}_{\{\tau > t\}} (J_t^1 + J_t^2),$$

where

$$J_t^1 := \mathbb{E}_{\mathbb{P}}\left( \int_t^\infty m_{u \wedge s} \, dF_u \,\Big|\, \mathcal{F}_t \right) = -\mathbb{E}_{\mathbb{P}}\left( \int_t^s m_u \, dG_u + \int_s^\infty m_s \, dG_u \,\Big|\, \mathcal{F}_t \right)$$

and

$$J_t^2 := \mathbb{E}_{\mathbb{P}}\left( \int_t^\infty m_{u \wedge s}^* \, dF_u \,\Big|\, \mathcal{F}_t \right) = -\mathbb{E}_{\mathbb{P}}\left( \int_t^s m_u^* \, dG_u + \int_s^\infty m_s^* \, dG_u \,\Big|\, \mathcal{F}_t \right).$$

For $J_t^1$, using Itô's formula, and taking into account the continuity of $G$ and the martingale property of $m$, we get (recall that $G_\infty = 1 - F_\infty = 0$)

$$J_t^1 = \mathbb{E}_{\mathbb{P}}\left( -\int_t^s m_{u-} \, dG_u - m_s(G_\infty - G_s) \,\Big|\, \mathcal{F}_t \right)$$

$$= \mathbb{E}_{\mathbb{P}}\left( G_t m_t - G_s m_s + \int_{]t,s]} G_u \, dm_u + \int_t^s d\langle G, m \rangle_u - m_s(G_\infty - G_s) \,\Big|\, \mathcal{F}_t \right)$$

$$= G_t m_t + \mathbb{E}_{\mathbb{P}}(\langle m, G \rangle_s \,|\, \mathcal{F}_t) - \langle m, G \rangle_t$$

$$= e^{-\Gamma_t} m_t + \langle m, F \rangle_t - \mathbb{E}_{\mathbb{P}}(\langle m, F \rangle_s \,|\, \mathcal{F}_t).$$

On the other hand, since $m^*$ is a continuous process of finite variation, another application of Itô's formula yields

$$J_t^2 = \mathbb{E}_{\mathbb{P}}\left( -\int_t^s m_{u-}^* \, dG_u - m_s^*(G_\infty - G_s) \,\Big|\, \mathcal{F}_t \right)$$

$$= \mathbb{E}_{\mathbb{P}}\left( G_t m_t^* - G_s m_s^* + \int_t^s G_u \, dm_u^* - m_s^*(G_\infty - G_s) \,\Big|\, \mathcal{F}_t \right)$$

$$= e^{-\Gamma_t} m_t^* - \langle m, F \rangle_t + \mathbb{E}_{\mathbb{P}}(\langle m, F \rangle_s \,|\, \mathcal{F}_t),$$

where the last equality is a consequence of the definition of $m^*$. Upon simplification, we obtain, for any $t \leq s$,

$$\mathbb{E}_{\mathbb{P}}(\bar{m}_\tau \,|\, \mathcal{G}_t) = \mathbb{1}_{\{\tau \leq t\}} \hat{m}_\tau + \mathbb{1}_{\{\tau > t\}} (m_t + m_t^*) = \hat{m}_t.$$

This completes the proof of part (ii). The proof of part (iii) is similar. The bounded submartingale $F$ is no longer assumed to follow a continuous process, but, in view of the continuity of $m$, we have $[m, F] = \langle m, F \rangle$.    □

### 5.1.3 Martingales Associated with the Hazard Process $\Gamma$

In this section, we assume that (G.1) is valid. The next result is a generalization of Corollary 4.1.2 and Lemma 4.1.2. Let us stress that the case when $\mathbb{F} = \mathbb{G}$ is not covered by Lemma 5.1.7.

**Lemma 5.1.7.** *The process $L$, given by the formula*

$$L_t := \mathbb{1}_{\{\tau>t\}}e^{\Gamma_t} = (1 - H_t)e^{\Gamma_t} = \frac{1 - H_t}{1 - F_t}, \qquad (5.24)$$

*follows a $\mathbb{G}$-martingale. Furthermore, for any bounded $\mathbb{F}$-martingale $m$, the product $Lm$ is a $\mathbb{G}$-martingale. If, in addition, $m$ follows also a $\mathbb{G}$-martingale then the quadratic covariation $[L, m]$, which equals*

$$[L, m]_t := L_t m_t - L_0 m_0 - \int_{]0,t]} L_{s-}\, dm_s - \int_{]0,t]} m_{s-}\, dL_s, \qquad (5.25)$$

*follows a $\mathbb{G}$-martingale.*

*Proof.* It is enough to check that we have, for any $t \le s$,

$$\mathbb{E}_{\mathbb{P}}(\mathbb{1}_{\{\tau>s\}}e^{\Gamma_s} \mid \mathcal{G}_t) = \mathbb{1}_{\{\tau>t\}}e^{\Gamma_t}.$$

In view of (5.11), this can be rewritten as follows:

$$\mathbb{1}_{\{\tau>t\}}e^{\Gamma_t}\,\mathbb{E}_{\mathbb{P}}(\mathbb{1}_{\{\tau>s\}}e^{\Gamma_s} \mid \mathcal{F}_t) = \mathbb{1}_{\{\tau>t\}}e^{\Gamma_t}.$$

To complete the proof of the first statement, it suffices to observe that

$$\mathbb{E}_{\mathbb{P}}(\mathbb{1}_{\{\tau>s\}}e^{\Gamma_s} \mid \mathcal{F}_t) = \mathbb{E}_{\mathbb{P}}(e^{\Gamma_s}\mathbb{E}_{\mathbb{P}}(\mathbb{1}_{\{\tau>s\}} \mid \mathcal{F}_s) \mid \mathcal{F}_t) = 1.$$

For the second part of the lemma, notice that in view of (5.2), for $t \le s$ we have

$$\mathbb{E}_{\mathbb{P}}(L_s m_s \mid \mathcal{G}_t) = \mathbb{E}_{\mathbb{P}}(\mathbb{1}_{\{\tau>t\}}L_s m_s \mid \mathcal{G}_t) = \mathbb{1}_{\{\tau>t\}}e^{\Gamma_t}\,\mathbb{E}_{\mathbb{P}}(\mathbb{1}_{\{\tau>s\}}e^{\Gamma_s}m_s \mid \mathcal{F}_t)$$

$$= \mathbb{1}_{\{\tau>t\}}e^{\Gamma_t}\,\mathbb{E}_{\mathbb{P}}(m_s e^{\Gamma_s}\mathbb{E}_{\mathbb{P}}(\mathbb{1}_{\{\tau>s\}} \mid \mathcal{F}_s) \mid \mathcal{F}_t) = (1 - H_t)e^{\Gamma_t}m_t = L_t m_t,$$

and thus $Lm$ follows a $\mathbb{G}$-martingale. The last statement is obvious. $\quad\square$

It is known that if the square-integrable $\mathbb{G}$-martingales $L$ and $m$ are mutually orthogonal (i.e., when the product $Lm$ follows a $\mathbb{G}$-martingale), then the predictable covariation $\langle L, m \rangle$ vanishes, and thus the quadratic covariation $[L, m]$ is a $\mathbb{G}$-martingale (see, for instance, Jacod (1979)).

Under the assumptions of Lemma 5.1.7, if $\Gamma$ is an increasing process (so that $L$ is a process of finite variation), we have $[L, m]_t = \sum_{u\le t} \Delta L_u \Delta m_u$. In this case, formula (5.25) can be rewritten as follows:

$$L_t m_t = L_0 m_0 + \int_{]0,t]} L_{s-}\, dm_s + \int_{]0,t]} m_s\, dL_s. \qquad (5.26)$$

In the next result, we deal with the continuous case; more precisely, we assume that $\Gamma$ is a continuous, increasing process. The following proposition is a direct counterpart of Propositions 4.2.1 and 4.2.2.

**Proposition 5.1.3.** *Assume that the* $\mathbb{F}$*-hazard process* $\Gamma$ *of* $\tau$ *follows a continuous, increasing process. Then the following statements are valid.*
(i) *The process* $\hat{M}_t = H_t - \Gamma_{t \wedge \tau}$ *follows a* $\mathbb{G}$*-martingale, specifically,*

$$\hat{M}_t = - \int_{]0,t]} e^{-\Gamma_u} \, dL_u. \tag{5.27}$$

*Furthermore,* $L_t = \mathcal{E}_t(-\hat{M})$, *that is,* $L$ *solves the linear integral equation*

$$L_t = 1 - \int_{]0,t]} L_{u-} \, d\hat{M}_u. \tag{5.28}$$

(ii) *If a bounded* $\mathbb{F}$*-martingale* $m$ *is also a* $\mathbb{G}$*-martingale*[1] *then the product* $\hat{M}m$ *is a* $\mathbb{G}$*-martingale; i.e.,* $\hat{M}$ *and* $m$ *are mutually orthogonal* $\mathbb{G}$*-martingales.*
(iii) *If* $m$ *is a bounded, predictable,* $\mathbb{F}$*-martingale then the product* $\hat{M}\tilde{m}$ *is a* $\mathbb{G}$*-martingale, where* $\tilde{m}$ *is the stopped process; that is,* $\tilde{m}_t = m_{t \wedge \tau}$ *for* $t \in \mathbb{R}_+$.

*Proof.* We shall first prove (i). The martingale property of $\hat{M}$ and equalities (5.27)–(5.28) can be shown by using the same arguments as those employed in the proof of Proposition 4.2.1, i.e., the integration by parts formula, combined with the definition of the $\mathbb{G}$-martingale $L$.

Let $g$ and $h$ be the two adapted processes, whose sample paths are right-continuous, with left-hand side limits. If $g$ and $h$ are of finite variation on $[0, t]$, then the following equality is valid (see Lemma 4.2.2)

$$g_t h_t = g_0 h_0 + \int_{]0,t]} g_u \, dh_u + \int_{]0,t]} h_{u-} \, dg_u,$$

where both integrals can be seen either as the Lebesgue-Stieltjes integrals or as the Itô integrals. In view of the assumed continuity of $\Gamma$, by taking $g_t = 1 - H_t$ and $h_t = e^{\Gamma_t}$, we obtain

$$L_t = (1 - H_t)e^{\Gamma_t} = 1 + \int_{]0,t]} e^{\Gamma_u} \big( (1 - H_u) \, d\Gamma_u - dH_u \big). \tag{5.29}$$

Consequently,

$$\hat{M}_t = H_t - \Gamma_{t \wedge \tau} = \int_{]0,t]} \big( dH_u - (1 - H_u) \, d\Gamma_u \big) = - \int_{]0,t]} e^{-\Gamma_u} \, dL_u,$$

and so, by virtue of Lemma 5.1.7, $\hat{M}$ is a $\mathbb{G}$-martingale. Notice also that (5.29) may be rewritten as follows:

$$L_t = 1 - \int_{]0,t]} e^{\Gamma_u} (1 - H_{u-}) \, d(H_u - \Gamma_{u \wedge \tau}) = 1 - \int_{]0,t]} L_{u-} \, d\hat{M}_u.$$

The proof of part (i) is thus completed.

---

[1] This holds, for instance, when the filtrations $\mathbb{F}$ and $\mathbb{G}$ have the so-called *martingale invariance property* under $\mathbb{P}$ (see Sect. 6.1.1).

Let us now prove (ii). By virtue of Lemma 5.1.7, $L$ and $m$ are orthogonal $\mathbb{G}$-martingales, and thus the quadratic covariation $[L, m]$ is also a $\mathbb{G}$-martingale. Consequently, using Itô's formula and (5.27), we get

$$
\begin{aligned}
\hat{M}_t m_t &= \hat{M}_0 m_0 + \int_{]0,t]} \hat{M}_{u-}\, dm_u + \int_{]0,t]} m_{u-}\, d\hat{M}_u + [\hat{M}, m]_t \\
&= \hat{M}_0 m_0 + \int_{]0,t]} \hat{M}_{u-}\, dm_u - \int_{]0,t]} m_{u-} e^{-\Gamma_u}\, dL_u - \int_{]0,t]} e^{-\Gamma_u}\, d[L, m]_u,
\end{aligned}
$$

where the integrands $L, m$ and $[L, m]$ follow $\mathbb{G}$-martingales.

To prove part (iii), we shall make use again of Itô's formula. If $m$ is a bounded, predictable, $\mathbb{F}$-martingale, then, by virtue of part (i) in Lemma 5.1.6, the stopped process $\tilde{m} = m^\tau$ is a $\mathbb{G}$-martingale. In view of Lemma 5.1.7, the product $Lm$ is a $\mathbb{G}$-martingale, and thus the stopped process $(Lm)^\tau = \tilde{L}\tilde{m}$, where we denote $\tilde{L}_t = L_{t \wedge \tau}$, is also a $\mathbb{G}$-martingale. Since

$$
[\tilde{L}, \tilde{m}]_t := \tilde{L}_t \tilde{m}_t - \tilde{L}_0 \tilde{m}_0 - \int_{]0,t]} \tilde{L}_{s-}\, d\tilde{m}_s - \int_{]0,t]} \tilde{m}_{s-}\, d\tilde{L}_s,
$$

we conclude that the quadratic covariation $[\tilde{L}, \tilde{m}]$ follows a $\mathbb{G}$-martingale. Consequently,

$$
\begin{aligned}
\hat{M}_t \tilde{m}_t &= \hat{M}_0 \tilde{m}_0 + \int_{]0,t]} \hat{M}_{u-}\, d\tilde{m}_u + \int_{]0,t]} \tilde{m}_{u-}\, d\hat{M}_u + [\hat{M}, \tilde{m}]_t \\
&= \hat{M}_0 \tilde{m}_0 + \int_{]0,t]} \hat{M}_{u-}\, d\tilde{m}_u - \int_{]0,t]} m_{u-} e^{-\Gamma_u}\, d\tilde{L}_u - \int_{]0,t]} e^{-\Gamma_u}\, d[L, \tilde{m}]_u \\
&= \hat{M}_0 \tilde{m}_0 + \int_{]0,t]} \hat{M}_{u-}\, d\tilde{m}_u - \int_{]0,t]} m_{u-} e^{-\Gamma_u}\, d\tilde{L}_u - \int_{]0,t]} e^{-\Gamma_u}\, d[\tilde{L}, \tilde{m}]_u,
\end{aligned}
$$

where we have used the well-known property of the quadratic covariation:

$$
[L, \tilde{m}]_t = [L, m^\tau]_t = [L, m]_{t \wedge \tau} = [L^\tau, m^\tau]_t = [\tilde{L}, \tilde{m}]_t.
$$

We conclude that the product $\hat{M}\tilde{m}$ is a $\mathbb{G}$-martingale. $\qquad\square$

**Corollary 5.1.4.** *For any bounded $\mathbb{F}$-predictable process $Z$, the following processes are $\mathbb{G}$-martingales:*

$$
V_t^1 = Z_\tau \mathbb{1}_{\{\tau \le t\}} - \int_0^{t \wedge \tau} Z_u\, d\Gamma_u = \int_0^t Z_u\, d\hat{M}_u, \tag{5.30}
$$

*and*

$$
V_t^2 = \exp\left(\mathbb{1}_{\{\tau \le t\}} Z_\tau\right) - \int_0^{t \wedge \tau} \left(e^{Z_u} - 1\right) d\Gamma_u. \tag{5.31}
$$

*Proof.* It is apparent that $V^1$ is a $\mathbb{G}$-martingale. The martingale property of $V^2$ is in turn an easy consequence of the martingale property of $V^1$. $\qquad\square$

*Remarks.* If the continuous process $\Gamma$ is not of finite variation, formula (5.29) becomes

$$L_t = (1 - H_t)e^{\Gamma_t} = 1 + \int_0^t e^{\Gamma_u}\big((1 - H_u)\,(d\Gamma_u + \tfrac{1}{2}\langle\Gamma\rangle_u) - dH_u\big)$$

and the process $\hat{M}$ is no longer a $\mathbb{G}$-martingale (but, obviously, the process $H_t - \Gamma_{t\wedge\tau} - \frac{1}{2}\langle\Gamma\rangle_{t\wedge\tau}$ is a $\mathbb{G}$-martingale).

### 5.1.4 Stochastic Intensity of a Random Time

Let us now consider the most widely used in practical applications case of an absolutely continuous $\mathbb{F}$-hazard process $\Gamma$. Assume that $\Gamma_t = \int_0^t \gamma_u\,du$ for some $\mathbb{F}$-progressively measurable process $\gamma$, referred to as the $\mathbb{F}$-*intensity* of a random time $\tau$, or simply, the *stochastic intensity* of $\tau$ (also known as the *hazard rate*). A continuous, $\mathbb{F}$-adapted process $\Gamma$ is manifestly $\mathbb{F}$-predictable, and thus there exists an $\mathbb{F}$-predictable modification of the $\mathbb{F}$-intensity $\gamma$ (see Lemma 1.36 in Chap. 1 of Jacod (1979) in this regard). Moreover, the $\mathbb{F}$-predictable version of the hazard rate is uniquely specified, $\mathbb{P}\otimes\ell$-a.e. on the product space $\Omega\times\mathbb{R}_+$ endowed with the $\sigma$-field of predictable sets.

By virtue of Proposition 5.1.3, the process $\hat{M}$, defined as

$$\hat{M}_t = H_t - \int_0^{t\wedge\tau} \gamma_u\,du = H_t - \int_0^t \mathbb{1}_{\{\tau\geq u\}}\gamma_u\,du,$$

follows a $\mathbb{G}$-martingale. The last property is frequently used in financial literature as the definition of a 'stochastic intensity' of a random time. In fact, it is not uncommon to refer to the $\mathbb{G}$-predictable process $\tilde{\lambda}$, defined as

$$\tilde{\lambda}_t = \mathbb{1}_{\{\tau\geq t\}}\gamma_t,$$

as the stochastic intensity of $\tau$. Under this convention, we have

$$\hat{M}_t = H_t - \int_0^t \tilde{\lambda}_u\,du, \quad \forall\,t \in \mathbb{R}_+,$$

so that the $\mathbb{G}$-intensity $\tilde{\lambda}$ is closely related to the concept of the $\mathbb{G}$-compensator of $H$ (for further details, we refer to Sect. 6.1.4). In this text, by a stochastic intensity of a random time $\tau$, we mean the $\mathbb{F}$-intensity, rather than the $\mathbb{G}$-intensity (unless $\tau$ is a $\mathbb{F}$-stopping time, so that $\mathbb{F} = \mathbb{G}$).

The next corollary clarifies the intuitive interpretation of $\mathbb{F}$-intensity $\gamma$ as the 'intensity of default, given the information flow $\mathbb{F}$.'

**Corollary 5.1.5.** *If the $\mathbb{F}$-hazard process $\Gamma$ of a random time $\tau$ is absolutely continuous, then for any $t \leq s$ we have*

$$\mathbb{P}\{\tau > s\,|\,\mathcal{G}_t\} = \mathbb{1}_{\{\tau>t\}}\,\mathbb{E}_{\mathbb{P}}\big(e^{-\int_t^s \gamma_u\,du}\,\big|\,\mathcal{F}_t\big)$$

*and*

$$\mathbb{P}\{t < \tau \leq s\,|\,\mathcal{G}_t\} = \mathbb{1}_{\{\tau>t\}}\,\mathbb{E}_{\mathbb{P}}\big(1 - e^{-\int_t^s \gamma_u\,du}\,\big|\,\mathcal{F}_t\big).$$

It is obvious that the $\mathbb{F}$-hazard function $\Gamma$ is not well defined when $\tau$ is an $\mathbb{F}$-stopping time, i.e., when $\mathbb{H} \subseteq \mathbb{F}$ (so that $\mathbb{G} = \mathbb{F}$). For this reason, Corollaries 5.1.1 and 5.1.5 cannot be directly applied to this situation. However, it appears that for $\tau$ from a certain class of $\mathbb{G}$-stopping times, we can find an increasing, $\mathbb{G}$-predictable process $\Lambda$ such that for any $t \leq s$ we have (for more details, see Sect. 6.1)

$$\mathbb{P}\{\tau > s \,|\, \mathcal{G}_t\} = \mathbb{1}_{\{\tau > t\}} \, \mathbb{E}_{\mathbb{P}}\big(e^{\Lambda_t - \Lambda_s} \,\big|\, \mathcal{G}_t\big).$$

If the process $\Lambda$ is absolutely continuous with respect to the Lebesgue measure, i.e., if $\Lambda_t = \int_0^t \lambda_u \, du$ for some $\mathbb{G}$-progressively measurable process $\lambda$, then we have

$$\mathbb{P}\{\tau > s \,|\, \mathcal{G}_t\} = \mathbb{1}_{\{\tau > t\}} \, \mathbb{E}_{\mathbb{P}}\big(e^{-\int_t^s \lambda_u \, du} \,\big|\, \mathcal{G}_t\big).$$

## 5.2 Martingale Representation Theorems

Our next goal is to establish few versions of the martingale representation theorem for $\mathbb{G}$-martingales. We shall first deal with the general case. Subsequently, we shall examine the case of the Brownian filtration.

### 5.2.1 General Case

In this section, we consider a general set-up. In particular, we do not assume that the filtration $\mathbb{F}$ supports only continuous martingales. However, we postulate that the process $F$ is continuous and increasing.

If $F_t = \mathbb{P}\{\tau \leq t \,|\, \mathcal{F}_t\}$ is a continuous, increasing process, then the associated hazard process $\Gamma$ also follows a continuous, increasing process, since $\Gamma_t = -\ln(1 - F_t)$. As shown in Proposition 5.1.3, the process $\hat{M} = H_t - \Gamma_{t \wedge \tau}$ is a $\mathbb{G}$-martingale.

**Proposition 5.2.1.** *Assume that the $\mathbb{F}$-hazard process $\Gamma$ of $\tau$ follows an increasing continuous process. Let $Z$ be an $\mathbb{F}$-predictable process such that the random variable $Z_\tau$ is integrable. Then the $\mathbb{G}$-martingale $M_t^Z := \mathbb{E}_{\mathbb{P}}(Z_\tau | \mathcal{G}_t)$ admits the following decomposition*

$$M_t^Z = m_0 + \int_{]0,t]} e^{\Gamma_u} \, d\tilde{m}_u + \int_{]0,t]} (Z_u - D_u) \, d\hat{M}_u, \tag{5.32}$$

*where $\tilde{m}_t = m_{t \wedge \tau}$, and $m$ is an $\mathbb{F}$-martingale, namely,*

$$m_t = \mathbb{E}_{\mathbb{P}}\Big(\int_0^\infty Z_u e^{-\Gamma_u} \, d\Gamma_u \,\Big|\, \mathcal{F}_t\Big) = \mathbb{E}_{\mathbb{P}}\Big(\int_0^\infty Z_u \, dF_u \,\Big|\, \mathcal{F}_t\Big), \tag{5.33}$$

*hence, in particular, $m_0 = M_0^Z$. Moreover,*

$$D_t = \mathbb{E}_{\mathbb{P}}\Big(\int_t^\infty Z_u e^{\Gamma_t - \Gamma_u} \, d\Gamma_u \,\Big|\, \mathcal{F}_t\Big) = e^{\Gamma_t} \mathbb{E}_{\mathbb{P}}\Big(\int_t^\infty Z_u \, dF_u \,\Big|\, \mathcal{F}_t\Big).$$

*Proof.* By virtue of Corollary 5.1.3, we have

$$M_t^Z = \mathbb{E}_{\mathbb{P}}(Z_\tau \,|\, \mathcal{G}_t) = \mathbb{1}_{\{\tau \le t\}} Z_\tau + \mathbb{1}_{\{\tau > t\}} \mathbb{E}_{\mathbb{P}}\left( \int_t^\infty Z_u e^{\Gamma_t - \Gamma_u} d\Gamma_u \,\Big|\, \mathcal{F}_t \right)$$

$$= \mathbb{1}_{\{\tau \le t\}} Z_\tau + \mathbb{1}_{\{\tau > t\}} D_t.$$

In particular, we have $M_t^Z = M_{t \wedge \tau}^Z$ for every $t \in \mathbb{R}_+$. In view of our assumptions, $e^{\Gamma_t}$ is a continuous, increasing process. The Itô integration by parts formula yields (notice that $m$ is a $\mathbb{G}$-semimartingale)

$$D_t = e^{\Gamma_t} m_t - e^{\Gamma_t} \int_0^t Z_u e^{-\Gamma_u} d\Gamma_u$$

$$= m_0 + \int_{]0,t]} e^{\Gamma_u} dm_u + \int_0^t m_u e^{\Gamma_u} d\Gamma_u - \int_0^t Z_u d\Gamma_u$$

$$- \int_0^t e^{\Gamma_u} \int_0^u Z_v e^{-\Gamma_v} d\Gamma_v \, d\Gamma_u$$

which entails that

$$D_t = m_0 + \int_{]0,t]} e^{\Gamma_u} dm_u + \int_0^t (D_u - Z_u) \, d\Gamma_u.$$

Furthermore, since $D$ is a right-continuous process with left-hand limits, we have

$$\mathbb{1}_{\{\tau > t\}} D_t = m_0 + \int_{]0,t \wedge \tau]} dD_u - \mathbb{1}_{\{\tau \le t\}} D_\tau$$

$$= m_0 + \int_{]0,t \wedge \tau]} dD_u - \mathbb{1}_{\{\tau \le t\}} M_{\tau-}^Z.$$

We conclude that

$$M_t^Z = m_0 + \int_{]0,t \wedge \tau]} e^{\Gamma_u} dm_u + \int_0^{t \wedge \tau} (D_u - Z_u) \, d\Gamma_u + \mathbb{1}_{\{\tau \le t\}}(Z_\tau - D_\tau).$$

The last formula immediately yields (5.32). □

The following corollary to Proposition 5.2.1 appears to be useful in the study of the case of the Brownian filtration.

**Corollary 5.2.1.** *Under the assumptions of Proposition 5.2.1, if, in addition, $\Delta D_\tau = 0$ (or, equivalently, $\Delta m_\tau = 0$), then*

$$M_t^Z = m_0 + \int_{]0,t]} e^{\Gamma_u} d\tilde{m}_u + \int_{]0,t]} (Z_u - M_{u-}^Z) \, d\hat{M}_u, \qquad (5.34)$$

*so that*

$$M_t^Z = m_0 + \tilde{M}_t^Z + \hat{M}_t^Z,$$

*where the two $\mathbb{G}$-martingales $\tilde{M}^Z$ and $\hat{M}^Z$ are mutually orthogonal, i.e., the product $\tilde{M}^Z \hat{M}^Z$ is a $\mathbb{G}$-martingale.*

*Proof.* Obviously, $D_u = M_u^Z$ for $u < \tau$, and thus also $D_\tau = D_{\tau-} = M_{\tau-}^Z$. Consequently,

$$\mathbb{1}_{\{\tau > t\}} D_t = m_0 + \int_{]0, t \wedge \tau]} dD_u - \mathbb{1}_{\{\tau \le t\}} M_{\tau-}^Z.$$

Since $\Gamma$ is a continuous process, we conclude that (5.34) is valid.

For the second statement, notice that, by virtue of part (iii) in Proposition 5.1.3, the quadratic covariation $[\hat{M}, \tilde{m}]$ is a $\mathbb{G}$-martingale, and thus the process[2]

$$[\tilde{M}^Z, \hat{M}^Z]_t = \int_{]0,t]} e^{\Gamma_u} (Z_u - M_{u-}^Z) \, d\,[\hat{M}, \tilde{m}]_u$$

follows a $\mathbb{G}$-martingale. We conclude that $\tilde{M}^Z \hat{M}^Z$ is a $\mathbb{G}$-martingale.    □

*Remarks.* Suppose that the filtration $\mathbb{F}$ supports only continuous martingales (for instance, $\mathbb{F}$ is the natural filtration of some Brownian motion). Then, it is obvious that for the process $m$ given by formula (5.33), we have $\Delta m_\tau = 0$. Under the present assumptions, equality (5.34) represents the decomposition of the $\mathbb{G}$-martingale $M^Z$ into the sum of the continuous $\mathbb{G}$-martingale $\tilde{M}^Z$ and the discontinuous $\mathbb{G}$-martingale $\hat{M}^Z$, where $\tilde{M}^Z$ and $\hat{M}^Z$ are mutually orthogonal $\mathbb{G}$-martingales. Formula (5.32) yields the so-called *canonical decomposition* of $M^Z$: $\tilde{M}^Z$ is the continuous martingale part of $M^Z$, and $\hat{M}^Z$ is the purely discontinuous martingale part $M^Z$.

**Corollary 5.2.2.** *Assume that $\Gamma$ is a continuous, increasing process. Let $h : \mathbb{R}_+ \to \mathbb{R}$ be a Borel measurable function such that the random variable $h(\tau)$ is integrable. Then the $\mathbb{G}$-martingale $M^h$, given by the formula*

$$M_t^h = \mathbb{E}_{\mathbb{P}}(h(\tau)| \mathcal{G}_t), \quad \forall t \in \mathbb{R}_+,$$

*admits the following decomposition*

$$M_t^h = m_0 + \int_{]0,t]} e^{\Gamma_u} d\tilde{m}_u + \int_{]0,t]} (h(u) - g_u) \, d\hat{M}_u,$$

*where $\tilde{m}_t = m_{t \wedge \tau}$ and $m$ is an $\mathbb{F}$-martingale*

$$m_t = \mathbb{E}_{\mathbb{P}}\left( \int_0^\infty h(u) e^{-\Gamma_u} d\Gamma_u \, \Big| \, \mathcal{F}_t \right) = \mathbb{E}_{\mathbb{P}}\left( \int_0^\infty h(u) \, dF_u \, \Big| \, \mathcal{F}_t \right)$$

*and*

$$g_t = e^{\Gamma_t} \mathbb{E}_{\mathbb{P}}\left( \int_t^\infty h(u) e^{-\Gamma_u} d\Gamma_u \, \Big| \, \mathcal{F}_t \right) = e^{\Gamma_t} \mathbb{E}_{\mathbb{P}}\left( \int_t^\infty h(u) \, dF_u \, \Big| \, \mathcal{F}_t \right).$$

The following corollary to Proposition 5.2.1 was already established in the previous chapter (cf. Proposition 4.3.2).

---

[2] In fact, it is easy to see that, under the present assumptions, we have $[\hat{M}, \tilde{m}] = 0$.

**Corollary 5.2.3.** *Assume that the filtration* $\mathbb{F}$ *is trivial and the hazard functions* $\Gamma$ *of a random time* $\tau$ *is continuous. Let* $M_t^h := \mathbb{E}_{\mathbb{P}}(h(\tau) \,|\, \mathcal{H}_t)$ *for some bounded, Borel measurable function* $h : \mathbb{R} \to \mathbb{R}$. *Then*

$$M_t^h = M_0^h + \int_{]0,t]} (h(u) - g(u)) \, d\hat{M}_u,$$

*where the function* $g : \mathbb{R}_+ \to \mathbb{R}$ *equals*

$$g(t) = e^{\Gamma(t)} \, \mathbb{E}_{\mathbb{P}}\big(h(\tau) \mathbb{1}_{\{\tau > t\}}\big) = e^{\Gamma(t)} \int_t^\infty h(u) \, dF(u).$$

*Proof.* In view of (5.32), it is enough to check that $M_t^h = \tilde{h}(t)$ on $\{\tau > t\}$, where

$$\tilde{h}(t) := e^{\Gamma(t)} \, \mathbb{E}_{\mathbb{P}}\big(h(\tau) \mathbb{1}_{\{\tau > t\}}\big).$$

Since $\Gamma$ is continuous, we have

$$\mathbb{E}_{\mathbb{P}}\big(h(\tau) \mathbb{1}_{\{\tau > t\}}\big) = \int_t^\infty h(u) e^{-\Gamma(u)} \, d\Gamma(u).$$

The following representation is valid, provided that the hazard function $\Gamma$ is continuous (cf. (4.8))

$$M_t^h = \mathbb{E}_{\mathbb{P}}(h(\tau) \,|\, \mathcal{H}_t) = \mathbb{1}_{\{\tau \le t\}} h(\tau) + \mathbb{1}_{\{\tau > t\}} e^{\Gamma(t)} \int_t^\infty h(u) e^{-\Gamma(u)} \, d\Gamma(u).$$

By combining the last two formulae, we obtain the desired equality.    □

### 5.2.2 Case of a Brownian Filtration

In this section, we consider the case of the Brownian filtration, i.e., we assume that the reference filtration $\mathbb{F} = \mathbb{F}^W$ for some Brownian motion $W$. We postulate that the Brownian motion $W$ remains a (continuous) martingale (and thus a Brownian motion) with respect to the enlarged filtration $\mathbb{G}$. We assume, as usual, that Condition (G.1) is satisfied.

Let us fix $T > 0$. In the next result, we do not need to assume that $F$ is an increasing process. In other words, we do not postulate that the $\mathbb{G}$-martingale $L$, given by formula (5.24), is a process of finite variation.

**Proposition 5.2.2.** *For a* $\mathbb{G}$-*measurable and integrable random variable* $X$, *we define the* $\mathbb{G}$-*martingale* $M_t^X = \mathbb{E}_{\mathbb{P}}(X \,|\, \mathcal{G}_t)$, $t \in [0, T]$. *Then* $M^X$ *admits the following representation*

$$M_t^X = M_0^X + \int_0^t \xi_u^X \, dW_u + \int_{]0,t]} \tilde{\zeta}_u^X \, dL_u = M_0^X + \tilde{L}_t^X + \hat{L}_t^X,$$

*where* $\xi^X$ *and* $\tilde{\zeta}^X$ *are* $\mathbb{G}$-*predictable stochastic processes. Moreover, the* $\mathbb{G}$-*martingales* $\tilde{L}_t^X$ *and* $\hat{L}_t^X$ *are mutually orthogonal.*

*Proof.* It is clear that $\mathbb{E}_{\mathbb{P}}(\mathbb{E}_{\mathbb{P}}(X \mid \mathcal{G}_T) \mid \mathcal{G}_t) = \mathbb{E}_{\mathbb{P}}(X \mid \mathcal{G}_t)$. Thus, we may assume that $X$ is $\mathcal{G}_T$-measurable. Since $\mathcal{G}_T = \mathcal{H}_T \vee \mathcal{F}_T$, it suffices to consider a random variable $X$ of the form $X = (1 - H_s)Y$ for some fixed $s \leq T$ and some $\mathcal{F}_T$-measurable random variable $Y$. Notice that

$$X = (1 - H_s)Y = (1 - H_s)e^{\Gamma_s}\tilde{Y} = L_s \tilde{Y},$$

where $\tilde{Y} = e^{-\Gamma_s}Y$ is an $\mathcal{F}_T$-measurable, integrable random variable. We introduce the $\mathbb{F}$-martingale $U$ by setting, for $t \in [0, T]$,

$$U_t = \mathbb{E}_{\mathbb{P}}(\tilde{Y} \mid \mathcal{F}_t) = \mathbb{E}_{\mathbb{P}}(\tilde{Y}) + \int_0^t \xi_u \, dW_u,$$

where the second equality is a consequence of the martingale representation property of the Brownian filtration (so that $\xi$ is an $\mathbb{F}$-predictable stochastic process). By assumption, $W$ is also a $\mathbb{G}$-martingale, and thus the process $U$ is not only an $\mathbb{F}$-martingale, but also a (continuous) $\mathbb{G}$-martingale. Invoking Lemma 5.1.7, we deduce that the quadratic covariation $[L, U]$ follows a $\mathbb{G}$-martingale. Furthermore,

$$[L, U]_t = \langle L^c, U \rangle_t + \sum_{u \leq t} \Delta L_u \, \Delta U_u = \langle L^c, U \rangle_t,$$

where we write $L^c$ to denote the continuous martingale part of $L$. It turns out that $[L, U]_t = \langle L^c, U \rangle_t = 0$ for every $t \in [0, T]$ (recall that any continuous martingale of finite variation is constant). Clearly $U_T = \tilde{Y}$, and this implies $X = L_s U_T$. The Itô integration by parts formula yields

$$X = L_0 U_0 + \int_0^T L_{t-} \, dU_t + \int_{]0,s]} U_{t-} \, dL_t + [L, U]_s$$

$$= \mathbb{E}_{\mathbb{P}}(\tilde{Y}) + \int_0^T \xi_t L_{t-} \, dW_t + \int_{]0,T]} U_t \mathbb{1}_{[0,s]}(t) \, dL_t.$$

We conclude that the asserted formula holds, with the following processes: $\xi_t^X = \xi_t L_{t-}$ and $\tilde{\zeta}_t^X = U_t \mathbb{1}_{[0,s]}(t)$. It remains to show that the $\mathbb{G}$-martingales $\tilde{L}^X$ and $\hat{L}^X$ are mutually orthogonal. As a consequence of Lemma 5.1.7, we get $[L, W] = \langle L, W \rangle = 0$. The $\mathbb{G}$-martingale property of the product $\tilde{L}^X \hat{L}^X$ thus follows easily from the Itô formula. □

**Corollary 5.2.4.** *We preserve the assumptions of Proposition 5.2.2. In addition, we assume that the hazard process $\Gamma$ is continuous. Then for any $\mathbb{G}$-martingale $N$ we have*

$$N_t = N_0 + \int_0^t \xi_u^N \, dW_u + \int_{]0,t]} \zeta_u^N \, d\hat{M}_u = M_0 + \tilde{M}_t^N + \hat{M}_t^N,$$

*where $\xi^N$ and $\zeta^N$ are $\mathbb{G}$-predictable stochastic processes. The continuous $\mathbb{G}$-martingale $\tilde{M}^N$ and the purely discontinuous $\mathbb{G}$-martingale $\hat{M}^N$ are mutually orthogonal.*

*Proof.* Observe that part (i) in Proposition 5.1.3 yields

$$\int_{]0,t]} U_{u-}\, dL_u = -\int_{]0,t]} U_u L_{u-}\, d\hat{M}_u.$$

The desired formula now follows upon setting $\zeta_t^N = -U_t L_{t-}$. The mutual orthogonality of the two $\mathbb{G}$-martingales $\tilde{M}^N$ and $\hat{M}^N$ is also obvious.    □

In the next result, we no longer postulate that the Brownian motion $W$ remains a martingale with respect to the enlarged filtration $\mathbb{G}$. By virtue of part (i) in Lemma 5.1.6, if $F$ is a continuous, increasing process, then the stopped process $W_{t\wedge\tau}$ follows a $\mathbb{G}$-martingale. In general, in view of part (iii) in Lemma 5.1.6, we see that the process $\hat{W}$, given by the formula

$$\hat{W}_t = W_{t\wedge\tau} + \int_0^{t\wedge\tau} (1-F_u)^{-1}\, d\langle W, F\rangle_u,$$

follows a $\mathbb{G}$-martingale.

Let $Z$ be an $\mathbb{F}$-predictable process such that the random variable $Z_\tau$ is integrable. We define an auxiliary process $m$ by setting (cf. (5.33))

$$m_t = \mathbb{E}_{\mathbb{P}}\left(\int_0^\infty Z_u e^{-\Gamma_u}\, d\Gamma_u \,\Big|\, \mathcal{F}_t\right).$$

It is clear that the continuous $\mathbb{F}$-martingale $m$ admits the integral representation with respect to the underlying Brownian motion $W$, namely,

$$m_t = m_0 + \int_0^t \xi_u^Z\, dW_u,$$

for some $\mathbb{F}$-predictable process $\xi^Z$. The next result is an immediate corollary to Proposition 5.2.1.

**Corollary 5.2.5.** *Assume that the $\mathbb{F}$-hazard process $\Gamma$ of $\tau$ follows a continuous, increasing process. Let $Z$ be an $\mathbb{F}$-predictable process such that the random variable $Z_\tau$ is integrable. Then the $\mathbb{G}$-martingale $M_t^Z = \mathbb{E}_{\mathbb{P}}(Z_\tau|\mathcal{G}_t)$ is the sum of two mutually orthogonal $\mathbb{G}$-martingales, namely,*

$$M_t^Z = m_0 + \int_0^t e^{\Gamma_u}\xi_u^Z\, d\hat{W}_u + \int_{]0,t]}(Z_u - M_{u-}^Z)\, d\hat{M}_u. \tag{5.35}$$

It is worth noting that on the random interval $[0,\tau]$ the left-continuous $\mathbb{G}$-predictable process $M_{t-}^Z$ coincides with some $\mathbb{F}$-predictable process. Therefore, equality (5.35) means also that

$$M_t^Z = m_0 + \int_0^{t\wedge\tau} \xi_u\, dW_u + \int_{]0,t\wedge\tau]} \zeta_u\, d\hat{M}_u$$

for some $\mathbb{F}$-predictable processes $\xi$ and $\zeta$.

## 5.3 Change of a Probability Measure

In this section, we deal with an equivalent change of a probability measure. We make the following standing assumptions: (a) the filtration $\mathbb{F}$ is generated by a Brownian motion, (b) any $\mathbb{F}$-martingale is a $\mathbb{G}$-martingale under $\mathbb{P}$, and (c) the hazard process $\Gamma$ of $\tau$ is a continuous, increasing process.

Let us fix $T > 0$. For a probability measure $\mathbb{P}^*$, equivalent to $\mathbb{P}$ on $(\Omega, \mathcal{G}_T)$, we introduce the Radon-Nikodým density process $\eta_t$, $t \leq T$, by setting

$$\eta_t := \frac{d\mathbb{P}^*}{d\mathbb{P}}\Big|_{\mathcal{G}_t} = \mathbb{E}_{\mathbb{P}}(X \mid \mathcal{G}_t), \quad \mathbb{P}\text{-a.s.}, \tag{5.36}$$

where $X$ is a $\mathcal{G}_T$-measurable and integrable random variable, such that $\mathbb{P}\{X > 0\} = 1$ and $\mathbb{E}_{\mathbb{P}}(X) = 1$. In view of Corollary 5.2.4, the $\mathbb{G}$-martingale $\eta$ admits the following integral representation:

$$\eta_t = 1 + \int_0^t \xi_u \, dW_u + \int_{]0,t]} \zeta_u \, d\hat{M}_u,$$

where $\xi$ and $\zeta$ are $\mathbb{G}$-predictable stochastic processes. Since $\eta$ is a strictly positive martingale, it can be also shown (see Proposition 6.20 in Jacod (1979)) that the process of left-hand limits $\eta_{t-}$ is strictly positive as well. Thus, we may rewrite the last formula as follows:

$$\eta_t = 1 + \int_{]0,t]} \eta_{u-} \big( \beta_u \, dW_u + \kappa_u \, d\hat{M}_u \big), \tag{5.37}$$

where $\beta_t = \xi_t \eta_{t-}^{-1}$ and $\kappa_t = \zeta_t \eta_{t-}^{-1}$ are $\mathbb{G}$-predictable processes.

From Proposition 5.1.3, we already know that the process $\hat{M}_t = H_t - \Gamma_{t \wedge \tau}$ is a $\mathbb{G}$-martingale under $\mathbb{P}$. Our goal is to examine the martingale associated with the jump process $H$ under $\mathbb{P}^*$. Let us emphasize that conditions (a)–(c) given above are among the assumptions of the next two results.

**Proposition 5.3.1.** *Let $\mathbb{P}^*$ be a probability measure on $(\Omega, \mathcal{G}_T)$ equivalent to the reference probability measure $\mathbb{P}$. Assume that the Radon-Nikodým density of $\mathbb{P}^*$ with respect to $\mathbb{P}$ is given by (5.36) with $\eta$ satisfying (5.37). Then the process*

$$W_t^* = W_t - \int_0^t \beta_u \, du, \quad \forall t \in [0, T], \tag{5.38}$$

*follows a Brownian motion with respect to $\mathbb{G}$ under $\mathbb{P}^*$, and the process $M_t^*$, $t \in [0, T]$, given by the formula*

$$M_t^* = \hat{M}_t - \int_0^{t \wedge \tau} \kappa_u \, d\Gamma_u = H_t - \int_0^{t \wedge \tau} (1 + \kappa_u) \, d\Gamma_u, \tag{5.39}$$

*follows a $\mathbb{G}$-martingale under $\mathbb{P}^*$. Furthermore, the $\mathbb{G}$-martingales $W^*$ and $M^*$ are mutually orthogonal under $\mathbb{P}^*$.*

*Proof.* Notice first that for $t \leq T$ we have

$$
\begin{aligned}
d(\eta_t W_t^*) &= W_t^* \, d\eta_t + \eta_{t-} \, dW_t^* + d\,[W^*, \eta]_t \\
&= W_t^* \, d\eta_t + \eta_{t-} \, dW_t - \eta_{t-}\beta_t \, dt + \eta_{t-}\beta_t \, d\,[W, W]_t \\
&= W_t^* \, d\eta_t + \eta_{t-} \, dW_t .
\end{aligned}
$$

This shows that $\eta W^*$ is a $\mathbb{G}$-(local)-martingale under $\mathbb{P}$, so that $W^*$ follows a $\mathbb{G}$-(local)-martingale under $\mathbb{P}^*$. Since the quadratic variation of $W^*$ under $\mathbb{P}^*$ equals $[W^*, W^*]_t = t$ and $W^*$ is continuous, by virtue of Lévy's characterization of the Brownian motion (see, e.g., Theorem 3.16 in Karatzas and Shreve (1997)), it is clear that $W^*$ follows a Brownian motion under $\mathbb{P}^*$. Similarly, for $t \leq T$,

$$
\begin{aligned}
d(\eta_t M_t^*) &= M_t^* \, d\eta_t + \eta_{t-} \, dM_t^* + d\,[M^*, \eta]_t \\
&= M_t^* \, d\eta_t + \eta_{t-} \, d\hat{M}_t - \eta_{t-}\kappa_t \, d\Gamma_{t \wedge \tau} + \eta_{t-}\kappa_t \, dH_t ,
\end{aligned}
$$

hence, after rearrangements, we obtain

$$
d(\eta_t M_t^*) = M_t^* \, d\eta_t + \eta_{t-}(1 + \kappa_t) \, d\hat{M}_t .
$$

We conclude that the product $\eta M^*$ follows a $\mathbb{G}$-(local)-martingale under $\mathbb{P}$. This shows, of course, that $M^*$ is a $\mathbb{G}$-(local)-martingale under $\mathbb{P}^*$. To end the proof, it suffices to observe that the process $W^*$ is continuous and $M^*$ follows a process of finite variation. $\qquad\square$

**Corollary 5.3.1.** *Let $Y$ be a $\mathbb{G}$-martingale with respect to $\mathbb{P}^*$. Then $Y$ admits the following decomposition*

$$
Y_t = Y_0 + \int_0^t \xi_u^* \, dW_u^* + \int_{]0,t]} \zeta_u^* \, dM_u^* , \tag{5.40}
$$

*where $\xi^*$ and $\zeta^*$ are $\mathbb{G}$-predictable stochastic processes.*

*Proof.* Consider the process $\tilde{Y}_t$, $t \in [0, T]$, which is given by the formula

$$
\tilde{Y}_t = \int_{]0,t]} \eta_{u-}^{-1} \, d(\eta_u Y_u) - \int_{]0,t]} \eta_{u-}^{-1} Y_{u-} \, d\eta_u .
$$

It is clear that $\tilde{Y}$ is a $\mathbb{G}$-martingale under $\mathbb{P}$. The Itô formula yields

$$
\eta_{u-}^{-1} \, d(\eta_u Y_u) = dY_u + \eta_{u-}^{-1} Y_{u-} \, d\eta_u + \eta_{u-}^{-1} \, d\,[Y, \eta]_u ,
$$

and so

$$
Y_t = Y_0 + \tilde{Y}_t - \int_{]0,t]} \eta_{u-}^{-1} \, d\,[Y, \eta]_u . \tag{5.41}
$$

From Corollary 5.2.4, we know that for some $\mathbb{G}$-predictable processes $\tilde{\xi}$ and $\tilde{\zeta}$ we have

$$
\tilde{Y}_t = Y_0 + \int_0^t \tilde{\xi}_u \, dW_u + \int_{]0,t]} \tilde{\zeta}_u \, d\hat{M}_u . \tag{5.42}
$$

Consequently, we obtain

$$dY_t = \tilde{\xi}_t \, dW_t + \tilde{\zeta}_t \, d\hat{M}_t - \eta_{t-}^{-1} \, d\,[Y,\eta]_t$$
$$= \tilde{\xi}_t \, dW_t^* + \tilde{\zeta}_t (1 + \kappa_t)^{-1} \, dM_t^*,$$

because (5.37), when combined with (5.41)–(5.42), yield

$$\eta_{t-}^{-1} \, d\,[Y,\eta]_t = \tilde{\xi}_t \beta_t \, dt + \tilde{\zeta}_t \kappa_t (1 + \kappa_t)^{-1} \, dH_t.$$

To derive the last equality, we note in particular that, in view of (5.41), we have (we take into account the continuity of the hazard process $\Gamma$)

$$\Delta[Y,\eta]_t = \eta_{t-} \tilde{\zeta}_t \kappa_t \, dH_t - \kappa_t \Delta[Y,\eta]_t.$$

We conclude that $Y$ satisfies (5.40) with $\xi^* = \tilde{\xi}$ and $\zeta^* = \tilde{\zeta}(1 + \kappa)^{-1}$.    □

*Remarks.* Since the martingale $M^*$ is stopped at $\tau$, we may assume, without loss of generality, that processes $\kappa$ in (5.39) and $\zeta^*$ in (5.40) are $\mathbb{F}$-predictable. Likewise, if the processes $W^*$ and $Y$, introduced in Proposition 5.3.1 and Corollary 5.3.1 respectively, are stopped at $\tau$, then the processes $\beta$ and $\xi^*$ can also be chosen to be $\mathbb{F}$-predictable. To emphasize this feature, we shall write $\tilde{\kappa}$ and $\tilde{\beta}$ to denote the unique $\mathbb{F}$-predictable processes that coincide with $\kappa$ and $\beta$ on the random interval $[0, \tau[$.

**Hazard rate of $\tau$ under $\mathbb{P}^*$.** In Sect. 4.4, we have shown that if $\mathbb{P}^*$ is a probability measure, which is absolutely continuous with respect to the reference probability measure $\mathbb{P}$, then, under mild technical assumptions, it is possible to derive a simple, but universal, relationship between the hazard functions $\Gamma$ and $\Gamma^*$ of $\tau$ under $\mathbb{P}$ and $\mathbb{P}^*$, respectively. In case of hazard processes, this issue becomes much more involved.

In view of Proposition 5.3.1, it would be natural to conjecture, at least under the assumptions (a)–(c) stated at the beginning of the present section, that if a probability measure $\mathbb{P}^*$ is equivalent to $\mathbb{P}$ on $(\Omega, \mathcal{G}_T)$, then the $\mathbb{F}$-hazard process $\Gamma^*$ of $\tau$ under $\mathbb{P}^*$ is given by the following expression:

$$\Gamma_t^* = \int_0^t (1 + \tilde{\kappa}_u) \, d\Gamma_u, \quad \forall \, t \in [0, T].$$

In case of a random time $\tau$, which admits the $\mathbb{F}$-intensity process $\gamma$ under $\mathbb{P}$, it would be natural to expect that $\tau$ admits also the $\mathbb{F}$-intensity process, denoted by $\gamma^*$, under $\mathbb{P}^*$, and that the relationship $\gamma_t^* = (1 + \tilde{\kappa}_t)\gamma_t$ is satisfied for every $t \in [0, T]$.

The reader should thus be warned that things are not that simple, and the above relationships – which are clearly satisfied by the $\mathbb{F}$-martingale hazard processes $\Lambda$, formally introduced in the next chapter – are not necessarily valid for the $\mathbb{F}$-hazard processes. We shall examine this problem in some detail in Sect. 7.2.

# 6. Martingale Hazard Process

In Sect. 4.5, we have introduced the concept of the martingale hazard function of a random time and we have examined the connection between this concept and the notion of the hazard function. It appeared, that both notions coincide if and only if the cumulative distribution function of $\tau$, and thus also its hazard function, are continuous (see Proposition 4.5.1). In this sense, the martingale hazard function uniquely characterizes the unconditional probability distribution of a continuously distributed random time. On the other hand, we have shown in Sect. 5.1.3 (see Proposition 5.1.3) that if the $\mathbb{F}$-hazard process is continuous, the process $H_t - \Gamma_{t \wedge \tau}$ follows a $\mathbb{G}$-martingale. The main goal of this chapter is to extend the concept to the case of a non-trivial filtration, and to examine whether a continuous $\mathbb{F}$-martingale hazard process uniquely specifies the $\mathbb{F}$-conditional survival probabilities of a random time.

## 6.1 Martingale Hazard Process $\Lambda$

In this chapter, we assume that (G.1) is valid, so that $\mathbb{G} = \mathbb{H} \vee \mathbb{F}$. It should be stressed that the case when $\mathbb{H} \subseteq \mathbb{F}$ (i.e., $\mathbb{F} = \mathbb{G}$) is not excluded. This means that the situation when $\tau$ is an $\mathbb{F}$-stopping time is also covered by the results of this chapter. The concept of the $(\mathbb{F}, \mathbb{G})$-martingale hazard process is a direct counterpart of the notion of the martingale hazard function of $\tau$ (the latter can be seen as the $(\mathbb{F}^0, \mathbb{H})$-martingale hazard process of $\tau$).

**Definition 6.1.1.** An $\mathbb{F}$-predictable, right-continuous, increasing process $\Lambda$ is called a $(\mathbb{F}, \mathbb{G})$-*martingale hazard process* of a random time $\tau$ if and only if the process $\tilde{M}_t := H_t - \Lambda_{t \wedge \tau}$ follows a $\mathbb{G}$-martingale. In addition, $\Lambda_0 = 0$. If, in addition, $\Lambda_t = \int_0^t \lambda_u \, du$ the $\mathbb{F}$-progressively measurable non-negative process $\lambda$ is referred to as the $(\mathbb{F}, \mathbb{G})$-*martingale intensity process*.

Under (G.1), a random time $\tau$ and a reference filtration $\mathbb{F}$ uniquely specify the enlarged filtration $\mathbb{G}$ through $\mathbb{G} = \mathbb{H} \vee \mathbb{F}$. Thus, when (G.1) holds, we find it convenient to refer to the $(\mathbb{F}, \mathbb{G})$-martingale hazard process of $\tau$ as the $\mathbb{F}$-*martingale hazard process* of $\tau$.

We first examine the case when the $\mathbb{F}$-martingale hazard process $\Lambda$ can be expressed through a straightforward counterpart of formula (4.26). To this end, we introduce the following condition.

**Condition (F.1)** For any $t \in \mathbb{R}_+$, the $\sigma$-fields $\mathcal{F}_\infty$ and $\mathcal{H}_t$ are conditionally independent given $\mathcal{F}_t$ under $\mathbb{P}$; that is, for any bounded, $\mathcal{F}_\infty$-measurable random variable $\xi$ and any bounded, $\mathcal{H}_t$-measurable random variable $\eta$ we have

$$\mathbb{E}_{\mathbb{P}}(\xi\eta \,|\, \mathcal{F}_t) = \mathbb{E}_{\mathbb{P}}(\xi \,|\, \mathcal{F}_t)\mathbb{E}_{\mathbb{P}}(\eta \,|\, \mathcal{F}_t).$$

Let us emphasize that Condition (F.1) is satisfied when $\tau$ is constructed through the canonical method (see Sect. 6.5 and 8.2.1). Since $\mathcal{F}_t \subseteq \mathcal{F}_\infty$, we may restate Condition (F.1) as follows.

**Condition (F.1a)** For any $t \in \mathbb{R}_+$ and every $u \leq t$, the following equality holds: $\mathbb{P}\{\tau \leq u \,|\, \mathcal{F}_t\} = \mathbb{P}\{\tau \leq u \,|\, \mathcal{F}_\infty\}$.

The following condition will also be useful.

**Condition (F.2)** The process $F_t = \mathbb{P}\{\tau \leq t \,|\, \mathcal{F}_t\}$ admits a modification with increasing sample paths.

Under (F.1), we have $F_t = \mathbb{P}\{\tau \leq t \,|\, \mathcal{F}_t\} = \mathbb{P}\{\tau \leq t \,|\, \mathcal{F}_\infty\}$ for any $t \in \mathbb{R}_+$. It is thus clear that in this case $F$ admits a modification with increasing sample paths, so that (F.2) is valid. However, the process $F$ is not necessarily $\mathbb{F}$-predictable (e.g., when $\tau$ is an $\mathbb{F}$-stopping time that is not $\mathbb{F}$-predictable).

### 6.1.1 Martingale Invariance Property

We work in the following abstract set-up: we are given a probability space $(\Omega, \mathcal{G}, \mathbb{P})$ endowed with a filtration $\mathbb{G}$; a reference filtration $\mathbb{F}$ is an arbitrary sub-filtration of $\mathbb{G}$. The definition of martingale invariance property is classic. It is important to notice that this property is not necessarily preserved under an equivalent change of the underlying probability measure $\mathbb{P}$.

**Definition 6.1.2.** A filtration $\mathbb{F}$ has the *martingale invariance property* with respect to a filtration $\mathbb{G}$ if any $\mathbb{F}$-martingale follows also a $\mathbb{G}$-martingale.

**Condition (M.1)** Filtrations $\mathbb{F}$ and $\mathbb{G}$, with $\mathbb{F} \subseteq \mathbb{G}$, satisfy (M.1) (under $\mathbb{P}$) whenever $\mathbb{F}$ has the martingale invariance property with respect to $\mathbb{G}$.

The following condition appears to be equivalent to (M.1).

**Condition (M.2)** For any $t \in \mathbb{R}_+$, the $\sigma$-fields $\mathcal{F}_\infty$ and $\mathcal{G}_t$ are conditionally independent given $\mathcal{F}_t$ under $\mathbb{P}$.

By the definition of conditional independence of $\sigma$-fields, Condition (M.2) means that for any bounded, $\mathcal{F}_\infty$-measurable random variable $\xi$ and any bounded, $\mathcal{G}_t$-measurable random variable $\eta$ we have

$$\mathbb{E}_{\mathbb{P}}(\xi\eta \,|\, \mathcal{F}_t) = \mathbb{E}_{\mathbb{P}}(\xi \,|\, \mathcal{F}_t)\mathbb{E}_{\mathbb{P}}(\eta \,|\, \mathcal{F}_t).$$

Notice that Condition (M.2) can also be re-expressed in the following way.

**Condition (M.2a)** For any $t \in \mathbb{R}_+$, and any $s \geq t$ the $\sigma$-fields $\mathcal{F}_s$ and $\mathcal{G}_t$ are conditionally independent given the $\sigma$-field $\mathcal{F}_t$.

Since $\mathcal{F}_t \subseteq \mathcal{G}_t$ and $\mathcal{F}_t \subseteq \mathcal{F}_\infty$, each of the following two conditions is also equivalent to (M.2).

**Condition (M.2b)** For any $t \in \mathbb{R}_+$ and any bounded, $\mathcal{F}_\infty$-measurable random variable $\xi$ we have $\mathbb{E}_\mathbb{P}(\xi \,|\, \mathcal{G}_t) = \mathbb{E}_\mathbb{P}(\xi \,|\, \mathcal{F}_t)$.

**Condition (M.2c)** For any $t \in \mathbb{R}_+$, and any bounded, $\mathcal{G}_t$-measurable random variable $\eta$ we have $\mathbb{E}_\mathbb{P}(\eta \,|\, \mathcal{F}_\infty) = \mathbb{E}_\mathbb{P}(\eta \,|\, \mathcal{F}_t)$.

**Lemma 6.1.1.** *A filtration $\mathbb{F}$ has the martingale invariance property with respect to a filtration $\mathbb{G}$ if and only if Condition (M.2) is satisfied. Put another way, the conditions (M.1) and (M.2) are equivalent.*

*Proof.* Suppose first that (M.2) holds. Let $M$ be an arbitrary $\mathbb{F}$-martingale. Then for any $t \leq s$ we have (the first equality below follows from (M.2b))

$$\mathbb{E}_\mathbb{P}(M_s \,|\, \mathcal{G}_t) = \mathbb{E}_\mathbb{P}(M_s \,|\, \mathcal{F}_t) = M_t,$$

and thus $M$ is a $\mathbb{G}$-martingale. Conversely, suppose that every $\mathbb{F}$-martingale is a $\mathbb{G}$-martingale. We shall check that this implies (M.2b). To this end, for any fixed $t \leq s$ we consider an arbitrary set $A \in \mathcal{F}_\infty$. We introduce the $\mathbb{F}$-martingale $M_t := \mathbb{P}\{A \,|\, \mathcal{F}_t\}$, $t \in \mathbb{R}_+$. Since $M$ is also a $\mathbb{G}$-martingale, we obtain

$$\mathbb{P}\{A \,|\, \mathcal{G}_t\} = M_t = \mathbb{P}\{A \,|\, \mathcal{F}_t\}.$$

By standard arguments, this shows that (M.2b) is valid. □

Assume now that Condition (G.1) holds – that is, we have $\mathbb{G} = \mathbb{H} \vee \mathbb{F}$ for some filtration $\mathbb{H}$. Let us recall that we have also introduced Condition (F.1). Since $\mathcal{H}_t \subseteq \mathcal{G}_t$, it is apparent that (M.2) is stronger than (F.1). It appears that both conditions are in fact equivalent.

**Lemma 6.1.2.** *Conditions (F.1) and (M.1) are equivalent.*

*Proof.* We already know that conditions (M.1) and (M.2) are equivalent, and (M.2) is stronger than (F.1). It is enough to check that (F.1) implies (M.2). Condition (F.1) is equivalent to the following condition: for any bounded, $\mathcal{F}_\infty$-measurable random variable $\xi$ we have $\mathbb{E}_\mathbb{P}(\xi \,|\, \mathcal{H}_t \vee \mathcal{F}_t) = \mathbb{E}_\mathbb{P}(\xi \,|\, \mathcal{F}_t)$. Since $\mathcal{G}_t = \mathcal{H}_t \vee \mathcal{F}_t$, this immediately gives (M.2b). □

### 6.1.2 Evaluation of $\Lambda$: Special Case

In this section, we assume that (G.1) and (F.2) hold.

**Proposition 6.1.1.** *Assume that $F$ is an increasing, $\mathbb{F}$-predictable process. Then the process $\Lambda$ given by the formula*

$$\Lambda_t = \int_{]0,t]} \frac{dF_u}{1 - F_{u-}} = \int_{]0,t]} \frac{d\mathbb{P}\{\tau \leq u \,|\, \mathcal{F}_u\}}{1 - \mathbb{P}\{\tau < u \,|\, \mathcal{F}_u\}} \tag{6.1}$$

*is the $\mathbb{F}$-martingale hazard process of $\tau$.*

*Proof.* We need to check that the compensated process $H_t - \Lambda_{t \wedge \tau}$ follows a $\mathbb{G}$-martingale, where $\mathbb{G} = \mathbb{H} \vee \mathbb{F}$. Using (5.3), for $t < s$ we obtain

$$\mathbb{E}_{\mathbb{P}}(H_s - H_t \,|\, \mathcal{G}_t) = \mathbb{P}\{t < \tau \le s \,|\, \mathcal{G}_t\} = \mathbb{1}_{\{\tau > t\}} \frac{\mathbb{P}\{t < \tau \le s \,|\, \mathcal{F}_t\}}{\mathbb{P}\{\tau > t \,|\, \mathcal{F}_t\}}$$

$$= \mathbb{1}_{\{\tau > t\}} \frac{\mathbb{E}_{\mathbb{P}}(F_s \,|\, \mathcal{F}_t) - F_t}{1 - F_t}.$$

On the other hand, for the process $\Lambda$ given by (6.1) we obtain

$$\mathbb{E}_{\mathbb{P}}(\Lambda_{s \wedge \tau} - \Lambda_{t \wedge \tau} \,|\, \mathcal{G}_t) = \mathbb{E}_{\mathbb{P}}\left( \int_{]t \wedge \tau, s \wedge \tau]} \frac{dF_u}{1 - F_{u-}} \,\bigg|\, \mathcal{G}_t \right) = \mathbb{E}_{\mathbb{P}}(\mathbb{1}_{\{\tau > t\}} Y \,|\, \mathcal{G}_t),$$

where

$$Y := \int_{]t, s \wedge \tau]} \frac{dF_u}{1 - F_{u-}} = \mathbb{1}_{\{\tau > t\}} Y. \tag{6.2}$$

Furthermore, using (5.11), we get

$$\mathbb{E}_{\mathbb{P}}(\mathbb{1}_{\{\tau > t\}} Y \,|\, \mathcal{G}_t) = \mathbb{1}_{\{\tau > t\}} \frac{\mathbb{E}_{\mathbb{P}}(\mathbb{1}_{\{\tau > t\}} Y \,|\, \mathcal{F}_t)}{\mathbb{P}\{\tau > t \,|\, \mathcal{F}_t\}}.$$

It is thus enough to verify that for $I := \mathbb{E}_{\mathbb{P}}(\mathbb{1}_{\{\tau > t\}} Y \,|\, \mathcal{F}_t)$ we have:

$$I = \mathbb{E}_{\mathbb{P}}\left( \int_{]t, s \wedge \tau]} \frac{dF_u}{1 - F_{u-}} \,\bigg|\, \mathcal{F}_t \right) = \mathbb{E}_{\mathbb{P}}(F_s - F_t \,|\, \mathcal{F}_t). \tag{6.3}$$

To this end, notice that

$$I = \mathbb{E}_{\mathbb{P}}\left( \mathbb{1}_{\{\tau > s\}} \int_{]t, s]} \frac{dF_u}{1 - F_{u-}} + \mathbb{1}_{\{t < \tau \le s\}} \int_{]t, s \wedge \tau]} \frac{dF_u}{1 - F_{u-}} \,\bigg|\, \mathcal{F}_t \right)$$

$$= \mathbb{E}_{\mathbb{P}}\left( \mathbb{E}_{\mathbb{P}}\left( \mathbb{1}_{\{\tau > s\}} \int_{]t, s]} \frac{dF_u}{1 - F_{u-}} \,\bigg|\, \mathcal{F}_s \right) + \mathbb{1}_{\{t < \tau \le s\}} \int_{]t, s \wedge \tau]} \frac{dF_u}{1 - F_{u-}} \,\bigg|\, \mathcal{F}_t \right)$$

$$= \mathbb{E}_{\mathbb{P}}\left( (1 - F_s) \int_{]t, s]} \frac{dF_u}{1 - F_{u-}} + \int_{]t, s]} \int_{]t, u]} \frac{dF_v}{1 - F_{v-}} \, dF_u \,\bigg|\, \mathcal{F}_t \right)$$

$$= \mathbb{E}_{\mathbb{P}}\left( (1 - F_s)(\Lambda_s - \Lambda_t) + \int_{]t, s]} (\Lambda_u - \Lambda_t) \, dF_u \,\bigg|\, \mathcal{F}_t \right),$$

where the third equality is a consequence of (5.19) applied to the $\mathbb{F}$-predictable process $Z_s = \int_{]t, s]} (1 - F_{u-})^{-1} dF_u$. To conclude the proof, one may now argue along similar lines as in the proof of part (i) in Proposition 4.5.1. Under the present assumptions, $\Lambda$ and $F$ are processes of finite variation, so that their continuous martingale parts vanish identically. The product rule (cf. (4.29)):

$$\int_{]t, s]} \Lambda_u \, dF_u = \Lambda_s F_s - \Lambda_t F_t - \int_{]t, s]} F_{u-} \, d\Lambda_u \tag{6.4}$$

is thus the path-by-path version of the deterministic integration by parts formula of Lemma 4.2.2. $\qquad\square$

*Remarks.* Alternatively, to evaluate the conditional expectation

$$K := \mathbb{E}_{\mathbb{P}}\big(\Lambda_{s\wedge\tau} - \Lambda_{t\wedge\tau}\,|\,\mathcal{G}_t\big),$$

we can directly apply formula (5.18) of Corollary 5.1.3. It is enough to notice that

$$K = \mathbb{E}_{\mathbb{P}}\big(1\!\!1_{\{\tau>s\}}(\Lambda_s - \Lambda_t)\,|\,\mathcal{G}_t\big) + \mathbb{E}_{\mathbb{P}}\big(1\!\!1_{\{t<\tau\le s\}}\,\tilde{\Lambda}_\tau\,|\,\mathcal{G}_t\big),$$

where, for a fixed $t$, we write $\tilde{\Lambda}_u = (\Lambda_u - \Lambda_t)1\!\!1_{]t,\infty[}(u)$ (so that $\tilde{\Lambda}$ follows an $\mathbb{F}$-predictable process). Therefore, an application of (5.18) yields

$$\mathbb{E}_{\mathbb{P}}\big(1\!\!1_{\{t<\tau\le s\}}\,\tilde{\Lambda}_\tau\,|\,\mathcal{G}_t\big) = 1\!\!1_{\{\tau>t\}}e^{\Gamma_t}\,\mathbb{E}_{\mathbb{P}}\Big(\int_{]t,s]}(\Lambda_u - \Lambda_t)\,dF_u\,\Big|\,\mathcal{F}_t\Big).$$

On the other hand, (5.11) gives

$$\mathbb{E}_{\mathbb{P}}\big(1\!\!1_{\{\tau>s\}}(\Lambda_s - \Lambda_t)\,|\,\mathcal{G}_t\big) = 1\!\!1_{\{\tau>t\}}e^{\Gamma_t}\,\mathbb{E}_{\mathbb{P}}\big(1\!\!1_{\{\tau>s\}}\,(\Lambda_s - \Lambda_t)\,|\,\mathcal{F}_t\big).$$

Combining the above formulae, we obtain

$$K = 1\!\!1_{\{\tau>t\}}e^{\Gamma_t}\,\mathbb{E}_{\mathbb{P}}\Big(1\!\!1_{\{\tau>s\}}\,(\Lambda_s - \Lambda_t) + \int_{]t,s]}(\Lambda_u - \Lambda_t)\,dF_u\,\Big|\,\mathcal{F}_t\Big)$$

$$= 1\!\!1_{\{\tau>t\}}e^{\Gamma_t}\,\mathbb{E}_{\mathbb{P}}\Big((1 - F_s)(\Lambda_s - \Lambda_t) + \int_{]t,s]}(\Lambda_u - \Lambda_t)\,dF_u\,\Big|\,\mathcal{F}_t\Big),$$

where the last equality is derived by conditioning first with respect to the $\sigma$-field $\mathcal{F}_s$.

### 6.1.3 Evaluation of $\Lambda$: General Case

We maintain the assumption that (G.1) holds. On the other hand, we assume that either (F.2) is not satisfied (so that the process $F$ is not increasing) or (F.2) is valid, but the increasing process $F$ is not $\mathbb{F}$-predictable.

*Example 6.1.1.* For instance, a random time $\tau$ can be an $\mathbb{F}$-stopping time, which is not $\mathbb{F}$-predictable. If $\tau$ is an $\mathbb{F}$-stopping time, we have $F = H$, and the process $H$ is not $\mathbb{F}$-predictable, unless a stopping time $\tau$ is $\mathbb{F}$-predictable.

As the next result shows, the $\mathbb{F}$-martingale hazard process $\Lambda$ can still be found through a suitable modification of formula (6.1). In the next result, we do not need to assume that (F.2) holds. We shall write $\tilde{F}$ to denote the $\mathbb{F}$-*compensator* of the bounded $\mathbb{F}$-submartingale $F$. This means that $\tilde{F}$ is the unique $\mathbb{F}$-predictable, increasing process, with $\tilde{F}_0 = 0$, and such that the *compensated process* $U = F - \tilde{F}$ follows an $\mathbb{F}$-martingale. Let us recall that the existence and uniqueness of $\tilde{F}$ is an immediate consequence of the Doob-Meyer decomposition theorem.

*Remarks.* In some applications, the $\mathbb{F}$-stopping time $\tau$ is assumed to be *totally inaccessible* (cf. Dellacherie (1972)). In this case, the compensator $\tilde{F}$ of the increasing process $F = H$ is known to follow an $\mathbb{F}$-adapted process with continuous increasing sample paths.

**Lemma 6.1.3.** *Let $Z$ be a bounded, $\mathbb{F}$-predictable process. Then for any $t \leq s$*

$$\mathbb{E}_{\mathbb{P}}(\mathbb{1}_{\{t<\tau\leq s\}} Z_\tau \mid \mathcal{F}_t) = \mathbb{E}_{\mathbb{P}}\left(\int_{]t,s]} Z_u \, d\tilde{F}_u \,\Big|\, \mathcal{F}_t\right).$$

*Proof.* The martingale property of $U = F - \tilde{F}$ yields

$$\mathbb{E}_{\mathbb{P}}\left(\int_{]t,s]} Z_u \, d(F_u - \tilde{F}_u) \,\Big|\, \mathcal{F}_t\right) = \mathbb{E}_{\mathbb{P}}\left(\int_{]t,s]} Z_u \, dU_u \,\Big|\, \mathcal{F}_t\right) = 0.$$

It is thus enough to make use of (5.19).                                    □

**Proposition 6.1.2.** (i) *The $\mathbb{F}$-martingale hazard process of a random time $\tau$ is given by the formula*

$$\Lambda_t = \int_{]0,t]} \frac{d\tilde{F}_u}{1 - F_{u-}}. \tag{6.5}$$

(ii) *Assume that $\tilde{F}_t = \tilde{F}_{t\wedge\tau}$ for every $t \in \mathbb{R}_+$, i.e., the process $\tilde{F}$ is stopped at a random time $\tau$. Then the equality $\Lambda = \tilde{F}$ is valid.*

*Proof.* It is clear that the process $\Lambda$ given by (6.5) is $\mathbb{F}$-predictable. We thus need only to verify that the process $\tilde{M}_t = H_t - \Lambda_{t\wedge\tau}$ follows a $\mathbb{G}$-martingale. In the first part of the proof, we proceed along the same lines as in the proof of Proposition 6.1.1. We find that, in the present case, it is enough to show that for any $s \geq t$ (cf. (6.3))

$$\tilde{I} := \mathbb{E}_{\mathbb{P}}\left(\int_{]t,s\wedge\tau]} \frac{d\tilde{F}_u}{1 - F_{u-}} \,\Big|\, \mathcal{F}_t\right) = \mathbb{E}_{\mathbb{P}}(F_s - F_t \mid \mathcal{F}_t) = \mathbb{E}_{\mathbb{P}}(\tilde{F}_s - \tilde{F}_t \mid \mathcal{F}_t),$$

where the second equality is a consequence of the definition of $\tilde{F}$. We have

$$\tilde{I} = \mathbb{E}_{\mathbb{P}}\left(\mathbb{1}_{\{\tau>s\}} \int_{]t,s]} \frac{d\tilde{F}_u}{1 - F_{u-}} + \mathbb{1}_{\{t<\tau\leq s\}} \int_{]t,s\wedge\tau]} \frac{d\tilde{F}_u}{1 - F_{u-}} \,\Big|\, \mathcal{F}_t\right)$$

$$= \mathbb{E}_{\mathbb{P}}\left(\mathbb{E}_{\mathbb{P}}\left(\mathbb{1}_{\{\tau>s\}} \int_{]t,s]} \frac{d\tilde{F}_u}{1 - F_{u-}} \,\Big|\, \mathcal{F}_s\right) + \mathbb{1}_{\{t<\tau\leq s\}} \int_{]t,s\wedge\tau]} \frac{d\tilde{F}_u}{1 - F_{u-}} \,\Big|\, \mathcal{F}_t\right)$$

$$= \mathbb{E}_{\mathbb{P}}\left((1 - F_s) \int_{]t,s]} \frac{d\tilde{F}_u}{1 - F_{u-}} + \int_{]t,s]} \int_{]t,u]} \frac{d\tilde{F}_v}{1 - F_{v-}} \, d\tilde{F}_u \,\Big|\, \mathcal{F}_t\right)$$

$$= \mathbb{E}_{\mathbb{P}}\left((\Lambda_s - \Lambda_t)(1 - F_s) + \int_{]t,s]} (\Lambda_u - \Lambda_t) \, d\tilde{F}_u \,\Big|\, \mathcal{F}_t\right),$$

where the third equality follows from Lemma 6.1.3, combined with equality (5.19). Since $\Lambda$ is $\mathbb{F}$-predictable and $U$ is an $\mathbb{F}$-martingale, we obtain

$$\mathbb{E}_{\mathbb{P}}\left(\int_{]t,s]} (\Lambda_u - \Lambda_t) \, dU_u \,\Big|\, \mathcal{F}_t\right) = 0,$$

which in turn entails that

$$\tilde{I} = \mathbb{E}_{\mathbb{P}}\left((\Lambda_s - \Lambda_t)(1 - F_s) + \int_{]t,s]}(\Lambda_u - \Lambda_t)\,d\tilde{F}_u \,\Big|\, \mathcal{F}_t\right)$$

$$= \mathbb{E}_{\mathbb{P}}\left((\Lambda_s - \Lambda_t)(1 - F_s) + \int_{]t,s]}(\Lambda_u - \Lambda_t)\,d(F_u - U_u) \,\Big|\, \mathcal{F}_t\right)$$

$$= \mathbb{E}_{\mathbb{P}}\left((\Lambda_s - \Lambda_t)(1 - F_s) + \int_{]t,s]}(\Lambda_u - \Lambda_t)\,dF_u \,\Big|\, \mathcal{F}_t\right).$$

Our goal is to show that $\tilde{I} = \mathbb{E}_{\mathbb{P}}(\tilde{F}_s - \tilde{F}_t \,|\, \mathcal{F}_t)$. For this purpose, we observe that

$$\int_{]t,s]}(\Lambda_u - \Lambda_t)\,dF_u = -\Lambda_t(F_s - F_t) + \int_{]t,s]}\Lambda_u\,dF_u.$$

Since $\Lambda$ is a process of finite variation, Itô's product rule yields

$$\int_{]t,s]}\Lambda_u\,dF_u = \Lambda_s F_s - \Lambda_t F_t - \int_{]t,s]}F_{u-}\,d\Lambda_u. \tag{6.6}$$

Finally, it follows From (6.5) that

$$\int_{]t,s]}F_{u-}\,d\Lambda_u = \Lambda_s - \Lambda_t - \tilde{F}_s + \tilde{F}_t.$$

Combining the above formulae, we conclude that

$$(\Lambda_s - \Lambda_t)(1 - F_s) + \int_{]t,s]}(\Lambda_u - \Lambda_t)\,dF_u = \tilde{F}_s - \tilde{F}_t. \tag{6.7}$$

This completes the proof of part (i). We shall now prove part (ii). We assume that $\tilde{F}_{t\wedge\tau} = \tilde{F}_t$ for $t \in \mathbb{R}_+$, and thus the process $F_t - \tilde{F}_{t\wedge\tau}$ is an $\mathbb{F}$-martingale. We wish to show that if the process $H_t - \tilde{F}_{t\wedge\tau}$ follows a $\mathbb{G}$-martingale, that is, for any $t \le s$,

$$\mathbb{E}_{\mathbb{P}}(H_s - \tilde{F}_{s\wedge\tau} \,|\, \mathcal{G}_t) = H_t - \tilde{F}_{t\wedge\tau}$$

or, equivalently,

$$\mathbb{E}_{\mathbb{P}}(H_s - H_t \,|\, \mathcal{G}_t) = \mathbb{E}_{\mathbb{P}}(\tilde{F}_{s\wedge\tau} - \tilde{F}_{t\wedge\tau} \,|\, \mathcal{G}_t).$$

By virtue of (5.3), we have

$$\mathbb{E}_{\mathbb{P}}(H_s - H_t \,|\, \mathcal{G}_t) = (1 - H_t)\frac{\mathbb{E}_{\mathbb{P}}(H_s - H_t \,|\, \mathcal{F}_t)}{\mathbb{E}_{\mathbb{P}}(1 - H_t \,|\, \mathcal{F}_t)}. \tag{6.8}$$

On the other hand, for the random variable $\tilde{J} := \mathbb{E}_{\mathbb{P}}(\tilde{F}_{s\wedge\tau} - \tilde{F}_{t\wedge\tau} \,|\, \mathcal{G}_t)$, we obtain

$$\tilde{J} = \mathbb{E}_{\mathbb{P}}\left(\mathbb{1}_{\{\tau > t\}}(\tilde{F}_{s\wedge\tau} - \tilde{F}_{t\wedge\tau}) \,|\, \mathcal{G}_t\right) = (1 - H_t)\frac{\mathbb{E}_{\mathbb{P}}(\tilde{F}_{s\wedge\tau} - \tilde{F}_{t\wedge\tau} \,|\, \mathcal{F}_t)}{\mathbb{E}_{\mathbb{P}}(1 - H_t \,|\, \mathcal{F}_t)}$$

$$= (1 - H_t)\frac{\mathbb{E}_{\mathbb{P}}(F_s - F_t \,|\, \mathcal{F}_t)}{\mathbb{E}_{\mathbb{P}}(1 - H_t \,|\, \mathcal{F}_t)} = (1 - H_t)\frac{\mathbb{E}_{\mathbb{P}}(H_s - H_t \,|\, \mathcal{F}_t)}{\mathbb{E}_{\mathbb{P}}(1 - H_t \,|\, \mathcal{F}_t)},$$

where the second equality follows from (5.2), and the third is a consequence of our assumption that the process $F_t - \tilde{F}_{t\wedge\tau}$ is an $\mathbb{F}$-martingale.  $\square$

Under (F.1), the process $\tilde{F}$ is never stopped at $\tau$, unless $\tau$ is an F-stopping time. To show this assume, on the contrary, that $\tilde{F}_t = \tilde{F}_{t\wedge\tau}$. Under (F.1), the process $F_t - \tilde{F}_{t\wedge\tau}$ is not only an F-martingale, but also a G-martingale (see Lemmas 6.1.1–6.1.2). Since by virtue of part (ii) in Proposition 6.1.2 the process $H_t - \tilde{F}_{t\wedge\tau}$ is a G-martingale, we conclude that $H - F$ follows a G-martingale. In view of the definition of $F$, the last property reads, for $t \le s$,

$$\mathbb{E}_{\mathbb{P}}\big(H_s - \mathbb{E}_{\mathbb{P}}(H_s \,|\, \mathcal{F}_s) \,\big|\, \mathcal{G}_t\big) = H_t - \mathbb{E}_{\mathbb{P}}(H_t \,|\, \mathcal{F}_t)$$

or, equivalently,

$$\mathbb{E}_{\mathbb{P}}(H_s - H_t \,|\, \mathcal{G}_t) = \mathbb{E}_{\mathbb{P}}\big(\mathbb{E}_{\mathbb{P}}(H_s \,|\, \mathcal{F}_s) \,|\, \mathcal{G}_t\big) - \mathbb{E}_{\mathbb{P}}(H_t \,|\, \mathcal{F}_t) = I_1 - I_2. \qquad (6.9)$$

Under (F.1), we have (cf. (F.1a))

$$I_1 = \mathbb{E}_{\mathbb{P}}(\mathbb{P}\{\tau \le s \,|\, \mathcal{F}_s\} \,|\, \mathcal{F}_t \vee \mathcal{H}_t) = \mathbb{E}_{\mathbb{P}}(\mathbb{P}\{\tau \le s \,|\, \mathcal{F}_\infty\} \,|\, \mathcal{F}_t \vee \mathcal{H}_t)$$
$$= \mathbb{E}_{\mathbb{P}}(\mathbb{P}\{\tau \le s \,|\, \mathcal{F}_\infty\} \,|\, \mathcal{F}_t).$$

The last equality follows from the $\mathcal{F}_\infty$-measurability of the random variable $\mathbb{P}\{\tau \le s \,|\, \mathcal{F}_\infty\}$, combined with the fact that the $\sigma$-fields $\mathcal{F}_\infty$ and $\mathcal{H}_t$ are conditionally independent given $\mathcal{F}_t$. Consequently,

$$I_1 = \mathbb{E}_{\mathbb{P}}(\mathbb{E}_{\mathbb{P}}(H_s \,|\, \mathcal{F}_\infty) \,|\, \mathcal{F}_t) = \mathbb{E}_{\mathbb{P}}(H_s \,|\, \mathcal{F}_t).$$

We conclude that (6.9) can be rewritten as follows:

$$\mathbb{E}_{\mathbb{P}}(H_s - H_t \,|\, \mathcal{G}_t) = \mathbb{E}_{\mathbb{P}}(H_s \,|\, \mathcal{F}_t) - \mathbb{E}_{\mathbb{P}}(H_t \,|\, \mathcal{F}_t).$$

Finally, by applying (6.8) to the left-hand side of the last equality, we obtain

$$(1 - H_t) \frac{\mathbb{E}_{\mathbb{P}}(H_s - H_t \,|\, \mathcal{F}_t)}{\mathbb{E}_{\mathbb{P}}(1 - H_t \,|\, \mathcal{F}_t)} = \mathbb{E}_{\mathbb{P}}(H_s - H_t \,|\, \mathcal{F}_t).$$

Letting $s$ tend to $\infty$ in the last formula, we obtain $H_t = \mathbb{E}_{\mathbb{P}}(H_t \,|\, \mathcal{F}_t)$ or, more explicitly, $\mathbb{P}\{\tau \le t \,|\, \mathcal{F}_t\} = \mathbb{1}_{\{\tau \le t\}}$ for every $t \in \mathbb{R}_+$. We conclude that a random time $\tau$ is indeed an F-stopping time.

### 6.1.4 Uniqueness of a Martingale Hazard Process $\Lambda$

We shall first examine the relationship between the concept of an F-martingale hazard process $\Lambda$ of $\tau$ and the classic notion of the G-*compensator* (that is, the *dual predictable projection*) of the jump process $H$ associated with a random time $\tau$. For convenience, the compensator of the process $H$ is henceforth called the compensator of $\tau$.

**Definition 6.1.3.** A process $A$ is a G-*compensator* of $\tau$ if and only if the following conditions are satisfied: (i) $A$ is a G-predictable, right-continuous, increasing process, with $A_0 = 0$, (ii) the process $H - A$ is a G-martingale.

It is well known that for any random time $\tau$ and any filtration $\mathbb{G}$ such that $\tau$ is a $\mathbb{G}$-stopping time there exists a unique $\mathbb{G}$-compensator $A$ of $\tau$. Moreover, $A_t = A_{t \wedge \tau}$, i.e., the increasing process $A$ is in fact stopped at $\tau$. In the next auxiliary result, we shall deal with an arbitrary filtration $\mathbb{F}$, which, when combined with the natural filtration $\mathbb{H}$ of a $\mathbb{G}$-stopping time $\tau$, generates the enlarged filtration $\mathbb{G}$. Since both statements are classic, the proof is omitted.

**Lemma 6.1.4.** *Let $\mathbb{F}$ be an arbitrary filtration such that $\mathbb{G} = \mathbb{H} \vee \mathbb{F}$. Then:*
(i) *Let $A$ be a $\mathbb{G}$-predictable right-continuous increasing process satisfying $A_t = A_{t \wedge \tau}$. Then there exists an $\mathbb{F}$-predictable right-continuous increasing process $\Lambda$ such that $A_t = \Lambda_{t \wedge \tau}$.*
(ii) *Let $\Lambda$ be an $\mathbb{F}$-predictable right-continuous increasing process. Then $A_t = \Lambda_{t \wedge \tau}$ is a $\mathbb{G}$-predictable right-continuous increasing process.*

The next proposition summarizes the connections between the $\mathbb{G}$-compensator $A$ of $\tau$ and the $\mathbb{F}$-martingale hazard process $\Lambda$ of $\tau$. Once more $\mathbb{F}$ is an arbitrary filtration such that $\mathbb{G} = \mathbb{H} \vee \mathbb{F}$.

**Proposition 6.1.3.** (i) *Let $A$ be the $\mathbb{G}$-compensator of $\tau$. Then there exists an $\mathbb{F}$-martingale hazard process $\Lambda$ such that $A_t = \Lambda_{t \wedge \tau}$.*
(ii) *Let $\Lambda$ be an $\mathbb{F}$-martingale hazard process of $\tau$. Then the process $A_t = \Lambda_{t \wedge \tau}$ is the $\mathbb{G}$-compensator of $\tau$.*

*Proof.* The first (second, resp.) statement follows from part (i) (part (ii), resp.) in Lemma 6.1.4.  □

From the uniqueness of the $\mathbb{G}$-compensator, combined with part (ii) in Proposition 6.1.3, it follows that the $\mathbb{F}$-martingale hazard process is unique up to time $\tau$ in the following sense: if $\Lambda^1$ and $\Lambda^2$ are the two $\mathbb{F}$-martingale hazard processes of $\tau$, then the stopped processes coincide: $\Lambda^1_{t \wedge \tau} = \Lambda^2_{t \wedge \tau}$ for every $t \in \mathbb{R}_+$. To ensure some sort of uniqueness after $\tau$ of an $\mathbb{F}$-martingale hazard process, one needs to impose some additional restrictions.

Let $\tau$ be a $\mathbb{G}$-stopping time $\tau$ for some filtration $\mathbb{G}$. Then the sub-filtration $\mathbb{F}$ of $\mathbb{G}$ for which we have $\mathbb{G} = \mathbb{H} \vee \mathbb{F}$ is not uniquely specified. Assume that $\mathbb{G} = \mathbb{H} \vee \mathbb{F}^1 = \mathbb{H} \vee \mathbb{F}^2$, and denote by $\Lambda^i$ an $\mathbb{F}^i$-martingale hazard process of $\tau$. Then $\Lambda^1_{t \wedge \tau} = A_{t \wedge \tau} = \Lambda^2_{t \wedge \tau}$. It seems natural to search for the $\hat{\mathbb{F}}$-martingale hazard process, where $\hat{\mathbb{F}}$ is a 'minimal' filtration for which $\mathbb{G} = \mathbb{H} \vee \hat{\mathbb{F}}$.

## 6.2 Relationships Between Hazard Processes $\Gamma$ and $\Lambda$

Let us assume that the $\mathbb{F}$-hazard process $\Gamma$ is well defined (in particular, $\tau$ is not an $\mathbb{F}$-stopping time). We already know that under (G.1), for any $\mathcal{F}_s$-measurable random variable $Y$ we have (cf. (5.13))

$$\mathbb{E}_{\mathbb{P}}(\mathbb{1}_{\{\tau > s\}} Y \mid \mathcal{G}_t) = \mathbb{1}_{\{\tau > t\}} \mathbb{E}_{\mathbb{P}}(Y e^{\Gamma_t - \Gamma_s} \mid \mathcal{F}_t). \tag{6.10}$$

The natural question, which arises in this context is: can we substitute $\Gamma$ with the $\mathbb{F}$-martingale hazard function $\Lambda$ in the last formula? Of course, the answer is trivial when it is known that the equality $\Lambda = \Gamma$ is satisfied, for instance, when conditions (G.1) and (F.2) are fulfilled and the process $F$ is continuous. We are thus in a position the following result, which corresponds to parts (ii)-(iii) of Proposition 4.5.1.

**Proposition 6.2.1.** *Under* (G.1) *and* (F.2), *the following statements hold.*
(i) *If the increasing process* $F$ *is* $\mathbb{F}$-*predictable, but* $F$ *is not continuous, then the* $\mathbb{F}$-*martingale hazard process* $\Lambda$ *is also discontinuous process and we have*

$$e^{-\Gamma_t} = e^{-\Lambda_t^c} \prod_{0 < u \leq t} (1 - \Delta\Lambda_u),$$

*where we write* $\Lambda^c$ *to denote the continuous component of* $\Lambda$. *More explicitly,*
$\Lambda_t^c = \Lambda_t - \sum_{0 \leq u \leq t} \Delta\Lambda_u$ *for every* $t \in \mathbb{R}_+$.
(ii) *If the increasing process* $F$ *is continuous, then the* $\mathbb{F}$-*martingale hazard process* $\Lambda$ *is also continuous and*

$$\Gamma_t = \Lambda_t = -\ln(1 - F_t), \quad \forall t \in \mathbb{R}_+.$$

*If, in addition, the process* $\Lambda = \Gamma$ *is absolutely continuous then for an integrable* $\mathcal{F}_s$-*measurable random variable* $Y$ *we get*

$$\mathbb{E}_{\mathbb{P}}(\mathbb{1}_{\{\tau > s\}} Y \,|\, \mathcal{G}_t) = \mathbb{1}_{\{\tau > t\}} \mathbb{E}_{\mathbb{P}}\big(Y e^{-\int_t^s \lambda_u \, du} \,|\, \mathcal{F}_t\big).$$

*Proof.* The hazard process $\Gamma$ is discontinuous if and only if the process $F$ admits discontinuities. Under (G.1) and (F.2), by virtue of Proposition 6.1.1, we have

$$G_t = -\int_{]0,t]} G_{u-} \, d\Lambda_u.$$

Since $\Lambda$ is a process of finite variation, we obtain (cf. (4.24)–(4.25))

$$e^{-\Gamma_t} = G_t = e^{-\Lambda_t^c} \prod_{0 < u \leq t} (1 - \Delta\Lambda_u).$$

The second statement is an immediate consequence of part (i).    $\square$

The following result is a straightforward consequence of Corollary 5.1.3.

**Corollary 6.2.1.** *Suppose that* (G.1) *and* (F.2) *hold and* $F$ *is a continuous process. Let* $Y = h(\tau)$ *for some bounded, continuous function* $h : \mathbb{R}_+ \to \mathbb{R}$. *Then*

$$\mathbb{E}_{\mathbb{P}}(Y \,|\, \mathcal{G}_t) = \mathbb{1}_{\{\tau \leq t\}} h(\tau) + \mathbb{1}_{\{\tau > t\}} \mathbb{E}_{\mathbb{P}}\Big(\int_t^\infty h(u) e^{\Lambda_t - \Lambda_u} \, d\Lambda_u \,\Big|\, \mathcal{F}_t\Big).$$

*Let* $Z$ *be a bounded,* $\mathbb{F}$-*predictable process. Then for any* $t \leq s$

$$\mathbb{E}_{\mathbb{P}}(Z_\tau \mathbb{1}_{\{t < \tau \leq s\}} \,|\, \mathcal{G}_t) = \mathbb{1}_{\{\tau > t\}} \mathbb{E}_{\mathbb{P}}\Big(\int_t^s Z_u e^{\Lambda_t - \Lambda_u} \, d\Lambda_u \,\Big|\, \mathcal{F}_t\Big).$$

In some instances, the $\mathbb{F}$-martingale hazard process of a random time $\tau$ can be found through the martingale approach. The following question thus arises: does the continuity of the $\mathbb{F}$-martingale hazard process $\Lambda$ imply the equality $\Lambda = \Gamma$? The next result furnishes a partial answer to this question.

**Proposition 6.2.2.** *Under* (G.1) *and* (F.2), *assume that the filtration* $\mathbb{F}$ *supports only continuous martingales. If the* $\mathbb{F}$-*martingale hazard process* $\Lambda$ *is continuous, then the hazard process* $\Gamma$ *is also continuous and* $\Gamma = \Lambda$.

*Proof.* We know that the $\mathbb{F}$-martingale hazard process $\Lambda$ is given by (6.5). Therefore, if $\Lambda$ is continuous, then the process $\tilde{F}$ is continuous as well. Since the $\mathbb{F}$-martingale $U = F - \tilde{F}$ is necessarily continuous, it results that $F = U + \tilde{F}$ follows a continuous, increasing process. We conclude that $\Lambda$ is given by (6.1), so that $\Lambda_t = -\ln(1 - F_t) = \Gamma_t$. □

Let us state the following conjecture.

**Conjecture (A).** Under assumptions (G.1) and (F.2), if the $\mathbb{F}$-martingale hazard process $\Lambda$ of $\tau$ is continuous, then the equality $\Gamma = \Lambda$ holds.

The following counter-example – borrowed from Elliott et al. (2000) – shows that in general (more specifically, when Condition (F.2) fails to hold) the implication in Conjecture (A) is not valid.

*Example 6.2.1.* Let $W$ be a standard Brownian motion on $(\Omega, \mathbb{F}, \mathbb{P})$, where $\mathbb{F} = \mathbb{F}^W$ is the natural filtration of $W$. We define a random time $\tau$ on $(\Omega, \mathcal{F}_1, \mathbb{P})$ by setting: $\tau = \sup \{ t \leq 1 : W_t = 0 \}$. In words, $\tau$ is the last passage time to 0 before time 1 by the Brownian motion $W$. We set $\mathbb{G} = \mathbb{H} \vee \mathbb{F}$. Then the $\mathbb{F}$-hazard process of $\tau$ equals $\Gamma_t = -\ln(1 - F_t)$, where

$$F_t = \mathbb{P}\{\tau \leq t \,|\, \mathcal{F}_t\} = \tilde{N}\Big(\frac{|W_t|}{\sqrt{1-t}}\Big), \quad \tilde{N}(x) := \sqrt{\frac{2}{\pi}} \int_0^x e^{-u^2/2}\, du.$$

Let $L^0$ stand for the local time of $W$ at the origin (for the properties of local times, we refer to Karatzas and Shreve (1991) or Revuz and Yor (1991)). We claim that the $\mathbb{F}$-martingale hazard process of $\tau$ equals, for $t \in [0,1]$,

$$\Lambda_t = \sqrt{\frac{2}{\pi}} \int_0^t \frac{dL_s^0}{\sqrt{1-s}}.$$

To check this, we shall use the following result (see Yor (1997)).

**Proposition 6.2.3.** *For every* $t \in [0,1)$, *the* $\mathbb{F}$-*hazard process of* $\tau$ *equals*

$$F_t = \mathbb{P}\{\tau \leq t \,|\, \mathcal{F}_t\} = \tilde{N}\Big(\frac{|W_t|}{\sqrt{1-t}}\Big). \tag{6.11}$$

*The* $\mathbb{F}$-*compensator of* $F$ *equals*

$$\tilde{F}_t = \sqrt{\frac{2}{\pi}} \int_0^t \frac{dL_s^0}{\sqrt{1-s}},$$

*where* $L^0$ *is the local time at the origin of the Brownian motion* $W$.

*Proof.* For any fixed $t < 1$, the event $\{\tau \le t\}$ coincides with the event $\{d_t > 1\}$, where $d_t = \inf\{u \ge t : W_u = 0\}$. Let us quote the following equality (cf. Yor (1997))

$$d_t = t + \inf\{u \ge 0 : W_{u+t} - W_t = -W_t\} = t + \hat{\tau}_{-W_t} \stackrel{d}{=} t + \frac{W_t^2}{G^2}. \qquad (6.12)$$

We denote here $\hat{\tau}_b = \inf\{u \ge 0 : \hat{W}_u = b\}$, where $\hat{W}_u = W_{u+t} - W_t$, $u \ge 0$, is a Brownian motion independent of $\mathcal{F}_t^W$. Also, $G$ has a Gaussian law with mean 0 and variance 1, and $G$ is independent of $W_t$.

Standard calculations show that, for any $a \in \mathbb{R}$,

$$\mathbb{P}\left(\frac{a^2}{G^2} > 1 - t\right) = \tilde{N}\left(\frac{|a|}{\sqrt{1-t}}\right). \qquad (6.13)$$

The Itô-Tanaka formula, combined with the classic identity:

$$x\tilde{N}'(x) + \tilde{N}''(x) = 0,$$

lead to

$$\tilde{N}\left(\frac{|W_t|}{\sqrt{1-t}}\right) = \int_0^t \tilde{N}'\left(\frac{|W_s|}{\sqrt{1-s}}\right) d\left(\frac{|W_s|}{\sqrt{1-s}}\right) + \frac{1}{2}\int_0^t \tilde{N}''\left(\frac{|W_s|}{\sqrt{1-s}}\right) \frac{ds}{1-s}$$

$$= \int_0^t \tilde{N}'\left(\frac{|W_s|}{\sqrt{1-s}}\right) \frac{\text{sgn}(W_s)}{\sqrt{1-s}} dW_s + \int_0^t \tilde{N}'\left(\frac{|W_s|}{\sqrt{1-s}}\right) \frac{dL_s^0}{\sqrt{1-s}}.$$

Let us recall the well-known property of the Brownian local time: if $g : \mathbb{R} \to \mathbb{R}$ is a non-negative, Borel measurable function, then

$$\int_0^t g(W_s)\, dL_s^0 = g(0)L_t^0, \quad \forall t \in \mathbb{R}_+.$$

We conclude that

$$\tilde{N}\left(\frac{|W_t|}{\sqrt{1-t}}\right) = \int_0^t \tilde{N}'\left(\frac{|W_s|}{\sqrt{1-s}}\right) \frac{\text{sgn}(W_s)}{\sqrt{1-s}} dW_s + \sqrt{\frac{2}{\pi}}\int_0^t \frac{dL_s^0}{\sqrt{1-s}}.$$

But, in view of (6.12)–(6.13), we have

$$F_t = \mathbb{P}\{\tau \le t \mid \mathcal{F}_t\} = \mathbb{P}\{d_t > 1 \mid \mathcal{F}_t\} = \tilde{N}\left(\frac{|W_t|}{\sqrt{1-t}}\right),$$

hence, the $\mathbb{F}$-compensator of $F$ equals

$$\tilde{F}_t = \sqrt{\frac{2}{\pi}}\int_0^t \frac{dL_s^0}{\sqrt{1-s}}.$$

This ends the proof of the proposition. □

We continue an analysis of our example. Again using the property of support of the local time, specifically, the equality

$$L_t^0 = \int_0^t \mathbb{1}_{\{W_s=0\}} \, dL_s^0, \quad \forall t \in \mathbb{R}_+,$$

we find that $L_t^0 = L_{t \wedge \tau}^0$, and thus $\tilde{F}_t = \tilde{F}_{t \wedge \tau}$. In view of part (ii) in Proposition 6.1.2, the $\mathbb{F}$-martingale hazard process $\Lambda$ thus equals $\tilde{F}$. Furthermore, both $\Gamma$ and $\Lambda$ are continuous processes, but $\Lambda$ is increasing, while $\Gamma$ has non-zero continuous martingale part. We conclude that $\Gamma \neq \Lambda$.

Note that Condition (F.2) is not satisfied in the present set-up (hence (F.1) does not hold either). We conclude that when (F.2) fails to hold, the continuity of $\Gamma$ and $\Lambda$ does not necessarily imply the equality $\Gamma = \Lambda$. Notice that the $\mathbb{G}$-compensator $A$ of $H$, which satisfies $A_t = \Lambda_{t \wedge \tau}$, is also equal to $\tilde{F}$ since

$$\tilde{F}_t = \sqrt{\frac{2}{\pi}} \int_0^{t \wedge \tau} \frac{dL_s^0}{\sqrt{1-s}} = \Lambda_{t \wedge \tau}.$$

Finally, let us notice that the $\mathbb{F}$-martingale hazard process $\Lambda$ represents at the same time the $\hat{\mathbb{F}}$-martingale hazard process of $\tau$, where $\hat{\mathbb{F}}$ stands for the natural filtration of the process $|W_t|$ (it is well known that $\hat{\mathbb{F}}$ is a strict sub-filtration of $\mathbb{F}$). Likewise, for every $t$ we have

$$F_t = \mathbb{P}\{\tau \leq t \,|\, \mathcal{F}_t\} = \mathbb{P}\{\tau \leq t \,|\, \hat{\mathcal{F}}_t\} = \hat{F}_t,$$

so that $\Gamma = \hat{\Gamma}_t$.

## 6.3 Martingale Representation Theorem

We consider the following set-up: we are given a reference filtration $\mathbb{F}$ and the enlarged filtration $\mathbb{G} = \mathbb{F} \vee \mathbb{H}$, where the filtration $\mathbb{H}$ is generated by a random time $\tau$. In addition, we assume that the assumptions of Proposition 6.1.2 are valid so that $F$ follows an increasing $\mathbb{F}$-predictable process, and the $\mathbb{F}$-martingale hazard process $\Lambda$ of a random time $\tau$ is given by the formula

$$\Lambda_t = \int_{]0,t]} \frac{dF_u}{1 - F_{u-}}. \tag{6.14}$$

By virtue of the definition of the $\mathbb{F}$-martingale hazard process, the compensated process $\tilde{M}_t = H_t - \Lambda_{t \wedge \tau}$ follows a $\mathbb{G}$-martingale. In Lemma 5.1.7, we have proved that the process $L$, given by the formula

$$L_t := \mathbb{1}_{\{\tau > t\}} (1 - F_t)^{-1} = \frac{1 - H_t}{1 - F_t},$$

follows a $\mathbb{G}$-martingale. We shall now check that the following equality is valid:

$$dL_t = -(1 - F_t)^{-1} \, d\tilde{M}_t. \tag{6.15}$$

To this end, we observe that

$$A_t := \Lambda_{t \wedge \tau} = \int_{]0,t]} \frac{1 - H_{u-}}{1 - F_{u-}} \, dF_u = \int_{]0,t]} L_{u-} \, dF_u. \tag{6.16}$$

Moreover, since

$$1 - H_t = L_t(1 - F_t),$$

using Itô's lemma, we obtain (notice that $L$ is a process of finite variation)

$$dH_t = L_{t-} \, dF_t - (1 - F_t) \, dL_t = A_t - (1 - F_t) \, dL_t$$

The following result is a counterpart of Proposition 5.2.1.

**Proposition 6.3.1.** *Let $Z$ be an $\mathbb{F}$-predictable process such that the random variable $Z_\tau$ is integrable. Then the $\mathbb{G}$-martingale $M_t^Z := \mathbb{E}_{\mathbb{P}}(Z_\tau \,|\, \mathcal{G}_t)$ admits the following integral representation*

$$M_t^Z = m_0 + \int_{]0,t]} L_{t-} \, dm_u + \int_{]0,t]} (Z_u - D_u) \, d\tilde{M}_u, \tag{6.17}$$

*where $m$ stands for an $\mathbb{F}$-martingale, given by the formula*

$$m_t = \mathbb{E}_{\mathbb{P}} \left( \int_0^\infty Z_u \, dF_u \,\Big|\, \mathcal{F}_t \right),$$

*and*

$$D_t = (1 - F_t)^{-1} \mathbb{E}_{\mathbb{P}} \left( \int_t^\infty Z_u \, dF_u \,\Big|\, \mathcal{F}_t \right).$$

*Proof.* By virtue of Proposition 5.1.1, we have (cf. (5.18))

$$M_t^Z = \mathbb{E}_{\mathbb{P}}(Z_\tau \,|\, \mathcal{G}_t) = H_t Z_\tau + (1 - H_t) D_t = H_t Z_\tau + \hat{D}_t,$$

where

$$\hat{D}_t := (1 - H_t) D_t = L_t \left( m_t - \int_{]0,t]} Z_u \, dF_u \right).$$

Since $L$ is a process of finite variation, we obtain

$$\begin{aligned} d\hat{D}_t &= L_{t-}(dm_t - Z_t \, dF_t) + D_t(1 - F_t) \, dL_t \\ &= L_{t-} \, dm_t - Z_t \, dD_t + D_t(1 - F_t) \, dL_t \\ &= L_{t-} \, dm_t - Z_t \, dD_t - D_t \, d\tilde{M}_t, \end{aligned}$$

where we have used (6.15) and (6.16). Consequently,

$$dM_t^Z = Z_t \, dH_t + d\hat{D}_t = L_{t-} \, dm_t + (Z_t - D_t) \, d\tilde{M}_t.$$

This gives the desired expression (6.17). □

It is clear that $m_0 = M_0^Z$. Notice also that equality (6.17) can be rewritten as follows:

$$M_t^Z = m_0 + \int_{]0,t \wedge \tau]} (1 - F_{t-})^{-1} \, dm_u + \int_{]0,t]} (Z_u - D_u) \, d\tilde{M}_u. \tag{6.18}$$

## 6.4 Case of the Martingale Invariance Property

In the next result, we shall work directly with the $\mathbb{F}$-martingale hazard process. Therefore, Proposition 6.4.1 also covers the case when the $\mathbb{F}$-hazard process $\Gamma$ does not exist (for example, when $\tau$ is an $\mathbb{F}$-stopping time). It appears that a counterpart of formula (6.10), with $\Gamma$ replaced by $\Lambda$, is valid. However, we need to impose here a suitable continuity condition. The following proposition is essentially due to Duffie et al. (1996).

**Proposition 6.4.1.** *Let* $\mathbb{H} \vee \mathbb{F} \subseteq \mathbb{G}$. *Assume that Condition* (M.1) *is valid, and the* $\mathbb{F}$-*martingale hazard process* $\Lambda$ *of a random time* $\tau$ *is continuous. For a fixed* $s > 0$, *let* $Y$ *stand for an* $\mathcal{F}_s$-*measurable, integrable random variable.*
(i) *If the* (*right-continuous*) *process* $V$, *given by the formula*

$$V_t = \mathbb{E}_{\mathbb{P}}\big(Y e^{\Lambda_t - \Lambda_s} \,|\, \mathcal{F}_t\big), \quad \forall\, t \in [0, s], \tag{6.19}$$

*is continuous at* $\tau$, *i.e., if* $\Delta V_{s \wedge \tau} = V_{s \wedge \tau} - V_{(s \wedge \tau)-} = 0$, *then for any* $t < s$ *we have*

$$\mathbb{E}_{\mathbb{P}}(\mathbb{1}_{\{\tau > s\}} Y \,|\, \mathcal{G}_t) = \mathbb{1}_{\{\tau > t\}} \, \mathbb{E}_{\mathbb{P}}\big(Y e^{\Lambda_t - \Lambda_s} \,|\, \mathcal{F}_t\big).$$

(ii) *If the process* $V$, *given by the formula*

$$V_t = \mathbb{E}_{\mathbb{P}}\big(e^{\Lambda_t - \Lambda_s} \,|\, \mathcal{F}_t\big), \quad \forall\, t \in [0, s], \tag{6.20}$$

*is continuous at* $\tau$ *then for any* $t \le s$ *we have*

$$\mathbb{P}\{\tau > s \,|\, \mathcal{G}_t\} = \mathbb{1}_{\{\tau > t\}} \mathbb{E}_{\mathbb{P}}\big(e^{\Lambda_t - \Lambda_s} \,|\, \mathcal{F}_t\big). \tag{6.21}$$

*Proof.* It is clear that it suffices to prove part (i). We shall first check that

$$U_t := \mathbb{1}_{\{\tau > t\}} V_t = \mathbb{E}_{\mathbb{P}}\big(\Delta V_\tau \mathbb{1}_{\{t < \tau \le s\}} + \mathbb{1}_{\{\tau > s\}} Y \,|\, \mathcal{G}_t\big) \tag{6.22}$$

or, equivalently,

$$U_t = \mathbb{E}_{\mathbb{P}}\Big( \int_{]t, s]} \Delta V_u \, dH_u + \mathbb{1}_{\{\tau > s\}} Y \,\Big|\, \mathcal{G}_t \Big). \tag{6.23}$$

In view of (6.19), we have $V_t = e^{\Lambda_t} m_t$, where $m$ denotes an $\mathbb{F}$-martingale: $m_t := \mathbb{E}_{\mathbb{P}}\big(Y e^{-\Lambda_s} \,|\, \mathcal{F}_t\big)$ for $t \in [0, s]$. In view of our assumptions, $m$ also follows a $\mathbb{G}$-martingale. Using Itô's product rule, we obtain

$$dV_t = m_{t-} \, de^{\Lambda_t} + e^{\Lambda_t} \, dm_t = V_{t-} e^{-\Lambda_t} \, de^{\Lambda_t} + e^{\Lambda_t} \, dm_t. \tag{6.24}$$

On the other hand, another application of Itô's product rule yields

$$dU_t = (1 - H_{t-}) \, dV_t - V_{t-} \, dH_t - \Delta V_t \Delta H_t.$$

Combining the last equality with (6.24), we obtain

$$dU_t = (1 - H_{t-})\big(V_{t-} e^{-\Lambda_t} \, de^{\Lambda_t} + e^{\Lambda_t} \, dm_t\big) - V_{t-} \, dH_t - \Delta V_t \, dH_t,$$

so that, after rearranging,

$$dU_t = -\Delta V_t\, dH_t + dC_t. \qquad (6.25)$$

In the last formula, we write $C$ to denote a $\mathbb{G}$-martingale

$$dC_t = (1 - H_{t-})e^{\Lambda_t}\, dm_t + dD_t,$$

where in turn the $\mathbb{G}$-martingale $D$ equals

$$dD_t = -V_{t-}\big(dH_t - (1 - H_{t-})e^{-\Lambda_t}\, de^{\Lambda_t}\big) = -V_{t-}\, d(H_t - \Lambda_{t\wedge\tau}) = -V_{t-}\, d\tilde{M}_t.$$

Since obviously $U_s = \mathbb{1}_{\{\tau>s\}}Y$, equality (6.25) implies (6.23). If the process $V$ is continuous at $\tau$, then (6.22) yields

$$\mathbb{E}_{\mathbb{P}}(\mathbb{1}_{\{\tau>s\}}\, Y \,|\, \mathcal{G}_t) = \mathbb{1}_{\{\tau>t\}} V_t = \mathbb{1}_{\{\tau>t\}}\, \mathbb{E}_{\mathbb{P}}\big(Y e^{\Lambda_t - \Lambda_s} \,|\, \mathcal{F}_t\big).$$

This completes the proof.    □

### 6.4.1 Valuation of Defaultable Claims

Let an $\mathbb{F}$-adapted process $B$ be given by the formula

$$B_t = \exp\left(\int_0^t r_u\, du\right), \quad \forall t \in \mathbb{R}_+,$$

for some $\mathbb{F}$-progressively measurable, integrable (short-term rate) process $r$. It is clear that $B$, referred to as the (default-free) savings account, follows a continuous process of finite variation. For a $\mathbb{G}$-predictable process $Z$, and a $\mathcal{G}_T$-measurable random variable $X$, we define the value process $S$ by setting

$$S_t = B_t\, \mathbb{E}_{\mathbb{P}}\left(\int_{]t,T]} B_u^{-1} Z_u\, dH_u + B_T^{-1} X \mathbb{1}_{\{T<\tau\}} \,\Big|\, \mathcal{G}_t\right), \qquad (6.26)$$

where $Z$ and $X$ satisfy suitable integrability conditions. The next result, borrowed from Duffie et al. (1996), is a suitable extension of Proposition 6.4.1. For convenience, we postpone the proof of this result to Sect. 8.3.

**Proposition 6.4.2.** *Assume that Condition (M.1) is fulfilled, and a random time $\tau$ admits an absolutely continuous $\mathbb{F}$-martingale hazard function $\Lambda$. For an $\mathbb{F}$-predictable process $Z$ and an $\mathcal{F}_T$-measurable random variable $X$, we define the process $V_t$, $t \in [0, T]$, by setting*

$$V_t = \tilde{B}_t\, \mathbb{E}_{\mathbb{P}}\left(\int_t^T \tilde{B}_u^{-1} Z_u \lambda_u\, du + \tilde{B}_T^{-1} X \,\Big|\, \mathcal{F}_t\right), \qquad (6.27)$$

*where $\tilde{B}$ is the 'savings account' corresponding to the default-risk-adjusted short-term rate $R_t = r_t + \lambda_t$, specifically,*

$$\tilde{B}_t = \exp\left(\int_0^t (r_u + \lambda_u)\, du\right).$$

*Then*

$$\mathbb{1}_{\{\tau>t\}} V_t = B_t\, \mathbb{E}_{\mathbb{P}}\left(B_\tau^{-1}(Z_\tau + \Delta V_\tau)\mathbb{1}_{\{t<\tau\leq T\}} + B_T^{-1} X \mathbb{1}_{\{T<\tau\}} \,\Big|\, \mathcal{G}_t\right).$$

**Corollary 6.4.1.** *Let the processes $S$ and $V$ be defined by formulae (6.26) and (6.27), respectively. Then*

$$S_t = \mathbb{1}_{\{\tau > t\}}\left(V_t - B_t\, \mathbb{E}_{\mathbb{P}}\big(B_\tau^{-1}\mathbb{1}_{\{\tau \leq T\}}\Delta V_\tau \,\big|\, \mathcal{G}_t\big)\right).$$

*If, in addition, $\Delta V_\tau = 0$, then $S_t = \mathbb{1}_{\{\tau > t\}}V_t$ for every $t \in [0, T]$.*

**Conjecture (B).** Under (G.1) and (F.1), if $Z$ follows an $\mathbb{F}$-predictable process and $X$ is an $\mathcal{F}_T$-measurable random variable, then the continuity condition $\Delta V_\tau = 0$ is satisfied.

Let us observe that instead of verifying conjecture (B), to establish the equality $S_t = \mathbb{1}_{\{\tau > t\}}V_t$, which is a handy form of the valuation formula (6.26), it suffices to show that $\Lambda = \Gamma$ and to make use of the following result.

**Proposition 6.4.3.** *Assume that the conditions (G.1) and (F.1) are valid and a random time $\tau$ admits an absolutely continuous $\mathbb{F}$-martingale hazard function $\Lambda$. Let $Z$ be an $\mathbb{F}$-predictable process, and let $X$ be an $\mathcal{F}_T$-measurable random variable. If $\Gamma = \Lambda$ then $S_t = \mathbb{1}_{\{\tau > t\}}V_t$ for $t \leq T$, where the processes $S$ and $V$ are given by expressions (6.26) and (6.27), respectively.*

*Proof.* In view of (F.1), we have (for the first equality, see Lemma 5.1.4)

$$\mathbb{P}\{\tau \geq u \,|\, \mathcal{F}_\infty \vee \mathcal{H}_t\} = \mathbb{1}_{\{\tau > t\}}\frac{\mathbb{P}\{\tau \geq u \,|\, \mathcal{F}_\infty\}}{\mathbb{P}\{\tau \geq t \,|\, \mathcal{F}_\infty\}} = \mathbb{1}_{\{\tau > t\}}\frac{\mathbb{P}\{\tau \geq u \,|\, \mathcal{F}_u\}}{\mathbb{P}\{\tau \geq t \,|\, \mathcal{F}_t\}}$$

for any $u > t$. Put more explicitly,

$$\mathbb{P}\{\tau \geq u \,|\, \mathcal{F}_\infty \vee \mathcal{H}_t\} = \mathbb{1}_{\{\tau > t\}}e^{\Gamma_t - \Gamma_u}.$$

If $Z$ is an $\mathbb{F}$-predictable process and $X$ is an $\mathcal{F}_T$-measurable random variable, using the $\mathbb{G}$-martingale property of the compensated process $H_t - \Lambda_{t \wedge \tau}$, we obtain

$$S_t = B_t\, \mathbb{E}_{\mathbb{P}}\left(\int_t^T B_u^{-1}Z_u\lambda_u \mathbb{1}_{\{u \leq \tau\}}\,du + B_T^{-1}X\mathbb{1}_{\{T < \tau\}} \,\bigg|\, \mathcal{G}_t\right)$$

$$= B_t\, \mathbb{E}_{\mathbb{P}}\left(\int_t^T B_u^{-1}Z_u\lambda_u\,\mathbb{P}\{\tau \geq u \,|\, \mathcal{F}_\infty \vee \mathcal{H}_t\}\,du \,\bigg|\, \mathcal{G}_t\right)$$

$$\quad + B_t\, \mathbb{E}_{\mathbb{P}}\left(B_T^{-1}X\,\mathbb{P}\{\tau > T \,|\, \mathcal{F}_\infty \vee \mathcal{H}_t\} \,\bigg|\, \mathcal{G}_t\right)$$

$$= \mathbb{1}_{\{\tau > t\}}B_t\, \mathbb{E}_{\mathbb{P}}\left(\int_t^T B_u^{-1}Z_u\lambda_u e^{\Gamma_t - \Gamma_u}\,du \,\bigg|\, \mathcal{F}_t \vee \mathcal{H}_t\right)$$

$$\quad + B_t\, \mathbb{1}_{\{\tau > t\}}\mathbb{E}_{\mathbb{P}}\left(B_T^{-1}Xe^{\Gamma_t - \Gamma_T} \,\bigg|\, \mathcal{F}_t \vee \mathcal{H}_t\right)$$

$$= \mathbb{1}_{\{\tau > t\}}B_t\, \mathbb{E}_{\mathbb{P}}\left(\int_t^T B_u^{-1}Z_u\lambda_u e^{\Lambda_t - \Lambda_u}\,du \,\bigg|\, \mathcal{F}_t\right)$$

$$\quad + B_t\, \mathbb{1}_{\{\tau > t\}}\mathbb{E}_{\mathbb{P}}\left(B_T^{-1}Xe^{\Lambda_t - \Lambda_T} \,\bigg|\, \mathcal{F}_t\right),$$

where the last equality is an immediate consequence of (F.1) (see, for instance, condition (M.2b)). Since $\tilde{B}_t = B_t e^{\Lambda_t}$, the result follows. $\qquad\square$

## 6.4.2 Case of a Stopping Time

In this section, we assume that the random time $\tau$ is an $\mathbb{F}$-stopping time. In other words, we postulate that $\mathbb{H} \subseteq \mathbb{F}$ or, equivalently, that $\mathbb{F} = \mathbb{G}$. Then, conditions (F.1), (F.2) and (M.1) are trivially satisfied. On the other hand, it is clear that $F = H$, and thus the $\mathbb{F}$-hazard process $\Gamma$ of $\tau$ is not well defined.

Let us comment very briefly on the classification of stopping times (see, e.g., Dellacherie (1972)). If $\tau$ is a $\mathbb{G}$-predictable stopping time, we get the trivial equality $\Lambda = H$, and thus the concept of a $\mathbb{G}$-martingale hazard process of a $\mathbb{G}$-predictable stopping time is of no real use. If $\tau$ is a totally inaccessible $\mathbb{G}$-stopping time, the $\mathbb{G}$-compensator of the associated jump process $H$ follows a continuous process (see Theorem V.T40 in Dellacherie (1972)). Recall that the $\mathbb{G}$-compensator of $H$ is always stopped at $\tau$.

From the previous section, we know that the process $\Lambda$ can be used in the evaluation of certain conditional expectations, provided that a certain continuity condition is fulfilled. The following result, which covers the case of a totally inaccessible $\mathbb{G}$-stopping time, is an immediate consequence of Proposition 6.4.1.

**Corollary 6.4.2.** *Assume that $\tau$ is a $\mathbb{G}$-stopping time and the $\mathbb{G}$-martingale hazard process $\Lambda$ of $\tau$ is continuous. For a fixed $T > 0$, let $Y$ be a $\mathcal{G}_T$-measurable, integrable random variable. If the process $V_t$, $t \in [0, T]$, given by the formula*

$$V_t = \mathbb{E}_{\mathbb{P}}(Ye^{\Lambda_t - \Lambda_T} \mid \mathcal{G}_t), \qquad (6.28)$$

*is continuous at $\tau$ then for any $t < T$ we have*

$$\mathbb{E}_{\mathbb{P}}(\mathbb{1}_{\{\tau > T\}} Y \mid \mathcal{G}_t) = \mathbb{1}_{\{\tau > t\}} \mathbb{E}_{\mathbb{P}}(Ye^{\Lambda_t - \Lambda_T} \mid \mathcal{G}_t).$$

*Example 6.4.1.* Let $\tau$ be a random time, given on some probability space $(\Omega, \mathcal{G}, \mathbb{P})$, such that the cumulative distribution function $F$ of $\tau$ is continuous, and $\mathbb{P}\{\tau > t\} > 0$ for every $t \in \mathbb{R}_+$. Let us take $\mathbb{G} = \mathbb{H}$. Then $\tau$ is a totally inaccessible $\mathbb{G}$-stopping time and its $\mathbb{G}$-martingale hazard process $\Lambda$ equals

$$\Lambda_{t \wedge \tau} = \int_0^{t \wedge \tau} \frac{dF(u)}{1 - F(u)}.$$

It is thus clear that we have $\Lambda_t = \Gamma^0(t \wedge \tau) = \Lambda^0(t \wedge \tau)$, where $\Gamma^0$ ($\Lambda^0$, resp.) is the hazard function (the martingale hazard function, resp.) of $\tau$. Let us set, for $t \in \mathbb{R}_+$,

$$\Lambda_t = \int_0^t \frac{dF(u)}{1 - F(u)} = \Gamma^0(t) = \Lambda^0(t).$$

For $\Lambda$ given above and any fixed $T > 0$, the process $V$ associated with the random variable $Y = 1$ does not have a discontinuity at $\tau$. Thus, for arbitrary $0 \le t < s$ we have (recall that here $\mathcal{G}_t = \mathcal{H}_t$)

$$\mathbb{P}\{\tau > s \mid \mathcal{G}_t\} = \mathbb{1}_{\{\tau > t\}} \mathbb{E}_{\mathbb{P}}(e^{\Lambda_t - \Lambda_s} \mid \mathcal{G}_t) = \mathbb{1}_{\{\tau > t\}} \frac{1 - F(s)}{1 - F(t)}.$$

## 6.5 Random Time with a Given Hazard Process

We shall examine the standard construction of a random time $\tau$ associated with a given hazard process $\Phi$. It appears that in this method, the process $\Phi$ can be considered either as the $\mathbb{F}$-hazard process $\Gamma$, or as the $\mathbb{F}$-martingale hazard process $\Lambda$. Indeed, we shall show that the following properties are valid:

(i) $\Phi$ coincides with the $\mathbb{F}$-hazard process $\Gamma$ of $\tau$,
(ii) $\Phi$ is the $\mathbb{F}$-martingale hazard process of a random time $\tau$,
(iii) $\Phi$ is the $\mathbb{G}$-martingale hazard process of a a $\mathbb{G}$-stopping time $\tau$.

Let $\Phi$ be an $\mathbb{F}$-adapted, continuous, increasing process given on a filtered probability space $(\tilde{\Omega}, \mathbb{F}, \tilde{\mathbb{P}})$ such that $\Phi_0 = 0$ and $\Phi_\infty = +\infty$. For instance, $\Phi$ can be given by the formula

$$\Phi_t = \int_0^t \phi_u \, du, \quad \forall t \in \mathbb{R}_+, \tag{6.29}$$

where $\phi$ is a non-negative, $\mathbb{F}$-progressively measurable process. Our goal is to construct a random time $\tau$, on an enlarged probability space $(\Omega, \mathcal{G}, \mathbb{P})$, in such a way that $\Phi$ is an $\mathbb{F}$-(martingale) hazard process of $\tau$. To this end, we assume that $\xi$ is a random variable on some probability space[1] $(\hat{\Omega}, \hat{\mathcal{F}}, \hat{\mathbb{P}})$, with the uniform probability law on $[0, 1]$. We may take the product space $(\Omega = \tilde{\Omega} \times \hat{\Omega}, \mathcal{G} = \mathcal{F}_\infty \otimes \hat{\mathcal{F}})$ with $\mathbb{P} = \tilde{\mathbb{P}} \otimes \hat{\mathbb{P}}$ as an enlarged probability space. We define $\tau : (\Omega, \mathcal{G}, \mathbb{P}) \to \mathbb{R}_+$ by setting

$$\tau = \inf \{ t \in \mathbb{R}_+ : e^{-\Phi_t} \leq \xi \} = \inf \{ t \in \mathbb{R}_+ : \Phi_t \geq -\ln \xi \}.$$

As usual, we set $\mathcal{G}_t = \mathcal{H}_t \vee \mathcal{F}_t$ for every $t$, so that Condition (G.1) is satisfied.

*Remarks.* It is worth stressing that the random time $\tau$ constructed above is not a stopping time with respect to the filtration $\mathbb{F}$. Furthermore, $\tau$ is a totally inaccessible stopping time with respect to the enlarged filtration $\mathbb{G} = \mathbb{F} \vee \mathbb{H}$.

We shall now check that properties (i)-(iii) listed above are satisfied.

*Proof of* (i). We shall find the process $F_t = \mathbb{P}\{\tau \leq t \,|\, \mathcal{F}_t\}$. Since clearly $\{\tau > t\} = \{e^{-\Phi_t} > \xi\}$, we get

$$\mathbb{P}\{\tau > t \,|\, \mathcal{F}_\infty\} = e^{-\Phi_t}.$$

Consequently,

$$1 - F_t = \mathbb{P}\{\tau > t \,|\, \mathcal{F}_t\} = \mathbb{E}_{\mathbb{P}}\big(\mathbb{P}\{\tau > t \,|\, \mathcal{F}_\infty\} \,|\, \mathcal{F}_t\big) = e^{-\Phi_t},$$

and so $F$ is an $\mathbb{F}$-adapted continuous increasing process. In addition,

$$F_t = 1 - e^{-\Phi_t} = \mathbb{P}\{\tau \leq t \,|\, \mathcal{F}_\infty\} = \mathbb{P}\{\tau \leq t \,|\, \mathcal{F}_t\}. \tag{6.30}$$

We conclude that $\Phi$ coincides with the $\mathbb{F}$-hazard process $\Gamma$ of $\tau$ under $\mathbb{P}$.

---

[1] It is enough to assume that we may define on $(\Omega, \mathcal{G}, \mathbb{P})$ a random variable $\xi$, which is uniformly distributed on $[0, 1]$, and which is independent of the process $\Phi$ (we then set $\hat{\mathcal{F}} = \sigma(\xi)$).

*Proof of* (ii). We shall now check that $\Phi$ represents the $\mathbb{F}$-martingale hazard process $\Lambda$. This can be done either directly, or by establishing the equality $\Lambda = \Gamma$. Since the process $\Phi$ is continuous, to show that $\Lambda = \Gamma$, it is enough to check that Condition (F.1a) (or, equivalently, Condition (F.1)) holds, and to apply Corollary 6.2.1. Let us check that (F.1a) is valid. To this end, we fix $t$ and we consider an arbitrary $u \leq t$. Since for any $u \in \mathbb{R}_+$ we have

$$\mathbb{P}\{\tau \leq u \,|\, \mathcal{F}_\infty\} = 1 - e^{-\Phi_u}, \tag{6.31}$$

we obtain the desired property:

$$\mathbb{P}\{\tau \leq u \,|\, \mathcal{F}_t\} = \mathbb{E}_{\mathbb{P}}\big(\mathbb{P}\{\tau \leq u \,|\, \mathcal{F}_\infty\} \,|\, \mathcal{F}_t\big) = 1 - e^{-\Phi_u} = \mathbb{P}\{\tau \leq u \,|\, \mathcal{F}_\infty\}.$$

Alternatively, we may check directly that (F.1) holds. Since

$$\{\tau \leq s\} = \{\Phi_s \geq -\ln \xi\} \in \hat{\mathcal{F}} \vee \mathcal{F}_s,$$

it is clear that $\mathcal{F}_t \subseteq \mathcal{H}_t \vee \mathcal{F}_t \subseteq \hat{\mathcal{F}} \vee \mathcal{F}_t$. Thus, for any bounded, $\mathcal{F}_\infty$-measurable random variable $\xi$ we have

$$\mathbb{E}_{\mathbb{P}}(\xi \,|\, \mathcal{H}_t \vee \mathcal{F}_t) = \mathbb{E}_{\mathbb{P}}(\xi \,|\, \hat{\mathcal{F}} \vee \mathcal{F}_t) = \mathbb{E}_{\mathbb{P}}(\xi \,|\, \mathcal{F}_t), \tag{6.32}$$

where the second equality is a consequence of the independence of $\hat{\mathcal{F}}$ and $\mathcal{F}_\infty$. This shows that (F.1) holds.

We conclude that the $\mathbb{F}$-martingale hazard process $\Lambda$ of $\tau$ coincides with $\Gamma$, so that, by virtue of part (i): $\Phi_t = \Lambda_t = \Gamma_t = -\ln(1 - F_t)$. Furthermore, we know that Condition (F.1) is equivalent to (M.2), and thus, by virtue of Lemma 6.1.1, the martingale invariance property holds, i.e., any $\mathbb{F}$-martingale also follow a martingale with respect to $\mathbb{G}$.

*Proof of* (iii). Let us now check directly that $\Phi$ is an $\mathbb{F}$-martingale hazard process of a random time $\tau$. Since $\Phi$ is a $\mathbb{F}$-predictable process (and thus a $\mathbb{G}$-predictable process), we will show at the same time that $\Phi$ is also the $\mathbb{G}$-martingale hazard process of a $\mathbb{G}$-stopping time $\tau$. We need to verify that the compensated process $H_t - \Phi_{t \wedge \tau}$ follows a $\mathbb{G}$-martingale. Since, for any $t \leq s$,

$$\mathbb{E}_{\mathbb{P}}(H_s - H_t \,|\, \mathcal{G}_t) = \mathbb{E}_{\mathbb{P}}(\mathbb{1}_{\{t < \tau \leq s\}} \,|\, \mathcal{G}_t) = \mathbb{1}_{\{\tau > t\}} \, \mathbb{E}_{\mathbb{P}}(\mathbb{1}_{\{t < \tau \leq s\}} \,|\, \mathcal{G}_t),$$

by virtue of Lemma 5.1.2, we have

$$\mathbb{E}_{\mathbb{P}}(H_s - H_t \,|\, \mathcal{G}_t) = \mathbb{1}_{\{\tau > t\}} \, \frac{\mathbb{P}\{t < \tau \leq s \,|\, \mathcal{F}_t\}}{\mathbb{P}\{\tau > t \,|\, \mathcal{F}_t\}} \, .$$

Using (6.30), we obtain

$$\mathbb{P}\{t < \tau \leq s \,|\, \mathcal{F}_t\} = \mathbb{E}_{\mathbb{P}}(F_s \,|\, \mathcal{F}_t) - F_t,$$

and this in turn shows that

$$\mathbb{E}_{\mathbb{P}}(H_s - H_t \mid \mathcal{G}_t) = \mathbb{1}_{\{\tau>t\}} \frac{\mathbb{E}_{\mathbb{P}}(F_s \mid \mathcal{F}_t) - F_t}{1 - F_t}. \tag{6.33}$$

On the other hand, if we set $Y = \Phi_{s\wedge\tau} - \Phi_{t\wedge\tau}$, then, in view of part (i), we get (cf. (6.2))

$$Y = \mathbb{1}_{\{\tau>t\}} Y = \ln\left(\frac{1 - F_{s\wedge\tau}}{1 - F_{t\wedge\tau}}\right) = \int_{]t,s\wedge\tau]} \frac{dF_u}{1 - F_u}.$$

Using again (5.2), we obtain (for the last equality in the formula below, see (6.3))

$$\mathbb{E}_{\mathbb{P}}(Y \mid \mathcal{G}_t) = \mathbb{1}_{\{\tau>t\}} \frac{\mathbb{E}_{\mathbb{P}}(Y \mid \mathcal{F}_t)}{\mathbb{P}\{\tau > t \mid \mathcal{F}_t\}} = \mathbb{1}_{\{\tau>t\}} \frac{\mathbb{E}_{\mathbb{P}}\left(\int_{]t,s\wedge\tau]}(1 - F_u)^{-1} dF_u \mid \mathcal{F}_t\right)}{1 - F_t}$$

$$= \mathbb{1}_{\{\tau>t\}} \frac{\mathbb{E}_{\mathbb{P}}(F_s \mid \mathcal{F}_t) - F_t}{1 - F_t}.$$

We conclude that the process $H_t - \Phi_{t\wedge\tau}$ follows a $\mathbb{G}$-martingale.

Let us analyze the differences between statements (i) and (iii). In part (i), we consider $\Phi$ as an $\mathbb{F}$-hazard process of $\tau$, then using Corollary 5.1.1 we deduce that for any $\mathcal{F}_s$-measurable random variable $Y$

$$\mathbb{E}_{\mathbb{P}}(\mathbb{1}_{\{\tau>s\}} Y \mid \mathcal{G}_t) = \mathbb{1}_{\{\tau>t\}} \mathbb{E}_{\mathbb{P}}(Y e^{\Phi_t - \Phi_s} \mid \mathcal{F}_t). \tag{6.34}$$

In part (iii), $\Phi$ is considered as the $\mathbb{G}$-martingale hazard process then, in view of Corollary 6.4.2, for any $\mathcal{G}_s$-measurable random variable $Y$ such that the associated process $V$ is continuous at $\tau$ we obtain

$$\mathbb{E}_{\mathbb{P}}(\mathbb{1}_{\{\tau>s\}} Y \mid \mathcal{G}_t) = \mathbb{1}_{\{\tau>t\}} \mathbb{E}_{\mathbb{P}}(Y e^{\Phi_t - \Phi_s} \mid \mathcal{G}_t). \tag{6.35}$$

If $Y$ is actually $\mathcal{F}_s$-measurable then we have (see (6.32))

$$\mathbb{E}_{\mathbb{P}}(Y e^{\Phi_t - \Phi_s} \mid \mathcal{G}_t) = \mathbb{E}_{\mathbb{P}}(Y e^{\Phi_t - \Phi_s} \mid \mathcal{F}_t \vee \mathcal{H}_t) = \mathbb{E}_{\mathbb{P}}(Y e^{\Phi_t - \Phi_s} \mid \mathcal{F}_t).$$

It follows that the associated process $V$ is necessarily continuous at $\tau$, and formulae (6.34) and (6.35) coincide.

*Remarks.* Assume that the process $\Phi$ is absolutely continuous, it satisfies (6.29) for some process $\phi$. Then equality (6.33) can be rewritten as follows:

$$\mathbb{P}\{t < \tau \le s \mid \mathcal{G}_t\} = \mathbb{1}_{\{\tau>t\}} \mathbb{E}_{\mathbb{P}}\left(1 - e^{-\int_t^s \phi_u \, du} \mid \mathcal{F}_t\right). \tag{6.36}$$

Using (6.31), we find that the cumulative distribution function of a random time $\tau$ under $\mathbb{P}$ equals

$$F(t) = \mathbb{P}\{\tau \le t\} = 1 - \mathbb{E}_{\mathbb{P}}\left(e^{-\int_0^t \phi_u \, du}\right) = 1 - e^{-\int_0^t \gamma^0(u) \, du},$$

where we write $\gamma^0$ to denote the unique $\mathbb{F}^0$-intensity (that is, the intensity function) of $\tau$.

Let us conclude this section by mentioning that the construction of a random time described above can be extended to the case of a finite family of $\mathbb{F}$-conditionally independent random times (see Sect. 9.1.2).

## 6.6 Poisson Process and Conditional Poisson Process

Until now, we have focused our attention on the case of a single random time and the associated jump process. In some financial applications, we need to model a sequence of successive random times. Almost invariably, this is done by making use of the so-called $\mathbb{F}$-*conditional Poisson process*, also known as the *doubly stochastic Poisson process*. The general idea is quite similar to the canonical construction of a single random time, which was examined in the previous section. We start by assuming that we are given a stochastic process $\Phi$, to be interpreted as the *hazard process*, and we construct a jump process, with unit jump size, such that the probabilistic features of consecutive jump times are governed by the hazard process $\Phi$.

**Poisson process with constant intensity.** Let us first recall the definition and the basic properties of the (time-homogeneous) Poisson process $N$ with constant intensity $\lambda > 0$.

**Definition 6.6.1.** A process $N$ defined on a probability space $(\Omega, \mathcal{G}, \mathbb{P})$ is called the *Poisson process* with intensity $\lambda$ with respect to $\mathbb{G}$ if $N_0 = 0$ and for any $0 \leq s < t$ the following two conditions are satisfied:
(i) the increment $N_t - N_s$ is independent of the $\sigma$-field $\mathcal{G}_s$,
(ii) the increment $N_t - N_s$ has the Poisson law with parameter $\lambda(t - s)$; specifically, for any $k = 0, 1, \ldots$ we have

$$\mathbb{P}\{N_t - N_s = k \,|\, \mathcal{G}_s\} = \mathbb{P}\{N_t - N_s = k\} = \frac{\lambda^k (t-s)^k}{k!} \, e^{-\lambda(t-s)}.$$

The Poisson process of Definition 6.6.1 is termed *time-homogeneous*, since the probability law of the increment $N_{t+h} - N_{s+h}$ is invariant with respect to the shift $h \geq -s$. In particular, for arbitrary $s < t$ the probability law of the increment $N_t - N_s$ coincides with the law of the random variable $N_{t-s}$. Let us finally observe that, for every $0 \leq s < t$,

$$\mathbb{E}_{\mathbb{P}}(N_t - N_s \,|\, \mathcal{G}_s) = \mathbb{E}_{\mathbb{P}}(N_t - N_s) = \lambda(t - s). \tag{6.37}$$

We take a version of the Poisson process whose sample paths are, with probability 1, right-continuous stepwise functions with all jumps of size 1. Let us set $\tau_0 = 0$, and let us denote by $\tau_1, \tau_2, \ldots$ the $\mathbb{G}$-stopping times given as the random moments of the successive jumps of $N$. For any $k = 0, 1, \ldots$

$$\tau_{k+1} = \inf \{t > \tau_k : N_t \neq N_{\tau_k}\} = \inf \{t > \tau_k : N_t - N_{\tau_k} = 1\}.$$

One shows without difficulties that $\mathbb{P}\{\lim_{k \to \infty} \tau_k = \infty\} = 1$. It is convenient to introduce the sequence $\xi_k$, $k \in \mathbb{N}$ of non-negative random variables, where $\xi_k = \tau_k - \tau_{k-1}$ for every $k \in \mathbb{N}$. Let us quote the following well known result.

**Proposition 6.6.1.** *The random variables $\xi_k$, $k \in \mathbb{N}$ are mutually independent and identically distributed, with the exponential law with parameter $\lambda$, that is, for every $k \in \mathbb{N}$ we have*

$$\mathbb{P}\{\xi_k \leq t\} = \mathbb{P}\{\tau_k - \tau_k \leq t\} = 1 - e^{-\lambda t}, \quad \forall t \in \mathbb{R}_+.$$

Proposition 6.6.1 suggests a simple construction of a process $N$, which follows a time-homogeneous Poisson process with respect to its natural filtration $\mathbb{F}^N$. Suppose that the probability space $(\Omega, \mathcal{G}, \mathbb{P})$ is large enough to support a family of mutually independent random variables $\xi_k$, $k \in \mathbb{N}$ with the common exponential law with parameter $\lambda > 0$. We define the process $N$ on $(\Omega, \mathcal{G}, \mathbb{P})$ by setting: $N_t = 0$ if $\{t < \xi_1\}$ and, for any natural $k$,

$$N_t = k \quad \text{if and only if} \quad \sum_{i=1}^{k} \xi_i \leq t < \sum_{i=1}^{k+1} \xi_i.$$

It can checked that the process $N$ defined in this way is indeed a Poisson process with parameter $\lambda$, with respect to its natural filtration $\mathbb{F}^N$. The jump times of $N$ are, of course, the random times $\tau_k = \sum_{i=1}^{k} \xi_i$, $k \in \mathbb{N}$.

Let us recall some useful equalities that are not hard to establish through elementary calculations involving the Poisson law. For any $a \in \mathbb{R}$ and $0 \leq s < t$ we have

$$\mathbb{E}_{\mathbb{P}}\big(e^{ia(N_t - N_s)} \,\big|\, \mathcal{G}_s\big) = \mathbb{E}_{\mathbb{P}}\big(e^{ia(N_t - N_s)}\big) = e^{\lambda(t-s)(e^{ia}-1)},$$

and

$$\mathbb{E}_{\mathbb{P}}\big(e^{a(N_t - N_s)} \,\big|\, \mathcal{G}_s\big) = \mathbb{E}_{\mathbb{P}}\big(e^{a(N_t - N_s)}\big) = e^{\lambda(t-s)(e^a-1)}.$$

The next result is an easy consequence of (6.37) and the above formulae. The proof of the proposition is thus left to the reader.

**Proposition 6.6.2.** *The following stochastic processes follow $\mathbb{G}$-martingales.*
(i) *The compensated Poisson process $\hat{N}$ defined as*

$$\hat{N}_t := N_t - \lambda t.$$

(ii) *For any $k \in \mathbb{N}$, the compensated Poisson process stopped at $\tau_k$*

$$\hat{M}_t^k := N_{t \wedge \tau_k} - \lambda(t \wedge \tau_k).$$

(iii) *For any $a \in \mathbb{R}$, the exponential martingale $M^a$ given by the formula*

$$M_t^a := e^{aN_t - \lambda t(e^a - 1)} = e^{a\hat{N}_t - \lambda t(e^a - a - 1)}.$$

(iv) *For any fixed $a \in \mathbb{R}$, the exponential martingale $K^a$ given by the formula*

$$K_t^a := e^{iaN_t - \lambda t(e^{ia} - 1)} = e^{ia\hat{N}_t - \lambda t(e^{ia} - ia - 1)}.$$

*Remarks.* (i) For any $\mathbb{G}$-martingale $M$, defined on some filtered probability space $(\Omega, \mathbb{G}, \mathbb{P})$, and an arbitrary $\mathbb{G}$-stopping time $\tau$, the stopped process $M_t^\tau = M_{t \wedge \tau}$ necessarily follows a $\mathbb{G}$-martingale. Thus, the second statement of the proposition is an immediate consequence of the first, combined with the simple observation that each jump time $\tau_k$ is a $\mathbb{G}$-stopping time.
(ii) Consider the random time $\tau = \tau_1$, where $\tau_1$ is the time of the first jump of the Poisson process $N$. Then $N_{t \wedge \tau} = N_{t \wedge \tau_1} = H_t$, so that the process $\hat{M}^1$ introduced in part (ii) of the proposition coincides with the martingale $\hat{M}$ associated with $\tau$.

(iii) The property described in part (iii) of Proposition 6.6.2 characterizes the Poisson process in the following sense: if $N_0 = 0$ and for every $a \in \mathbb{R}$ the process $M^a$ is a $\mathbb{G}$-martingale, then $N$ follows the Poisson process with parameter $\lambda$. Indeed, the martingale property of $M^a$ yields

$$\mathbb{E}_\mathbb{P}\big(e^{a(N_t - N_s)} \,|\, \mathcal{G}_s\big) = e^{\lambda(t-s)(e^a - 1)}, \quad \forall 0 \le s < t.$$

By standard arguments, this implies that the random variable $N_t - N_s$ is independent of the $\sigma$-field $\mathcal{G}_s$, and has the Poisson law with parameter $\lambda(t-s)$. A similar remark applies to property (iv) in Proposition 6.6.2.

Let us consider the case of a Brownian motion $W$ and a Poisson process $N$ that are defined on a common filtered probability space $(\Omega, \mathbb{G}, \mathbb{P})$. In particular, for every $0 \le s < t$, the increment $W_t - W_s$ is independent of the $\sigma$-field $\mathcal{G}_s$, and has the Gaussian law $N(0, t - s)$. It might be useful to recall that for any real number $b$ the following processes follow martingales with respect to $\mathbb{G}$:

$$\hat{W}_t = W_t - t, \quad m_t^b = e^{bW_t - \frac{1}{2}b^2 t}, \quad k_t^b = e^{ibW_t + \frac{1}{2}b^2 t}.$$

The next result shows that a Brownian motion $W$ and a Poisson process $N$, with respect to a common filtration $\mathbb{G}$, are necessarily mutually independent.

**Proposition 6.6.3.** *Let a Brownian motion $W$ and a Poisson process $N$ be defined on a common filtered probability space $(\Omega, \mathbb{G}, \mathbb{P})$. Then the two processes $W$ and $N$ are mutually independent.*

*Proof.* Let us sketch the proof. For a fixed $a \in \mathbb{R}$ and any $t > 0$, we have

$$e^{iaN_t} = 1 + \sum_{0 < u \le t} (e^{iaN_t} - e^{iaN_{t-}}) = 1 + \int_{]0,t]} (e^{ia} - 1)e^{iaN_{u-}}\, dN_u,$$

$$= 1 + \int_{]0,t]} (e^{ia} - 1)e^{iaN_{u-}}\, d\hat{N}_u + \lambda \int_0^t (e^{ia} - 1)e^{iaN_{u-}}\, du.$$

On the other hand, for any $b \in \mathbb{R}$, the Itô formula yields

$$e^{ibW_t} = 1 + ib \int_0^t e^{ibW_u}\, dW_u - \frac{1}{2}b^2 \int_0^t e^{ibW_u}\, du.$$

The continuous martingale part of the compensated Poisson process $\hat{N}$ is identically equal to 0 (since $\hat{N}$ is a process of finite variation), and obviously the processes $\hat{N}$ and $W$ have no common jumps. Thus, using the Itô product rule for semimartingales, we obtain

$$e^{i(aN_t + bW_t)} = 1 + ib \int_0^t e^{i(aN_u + bW_u)}\, dW_u - \frac{1}{2}b^2 \int_0^t e^{i(aN_u + bW_u)}\, du$$

$$+ \int_{]0,t]} (e^{ia} - 1)e^{i(aN_{u-} + bW_u)}\, d\hat{N}_u + \lambda \int_0^t (e^{ia} - 1)e^{i(aN_u + bW_u)}\, du.$$

Let us denote $f_{a,b}(t) = \mathbb{E}_{\mathbb{P}}(e^{i(aN_t+bW_t)})$. By taking the expectations of both sides of the last equality, we get

$$f_{a,b}(t) = 1 + \lambda \int_0^t (e^{ia} - 1)f_{a,b}(u)\,du - \frac{1}{2}b^2 \int_0^t f_{a,b}(u)\,du.$$

By solving the last equation, we obtain, for arbitrary $a, b \in \mathbb{R}$,

$$\mathbb{E}_{\mathbb{P}}(e^{i(aN_t+bW_t)}) = f_{a,b}(t) = e^{\lambda t(e^{ia}-1)}e^{-\frac{1}{2}b^2 t} = \mathbb{E}_{\mathbb{P}}(e^{iaN_t})\mathbb{E}_{\mathbb{P}}(e^{ibW_t}).$$

We conclude that for any $t \in \mathbb{R}_+$ the random variables $W_t$ and $N_t$ are mutually independent under $\mathbb{P}$.

In the second step, we fix $0 < t < s$, and we consider the following expectation, for arbitrary real numbers $a_1, a_2, b_1$ and $b_2$,

$$f(t,s) := \mathbb{E}_{\mathbb{P}}(e^{i(a_1 N_t+a_2 N_s+b_1 W_t+b_2 W_s)}).$$

Let us denote $\tilde{a}_1 = a_1 + a_2$ and $\tilde{b}_1 = b_1 + b_2$. Then

$$
\begin{aligned}
f(t,s) &= \mathbb{E}_{\mathbb{P}}(e^{i(a_1 N_t+a_2 N_s+b_1 W_t+b_2 W_s)}) \\
&= \mathbb{E}_{\mathbb{P}}(\mathbb{E}_{\mathbb{P}}(e^{i(\tilde{a}_1 N_t+a_2(N_s-N_t)+\tilde{b}_1 W_t+b_2(W_s-W_t))}\,|\,\mathcal{G}_t)) \\
&= \mathbb{E}_{\mathbb{P}}(e^{i(\tilde{a}_1 N_t+\tilde{b}_1 W_t)}\mathbb{E}_{\mathbb{P}}(e^{i(a_2(N_s-N_t)+b_2(W_s-W_t))}\,|\,\mathcal{G}_t)) \\
&= \mathbb{E}_{\mathbb{P}}(e^{i(\tilde{a}_1 N_t+\tilde{b}_1 W_t)}\mathbb{E}_{\mathbb{P}}(e^{i(a_2 N_{t-s}+b_2 W_{t-s})})) \\
&= f_{a_1,b_1}(t-s)\,\mathbb{E}_{\mathbb{P}}(e^{i(\tilde{a}_1 N_t+\tilde{b}_1 W_t)}) \\
&= f_{a_1,b_1}(t-s)f_{\tilde{a}_1,\tilde{b}_1}(t),
\end{aligned}
$$

where we have used, in particular, the independence of the increment $N_t - N_s$ (and $W_t - W_s$) of the $\sigma$-field $\mathcal{G}_t$, and the time-homogeneity of $N$ and $W$. By setting $b_1 = b_2 = 0$ in the last formula, we obtain

$$\mathbb{E}_{\mathbb{P}}(e^{i(a_1 N_t+a_2 N_s)}) = f_{a_1,0}(t-s)f_{\tilde{a}_1,0}(t),$$

while the choice of $a_1 = a_2 = 0$ yields

$$\mathbb{E}_{\mathbb{P}}(e^{i(b_1 W_t+b_2 W_s)}) = f_{0,b_1}(t-s)f_{0,\tilde{b}_1}(t).$$

It is not difficult to check that

$$f_{a_1,b_1}(t-s)f_{\tilde{a}_1,\tilde{b}_1}(t) = f_{a_1,0}(t-s)f_{\tilde{a}_1,0}(t)f_{0,b_1}(t-s)f_{0,\tilde{b}_1}(t).$$

We conclude that for any $0 \leq t < s$ and arbitrary $a_1, a_2, b_1, b_2 \in \mathbb{R}$:

$$\mathbb{E}_{\mathbb{P}}(e^{i(a_1 N_t+a_2 N_s+b_1 W_t+b_2 W_s)}) = \mathbb{E}_{\mathbb{P}}(e^{i(a_1 N_t+a_2 N_s)})\mathbb{E}_{\mathbb{P}}(e^{i(b_1 W_t+b_2 W_s)}).$$

This means that the random variables $(N_t, N_s)$ and $(W_t, W_s)$ are mutually independent. By proceeding along the same lines, one may check that the random variables $(N_{t_1}, \ldots, N_{t_n})$ and $(W_{t_1}, \ldots, W_{t_n})$ are mutually independent for any $n \in \mathbb{N}$ and for any choice of $0 \leq t_1 < \cdots < t_n$.  $\square$

Let us now examine the behavior of the Poisson process under a specific equivalent change of the underlying probability measure. For a fixed $T > 0$, we introduce a probability measure $\mathbb{P}^*$ on $(\Omega, \mathcal{G}_T)$ by setting

$$\frac{d\mathbb{P}^*}{d\mathbb{P}}\bigg|_{\mathcal{G}_T} = \eta_T, \quad \mathbb{P}\text{-a.s.,} \tag{6.38}$$

where the Radon-Nikodým density process $\eta_t$, $t \in [0, T]$, satisfies

$$d\eta_t = \eta_{t-}\kappa\, d\hat{N}_t, \quad \eta_0 = 1, \tag{6.39}$$

for some constant $\kappa > -1$. Since $Y := \kappa\hat{N}$ is a process of finite variation, we know from Lemma 4.4.1 that (6.39) admits a unique solution, denoted as $\mathcal{E}_t(Y)$ or $\mathcal{E}_t(\kappa\hat{N})$; it can be seen as a special case of the Doléans (or stochastic) exponential. By solving (6.39) path-by-path, we obtain

$$\eta_t = \mathcal{E}_t(\kappa\hat{N}) = e^{Y_t} \prod_{0 < u \le t} (1 + \Delta Y_u)e^{-\Delta Y_u} = e^{Y_t^c} \prod_{0 < u \le t} (1 + \Delta Y_u),$$

where $Y_t^c := Y_t - \sum_{0 < u \le t} \Delta Y_u$ is the path-by-path continuous part of $Y$. Direct calculations show that

$$\eta_t = e^{-\kappa\lambda t} \prod_{0 < u \le t} (1 + \kappa\Delta N_u) = e^{-\kappa\lambda t}(1 + \kappa)^{N_t} = e^{N_t \ln(1+\kappa) - \kappa\lambda t},$$

where the last equality holds if $\kappa > -1$. Upon setting $a = \ln(1 + \kappa)$ in part (iii) of Proposition 6.6.2, we get $M^a = \eta$; this confirms that the process $\eta$ follows a $\mathbb{G}$-martingale under $\mathbb{P}$. We have thus proved the following result.

**Lemma 6.6.1.** *Assume that $\kappa > -1$. The unique solution $\eta$ to the SDE (6.39) follows an exponential $\mathbb{G}$-martingale under $\mathbb{P}$. Specifically,*

$$\eta_t = e^{N_t \ln(1+\kappa) - \kappa\lambda t} = e^{\hat{N}_t \ln(1+\kappa) - \lambda t(\kappa - \ln(1+\kappa))} = M_t^a, \tag{6.40}$$

*where $a = \ln(1+\kappa)$. In particular, the random variable $\eta_T$ is strictly positive, $\mathbb{P}$a.s. and $\mathbb{E}_\mathbb{P}(\eta_T) = 1$. Furthermore, the process $M^a$ solves the following SDE:*

$$dM_t^a = M_{t-}^a (e^a - 1)\, d\hat{N}_t, \quad M_0^a = 1. \tag{6.41}$$

We are in the position to establish the well-known result, which states that under $\mathbb{P}^*$ the process $N_t$, $t \in [0, T]$, follows a Poisson process with the constant intensity $\lambda^* = (1 + \kappa)\lambda$.

**Proposition 6.6.4.** *Assume that under $\mathbb{P}$ a process $N$ is a Poisson process with intensity $\lambda$ with respect to the filtration $\mathbb{G}$. Suppose that the probability measure $\mathbb{P}^*$ is defined on $(\Omega, \mathcal{G}_T)$ through (6.38) and (6.39) for some $\kappa > -1$.*
*(i) The process $N_t$, $t \in [0, T]$, follows a Poisson process under $\mathbb{P}^*$ with respect to $\mathbb{G}$ with the constant intensity $\lambda^* = (1 + \kappa)\lambda$.*
*(ii) The compensated process $N_t^*$, $t \in [0, T]$, defined as*

$$N_t^* = N_t - \lambda^* t = N_t - (1 + \kappa)\lambda t = \hat{N}_t - \kappa\lambda t,$$

*follows a $\mathbb{P}^*$-martingale with respect to $\mathbb{G}$.*

*Proof.* From remark (iii) after Proposition 6.6.2, we know that it suffices to find $\lambda^*$ such that, for any fixed $b \in \mathbb{R}$, the process $\tilde{M}^b$, given as

$$\tilde{M}^b_t := e^{bN_t - \lambda^* t(e^b - 1)}, \quad \forall t \in [0, T], \tag{6.42}$$

follows a $\mathbb{G}$-martingale under $\mathbb{P}^*$. By standard arguments, the process $\tilde{M}^b$ is a $\mathbb{P}^*$-martingale if and only if the product $\tilde{M}^b \eta$ is a martingale under the original probability measure $\mathbb{P}$. But in view of (6.40), we have

$$\tilde{M}^b_t \eta_t = \exp\left(N_t(b + \ln(1 + \kappa)) - t(\kappa\lambda + \lambda^*(e^b - 1))\right).$$

Let us write $a = b + \ln(1 + \kappa)$. Since $b$ is an arbitrary real number, so is $a$. Then, by virtue of part (iii) in Proposition 6.6.2, we necessarily have

$$\kappa\lambda + \lambda^*(e^b - 1) = \lambda(e^a - 1).$$

After simplifications, we conclude that, for any fixed real number $b$, the process $\tilde{M}^b$ defined by (6.42) is a $\mathbb{G}$-martingale under $\mathbb{P}^*$ if and only if $\lambda^* = (1 + \kappa)\lambda$. In other words, the intensity $\lambda^*$ of $N$ under $\mathbb{P}^*$ satisfies $\lambda^* = (1 + \kappa)\lambda$. Also the second statement is clear. $\qquad\square$

*Remarks.* Assume that $\mathbb{G} = \mathbb{F}^N$, i.e., the filtration $\mathbb{G}$ is generated by some Poisson process $N$. Then any strictly positive $\mathbb{G}$-martingale $\eta$ under $\mathbb{P}$ is known to satisfy (6.39) for some $\mathbb{G}$-predictable process $\kappa$.

Assume that $W$ is a Brownian motion and $N$ follows a Poisson process under $\mathbb{P}$ with respect to $\mathbb{G}$. Let $\eta$ satisfy

$$d\eta_t = \eta_{t-}(\beta_t \, dW_t + \kappa \, d\hat{N}_t), \quad \eta_0 = 1, \tag{6.43}$$

for some $\mathbb{G}$-predictable stochastic process $\beta$ and some constant $\kappa > -1$. A simple application of the Itô's product rule shows that if processes $\eta^1$ and $\eta^2$ satisfy:

$$d\eta^1_t = \eta^1_{t-}\beta_t \, dW_t, \quad d\eta^2_t = \eta^2_{t-}\kappa \, d\hat{N}_t,$$

then the product $\eta_t := \eta^1_t \eta^2_t$ satisfies (6.43). Taking the uniqueness of solutions to the linear SDE (6.43) for granted, we conclude that the unique solution to this SDE is given by the expression:

$$\eta_t = \exp\left(\int_0^t \beta_u \, dW_u - \frac{1}{2}\int_0^t \beta_u^2 \, du\right)\exp\left(N_t \ln(1 + \kappa) - \kappa\lambda t\right). \tag{6.44}$$

The proof of the next result is left to the reader as exercise.

**Proposition 6.6.5.** *Let the probability $\mathbb{P}^*$ be given by (6.38) and (6.44) for some constant $\kappa > -1$ and a $\mathbb{G}$-predictable process $\beta$, such that $\mathbb{E}_{\mathbb{P}}(\eta_T) = 1$.*
*(i) The process $W^*_t = W_t - \int_0^t \beta_u \, du, \, t \in [0, T]$, follows a Brownian motion under $\mathbb{P}^*$, with respect to the filtration $\mathbb{G}$.*
*(ii) The process $N_t, \, t \in [0, T]$, follows a Poisson process with the constant intensity $\lambda^* = (1 + \kappa)\lambda$ under $\mathbb{P}^*$, with respect to the filtration $\mathbb{G}$.*
*(iii) Processes $W^*$ and $N$ are mutually independent under $\mathbb{P}^*$.*

**Poisson process with deterministic intensity.** Let $\lambda : \mathbb{R}_+ \to \mathbb{R}_+$ be any non-negative, locally integrable function such that $\int_0^\infty \lambda(u)\,du = \infty$. By definition, the process $N$ (with $N_0 = 0$) is the Poisson process with *intensity function* $\lambda$ if for every $0 \le s < t$ the increment $N_t - N_s$ is independent of the $\sigma$-field $\mathcal{G}_s$, and has the Poisson law with parameter $\Lambda(t) - \Lambda(s)$, where the *hazard function* $\Lambda$ equals $\Lambda(t) = \int_0^t \lambda(u)\,du$.

More generally, let $\Lambda : \mathbb{R}_+ \to \mathbb{R}_+$ be a right-continuous, increasing function with $\Lambda(0) = 0$ and $\Lambda(\infty) = \infty$. The Poisson process with the hazard function $\Lambda$ satisfies, for every $0 \le s < t$ and every $k = 0, 1, \ldots$:

$$\mathbb{P}\{N_t - N_s = k \,|\, \mathcal{G}_s\} = \mathbb{P}\{N_t - N_s = k\} = \frac{(\Lambda(t) - \Lambda(s))^k}{k!} \, e^{-(\Lambda(t) - \Lambda(s))}.$$

*Example 6.6.1.* The most convenient and widely used method of constructing a Poisson process with a hazard function $\Lambda$ runs as follows: we take a Poisson process $\tilde{N}$ with the constant intensity $\lambda = 1$, with respect to some filtration $\tilde{\mathbb{G}}$, and we define the time-changed process $N_t := \tilde{N}_{\Lambda(t)}$. The process $N$ is easily seen to follow a Poisson process with the hazard function $\Lambda$, with respect to the time-changed filtration $\mathbb{G}$, where $\mathcal{G}_t = \tilde{\mathcal{G}}_{\Lambda(t)}$ for every $t \in \mathbb{R}_+$.

Since for arbitrary $0 \le s < t$

$$\mathbb{E}_{\mathbb{P}}(N_t - N_s \,|\, \mathcal{G}_s) = \mathbb{E}_{\mathbb{P}}(N_t - N_s) = \Lambda(t) - \Lambda(s),$$

it is clear that the compensated Poisson process $\hat{N}_t = N_t - \Lambda(t)$ follows a $\mathbb{G}$-martingale under $\mathbb{P}$. A suitable generalization of Proposition 6.6.3 shows that a Poisson process with the hazard function $\Lambda$ and a Brownian motion with respect to $\mathbb{G}$ follow mutually independent processes under $\mathbb{P}$. The proof of the next lemma relies on a direct application of the Itô formula, and so it is omitted.

**Lemma 6.6.2.** *Let $Z$ be an arbitrary bounded, $\mathbb{G}$-predictable process. Then the process $M^Z$, given by the formula*

$$M_t^Z = \exp\left(\int_{]0,t]} Z_u \, dN_u - \int_0^t (e^{Z_u} - 1)\, d\Lambda(u)\right),$$

*follows a $\mathbb{G}$-martingale under $\mathbb{P}$. Moreover, $M^Z$ is the unique solution to the SDE*

$$dM_t^Z = M_{t-}^Z(e^{Z_t} - 1)\, d\hat{N}_t, \quad M_0^Z = 1.$$

In case of a Poisson process with intensity function $\lambda$, it can be easily deduced from Lemma 6.6.2 that, for any (Borel measurable) function $\kappa : \mathbb{R}_+ \to (-1, \infty)$, the process

$$\zeta_t = \exp\left(\int_{]0,t]} \ln(1 + \kappa(u))\, dN_u - \int_0^t \kappa(u)\lambda(u)\, du\right)$$

is the unique solution to the SDE

$$d\zeta_t = \zeta_{t-}\kappa(t)\, d\hat{N}_t, \quad \eta_0 = 1.$$

Using similar arguments as in the case of constant $\kappa$, one can show that the unique solution to the SDE

$$d\eta_t = \eta_{t-}\left(\beta_t\, dW_t + \kappa(t)\, d\hat{N}_t\right), \quad \eta_0 = 1,$$

is given by the following expression:

$$\eta_t = \zeta_t \exp\left(\int_0^t \beta_u\, dW_u - \frac{1}{2}\int_0^t \beta_u^2\, du\right). \tag{6.45}$$

The next result generalizes Proposition 6.6.5. Again, the proof is left to the reader.

**Proposition 6.6.6.** *Let $\mathbb{P}^*$ be a probability measure equivalent to $\mathbb{P}$ on $(\Omega, \mathcal{G}_T)$, such that the density process $\eta$ in (6.38) is given by (6.45). Then, under $\mathbb{P}^*$ and with respect to $\mathbb{G}$:*
*(i) the process $W_t^* = W_t - \int_0^t \beta_u\, du$, $t \in [0, T]$, follows a Brownian motion,*
*(ii) the process $N_t$, $t \in [0, T]$, follows a Poisson process with the intensity function $\lambda^*(t) = 1 + \kappa(t)\lambda(t)$,*
*(iii) processes $W^*$ and $N$ are mutually independent under $\mathbb{P}^*$.*

**Conditional Poisson process.** We start by assuming that we are given a filtered probability space $(\Omega, \mathbb{G}, \mathbb{P})$ and a certain sub-filtration $\mathbb{F}$ of $\mathbb{G}$. Let $\Phi$ be an $\mathbb{F}$-adapted, right-continuous, increasing process, with $\Phi_0 = 0$ and $\Phi_\infty = \infty$. We refer to $\Phi$ as the *hazard process*. In some cases, we have $\Phi_t = \int_0^t \phi_u\, du$ for some $\mathbb{F}$-progressively measurable process $\phi$ with locally integrable sample paths. Then the process $\phi$ is called the *intensity process*. We are in a position to state the definition of the $\mathbb{F}$-conditional Poisson process associated with $\Phi$. Slightly different, but essentially equivalent, definition of a conditional Poisson process (also known as the doubly stochastic Poisson process) can be found in Brémaud (1981) and Last and Brandt (1995).

**Definition 6.6.2.** A process $N$ defined on a probability space $(\Omega, \mathbb{G}, \mathbb{P})$ is called the $\mathbb{F}$-*conditional Poisson process* with respect to $\mathbb{G}$, associated with the hazard process $\Phi$, if for any $0 \le s < t$ and every $k = 0, 1, \ldots$

$$\mathbb{P}\{N_t - N_s = k\,|\,\mathcal{G}_s \vee \mathcal{F}_\infty\} = \frac{(\Phi_t - \Phi_s)^k}{k!}\, e^{-(\Phi_t - \Phi_s)}, \tag{6.46}$$

where $\mathcal{F}_\infty = \sigma(\mathcal{F}_u : u \in \mathbb{R}_+)$.

At the intuitive level, if a particular sample path $\Phi_\cdot(\omega)$ of the hazard process is known, the process $N$ has exactly the same properties as the Poisson process with respect to $\mathbb{G}$ with the (deterministic) hazard function $\Phi_\cdot(\omega)$. In particular, it follows from (6.46) that

$$\mathbb{P}\{N_t - N_s = k\,|\,\mathcal{G}_s \vee \mathcal{F}_\infty\} = \mathbb{P}\{N_t - N_s = k\,|\,\mathcal{F}_\infty\},$$

i.e., conditionally on the $\sigma$-field $\mathcal{F}_\infty$ the increment $N_t - N_s$ is independent of the $\sigma$-field $\mathcal{G}_s$.

Similarly, for any $0 \leq s < t \leq u$ and every $k = 0, 1, \ldots$, we have

$$\mathbb{P}\{N_t - N_s = k \,|\, \mathcal{G}_s \vee \mathcal{F}_u\} = \frac{(\varPhi_t - \varPhi_s)^k}{k!} \, e^{-(\varPhi_t - \varPhi_s)}. \qquad (6.47)$$

In other words, conditionally on the $\sigma$-field $\mathcal{F}_u$ the process $N_t$, $t \in [0, u]$, behaves like a Poisson process with the hazard function $\varPhi$. Finally, for any $n \in \mathbb{N}$, any non-negative integers $k_1, \ldots, k_n$, and arbitrary non-negative real numbers $s_1 < t_1 \leq s_2 < t_2 \leq \ldots \leq s_n < t_n$ we have

$$\mathbb{P}\Big( \bigcap_{i=1}^{n} \{N_{t_i} - N_{s_i} = k_i\} \Big) = \mathbb{E}_{\mathbb{P}}\Big( \prod_{i=1}^{n} \frac{(\varPhi_{t_i} - \varPhi_{s_i})^{k_i}}{k_i!} \, e^{-(\varPhi_{t_i} - \varPhi_{s_i})} \Big).$$

Let us notice that in all conditional expectations above, the reference filtration $\mathbb{F}$ can be replaced by the filtration $\mathbb{F}^{\varPhi}$ generated by the hazard process. In fact, an $\mathbb{F}$-conditional Poisson process with respect to $\mathbb{G}$ follows also a conditional Poisson process with respect to the filtrations: $\mathbb{F}^N \vee \mathbb{F}$ and $\mathbb{F}^N \vee \mathbb{F}^{\varPhi}$ (with the same hazard process).

We shall henceforth postulate that $\mathbb{E}_{\mathbb{P}}(\varPhi_t) < \infty$ for every $t \in \mathbb{R}_+$.

**Lemma 6.6.3.** *The compensated process $\hat{N}_t = N_t - \varPhi_t$ follows a martingale with respect to $\mathbb{G}$.*

*Proof.* It is enough to notice that, for arbitrary $0 \leq s < t$,

$$\mathbb{E}_{\mathbb{P}}(N_t - \varPhi_t \,|\, \mathcal{G}_s) = \mathbb{E}_{\mathbb{P}}(\mathbb{E}_{\mathbb{P}}(N_t - \varPhi_t \,|\, \mathcal{G}_s \vee \mathcal{F}_{\infty}) \,|\, \mathcal{G}_s) = \mathbb{E}_{\mathbb{P}}(N_s - \varPhi_s \,|\, \mathcal{G}_s) = N_s - \varPhi_s,$$

where in the second equality we have used the property of a Poisson process with deterministic hazard function. □

Given the two filtrations $\mathbb{F}$ and $\mathbb{G}$ and the hazard process $\varPhi$, it is not obvious whether we may find a process $N$, which would satisfy Definition 6.6.2. To provide a simple construction of a conditional Poisson process, we assume that the underlying probability space $(\Omega, \mathcal{G}, \mathbb{P})$, endowed with a reference filtration $\mathbb{F}$, is sufficiently large to accommodate for the following stochastic processes: a Poisson process $\tilde{N}$ with the constant intensity $\lambda = 1$ and an $\mathbb{F}$-adapted hazard process $\varPhi$. In addition, we postulate that the Poisson process $\tilde{N}$ is independent of the filtration $\mathbb{F}$

*Remark.* Given a filtered probability space $(\Omega, \mathbb{F}, \mathbb{P})$, it is always possible to enlarge it in such a way that there exists a Poisson process $\tilde{N}$ with $\lambda = 1$, independent of the filtration $\mathbb{F}$, and defined on the enlarged space.

Under the present assumptions, for every $0 \leq s < t$, any $u \in \mathbb{R}_+$, and any non-negative integer $k$, we have

$$\mathbb{P}\{\tilde{N}_t - \tilde{N}_s = k \,|\, \mathcal{F}_{\infty}\} = \mathbb{P}\{\tilde{N}_t - \tilde{N}_s = k \,|\, \mathcal{F}_u\} = \mathbb{P}\{\tilde{N}_t - \tilde{N}_s = k\}$$

and

$$\mathbb{P}\{\tilde{N}_t - \tilde{N}_s = k \,|\, \mathcal{F}_s^{\tilde{N}} \vee \mathcal{F}_s\} = \mathbb{P}\{\tilde{N}_t - \tilde{N}_s = k\} = \frac{(t - s)^k}{k!} \, e^{-(t-s)}.$$

The next result describes an explicit construction of a conditional Poisson process. This construction is based on a random time change associated with the increasing process $\Phi$.

**Proposition 6.6.7.** *Let $\tilde{N}$ be a Poisson process with the constant intensity $\lambda = 1$, independent of a reference filtration $\mathbb{F}$, and let $\Phi$ be an $\mathbb{F}$-adapted, right-continuous, increasing process. Then the process $N_t = \tilde{N}_{\Phi_t}$, $t \in \mathbb{R}_+$, follows the $\mathbb{F}$-conditional Poisson process with the hazard process $\Phi$ with respect to the filtration $\mathbb{G} = \mathbb{F}^N \vee \mathbb{F}$.*

*Proof.* Since $\mathcal{G}_s \vee \mathcal{F}_\infty = \mathcal{F}_s^N \vee \mathcal{F}_\infty$, it suffices to check that

$$\mathbb{P}\{N_t - N_s = k \,|\, \mathcal{F}_s^N \vee \mathcal{F}_\infty\} = \frac{(\Phi_t - \Phi_s)^k}{k!} \, e^{-(\Phi_t - \Phi_s)}$$

or, equivalently,

$$\mathbb{P}\{\tilde{N}_{\Phi_t} - \tilde{N}_{\Phi_s} = k \,|\, \mathcal{F}_{\Phi_s}^{\tilde{N}} \vee \mathcal{F}_\infty\} = \frac{(\Phi_t - \Phi_s)^k}{k!} \, e^{-(\Phi_t - \Phi_s)}.$$

The last equality follows from the assumed independence of $\tilde{N}$ and $\mathbb{F}$.    □

*Remark.* Within the setting of Proposition 6.6.7, any $\mathbb{F}$-martingale is also a $\mathbb{G}$-martingale, so that Condition (M.1) is satisfied.

The total number of jumps of the conditional Poisson process is obviously unbounded with probability 1. In some financial models (see, e.g., Lando (1998) or Duffie and Singleton (1999)), only the properties of the first jump are relevant, though. There exist many ways of constructing the conditional Poisson process, but Condition (F.1) is always satisfied by the first jump of such a process, since it follows directly from Definition 6.6.2. In effect, if we denote $\tau = \tau_1$, then for any $t \in \mathbb{R}_+$ and $u \geq t$ we have (cf. Condition (F1.a) of Sect. 6.1)

$$\mathbb{P}\{\tau \leq t \,|\, \mathcal{F}_u\} = \mathbb{P}\{N_t \geq 1 \,|\, \mathcal{F}_u\} = \mathbb{P}\{N_t - N_0 \geq 1 \,|\, \mathcal{G}_0 \vee \mathcal{F}_u\} = \mathbb{P}\{\tau \leq u \,|\, \mathcal{F}_\infty\},$$

where the last equality follows from (6.47). It is also clear, once more by (6.47), that $\mathbb{P}\{\tau \leq t \,|\, \mathcal{F}_u\} = e^{-\Phi_u}$ for every $0 \leq t \leq u$.

*Example 6.6.2. Cox process.* In some applications, it is natural to consider a special case of an $\mathbb{F}$-conditional Poisson process, with the filtration $\mathbb{F}$ generated by a certain stochastic process, representing the *state variables*. To be more specific, on considers a conditional Poisson process with the intensity process $\phi$ given as $\phi_t = g(t, Y_t)$, where $Y$ is an $\mathbb{R}^d$-valued stochastic process independent of the Poisson process $\tilde{N}$, and $g : \mathbb{R}_+ \times \mathbb{R}^d \to \mathbb{R}_+$ is a (continuous) function. The reference filtration $\mathbb{F}$ is typically chosen to be the natural filtration of the process $Y$; that is, we take $\mathbb{F} = \mathbb{F}^Y$. In such a case, the resulting $\mathbb{F}$-conditional Poisson process is referred[2] to as the *Cox process* associated with the state variables process $Y$, and the intensity function $g$.

---

[2] It should be acknowledged that the terminology in this area is not uniform across various sources.

Our last goal is to examine the behavior of an $\mathbb{F}$-conditional Poisson process $N$ under an equivalent change of a probability measure. Let us assume, for the sake of simplicity, that the hazard process $\Phi$ is continuous, and the reference filtration $\mathbb{F}$ is generated by a process $W$, which follows a Brownian motion with respect to $\mathbb{G}$. For a fixed $T > 0$, we define the probability measure $\mathbb{P}^*$ on $(\Omega, \mathcal{G}_T)$ by setting

$$\frac{d\mathbb{P}^*}{d\mathbb{P}}\bigg|_{\mathcal{G}_T} = \eta_T, \quad \mathbb{P}\text{-a.s.,} \tag{6.48}$$

where the Radon-Nikodým density process $\eta_t$, $t \in [0, T]$, solves the SDE

$$d\eta_t = \eta_{t-}\big(\beta_t \, dW_t + \kappa_t \, d\hat{N}_t\big), \quad \eta_0 = 1, \tag{6.49}$$

for some $\mathbb{G}$-predictable processes $\beta$ and $\kappa$ such that $\kappa > -1$ and $\mathbb{E}_{\mathbb{P}}(\eta_T) = 1$. An application of Itô's product rule shows that the unique solution to (6.49) is equal to the product $\nu_t \zeta_t$, where $d\nu_t = \nu_t \beta_t \, dW_t$ and $d\zeta_t = \zeta_{t-}\kappa_t \, d\hat{N}_t$, with $\nu_0 = \zeta_0 = 1$. The solutions to the last two equations are

$$\nu_t = \exp\left(\int_0^t \beta_u \, dW_u - \frac{1}{2}\int_0^t \beta_u^2 \, du\right)$$

and

$$\zeta_t = \exp\left(U_t\right) \prod_{0 < u \le t} (1 + \Delta U_u)\exp\left(-\Delta U_u\right),$$

respectively, where we denote $U_t = \int_{]0,t]} \kappa_u \, d\hat{N}_u$. It is useful to observe that $\zeta$ admits the following representations:

$$\zeta_t = \exp\left(-\int_0^t \kappa_u \, d\Phi_u\right) \prod_{0 < u \le t} (1 + \kappa_u \Delta N_u),$$

and

$$\zeta_t = \exp\left(\int_{]0,t]} \ln(1 + \kappa_u) \, dN_u - \int_0^t \kappa_u \, d\Phi_u\right).$$

The following result is a counterpart of Proposition 5.3.1.

**Proposition 6.6.8.** *Let the Radon-Nikodým density of $\mathbb{P}^*$ with respect to $\mathbb{P}$ be given by (6.48)–(6.49). Then the process $W_t^* = W_t - \int_0^t \beta_u \, du$, $t \in [0, T]$, follows a Brownian motion with respect to $\mathbb{G}$ under $\mathbb{P}^*$, and the process*

$$N_t^* = \hat{N}_t - \int_0^t \kappa_u \, d\Phi_u = N_t - \int_0^t (1 + \kappa_u) \, d\Phi_u, \quad \forall\, t \in [0, T], \tag{6.50}$$

*follows a $\mathbb{G}$-martingale under $\mathbb{P}^*$. If, in addition, the process $\kappa$ is $\mathbb{F}$-adapted, then the process $N$ follows under $\mathbb{P}^*$ an $\mathbb{F}$-conditional Poisson process with respect to $\mathbb{G}$, and the hazard process of $N$ under $\mathbb{P}^*$ equals*

$$\Phi_t^* = \int_0^t (1 + \kappa_u) \, d\Phi_u.$$

# 7. Case of Several Random Times

In this chapter, we assume throughout that we are given a finite collection $\tau_1, \ldots, \tau_n$ of random times, defined on a common probability space $(\Omega, \mathcal{G}, \mathbb{P})$ endowed with a filtration $\mathbb{F}$. We define the family of jump processes $H^i$, $i = 1, \ldots, n$ by setting $H_t^i = \mathbb{1}_{\{\tau_i \leq t\}}$ and we write $\mathbb{H}^i$ to denote the filtration generated by the jump process $H^i$. Let us introduce the enlarged filtration $\mathbb{G}$ by setting $\mathbb{G} = \mathbb{H}^1 \vee \ldots \vee \mathbb{H}^n \vee \mathbb{F}$. One of our goals is to examine the relationship between the $(\mathbb{F}, \mathbb{G})$-martingale hazard processes of random times $\tau_1, \ldots, \tau_n$, and the $(\mathbb{F}, \mathbb{G})$-martingale hazard process of their minimum, i.e., of the random time $\tau = \min(\tau_1, \ldots, \tau_n)$.

Sect. 7.1 deals with the basic questions associated with the minimum of several random times. Although the martingale hazard process of the minimum of a finite collection of random times is easily obtained, the important issue whether it can be used as a tool to evaluate conditional expectations appears to be rather difficult, and only partial answers to this question are known. Subsequently, we establish a version of the martingale representation theorem in case when the reference $\mathbb{F}$ is generated by a Brownian motion. Sect. 7.2 is devoted to the study of the properties of hazard processes under an equivalent change of the underlying probability measure. Finally, in Sect. 7.3 we provide a thorough analysis of the important counter-example due to Kusuoka (1999). Results presented in this chapter are mainly drawn from Duffie (1998a), Kusuoka (1999) and Jeanblanc and Rutkowski (2000b).

## 7.1 Minimum of Several Random Times

We shall examine the following problem: given a finite family of random times $\tau_i$, $i = 1, \ldots, n$ and the associated hazard processes $\Gamma^i$, $i = 1, \ldots, n$, we search for the hazard process of the minimum of $\tau_1, \ldots, \tau_n$, that is, of the random time $\tau = \tau_1 \wedge \ldots \wedge \tau_n$. It should be made clear that this problem cannot be solved in such a generality – that is, without the exact knowledge of the joint probability law of $(\tau_1, \ldots, \tau_n)$. In effect, the solution heavily depends on specific assumptions on random times, as well as on the choice of filtrations.

### 7.1.1 Hazard Function

We will first focus on the calculation of the hazard function of the minimum of a finite family of mutually independent random times (for the case of conditionally independent random time, see Sect. 9.1.2).

**Lemma 7.1.1.** *Let $\tau_i$, $i = 1, \ldots, n$ be $n$ random times defined on a common probability space $(\Omega, \mathcal{G}, \mathbb{P})$. Assume that $\tau_i$ admits the hazard function $\Gamma_i$. If $\tau_i$, $i = 1, \ldots, n$ are mutually independent random variables, then the hazard function $\Gamma$ of $\tau$ is equal to the sum of hazard functions $\Gamma_i$, $i = 1, \ldots, n$.*

*Proof.* For any $t \in \mathbb{R}_+$ we have

$$e^{-\Gamma(t)} = 1 - F(t) = \mathbb{P}\{\tau > t\} = \mathbb{P}\{\min(\tau_1, \ldots, \tau_n) > t\} = \prod_{i=1}^{n} \mathbb{P}\{\tau_i > t\}$$

$$= \prod_{i=1}^{n}(1 - F_i(t)) = \prod_{i=1}^{n} e^{-\Gamma_i(t)} = e^{-\sum_{i=1}^{n} \Gamma_i(t)}. \qquad \square$$

Let us examine the case of continuous distribution functions $F_i$, $i = 1, \ldots, n$. In this case, we also get $\Lambda(t) = \sum_{i=1}^{n} \Lambda_i(t)$. In particular, if each random time $\tau_i$, $i = 1, \ldots, n$ admits the intensity function $\gamma_i(t) = \lambda_i(t) = f_i(t)(1 - F_i(t))^{-1}$, then the process

$$H_t - \sum_{i=1}^{n} \int_0^{t \wedge \tau} \gamma_i(u)\, du = \mathbb{1}_{\{\tau \le t\}} - \sum_{i=1}^{n} \int_0^{t \wedge \tau} \lambda_i(u)\, du$$

follows an $\mathbb{H}$-martingale, where $\mathbb{H} = \mathbb{H}^1 \vee \ldots \vee \mathbb{H}^n$. Conversely, if the hazard function of $\tau$ satisfies $\Lambda(t) = \Gamma(t) = \sum_{i=1}^{n} \Gamma_i(t) = \sum_{i=1}^{n} \Lambda_i(t)$, for every $t \in \mathbb{R}_+$, then $\mathbb{P}\{\tau_1 > t, \ldots, \tau_n > t\} = \prod_{i=1}^{n} \mathbb{P}\{\tau_i > t\}$ for every $t \in \mathbb{R}_+$.

### 7.1.2 Martingale Hazard Process

In this section, we assume that $\tau_i$, $i = 1, \ldots, n$ are random times such that $\mathbb{P}\{\tau_i = \tau_j\} = 0$ for $i \ne j$. The next lemma is borrowed from Duffie (1998a).

**Lemma 7.1.2.** *The $(\mathbb{F}, \mathbb{G})$-martingale hazard process of $\tau = \tau_1 \wedge \ldots \wedge \tau_n$ is equal to the sum of $(\mathbb{F}, \mathbb{G})$-martingale hazard processes $\Lambda^i$, i.e., $\Lambda = \sum_{i=1}^{n} \Lambda^i$. If $\Lambda$ is a continuous process, the process $\tilde{L}_t := (1 - H_t)e^{\Lambda_t}$ is a $\mathbb{G}$-martingale.*

*Proof.* We know that for any $i = 1, \ldots, n$ the process $\tilde{M}_t^i := H_t^i - \Lambda_{t \wedge \tau_i}^i$ is a $\mathbb{G}$-martingale. Since the random times $\tau_i$, $i = 1 \ldots, n$ are $\mathbb{G}$-stopping times, $\tau$ is a $\mathbb{G}$-stopping time as well. Consequently, by the well-known property of martingales, for any fixed $i = 1, \ldots n$, the stopped process

$$(\tilde{M}_t^i)^\tau = H_{t \wedge \tau}^i - \Lambda_{t \wedge \tau_i \wedge \tau}^i = H_{t \wedge \tau}^i - \Lambda_{t \wedge \tau}^i$$

also follows a $\mathbb{G}$-martingale.

Since $\mathbb{P}\{\tau_i = \tau_j\} = 0$ for $i \neq j$, we have $\sum_{i=1}^n H_{t \wedge \tau}^i = H_t = \mathbb{1}_{\{\tau \leq t\}}$, so that the process

$$\tilde{M}_t := H_t - \sum_{i=1}^n \Lambda_{t \wedge \tau}^i = \sum_{i=1}^n (\tilde{M}_t^i)^\tau$$

follows a $\mathbb{G}$-martingale, as the sum of $\mathbb{G}$-martingales. We conclude that the $(\mathbb{F}, \mathbb{G})$-martingale hazard process $\Lambda$ of $\tau$ satisfies: $\Lambda_t = \sum_{i=1}^n \Lambda_t^i$ for $t \in \mathbb{R}_+$. The second statement is an easy consequence of Itô's formula, which gives

$$\tilde{L}_t = 1 - \int_{]0,t]} \tilde{L}_{u-} \, d\tilde{M}_u. \tag{7.1}$$

This ends the proof.    $\square$

The striking feature of Lemma 7.1.2 is that the $(\mathbb{F}, \mathbb{G})$-martingale hazard process of $\tau$ can be easily found, even in case when the joint probability law of random times $\tau_1, \ldots, \tau_n$ is not known. It should thus be observed that in order to make effective use of the notion of an $(\mathbb{F}, \mathbb{G})$-martingale hazard process $\Lambda$, we need to show, in addition, that $\Lambda$ actually possesses the desired properties. For instance, it would be important to know whether the equality

$$\mathbb{P}\{\tau > s \mid \mathcal{G}_t\} = \mathbb{1}_{\{\tau > t\}} \mathbb{E}_{\mathbb{P}}(e^{\Lambda_t - \Lambda_s} \mid \mathcal{F}_t)$$

is valid for every $t \leq s$. One can also ask, more generally, whether we have

$$\mathbb{E}_{\mathbb{P}}(\mathbb{1}_{\{\tau > s\}} Y \mid \mathcal{G}_t) = \mathbb{1}_{\{\tau > t\}} \mathbb{E}_{\mathbb{P}}(Y e^{\Lambda_t - \Lambda_s} \mid \mathcal{F}_t)$$

for any bounded, $\mathcal{F}_s$-measurable random variable $Y$.

From now on, we shall assume that the martingale invariance property (M.1) is satisfied by the two filtrations $\mathbb{F}$ and $\mathbb{G}$. Combining Lemma 7.1.2 with part (ii) in Proposition 6.4.1, we get immediately the following result.

**Proposition 7.1.1.** *Assume that each random time $\tau_i$ admits a continuous $(\mathbb{F}, \mathbb{G})$-martingale hazard process $\Lambda^i$. Let us set $\Lambda = \sum_{i=1}^n \Lambda^i$, and let $Y$ be a bounded, $\mathcal{F}_s$-measurable random variable. If the process $V$, defined as*

$$V_t = \mathbb{E}_{\mathbb{P}}(Y e^{\Lambda_t - \Lambda_s} \mid \mathcal{F}_t), \quad \forall t \in [0, s],$$

*is continuous at $\tau$, then for any $t < s$ we have*

$$\mathbb{E}_{\mathbb{P}}(\mathbb{1}_{\{\tau > s\}} Y \mid \mathcal{G}_t) = \mathbb{1}_{\{\tau > t\}} \mathbb{E}_{\mathbb{P}}(Y e^{\Lambda_t - \Lambda_s} \mid \mathcal{F}_t).$$

In case when processes $\Lambda^i$ are absolutely continuous, we have

$$\mathbb{E}_{\mathbb{P}}(\mathbb{1}_{\{\tau > s\}} Y \mid \mathcal{G}_t) = \mathbb{1}_{\{\tau > t\}} \mathbb{E}_{\mathbb{P}}\left(Y e^{-\sum_{i=1}^n \int_t^s \lambda_u^i \, du} \mid \mathcal{F}_t\right).$$

At first glance, Proposition 7.1.1 might seem to be a strong and useful result, since it covers the case of independent and dependent random times. Notice, however, that the assumptions in Proposition 7.1.1 are rather stringent, and thus the number of circumstances when Proposition 7.1.1 can be effectively applied is in fact rather limited. The most widely used in practice case is examined in Sect. 9.1.2, in which we construct a finite family of random times conditionally independent with respect to the reference filtration $\mathbb{F}$.

### 7.1.3 Martingale Representation Theorem

In this section, we consider the case when the reference filtration $\mathbb{F}$ is the Brownian filtration, i.e., we assume that $\mathbb{F} = \mathbb{F}^W$ for some Brownian motion $W$. We postulate that $W$ remains a martingale (and thus, a Brownian motion) with respect to the enlarged filtration $\mathbb{G} = \mathbb{H}^1 \vee \ldots \vee \mathbb{H}^n \vee \mathbb{F}$. In view of the martingale representation property of the Brownian filtration, this means that any $\mathbb{F}$-local martingale is also a local martingale with respect to $\mathbb{G}$ (or, indeed, with respect to any filtration $\tilde{\mathbb{F}}$ such that $\mathbb{F} \subseteq \tilde{\mathbb{F}} \subseteq \mathbb{G}$), and so Condition (M.1) is satisfied. It is worth noting that the case when $\mathbb{F}$ is a trivial filtration is also covered by the results of this section, though.

One of our goals is to generalize the martingale representation property established in Corollary 5.2.4 (see also Proposition 5.2.2). Recall that in Corollary 5.2.4 we have assumed that the $\mathbb{F}$-hazard process $\Gamma$ of a random time $\tau$ is an increasing continuous process. Also, by virtue of Proposition 6.2.1, under the assumptions of Corollary 5.2.4 we have $\Gamma = \Lambda$; that is, the $\mathbb{F}$-hazard process $\Gamma$ and the $(\mathbb{F}, \mathbb{G})$-martingale hazard process $\Lambda$ coincide.

In the present set-up, we prefer to make assumptions directly about the $(\mathbb{F}, \mathbb{G})$-martingale hazard processes $\Lambda^i$ of the random times $\tau_i$, $i = 1, \ldots, n$. We assume throughout that the processes $\Lambda^i$, $i = 1, \ldots, n$ are continuous. Recall that by virtue of the definition of the $(\mathbb{F}, \mathbb{G})$-martingale hazard process $\Lambda^i$ of a random time $\tau_i$, for each $i = 1, \ldots, n$ the compensated jump process

$$\tilde{M}_t^i = H_t^i - \Lambda_{t \wedge \tau_i}^i$$

follows a $\mathbb{G}$-martingale. It is thus easy to see that the process

$$\tilde{L}_t^i = (1 - H_t^i) e^{\Lambda_t^i}$$

also follows a $\mathbb{G}$-martingale, since clearly (cf. (7.1))

$$\tilde{L}_t^i = 1 - \int_{]0,t]} \tilde{L}_{u-}^i \, d\tilde{M}_u^i. \tag{7.2}$$

One also checks without difficulty that $\tilde{L}^i$ and $\tilde{L}^j$ are mutually orthogonal $\mathbb{G}$-martingales for any $i \neq j$ (a similar remark applies to $\tilde{M}^i$ and $\tilde{M}^j$).

For any fixed $k \in \{0, \ldots, n\}$, we introduce an auxiliary sub-filtration $\tilde{\mathbb{G}}^k = \mathbb{H}^1 \vee \ldots \vee \mathbb{H}^k \vee \mathbb{F}$. Since $k$ is fixed, we shall write $\tilde{\mathbb{G}}$ rather than $\tilde{\mathbb{G}}^k$ in the sequel. Obviously $\tilde{\mathbb{G}} = \mathbb{G}$ when $k = n$, and, by convention, $\tilde{\mathbb{G}} = \mathbb{F}$ when $k = 0$. Obviously, for any fixed $k$ and arbitrary $1 \leq i \leq k$ processes $\tilde{L}^i$ and $\tilde{M}^i$ are $\tilde{\mathbb{G}}$-adapted. In addition, $\tilde{L}^i$ and $\tilde{L}^j$ are mutually orthogonal $\tilde{\mathbb{G}}$-martingales for arbitrary $1 \leq i, j \leq k$ such that $i \neq j$. A trivial modification of Lemma 7.1.2 shows that the $(\mathbb{F}, \tilde{\mathbb{G}})$-martingale hazard process of the random time $\tilde{\tau} := \tau_1 \wedge \ldots \wedge \tau_k$ equals $\tilde{\Lambda} = \sum_{i=1}^k \Lambda^i$. In other words, the process $\tilde{H}_t - \sum_{i=1}^k \Lambda_{t \wedge \tilde{\tau}}^i$ is a $\tilde{\mathbb{G}}$-martingale, where we set $\tilde{H}_t = \mathbb{1}_{\{\tilde{\tau} \leq t\}}$.

**Proposition 7.1.2.** *Assume that the* $\mathbb{F}$*-Brownian motion* $W$ *remains a Brownian motion with respect to the enlarged filtration* $\mathbb{G} = \mathbb{H}^1 \vee \ldots \vee \mathbb{H}^n \vee \mathbb{F}$. *Let* $Y$ *be a bounded,* $\mathcal{F}_T$*-measurable random variable, and let* $\tilde{\tau} = \tau_1 \wedge \ldots \wedge \tau_k$. *Then for* $t \leq s \leq T$ *we have*

$$\mathbb{E}_{\mathbb{P}}(\mathbb{1}_{\{\tilde{\tau}>s\}} Y \,|\, \mathcal{G}_t) = \mathbb{E}_{\mathbb{P}}(\mathbb{1}_{\{\tilde{\tau}>s\}} Y \,|\, \tilde{\mathcal{G}}_t) = \mathbb{1}_{\{\tilde{\tau}>t\}} \mathbb{E}_{\mathbb{P}}(Y e^{\tilde{A}_t - \tilde{A}_s} \,|\, \mathcal{F}_t).$$

*For every* $t \leq s$ *we have*

$$\mathbb{P}\{\tilde{\tau} > s \,|\, \mathcal{G}_t\} = \mathbb{P}\{\tilde{\tau} > s \,|\, \tilde{\mathcal{G}}_t\} = \mathbb{1}_{\{\tilde{\tau}>t\}} \mathbb{E}_{\mathbb{P}}(e^{\tilde{A}_t - \tilde{A}_s} \,|\, \mathcal{F}_t),$$

*in particular, for* $\tau = \tau_1 \wedge \ldots \wedge \tau_n$ *we have, for every* $t \leq s$,

$$\mathbb{P}\{\tau > s \,|\, \mathcal{G}_t\} = \mathbb{1}_{\{\tau>t\}} \mathbb{E}_{\mathbb{P}}(e^{A_t - A_s} \,|\, \mathcal{F}_t),$$

*where* $A = \sum_{i=1}^n A^i$.

*Proof.* We fix $s \leq T$. For every $t \in [0, T]$, we set

$$\tilde{Y}_t = \mathbb{E}_{\mathbb{P}}(Y e^{-\tilde{A}_s} \,|\, \mathcal{F}_t).$$

Let the process $U$ be given by the formula, for every $t \in [0, T]$,

$$U_t = (1 - \tilde{H}_{t \wedge s}) e^{\tilde{A}_{t \wedge s}} = \prod_{i=1}^k \tilde{L}^i_{t \wedge s}. \tag{7.3}$$

Under the present assumptions, the process $\tilde{Y}$ is a continuous $\mathbb{G}$-martingale, and thus also a $\tilde{\mathbb{G}}$-martingale. The process $U$, which is manifestly of finite variation, is also a $\tilde{\mathbb{G}}$-martingale as the product of mutually orthogonal $\tilde{\mathbb{G}}$-martingales $\tilde{L}^1, \ldots, \tilde{L}^k$ (stopped at $s$). Thus, the product $U\tilde{Y}$ is a $\tilde{\mathbb{G}}$-martingale. This in turn yields, for any $t \leq s$

$$\mathbb{E}_{\mathbb{P}}(\mathbb{1}_{\{\tilde{\tau}>s\}} Y \,|\, \tilde{\mathcal{G}}_t) = \mathbb{E}_{\mathbb{P}}(U_T \tilde{Y}_T \,|\, \tilde{\mathcal{G}}_t) = U_t \tilde{Y}_t = (1 - \tilde{H}_t) e^{\tilde{A}_t} \mathbb{E}_{\mathbb{P}}(Y e^{-\tilde{A}_s} \,|\, \mathcal{F}_t)$$

as expected. It is also clear that we may replace the filtration $\tilde{\mathbb{G}}$ by $\mathbb{G}$ in the above reasoning. □

For a fixed $k \in \{0, \ldots, n-1\}$, we introduce an auxiliary sub-filtration $\tilde{\mathbb{F}}^k := \mathbb{H}^{k+1} \vee \ldots \vee \mathbb{H}^n \vee \mathbb{F}$, and we write briefly $\tilde{\mathbb{F}} = \tilde{\mathbb{F}}^k$. The next result generalizes Proposition 5.2.2. Recall that we have assumed that the $\mathbb{F}$-martingale hazard process $A^i$ is continuous for any $i = 1, \ldots, n$.

**Proposition 7.1.3.** *Assume that the Brownian motion* $W$ *remains a Brownian motion with respect to* $\mathbb{G}$. *Let* $X$ *be a bounded,* $\tilde{\mathcal{F}}_T$*-measurable random variable. Then the* $\tilde{\mathbb{F}}$*-martingale* $M$, *which equals* $M_t = \mathbb{E}_{\mathbb{P}}(X \,|\, \tilde{\mathcal{F}}_t)$, $t \in [0, T]$, *has the following (unique) integral representation*

$$M_t = M_0 + \int_0^t \xi_u \, dW_u + \sum_{i=k+1}^n \int_{]0,t]} \zeta^i_u \, d\tilde{M}^i_u, \tag{7.4}$$

*where* $\xi$ *and* $\zeta^i$ *for* $i = k+1, \ldots, n$ *follow* $\tilde{\mathbb{F}}$*-predictable stochastic processes.*

*Proof.* The demonstration is similar to the proof of Proposition 5.2.2. We begin by noticing that it is enough to consider a random variable $X$ of the following form:

$$X = Y \prod_{j=1}^{r} (1 - H_{s_j}^{i_j})$$

for some $r \leq n - k$, where $0 < s_1 < \cdots < s_r \leq T$, $k + 1 \leq i_1 < \cdots < i_r \leq n$, and where $Y$ is assumed to be a bounded, $\mathcal{F}_T$-measurable random variable. We introduce the following auxiliary $\mathbb{F}$-martingale:

$$\tilde{Y}_t = \mathbb{E}_{\mathbb{P}}\Big(Y \exp\Big(-\sum_{i=1}^{r} \Lambda_{s_i}^{i_j}\Big)\Big|\mathcal{F}_t\Big).$$

Because $\mathbb{F}$ is generated by a Brownian motion $W$, invoking the martingale representation property of the Brownian filtration, we conclude that $\tilde{Y}$ follows a continuous process, admitting the following integral representation:

$$\tilde{Y}_t = \tilde{Y}_0 + \int_0^t \tilde{\xi}_u \, dW_u, \quad \forall t \in [0, T],$$

for some $\mathbb{F}$-predictable process $\tilde{\xi}$. By assumption, $W$ remains a martingale with respect to $\mathbb{G}$, and so $\tilde{Y}$ is a $\mathbb{G}$-martingale as well.[1] As a continuous $\mathbb{G}$-martingale, $\tilde{Y}$ is orthogonal to each $\mathbb{G}$-martingale of finite variation $\tilde{M}^i$. Using Itô's formula and (7.2), we obtain

$$Y \prod_{j=1}^{r}(1 - H_{s_j}^{i_j}) = \tilde{Y}_T \prod_{j=1}^{r} \tilde{L}_{s_j}^{i_j}$$

$$= \tilde{Y}_0 + \int_0^T \prod_{j=1}^{r} \tilde{L}_{(u \wedge s_j)-}^{i_j} \, d\tilde{Y}_u - \sum_{l=1}^{r} \int_{]0, s_l]} \tilde{Y}_{u-} \prod_{j=1}^{r} \tilde{L}_{(u \wedge s_j)-}^{i_j} \, d\tilde{M}_u^{i_l}.$$

The last formula leads to (7.4). The uniqueness follows from the mutual orthogonality of integrals in (7.4).                                                    □

If the random variable $X$ is merely $\mathcal{G}_T$-measurable, we may still apply Proposition 7.1.3 to the $\tilde{\mathbb{F}}$-martingale $M_t = \mathbb{E}_{\mathbb{P}}(X \,|\, \tilde{\mathcal{F}}_t)$, since clearly $M_t = \mathbb{E}_{\mathbb{P}}(\hat{X} \,|\, \tilde{\mathcal{F}}_t)$, where $\hat{X} := \mathbb{E}_{\mathbb{P}}(X \,|\, \tilde{\mathcal{F}}_T)$ is an $\tilde{\mathcal{F}}_T$-measurable random variable. This shows that representation (7.4) holds indeed for any $\tilde{\mathbb{F}}$-martingale.

It is also interesting to observe that in Proposition 7.1.2 we may substitute the Brownian filtration $\mathbb{F}$ with the filtration $\bar{\mathbb{F}} := \mathbb{H}^{k+1} \vee \ldots \vee \mathbb{H}^n \vee \mathbb{F}$. First, it is clear that $\tilde{\Lambda} = \sum_{i=1}^{k} \Lambda^i$ is also the $(\bar{\mathbb{F}}, \mathbb{G})$-martingale hazard process of a random time $\tilde{\tau}$. Second, Proposition 7.1.3 shows that the process

$$\hat{Y}_t := \mathbb{E}_{\mathbb{P}}\big(Y e^{-\tilde{\Lambda}_s} \,\big|\, \tilde{\mathcal{F}}_t\big), \quad \forall t \in [0, T],$$

where $Y$ is a bounded, $\tilde{\mathcal{F}}_T$-measurable random variable, admits the following

---

[1] Since $\tilde{Y}$ is manifestly $\bar{\mathbb{F}}$-adapted, it also follows a martingale with respect to $\bar{\mathbb{F}}$.

integral representation:

$$\hat{Y}_t = \hat{Y}_0 + \int_0^t \xi_u \, dW_u + \sum_{i=k+1}^{n} \int_{]0,t]} \zeta_u^i \, d\tilde{M}_u^i,$$

where $\xi$ and $\zeta^i$, $i = k+1, \ldots, n$ are $\mathbb{F}$-predictable processes. Therefore, $\hat{Y}$ is a $\mathbb{G}$-martingale orthogonal to the $\mathbb{G}$-martingale $U$ given by formula (7.3). Arguing as in the proof of Proposition 7.1.2, we obtain the following result.

**Corollary 7.1.1.** *Let $Y$ be a bounded, $\tilde{\mathcal{F}}_T$-measurable random variable, and let $\tilde{\tau} = \tau_1 \wedge \ldots \wedge \tau_k$. Then for every $t \le s \le T$ we have*

$$\mathbb{E}_{\mathbb{P}}(\mathbb{1}_{\{\tilde{\tau} > s\}} Y \,|\, \mathcal{G}_t) = \mathbb{1}_{\{\tilde{\tau} > t\}} \mathbb{E}_{\mathbb{P}}(Y e^{\tilde{\Lambda}_t - \tilde{\Lambda}_s} \,|\, \tilde{\mathcal{F}}_t).$$

*In particular, for every $t \le s$ we have*

$$\mathbb{P}\{\tilde{\tau} > s \,|\, \mathcal{G}_t\} = \mathbb{1}_{\{\tilde{\tau} > t\}} \mathbb{E}_{\mathbb{P}}(e^{\tilde{\Lambda}_t - \tilde{\Lambda}_s} \,|\, \tilde{\mathcal{F}}_t).$$

Of course, the boundedness of $X$ ($Y$, resp.) in Proposition 7.1.3 (in Corollary 7.1.1, resp.) is not a necessary condition, so that it can be relaxed.

## 7.2 Change of a Probability Measure

In this section, in which we follow Kusuoka (1999), results of Sect. 5.3 are extended to the case of several random times. We preserve the assumptions of Sect. 7.1.3. In particular, the filtration $\mathbb{F}$ is generated by a Brownian motion $W$, which is also a $\mathbb{G}$-martingale (the case of a trivial filtration $\mathbb{F}$ is also covered by the results of this section, though), and the processes $\Lambda^i$ are continuous. For a fixed $T > 0$, we shall examine the properties of $\tilde{\tau}$ under a probability measure $\mathbb{P}^*$, which is equivalent to $\mathbb{P}$ on $(\Omega, \mathcal{G}_T)$. To this end, we introduce the associated $\mathbb{G}$-martingale $\eta$ by setting, for $t \in [0, T]$,

$$\eta_t := \frac{d\mathbb{P}^*}{d\mathbb{P}} \Big|_{\mathcal{G}_t} = \mathbb{E}_{\mathbb{P}}(X \,|\, \mathcal{G}_t), \quad \mathbb{P}\text{-a.s.},$$

where $X$ stands for an arbitrary $\mathcal{G}_T$-measurable random variable, such that $\mathbb{P}\{X > 0\} = 1$ and $\mathbb{E}_{\mathbb{P}}(X) = 1$. By virtue of Proposition 7.1.3, the Radon-Nikodým density process $\eta$ introduced above has the following integral representation:

$$\eta_t = 1 + \int_0^t \xi_u \, dW_u + \sum_{i=1}^{n} \int_{]0,t]} \zeta_u^i \, d\tilde{M}_u^i,$$

where $\xi$ and $\zeta^i$ for $i = 1, \ldots, n$ are $\mathbb{G}$-predictable stochastic processes. It can also be easily checked that $\eta$ follows a strictly positive stochastic process. Hence, we may rewrite the last formula as follows:

$$\eta_t = 1 + \int_{]0,t]} \eta_{u-} \Big( \beta_u \, dW_u + \sum_{i=1}^{n} \kappa_u^i \, d\tilde{M}_u^i \Big), \tag{7.5}$$

where $\beta$ and $\kappa^i > -1$ for $i = 1, \ldots, n$ are $\mathbb{G}$-predictable processes.

The next result extends Proposition 5.3.1. The proof of Proposition 7.2.1 relies on similar arguments as the proof of Proposition 5.3.1, and thus it is left to the reader.

**Proposition 7.2.1.** *Let* $\mathbb{P}^*$ *be a probability measure equivalent to* $\mathbb{P}$ *on* $(\Omega, \mathcal{G}_T)$. *Suppose that the Radon-Nikodým density process* $\eta$ *of* $\mathbb{P}^*$ *with respect to* $\mathbb{P}$ *is given by formula (7.5). Then the process*

$$W_t^* = W_t - \int_0^t \beta_u \, du, \quad \forall t \in [0, T],$$

*follows a* $\mathbb{G}$-*Brownian motion under* $\mathbb{P}^*$, *and for each* $i = 1, \ldots, n$ *the process* $M_t^{i*}$, $t \in [0, T]$, *given by the formula*

$$M_t^{i*} := \tilde{M}_t^i - \int_{]0, t \wedge \tau_i]} \kappa_u^i \, d\Lambda_u^i = H_t^i - \int_{]0, t \wedge \tau_i]} (1 + \kappa_u^i) \, d\Lambda_u^i, \qquad (7.6)$$

*is a* $\mathbb{G}$-*martingale orthogonal to* $W^*$ *under* $\mathbb{P}^*$. *Moreover, processes* $M^{i*}$ *and* $M^{j*}$, $i \neq j$ *are mutually orthogonal* $\mathbb{G}$-*martingales under* $\mathbb{P}^*$.

Although, by virtue of the last result, the process $M^{i*}$ follows a $\mathbb{G}$-martingale under $\mathbb{P}^*$, it should be stressed that the process $\int_{]0, t]} (1 + \kappa_u^i) \, d\Lambda_u^i$ does not necessarily represent the $(\mathbb{F}, \mathbb{G})$-martingale hazard process of $\tau_i$ under $\mathbb{P}^*$, since it is not $\mathbb{F}$-adapted but merely $\mathbb{G}$-adapted, in general. To circumvent this difficulty, we shall choose a suitable version of the process $\kappa^i$. For any fixed $i$, we take a process $\kappa^{i*}$, which coincides with $\kappa^i$ on a random interval $[0, \tau_i]$, and which is predictable with respect to the enlarged filtration $\mathbb{F}^{i*}$, which is given as:

$$\mathbb{F}^{i*} = \mathbb{H}^1 \vee \ldots \vee \mathbb{H}^{i-1} \vee \mathbb{H}^{i+1} \vee \ldots \mathbb{H}^n \vee \mathbb{F}.$$

Then, it is rather obvious that the process

$$M_t^{i*} := H_t^i - \int_{]0, t \wedge \tau_i]} (1 + \kappa_u^i) \, d\Lambda_u^i = H_t^i - \int_{]0, t \wedge \tau_i]} (1 + \kappa_u^{i*}) \, d\Lambda_u^i$$

follows a $\mathbb{G}$-martingale under $\mathbb{P}^*$. We conclude that, for any fixed $i$, the process $\Lambda_t^{i*}$, $t \in [0, T]$, given by the formula

$$\Lambda_t^{i*} = \int_{]0, t]} (1 + \kappa_u^{i*}) \, d\Lambda_u^i$$

represents the $(\mathbb{F}^{i*}, \mathbb{G})$-martingale hazard process of $\tau_i$ under $\mathbb{P}^*$. This does not mean, however, that the following equality is valid for $s \leq t \leq T$

$$\mathbb{P}^*\{\tau_i > s \mid \mathcal{G}_t\} = \mathbb{1}_{\{\tau_i > t\}} \mathbb{E}_{\mathbb{P}^*}\left(e^{\Lambda_t^{i*} - \Lambda_s^{i*}} \mid \mathcal{F}_t^{i*}\right).$$

We prefer to examine the last issue in a slightly more general framework. For a fixed $k \leq n$, we consider the random time $\tilde{\tau} = \tau_1 \wedge \ldots \wedge \tau_k$. Since the order of random times is not essential here, the analysis presented in the sequel also covers the case of a single random time $\tau_i$, for any choice of $i = 1, \ldots, n$.

As in Sect. 7.1.3, we introduce an auxiliary filtration $\tilde{\mathbb{F}} = \mathbb{H}^{k+1} \vee \ldots \vee \mathbb{H}^n \vee \mathbb{F}$. For any $i = 1, \ldots, n$, we denote by $\tilde{\kappa}^i$ ($\tilde{\beta}$, resp.) the $\tilde{\mathbb{F}}$-predictable process such that $\tilde{\kappa}^i = \kappa^i$ ($\tilde{\beta} = \beta$, resp.) on the random set $[0, \tilde{\tau}]$. Let us set

$$\tilde{W}_t^* = W_t - \int_0^t \tilde{\beta}_u \, du,$$

and

$$\tilde{M}_t^{i*} = H_t^i - \int_{]0, t \wedge \tau_i]} (1 + \tilde{\kappa}_u^i) \, d\Lambda_u^i$$

for any $i = 1, \ldots, n$. Notice that the processes $\tilde{W}^*$ and $\tilde{M}^{i*}$ follow $\mathbb{G}$-martingales under $\mathbb{P}^*$, provided that they are stopped at the random time $\tilde{\tau}$ (since clearly $\tilde{W}_{t \wedge \tilde{\tau}}^* = W_{t \wedge \tilde{\tau}}^*$ and $\tilde{M}_{t \wedge \tilde{\tau}}^{i*} = M_{t \wedge \tilde{\tau}}^{i*}$). Thus, letting $\tilde{H}_t = \mathbb{1}_{\{\tilde{\tau} \le t\}}$, the process

$$\tilde{H}_t - \sum_{i=1}^k \int_{]0, t \wedge \tilde{\tau}]} (1 + \tilde{\kappa}_u^i) \, d\Lambda_u^i = \sum_{i=1}^k M_{t \wedge \tilde{\tau}}^{i*}$$

also follows a $\mathbb{G}$-martingale. This shows that the $\tilde{\mathbb{F}}$-predictable process $\Lambda^*$, defined as

$$\Lambda_t^* = \sum_{i=1}^k \int_{]0, t]} (1 + \tilde{\kappa}_u^i) \, d\Lambda_u^i, \tag{7.7}$$

represents the $(\tilde{\mathbb{F}}, \mathbb{G})$-martingale hazard process of the random time $\tilde{\tau}$ under an equivalent probability measure $\mathbb{P}^*$. In view of Corollary 7.1.1, it is tempting to conjecture that for any bounded, $\tilde{\mathcal{F}}_T$-measurable random variable $Y$ and arbitrary $t \le s \le T$ we have

$$\mathbb{E}_{\mathbb{P}^*}(\mathbb{1}_{\{\tilde{\tau} > s\}} Y \,|\, \mathcal{G}_t) = \mathbb{1}_{\{\tilde{\tau} > t\}} \mathbb{E}_{\mathbb{P}^*}\big(Y e^{\Lambda_t^* - \Lambda_s^*} \,\big|\, \tilde{\mathcal{F}}_t\big). \tag{7.8}$$

It appears that if we wish the last equality to hold, we need to substitute the probability measure $\mathbb{P}^*$ on the right-hand side of (7.8) with some related probability measure. To this end, we introduce the following stochastic processes $\hat{\eta}^l$, $l = 1, 2, 3$:

$$\hat{\eta}_t^1 = 1 + \int_{]0, t]} \hat{\eta}_{u-}^1 \Big( \tilde{\beta}_u \, dW_u + \sum_{i=k+1}^n \tilde{\kappa}_u^i \, d\tilde{M}_u^i \Big), \tag{7.9}$$

$$\hat{\eta}_t^2 = 1 + \int_{]0, t]} \hat{\eta}_{u-}^2 \Big( \tilde{\beta}_u \, dW_u + \sum_{i=1}^n \tilde{\kappa}_u^i \, d\tilde{M}_u^i \Big),$$

and

$$\hat{\eta}_t^3 = 1 + \int_{]0, t]} \hat{\eta}_{u-}^3 \Big( \tilde{\beta}_u \, dW_u + \sum_{i=1}^k \kappa_u^i \, d\tilde{M}_u^i + \sum_{i=k+1}^n \tilde{\kappa}_u^i \, d\tilde{M}_u^i \Big).$$

Processes $\hat{\eta}^l$, $l = 1, 2, 3$ will play the role of the Radon-Nikodým density processes of some probability measures equivalent to $\mathbb{P}$ on $(\Omega, \mathcal{G}_T)$.

On one hand, we note that the process $\hat{\eta}^1$ is $\mathbb{F}$-adapted (since, in particular, each process $\tilde{M}^i$ is adapted to the filtration $\mathbb{H}^i \vee \mathbb{F}$). On the other hand, the processes $\hat{\eta}^2$ and $\hat{\eta}^3$ are $\mathbb{G}$-adapted, but they are not necessarily $\tilde{\mathbb{F}}$-adapted. To overcome this deficiency, we introduce their $\tilde{\mathbb{F}}$-adapted (RCLL) modifications by formally setting $\tilde{\eta}_t^l = \mathbb{E}_{\mathbb{P}}(\hat{\eta}_T^l \mid \tilde{\mathcal{F}}_t)$ for $l = 2, 3$ and every $t \in [0, T]$.

**Lemma 7.2.1.** *We have*

$$\tilde{\eta}_t^l = 1 + \int_{]0,t]} \tilde{\eta}_{u-}^l \left( \tilde{\beta}_u \, dW_u + \sum_{i=k+1}^{n} \tilde{\kappa}_u^i \, d\tilde{M}_u^i \right). \tag{7.10}$$

*Proof.* For $l = 1$, formula (7.10) coincides with (7.9). For $l = 2, 3$, it follows from the uniqueness of the martingale representation property (see Proposition 7.1.3), combined with the mutual orthogonality of integrals in the definition of processes $\hat{\eta}^2$ and $\hat{\eta}^3$. □

We define a probability measure $\mathbb{P}_l$ on $(\Omega, \mathcal{G}_T)$ by setting, for $l = 1, 2, 3$ and every $t \in [0, T]$,

$$\hat{\eta}_t^l := \left. \frac{d\mathbb{P}_l}{d\mathbb{P}} \right|_{\mathcal{G}_t}, \quad \mathbb{P}\text{-a.s.}, \tag{7.11}$$

It is thus clear that

$$\tilde{\eta}_t^l = \mathbb{E}_{\mathbb{P}}(\hat{\eta}_T^l \mid \tilde{\mathcal{F}}_t) = \left. \frac{d\mathbb{P}_l}{d\mathbb{P}} \right|_{\tilde{\mathcal{F}}_t}, \quad \mathbb{P}\text{-a.s.}$$

The following counterpart of Corollary 7.1.1 is due to Kusuoka (1999).

**Proposition 7.2.2.** *Let $Y$ be a bounded, $\tilde{\mathcal{F}}_T$-measurable random variable. Then for any $t \le s \le T$ and $l = 1, 2, 3$ we have*

$$\mathbb{E}_{\mathbb{P}^*}(\mathbb{1}_{\{\tilde{\tau}>s\}} Y \mid \mathcal{G}_t) = \mathbb{1}_{\{\tilde{\tau}>t\}} \mathbb{E}_{\mathbb{P}_l}\left(Y e^{\Lambda_t^* - \Lambda_s^*} \mid \tilde{\mathcal{F}}_t\right),$$

*where the process $\Lambda^*$ is given by formula (7.7). In particular, we have*

$$\mathbb{P}^*\{\tilde{\tau} > s \mid \mathcal{G}_t\} = \mathbb{1}_{\{\tilde{\tau}>t\}} \mathbb{E}_{\mathbb{P}_l}\left(e^{\Lambda_t^* - \Lambda_s^*} \mid \tilde{\mathcal{F}}_t\right).$$

The proof of Proposition 7.2.2 parallels the demonstration of Proposition 7.1.2 (see also remarks preceding Corollary 7.1.1). We need first to establish some preliminary results. The next lemma provides a counterpart of the integral representation (7.4).

**Lemma 7.2.2.** *Let $Y$ be an $\tilde{\mathbb{F}}$-martingale under $\mathbb{P}_l$ for some $l \in \{1, 2, 3\}$. Then there exist $\tilde{\mathbb{F}}$-predictable processes $\tilde{\xi}$ and $\tilde{\zeta}^i$, $i = k+1, \ldots, n$ such that*

$$Y_t = Y_0 + \int_0^t \tilde{\xi}_u \, d\tilde{W}_u^* + \sum_{i=k+1}^{n} \int_{]0,t]} \tilde{\zeta}_u^i \, d\tilde{M}_u^{i*}. \tag{7.12}$$

*Proof.* The proof combines the calculations already employed in the proof of Corollary 5.3.1 with the martingale representation property under the original probability measure $\mathbb{P}$, which was previously established in Proposition 7.1.3.

We fix $l$, and we write $\tilde{\eta}_t = \mathbb{E}_{\mathbb{P}}(\eta_T^l \mid \tilde{\mathcal{F}}_t)$ (recall that $\tilde{\eta}_t = \hat{\eta}_t^1$ for $l = 1$). Let us introduce an auxiliary process $\tilde{Y}$, by setting, for $t \in [0, T]$,

$$\tilde{Y}_t = \int_{]0,t]} \tilde{\eta}_{u-}^{-1} \, d(\tilde{\eta}_u Y_u) - \int_{]0,t]} \tilde{\eta}_{u-}^{-1} Y_{u-} \, d\tilde{\eta}_u.$$

The process $\tilde{Y}$ follows an $\tilde{\mathbb{F}}$-martingale under $\mathbb{P}$. Since Itô's formula yields

$$\tilde{\eta}_{u-}^{-1} \, d(\tilde{\eta}_u Y_u) = dY_u + \tilde{\eta}_{u-}^{-1} Y_{u-} \, d\tilde{\eta}_u + \tilde{\eta}_{u-}^{-1} \, d\,[Y, \tilde{\eta}]_u,$$

the process $Y$ admits the following representation:

$$Y_t = Y_0 + \tilde{Y}_t - \int_{]0,t]} \tilde{\eta}_{u-}^{-1} \, d\,[Y, \tilde{\eta}]_u.$$

On the other hand, invoking Proposition 7.1.3, we obtain the following integral representation for the process $\tilde{Y}$

$$\tilde{Y}_t = \int_0^t \xi_u \, dW_u + \sum_{i=k+1}^n \int_{]0,t]} \zeta_u^i \, d\tilde{M}_u^i,$$

where $\xi$ and $\zeta^i$, $i = 1, \ldots, k$ are $\tilde{\mathbb{F}}$-predictable processes. Consequently, we have

$$dY_t = \xi_t \, dW_t + \sum_{i=k+1}^n \zeta_t^i \, d\tilde{M}_t^i - \tilde{\eta}_{t-}^{-1} \, d\,[Y, \tilde{\eta}]_t, \qquad (7.13)$$

and, finally,

$$dY_t = \xi_t \, d\tilde{W}_t^* + \sum_{i=k+1}^n \zeta_t^i (1 + \tilde{\kappa}_t^i)^{-1} \, d\tilde{M}_t^{i*}.$$

To derive the last equality from (7.13), it is enough to observe that (7.10) combined with (7.6) yield

$$\tilde{\eta}_{t-}^{-1} \, d\,[Y, \tilde{\eta}]_t = \xi_t \tilde{\beta}_t \, dt + \sum_{i=k+1}^n \zeta_t^i \tilde{\kappa}_t^i (1 + \tilde{\kappa}_t^i)^{-1} \, dH_t^i,$$

where the last equality follows in turn from the following relationship:

$$\Delta[Y, \tilde{\eta}]_t = \tilde{\eta}_{t-} \sum_{i=k+1}^n \left( \zeta_t^i \tilde{\kappa}_t^i - \tilde{\kappa}_t^i \tilde{\eta}_{t-}^{-1} \Delta[Y, \tilde{\eta}]_t \right) dH_t^i.$$

We conclude that $Y$ satisfies equality (7.12) with the following processes:

$$\tilde{\xi}_t = \xi_t, \quad \tilde{\zeta}_t^i = \zeta_t^i (1 + \tilde{\kappa}_t^i)^{-1}$$

for $i = k+1, \ldots, n$. $\qquad \qquad \square$

**Corollary 7.2.1.** *Let* $Y$ *be a bounded,* $\tilde{\mathcal{F}}_T$*-measurable random variable. For a fixed* $s \leq T$, *we define the process* $\hat{Y}$ *by setting*

$$\hat{Y}_t = \mathbb{E}_{\mathbb{P}_l}(Y e^{-A_s^*} \,|\, \tilde{\mathcal{F}}_t), \quad \forall t \in [0, T]. \tag{7.14}$$

*Then the process* $\hat{Y}$ *has the following integral representation under* $\mathbb{P}_l$ :

$$\hat{Y}_t = \hat{Y}_0 + \int_0^t \hat{\xi}_u \, d\tilde{W}_u^* + \sum_{i=k+1}^{n} \int_{]0,t]} \hat{\zeta}_u^i \, d\tilde{M}_u^{i*}, \tag{7.15}$$

*where* $\hat{\xi}$ *and* $\hat{\zeta}^i$, $i = k+1, \ldots, n$ *are* $\tilde{\mathbb{F}}$*-predictable processes. The stopped process* $\hat{Y}_{t \wedge \tilde{\tau}}$ *follows a* $\mathbb{G}$*-martingale orthogonal under* $\mathbb{P}^*$ *to the* $\mathbb{G}$*-martingales* $M^{i*}$, $i = 1, \ldots, k$.

*Proof.* It is enough to apply Lemma 7.2.2, to notice that the stopped process $\hat{Y}_{t \wedge \tilde{\tau}}$ satisfies

$$\hat{Y}_{t \wedge \tilde{\tau}} = \hat{Y}_0 + \int_0^t \hat{\xi}_u \, d\tilde{W}_{u \wedge \tilde{\tau}}^* + \sum_{i=k+1}^{n} \int_{]0,t]} \hat{\zeta}_u^i \, d\tilde{M}_{u \wedge \tilde{\tau}}^{i*},$$

and to recall that $\tilde{W}_{t \wedge \tilde{\tau}}^* = W_{t \wedge \tilde{\tau}}^*$ and $\tilde{M}_{t \wedge \tilde{\tau}}^{i*} = M_{t \wedge \tilde{\tau}}^{i*}$. ☐

We are in a position to proceed to the proof of Proposition 7.2.2.

*Proof of Proposition 7.2.2.* For a fixed $s \leq T$, let $\hat{Y}$ be the process defined through formula (7.14). Furthermore, let $U$ be the process given by the following expression (notice that the process $U$ is stopped at $\tilde{\tau} \wedge s$):

$$U_t = (1 - \tilde{H}_{t \wedge s}) e^{A_{t \wedge s}^*} = (1 - \tilde{H}_{t \wedge s}) \prod_{i=1}^{k} e^{\int_0^{t \wedge s}(1 + \tilde{\kappa}_u^i) dA_u^i}$$

$$= \prod_{i=1}^{k}(1 - H_{t \wedge s}^i) e^{\int_0^{t \wedge s}(1 + \kappa_u^i) dA_u^i} = \prod_{i=1}^{k} L_{t \wedge s}^{i*},$$

where we denote

$$L_t^{i*} = (1 - H_t^i) e^{\int_0^t (1 + \kappa_u^i) dA_u^i}.$$

It is useful to observe that (cf. (7.2))

$$L_t^{i*} = 1 - \int_{]0,t]} L_{u-}^{i*} \, dM_u^{i*}. \tag{7.16}$$

In view of (7.15), the above representation of the process $U$, and equation (7.16), we conclude that the two processes, $U$ and $\hat{Y}_{t \wedge \tilde{\tau}}$, follow mutually orthogonal $\mathbb{G}$-martingales under $\mathbb{P}^*$. Thus,

$$\mathbb{E}_{\mathbb{P}^*}(\mathbb{1}_{\{\tilde{\tau} > s\}} Y \,|\, \mathcal{G}_t) = \mathbb{E}_{\mathbb{P}^*}(U_T \hat{Y}_T \,|\, \mathcal{G}_t) = U_t \hat{Y}_t = (1 - \tilde{H}_t) e^{A_t^*} \mathbb{E}_{\mathbb{P}_l}(Y e^{-A_s^*} \,|\, \tilde{\mathcal{F}}_t).$$

The last expression yields the needed formulae. ☐

## 7.3 Kusuoka's Counter-Example

This section provides an analysis of an counter-example, due to Kusuoka (1999), which shows that formula (7.8) may fail to hold, in general. Assume that under the original probability measure $\mathbb{P}$ the random times $\tau_i$, $i = 1, 2$ are mutually independent random variables, with exponential laws with parameters $\lambda_1$ and $\lambda_2$, respectively. The joint probability law of the pair $(\tau_1, \tau_2)$ under $\mathbb{P}$ has the density function

$$f(x, y) = \lambda_1 \lambda_2 \, e^{-(\lambda_1 x + \lambda_2 y)}, \quad \forall\, (x, y) \in \mathbb{R}_+^2.$$

We shall examine these random times under a particular equivalent change of probability measure $\mathbb{P}^*$. It will appear that under $\mathbb{P}^*$ the intensity of the random time $\tau_1$ jumps from the original value $\lambda_1$ to the pre-determined value $\alpha_1$ as soon as $\tau_2$ occurs. The behavior of the intensity of $\tau_2$ is analogous. Such a specification of the stochastic intensity of dependent random times appears in a natural way in certain practical applications related to the valuation of defaultable claims with counterparty risk. Let us observe that the reference filtration $\mathbb{F}$ is trivial in Kusuoka's example.

Let $\alpha_1$ and $\alpha_2$ be strictly positive real numbers. For a fixed $T > 0$,[2] we introduce an equivalent probability measure $\mathbb{P}^*$ on $(\Omega, \mathcal{G})$ by setting

$$\frac{d\mathbb{P}^*}{d\mathbb{P}} = \eta_T, \quad \mathbb{P}\text{-a.s.,}$$

where the Radon-Nikodým density process $\eta_t$, $t \in [0, T]$, satisfies

$$\eta_t = 1 + \sum_{i=1}^{2} \int_{]0,t]} \eta_{u-} \kappa_u^i \, d\tilde{M}_u^i,$$

for processes $\kappa^1$ and $\kappa^2$ defined as follows:

$$\kappa_t^1 = \mathbb{1}_{\{\tau_2 < t\}} \left( \frac{\alpha_1}{\lambda_1} - 1 \right), \quad \kappa_t^2 = \mathbb{1}_{\{\tau_1 < t\}} \left( \frac{\alpha_2}{\lambda_2} - 1 \right).$$

The process $\kappa^1$ ($\kappa^2$, resp.) is obviously $\mathbb{H}^2$-predictable ($\mathbb{H}^1$-predictable, resp.) For the sake of our further purposes, it is useful to notice that $\eta_T = \eta_T^1 \eta_T^2$, where for every $t \in [0, T]$ and $i = 1, 2$ we have

$$\eta_t^i = 1 + \int_{]0,t]} \eta_{u-}^i \kappa_u^i \, d\tilde{M}_u^i.$$

More explicitly,

$$\eta_t^1 = \mathbb{1}_{\{\tau_1 \leq \tau_2\}} + \mathbb{1}_{\{\tau_2 < t \leq \tau_1\}} + \mathbb{1}_{\{t \leq \tau_2 < \tau_1\}} e^{-(\alpha_1 - \lambda_1)(t - \tau_2)}$$
$$+ \mathbb{1}_{\{\tau_2 < \tau_1 < t\}} \frac{\alpha_1}{\lambda_1} e^{-(\alpha_1 - \lambda_1)(\tau_1 - \tau_2)},$$

and a similar formula is valid for the process $\eta^2$.

---

[2] We follow here the convention adopted in Kusuoka (1999). It is important to observe that we may take here $T = \infty$.

It is easy to see that $\Lambda_t^{i*} = \int_0^t \lambda_u^{i*}\, du$ for $i = 1, 2$, where the processes $\lambda_t^{i*}$, $t \in [0, T]$, $i = 1, 2$ are given by the following expressions:

$$\lambda_t^{*1} = \lambda_1(1 - H_t^2) + \alpha_1 H_t^2 = \lambda_1 \mathbb{1}_{\{\tau_2 > t\}} + \alpha_1 \mathbb{1}_{\{\tau_2 \leq t\}},$$

and

$$\lambda_t^{*2} = \lambda_2(1 - H_t^1) + \alpha_2 H_t^1 = \lambda_2 \mathbb{1}_{\{\tau_1 > t\}} + \alpha_2 \mathbb{1}_{\{\tau_1 \leq t\}}.$$

Put another way, the compensated jump processes, for $t \in [0, T]$,

$$H_t^1 - \int_0^{t \wedge \tau_1} \left( \lambda_1 \mathbb{1}_{\{\tau_2 > u\}} + \alpha_1 \mathbb{1}_{\{\tau_2 \leq u\}} \right) du$$

and

$$H_t^2 - \int_0^{t \wedge \tau_2} \left( \lambda_2 \mathbb{1}_{\{\tau_1 > u\}} + \alpha_2 \mathbb{1}_{\{\tau_1 \leq u\}} \right) du$$

follow martingales under $\mathbb{P}^*$ with respect to the joint filtration $\mathbb{G} = \mathbb{H}^1 \vee \mathbb{H}^2$. In view of the assumed symmetry, it is enough to consider the random time $\tilde{\tau} = \tau_1$ (i.e., we take $n = 2$ and $k = 1$). Notice that, in the present setup, we have $\tilde{\kappa}_t^2 = 0$, since obviously $\kappa_t^2 = 0$ on the random interval $[0, \tau_1]$. Thus, the probability measure $\mathbb{P}_1$ given by formulae (7.9)–(7.11) coincides with the original probability measure $\mathbb{P}$. It is useful to notice the process $\kappa^1$ is left-continuous and $\mathbb{H}^2$-adapted, so that it is $\mathbb{H}^2$-predictable. This implies that $\tilde{\kappa}_t^1 = \kappa_t^1$ for every $t \in [0, T]$. Consequently, the probability measures $\mathbb{P}_2$ and $\mathbb{P}_3$ coincide with the probability measure $\mathbb{P}_1^*$ given by formula (7.21) below. Our goal is to find an explicit formula for the conditional expectation of survival:

$$\mathbb{P}^*\{\tau_1 > s \,|\, \mathcal{H}_t^1 \vee \mathcal{H}_t^2\}$$

for every $t \leq s \leq T$. To be more specific, we shall verify that the following equalities are valid:

$$\mathbb{P}^*\{\tau_1 > s \,|\, \mathcal{H}_t^1 \vee \mathcal{H}_t^2\} = \mathbb{1}_{\{\tau_1 > t\}} \mathbb{E}_{\mathbb{P}_1}\left( e^{\Lambda_t^{1*} - \Lambda_s^{1*}} \,|\, \mathcal{H}_t^2 \right)$$

and

$$\mathbb{P}^*\{\tau_1 > s \,|\, \mathcal{H}_t^1 \vee \mathcal{H}_t^2\} = \mathbb{1}_{\{\tau_1 > t\}} \mathbb{E}_{\mathbb{P}}\left( e^{\Lambda_t^{1*} - \Lambda_s^{1*}} \,|\, \mathcal{H}_t^2 \right).$$

In fact, the second equality above is an obvious consequence of the first, combined with the equality $\mathbb{P}_1 = \mathbb{P}$. We shall also check directly that:

$$\mathbb{P}^*\{\tau_1 > s \,|\, \mathcal{H}_t^1 \vee \mathcal{H}_t^2\} \neq \mathbb{1}_{\{\tau_1 > t\}} \mathbb{E}_{\mathbb{P}^*}\left( e^{\Lambda_t^{1*} - \Lambda_s^{1*}} \,|\, \mathcal{H}_t^2 \right).$$

This leads to an important conclusion that $\Lambda^{*1}$ does not represent the $\mathbb{H}^2$-hazard process of $\tau_1$ under the equivalent probability measure $\mathbb{P}^*$ (as we shall see later on, the $\mathbb{H}^2$-hazard process of $\tau_1$ under $\mathbb{P}^*$ has a discontinuity at $\tau_2$). We shall also examine the validity of the martingale invariance property (M.1). In the present setting, $\mathbb{G} = \mathbb{H}^1 \vee \mathbb{H}^2$ and the role of the reference filtration $\mathbb{F}$ is played by the filtration $\mathbb{H}^2$ generated by $\tau_2$. In view of the postulated mutual independence of random times $\tau_1$ and $\tau_2$ under $\mathbb{P}$, it is clear that Condition (M.1) is clearly satisfied under the original probability measure $\mathbb{P}$. We shall show that this condition fails to hold under the equivalent probability measure $\mathbb{P}^*$.

**Unconditional probability distribution of $\tau_1$ under $\mathbb{P}^*$.** We find it convenient to first derive the unconditional law of $\tau_1$ under $\mathbb{P}^*$. For simplicity of exposition, we assume throughout that $\lambda_1+\lambda_2-\alpha_1 \neq 0$ and $\lambda_1+\lambda_2-\alpha_2 \neq 0$. Notice that the marginal probability density function $f^*_{\tau_1}$ of the random time $\tau_1$ under $\mathbb{P}^*$ equals

$$f^*_{\tau_1}(t) = \int_0^t \lambda_1\lambda_2 \frac{\alpha_1}{\lambda_1} e^{-(\alpha_1-\lambda_1)(t-y)} e^{-(\lambda_1 t+\lambda_2 y)}\, dy$$

$$+ \int_t^T \lambda_1\lambda_2 \frac{\alpha_2}{\lambda_2} e^{-(\alpha_2-\lambda_2)(y-t)} e^{-(\lambda_1 t+\lambda_2 y)}\, dy$$

$$+ \int_t^\infty \lambda_1\lambda_2 e^{-(\alpha_2-\lambda_2)(T-t)} e^{-(\lambda_1 t+\lambda_2 y)}\, dy$$

$$= \frac{1}{\lambda_1+\lambda_2-\alpha_1} \left( \alpha_1\lambda_2 e^{-\alpha_1 t} + (\lambda_1-\alpha_1)(\lambda_1+\lambda_2)e^{-(\lambda_1+\lambda_2)t} \right)$$

for every $t \leq T$. For $t > T$ we have

$$f^*_{\tau_1}(t) = \int_0^T \lambda_1\lambda_2 e^{-(\alpha_1-\lambda_1)(T-y)} e^{-(\lambda_1 t+\lambda_2 y)}\, dy$$

$$+ \int_T^\infty \lambda_1\lambda_2 e^{-(\lambda_1 t+\lambda_2 y)}\, dy$$

$$= \frac{\lambda_1 e^{-\lambda_1 t}}{\lambda_1+\lambda_2-\alpha_1} \left( \lambda_2 e^{-(\alpha_1-\lambda_1)T} + (\lambda_1-\alpha_1)e^{-\lambda_2 T} \right),$$

which entails that, for any $s \in [0, T]$,

$$\mathbb{P}^*\{\tau_1 > s\} = \frac{1}{\lambda_1+\lambda_2-\alpha_1} \left( \lambda_2 e^{-\alpha_1 s} + (\lambda_1-\alpha_1)e^{-(\lambda_1+\lambda_2)s} \right). \qquad (7.17)$$

For $s > T$ we have

$$\mathbb{P}^*\{\tau_1 > s\} = \frac{e^{-\lambda_1 s}}{\lambda_1+\lambda_2-\alpha_1} \left( \lambda_2 e^{-(\alpha_1-\lambda_1)T} + (\lambda_1-\alpha_1)e^{-\lambda_2 T} \right). \qquad (7.18)$$

**Conditional probability distribution of $\tau_1$ under $\mathbb{P}^*$.** Our next aim is to derive explicit formulae for the joint probability distribution of $(\tau_1, \tau_2)$ and for the conditional probability $I := \mathbb{P}^*\{\tau_1 > s \,|\, \mathcal{H}_t^1 \vee \mathcal{H}_t^2\}$.

**Lemma 7.3.1.** *For every $t \leq s \leq T$ we have*

$$\mathbb{P}^*\{\tau_1 > s, \tau_2 > t\} = \mathbb{P}^*\{\tau_1 > s\} - \mathbb{P}^*\{\tau_1 > s, \tau_2 \leq t\}$$

$$= \frac{\lambda_1-\alpha_1}{\lambda_1+\lambda_2-\alpha_1} e^{-(\lambda_1+\lambda_2)s} + \frac{\lambda_2}{\lambda_1+\lambda_2-\alpha_1} e^{-\alpha_1 s-(\lambda_1+\lambda_2-\alpha_1)t}$$

*and*

$$I = \mathbb{1}_{\{\tau_1>t,\tau_2>t\}} \frac{1}{\lambda_1+\lambda_2-\alpha_1} \left( \lambda_2 e^{-\alpha_1(s-t)} + (\lambda_1-\alpha_1)e^{-(\lambda_1+\lambda_2)(s-t)} \right)$$

$$+ \mathbb{1}_{\{\tau_2 \leq t<\tau_1\}} e^{-\alpha_1(s-t)}.$$

*Proof.* In view of results of Sect. 5.1.1, for any $t \leq s$ we have

$$I = \mathbb{P}^*\{\tau_1 > s \mid \mathcal{H}_t^1 \vee \mathcal{H}_t^2\} = (1 - H_t^1) \frac{\mathbb{P}^*\{\tau_1 > s \mid \mathcal{H}_t^2\}}{\mathbb{P}^*\{\tau_1 > t \mid \mathcal{H}_t^2\}}, \tag{7.19}$$

where in turn

$$\mathbb{P}^*\{\tau_1 > s \mid \mathcal{H}_t^2\} = (1 - H_t^2) \frac{\mathbb{P}^*\{\tau_1 > s, \tau_2 > t\}}{\mathbb{P}^*\{\tau_2 > t\}} + H_t^2 \mathbb{P}^*\{\tau_1 > s \mid \tau_2\}.$$

Combining the last formula with (7.19), we obtain

$$I = (1 - H_t^1)(1 - H_t^2) \frac{\mathbb{P}^*\{\tau_1 > s, \tau_2 > t\}}{\mathbb{P}^*\{\tau_1 > t, \tau_2 > t\}} + (1 - H_t^1)H_t^2 \frac{\mathbb{P}^*\{\tau_1 > s \mid \tau_2\}}{\mathbb{P}^*\{\tau_1 > t \mid \tau_2\}}$$

or, more explicitly,

$$I = \mathbb{1}_{\{\tau_1 > t, \tau_2 > t\}} \frac{\mathbb{P}^*\{\tau_1 > s, \tau_2 > t\}}{\mathbb{P}^*\{\tau_1 > t, \tau_2 > t\}} + \mathbb{1}_{\{\tau_2 \leq t < \tau_1\}} \frac{\mathbb{P}^*\{\tau_1 > s \mid \tau_2\}}{\mathbb{P}^*\{\tau_1 > t \mid \tau_2\}}$$

$$= \mathbb{1}_{\{\tau_1 > t, \tau_2 > t\}} I_1 + \mathbb{1}_{\{\tau_2 \leq t < \tau_1\}} I_2.$$

To evaluate $I_1$, observe that

$$\mathbb{P}^*\{\tau_1 > s, \tau_2 \leq t\} = \int_s^T \int_0^t \lambda_1 \lambda_2 \frac{\alpha_1}{\lambda_1} e^{-(\alpha_1 - \lambda_1)(x-y)} e^{-(\lambda_1 x + \lambda_2 y)} \, dx \, dy$$

$$+ \int_T^\infty \int_0^t \lambda_1 \lambda_2 e^{-(\alpha_1 - \lambda_1)(T-y)} e^{-(\lambda_1 x + \lambda_2 y)} \, dx \, dy$$

$$= \frac{\lambda_2 e^{-\alpha_1 s}}{\lambda_1 + \lambda_2 - \alpha_1} \left(1 - e^{-(\lambda_1 + \lambda_2 - \alpha_1)t}\right).$$

Combining the last formula with (7.17), we obtain the first asserted formula.

$$\mathbb{P}^*\{\tau_1 > s, \tau_2 > t\} = \mathbb{P}^*\{\tau_1 > s\} - \mathbb{P}^*\{\tau_1 > s, \tau_2 \leq t\}$$

$$= \frac{\lambda_1 - \alpha_1}{\lambda_1 + \lambda_2 - \alpha_1} e^{-(\lambda_1 + \lambda_2)s} + \frac{\lambda_2}{\lambda_1 + \lambda_2 - \alpha_1} e^{-\alpha_1 s - (\lambda_1 + \lambda_2 - \alpha_1)t}.$$

Consequently,

$$I_1 = \frac{\mathbb{P}^*\{\tau_1 > s, \tau_2 > t\}}{\mathbb{P}^*\{\tau_1 > t, \tau_2 > t\}}$$

$$= \frac{1}{\lambda_1 + \lambda_2 - \alpha_1} \left(\lambda_2 e^{-\alpha_1(s-t)} + (\lambda_1 - \alpha_1)e^{-(\lambda_1 + \lambda_2)(s-t)}\right).$$

To establish the formula for $I$, it thus remains to find $I_2$. To this end, it is enough to check that $I_3 := \mathbb{1}_{\{\tau_2 \leq t\}} \mathbb{P}^*\{\tau_1 > s \mid \tau_2\}$ equals, for arbitrary $t \leq s \leq T$,

$$I_3 = \mathbb{1}_{\{\tau_2 \leq t\}} \frac{(\lambda_1 + \lambda_2 - \alpha_2)\lambda_2 e^{-\alpha_1(s-\tau_2)}}{\lambda_1 \alpha_2 e^{(\lambda_1 + \lambda_2 - \alpha_2)\tau_2} + (\lambda_2 - \alpha_2)(\lambda_1 + \lambda_2)}. \tag{7.20}$$

In effect, the last formula immediately yields

$$\mathbb{1}_{\{\tau_2 \le t < \tau_1\}} \frac{\mathbb{P}^* \{\tau_1 > s \mid \tau_2\}}{\mathbb{P}^* \{\tau_1 > t \mid \tau_2\}} = \mathbb{1}_{\{\tau_2 \le t < \tau_1\}} e^{-\alpha_1(s-t)},$$

which is the desired result. To evaluate $I_3$, we may, for instance, notice that for any $u \le s$

$$\mathbb{P}^* \{\tau_1 > s \mid \tau_2 = u\} = \frac{1}{f_{\tau_2}^*(u)} \int_s^T \frac{\alpha_1}{\lambda_1} e^{-(\alpha_1 - \lambda_1)(x-u)} f(x, u) \, dx$$

$$+ \frac{1}{f_{\tau_2}^*(u)} \int_T^\infty e^{-(\alpha_1 - \lambda_1)(T-u)} f(x, u) \, dx$$

or, more explicitly,

$$\mathbb{P}^* \{\tau_1 > s \mid \tau_2 = u\} = \frac{(\lambda_1 + \lambda_2 - \alpha_2) \lambda_2 e^{-(\lambda_1 + \lambda_2 - \alpha_1) u} e^{-\alpha_1 s}}{\lambda_1 \alpha_2 e^{-\alpha_2 u} + (\lambda_2 - \alpha_2)(\lambda_1 + \lambda_2) e^{-(\lambda_1 + \lambda_2) u}}.$$

The last formula yields (7.20), upon simplification. An alternative (somewhat lengthy) way of deriving the expression for $I_3$ relies on the direct use of the Bayes formula:

$$\mathbb{1}_{\{\tau_2 \le t\}} \mathbb{P}^* \{\tau_1 > s \mid \tau_2\} = \mathbb{1}_{\{\tau_2 \le t\}} \frac{\mathbb{E}_{\mathbb{P}}(\eta_s \mathbb{1}_{\{\tau_1 > s\}} \mid \tau_2)}{\mathbb{E}_{\mathbb{P}}(\eta_s \mid \tau_2)}.$$

In this method, it suffices to check that for arbitrary $t \le s \le T$ we have

$$\mathbb{1}_{\{\tau_2 \le t\}} \mathbb{E}_{\mathbb{P}}(\eta_s \mathbb{1}_{\{\tau_1 > s\}} \mid \tau_2) = \mathbb{1}_{\{\tau_2 \le t\}} e^{-\alpha_1 s} e^{-(\lambda_1 - \alpha_1) \tau_2}$$

and

$$\mathbb{1}_{\{\tau_2 \le t\}} \mathbb{E}_{\mathbb{P}}(\eta_s \mid \tau_2) = \mathbb{1}_{\{\tau_2 \le t\}} \frac{f_{\tau_2}^*(\tau_2)}{f_{\tau_2}(\tau_2)},$$

where $f_{\tau_2}(u) = \lambda_2 e^{-\lambda_2 u}$. The details are omitted. □

*Remarks.* Observe that in order to find $\mathbb{P}^* \{\tau_1 > s, \tau_2 > t\}$ for $t \le s \le T$, it suffices in fact to notice that

$$J := \mathbb{P}^* \{\tau_1 > s, \tau_2 > t\} = \mathbb{E}_{\mathbb{P}}(\eta_T \mathbb{1}_{\{\tau_1 > s, \tau_2 > t\}}) = \mathbb{E}_{\mathbb{P}}(\eta_s \mathbb{1}_{\{\tau_1 > s, \tau_2 > t\}})$$

and

$$\eta_s \mathbb{1}_{\{\tau_1 > s, \tau_2 > t\}} = \eta_s^1 \mathbb{1}_{\{\tau_1 > s, \tau_2 > t\}} = \mathbb{1}_{\{\tau_1 > s, \tau_2 > s\}} + \mathbb{1}_{\{t < \tau_2 < s < \tau_1\}} e^{-(\alpha_1 - \lambda_1)(s-\tau_2)}.$$

Therefore,

$$J = \int_s^\infty \int_s^\infty \lambda_1 \lambda_2 e^{-(\lambda_1 x + \lambda_2 y)} \, dx \, dy$$

$$+ \int_s^\infty \int_t^s \lambda_1 \lambda_2 e^{-(\alpha_1 - \lambda_1)(s-y)} e^{-(\lambda_1 x + \lambda_2 y)} \, dx \, dy$$

$$= \frac{\lambda_1 - \alpha_1}{\lambda_1 + \lambda_2 - \alpha_1} e^{-(\lambda_1 + \lambda_2) s} + \frac{\lambda_2}{\lambda_1 + \lambda_2 - \alpha_1} e^{-\alpha_1 s - (\lambda_1 + \lambda_2 - \alpha_1) t}.$$

Let us introduce a probability measure $\mathbb{P}_1^*$ by setting

$$\frac{d\mathbb{P}_1^*}{d\mathbb{P}} = \eta_T^1, \quad \mathbb{P}\text{-a.s.} \tag{7.21}$$

Observe that the marginal probability density function $\tilde{f}_{\tau_1}$ of $\tau_1$ under $\mathbb{P}_1^*$ coincides with $f_{\tau_1}^*$, because for every $t \leq T$ we have

$$
\begin{aligned}
\tilde{f}_{\tau_1}(t) &= \int_0^t \lambda_1 \lambda_2 \frac{\alpha_1}{\lambda_1} e^{-(\alpha_1-\lambda_1)(t-y)} e^{-(\lambda_1 t + \lambda_2 y)} \, dx dy \\
&\quad + \int_t^\infty \lambda_1 \lambda_2 e^{-(\lambda_1 t + \lambda_2 y)} \, dx dy \\
&= \frac{1}{\lambda_1 + \lambda_2 - \alpha_1} \left( \alpha_1 \lambda_2 e^{-\alpha_1 t} + (\lambda_1 - \alpha_1)(\lambda_1 + \lambda_2) e^{-(\lambda_1 + \lambda_2)t} \right).
\end{aligned}
$$

It is also obvious that $\tilde{f}_{\tau_1} = f_{\tau_1}^*$ for $t > T$. In fact, one may also easily deduce from the calculations in the proof of Lemma 7.3.1 and the last remarks that:

$$
\begin{aligned}
I &= \mathbb{P}^*\{\tau_1 > s \mid \mathcal{H}_t^1 \vee \mathcal{H}_t^2\} = \mathbb{P}_1^*\{\tau_1 > s \mid \mathcal{H}_t^1 \vee \mathcal{H}_t^2\} \\
&= \mathbb{1}_{\{\tau_1 > t, \tau_2 > t\}} \frac{\mathbb{P}_1^*\{\tau_1 > s, \tau_2 > t\}}{\mathbb{P}_1^*\{\tau_1 > t, \tau_2 > t\}} + \mathbb{1}_{\{\tau_2 \leq t < \tau_1\}} \frac{\mathbb{P}_1^*\{\tau_1 > s \mid \tau_2\}}{\mathbb{P}_1^*\{\tau_1 > t \mid \tau_2\}}.
\end{aligned}
$$

**Intensity of $\tau_1$ under $\mathbb{P}^*$.** We shall now focus on the intensity process of $\tau_1$ under $\mathbb{P}^*$. We have

$$\Lambda_t^{1*} = \int_0^t \left( \lambda_1 \mathbb{1}_{\{\tau_2 > u\}} + \alpha_1 \mathbb{1}_{\{\tau_2 \leq u\}} \right) du = \lambda_1 (t \wedge \tau_2) + \alpha_1 (t \vee \tau_2 - \tau_2). \tag{7.22}$$

The first equality in next result is merely a special case of Proposition 7.2.2. Note that equality (7.23) shows that $\Lambda^{1*}$ can be used to evaluate the conditional probability of survival. Inequality (7.24) makes it clear that the process $\Lambda^{1*}$ is not the $\mathbb{H}^2$-hazard process of $\tau_1$ under $\mathbb{P}^*$, though.

**Proposition 7.3.1.** *Let us set $I = \mathbb{P}^*\{\tau_1 > s \mid \mathcal{H}_t^1 \vee \mathcal{H}_t^2\}$. Then for every $t < s \leq T$ we have*

$$
\begin{aligned}
I &= \mathbb{1}_{\{\tau_1 > t\}} \mathbb{E}_{\mathbb{P}}(e^{\Lambda_t^{1*} - \Lambda_s^{1*}} \mid \mathcal{H}_t^2) = \mathbb{1}_{\{\tau_1 > t\}} \mathbb{E}_{\mathbb{P}_1^*}(e^{\Lambda_t^{1*} - \Lambda_s^{1*}} \mid \mathcal{H}_t^2) \\
&= \mathbb{1}_{\{\tau_1 > t\}} \mathbb{E}_{\mathbb{P}_1}(e^{\Lambda_t^{1*} - \Lambda_s^{1*}} \mid \mathcal{H}_t^2) \tag{7.23}
\end{aligned}
$$

*and*

$$\mathbb{P}^*\{\tau_1 > s \mid \mathcal{H}_t^1 \vee \mathcal{H}_t^2\} \neq \mathbb{1}_{\{\tau_1 > t\}} \mathbb{E}_{\mathbb{P}^*}(e^{\Lambda_t^{1*} - \Lambda_s^{1*}} \mid \mathcal{H}_t^2). \tag{7.24}$$

*Proof.* Recall that an explicit formula for the conditional probability $I$ was already found in Lemma 7.3.1. We are going to check that $I = \tilde{I}$, where

$$\tilde{I} := \mathbb{1}_{\{\tau_1 > t\}} \mathbb{E}_{\mathbb{P}}(e^{\Lambda_t^{1*} - \Lambda_s^{1*}} \mid \mathcal{H}_t^2).$$

To this end, it is enough to verify that (see Lemma 7.3.1)

$$\tilde{I} = \mathbb{1}_{\{\tau_2 > t\}} \frac{1}{\lambda_1 + \lambda_2 - \alpha_1} \left( \lambda_2 e^{-\alpha_1(s-t)} + (\lambda_1 - \alpha_1) e^{-(\lambda_1 + \lambda_2)(s-t)} \right)$$
$$+ \mathbb{1}_{\{\tau_2 \le t\}} e^{-\alpha_1(s-t)}.$$

If we denote $Y = e^{\Lambda_t^{1*} - \Lambda_s^{1*}}$, then the general formula yields

$$\mathbb{E}_{\mathbb{P}}\left( e^{\Lambda_t^{1*} - \Lambda_s^{1*}} \mid \mathcal{H}_t^2 \right) = \mathbb{1}_{\{\tau_2 > t\}} \frac{\mathbb{E}_{\mathbb{P}}(Y \mathbb{1}_{\{\tau_2 > t\}})}{\mathbb{P}\{\tau_2 > t\}} + \mathbb{1}_{\{\tau_2 \le t\}} \mathbb{E}_{\mathbb{P}}(Y \mid \tau_2).$$

Standard calculations show that

$$\mathbb{E}_{\mathbb{P}}(Y \mathbb{1}_{\{\tau_2 > t\}}) = \mathbb{E}_{\mathbb{P}}\left( \mathbb{1}_{\{\tau_2 > t\}} e^{\lambda_1(t - s \wedge \tau_2) + \alpha_1(\tau_2 - s \vee \tau_2)} \right)$$
$$= \int_t^s e^{\lambda_1(t-u) + \alpha_1(u-s)} \lambda_2 e^{-\lambda_2 u} \, du + \int_s^\infty e^{\lambda_1(t-s)} \lambda_2 e^{-\lambda_2 u} \, du$$
$$= \frac{\lambda_2 e^{\lambda_1 t - \alpha_1 s}}{\lambda_1 + \lambda_2 - \alpha_1} \left( e^{-(\lambda_1 + \lambda_2 - \alpha_1)t} - e^{-(\lambda_1 + \lambda_2 - \alpha_1)s} \right) + e^{\lambda_1 t} e^{-(\lambda_1 + \lambda_2)s},$$

and, of course, $\mathbb{P}\{\tau_2 > t\} = e^{-\lambda_2 t}$. Consequently, we obtain

$$\frac{\mathbb{E}_{\mathbb{P}}(Y \mathbb{1}_{\{\tau_2 > t\}})}{\mathbb{P}\{\tau_2 > t\}} = \frac{1}{\lambda_1 + \lambda_2 - \alpha_1} \left( \lambda_2 e^{-\alpha_1(s-t)} + (\lambda_1 - \alpha_1) e^{-(\lambda_1 + \lambda_2)(s-t)} \right),$$

as expected. Furthermore, it follows from (7.22) that

$$\mathbb{1}_{\{\tau_2 \le t\}} \mathbb{E}_{\mathbb{P}}(e^{\Lambda_t^{1*} - \Lambda_s^{1*}} \mid \tau_2) = \mathbb{1}_{\{\tau_2 \le t\}} e^{-\alpha_1(s-t)}.$$

This ends the proof of the first equality in (7.23). The second equality in (7.23) follows from the above calculations and the fact that the probability distribution of $\tau_1$ under $\mathbb{P}_1^*$ is identical with its probability distribution under $\mathbb{P}^*$. Finally, the last equality in (7.23) is trivial since $\mathbb{P}_1 = \mathbb{P}$.

We shall examine (7.24) for $t = 0$ (the general case is left to the reader). We wish to show that for $s \le T$

$$\mathbb{P}^*\{\tau_1 > s\} \ne \mathbb{E}_{\mathbb{P}^*}(e^{-\Lambda_s^{1*}}), \tag{7.25}$$

where the left-hand side is given by (7.17). We have

$$\mathbb{E}_{\mathbb{P}^*}\left( e^{-\Lambda_s^{1*}} \right) = \mathbb{E}_{\mathbb{P}^*}\left( e^{-\lambda_1(s \wedge \tau_2) - \alpha_1(s \vee \tau_2 - \tau_2)} \right)$$
$$= \int_0^s e^{-\lambda_1 u - \alpha_1(s-u)} f_{\tau_2}^*(u) \, du + \int_s^\infty e^{-\lambda_1 s} f_{\tau_2}^*(u) \, du.$$

Consequently,

$$\mathbb{E}_{\mathbb{P}^*}\left( e^{-\Lambda_s^{1*}} \right) = \int_0^s e^{-\lambda_1 u - \alpha_1(s-u)} f_{\tau_2}^*(u) \, du + e^{-\lambda_1 s} \mathbb{P}^*\{\tau_2 > s\},$$

where (cf. Sect. 7.3)

$$f_{\tau_2}^*(u) = \frac{1}{\lambda_1 + \lambda_2 - \alpha_2}\left(\alpha_2\lambda_1 e^{-\alpha_2 u} + (\lambda_2 - \alpha_2)(\lambda_1 + \lambda_2)e^{-(\lambda_1+\lambda_2)u}\right)$$

for $u \le s \le T$, and

$$\mathbb{P}^*\{\tau_2 > s\} = \frac{1}{\lambda_1 + \lambda_2 - \alpha_2}\left(\lambda_1 e^{-\alpha_2 s} + (\lambda_2 - \alpha_2)e^{-(\lambda_1+\lambda_2)s}\right).$$

Straightforward calculations yield

$$\mathbb{E}_{\mathbb{P}^*}\left(e^{-\Lambda_s^{1*}}\right) = \frac{1}{\lambda_1 + \lambda_2 - \alpha_2}\Bigg\{ \frac{\lambda_1\alpha_2}{\lambda_1 - \alpha_1 + \alpha_2}\left(e^{-\alpha_1 s} - e^{-(\lambda_1+\alpha_2)s}\right)$$
$$+ \frac{(\lambda_2 - \alpha_2)(\lambda_1 + \lambda_2)}{2\lambda_1 + \lambda_2 - \alpha_1}\left(e^{-\alpha_1 s} - e^{-(2\lambda_1+\lambda_2)s}\right)$$
$$+ \left(\lambda_1 e^{-(\lambda_1+\alpha_2)s} + (\lambda_2 - \alpha_2)e^{-(2\lambda_1+\lambda_2)s}\right)\Bigg\},$$

which shows, when combined with (7.17), that inequality (7.25) holds.    □

If $\lambda_i = \alpha_i$ for $i = 1, 2$, then the last formula yields, as expected, $\mathbb{E}_{\mathbb{P}^*}(e^{-\Lambda_s^{1*}}) = e^{-\lambda_1 s} = \mathbb{P}\{\tau_1 > s\}$. In fact, the last equalities are also valid when $\lambda_2 \ne \alpha_2$, but $\lambda_1 = \alpha_1$. Let us now assume that $\lambda_2 = \alpha_2$, but $\lambda_1 \ne \alpha_1$ (this corresponds to the equality: $\mathbb{P}^* = \mathbb{P}_1^*$). Then we get

$$\mathbb{E}_{\mathbb{P}_1^*}\left(e^{-\Lambda_s^{1*}}\right) = \frac{1}{\lambda_1 + \lambda_2 - \alpha_1}\left(\lambda_2 e^{-\alpha_1 s} + (\lambda_1 - \alpha_1)e^{-(\lambda_1+\lambda_2)s}\right)$$
$$= \mathbb{P}_1^*\{\tau_1 > s\} = \mathbb{P}^*\{\tau_1 > s\}.$$

This coincides with the second equality in (7.23), in the special case of $t = 0$.

### 7.3.1 Validity of Condition (F.2)

We shall now examine the validity of Condition (F.2) under $\mathbb{P}^*$. In the present context, we consider the random time $\tau = \tau_1$ and we take: $\mathbb{F} = \mathbb{H}^2$ and $t \le T$. Thus, Condition (F.2) now reads as follows.

**Condition (F.2)** The process $F_t = \mathbb{P}^*\{\tau_1 \le t \,|\, \mathcal{H}_t^2\}$, $t \in [0, T]$, admits a modification with increasing sample paths.

Let us denote $G_t = 1 - F_t = \mathbb{P}^*\{\tau_1 > t \,|\, \mathcal{H}_t^2\}$. From the proof of Lemma 7.3.1, it follows that

$$G_t = (1 - H_t^2)\frac{\mathbb{P}^*\{\tau_1 > t, \tau_2 > t\}}{\mathbb{P}^*\{\tau_2 > t\}} + H_t^2\,\mathbb{P}^*\{\tau_1 > t \,|\, \tau_2\},$$

where

$$\mathbb{P}^*\{\tau_1 > t, \tau_2 > t\} = \frac{\lambda_1 - \alpha_1}{\lambda_1 + \lambda_2 - \alpha_1}e^{-(\lambda_1+\lambda_2)t}$$
$$+ \frac{\lambda_2}{\lambda_1 + \lambda_2 - \alpha_1}e^{-\alpha_1 t - (\lambda_1+\lambda_2-\alpha_1)t}$$
$$= e^{-(\lambda_1+\lambda_2)t}.$$

and (cf. (7.17)–(7.18))

$$\mathbb{P}^*\{\tau_2 > t\} = \frac{1}{\lambda_1 + \lambda_2 - \alpha_2}\left(\lambda_1 e^{-\alpha_2 t} + (\lambda_2 - \alpha_2)e^{-(\lambda_1+\lambda_2)t}\right)$$

for $t \leq T$, whereas

$$\mathbb{P}^*\{\tau_2 > t\} = \frac{e^{-\lambda_2 t}}{\lambda_1 + \lambda_2 - \alpha_2}\left(\lambda_1 e^{-(\alpha_2-\lambda_2)T} + (\lambda_2 - \alpha_2)e^{-\lambda_1 T}\right)$$

for $t > T$. On the other hand, for $u \leq t \leq T$ we have (cf. (7.20))

$$\mathbb{P}^*\{\tau_1 > t \,|\, \tau_2 = u\} = \frac{(\lambda_1 + \lambda_2 - \alpha_2)\lambda_2 e^{-\alpha_1(t-u)}}{\lambda_1\alpha_2 e^{(\lambda_1+\lambda_2-\alpha_2)u} + (\lambda_2 - \alpha_2)(\lambda_1 + \lambda_2)}.$$

Similar calculations yield

$$\mathbb{P}^*\{\tau_1 > t \,|\, \tau_2 = u\} = \frac{(\lambda_1 + \lambda_2 - \alpha_2)\lambda_2 e^{-(\alpha_1-\lambda_1)T}e^{\alpha_1 u - \lambda_1 t}}{\lambda_1\alpha_2 e^{(\lambda_1+\lambda_2-\alpha_2)u} + (\lambda_2 - \alpha_2)(\lambda_1 + \lambda_2)}$$

for $u \leq T < t$, and finally

$$\mathbb{P}^*\{\tau_1 > t \,|\, \tau_2 = u\} = \frac{(\lambda_1 + \lambda_2 - \alpha_2)e^{\lambda_2 u - \lambda_1 t}}{\lambda_1\lambda_2 e^{-(\alpha_2-\lambda_2)T} + (\lambda_2 - \alpha_2)e^{-\lambda_1 T}}$$

for $T < u \leq t$. Combining the above formulae, we obtain

$$G_t = \mathbb{1}_{\{t<\tau_2\leq T\}}\frac{c}{\lambda_1 e^{ct} + \lambda_2 - \alpha_2} + \mathbb{1}_{\{T<t<\tau_2\}}\frac{ce^{\lambda_1(T-t)}}{\lambda_1 e^{cT} + \lambda_2 - \alpha_2}$$
$$+ \mathbb{1}_{\{\tau_2=u\leq t\leq T\}}\frac{c\lambda_2 e^{\alpha_1(u-t)}}{\lambda_1\alpha_2 e^{cu} + (\lambda_2 - \alpha_2)(\lambda_1 + \lambda_2)}$$
$$+ \mathbb{1}_{\{\tau_2=u\leq T<t\}}\frac{c\lambda_2 e^{(\lambda_1-\alpha_1)T}e^{\alpha_1 u - \lambda_1 t}}{\lambda_1\alpha_2 e^{cu} + (\lambda_2 - \alpha_2)(\lambda_1 + \lambda_2)}$$
$$+ \mathbb{1}_{\{T<\tau_2=u\leq t\}}\frac{ce^{\lambda_1 T}e^{\lambda_2 u - \lambda_1 t}}{\lambda_1\lambda_2 e^{cT} + (\lambda_2 - \alpha_2)},$$

where we denote $c = \lambda_1 + \lambda_2 - \alpha_2$. In particular, for every $t \in [0, T]$ we have

$$G_t = \mathbb{1}_{\{t<\tau_2\leq T\}}\frac{c}{\lambda_1 e^{ct} + \lambda_2 - \alpha_2} + \mathbb{1}_{\{\tau_2\leq t\leq T\}}\frac{c\lambda_2 e^{-\alpha_1(t-\tau_2)}}{\lambda_1\alpha_2 e^{c\tau_2} + (\lambda_2 - \alpha_2)(\lambda_1 + \lambda_2)}.$$

Since both terms on the right-hand side of the last formula can be shown to follow decreasing functions, it is enough to examine the jump at $\tau_2$, which equals

$$\Delta = \frac{c}{\lambda_1 e^{c\tau_2} + \lambda_2 - \alpha_2} - \frac{c\lambda_2}{\lambda_1\alpha_2 e^{c\tau_2} + (\lambda_2 - \alpha_2)(\lambda_1 + \lambda_2)}.$$

Straightforward calculations show that $\Delta \leq 0$ if and only if $\lambda_2 \leq \alpha_2$.

### 7.3.2 Validity of Condition (M.1)

Recall that in the present set-up we have $\mathbb{G} = \mathbb{H}^1 \vee \mathbb{H}^2$. Suppose that we choose $\tau = \tau_1$ as the reference random time. This means that the filtration $\mathbb{H}^2$ generated by $\tau_2$ is chosen to play the role of the reference filtration $\mathbb{F}$. Therefore, Condition (M.1) introduced in Sect. 6.1.1 can be restated as follows.

**Condition (M.1)** An arbitrary $\mathbb{H}^2$-martingale under $\mathbb{P}^*$ also follows a $\mathbb{G}$-martingale under $\mathbb{P}^*$.

In Sect. 6.1.1, we have shown that Condition (M.1) is equivalent to Condition (M2.b). In the present setting, the latter condition takes the following form.

**Condition (M.2b)** For any bounded, $\mathcal{H}^2_\infty$-measurable random variable $\xi$, the equality

$$\mathbb{E}_{\mathbb{P}^*}(\xi \,|\, \mathcal{H}^1_t \vee \mathcal{H}^2_t) = \mathbb{E}_{\mathbb{P}^*}(\xi \,|\, \mathcal{H}^2_t)$$

is valid for arbitrary $t \in \mathbb{R}_+$.

Using the calculations from the preceding sections, we shall now verify that the last condition is not satisfied in Kusuoka's example. To this end, we may take, for instance, arbitrary two times $t < s \le T$ and the $\mathcal{H}^2_\infty$-measurable random variable $\xi = 1\!\!1_{\{\tau_2 > s\}}$. By applying a suitable modification of formula (7.17), we obtain

$$\mathbb{E}_{\mathbb{P}^*}(\xi \,|\, \mathcal{H}^2_t) = \mathbb{P}^*\{\tau_2 > s \,|\, \mathcal{H}^2_t\} = 1\!\!1_{\{\tau_2 > t\}} \frac{\mathbb{P}^*\{\tau_2 > s\}}{\mathbb{P}^*\{\tau_2 > t\}}$$

$$= 1\!\!1_{\{\tau_2 > t\}} \frac{\lambda_1 e^{-\alpha_2 s} + (\lambda_2 - \alpha_2) e^{-(\lambda_1 + \lambda_2)s}}{\lambda_1 e^{-\alpha_2 t} + (\lambda_2 - \alpha_2) e^{-(\lambda_1 + \lambda_2)t}}.$$

On the other hand, Lemma 7.3.1 yields

$$\mathbb{E}_{\mathbb{P}^*}(\xi \,|\, \mathcal{H}^1_t \vee \mathcal{H}^2_t) = \mathbb{P}^*\{\tau_2 > s \,|\, \mathcal{H}^1_t \vee \mathcal{H}^2_t\}$$

$$= 1\!\!1_{\{\tau_1 > t, \tau_2 > t\}} \frac{\mathbb{P}^*\{\tau_1 > t, \tau_2 > s\}}{\mathbb{P}^*\{\tau_1 > t, \tau_2 > t\}} + 1\!\!1_{\{\tau_1 \le t < \tau_2\}} \frac{\mathbb{P}^*\{\tau_2 > s \,|\, \tau_1\}}{\mathbb{P}^*\{\tau_2 > t \,|\, \tau_1\}}$$

$$= 1\!\!1_{\{\tau_1 > t, \tau_2 > t\}} \frac{1}{\lambda_1 + \lambda_2 - \alpha_2} \left( \lambda_1 e^{-\alpha_2(s-t)} + (\lambda_2 - \alpha_2) e^{-(\lambda_1 + \lambda_2)(s-t)} \right)$$

$$+ 1\!\!1_{\{\tau_1 \le t < \tau_1\}} e^{-\alpha_2(s-t)}.$$

It is thus clear that for any $s > t$ we have

$$\mathbb{E}_{\mathbb{P}^*}(\xi \,|\, \mathcal{H}^1_t \vee \mathcal{H}^2_t) \ne \mathbb{E}_{\mathbb{P}^*}(\xi \,|\, \mathcal{H}^2_t).$$

We conclude that the martingale invariance property (M.1) fails to hold under the equivalent probability measure $\mathbb{P}^*$. Let us make a trivial observation that, due to the assumed independence of random times $\tau_1$ and $\tau_2$ under $\mathbb{P}$, Condition (M.1) is satisfied under the original probability measure $\mathbb{P}$.

# Part III

# Reduced-Form Modeling

# 8. Intensity-Based Valuation of Defaultable Claims

As already mentioned in Sect. 1.4, existing approaches to the modeling of credit (or default) risk may be divided into two broad classes: *structural models* and *reduced-form models*. In the former approach, the total value of the firm's assets is directly used to determine the default event, which occurs when the firm's value falls through some boundary. It results that here the default time is a *predictable* stopping time with respect to the reference filtration modeling the information flow available to the traders. This means that the random time of default is announced by an increasing sequence of stopping times. By contrast, in the latter approach, the firm's value process either is not modeled at all, or it plays only an auxiliary role of a state variable. The default time is modeled as a stopping time that is not predictable; the default event thus arrives as a total surprise. Formally, the random time of default event is given here as a *totally inaccessible* stopping time, in the terminology of the general theory of stochastic processes (see, e.g., Dellacherie (1972) or Jacod and Shiryaev (1987)). The main tool in this approach is an exogenous specification of the conditional probability of default, given that default has not yet occurred. Since in most cases this is done by means of the hazard rate (or intensity) of default, reduced-form models are also commonly known as *hazard rate models* or *intensity-based models*. From a long list of papers devoted to the reduced-form methodology, let us mention a few: Artzner and Delbaen (1995), Jarrow and Turnbull (1995), Duffie et al. (1996), Duffie and Singleton (1997, 1999), Lando (1998), Schlögl (1998), Schönbucher (1998b), Wong (1998), Elliott et al. (2000), and Bélanger et al. (2001).

Some authors, including Madan and Unal (1998, 2000) and Davydov et al. (2000), combine the basic ideas from both abovementioned approaches, by postulating that the hazard rate of default event is directly linked to the current value of the firm's assets (or the firm's equity). Reduced-form models with this specific feature are referred to as *hybrid models*. In this set-up, the default time is still a totally inaccessible stopping time, but the likelihood of default may grow rapidly when the total value of the firm's assets (or the value of the firm's equity) approaches some barrier. A more involved way of producing a hybrid model was examined by Duffie and Lando (2001), who consider a firm's value model with incomplete accounting data; results of this paper are not examined in the present text, though.

## 8.1 Defaultable Claims

In this section, we present basic results that can be obtained through the intensity-based approach to the valuation of defaultable claims. We assume that we are given the underlying probability space $(\Omega, \mathcal{G}, \mathbb{Q}^*)$, endowed with the filtration $\mathbb{F} = (\mathcal{F}_t)_{t \geq 0}$ (of course, $\mathcal{F}_t \subseteq \mathcal{G}$ for any $t$). The probability measure $\mathbb{Q}^*$ is interpreted as a *spot martingale measure* for our model of securities market; the real-world probability measure will be denoted by $\mathbb{Q}$. All processes introduced below are defined on the probability space $(\Omega, \mathcal{G}, \mathbb{Q}^*)$.

As in Chap. 2, we formally identify a *defaultable claim* with a quintuple $DCT = (X, A, \tilde{X}, Z, \tau)$. The *default time* $\tau$ is an arbitrary non-negative random variable, which is defined on the underlying probability space $(\Omega, \mathcal{G}, \mathbb{Q}^*)$; in particular, $\mathbb{Q}^*\{\tau < +\infty\} = 1$. For the sake of convenience, we shall usually assume that $\mathbb{Q}^*\{\tau = 0\} = 0$ and $\mathbb{Q}^*\{\tau > t\} > 0$ for every $t \in \mathbb{R}_+$. For a given default time $\tau$, we introduce the associated jump process $H$ by setting $H_t = \mathbb{1}_{\{\tau \leq t\}}$ for $t \in \mathbb{R}_+$. We shall refer to $H$ as the *default process*. It is obvious that $H$ is a right-continuous process. Let $\mathbb{H}$ be the filtration generated by the process $H$ – i.e., $\mathcal{H}_t = \sigma(H_u : u \leq t) = \sigma(\{\tau \leq u\} : u \leq t)$.

An essential role is played by the enlarged filtration $\mathbb{G} = \mathbb{H} \vee \mathbb{F}$. By definition, for every $t$ we set $\mathcal{G}_t = \mathcal{H}_t \vee \mathcal{F}_t = \sigma(\mathcal{H}_t, \mathcal{F}_t)$. It should be emphasized that the *default time* $\tau$ is not necessarily a stopping time with respect to the filtration $\mathbb{F}$. On the other hand, $\tau$ is, of course, a stopping time with respect to the filtration $\mathbb{G}$. In most intensity-based models, the underlying filtration $\mathbb{G}$ encompasses a certain Brownian filtration $\mathbb{F}$; $\mathbb{G}$ is usually strictly larger than $\mathbb{F}$, though. In this case, the default time is usually modeled in such a way that it is not a predictable stopping time with respect to the filtration $\mathbb{G}$. Recall that if $\tau$ is a stopping time with the Brownian filtration $\mathbb{F}$, then it is necessarily a predictable stopping time.

The *short-term interest rate* process $r$ follows an $\mathbb{F}$-progressively measurable process, such that the savings account $B$, given by the usual expression:

$$B_t = \exp\left(\int_0^t r_u \, du\right), \quad \forall t \in \mathbb{R}_+,$$

is well defined.

Similarly as in Chap. 2, we introduce the following random variables and processes that specify the cash flows associated with a defaultable claim:
- the *promised contingent claim* $X$, representing the payoff received by the owner of the claim at time $T$, if there was no default prior to or at time $T$,
- the process $A$ representing the *promised dividends* – that is, the stream of (continuous or discrete) cash flows received by the owner of the claim prior to default,
- the *recovery process* $Z$, representing the recovery payoff at the time of default, if default occurs prior to or at the maturity date $T$,
- the *recovery claim* $\tilde{X}$, representing the recovery payoff at time $T$, if default occurs prior to or at the maturity date $T$.

We shall postulate throughout the text that the processes $Z$ and $A$ are predictable with respect to the reference filtration $\mathbb{F}$, and that the random variables $X$ and $\tilde{X}$ are $\mathcal{F}_T$-measurable. By assumption, the promised dividends process $A$ follows a process of finite variation, with $A_0 = 0$. As usual, the sample paths of all processes are assumed to be right-continuous functions, with finite left-hand limits, with probability one. We shall assume without mentioning that all the above random objects satisfy suitable integrability conditions that are needed for evaluating the functionals introduced later on.

### 8.1.1 Risk-Neutral Valuation Formula

We place ourselves within the framework of an arbitrage-free financial market model. Specifically, we postulate that the underlying probability measure $\mathbb{Q}^*$ is the *spot martingale measure* (or the *risk-neutral probability*), meaning that the price process of any tradeable security, which pays no coupons or dividends, necessarily follows a $\mathbb{G}$-martingale under $\mathbb{Q}^*$, when discounted by the savings account $B$. Let us first recall the definitions of the dividend process and the price process of a defaultable claim (see Definition 2.1.1 in Chap. 2).

**Definition 8.1.1.** The *dividend process* $D$ of a defaultable claim $DCT = (X, A, \tilde{X}, Z, \tau)$ equals

$$D_t = X^d(T)\mathbb{1}_{[T,\infty[}(t) + \int_{]0,t]} (1 - H_u)\, dA_u + \int_{]0,t]} Z_u\, dH_u, \qquad (8.1)$$

where $X^d(T) = X\mathbb{1}_{\{\tau > T\}} + \tilde{X}\mathbb{1}_{\{\tau \leq T\}}$.

The next definition mimics Definition 2.1.2 of the *price process* (or the *value process*) of a defaultable claim. Expression (8.2) is henceforth referred to as the *risk-neutral valuation formula*.

**Definition 8.1.2.** The (ex-dividend) *price process* $X^d(\cdot, T)$ of a defaultable claim $DCT = (X, A, \tilde{X}, Z, \tau)$, which settles at time $T$, is given as

$$X^d(t, T) = B_t\, \mathbb{E}_{\mathbb{Q}^*}\left( \int_{]t,T]} B_u^{-1}\, dD_u \,\Big|\, \mathcal{G}_t \right), \qquad \forall\, t \in [0, T). \qquad (8.2)$$

In addition, we set $X^d(T, T) = X^d(T)$.

Before presenting the no-arbitrage arguments supporting Definition 8.1.2, let us consider a few special cases of the risk-neutral valuation formula (8.2). For the sake of brevity, we shall write $S_t^0 = X^d(t, T)$. Combining (8.1) with (8.2), we obtain

$$S_t^0 = B_t\, \mathbb{E}_{\mathbb{Q}^*}\left( \int_{]t,T]} B_u^{-1}(1 - H_u)\, dA_u + \int_{]t,T]} B_u^{-1} Z_u\, dH_u + B_T^{-1} X^d(T) \,\Big|\, \mathcal{G}_t \right).$$

where, as before, we denote

$$X^d(T) = \tilde{X}\mathbb{1}_{\{\tau \leq T\}} + X\mathbb{1}_{\{\tau > T\}} = \tilde{X}H_T + X(1 - H_T).$$

If the claim does not pay any dividends prior to default – that is, if $A \equiv 0$, and if $\tilde{X} = 0$, the risk-neutral valuation formula simplifies to:

$$S_t^0 = B_t\, \mathbb{E}_{\mathbb{Q}^*}\left(B_\tau^{-1}Z_\tau\mathbb{1}_{\{t < \tau \leq T\}} + B_T^{-1}X\mathbb{1}_{\{\tau > T\}}\,\Big|\,\mathcal{G}_t\right). \tag{8.3}$$

It is apparent that in this case $S_t^0 = 0$ on the set $\{\tau \leq t\}$, and so

$$S_t^0 = \mathbb{1}_{\{\tau > t\}}B_t\, \mathbb{E}_{\mathbb{Q}^*}\left(B_\tau^{-1}Z_\tau\mathbb{1}_{\{t < \tau \leq T\}} + B_T^{-1}X\mathbb{1}_{\{\tau > T\}}\,\Big|\,\mathcal{G}_t\right). \tag{8.4}$$

It should be stressed that we do not postulate here that a defaultable claim is attainable. In fact, within the framework of the intensity-based approach, a defaultable claim typically cannot be duplicated by trading in default-free securities, so that the standard arguments based on the existence of a replicating strategy do not apply in this setting. On the other hand, the valuation formula (8.2) can be supported by suitable no-arbitrage arguments, along similar lines[1] as in Sect. 2.1. Let us briefly recall these arguments.

To this end, we assume that $S^1, \ldots, S^n$ are price processes of $n$ non-dividend paying primary assets in our market model, with $S^n = B$. We do not need to be more specific about the nature of primary assets here. It suffices to assume that the savings account $B$ is well-defined. Let the $0^{th}$ asset correspond to the defaultable claim so that $S_t^0 = X^d(t, T)$. We write $\phi = (\phi^0, \ldots, \phi^k)$ to denote an $\mathbb{G}$-predictable process representing a trading strategy. The *wealth process* $U(\phi)$ of a strategy $\phi$ is given by the formula

$$U_t(\phi) = \sum_{i=0}^{k} \phi_t^i S_t^i, \quad \forall t \in [0, T].$$

A strategy $\phi$ is called *self-financing*, provided that $U_t(\phi) = U_0(\phi) + G_t(\phi)$ for every $t \in [0, T]$, where the *gains process* $G(\phi)$ is defined as follows

$$G_t(\phi) := \int_{]0,t]} \phi_u^0\, dD_u + \sum_{i=0}^{k} \int_{]0,t]} \phi_u^i\, dS_u^i.$$

The following result is merely a reformulation of Corollary 2.1.1.

**Proposition 8.1.1.** *For any self-financing trading strategy* $\phi = (\phi^0, \ldots, \phi^k)$, *the discounted wealth process* $\tilde{U}_t(\phi) = B_t^{-1}U_t(\phi)$, $t \in [0, T]$, *follows a local martingale under* $\mathbb{Q}^*$ *with respect to the filtration* $\mathbb{G}$.

It is customary to restrict the class of trading strategies, by postulating that the discounted wealth of an *admissible* strategy follows a martingale under $\mathbb{Q}^*$ (to this end, it suffices, for instance, to consider only strategies with non-negative wealth processes). Proposition 8.1.1 shows that if the original securities market model is arbitrage-free, and the ex-dividend price process of an additional security (i.e., of a defaultable claim) is given by Definition 8.1.2, then the arbitrage-free feature of the securities market model is preserved.

---

[1] Of course, the probability $\mathbb{P}^*$ should be substituted with $\mathbb{Q}^*$ in Sect. 2.1.

## 8.2 Valuation via the Hazard Process

Our first goal is to recall the definition and the basic properties of the F-hazard process $\Gamma$ of a random time (for more details, the reader is referred to Sect. 5.1). Before stating the definition of the hazard process, let us quote the following useful formula, established in Lemma 5.1.2,

$$\mathbb{Q}^*\{t < \tau \le T \,|\, \mathcal{G}_t\} = \mathbb{1}_{\{\tau > t\}} \frac{\mathbb{Q}^*\{t < \tau \le T \,|\, \mathcal{F}_t\}}{\mathbb{Q}^*\{\tau > t \,|\, \mathcal{F}_t\}}. \tag{8.5}$$

Let us denote $F_t = \mathbb{Q}^*\{\tau \le t \,|\, \mathcal{F}_t\}$. We shall postulate throughout that $F_0 =$ and that the inequality $F_t < 1$ is valid for every $t \in \mathbb{R}_+$. The *survival process* $G$ of the random time $\tau$ with respect to the reference filtration $\mathbb{F}$ equals

$$G_t := 1 - F_t = \mathbb{Q}^*\{\tau > t \,|\, \mathcal{F}_t\}, \quad \forall t \in \mathbb{R}_+.$$

Since $\{\tau \le t\} \subseteq \{\tau \le s\}$, for any $0 \le t \le s$ we have:

$$\mathbb{E}_{\mathbb{Q}^*}(F_s \,|\, \mathcal{F}_t) = \mathbb{E}_{\mathbb{Q}^*}(\mathbb{Q}^*\{\tau \le s \,|\, \mathcal{F}_s\} \,|\, \mathcal{F}_t)$$
$$= \mathbb{Q}^*\{\tau \le s \,|\, \mathcal{F}_t\} \ge \mathbb{Q}^*\{\tau \le t \,|\, \mathcal{F}_t\} = F_t,$$

and so the process $F$ (the survival process $G$, resp.) follows a bounded, non-negative $\mathbb{F}$-submartingale ($\mathbb{F}$-supermartingale, resp.) under $\mathbb{Q}^*$. The hazard process of the default time, given the flow of information represented by the filtration $\mathbb{F}$, is formally introduced through the following definition.

**Definition 8.2.1.** The $\mathbb{F}$-*hazard process* of $\tau$ under $\mathbb{Q}^*$, denoted by $\Gamma$, is defined through the formula $1 - F_t = e^{-\Gamma_t}$ or, equivalently,

$$\Gamma_t := -\ln G_t = -\ln(1 - F_t), \quad \forall t \in \mathbb{R}_+.$$

Since obviously $G_0 = 1$, it is clear that $\Gamma_0 = 0$. In view of the equality $\mathbb{Q}^*\{\tau < +\infty\} = 1$, it is also easy to see that $\lim_{t \to \infty} \Gamma_t = \infty$. For the sake of conciseness, we shall refer to $\Gamma$ as the $\mathbb{F}$-hazard process of $\tau$, rather than the $\mathbb{F}$-hazard process of $\tau$ under $\mathbb{Q}^*$. If no risk of ambiguity arises, we shall simply call it the hazard process of $\tau$. Combining formula (8.5) with the definition of the hazard process, we obtain

$$\mathbb{Q}^*\{t < \tau \le T \,|\, \mathcal{G}_t\} = \mathbb{1}_{\{\tau > t\}} e^{\Gamma_t} \mathbb{E}_{\mathbb{Q}^*}\left(e^{-\Gamma_t} - e^{-\Gamma_T} \,\middle|\, \mathcal{F}_t\right)$$
$$= \mathbb{1}_{\{\tau > t\}} \mathbb{E}_{\mathbb{Q}^*}\left(1 - e^{\Gamma_t - \Gamma_T} \,\middle|\, \mathcal{F}_t\right).$$

It is evident that the hazard process $\Gamma$ is continuous if and only if the submartingale $F$, and thus also the supermartingale $G$, follow continuous processes. Assume, in addition, that the sample paths of $F$ are non-decreasing functions; this amounts to postulating that the martingale part of $F$ vanishes. We adopt the widely used convention of calling such a process an *increasing continuous process*. In this case, the hazard process $\Gamma$ of $\tau$ also follows an increasing continuous process.

Our next goal is to study the following question. Suppose the $\mathbb{F}$-hazard process $\Gamma$ is $\hat{\mathbb{F}}$-adapted, for some sub-filtration $\hat{\mathbb{F}}$ of $\mathbb{F}$. Does this mean that $\Gamma$ also represents the $\hat{\mathbb{F}}$-hazard process of $\tau$? It appears that the answer to this question is positive, as the following result shows.

**Lemma 8.2.1.** *Assume that the $\mathbb{F}$-hazard process of $\tau$ under $\mathbb{Q}^*$ follows an $\hat{\mathbb{F}}$-adapted process for some sub-filtration $\hat{\mathbb{F}}$ of $\mathbb{F}$. Then $\Gamma$ is also the $\hat{\mathbb{F}}$-hazard process of $\tau$ under $\mathbb{Q}^*$.*

*Proof.* The proof is elementary. Let us set $\hat{G}_t = \mathbb{Q}^*\{\tau > t \,|\, \hat{\mathcal{F}}_t\}$, and let us denote by $\hat{\Gamma}$ the process, which satisfies $\hat{G}_t = e^{-\hat{\Gamma}_t}$. Under the assumptions of the lemma, the process $G_t = e^{-\Gamma_t}$ is $\hat{\mathbb{F}}$-adapted, and so

$$\hat{G}_t = \mathbb{Q}^*\{\tau > t \,|\, \hat{\mathcal{F}}_t\} = \mathbb{E}_{\mathbb{Q}^*}(\mathbb{Q}^*\{\tau > t \,|\, \mathcal{F}_t\} \,|\, \hat{\mathcal{F}}_t) = \mathbb{Q}^*\{\tau > t \,|\, \mathcal{F}_t\} = G_t.$$

This immediately implies the equality $\hat{\Gamma} = \Gamma$.    □

**Stochastic intensity.** In most of the recently developed reduced-form models of credit risk, the hazard process $\Gamma$ of a default time is postulated to have absolutely continuous sample paths (with respect to the Lebesgue measure on $\mathbb{R}_+$). Specifically, it is assumed that the hazard process $\Gamma$ of $\tau$ admits the following integral representation

$$\Gamma_t = \int_0^t \gamma_u \, du, \quad \forall t \in \mathbb{R}_+,$$

for some non-negative, $\mathbb{F}$-progressively measurable stochastic process $\gamma$, with integrable sample paths. In addition, we assume that $\int_0^\infty \gamma_u \, du = \infty$, $\mathbb{Q}^*$-a.s. The process $\gamma$ is called the $\mathbb{F}$-*hazard rate* or the $\mathbb{F}$-*intensity* of $\tau$. It is also customary to refer to $\gamma$ as the *stochastic intensity* of $\tau$, especially when the choice of the reference filtration $\mathbb{F}$ is clear from the context.

In terms of the stochastic intensity of a default time, the conditional probability of the default event $\{t < \tau \leq T\}$, given the information $\mathcal{G}_t$ available at time $t$, equals

$$\mathbb{Q}^*\{t < \tau \leq T \,|\, \mathcal{G}_t\} = \mathbb{1}_{\{\tau > t\}} \mathbb{E}_{\mathbb{Q}^*}\left(1 - e^{-\int_t^T \gamma_u \, du} \,\Big|\, \mathcal{F}_t\right). \tag{8.6}$$

Since the event $\{\tau \leq t\}$ manifestly belongs to the $\sigma$-field $\mathcal{G}_t$, we also have

$$\mathbb{Q}^*\{\tau \leq T \,|\, \mathcal{G}_t\} = \mathbb{1}_{\{\tau \leq t\}} + \mathbb{1}_{\{\tau > t\}} \mathbb{E}_{\mathbb{Q}^*}\left(1 - e^{-\int_t^T \gamma_u \, du} \,\Big|\, \mathcal{F}_t\right).$$

Since the event $\{\tau > t\}$ also belongs to $\mathcal{G}_t$, we obtain

$$\mathbb{Q}^*\{t < \tau \leq T \,|\, \mathcal{G}_t\} + \mathbb{Q}^*\{\tau > T \,|\, \mathcal{G}_t\} = \mathbb{Q}^*\{\tau > t \,|\, \mathcal{G}_t\} = \mathbb{1}_{\{\tau > t\}},$$

so that the conditional probability of the non-default event $\{\tau > T\}$ equals

$$\mathbb{Q}^*\{\tau > T \,|\, \mathcal{G}_t\} = \mathbb{1}_{\{\tau > t\}} \mathbb{E}_{\mathbb{Q}^*}\left(e^{-\int_t^T \gamma_u \, du} \,\Big|\, \mathcal{F}_t\right). \tag{8.7}$$

**Intensity function.** In some instances, the intensity of a default time is non-random; in such cases, it is referred to as the *intensity function* of $\tau$ (the properties of this function were examined at some length in Chap. 4). The concept of intensity function appears, for instance, when the trivial filtration is chosen as the reference filtration $\mathbb{F}$, so that $\mathbb{G} = \mathbb{H}$. To emphasize the deterministic character of the hazard function, we shall write $\gamma(t)$, rather than $\gamma_t$, and so formulae (8.6)–(8.7) become

$$\mathbb{Q}^*\{t < \tau < T \,|\, \mathcal{H}_t\} = \mathbb{1}_{\{\tau > t\}}\left(1 - e^{-\int_t^T \gamma(u)du}\right), \qquad (8.8)$$

and

$$\mathbb{Q}^*\{\tau > T \,|\, \mathcal{H}_t\} = \mathbb{1}_{\{\tau > t\}} e^{-\int_t^T \gamma(u)du}, \qquad (8.9)$$

respectively. Recall that $\mathcal{H}_t = \sigma(H_u : u \leq t) = \sigma(\{\tau \leq u\} : u \leq t)$, and thus $\mathbb{H} = (\mathcal{H}_t)_{t \geq 0}$ is the natural filtration of the random time $\tau$. The assumption that the filtration $\mathbb{H}$ models the flow of information available to a trader amounts to saying that he has no access to the market data other than the occurrence of the default time $\tau$.

In some more general circumstances – for instance, when the default time $\tau$ is independent of a (non-trivial) filtration $\mathbb{F}$ – it has a deterministic intensity with respect to $\mathbb{F}$, and equalities (8.8)–(8.9) remain valid with the $\sigma$-field $\mathcal{H}_t$ replaced by a strictly larger $\sigma$-field $\mathcal{G}_t = \mathcal{H}_t \vee \mathcal{F}_t$.

### 8.2.1 Canonical Construction of a Default Time

We shall now briefly describe the most commonly used construction of a default time associated with a given hazard process $\Gamma$. For a more detailed analysis of this procedure, the interested reader is referred to Sect. 6.5. It should be stressed that the random time obtained through this particular method – which will be called the *canonical construction* in what follows – has certain specific features that are not necessarily shared by all random times with a given $\mathbb{F}$-hazard process $\Gamma$. In other words, the knowledge of the $\mathbb{F}$-hazard process of the default time $\tau$ does not uniquely specify all properties of $\tau$ with respect to the reference filtration $\mathbb{F}$.

We assume that we are given an $\mathbb{F}$-adapted, right-continuous, increasing process $\Gamma$ defined on a filtered probability space $(\tilde{\Omega}, \mathbb{F}, \mathbb{P}^*)$. As usual, we assume that $\Gamma_0 = 0$ and $\Gamma_\infty = +\infty$. In many instances, $\Gamma$ is given by the equality

$$\Gamma_t = \int_0^t \gamma_u \, du, \quad \forall t \in \mathbb{R}_+,$$

for some non-negative, $\mathbb{F}$-progressively measurable intensity process $\gamma$.

To construct a random time $\tau$ such that $\Gamma$ is the $\mathbb{F}$-hazard process of $\tau$, we need to enlarge the underlying probability space $\tilde{\Omega}$. This also means that $\Gamma$ is not the $\mathbb{F}$-hazard process of $\tau$ under $\mathbb{P}^*$, but rather the $\mathbb{F}$-hazard process of $\tau$ under a suitable extension $\mathbb{Q}^*$ of the probability measure $\mathbb{P}^*$.

Let $\xi$ be a random variable defined on some probability space $(\hat{\Omega}, \hat{\mathcal{F}}, \hat{\mathbb{Q}})$, uniformly distributed on the interval $[0, 1]$ under $\hat{\mathbb{Q}}$. We consider the product space $\Omega = \tilde{\Omega} \times \hat{\Omega}$, endowed with the product $\sigma$-field $\mathcal{G} = \mathcal{F}_\infty \otimes \hat{\mathcal{F}}$ and the product probability measure $\mathbb{Q}^* = \mathbb{P}^* \otimes \hat{\mathbb{Q}}$. The latter equality means that for arbitrary events $A \in \mathcal{F}_\infty$ and $B \in \hat{\mathcal{F}}$ we have $\mathbb{Q}^*\{A \times B\} = \mathbb{P}^*\{A\}\hat{\mathbb{Q}}\{B\}$.

*Remarks.* An alternative way of achieving basically the same goal relies on postulating that the underlying probability space $(\tilde{\Omega}, \mathbb{F}, \mathbb{P}^*)$ is sufficiently rich to support a random variable $\xi$, uniformly distributed on the interval $[0, 1]$, and independent of the filtration $\mathbb{F}$ under $\mathbb{P}^*$. In this version of the canonical construction, $\Gamma$ represents the $\mathbb{F}$-hazard process of $\tau$ under $\mathbb{P}^*$.

We define the random time $\tau : \Omega \to \mathbb{R}_+$ by setting

$$\tau = \inf\{t \in \mathbb{R}_+ : e^{-\Gamma_t} \le \xi\} = \inf\{t \in \mathbb{R}_+ : \Gamma_t \ge \eta\}, \qquad (8.10)$$

where the random variable $\eta = -\ln \xi$ has a unit exponential law under $\mathbb{Q}^*$. It is not difficult to find the process $F_t = \mathbb{Q}^*\{\tau \le t \,|\, \mathcal{F}_t\}$. Indeed, since clearly $\{\tau > t\} = \{\xi < e^{-\Gamma_t}\}$ and the random variable $\Gamma_t$ is $\mathcal{F}_\infty$-measurable, we obtain

$$\mathbb{Q}^*\{\tau > t \,|\, \mathcal{F}_\infty\} = \mathbb{Q}^*\{\xi < e^{-\Gamma_t} \,|\, \mathcal{F}_\infty\} = \hat{\mathbb{Q}}\{\xi < e^x\}_{x=\Gamma_t} = e^{-\Gamma_t}. \quad (8.11)$$

Consequently, we have

$$1 - F_t = \mathbb{Q}^*\{\tau > t \,|\, \mathcal{F}_t\} = \mathbb{E}_{\mathbb{Q}^*}\big(\mathbb{Q}^*\{\tau > t \,|\, \mathcal{F}_\infty\} \,|\, \mathcal{F}_t\big) = e^{-\Gamma_t}, \qquad (8.12)$$

and so $F$ is an $\mathbb{F}$-adapted, right-continuous, increasing process. It is also clear that the process $\Gamma$ represents the $\mathbb{F}$-hazard process of $\tau$ under $\mathbb{Q}^*$. As an immediate consequence of (8.11) and (8.12), we obtain the following interesting property of the canonical construction of the default time:

$$\mathbb{Q}^*\{\tau \le t \,|\, \mathcal{F}_\infty\} = \mathbb{Q}^*\{\tau \le t \,|\, \mathcal{F}_t\}, \quad \forall t \in \mathbb{R}_+. \qquad (8.13)$$

Let us now analyze some important consequences of (8.13). First, we obtain

$$\mathbb{Q}^*\{\tau \le t \,|\, \mathcal{F}_\infty\} = \mathbb{Q}^*\{\tau \le t \,|\, \mathcal{F}_u\} = \mathbb{Q}^*\{\tau \le t \,|\, \mathcal{F}_t\} = e^{-\Gamma_t} \qquad (8.14)$$

for arbitrary two dates $0 \le t \le u$. Notice that only the last equality in (8.14) is necessarily satisfied by the $\mathbb{F}$-hazard process $\Gamma$ of $\tau$; the first two equalities are additional features of the canonical construction of $\tau$, meaning that they are not necessarily valid in a general set-up.

Equality (8.14) entails the conditional independence under $\mathbb{Q}^*$ of the $\sigma$-fields $\mathcal{H}_t$ and $\mathcal{F}_t$, given the $\sigma$-field $\mathcal{F}_\infty$. Such a property of the two filtrations $\mathbb{H}$ and $\mathbb{F}$ was introduced in Sect. 6.1, where it was termed Condition (F.1). By virtue of Lemma 6.1.2, Condition (F.1) is equivalent to Condition (M.1), which in turn can be stated as follows: an arbitrary $\mathbb{F}$-martingale also follows a $\mathbb{G}$-martingale under $\mathbb{Q}^*$. We refer to Sect. 6.1 for a thorough discussion of this condition, which was previously studied by, among others, Brémaud and Yor (1978), Dellacherie and Meyer (1978a), Mazziotto and Szpirglas (1979), Kusuoka (1999), and Elliott et al. (2000).

By virtue of the analysis in Sect. 6.5 (or, by virtue of Corollary 6.2.1), we have the following lemma.

**Lemma 8.2.2.** *Assume that the process $\Gamma$ is continuous. Then the $(\mathbb{F}, \mathbb{G})$-martingale hazard process $\Lambda$ of the random time $\tau$, given by (8.10), coincides with the $\mathbb{F}$-hazard process $\Gamma$ of $\tau$.*

*Remarks.* In most credit risk models, the reference filtration $\mathbb{F}$ is generated by the process $W$ that follows a Brownian motion under $\mathbb{P}^*$. In view of the martingale invariance property, the canonical construction ensures that the Brownian motion process $W$ remains a continuous martingale (and thus a Brownian motion) under the extended probability measure $\mathbb{Q}^*$ and with respect to the enlarged filtration $\mathbb{G}$. Let us stress again that $\mathbb{Q}^*\{A \times \hat{\Omega}\} = \mathbb{P}^*\{A\}$ for any event $A \in \mathcal{F}_\infty$; that is, the restriction of the probability measure $\mathbb{Q}^*$ to the $\sigma$-field $\mathcal{F}_\infty$ coincides with $\mathbb{P}^*$.

*Example 8.2.1. Deterministic hazard process.* Let us assume that the underlying filtration $\mathbb{F}$ is non-trivial, but the $\mathbb{F}$-hazard process $\Gamma$ is postulated to follow a deterministic function; that is, the $\mathbb{F}$-hazard process equals $\Gamma$ for some function $\Gamma : \mathbb{R}_+ \to \mathbb{R}_+$. Assume that the default time $\tau$ is defined as before – i.e.,

$$\tau = \inf \{\, t \in \mathbb{R}_+ : e^{-\Gamma(t)} \leq \xi \,\}.$$

We claim that the default process $H$ is independent of the filtration $\mathbb{F}$ or, equivalently, that the filtration $\mathbb{H}$ generated by the default process $H$ is independent of the filtration $\mathbb{F}$ under $\mathbb{Q}^*$. It suffices to check that we have, for any fixed $t \in \mathbb{R}_+$ and arbitrary $0 \leq u \leq t$,

$$\mathbb{Q}^*\{\tau \leq u \,|\, \mathcal{F}_t\} = \mathbb{Q}^*\{\tau \leq u\}. \tag{8.15}$$

Equality (8.15) easily follows from (8.14). In effect, we have

$$\mathbb{Q}^*\{\tau \leq u \,|\, \mathcal{F}_t\} = \mathbb{Q}^*\{\tau \leq u \,|\, \mathcal{F}_u\} = 1 - e^{-\Gamma(u)} = \mathbb{Q}^*\{\tau \leq u\},$$

where the last equality is a consequence of the assumption that the hazard process is deterministic.

If the default process $H$ is independent of the filtration $\mathbb{F}$ then any $\mathbb{F}$-adapted process $Y$ is independent of $H$ under $\mathbb{Q}^*$. In particular, since the short-term rate $r$ follows an $\mathbb{F}$-adapted process, processes $H$ and $r$ are mutually independent under $\mathbb{Q}^*$ when the $\mathbb{F}$-hazard process of $\tau$ is deterministic, and the default time $\tau$ is constructed through the canonical approach

*Example 8.2.2. State variables.* In some financial models, it is assumed that the reference filtration $\mathbb{F}$ is generated by some stochastic process, $Y$ say. More specifically, the $\mathbb{F}$-intensity of the default time is given by the equality

$$\Gamma_t = \int_0^t g(u, Y_u)\, du, \quad \forall\, t \in \mathbb{R}_+,$$

for some function $g : \mathbb{R}_+ \times \mathcal{Y} \to \mathbb{R}_+$ satisfying mild technical assumptions, where $\mathcal{Y}$ denotes the state space for the process $Y$ (typically, $\mathcal{Y} = \mathbb{R}^d$).

### 8.2.2 Integral Representation of the Value Process

Our next goal is to establish a convenient representation for the pre-default value of a defaultable claim in terms of the hazard process $\Gamma$ of the default time. For the sake of conciseness, we denote

$$I_t(A) = B_t \, \mathbb{E}_{\mathbb{Q}^*} \left( \int_{]t,T]} B_u^{-1} (1 - H_u) \, dA_u \,\Big|\, \mathcal{G}_t \right)$$

and

$$J_t(Z) = B_t \, \mathbb{E}_{\mathbb{Q}^*} \left( \mathbb{1}_{\{t < \tau \le T\}} B_\tau^{-1} Z_\tau \,\Big|\, \mathcal{G}_t \right).$$

We shall also write

$$\tilde{K}_t = B_t \, \mathbb{E}_{\mathbb{Q}^*} \left( B_T^{-1} \tilde{X} \mathbb{1}_{\{\tau \le T\}} \,\big|\, \mathcal{G}_t \right), \quad K_t = B_t \, \mathbb{E}_{\mathbb{Q}^*} \left( B_T^{-1} X \mathbb{1}_{\{T < \tau\}} \,\big|\, \mathcal{G}_t \right).$$

It is thus clear that $S_t^0 = I_t(A) + J_t(Z) + \tilde{K}_t + K_t$. The following proposition is a rather straightforward corollary to the results established in Chap. 5. Let us stress that we do not need to assume here that the default time $\tau$ was constructed through the canonical method.

**Proposition 8.2.1.** *The pre-default value process $S_t^0$ of a defaultable claim $(X, A, 0, Z, \tau)$ admits the following representation for $t \in [0, T]$*

$$S_t^0 = \mathbb{1}_{\{\tau > t\}} G_t^{-1} B_t \, \mathbb{E}_{\mathbb{Q}^*} \left( \int_{]t,T]} B_u^{-1} (G_u \, dA_u - Z_u \, dG_u) + G_T B_T^{-1} X \,\Big|\, \mathcal{F}_t \right).$$

*If the survival process $G$, and thus also the hazard process $\Gamma$, are continuous, then*

$$S_t^0 = \mathbb{1}_{\{\tau > t\}} B_t \, \mathbb{E}_{\mathbb{Q}^*} \left( \int_{]t,T]} B_u^{-1} e^{\Gamma_t - \Gamma_u} (dA_u + Z_u \, d\Gamma_u) + B_T^{-1} X e^{\Gamma_t - \Gamma_T} \,\Big|\, \mathcal{F}_t \right).$$

*Proof.* Since $\tilde{X} = 0$, it is obvious that $\tilde{K}_t = 0$ for $t \in [0, T]$, and so the value process satisfies: $S_t^0 = I_t(A) + J_t(Z) + K_t$. By applying Proposition 5.1.2 to the process of finite variation $\int_{]0,t]} B_u^{-1} \, dA_u$, we obtain

$$I_t(A) = \mathbb{1}_{\{\tau > t\}} G_t^{-1} B_t \, \mathbb{E}_{\mathbb{Q}^*} \left( \int_{]t,T]} B_u^{-1} G_u \, dA_u \,\Big|\, \mathcal{F}_t \right)$$

or, equivalently,

$$I_t(A) = \mathbb{1}_{\{\tau > t\}} B_t \, \mathbb{E}_{\mathbb{Q}^*} \left( \int_{]t,T]} B_u^{-1} e^{\Gamma_t - \Gamma_u} \, dA_u \,\Big|\, \mathcal{F}_t \right).$$

Furthermore, formula (5.18) of Proposition 5.1.1 yields

$$J_t(Z) = -\mathbb{1}_{\{\tau > t\}} G_t^{-1} B_t \, \mathbb{E}_{\mathbb{Q}^*} \left( \int_{]t,T]} B_u^{-1} Z_u \, dG_u \,\Big|\, \mathcal{F}_t \right).$$

If, in addition, the survival process $G$ is a continuous (and thus a decreasing) process, the hazard process $\Gamma$ is an increasing continuous process, and

$$J_t(Z) = \mathbb{1}_{\{\tau>t\}} B_t \, \mathbb{E}_{\mathbb{Q}^*} \left( \int_t^T B_u^{-1} e^{\Gamma_t - \Gamma_u} Z_u \, d\Gamma_u \, \Big| \, \mathcal{F}_t \right).$$

Finally, it follows from (5.11) that

$$K_t = \mathbb{1}_{\{\tau>t\}} G_t^{-1} B_t \, \mathbb{E}_{\mathbb{Q}^*} (\mathbb{1}_{\{\tau>T\}} B_T^{-1} X \, | \, \mathcal{F}_t). \tag{8.16}$$

Since the random variables $X$ and $B_T$ are $\mathcal{F}_T$-measurable, we also have (see (5.13))

$$K_t = \mathbb{1}_{\{\tau>t\}} G_t^{-1} B_t \, \mathbb{E}_{\mathbb{Q}^*} (G_T B_T^{-1} X \, | \, \mathcal{F}_t) = \mathbb{1}_{\{\tau>t\}} B_t \, \mathbb{E}_{\mathbb{Q}^*} (B_T^{-1} X e^{\Gamma_t - \Gamma_T} \, | \, \mathcal{F}_t).$$

Both formulae of the proposition are obtained upon summation. □

**Corollary 8.2.1.** *Assume that the $\mathbb{F}$-hazard process $\Gamma$ follows a continuous process of finite variation. Then the pre-default value of a defaultable claim $(X, A, 0, Z, \tau)$ coincides with the pre-default value of a defaultable claim $(X, \hat{A}, 0, 0, \tau)$, where $\hat{A}_t = A_t + \int_0^t Z_u \, d\Gamma_u$.*

*Remarks.* We have omitted in Proposition 8.2.1 the recovery payoff $\tilde{X}$, since the expression based on the hazard process of the default time does not easily cover the case of a general $\mathcal{F}_T$-measurable random variable. However, in the special case when $\tilde{X} = \delta$ for some constant $\delta$, it suffices to substitute $\tilde{X}$ with an equivalent payoff $\delta B(\tau, T)$ at time of default.

Let us return to the case of the default time that admits the stochastic intensity $\gamma$. The second formula of Proposition 8.2.1 now takes the following form

$$S_t^0 = \mathbb{1}_{\{\tau>t\}} \, \mathbb{E}_{\mathbb{Q}^*} \left( \int_{]t,T]} e^{-\int_t^u (r_v + \gamma_v) dv} (dA_u + \gamma_u Z_u \, du) \, \Big| \, \mathcal{F}_t \right)$$

$$+ \, \mathbb{1}_{\{\tau>t\}} \, \mathbb{E}_{\mathbb{Q}^*} \left( e^{-\int_t^T (r_v + \gamma_v) dv} X \, \Big| \, \mathcal{F}_t \right).$$

To get a more concise representation for the last expression, we introduce the *default-risk-adjusted interest rate* $\tilde{r} = r + \gamma$ and the associated *default-risk-adjusted savings account* $\tilde{B}$, given by the formula

$$\tilde{B}_t = \exp \left( \int_0^t \tilde{r}_u \, du \right), \quad \forall \, t \in \mathbb{R}_+. \tag{8.17}$$

Although $\tilde{B}$ does not represent the price of a tradeable security, it has similar features as the savings account $B$; in particular, $\tilde{B}$ also follows an $\mathbb{F}$-adapted, continuous process of finite variation. In terms of the process $\tilde{B}$, we have

$$S_t^0 = \mathbb{1}_{\{\tau>t\}} \tilde{B}_t \, \mathbb{E}_{\mathbb{Q}^*} \left( \int_{]t,T]} \tilde{B}_u^{-1} \, dA_u + \int_t^T \tilde{B}_u^{-1} Z_u \gamma_u \, du + \tilde{B}_T^{-1} X \, \Big| \, \mathcal{F}_t \right). \tag{8.18}$$

It is noteworthy that the default time $\tau$ does not appear explicitly in the conditional expectation on the right-hand side of (8.18).

### 8.2.3 Case of a Deterministic Intensity

For the sake of simplicity, we shall assume in this section that:
– the default time admits the intensity function $\gamma$ with respect to $\mathbb{F}$,
– the continuously compounded interest rate $r$ is deterministic.
In view of the latter assumption, at time $t$ the price of a unit default-free
zero-coupon bond of maturity $T$ equals

$$B(t,T) = e^{-\int_t^T r(v)\,dv}, \quad \forall t \in [0,T].$$

Our goal is to derive some integral representations for the pre-default values
of simple defaultable claims. We take $A \equiv 0$, $\tilde{X} = 0$ and $Z_\tau = h(\tau)$ for some
continuous function $h : \mathbb{R}_+ \to \mathbb{R}$. If, in addition, the promised payoff $X$ is
non-random, the pre-default value of the claim equals

$$S_t^0 = \mathbb{1}_{\{\tau > t\}} B_t \left( \int_t^T e^{-\int_t^u \gamma(v)\,dv} B_u^{-1} \gamma(u) h(u)\,du + B_T^{-1} X e^{-\int_t^T \gamma(v)\,dv} \right)$$

or, equivalently,

$$S_t^0 = \mathbb{1}_{\{\tau > t\}} \left( \int_t^T e^{-\int_t^u \tilde{r}(v)\,dv} \gamma(u) h(u)\,du + X e^{-\int_t^T \tilde{r}(v)\,dv} \right), \tag{8.19}$$

where $\tilde{r}(v) = r(v) + \gamma(v)$.

*Remarks.* Let us again stress that $S_t^0$ represents only the pre-default value of
a defaultable claim. At any date $t$, the discounted payoff of the defaultable
claim introduced above is given by the following expression

$$Y_t = \mathbb{1}_{\{\tau \le T\}} h(\tau) e^{-\int_t^\tau r(v)\,dv} + \mathbb{1}_{\{\tau > T\}} X e^{-\int_t^T r(v)\,dv}.$$

Thus, the 'full' value at time $t$ of a defaultable claim equals

$$\mathbb{E}_{\mathbb{Q}^*}(Y_t \mid \mathcal{H}_t) = \mathbb{1}_{\{\tau > t\}} \left( \int_t^T e^{-\int_t^u \tilde{r}(v)\,dv} h(u) \gamma(u)\,du + X e^{-\int_t^T \tilde{r}(v)\,dv} \right)$$

$$+ \mathbb{1}_{\{\tau \le t\}} h(\tau) e^{\int_\tau^t r(v)\,dv}.$$

The additional third term in the last formula represents the current value of
the recovery cash flow $h(\tau)$ received by the owner of the claim at the time of
default, and reinvested in the savings account.

*Example 8.2.3.* Let us consider some examples of corporate zero-coupon
bonds with maturity date $T$ that are subject to various recovery schemes.
In the next section, we shall study these schemes in the context of general
contingent claims. In all cases examined below, the pre-default value of a
corporate bond appears to be proportional to the bond's face value, $L$. In
what follows, when referring to the pre-default values of corporate bonds, we
shall usually set $L = 1$ and we shall suppress $L$ from the notation.

**Zero recovery.** Let us first consider a corporate zero-coupon bond with *zero recovery* at default. This corresponds to the choice of $h = 0$ and $X = L = 1$ in (8.19). Denoting by $D^0(t, T)$ the pre-default value at time $t$ of such a bond, for every $t \in [0, T]$ we obtain:

$$D^0(t, T) = \mathbb{1}_{\{\tau > t\}} e^{-\int_t^T (r(v) + \gamma(v)) dv} = \mathbb{1}_{\{\tau > t\}} B(t, T) e^{-\int_t^T \gamma(v) dv}.$$

Under the zero recovery scheme, the corporate bond becomes, of course, valueless as soon as the default occurs.

**Fractional recovery of par value.** Let us assume that the recovery function $h$ satisfies $h = \delta L = \delta$ for some constant recovery coefficient $0 \le \delta \le 1$. The corresponding recovery scheme is aptly termed the *fractional recovery of par value*. The pre-default value at time $t$ of a corporate bond that is subject to this recovery scheme, denoted by $\tilde{D}^\delta(t, T)$, equals

$$\tilde{D}^\delta(t, T) = \mathbb{1}_{\{\tau > t\}} \left( \int_t^T e^{-\int_t^u \tilde{r}(v) dv} \delta \gamma(u) \, du + e^{-\int_t^T \tilde{r}(v) dv} \right).$$

Notice that $\tilde{D}^\delta(t, T)$ represents the value before default of a corporate bond that pays at time of default a constant payoff proportional to the bond's face value, in case the bond defaults before or at the bond's maturity date $T$. However, it is clear that constant coefficient $\delta$ can be replaced by some function $\delta(t)$ of time (the same remark applies to the next recovery scheme).

**Fractional recovery of Treasury value.** Let us finally assume that the recovery function equals

$$h(\tau) = \delta L e^{-\int_\tau^T r(v) dv} = \delta e^{-\int_\tau^T r(v) dv}. \tag{8.20}$$

The above specification of the recovery function describes a corporate zero-coupon bond with the so-called *fractional recovery of Treasury value*. Indeed, since the payoff $h(\tau)$ can be invested in the savings account, we may formally postulate that the bond pays the constant payoff $\delta$ at maturity $T$ if default occurs before maturity (otherwise, it pays the nominal value $L = 1$). We may thus equally well postulate that the recovery payoff at default equals

$$h(\tau) = \delta B(\tau, T). \tag{8.21}$$

Under the present assumptions, the two alternative specifications, (8.20) and (8.21), yield an identical pre-default value of a corporate bond with the fractional recovery of Treasury value. It is thus interesting to notice that the latter specification is much more convenient if random character of interest rates is taken into account. Indeed, in all models the current value bond price $B(\tau, T)$ is known at time $\tau$, and thus one can always define the recovery process $Z$ by setting $Z_t = \delta B(t, T)$ for every $t \in [0, T]$. On the other hand, the right-hand side of (8.20) is not observed at time $\tau$ under the uncertainty of interest rates.

Let us denote by $D^\delta(t,T)$ the pre-default value of a unit corporate bond with the fractional recovery of Treasury value. By plugging (8.20) into the general formula (8.19), we obtain

$$D^\delta(t,T) = \mathbb{1}_{\{\tau>t\}} \left( \int_t^T e^{-\int_t^T r(v)dv} e^{-\int_t^u \gamma(v)dv} \delta\gamma(u)\,du + e^{-\int_t^T \tilde{r}(v)dv} \right),$$

that is,

$$D^\delta(t,T) = \mathbb{1}_{\{\tau>t\}} B(t,T) \left\{ \delta\left(1 - e^{-\int_t^T \gamma(v)\,dv}\right) + e^{-\int_t^T \gamma(v)\,dv} \right\}.$$

We end this example by noticing that the pre-default value $D^\delta(t,T)$ of a corporate bond with the fractional recovery of Treasury value can also be expressed as follows (see (8.6))

$$D^\delta(t,T) = B(t,T) \left( \delta\, \mathbb{Q}^* \{ t < \tau \le T \,|\, \mathcal{G}_t \} + \mathbb{Q}^* \{ \tau > T \,|\, \mathcal{G}_t \} \right).$$

It is worth stressing that the last representation, though apparently universal, in fact hinges on the non-random character of interest rates postulated in this section. In a more general setting, we need to impose some further assumptions, as well as to substitute the spot martingale measure $\mathbb{Q}^*$ with the associated forward martingale measure $\mathbb{Q}_T$.

### 8.2.4 Implied Probabilities of Default

Simple valuation formulae based on the intensity function are frequently used by practitioners in order to calibrate the model (see, e.g., Li (1998b)). The basic idea is to derive the default probabilities implicit in market quotes of traded defaultable securities (corporate bonds, default swaps, etc.), and to subsequently use these probabilities to value defaultable securities that are not quoted in the market. It is apparent that such an approach to model's calibration parallels the widely popular method of using implied volatilities of publicly traded (or at least liquid) options to value these exotic options for which the market quotes are either not available, or not reliable. Typically, it is postulated that:

- the interest rate process and the default process are mutually independent under the spot martingale measure $\mathbb{Q}^*$,
- the default can only be observed at some date from a given finite collection of dates $0 < T_1 < \cdots < T_n = T^*$, for some horizon date $T^*$,
- we are given the default-free term structure of interest rates, formally identified here with the prices $B(0, T_i)$, $i = 1, \ldots, n$ of zero-coupon Treasury bonds.

The first assumption means that we are interested in finding the intensity function, as opposed to the intensity process with respect to some non-trivial filtration $\mathbb{F}$. The second means that we are not interested in the exact behavior of the intensity function between the 'observed default dates,' so that we may adopt a convention that the intensity function is constant between each two dates $T_i$ and $T_{i+1}$.

In view of the preceding discussion, we may, and do, assume that the intensity function $\lambda : [0, T^*] \to \mathbb{R}_+$ satisfies $\lambda(t) = \sum_{i=1}^{n} \alpha_i \mathbb{1}_{[T_{i-1}, T_i[}(t)$ for some positive constants $\alpha_i$, $i = 0, \dots, n-1$ (we set $T_0 = 0$). This in turn amounts to postulate that for every $j = 1, \dots, n$

$$q_j^* = G(T_{j-1}) - G(T_j) = \exp\left(-\sum_{i=1}^{j-1} \alpha_i(T_i - T_{i-1})\right) - \exp\left(-\sum_{i=1}^{j} \alpha_i(T_i - T_{i-1})\right)$$

where

$$q_j^* := \mathbb{Q}^*\{T_{j-1} < \tau \le T_j\}, \quad G(T_j) = \mathbb{Q}^*\{\tau > T_j\}.$$

Notice that in general the inequality $\sum_{j=1}^{n} q_j^* = \mathbb{Q}^*\{\tau \le T_n\} \le 1$ is valid. In other words, we do not need to assume that the default will definitely happen before or at the horizon date $T^*$.

*Example 8.2.4.* Let us assume, for instance, that our goal is to calibrate the model to market quotes of a family of default swaps. For simplicity, we assume that $T_1, \dots, T_n$ are payment dates, and $T_n$ is the maturity of the contract. Under the independence assumption, the present value at time 0 of the default payment leg is

$$I_1 = \sum_{i=1}^{n} B(0, T_i) X_{T_i} q_i^* = \sum_{i=1}^{n} B(0, T_i) X_{T_i} \big(G(T_{i-1}) - G(T_i)\big),$$

where the (non-random) payoffs $X_{T_i}$ are typically expressed either as some fixed amounts, or as a percentage of the present value of the future coupons and the face value of the underlying bond discounted at the risk-free rate.

The premium payment leg is defined as a stream of fixed cash flows $\kappa$ that are paid until the maturity of the contract or until default, whichever comes first. The present value of these cash flows at time 0 equals

$$I_2 = B_0 \, \mathbb{E}_{\mathbb{Q}^*}\left(\sum_{i=1}^{n} \mathbb{1}_{\{T_{i-1} < \tau \le T_i\}} \sum_{j=1}^{i} B_{T_j}^{-1} \kappa \,\Big|\, \mathcal{G}_0\right)$$

$$= \kappa \sum_{i=1}^{n} q_i^* \sum_{j=1}^{i} B(0, T_j) = \kappa \sum_{i=1}^{n} B(0, T_i) \sum_{j=1}^{i} q_j^*$$

$$= \kappa \sum_{i=1}^{n} B(0, T_i) \mathbb{Q}^*\{\tau \le T_i\} = \kappa \sum_{i=1}^{n} B(0, T_i)(1 - G(T_i)).$$

Given a portfolio of default swaps and their market quotes, we may search for the values of $\alpha_i$, $i = 0, \dots, n-1$. The calibration procedure relies on solving non-linear equation of the form $I_1 = I_2$. In principle, the quoted default swaps should be repriced correctly within the calibrated model. Using this approach, we may not only value new issues of contracts that are exposed to the default risk of the underlying entity, but we may also mark to market outstanding default swaps and other defaultable contracts.

### 8.2.5 Exogenous Recovery Rules

We shall now return to the case of a defaultable claim $DCT = (X, A, 0, Z, \tau)$. In Sect. 8.2.3, we have briefly presented several alternative recovery rules in the context of intensity-based valuation of a corporate bond. As expected, these schemes can be extended to the case of an arbitrary defaultable claim. We shall now examine these extensions in some detail.

**Fractional recovery of par value.** We need to assume here that the par value (or the face value) of a defaultable claim is well defined. Denoting by $L$ the constant representing the claim's par value and by $\delta$ the claim's recovery rate, we set $Z_t = \delta L$. Therefore, the pre-default value, denoted by $\tilde{D}_t^\delta$, equals

$$\tilde{D}_t^\delta = B_t \, \mathbb{E}_{\mathbb{Q}^*} \bigg( \int_{]t,T]} B_u^{-1}(1-H_u) \, dA_u + \int_{]t,T]} B_u^{-1} \delta L \, dH_u + B_T^{-1} X \mathbb{1}_{\{\tau > T\}} \bigg| \mathcal{G}_t \bigg).$$

Consequently, by virtue of Proposition 8.2.1,

$$\tilde{D}_t^\delta = \mathbb{1}_{\{\tau > t\}} G_t^{-1} B_t \, \mathbb{E}_{\mathbb{Q}^*} \bigg( \int_{]t,T]} B_u^{-1}(G_u \, dA_u - \delta L \, dG_u) + G_T B_T^{-1} X \bigg| \mathcal{F}_t \bigg),$$

where $G$ is the survival process of $\tau$ with respect to the reference filtration $\mathbb{F}$. In the case of a continuous survival process $G$, the last formula yields

$$\tilde{D}_t^\delta = \mathbb{1}_{\{\tau > t\}} B_t \, \mathbb{E}_{\mathbb{Q}^*} \bigg( \int_{]t,T]} B_u^{-1} e^{\Gamma_t - \Gamma_u} (dA_u + \delta L \, d\Gamma_u) + B_T^{-1} X e^{\Gamma_t - \Gamma_T} \bigg| \mathcal{F}_t \bigg),$$

where $\Gamma_t = -\ln G_t$ is the $\mathbb{F}$-hazard process of the default time.

*Example 8.2.5.* Let us first assume that $A \equiv 0$. We shall write $U(t, T)$ to denote the price of a *digital default put* – that is, a default-risk sensitive security, which pays one unit of cash at time $\tau$ if default occurs prior to or at $T$, and pays zero otherwise. Formally, a digital default put corresponds to a defaultable claim of the form $(0, 0, 0, 1, \tau)$. We have $U(t, T) = \tilde{D}^1(t, T) - D^0(t, T)$ or, more explicitly,

$$U(t, T) = B_t \, \mathbb{E}_{\mathbb{Q}^*}(B_\tau^{-1} \mathbb{1}_{\{\tau \le T\}} \,|\, \mathcal{G}_t).$$

Let $\mathbb{Q}_T$ be the $T$-forward martingale measure, associated with $\mathbb{Q}^*$ through the formula

$$\frac{d\mathbb{Q}_T}{d\mathbb{Q}^*} = \frac{1}{B(0, T) B_T}, \quad \mathbb{Q}^*\text{-a.s.}$$

Using the abstract Bayes rule, we obtain the following representation for the price of a defaultable claim in terms of the forward martingale measure

$$\tilde{D}_t^\delta = \mathbb{1}_{\{\tau > t\}} \delta L S(t, T) + \mathbb{1}_{\{\tau > t\}} B(t, T) \, \mathbb{E}_{\mathbb{Q}_T} \left( X e^{\Gamma_t - \Gamma_T} \,|\, \mathcal{F}_t \right).$$

Notice that the hazard process of $\tau$ is not affected by the change of probability measure from $\mathbb{Q}^*$ to $\mathbb{Q}_T$. If the promised dividends process $A$ does not vanish, we need to add an extra term on the right-hand side of the last equality.

As already observed in Corollary 8.2.1, if the $\mathbb{F}$-hazard process $\Gamma$ follows a continuous process of finite variation, we may set $Z \equiv 0$ and substitute the promised dividends process $A$ with the process $\hat{A}_t = A_t + \delta L\Gamma_t$. In other words, from the point of view of arbitrage-free valuation the two defaultable claims $(X, A, \tilde{X}, \delta L, \tau)$ and $(X, A + \delta L\Gamma, \tilde{X}, 0, \tau)$ are essentially equivalent if $\Gamma$ is a continuous process of finite variation.

Finally, if the default time $\tau$ admits the $\mathbb{F}$-intensity process $\gamma$, then we have (cf. (8.18))

$$\tilde{D}_t^\delta = \mathbb{1}_{\{\tau > t\}} \tilde{B}_t \, \mathbb{E}_{\mathbb{Q}^*} \left( \int_{]t,T]} \tilde{B}_u^{-1} \, dA_u + \delta L \int_t^T \tilde{B}_u^{-1} \gamma_u \, du + \tilde{B}_T^{-1} X \,\Big|\, \mathcal{F}_t \right),$$

where the default-risk-adjusted savings account $\tilde{B}$ is given by (8.17). If, in addition, the sample paths of the process $A$ are absolutely continuous functions: $A_t = \int_0^t a_u \, du$, then

$$\tilde{D}_t^\delta = \mathbb{1}_{\{\tau > t\}} \tilde{B}_t \, \mathbb{E}_{\mathbb{Q}^*} \left( \int_t^T \tilde{B}_u^{-1} (a_u + \delta L\gamma_u) \, du + \tilde{B}_T^{-1} X \,\Big|\, \mathcal{F}_t \right)$$

$$= \mathbb{1}_{\{\tau > t\}} \tilde{B}_t \, \mathbb{E}_{\mathbb{Q}^*} \left( \int_t^T \tilde{B}_u^{-1} (a_u \gamma_u^{-1} + \delta L) \gamma_u \, du + \tilde{B}_T^{-1} X \,\Big|\, \mathcal{F}_t \right),$$

where the last equality holds, provided that $\gamma > 0$. We may here choose, without loss of generality, $\mathbb{F}$-predictable versions of processes $a$ and $\gamma$. In view of the considerations above, we are in a position to state the following corollary, which furnishes still another equivalent representation of a defaultable claim with fractional recovery of par value.

**Corollary 8.2.2.** *Assume that $A_t = \int_0^t a_u \, du$ and $\Gamma_t = \int_0^t \gamma_u \, du$ with $\gamma > 0$. Then a defaultable claim $(X, A, \tilde{X}, \delta L, \tau)$ is equivalent to a defaultable claim $(X, 0, \tilde{X}, \hat{Z}, \tau)$, where $\hat{Z}_t = \delta L + a_t \gamma_t^{-1}$.*

**Fractional recovery of no-default value.** In case of a general contingent claim, the counterpart of the fractional recovery of Treasury value scheme is referred to as the *fractional recovery of no-default value*. In this scheme, it is assumed that the owner of a defaultable claim receives at time of default a fixed fraction of a market value of an equivalent non-defaultable security. By definition, the *no-default value* (also known as the *Treasury value*) of a defaultable claim $(X, A, \tilde{X}, Z, \tau)$ is equal to the expected discounted value of the promised dividends $A$ and the promised contingent claim $X$, specifically:

$$U_t = B_t \, \mathbb{E}_{\mathbb{Q}^*} \left( \int_{[t,T]} B_u^{-1} \, dA_u + B_T^{-1} X \,\Big|\, \mathcal{G}_t \right). \tag{8.22}$$

Notice that $U$ includes also the dividends paid at time $t$. When valuing a defaultable claim $(X, A, \tilde{X}, Z, \tau)$ with fractional recovery of no-default value, we set $\tilde{X} = 0$ and $Z_t = \delta U_t$, where $U$ is given by the last formula.

Put more explicitly, the pre-default value equals

$$D_t^\delta = B_t \, \mathbb{E}_{\mathbb{Q}^*} \left( \int_{]t,T]} B_u^{-1}(1-H_u) \, dA_u + \int_{]t,T]} B_u^{-1}\delta U_u \, dH_u + B_T^{-1} X \, \mathbb{1}_{\{\tau>T\}} \, \Big| \, \mathcal{G}_t \right).$$

**Proposition 8.2.2.** *For any $t < T$ we have $D_t^\delta = (1 - \delta)\tilde{D}_t^0 + \mathbb{1}_{\{\tau>t\}}\delta\tilde{U}_t$, where the process $\tilde{D}_t^0$, which equals*

$$\tilde{D}_t^0 = \mathbb{1}_{\{\tau>t\}} G_t^{-1} B_t \, \mathbb{E}_{\mathbb{Q}^*} \left( \int_{]t,T]} B_u^{-1} G_u \, dA_u + G_T B_T^{-1} X \, \Big| \, \mathcal{F}_t \right),$$

*represents the pre-default value of a defaultable claim $(X, A, 0, 0, \tau)$ with zero recovery and $\tilde{U}_t$ is given by*

$$\tilde{U}_t = B_t \, \mathbb{E}_{\mathbb{Q}^*} \left( \int_{]t,T]} B_u^{-1} \, dA_u + B_T^{-1} X \, \Big| \, \mathcal{G}_t \right). \qquad (8.23)$$

*Proof.* We shall sketch the proof. Since manifestly

$$D_t^\delta = \tilde{D}_t^0 + \delta B_t \, \mathbb{E}_{\mathbb{Q}^*} \left( \int_{]t,T]} B_u^{-1} U_u \, dH_u \, \Big| \, \mathcal{G}_t \right),$$

it suffices to show that the following equality is valid:

$$\mathbb{1}_{\{\tau>t\}}\tilde{U}_t = B_t \, \mathbb{E}_{\mathbb{Q}^*} \left( \int_{]t,T]} B_u^{-1}(1 - H_u) \, dA_u + B_T^{-1} X \mathbb{1}_{\{\tau>T\}} \, \Big| \, \mathcal{G}_t \right) + J,$$

where we have set

$$J := B_t \, \mathbb{E}_{\mathbb{Q}^*} \left( \int_{]t,T]} B_u^{-1} U_u \, dH_u \, \Big| \, \mathcal{G}_t \right).$$

But

$$J = B_t \, \mathbb{E}_{\mathbb{Q}^*} \left( \int_{]t,T]} \mathbb{E}_{\mathbb{Q}^*} \left( \int_{[u,T]} B_v^{-1} \, dA_v + B_T^{-1} X \, \Big| \, \mathcal{G}_u \right) dH_u \, \Big| \, \mathcal{G}_t \right)$$

$$= B_t \, \mathbb{E}_{\mathbb{Q}^*} \left( \int_{]t,T]} \left( \int_{[u,T]} B_v^{-1} \, dA_v + B_T^{-1} X \right) dH_u \, \Big| \, \mathcal{G}_t \right)$$

$$= B_t \, \mathbb{E}_{\mathbb{Q}^*} \left( \mathbb{1}_{\{\tau>t\}} \int_{]t,T]} B_u^{-1} H_u \, dA_u + B_T^{-1} X \mathbb{1}_{\{t<\tau\leq T\}} \, \Big| \, \mathcal{G}_t \right),$$

where we have used, in particular, Fubini's theorem. $\qquad \square$

*Remarks.* (i) In formulae (8.22) and (8.23), the conditioning with respect to the $\sigma$-field $\mathcal{G}_t$ may be replaced by conditioning with respect to $\mathcal{F}_t$.
(ii) Bélanger et al. (2001) derive several interesting relationships between values of defaultable claims that are subject to alternative recovery schemes.
(iii) The so-called *endogenous recovery schemes* – the *fractional recovery of market value rule*, in particular – are examined in the next section.

## 8.3 Valuation via the Martingale Approach

We shall now present the martingale approach to the intensity-based valuation of defaultable claims. We make the following standing assumption.

**Assumption (D)** The default process $H_t = \mathbb{1}_{\{\tau \le t\}}$ admits the $\mathbb{F}$-martingale intensity process $\lambda$ under the spot martingale measure $\mathbb{Q}^*$.

Recall that the $\mathbb{F}$-*martingale intensity* $\lambda$ of $\tau$ is an $\mathbb{F}$-progressively measurable process such that the compensated process $\hat{M}$, given by the formula

$$\hat{M}_t := H_t - \int_0^{t \wedge \tau} \lambda_u \, du = H_t - \int_0^t \tilde{\lambda}_u \, du, \quad \forall t \in \mathbb{R}_+, \tag{8.24}$$

follows a $\mathbb{G}$-martingale under $\mathbb{Q}^*$ (cf. Definition 6.1.1). Notice that for the sake of brevity we have denoted $\tilde{\lambda}_t := \mathbb{1}_{\{\tau \ge t\}} \lambda_t$. Of course, in general the process $\tilde{\lambda}$ is not necessarily $\mathbb{F}$-adapted.

*Remarks.* The reference filtration $\mathbb{F}$ is usually strictly smaller than $\mathbb{G}$. The case of an $\mathbb{F}$-stopping time – that is, the case when $\mathbb{H} \subseteq \mathbb{F}$ – is not excluded a priori, though. The results below have thus the potential to also cover these situations when the $\mathbb{F}$-hazard process $\Gamma$ is not well defined (let us recall that the last feature depends on the choice of the reference filtration $\mathbb{F}$).

The next result provides an alternative representation for the price process of a defaultable claim $DCT = (X, A, \tilde{X}, Z, \tau)$. It appears that the integration with respect to the jump process $H$ can be substituted with the integration with respect to the associated intensity measure $\tilde{\lambda}_t \, dt$. As before, for the sake of brevity, we shall denote the pre-default value by $S_t^0$, and we restrict our attention to the case of $A \equiv 0$ and $\tilde{X} = 0$. As apparent from the previous section, the case of a more general defaultable claim can be handled by similar techniques.

**Proposition 8.3.1.** *The pre-default value process $S_t^0$ of a defaultable claim $DCT = (X, 0, 0, Z, \tau)$ admits the following representations:*

$$S_t^0 = B_t \, \mathbb{E}_{\mathbb{Q}^*} \left( \int_t^T B_u^{-1} Z_u \tilde{\lambda}_u \, du + B_T^{-1} X \mathbb{1}_{\{\tau > T\}} \, \Big| \, \mathcal{G}_t \right) \tag{8.25}$$

*and*

$$S_t^0 = \mathbb{E}_{\mathbb{Q}^*} \left( \int_t^T \left( Z_u \tilde{\lambda}_u - r_u S_u^0 \right) du + X \mathbb{1}_{\{\tau > T\}} \, \Big| \, \mathcal{G}_t \right). \tag{8.26}$$

*Proof.* The first representation follows from (8.2), combined with the equality

$$\mathbb{E}_{\mathbb{Q}^*} \left( \int_{]t,T]} B_u^{-1} Z_u \, dH_u \, \Big| \, \mathcal{G}_t \right) = \mathbb{E}_{\mathbb{Q}^*} \left( \int_{]t,T]} B_u^{-1} Z_u \left( d\hat{M}_u + \tilde{\lambda}_u \, du \right) \Big| \, \mathcal{G}_t \right),$$

which in turn is an immediate consequence of (8.24). To establish the second representation, it is enough to rewrite (8.25) as follows:

$$S_t^0 = B_t \left( \tilde{M}_t - \int_0^t B_u^{-1} Z_u \tilde{\lambda}_u \, du \right), \tag{8.27}$$

where

$$\tilde{M}_t := \mathbb{E}_{\mathbb{Q}^*} \left( \int_0^T B_u^{-1} Z_u \tilde{\lambda}_u \, du + B_T^{-1} X \mathbb{1}_{\{T < \tau\}} \,\Big|\, \mathcal{G}_t \right).$$

Applying Itô's formula to (8.27), we obtain the equality

$$dS_t^0 = (r_t S_t^0 - Z_t \tilde{\lambda}_t) \, dt + B_t \, d\tilde{M}_t,$$

which entails that

$$\mathbb{E}_{\mathbb{Q}^*}(S_T^0 \,|\, \mathcal{G}_t) = S_t^0 + \mathbb{E}_{\mathbb{Q}^*} \left( \int_t^T (r_u S_u^0 - Z_u \tilde{\lambda}_u) \, du \,\Big|\, \mathcal{G}_t \right).$$

Since clearly $S_T^0 = X \mathbb{1}_{\{T < \tau\}}$, the last equality yields (8.26).    □

The next result – due to Duffie et al. (1996) – plays a crucial role in the martingale approach to the valuation of defaultable securities (the proof of Proposition 8.3.2 is based on the same arguments as that of Proposition 6.4.1; we provide it here for the reader's convenience). As before, $\tilde{B}$ stands for the artificial 'savings account' corresponding to the default-risk-adjusted interest rate $\tilde{r}_t = r_t + \lambda_t$, i.e.,

$$\tilde{B}_t = \exp \left( \int_0^t (r_u + \lambda_u) \, du \right). \tag{8.28}$$

**Proposition 8.3.2.** *Assume that we are given an* $\mathbb{F}$*-predictable process* $Z$ *and an* $\mathcal{F}_T$*-measurable random variable* $X$*. We define the process* $V$ *by setting*

$$V_t = \tilde{B}_t \, \mathbb{E}_{\mathbb{Q}^*} \left( \int_t^T \tilde{B}_u^{-1} Z_u \lambda_u \, du + \tilde{B}_T^{-1} X \,\Big|\, \mathcal{G}_t \right). \tag{8.29}$$

*Then the process* $U_t := \mathbb{1}_{\{\tau > t\}} V_t$ *satisfies*

$$U_t = B_t \, \mathbb{E}_{\mathbb{Q}^*} \left( B_\tau^{-1} (Z_\tau + \Delta V_\tau) \mathbb{1}_{\{t < \tau \le T\}} + B_T^{-1} X \mathbb{1}_{\{T < \tau\}} \,\Big|\, \mathcal{G}_t \right) \tag{8.30}$$

*or, equivalently,*

$$U_t = B_t \, \mathbb{E}_{\mathbb{Q}^*} \left( \int_{]t,T]} B_u^{-1} (Z_u + \Delta V_u) \, dH_u + B_T^{-1} X \mathbb{1}_{\{T < \tau\}} \,\Big|\, \mathcal{G}_t \right). \tag{8.31}$$

*Proof.* We shall establish (8.31). In view of (8.29), we have

$$V_t = \tilde{B}_t \left( N_t - \int_0^t \tilde{B}_u^{-1} Z_u \lambda_u \, du \right), \tag{8.32}$$

where, under suitable integrability conditions, the process $N$, given by

$$N_t := \mathbb{E}_{\mathbb{Q}^*} \left( \int_0^T \tilde{B}_u^{-1} Z_u \lambda_u \, du + \tilde{B}_T^{-1} X \,\Big|\, \mathcal{G}_t \right), \tag{8.33}$$

follows a $\mathbb{G}$-martingale. Using Itô's product rule, we obtain

$$dV_t = r_t V_t \, dt - (Z_t - V_{t-}) \lambda_t \, dt + \tilde{B}_t \, dN_t. \tag{8.34}$$

Notice that $U_t = \tilde{H}_t V_t$, where $\tilde{H}_t = 1 - H_t = \mathbb{1}_{\{\tau > t\}}$. Since $\tilde{H}$ follows a process of finite variation, an application of Itô's product rule yields

$$dU_t = d(\tilde{H}_t V_t) = \tilde{H}_{t-}\, dV_t + V_{t-}\, d\tilde{H}_t + \Delta V_t \Delta \tilde{H}_t.$$

In view of (8.34) and equality $\tilde{\lambda}_t = \lambda_t \mathbb{1}_{\{t \leq \tau\}}$, we also have

$$dU_t = \tilde{H}_{t-}\big(r_t V_t\, dt - (Z_t - V_{t-})\tilde{\lambda}_t\, dt + \tilde{B}_t\, dN_t\big) + V_{t-}\, d\tilde{H}_t + \Delta V_t \Delta \tilde{H}_t.$$

Rearranging and noticing that $\Delta \tilde{H}_t = -\Delta H_t$, we obtain

$$dU_t = r_t U_t\, dt - (Z_t + \Delta V_t)\, dH_t + d\tilde{N}_t, \tag{8.35}$$

where $\tilde{N}$ satisfies

$$d\tilde{N}_t = \tilde{H}_{t-}\tilde{B}_t\, dN_t + (Z_t - V_{t-})\, d\hat{M}_t.$$

Since $U_T = X\mathbb{1}_{\{T < \tau\}}$, equation (8.35) leads to expression (8.31), provided that the $\mathbb{G}$-local martingale $\tilde{N}$ follows a 'true' martingale under $\mathbb{Q}^*$.    □

In view of the next corollary, it is natural to refer to the process $V$ given by (8.29) as the *pre-default value* of a defaultable claim $X$. To derive formula (8.36) below, it suffices to combine (8.3) with (8.30).

**Corollary 8.3.1.** *Let the processes $S^0$ and $V$ be defined by (8.2) and (8.29), respectively. Then*

$$S_t^0 = U_t - B_t\, \mathbb{E}_{\mathbb{Q}^*}\big(B_\tau^{-1}\mathbb{1}_{\{t < \tau \leq T\}}\Delta V_\tau \,\big|\, \mathcal{G}_t\big). \tag{8.36}$$

*If $\Delta V_\tau = 0$ or, more generally,*

$$\mathbb{E}_{\mathbb{Q}^*}\big(B_\tau^{-1}\mathbb{1}_{\{t < \tau \leq T\}}\Delta V_\tau \,\big|\, \mathcal{G}_t\big) = 0 \tag{8.37}$$

*then $S_t^0 = U_t = \mathbb{1}_{\{\tau > t\}}V_t$ for every $t \in [0, T]$. More explicitly,*

$$S_t^0 = \mathbb{1}_{\{\tau > t\}}\tilde{B}_t\, \mathbb{E}_{\mathbb{Q}^*}\Big(\int_t^T \tilde{B}_u^{-1}Z_u\lambda_u\, du + \tilde{B}_T^{-1}X \,\Big|\, \mathcal{G}_t\Big). \tag{8.38}$$

For the validity of Propositions 8.3.2 and 8.3.1, as well as Corollary 8.3.1, it is enough to assume that the recovery process $Z$ is $\mathbb{G}$-predictable, and $X$ is a $\mathcal{G}_T$-measurable random variable. The continuity condition (8.37) seems to be rather difficult to verify in a general set-up. It can be established, however, if certain additional restrictions are imposed on the underlying filtrations $\mathbb{F}$ and $\mathbb{G}$. In addition, we need to restrict our attention to the case of $\mathbb{F}$-predictable processes $B$ and $Z$, and an $\mathcal{F}_T$-measurable random variable $X$.

*Remarks.* We already know that the martingale condition (M.1), recalled in the next section, is satisfied when the default time $\tau$ is modeled either as the first jump time of an $\mathbb{F}$-conditional Poisson process (see Sect. 6.6) or through the canonical construction (see Sect. 8.2.1). But in these cases, the $\mathbb{F}$-hazard process $\Gamma$ of $\tau$ is well defined, and so expression (8.41) easily follows from the general formula (8.18), so that the martingale approach is not needed.

## 8.3.1 Martingale Hypotheses

Conditions that we are going to recall here were already employed several times in this text (see, for instance, Sect. 6.1 or Sect. 8.2.1). Let $\mathbb{F} \subseteq \mathbb{G}$ be two filtrations on a common probability space $(\Omega, \mathcal{G}, \mathbb{Q}^*)$. We say that $\mathbb{F}$ has the *martingale invariance property* with respect to $\mathbb{G}$ under $\mathbb{Q}^*$ if every $\mathbb{F}$-martingale is also a $\mathbb{G}$-martingale. This property of the pair $(\mathbb{F}, \mathbb{G})$ of filtrations was referred to as Condition (M.1) in Sect. 6.1. In the present framework, we have $\mathbb{G} = \mathbb{H} \vee \mathbb{F}$, where the filtration $\mathbb{H}$ is generated by the default process. We may thus also introduce the following conditions.

**Condition (M.2)** For any $t \in \mathbb{R}_+$, the $\sigma$-fields $\mathcal{F}_\infty$ and $\mathcal{G}_t$ are conditionally independent given $\mathcal{F}_t$ under $\mathbb{Q}^*$.

**Condition (F.1)** For any $t \in \mathbb{R}_+$, the $\sigma$-fields $\mathcal{F}_\infty$ and $\mathcal{H}_t$ are conditionally independent given $\mathcal{F}_t$ under $\mathbb{Q}^*$.

It is known that Condition (F.1) admits the following equivalent form (see Condition (F1.a) in Sect. 6.1):

$$\mathbb{Q}^*\{\tau \leq u \,|\, \mathcal{F}_t\} = \mathbb{Q}^*\{\tau \leq u \,|\, \mathcal{F}_\infty\}, \quad \forall \, 0 \leq u \leq t. \tag{8.39}$$

Moreover, by virtue of Lemmas 6.1.1 and 6.1.2, Conditions (M.1), (M.2) and (F.1) are equivalent when $\mathbb{G} = \mathbb{H} \vee \mathbb{F}$.

We claim that under (M.1), conditioning with respect to $\mathcal{G}_t$ in (8.29) may be substituted with conditioning with respect to $\mathcal{F}_t$. In other words, we claim that the pre-default value $V$, defined by (8.29), also satisfies

$$V_t = \tilde{B}_t \, \mathbb{E}_{\mathbb{Q}^*} \left( \int_t^T \tilde{B}_u^{-1} Z_u \lambda_u \, du + \tilde{B}_T^{-1} X \,\Big|\, \mathcal{F}_t \right). \tag{8.40}$$

To simplest way to establish the last equality, is to use a still another condition, which is also known to be equivalent to (M.2) (and thus to (M.1)).

**Condition (M.2b)** For any $t \in \mathbb{R}_+$, and any bounded (or $\mathbb{Q}^*$-integrable), $\mathcal{F}_\infty$-measurable random variable $\xi$ we have $\mathbb{E}_{\mathbb{Q}^*}(\xi \,|\, \mathcal{G}_t) = \mathbb{E}_{\mathbb{Q}^*}(\xi \,|\, \mathcal{F}_t)$.

Indeed, since both $B$ and $Z$ follow $\mathbb{F}$-adapted processes, and $X$ is an $\mathcal{F}_T$-measurable random variable, (M2.b) combined with (8.29) immediately yield (8.40). An alternative way to derive (8.40) would rely on a simple modification of the proof of Proposition 8.3.2.

If the process $V$ given by (8.40) satisfies $\Delta V_\tau = 0$, Proposition 8.3.2 now yields

$$S_t^0 = \mathbb{1}_{\{\tau > t\}} \tilde{B}_t \, \mathbb{E}_{\mathbb{Q}^*} \left( \int_t^T \tilde{B}_u^{-1} Z_u \lambda_u \, du + \tilde{B}_T^{-1} X \,\Big|\, \mathcal{F}_t \right). \tag{8.41}$$

In some cases – for instance, when the filtration $\mathbb{F}$ is generated by a Brownian motion under $\mathbb{Q}^*$ – the continuity of $V$ is trivial. In most cases, however, it seems more convenient to derive formula (8.41) using standard results on intensities of random times, rather than to rely on Proposition 8.3.2.

### 8.3.2 Endogenous Recovery Rules

Assume now that the recovery process $Z$ is not an exogenously specified process, but rather some non-anticipating functional of the value process $S^0$. Using the generic notation $Z_t = p(S^0_{t-}, t)$, we conclude that the pre-default value process $S^0$ is given as a solution to the *backward stochastic differential equation* of the form (as before, we set $A \equiv 0$ and $\tilde{X} = 0$)

$$S^0_t = B_t \, \mathbb{E}_{\mathbb{Q}^*} \left( \int_{]t,T]} B_u^{-1} p(S^0_{u-}, u) \, dH_u + B_T^{-1} X \, \Big| \, \mathcal{G}_t \right).$$

To simplify the exposition, we shall work under the martingale invariance hypothesis (M.1).

**Fractional recovery of market value.** Following Duffie and Singleton (1997, 1999), we first assume that the recovery process satisfies $Z_t = K_t S^0_{t-}$, where $K$ is a given $\mathbb{F}$-predictable process. Since $S^0$ represents the market value of a defaultable claim, and $K_t$ can be interpreted as the recovery rate, it is natural to refer to this scheme as the *fractional recovery of market value*. The following result furnishes a convenient representation for the pre-default value process $V$.

**Proposition 8.3.3.** *Let $V$ be a solution to (8.40) with $Z_t = K_t V_{t-}$ for some $\mathbb{F}$-predictable process $K$ or, more explicitly,*

$$V_t = \tilde{B}_t \, \mathbb{E}_{\mathbb{Q}^*} \left( \int_t^T \tilde{B}_u^{-1} K_u V_u \lambda_u \, du + \tilde{B}_T^{-1} X \, \Big| \, \mathcal{F}_t \right). \tag{8.42}$$

*Then $V_t$ is given by the formula*

$$V_t = \hat{B}_t \, \mathbb{E}_{\mathbb{Q}^*} \left( \hat{B}_T^{-1} X \, \big| \, \mathcal{F}_t \right), \tag{8.43}$$

*where $\hat{B}$ equals*

$$\hat{B}_t = \exp \left( \int_0^t (r_u + (1 - K_u)\lambda_u) \, du \right).$$

*Proof.* We proceed along the same lines as in the proof of Proposition 8.3.2, hence, we shall merely sketch the proof. In view of (8.34), with the process $N$ given as

$$N_t = \mathbb{E}_{\mathbb{Q}^*} \left( \int_0^T \tilde{B}_u^{-1} K_u V_u \lambda_u \, du + \tilde{B}_T^{-1} X \, \Big| \, \mathcal{F}_t \right),$$

we obtain

$$dV_t = V_t(r_t + \lambda_t) \, dt - K_t V_t \lambda_t \, dt + \tilde{B}_t \, dN_t$$

or, equivalently,

$$dV_t = V_t(r_t + (1 - K_t)\lambda_t) \, dt + \tilde{B}_t \, dN_t.$$

This immediately yields (8.43), provided that the $\mathbb{F}$-local martingale associated with last term follows in fact a martingale under $\mathbb{Q}^*$.    □

Proposition 8.3.3 shows that equation (8.42) admits a unique solution. In addition, it is apparent that the process $U_t := \mathbb{1}_{\{\tau > t\}} V_t$ satisfies (cf. (8.30))

$$U_t = B_t \, \mathbb{E}_{\mathbb{Q}^*}\left(B_\tau^{-1}(K_\tau V_{\tau-} + \Delta V_\tau)\mathbb{1}_{\{t < \tau \le T\}} + B_T^{-1} X \mathbb{1}_{\{T < \tau\}} \,\Big|\, \mathcal{G}_t\right).$$

Since process $Z_t = K_t V_{t-}$ is $\mathbb{F}$-predictable and clearly $V_{\tau-} = U_{\tau-}$, by combining Propositions 8.3.2 and 8.3.3 we obtain the following result.

**Corollary 8.3.2.** *Let $V$ be given by (8.42) for some $\mathbb{F}$-predictable process $K$. Assume that $\Delta V_\tau = 0$. Then the process $U_t = \mathbb{1}_{\{\tau > t\}} V_t$ satisfies*

$$U_t = B_t \, \mathbb{E}_{\mathbb{Q}^*}\left(B_\tau^{-1} K_\tau U_{\tau-}\mathbb{1}_{\{t < \tau \le T\}} + B_T^{-1} X \mathbb{1}_{\{T < \tau\}} \,\Big|\, \mathcal{G}_t\right). \tag{8.44}$$

*In this case, we have*

$$S_t^0 = \mathbb{1}_{\{\tau > t\}} \hat{B}_t \, \mathbb{E}_{\mathbb{Q}^*}\left(\hat{B}_T^{-1} X \,\big|\, \mathcal{F}_t\right). \tag{8.45}$$

In view of Corollary 8.3.2, the process $U$ satisfies equation (8.44) – i.e., the implicit definition of the price process $S^0$ of a defaultable claim. Of course, we have not established the uniqueness of solutions to the backward SDE (BSDE, for short)

$$S_t^0 = B_t \, \mathbb{E}_{\mathbb{Q}^*}\left(B_\tau^{-1} K_\tau S_{\tau-}^0 \mathbb{1}_{\{t < \tau \le T\}} + B_T^{-1} X \mathbb{1}_{\{T < \tau\}} \,\Big|\, \mathcal{G}_t\right),$$

but we have taken the uniqueness for granted and we have shown that this equation admits a solution. However, the uniqueness of solutions to our equation can be deduced from standard results on BSDEs. To this end, one might use the following equivalent representation for the last equation (cf. (8.26)):

$$S_t^0 = \mathbb{E}_{\mathbb{Q}^*}\left(\int_t^T S_u^0(K_u \tilde{\lambda}_u - r_u)\, du + X \mathbb{1}_{\{T < \tau\}} \,\Big|\, \mathcal{G}_t\right).$$

For the introduction to the theory of BSDEs, and various applications of BSDEs in stochastic optimal control and mathematical finance, we refer to Pardoux and Peng (1990), Duffie and Epstein (1992), Cvitanić and Karatzas (1993), Peng (1993), El Karoui and Quenez (1997a, 1997b), El Karoui et al. (1997), as well as to the monograph by Ma and Yong (1999).

*Example 8.3.1.* Let us consider the special case of a constant process $K_t = \delta$, where $\delta$ is a real number from the interval $[0, 1]$. In view of the definition of the price process $S^0$, we have (cf. (8.4))

$$S_t^0 = \mathbb{1}_{\{\tau > t\}} B_t \, \mathbb{E}_{\mathbb{Q}^*}\left(B_\tau^{-1} \delta S_{\tau-}^0 \mathbb{1}_{\{t < \tau \le T\}} + B_T^{-1} X \mathbb{1}_{\{T < \tau\}} \,\Big|\, \mathcal{G}_t\right).$$

By virtue of Proposition 8.3.3, the pre-default value process $V$, which equals

$$V_t = \tilde{B}_t \, \mathbb{E}_{\mathbb{Q}^*}\left(\int_t^T \tilde{B}_u^{-1} \delta V_u \lambda_u \, du + \tilde{B}_T^{-1} X \,\Big|\, \mathcal{F}_t\right),$$

admits also the following simple representation:

$$V_t = \mathbb{E}_{\mathbb{Q}^*}\left(Xe^{-\int_t^T (r_u+(1-\delta)\lambda_u)du}\,\Big|\,\mathcal{F}_t\right).$$

Consequently, under the assumption that $\Delta V_\tau = 0$, we obtain

$$S_t^0 = \mathbb{1}_{\{\tau>t\}}\,\mathbb{E}_{\mathbb{Q}^*}\left(Xe^{-\int_t^T (r_u+(1-\delta)\lambda_u)du}\,\Big|\,\mathcal{F}_t\right).$$

In the special case of zero recovery (i.e., when $\delta = 0$) the last formula can be seen as a particular case of (8.41) with $Z \equiv 0$ (or as a counterpart of (8.18) with $A \equiv 0$ and $Z \equiv 0$). For the case of full recovery (i.e., when $\delta = 1$) we get

$$S_t^0 = \mathbb{1}_{\{\tau>t\}}\,\mathbb{E}_{\mathbb{Q}^*}\left(Xe^{-\int_t^T r_u\,du}\,\Big|\,\mathcal{F}_t\right),$$

which is, of course, the price of a non-defaultable claim $X$ (the presence of the indicator function $\mathbb{1}_{\{\tau>t\}}$ is merely the result of the convention adopted in the definition of the pre-default value $S^0$).

**General recovery rule.** Since in practice the recovery payoff is not necessarily a linear function of the market value of a contract, it is natural to extend the approach presented in the preceding paragraph by postulating that the recovery process satisfies $Z_t = p(S_{t-}^0, t)$, where, for instance, the function $p : \mathbb{R} \times \mathbb{R}_+ \to \mathbb{R}$ is Lipschitz continuous with respect to the first variable, and satisfies $p(0, t) = 0$. In such a case, one cannot expect to derive a quasi-explicit representation similar to (8.45). In fact, we merely have the following result, which again is a consequence of Proposition 8.3.2. Let us stress once more that the existence and uniqueness of solutions to (8.47) and (8.49) is taken for granted. Corollary 8.3.3 establishes the equivalence between various representations of the unique solution $S^0$ to equation (8.46).

**Corollary 8.3.3.** *Let $S^0$ be the unique solution to the backward SDE*

$$S_t^0 = B_t\,\mathbb{E}_{\mathbb{Q}^*}\left(B_\tau^{-1}p(S_{\tau-}^0, \tau)\mathbb{1}_{\{t<\tau\leq T\}} + B_T^{-1}X\mathbb{1}_{\{T<\tau\}}\,\Big|\,\mathcal{G}_t\right) \qquad (8.46)$$

*or, equivalently, to the equation (cf. (8.26))*

$$S_t^0 = \mathbb{E}_{\mathbb{Q}^*}\left(\int_t^T (p(S_u^0, u)\tilde{\lambda}_u - r_u S_u^0)\,du + X\mathbb{1}_{\{T<\tau\}}\,\Big|\,\mathcal{G}_t\right). \qquad (8.47)$$

*Let $V$ be the unique solution to the backward SDE*

$$V_t = \tilde{B}_t\,\mathbb{E}_{\mathbb{Q}^*}\left(\int_t^T \tilde{B}_u^{-1}p(V_u, u)\lambda_u\,du + \tilde{B}_T^{-1}X\,\Big|\,\mathcal{F}_t\right) \qquad (8.48)$$

*or, equivalently, to the equation*

$$V_t = \mathbb{E}_{\mathbb{Q}^*}\left(\int_t^T (p(V_u, u)\lambda_u - (r_u + \lambda_u)V_u)\,du + X\,\Big|\,\mathcal{F}_t\right). \qquad (8.49)$$

*If $\Delta V_\tau = 0$, then $S_t^0 = U_t := \mathbb{1}_{\{\tau>t\}}V_t$. Otherwise, $S^0$ is given by (8.36).*

## 8.4 Hedging of Defaultable Claims

Our exposition is largely based on Blanchet-Scalliet and Jeanblanc (2001); related results can also be found in Wong (1998) and Bélanger et al. (2001). The line of arguments can be summarized as follows. First, we assume that we are given an arbitrage-free model of default-free securities. Next, we introduce a default time, which is typically defined on an enlarged probability space. We postulate that zero-coupon corporate bonds are priced through the risk-neutral valuation formula under the extended spot martingale measure $\mathbb{Q}^*$. In this way, we obtain the price process of a defaultable bond with a given maturity, or, more generally, price processes for a family of defaultable bonds with differing maturities.

For an arbitrary defaultable claim from a certain class, we show the existence of a replicating strategy based on continuous trading in default-free securities and defaultable bonds. Finally, we conclude that any defaultable claim can be valued through the risk-neutral valuation formula.

We make the standard assumption that the market for default-free and defaultable securities is frictionless, with continuous trading over a finite time interval $[0, T^*]$, where $T^* > 0$ is a finite horizon date. Typically, the primary securities are default-free zero-coupon bonds with differing maturities; shares and the savings account can also be included in the model. In fact, we do not need to specify the primary assets or other features of the default-free market. For further purposes, it suffices to postulate the existence of a replicating strategy for any contingent claim in the default-free market. In other words, we assume that this market is complete. We shall also make the following standing assumptions regarding the market of default-free securities and the specification of the default time $\tau$:

- the uncertainty in the default-free securities market is modeled through a reference filtration $\mathbb{F}$, on the underlying probability space $(\tilde{\Omega}, \mathcal{F}, \mathbb{P})$,
- the default-free securities market is arbitrage-free, specifically, there exists a unique spot martingale measure $\mathbb{P}^*$, equivalent to $\mathbb{P}$ on $(\tilde{\Omega}, \mathcal{F}_{T^*})$,
- the default time $\tau$ is a random time on the enlarged probability space $(\Omega, \mathcal{G}, \mathbb{Q}^*)$, for some (fixed) probability measure $\mathbb{Q}^*$,
- the enlarged filtration $\mathbb{G}$ satisfies $\mathbb{G} = \mathbb{F} \vee \mathbb{H}$,
- the restriction of the probability measure $\mathbb{Q}^*$ to the $\sigma$-field $\mathcal{F}_{T^*}$ coincides with $\mathbb{P}^*$.

Let us mention that the completeness of the reference default-free market, and thus also the uniqueness of the spot martingale measure $\mathbb{P}^*$, are not essential; these assumptions are made to simplify the exposition. For the sake of simplicity, we also impose the following technical conditions:

- any $\mathbb{F}$-martingale under $\mathbb{Q}^*$ also follows a $\mathbb{G}$-martingale under $\mathbb{Q}^*$, that is, the martingale invariance property (M.1) holds,
- the reference filtration is such that any $\mathbb{F}$-martingale is continuous,
- the $\mathbb{F}$-hazard process $\Gamma$ of $\tau$ is continuous.

*Remarks.* Blanchet-Scalliet and Jeanblanc (2001) argue that the martingale invariance property (M.1) (cf. Sect. 8.3.1) is in fact a natural requirement in some approaches to default risk modeling. Let us briefly report their arguments. Assume that the default-free model is arbitrage-free and complete, so that any $\mathcal{F}_{T^*}$-measurable, $\mathbb{P}^*$-integrable random variable $X$ admits a self-financing replicating strategy. Then the $\mathbb{F}$-martingale $\mathbb{E}_{\mathbb{P}^*}(XB_{T^*}^{-1} \,|\, \mathcal{F}_t)$ represents the discounted price process for $X$. Suppose that any $\mathcal{F}_{T^*}$-measurable contingent claim $X$ is also tradeable in the extended arbitrage-free market, equipped with the enlarged filtration $\mathbb{G}$ and an extended spot martingale measure $\mathbb{Q}^*$. It then clearly represents an attainable contingent claim in the extended arbitrage-free market, and it admits the same price and the same replicating strategy in both markets. We conclude that the equality $\mathbb{E}_{\mathbb{Q}^*}(XB_{T^*}^{-1} \,|\, \mathcal{G}_t) = \mathbb{E}_{\mathbb{P}^*}(XB_{T^*}^{-1} \,|\, \mathcal{F}_t)$ necessarily holds, so that any $\mathbb{F}$-martingale under $\mathbb{Q}^*$ is also a $\mathbb{G}$-martingale under $\mathbb{Q}^*$. The last equality also shows that the restriction of $\mathbb{Q}^*$ to $\mathcal{F}_{T^*}$ coincides with $\mathbb{P}^*$.

**Case of zero recovery.** Our goal is to examine replication of defaultable claims with zero recovery. In this case, it is natural to assume that defaultable securities, which will be used to hedge defaultable claims are also subject to the zero recovery scheme. Recall that we assume that $\mathbb{Q}^*$ is the spot martingale measure for the enlarged market model, in which not only default-free securities, but also defaultable securities, are traded. In particular, we postulate that the price process of any defaultable bond of any maturity $T \leq T^*$ satisfies

$$D^0(t,T) = B_t \, \mathbb{E}_{\mathbb{Q}^*}(B_T^{-1} \mathbb{1}_{\{\tau > T\}} \,|\, \mathcal{G}_t) = B(t,T) \, \mathbb{Q}_T\{\tau > T \,|\, \mathcal{G}_t\}. \qquad (8.50)$$

Before proceeding further, we shall first recall some basic facts about the martingale associated with a random time. As usual, we set $H_t = \mathbb{1}_{\{\tau \leq t\}}$. By virtue of Lemma 5.1.7, the process $L_t = \mathbb{1}_{\{\tau > t\}} e^{\Gamma_t} = (1 - H_t) e^{\Gamma_t}$ follows a $\mathbb{G}$-martingale under $\mathbb{Q}^*$. From Proposition 5.1.3, we know that

$$L_t = 1 - \int_{]0,t]} L_{u-} \, d\hat{M}_u, \qquad (8.51)$$

so that $dL_t = -L_{t-} \, d\hat{M}_t$. Finally, as shown in Proposition 5.1.3, the compensated process $\hat{M} = H_t - \Gamma_{t \wedge \tau}$ also follows a $\mathbb{G}$-martingale.[2] Our first goal is to derive the dynamics for the price process of a defaultable bond. To achieve this, we introduce a strictly positive $\mathbb{F}$-martingale $m$ by setting

$$m_t = \mathbb{E}_{\mathbb{Q}^*}(B_T^{-1} e^{-\Gamma_T} \,|\, \mathcal{F}_t) = \mathbb{E}_{\mathbb{P}^*}(B_T^{-1} e^{-\Gamma_T} \,|\, \mathcal{F}_t).$$

Observe that $m$ follows an $\mathbb{F}$-martingale not only under $\mathbb{Q}^*$, but also under $\mathbb{P}^*$ (since the restriction of $\mathbb{Q}^*$ to the $\sigma$-field $\mathcal{F}_{T^*}$ coincides with $\mathbb{P}^*$). The next result is borrowed from Blanchet-Scalliet and Jeanblanc (2001).

---

[2] The continuity of $\Gamma$ is essential here. Blanchet-Scalliet and Jeanblanc (2001) also examine the case of a discontinuous hazard process $\Gamma$.

**Proposition 8.4.1.** *Let the price process $D^0(t, T)$ be given by (8.50). Then*

$$dD^0(t, T) = D^0(t-, T)(r_t \, dt - d\hat{M}_t) + B_t L_{t-} \, dm_t. \qquad (8.52)$$

*Proof.* Let us denote $Z^0(t, T) = D^0(t, T)B_t^{-1}$. The process $Z^0(t, T)$ follows a $\mathbb{G}$-martingale under $\mathbb{Q}^*$ and

$$Z^0(t, T) = \mathbb{1}_{\{\tau > t\}} e^{\Gamma_t} \, \mathbb{E}_{\mathbb{Q}^*} (B_T^{-1} e^{-\Gamma_T} \mid \mathcal{F}_t) = L_t m_t.$$

Since the $\mathbb{F}$-martingale $m$ is continuous, and $L$ follows a process of finite variation, an application of Itô's product rule yields

$$dZ^0(t, T) = L_{t-} \, dm_t + m_t \, dL_t = L_{t-} \, dm_t - Z^0(t-, T) \, d\hat{M}_t. \qquad (8.53)$$

By applying Itô's formula to the product $B_t Z^0(t, T)$, we obtain (8.52).    □

From now on, we assume that the defaultable bond, with the price process $D^0(t, T)$, is included in the class of traded securities. Our next goal is to show that any contingent claim with zero promised dividends, zero recovery and maturity date $T$ can be replicated through continuous trading in default-free securities and the defaultable bond. Thus, its arbitrage price is given by the risk-neutral valuation formula (8.1). For this purpose, we shall proceed as follows. First, we assume that the price process of a defaultable claim is actually given by (8.1). Subsequently, we shall provide a suitable martingale representation theorem, which in turn will yield the replicating strategy.

Formally, we consider a defaultable claim of a simple form $(X, 0, 0, 0, \tau)$, which settles at time $T$. As usual, the promised payoff $X$ is assumed to be an $\mathcal{F}_T$-measurable random variable, integrable with respect to $\mathbb{P}^*$. We postulate that its price is given by the risk-neutral valuation formula (8.3). Let $\tilde{S}_t^0 = S_t^0/B_t$ be the discounted pre-default value of this claim, that is,

$$\tilde{S}_t^0 = \mathbb{E}_{\mathbb{Q}^*} (B_T^{-1} X \mathbb{1}_{\{\tau > T\}} \mid \mathcal{G}_t).$$

Since $\tilde{S}_t^0 = 0$ on the set $\{\tau \geq t\}$, and we assume here zero-recovery rule, it is clear that $\tilde{S}_t^0$ in fact represents the discounted value of a defaultable claim not only strictly prior to default, but indeed for any $t \in [0, T]$.

Using, for instance, Proposition 8.2.1 with $A \equiv 0$ and $Z \equiv 0$, we obtain

$$\tilde{S}_t^0 = \mathbb{1}_{\{\tau > t\}} G_t^{-1} \, \mathbb{E}_{\mathbb{Q}^*} (G_T B_T^{-1} X \mid \mathcal{F}_t), \quad \forall t \in [0, T].$$

Consequently, we have

$$\tilde{S}_t^0 = \mathbb{1}_{\{\tau > t\}} e^{\Gamma_t} \, \mathbb{E}_{\mathbb{Q}^*} (B_T^{-1} e^{-\Gamma_T} X \mid \mathcal{F}_t) = L_t m_t^X,$$

where $m^X$ is the $\mathbb{F}$-martingale under $\mathbb{Q}^*$ (and so also under $\mathbb{P}^*$), defined as

$$m_t^X = \mathbb{E}_{\mathbb{Q}^*} (B_T^{-1} e^{-\Gamma_T} X \mid \mathcal{F}_t) = \mathbb{E}_{\mathbb{P}^*} (B_T^{-1} e^{-\Gamma_T} X \mid \mathcal{F}_t). \qquad (8.54)$$

Let us first state the martingale representation theorem, which is an immediate consequence of Proposition 5.2.1 applied to the $\mathbb{F}$-predictable process $Z_t = X \mathbb{1}_{\{t > T\}}$ (a direct proof is also provided below).

**Lemma 8.4.1.** *The $\mathbb{G}$-martingale $\tilde{S}^0$ admits the integral representation*

$$\tilde{S}_t^0 = \tilde{S}_0^0 + \int_0^{t \wedge \tau} e^{\Gamma_u} dm_u^X - \int_{]0,t \wedge \tau]} e^{\Gamma_u} m_u^X \, d\hat{M}_u.$$

*Proof.* We know that $\tilde{S}_t^0 = L_t m_t^X$, where $L$ and $m^X$ are given by (8.51) and (8.54), respectively. In view of the continuity of the $\mathbb{F}$-martingale $m^X$, the Itô product rule gives

$$\tilde{S}_t^0 = \tilde{S}_0^0 + \int_0^t L_{u-} dm_u^X - \int_{]0,t]} e^{\Gamma_u} m_u^X \, d\hat{M}_u.$$

Since $\hat{M}$ is stopped at $\tau$ and $L_t = \mathbb{1}_{\{\tau > t\}} e^{\Gamma_t}$, this yields the formula. $\qquad \square$

Recall that we have assumed that the default-free market is complete. Therefore the two $\mathcal{F}_T$-measurable contingent claims $Y_1 = e^{-\Gamma_T}$ and $Y_2 = Xe^{-\Gamma_T}$ admit replicating strategies in this market. Without loss of generality, we may thus consider their price processes as primary securities in what follows. Specifically, we shall use these processes in our construction of a replicating strategy for a defaultable claim.

**Proposition 8.4.2.** *Let us denote $\zeta_t^X = m_t^X m_t^{-1}$. On the set $\{t \le \tau\}$, the replicating strategy for the discounted price process $\tilde{S}^0$ equals*

$$\phi_t^0 = \zeta_t^X, \quad \phi_t^1 = e^{\Gamma_t} \zeta_t^X, \quad \phi_t^2 = e^{\Gamma_t},$$

*where the hedging instruments are: the discounted price process $Z^0(t,T)$ of the $T$-maturity zero-coupon bond with zero recovery and the discounted price processes of default-free claims $Y_1 = e^{-\Gamma_T}$ and $Y_2 = Xe^{-\Gamma_T}$. On the set $\{t > \tau\}$, the replicating strategy is identically equal to zero.*

*Proof.* Recall that discounted price processes of default-free claims $Y_1 = e^{-\Gamma_T}$ and $Y_2 = Xe^{-\Gamma_T}$ are denoted by $m$ and $m^X$, respectively. From (8.53), we obtain

$$dZ^0(t,T) - L_{t-} dm_t = -L_{t-} m_t \, d\hat{M}_t = -e^{\Gamma_t} m_t \, d\hat{M}_t.$$

Combining the last equality with Lemma 8.4.1, we obtain

$$\tilde{S}_t^0 = \tilde{S}_0^0 + \int_0^{t \wedge \tau} e^{\Gamma_u} dm_u^X - \int_{]0,t \wedge \tau]} e^{\Gamma_u} m_u^X \, d\hat{M}_u$$

$$= = \tilde{S}_0^0 + \int_{]0,t \wedge \tau]} \zeta_u^X dZ^0(u,T) - \int_0^{t \wedge \tau} e^{\Gamma_u} \zeta_u^X \, dm_u + \int_0^{t \wedge \tau} e^{\Gamma_u} dm_u^X,$$

where we denote $\zeta_u^X = m_u^X m_u^{-1}$. This completes the proof. $\qquad \square$

Of course, the (self-financing) replicating strategy for a defaultable claim $(X, 0, 0, 0, \tau)$ also involves a component representing an investment in the savings account. We conclude that, for every $t \in [0, T]$ the cost of replication of this claim equals,

$$S_t^0 = B_t \, \mathbb{E}_{\mathbb{Q}^*}(B_T^{-1} X \mathbb{1}_{\{\tau > T\}} \mid \mathcal{G}_t).$$

## 8.5 General Reduced-Form Approach

Wong (1998) and Bélanger et al. (2001) propose to unify the structural and the reduced-form approaches by constructing a fairly general framework for the credit risk modeling, which encompasses most classic models of the default time. We start with a complete probability space $(\Omega, \mathcal{F}, \mathbb{Q}^*)$, with a $d$-dimensional standard Brownian motion $W_t^*$, $t \in [0, T^*]$, for some fixed horizon date $T^*$. The filtration $\mathbb{F}$ is the usual $\mathbb{Q}^*$-augmented filtration generated by $W^*$, and the short-term interest rate $r$ is assumed to follow an $\mathbb{F}$-progressively measurable process, with integrable sample paths.

The main step is the construction of the default time $\tau$; it generalizes the canonical construction described in Sect. 6.5 and 8.2.1. First, we introduce a non-decreasing, $\mathbb{F}$-predictable process $\Psi$ whose sample paths are right-continuous with left-hand limits, with $\Psi_0 = 0$. In addition, we assume that the underlying probability space $(\Omega, \mathcal{F}, \mathbb{Q}^*)$ supports a strictly positive random variable $\eta$, independent of the $\sigma$-field $\mathcal{F}_{T^*}$, with the (right-continuous) cumulative distribution function $\hat{F}$: $\hat{F}(x) = \mathbb{Q}^*\{\eta \leq x\}$ for $x \in \mathbb{R}$.

As expected, the random time $\tau$ is defined on $(\Omega, \mathcal{F}, \mathbb{Q}^*)$ by setting (cf. Sect. 6.5 and 8.2.1)

$$\tau = \inf \{ t \in [0, T^*] : \Psi_t \geq \eta \}. \tag{8.55}$$

with the usual convention that $\inf \emptyset = \infty$. We are now going to show that this construction indeed covers most models encountered in financial literature.

*Example 8.5.1. Structural approach.* Suppose that an $\mathbb{F}$-predictable process $V$ models the evolution of the firm's value, and default occurs if $V$ falls to some deterministic boundary $\bar{v} : [0, T^*] \to \mathbb{R}$ with $\bar{v}(0) < V_0$, specifically,

$$\tau = \inf \{ t \in [0, T^*] : V_t \leq \bar{v}(t) \}.$$

To obtain this random time in the present framework, it is enough to take $\eta = V_0 - \bar{v}(0)$ and

$$\Psi_t = \sup_{0 \leq s \leq t} (V_0 - V_s + \bar{v}(s) - \bar{v}(0)).$$

*Example 8.5.2. Intensity-based approach.* Let us take $\Psi_t = \int_0^t \gamma_u \, du$ for some $\mathbb{F}$-progressively measurable process $\gamma$ with integrable sample paths. In addition, we assume that an auxiliary random variable $\eta$ has a unit exponential law under $\mathbb{Q}^*$. Then Wong's approach reduces to the canonical construction of a default time $\tau$ with $\mathbb{F}$-intensity $\gamma$ that was presented in Sect. 8.2.1.

*Example 8.5.3. Ratings-based approach.* Bélanger et al. (2001) point out that a simple model for credit migrations can be obtained within their setting by proceeding as follows: let $\mathcal{K} = \{1, \ldots, K\}$ be the set of possible ratings, with $K$ representing default, and let $\tilde{C}_n, n \in \mathbb{N}^*$ be a Markov chain, independent of $\mathcal{F}_{T^*}$, with state space $\mathcal{K}$ and some transition matrix $\tilde{P}$.

To construct a migration process $C$, we postulate the existence of a sequence $\eta_n$, $n \in \mathbb{N}$, of positive random variables independent of $\mathcal{F}_{T^*}$ and $\bar{C}$, and independent of one another. We assume that the initial rating is $C_0 = \bar{C}_0$, and we set $\tau_0 = 0$. For any $n \in \mathbb{N}$, the jump time $\tau_n$ is defined recursively by the formula

$$\tau_n = \inf \left\{ t \in [0, T^*] : t \geq \tau_{n-1} \text{ and } \Psi_t - \Psi_{\tau_{n-1}} \geq \eta_n \right\}.$$

Finally, we set $C_t = \bar{C}_{n-1}$ for $\tau_{n-1} \leq t < \tau_n$ for every $n \in \mathbb{N}$. Using the terminology of Sect. 11.2.1, we may say that the discrete-time Markov chain $\bar{C}$ is embedded in the continuous-time migration process $C$.

*Remarks.* In practice, the intensities of credit migrations are also state-dependent – that is, they depend on the current rating of a firm. Therefore, it seems more natural to introduce a family of processes $\Psi^i$, $i \in \mathcal{K}$, rather than a single process $\Psi$, and to define the sequence of jump times specifying the timing of credit migrations as follows:

$$\tau_n = \inf \left\{ t \in [0, T^*] : t \geq \tau_{n-1} \text{ and } \Psi^i_t - \Psi^i_{\tau_{n-1}} \geq \eta_n \right\}$$

on the set $\{C_{\tau_{n-1}} = i\}$. The last construction is similar to the construction of a continuous-time $\mathbb{F}$-conditionally Markov chain presented in Sect. 11.3. However, since the discrete-time Markov chain $\bar{C}$ is exogenously specified in the present example, the migration process $C$ does not necessarily follow an $\mathbb{F}$-conditionally Markov chain.

*Example 8.5.4. Fixed payment dates.* It is also possible to postulate that the default may occur only on some pre-determined dates (for instance, the coupon payment dates of a corporate bond). To this end, it suffices to take a piecewise constant function $\Psi$ with jumps at these dates.

Let us now examine the $\mathbb{F}$-hazard process of the random time $\tau$ given by expression (8.55). Since $\{\tau \leq t\} = \{\eta \leq \Psi_t\}$ and $\eta$ is independent of the $\sigma$-field $\mathcal{F}_{T^*}$, we obtain (cf. Sect. 5.1)

$$F_t := \mathbb{Q}^* \{\tau \leq t \mid \mathcal{F}_t\} = \mathbb{Q}^* \{\tau \leq t \mid \mathcal{F}_{T^*}\} = \hat{F}(\Psi_t),$$

and so the $\mathbb{F}$-survival process equals $G_t = 1 - \hat{F}(\Psi_t)$ for every $t \in [0, T^*]$. Also, let us set

$$\tau^* = \inf \left\{ t \in [0, T^*] : \hat{F}(\Psi_t) = 1 \right\} = \inf \left\{ t \in [0, T^*] : G_t = 0 \right\}.$$

The $\mathbb{F}$-hazard process $\Gamma$ of $\tau$ is defined on the random interval $[0, \tau^*[$ through the usual formula: $\Gamma_t = -\ln G_t$ (see Definition 5.1.1). Since $\Psi_0 = 0$ we have $\Gamma_0 = 0$; moreover, $\Gamma > 0$ for $0 \leq t < \tau^*$. Let us set, for every $t \in [0, T^*]$,

$$\Lambda_t = \int_{]0,t]} \mathbb{1}_{\{F_{u-} < 1\}} \frac{dF_u}{1 - F_{u-}} = -\int_{]0,t]} \mathbb{1}_{\{G_{u-} > 0\}} \frac{dG_u}{G_{u-}}, \qquad (8.56)$$

where by convention $F_{0-} = 1 - G_{0-} = 0$.

Then $dG_t = -G_{t-}\, d\Lambda_t$ or, equivalently,

$$G_t = \exp(-\Lambda_t^c) \prod_{0<u\leq t} (1 - \Delta\Lambda_u), \quad \forall t \in [0, T^*], \tag{8.57}$$

where $\Delta\Lambda_u = \Lambda_u - \Lambda_{u-}$ and the continuous process $\Lambda^c$, given by the equality $\Lambda_t^c = \Lambda_t - \sum_{0<u\leq t} \Delta\Lambda_u$, $\forall t \in [0, T^*]$, is the path-by-path continuous part of $\Lambda$.

*Remarks.* (i) Recall that we have derived a slightly less general representation for the $\mathbb{F}$-martingale hazard process $\Lambda$ in Sect. 6.1.2, where we have assumed that the inequality $G_t > 0$ is valid for every $t \in \mathbb{R}_+$. Let us emphasize that conditions (G.1) and (F.2) of Chap. 5 are easily seen to be satisfied in the present construction of the random time $\tau$.

(ii) In the special case when the auxiliary random variable $\eta$ is exponentially distributed under $\mathbb{Q}^*$, we have $\hat{F}(x) = 1 - e^{-x}$, and thus $G_t > 0$ for every $t$. In this case, we have $\tau^* = \infty$, so that the indicator functions $\mathbb{1}_{\{F_{u-}<1\}}$ and $\mathbb{1}_{\{G_{u-}>0\}}$ in (8.56) may be omitted.

(iii) Let us now assume that $\tau^* \leq T^*$ with positive probability. If the process $G$ jumps to 0 at $\tau^*$, then $\Lambda_t = \Lambda_{\tau^*-} + 1$ for $\tau^* \leq t \leq T^*$. On the other hand, if $G_{\tau^*-} = 0$ then, by virtue of (8.57), we have $\Lambda_t = \Lambda_{\tau^*-} = \infty$ for $\tau^* \leq t \leq T^*$.

Wong (1998) and Bélanger et al. (2001) provide a detailed study of defaultable claims that are subject to alternative recovery rules, introduced in Sect. 8.2.5. In particular, they derive several interesting relationships between prices of defaultable contingent claim under different conventions.

Another topic examined in these papers is the hedging of credit derivatives. For this purpose, they establish a suitable version of the martingale representation theorem and they derive the dynamics of the value process of a defaultable bond (see Theorems 5.1 and 6.1 in Bélanger et al. (2001)); their results may be seen as generalizations of Propositions 8.4.1–8.4.2. Let us also mention in this regard that Greenfield (2000) construct hedges for some credit derivatives in a reduced-form model, under an additional assumption that the market and credit risks are independent under the risk-neutral probability measure $\mathbb{Q}^*$. Finally, Wong (1998) and Bélanger et al. (2001) examine the compatibility of their abstract approach and the HJM-type methodology for the modeling of default-free and defaultable term structures. We postpone the study of this issue to Chap. 13.

It is not difficult to check that most results in Wong (1998) and Bélanger et al. (2001) become in fact trivial, when specified to the case of structural approach, as described in Example 8.5.1. For this reason, in our opinion, it would be more appropriate to term this approach a 'general reduced-form methodology,' rather than a 'unified model of credit risk.' [3]

---

[3] *Added in proof:* The paper Bélanger et al. (2001) will appear in *Mathematical Finance* under a new title "A General Framework for Pricing Credit Risk."

# 8.6 Reduced-Form Models with State Variables

In this section, we survey results from Lando (1998), Duffie and Singleton (1994, 1997, 1999), Madan and Unal (1998, 2000), and Davydov et al. (2000). We place ourselves within the general framework of Sect. 8.3, and we take as given an arbitrage-free setting in which all securities are priced in terms of some short-term rate process $r$ and equivalent spot martingale measure $\mathbb{Q}^*$. In order to make the model of Sect. 8.3 analytically more tractable, one needs to impose additional conditions on the default time $\tau$ – more specifically, on the hazard-rate process $\lambda$ associated with $\tau$. The *state-variables process* $Y$ with Markovian features is used precisely for this purpose.

### 8.6.1 Lando's Approach

We assume that we are given a $k$-dimensional stochastic process $Y$ defined on the underlying filtered probability space $(\Omega, \mathbb{F}, \mathbb{Q}^*)$. We postulate that $Y$ is $\mathbb{F}$-adapted and follows an $\mathbb{F}$-Markov process under the spot martingale measure $\mathbb{Q}^*$. The process $Y$ is assumed to model the dynamics of 'state variables' that underpin the evolution of all other variables in our model of the economy. We postulate that the default time $\tau$ is the first jump time of a *Cox process* with the intensity of the form $\lambda_t = \lambda(Y_t)$, for some function $\lambda : \mathbb{R}^k \to \mathbb{R}_+$.[4] It is thus clear that the intensity of $\tau$ is an $\mathbb{F}$-adapted stochastic process.

*Remarks.* In Chap. 11 below, we examine the concept of a $\mathbb{G}$-Markov chain. The definition of a $\mathbb{G}$-Markov chain can be extended in a rather straightforward manner to the case of a general (i.e., not necessarily countable) state space, resulting in the definition of a $\mathbb{G}$-Markov process, which takes values in an arbitrary measurable space. In the present set-up, we deal with a special case of an $\mathbb{R}^k$-valued $\mathbb{F}$-Markov process for some pre-specified reference filtration $\mathbb{F}$. In some circumstances, however, one may choose $\mathbb{F} = \mathbb{F}^Y$, where $\mathbb{F}^Y$ is the filtration generated by $Y$.

The canonical construction of the default time $\tau$ with these properties can be achieved as follows. Let $\mathbb{F}$ be some filtration, such that the process $Y$ is $\mathbb{F}$-adapted, and let $\eta$ be a random variable, independent of $\mathbb{F}$, with a unit exponential probability law under $\mathbb{Q}^*$. Of course, $\eta$ and $Y$ are given on a common probability space $(\Omega, \mathcal{G}, \mathbb{Q}^*)$, and thus a suitable enlargement of the underlying probability space may be required to produce $\eta$. To define the default time $\tau$, which is the first jump of the corresponding Cox process, it suffices to set

$$\tau = \inf \left\{ t \in \mathbb{R}_+ : \int_0^t \lambda(Y_u) \, du \geq \eta \right\}. \tag{8.58}$$

It is important to notice that the martingale condition (M.1) (or, equivalently, (M.2)) of Sect. 8.3.1 is satisfied for this construction of the default time $\tau$.

---

[4] See Sect. 6.6, where the notion of a *Cox process* is related to the notion of a conditional Poisson process.

To take a full advantage from the above specification of the default time $\tau$ in terms of state variables, we also assume that:

- the promised payoff $X$ of a defaultable contingent claim with the settlement date $T$ is represented by an $\mathcal{F}_T$-measurable random variable,
- the recovery process $Z$ is $\mathbb{F}$-predictable,
- the short-term interest rate process satisfies $r_t = r(Y_t)$ for some function $r : \mathbb{R}^k \to \mathbb{R}$,

Of course, the latter postulate agrees with our interpretation of $Y$ as the process of state variables. Under this set of assumptions, in all previously established formulae in which the default time $\tau$ does not appear explicitly, but only through its $\mathbb{F}$-hazard-rate process $\lambda_t = \lambda(Y_t)$, we may substitute the conditional expectation with respect to $\mathcal{G}_t$ with the conditioning relative to the $\sigma$-field $\mathcal{F}_t$. For instance, making use of the present assumptions and expression (8.41), we obtain the following generic valuation formula

$$S_t = \mathbb{1}_{\{\tau>t\}} \mathbb{E}_{\mathbb{Q}^*} \left( \int_t^T e^{-\int_t^u R(Y_v)\,dv} Z_u \lambda(Y_u)\,du + e^{-\int_t^T R(Y_v)\,dv} X \,\Big|\, \mathcal{F}_t \right),$$

where $R(Y_u) = r(Y_u) + \lambda(Y_u)$. Let us notice that the above formula is a direct consequence of equality (8.38), combined with the observation that $\mathcal{F}_t \subseteq \mathcal{G}_t \subseteq \mathcal{F}_t \vee \sigma(\eta)$, where, by assumption, the $\sigma$-fields $\mathcal{F}_T$ and $\sigma(\eta)$ are mutually independent.

As shown by Lando (1998), this formula can also be derived directly – that is, without making use of the pre-default value process $V$ (in other words, directly using Proposition 8.3.1, rather than a suitable version of Corollary 8.3.1). The next result provides a direct proof of the generic valuation formula in Lando's setting.

**Proposition 8.6.1.** *Let the default time $\tau$ be given by (8.58). Then we have*

$$S_t = \mathbb{1}_{\{\tau>t\}} \tilde{B}_t \, \mathbb{E}_{\mathbb{Q}^*} \left( \int_t^T \tilde{B}_u^{-1} Z_u \lambda(Y_u)\,du + \tilde{B}_T^{-1} X \,\Big|\, \mathcal{F}_t \right),$$

*where $\tilde{B}$ is the default-risk-adjusted savings account*

$$\tilde{B}_t = \exp\left( \int_0^t \big(r(Y_u) + \lambda(Y_u)\big)\,du \right).$$

*Proof.* The proof of the proposition is motivated by the demonstration of Proposition 3.1 in Lando (1998). We begin by noticing that by virtue of (8.58), for any two dates $0 \le t \le u \le T$ we have

$$\mathbb{Q}^*\{\tau > u \,|\, \mathcal{F}_T \vee \mathcal{H}_t\} = \begin{cases} \exp\left( -\int_t^u \lambda(Y_v)\,dv \right), & \text{on the set } \{\tau > t\}, \\ 0, & \text{on the set } \{\tau \le t\}, \end{cases}$$

where, as usual, $\mathcal{H}_t = \sigma(H_u : u \le t)$ is the natural filtration of the default process $H_t = \mathbb{1}_{\{\tau \le t\}}$.

Therefore (cf. (8.25)),

$$S_t = B_t \, \mathbb{E}_{\mathbb{Q}^*} \left( \int_t^T B_u^{-1} Z_u \lambda(Y_u) \mathbb{1}_{\{u \leq \tau\}} \, du + B_T^{-1} X \mathbb{1}_{\{\tau > T\}} \,\Big|\, \mathcal{G}_t \right)$$

$$= B_t \, \mathbb{E}_{\mathbb{Q}^*} \left( \int_t^T B_u^{-1} Z_u \lambda(Y_u) \, \mathbb{Q}^* \{\tau \geq u \,|\, \mathcal{F}_T \vee \mathcal{H}_t\} \, du \,\Big|\, \mathcal{G}_t \right)$$

$$+ \, B_t \, \mathbb{E}_{\mathbb{Q}^*} \left( B_T^{-1} X \, \mathbb{Q}^* \{\tau > T \,|\, \mathcal{F}_T \vee \mathcal{H}_t\} \,\Big|\, \mathcal{G}_t \right)$$

$$= \mathbb{1}_{\{\tau > t\}} B_t \, \mathbb{E}_{\mathbb{Q}^*} \left( \int_t^T B_u^{-1} Z_u \lambda(Y_u) \exp\left( -\int_t^u \lambda(Y_v) \, dv \right) du \,\Big|\, \mathcal{G}_t \right)$$

$$+ \, \mathbb{1}_{\{\tau > t\}} B_t \, \mathbb{E}_{\mathbb{Q}^*} \left( B_T^{-1} X \exp\left( -\int_t^T \lambda(Y_v) \, dv \right) \,\Big|\, \mathcal{G}_t \right)$$

$$= \mathbb{1}_{\{t < \tau\}} \tilde{B}_t \, \mathbb{E}_{\mathbb{Q}^*} \left( \int_t^T \tilde{B}_u^{-1} Z_u \lambda(Y_u) \, du + \tilde{B}_T^{-1} X \,\Big|\, \mathcal{G}_t \right).$$

We now wish to substitute $\mathcal{G}_t$ with $\mathcal{F}_t$ in the last expression. First, we observe that conditioning with respect to $\mathcal{G}_t$ in our case coincides with conditioning with respect to $\mathcal{F}_t \vee \mathcal{H}_t \subseteq \mathcal{F}_t \vee \sigma(\eta)$. Furthermore, the random variable $\eta$ is independent of $\mathcal{F}_\infty$, so that the $\sigma$-fields $\mathcal{F}_\infty$ and $\mathcal{H}_t$ are conditionally independent with respect to $\mathcal{F}_t$. Since the random variable under the sign of the conditional expectation above is measurable with respect to $\mathcal{F}_T \subset \mathcal{F}_\infty$, to complete the proof, it suffices to apply Condition (M2.b) of Sect. 8.3.1. $\square$

Notice that Proposition 8.6.1, combined with Corollary 8.3.1, make it clear that the jump $\Delta V_\tau$ plays no role in the present set-up. Indeed, it shows that we always have $S_t = \mathbb{1}_{\{\tau > t\}} V_t$, where the process $V$ is given by (8.29). Hence, combining the definition of the price process of a defaultable claim with (8.30), we find that, under the present assumptions, the pre-default process $V$ of any defaultable claim $(X, Z, \tau)$ necessarily satisfies (cf. (8.37))

$$\mathbb{E}_{\mathbb{Q}^*} \left( B_\tau^{-1} \Delta V_\tau \mathbb{1}_{\{t < \tau \leq T\}} \,\big|\, \mathcal{G}_t \right) = 0, \quad \forall t \in [0, T].$$

## 8.6.2 Duffie and Singleton Approach

Duffie and Singleton (1994, 1997, 1999) develop an econometric model of defaultable term structures, by postulating that:
- the evolution of relevant economic factors is modeled through a state-variables process $Y$, which follows an $\mathbb{F}$-Markov process under the spot martingale measure $\mathbb{Q}^*$,
- the promised contingent claim has the form $X = g(Y_T)$ for some measurable function $g : \mathbb{R}^k \to \mathbb{R}$, the promised dividends process $A \equiv 0$,
- the defaultable claim in question is subject to the fractional recovery of market value scheme with some $\mathbb{F}$-predictable process $K$ (see Sect. 8.3.2),
- the default-risk-adjusted short-term interest rate process $R$ is given by the equality $R_t = r_t + (1 - K_t)\lambda_t = \rho(Y_t)$ (cf. Proposition 8.3.3), where $\rho : \mathbb{R}^k \to \mathbb{R}$ is a measurable function.

The following result – borrowed from Duffie and Singleton (1999) – provides a quasi-explicit expression for the pre-default value of a defaultable claim of the form described above. For the definition and properties of the pre-default value process $V$ in the context of the fractional recovery of market value scheme, see Sect. 8.3.

**Proposition 8.6.2.** *Under the present assumptions, the pre-default value process $V$ satisfies*

$$V_t = \mathbb{E}_{\mathbb{Q}^*}\left\{ \exp\left( -\int_t^T \rho(Y_u)\, du \right) g(Y_T) \,\Big|\, Y_t \right\}. \tag{8.59}$$

*Proof.* To establish (8.59), it suffices to combine the reasoning used in the proof of Proposition 8.3.3 with the proof of Proposition 8.6.1, and to employ the F-Markov property of $Y$.  □

Duffie and Singleton (1999) first consider the case when the state-variables process $Y$ follows a non-degenerate diffusion process. More specifically,

$$dY_t = \mu(Y_t)\, dt + \sigma(Y_t)\, dW_t^*, \tag{8.60}$$

where $W^*$ is a $d$-dimensional standard Brownian motion under $\mathbb{Q}^*$, and the coefficients $\mu$ and $\sigma$, with values in $\mathbb{R}^k$ and the space of $k \times d$ matrices, respectively, are sufficiently regular to ensure the existence of a unique, global, strong solution to (8.60). They argue that in this case the process $V$ given by (8.29) satisfies $\Delta V_\tau = 0$, so that $S_t = \mathbb{1}_{\{\tau > t\}} V_t$. Subsequently, they also examine the case of a jump-diffusion state variables process $Y$; again, the no-jump at default condition is satisfied by the associated process $V$. In both cases, a suitable (integro-)differential equation satisfied by the value function of a defaultable claim can be used to find the price of a defaultable claim, usually through a suitable numerical procedure.

Duffie and Singleton (1999) provide a financial interpretation, in which the process $Y$ models both the firm-specific and macroeconomic variables, which influence the hazard rate process $\lambda$ and the fractional loss process $1 - K$, resulting in the *risk-neutral mean loss rate process* $(1 - K)\lambda$. In fact, they also consider a more general case when the adjusted short-term rate equals $R_t = r_t + (1 - K_t)\lambda_t + l_t$, where the process $l$ is meant to represents an additional component of the short-term credit spread, due to the presence of the *liquidity risk*. In this way, they are able to formally include the liquidity risk in their model, without being forced to analyze in detail the nature of this component of the overall risk. Formally, we thus have $R_t = r_t + r_t^d + r_t^l$, where $r_t, r_t^d$ and $r_t^l$ represent the interest rate (market) risk, the default risk of a given financial instrument, and the liquidity risk associated with this instrument, respectively. Finally, they postulate that the adjusted short-term rate $R$ may also depend on the current value of the defaultable claim; more specifically, they assume that $R_t = \rho(Y_t, U_t)$, where $U$ stands for the left-continuous version of the pre-default value process of a defaultable claim.

Under the latter assumption, formula (8.59) becomes

$$J(Y_t, t) = \mathbb{E}_{\mathbb{Q}^*}\left\{ \exp\left( - \int_t^T \rho(Y_u, J(Y_u, u))\, du \right) g(Y_T) \,\Big|\, Y_t \right\}.$$

In the special case of interest rate swaps with one- or two-sided default risk, the last equation was treated numerically by Duffie and Huang (1996) and Huge and Lando (1999).

Duffie and Singleton (1999) give some arguments supporting the use of the fractional recovery of market value, as opposed to the fractional recovery of Treasury value and the fractional recovery of par value (recall that various forms of recovery at default were examined in some detail in Sect. 8.2.5). They argue correctly that in choosing the most convenient version of the recovery rule, apart from striving to create a computationally efficient model, it is essential to take into account the legal structure of the instrument to be priced. For instance, the fractional recovery of market value is well suited to the case of market-to-market defaultable swaps, as described in Sect. 14.4 and 14.5. On the other hand, for the case of corporate bonds, the choice of the fractional recovery of par value, possibly with a random recovery rate, seems to be the closest to the real-world market conventions. The interested reader may consult the original paper for a detailed discussion of these practically important issues.

Let us end this section by presenting briefly the two particular examples of econometric models for defaultable term structure, implemented by Duffie and Singleton (1997, 1999) and Dai and Singleton (2000).

*Example 8.6.1. Square-root diffusion model of Y.* As a first example, consider a model in which

$$r_t = \alpha_0 + \alpha_1 Y_t^1 + \alpha_2 Y_t^2 + \alpha_3 Y_t^3, \tag{8.61}$$

and the short-term credit spread $s_t = R_t - r_t$ satisfies

$$s_t = \gamma_0 + \gamma_1 Y_t^1 + \gamma_2 Y_t^2 + \gamma_3 Y_t^3, \tag{8.62}$$

for some constants $\alpha_i, \gamma_i,\ i = 0, \ldots, 3$. In addition, assume that the three-dimensional process $Y_t = (Y_t^1, Y_t^2, Y_t^3)$ is governed by the SDE

$$dY_t = \kappa(\theta - Y_t)\, dt + \sqrt{S(t)}\, dW_t^*, \tag{8.63}$$

where $\kappa$ is a $3 \times 3$ matrix with positive diagonal and nonpositive off-diagonal entries, $\theta$ belongs to $\mathbb{R}_+^3$, and $S(t)$ stands for the $3 \times 3$ diagonal matrix with diagonal elements $Y_t^1, Y_t^2$ and $Y_t^3$. Of course, in this case $W^*$ is a three-dimensional standard Brownian motion under $\mathbb{Q}^*$. In case of a diagonal matrix $\kappa$, processes $Y^1, Y^2$ and $Y^3$ are mutually independent square-root diffusions under $\mathbb{Q}^*$, and they are easily recognized as the classic Cox-Ingersoll-Ross-type models (cf. Cox et al. (1985a, 1985b)). Duffee (1999) performed an empirical study of such a model, under the additional assumptions that: $\alpha_0 = -1$ and $\alpha_3 = 0$.

*Example 8.6.2. Model with flexible correlation structure for* $(r, s)$. Assume first that $\kappa$ is a diagonal matrix. Then the short-term rate $r$ and the short-term credit spread $s$ cannot be negatively correlated over infinitesimally small time intervals, unless some of the coefficients in (8.61)–(8.62) are negative. But this in turn implies that the short-term rate process and/or the hazard-rate process may take on negative values, which is clearly an undesirable property. More generally, as noted by Dai and Singleton (2000), any well-defined model of Example 8.6.1 does not allow for negative correlation among any of the state variables, as the off-diagonal elements of $\kappa$ must be nonpositive.

The next example is aimed to overcome this drawback – that is, to gain more flexibility in the correlation structure among the state variables. This example is in fact a modification of the previous one. Instead of (8.63), we now postulate that

$$dY_t = \kappa(\theta - Y_t)\, dt + \Sigma \sqrt{S(t)}\, dW_t^*, \qquad (8.64)$$

where $\Sigma$ is a $3 \times 3$ matrix, and

$$S_{11}(t) = Y_t^1, \quad S_{22}(t) = \beta_{22} Y_t^2, \quad S_{33}(t) = b_3 + \beta_{31} Y_t^1 + \beta_{32} Y_t^2$$

with strictly positive coefficients $\beta_{ij}$. Finally, let $r$ be given by (8.61), and let $s$ satisfy (8.62) with $\gamma_3 = 0$, all other coefficients in (8.61)–(8.62) being now strictly positive constants. Following Dai and Singleton (2000), let us take

$$\kappa = \begin{pmatrix} \kappa_{11} & \kappa_{12} & 0 \\ \kappa_{21} & \kappa_{22} & 0 \\ 0 & 0 & \kappa_{33} \end{pmatrix}, \quad \Sigma = \begin{pmatrix} 1 & 0 & 0 \\ 0 & 1 & 0 \\ \sigma_{31} & \sigma_{32} & 1 \end{pmatrix},$$

where the off-diagonal entries of $\kappa$ are nonpositive. It is clear that under the present assumptions the short-term credit spread $s$ follows a strictly positive process. In addition, by making a judicious choice of the signs of $\sigma_{31}$ and $\sigma_{32}$, we may introduce either a positive or a negative short-term correlation between the increments of $r$ and $s$. Duffie and Singleton (1999) further specialize this model, by postulating that $r_t = \alpha_0 + Y_t^2 + Y_t^3$, $S_{33}(t) = b_3 + \beta_{32} Y_t^2$, and the matrices $\kappa$ and $\Sigma$ satisfy

$$\kappa = \begin{pmatrix} \kappa_{11} & \kappa_{12} & 0 \\ 0 & \kappa_{22} & 0 \\ 0 & 0 & \kappa_{33} \end{pmatrix}, \quad \Sigma = \begin{pmatrix} 1 & 0 & 0 \\ 0 & 1 & 0 \\ 0 & \sigma_{32} & 1 \end{pmatrix}.$$

With these restrictions, we obtain for $r$ a two-factor affine model with state variables $Y^2$ and $Y^3$, which can be estimated independently of the credit-risk spread $s$. In other words, the parameters of the Treasury bond model can be estimated without using corporate bond data.

*Remarks.* An alternative way of introducing negative correlations among the state variables was proposed by Duffie and Liu (2001), who assumed that the short-term rate process $r$ is an affine function of the squared Gaussian process of state variables. We do not go into details here. Let us only mention that they mainly focus on the pricing of floating-rate corporate debt.

### 8.6.3 Hybrid Methodologies

A so-called *hybrid approach* can be seen as a variant of the intensity-based modeling with state variables. In this approach, the time of default in modeled in terms of the stochastic intensity, but the conditional probability of default is directly related to the current level of the firm's value (or the current market value of the firm's equity). In this sense, a hybrid approach combines ideas from both the intensity-based methodology and the classic value-of-the-firm approach.

**Madan and Unal (1998) approach.** Madan and Unal (1998) consider the discounted equity value (including reinvested dividends) process $E_t^* = B_t^{-1} E_t$ as the unique Markovian state variable in their intensity-based model. They postulate that the hazard rate of default equals $\lambda_t = \lambda(E_t^*, t)$ for some function $\lambda : \mathbb{R}_+ \to \mathbb{R}_+$. The process $E^*$ is assumed to follow a diffusion process, specifically

$$dE_t^* = \sigma(E_t^*, t) \, dW_t^*, \quad E_0^* > 0, \tag{8.65}$$

under the spot martingale measure $\mathbb{P}^*$. We assume, of course, that the SDE (8.65) admits a unique, strong, global solution $E^*$. Naturally, the solution $E^*$ follows a Markov process under $\mathbb{P}^*$. We may assume that the process $E^*$ takes on strictly positive values: $E_t^* > 0$ for every $t \in [0, T^*]$. As usual, the default time $\tau$ is given by the canonical construction, so that it is defined on an enlarged probability space $(\Omega, \mathbb{G}, \mathbb{Q}^*)$ (see, e.g., formula (8.58)). Let us notice that the $\mathbb{P}^*$-standard Brownian motion $W^*$ remains a standard Brownian motion under $\mathbb{Q}^*$.

To avoid making a particular choice of a default-free term structure model, Madan and Unal (1998) focus on the futures price of a corporate bond. As it is well known,[5] the futures price $\pi^f(X)$ of a contingent claim $X$, for the settlement date $T$, is given by the conditional expectation under the spot martingale measure: $\pi_t^f(X) = \mathbb{E}_{\mathbb{Q}^*}(X \,|\, \mathcal{G}_t)$ for $t \in [0, T]$. In our case, the futures price $D^f(t, T)$ of a defaultable zero-coupon bond with zero recovery equals $D^f(t, T) = \mathbb{Q}^* \{\tau > T \,|\, \mathcal{G}_t\}$ for $t \in [0, T]$. More explicitly,

$$D^f(t, T) = \mathbb{1}_{\{\tau > t\}} \, \mathbb{E}_{\mathbb{P}^*} \left( e^{-\int_t^T \lambda(E_u^*, u) du} \,\middle|\, \mathcal{F}_t \right) = \mathbb{1}_{\{\tau > t\}} v(E_t^*, t)$$

for some function $v : \mathbb{R}_+ \to \mathbb{R}_+$. By virtue of (8.65) and the Feynman-Kac theorem (see Karatzas and Shreve (1991)), the function $v$ satisfies, under mild technical assumptions, the following pricing PDE

$$v_t(x, t) + \tfrac{1}{2}\sigma^2(x, t) v_{xx}(x, t) - \lambda(x, t) v(x, t) = 0$$

subject to the terminal condition $v(x, T) = 1$. For the sake of notational simplicity, we have assumed here that the process $W^*$ is one-dimensional.

---

[5] See, for instance, Duffie and Stanton (1992) or Sect. 3.2 and Sect. 15.2 in Musiela and Rutkowski (1997a).

*Example 8.6.3.* As a judicious choice of time-homogeneous intensity function $\lambda : \mathbb{R}_+ \to \mathbb{R}_+$, Madan and Unal (1998) propose to take the function

$$\lambda(x) = c(\ln(x/\bar{v}))^{-2}, \tag{8.66}$$

where $c$ and $\bar{v}$ are strictly positive constants. It is interesting to notice that the stochastic intensity $\lambda_t = \lambda(E_t^*)$ tends to infinity, when the discounted equity value $E_t^*$ approaches, either from above or from below, the critical level $\bar{v}$. In addition, we assume the dynamics of $E^*$ are:

$$dE_t^* = \sigma E_t^* \, dW_t^*, \quad E_0^* > 0, \tag{8.67}$$

in other words, we take $\sigma(x, t) = \sigma x$ for some constant volatility coefficient $\sigma$. Madan and Unal (1998) show that under assumptions (8.66)–(8.67) the futures price of a corporate bond equals $D^f(t, T) = G_\nu(h(E_t^*, T - t))$, where the parameter $\nu$ satisfies $\nu(\nu + 1) = 2c\sigma^{-2}$ and

$$h(x, t) = \frac{2\sigma^2 t}{\left( \ln(x/\bar{v}) - \sigma^2 t/2 \right)^2}.$$

For a fixed value of the parameter $\nu$, the function $G_\nu : \mathbb{R}_+ \to \mathbb{R}$ satisfies the second-order ODE:

$$x^2 G_\nu''(x) + \left( \frac{3}{2}x - 1 \right) G_\nu'(x) - \frac{\nu(\nu + 1)}{4} G_\nu(x) = 0$$

with the initial conditions $G_\nu(0) = 1$ and $G_\nu'(0) = -\nu(\nu + 1)/4$. As shown by Madan and Unal (1998), the quasi-explicit valuation formula above may serve to produce estimates of parameters of the hazard rate process, based on the observed market yields on defaultable bonds.

As a second important ingredient of their model, Madan and Unal (1998) take the magnitude of losses at default represented by a random recovery rate, which is termed a *payout rate* therein. They discuss a method for estimating the mean, and possibly also higher moments, of the conditional law of recovery rate, based on the observed prices of credit-risk sensitive instruments.

For the special case of senior and junior debts, they notice (following Black and Cox (1996)) that the payoffs of these instruments are equivalent to a put option and a call option, respectively, written on the firm's average recovery rate. Using this observation, they develop a procedure for estimating the risk-neutral probability law of the payout ratio. To this end, they postulate that the probability law in question belongs to the family of Beta distributions, with the following probability density function:

$$f(y; \alpha, \beta) = \frac{\Gamma(\alpha + \beta)}{\Gamma(\alpha)\Gamma(\beta)} \, y^{\alpha-1}(1 - y)^{\beta-1}, \quad \forall y \in (0, 1),$$

where $\Gamma$ is the Gamma function and $\alpha, \beta$ are positive parameters. The interested reader is referred to the original paper for a detailed description of the maximum likelihood estimation procedure for these parameters.

**Madan and Unal (2000) approach.** Madan and Unal (2000) start by postulating that at a random time the firm faces the payment of a random loss amount $\xi$, with the risk-neutral cumulative distribution function $\hat{F}$ (the associated probability density function will be henceforth denoted by $\hat{f}$). As usual, the equity value of the firm equals the value of its assets less its liabilities. Specifically, the firm's structure is modeled through the equality

$$E_t = V_t + g(r_t, t) - l(r_t, t),$$

where $V_t$ represents the (interest-rate insensitive) market value of cash assets held by the firm, $g(r_t, t)$ stands for the (interest-rate sensitive) market value of other assets, and $l(r_t, t)$ is the (interest-rate sensitive) value of the future firm's liabilities, discounted at the risk-free Treasury rate. It is interesting to notice that the sensitivity of equity to interest rates can be positive or negative here; this feature is referred to as a positive or negative *duration gap* in financial literature. At the intuitive level, default occurs if the loss is larger than the equity $E_t$ in place, that is, when $\xi > E_t$ and a payment in the amount $\xi$ is due. To account for the last feature, Madan and Unal (2000) formally postulate that the intensity of default equals

$$\lambda_t = \lambda(V_t, r_t) = \bar{\lambda}\big(1 - \hat{F}(E_t)\big), \tag{8.68}$$

where the constant $\bar{\lambda} > 0$ represents the hazard rate of loss event. This means that the loss arrival is governed by a Poisson process with a constant parameter $\bar{\lambda}$, the default intensity is proportional to $\bar{\lambda}$, but it is a decreasing function of the current level of the firms' equity. For the sake of analytical tractability, Madan and Unal (2000) consider a first-order approximation of (8.68) around the reference levels $\ln V_0$ and $r_0$ of the logarithm of cash assets and interest rates. Specifically, they set

$$\lambda_t = \lambda_0 - \bar{\lambda}\hat{f}(E_0)\big(V_0 \ln(V_t/V_0) + (g_r(r_0, 0) - l_r(r_0, 0))(r_t - r_0)\big),$$

so that $\lambda_t = \beta_0 + \beta_1 \ln V_t + \beta_2 r_t$ for some constant parameters $\beta_0, \beta_1$ and $\beta_2$.

*Example 8.6.4.* For the sake of analytical tractability, Madan and Unal (2000) further specify their model by assuming that the loss level $\xi$ is exponentially distributed with a mean level $\mu > 0$, i.e., $\hat{F}(y) = 1 - \exp(-y/\mu)$ for $y \in \mathbb{R}_+$, the value of cash assets is governed by the SDE:

$$dV_t = V_t\big(r_t\, dt + \sigma_V\, dW_t^*\big),$$

and the default-free short-term rate follows Vasicek's model (cf. (2.44)):

$$dr_t = (a - br_t)\, dt + \sigma_r\, d\tilde{W}_t.$$

Both equations above hold under the spot martingale measure $\mathbb{Q}^*$, and the Brownian motions $\tilde{W}$ and $W^*$ are mutually dependent, with constant instantaneous correlation coefficient $\rho_{Vr}$. Under this set of assumptions, they derive a closed-form solution for the price of a corporate bond that is subject to the fractional recovery of Treasury value with a constant recovery rate. Unfortunately, their valuation formula is too long to be reproduced here.

**Davydov, Linetsky and Lotz results.** Following some ideas from Linet-sky (1997) and Lotz (1998), Davydov et al. (2000) propose to modify the Madan and Unal (2000) approach. They postulate that, under the spot mar-tingale measure $\mathbb{Q}^*$, the equity process satisfies

$$dE_t = E_t\left(r\,dt + \sigma\,dW_t^*\right)$$

for a constant interest rate $r$. Consequently, for every $t \in [0, T]$, the condi-tional probability of no-default equals

$$\mathbb{Q}^*\{\tau > T \mid \mathcal{G}_t\} = \mathbb{1}_{\{\tau > t\}}\,\mathbb{E}_{\mathbb{P}^*}\left(e^{-\bar{\lambda}\int_t^T (1-\hat{F}(E_u))\,du}\,\Big|\,\mathcal{F}_t\right).$$

For $t = 0$, by applying Girsanov's theorem, we obtain

$$\mathbb{Q}^*\{\tau > T\} = e^{-\eta^2 T/2}\,\mathbb{E}_{\mathbb{P}^*}\left(h(W_T^*)e^{-\bar{\lambda}\int_0^T k(W_u^*)\,du}\right),$$

where we denote

$$\eta = \sigma^{-2}(r - \sigma^2/2), \quad h(x) = e^{\eta x}, \quad k(x) = 1 - \hat{F}(E_0 e^{\sigma x}).$$

For a fixed $s > 0$, we introduce the Laplace transform $u : \mathbb{R} \to \mathbb{R}$ by setting

$$u(x) = \mathbb{E}_{\mathbb{P}^*}\left(\int_0^\infty h(W_T^*)e^{-\int_0^T (s+\bar{\lambda}k(W_u^*))\,du}\,dT\,\Big|\,W_0^* = x\right).$$

From the Feynman-Kac formula (see p. 267 in Karatzas and Shreve (1991)), it results that $u$ is the unique bounded solution to the following second-order ODE:

$$\tfrac{1}{2}u''(x) - (s + \bar{\lambda}k(x))u(x) = -h(x).$$

It is well known that the solution to the last equation satisfies

$$u(x) = \int_{-\infty}^\infty G(x, y)h(y)\,dy,$$

where in turn the Green function $G(x, y)$ solves the following inhomogeneous equation (notice that $\delta_0$ stands here for Dirac's delta function)

$$\tfrac{1}{2}G_{xx}(x, y) - (s + \bar{\lambda}k(x))G(x, y) = -\delta_0(x - y)$$

subject to the boundary conditions at $x = y$:

$$\lim_{\epsilon \to 0^+}\left(G(y - \epsilon, y) - G(y + \epsilon, y)\right) = 0,$$
$$\lim_{\epsilon \to 0^+}\left(G_x(y - \epsilon, y) - G_x(y + \epsilon, y)\right) = 2,$$

and asymptotic boundary condition at infinity: $\lim_{x \to \pm\infty} G(x, y) = 0$. Let $\psi$ ($\phi$, resp.) be a strictly increasing (a strictly decreasing, resp.) solution to the homogeneous equation

$$\tfrac{1}{2}u''(x) - (s + \bar{\lambda}k(x))u(x) = 0.$$

Then
$$G(x, y) = \frac{2}{w(x)} \begin{cases} \psi(x)\phi(y), & \text{if } x \le y, \\ \phi(x)\psi(y), & \text{if } x \ge y, \end{cases}$$
where $w(x) := \psi'(x)\phi(x) - \psi(x)\phi'(x)$ stands for the Wronskian. For the special case of $h(y) = e^{\eta y}$ and $x = 0$, we obtain

$$u(0) = \frac{2}{w(0)} \left( \phi(0) \int_{-\infty}^{0} e^{\eta y} \psi(y) \, dy + \psi(0) \int_{0}^{\infty} e^{\eta y} \phi(y) \, dy \right).$$

*Example 8.6.5.* In addition, let us assume that the loss size is constant: $\xi = K$ for a strictly positive constant $K$. Then $1 - \hat{F}(x) = \mathbb{1}_{(-\infty, K)}(x)$, and

$$\lambda_t = \bar{\lambda} \mathbb{1}_{\{E_t < K\}} = \begin{cases} \bar{\lambda}, & \text{if } E_t < K, \\ 0, & \text{if } E_t \ge K. \end{cases}$$

Consequently, the hazard process $\Lambda$ equals

$$\Lambda_t = \int_0^t \lambda_u \, du = \bar{\lambda} \int_0^t \mathbb{1}_{\{E_u < K\}} \, du,$$

so that it is proportional to the occupation time of the equity process $E$ below the fixed level $K$. Furthermore,

$$\mathbb{Q}^* \{\tau > T\} = e^{-\eta^2 T/2} \, \mathbb{E}_{\mathbb{P}^*} \left( e^{\eta W_T^* - \bar{\lambda} \int_0^T \mathbb{1}_{\{W_u^* < -\varsigma\}} \, du} \right),$$

where $\varsigma = \sigma^{-1} \ln(E_0/K)$. A closed-form solution for the probability of no-default $\mathbb{Q}^* \{\tau > T\}$ was derived by Linetsky (1999), who dealt with the so-called *step options* (a type of a knock-out options, with knockout based on the amount of time spent by the underlying asset below some level).

*Example 8.6.6.* Davydov et al. (2000) observe that the first-order approximation employed by Madan and Unal (2000) may yield negative values for the hazard rate $\lambda$. To overcome this deficiency, they take the positive part of the first-order approximation as the intensity of default. Since for $g \equiv l \equiv 0$, Madan and Unal (2000) specification of $\lambda$ reduces to

$$\lambda_t = \bar{\lambda} \left( 1 - \hat{F}(E_0) - E_0 \hat{f}(E_0) \ln(E_t/E_0) \right),$$

Davydov et al. (2000) postulate that

$$\lambda_t = \bar{\lambda} \left( 1 - \hat{F}(E_0) - E_0 \hat{f}(E_0) \ln(E_t/E_0) \right)^+.$$

With this specification of default intensity, we obtain

$$\mathbb{Q}^* \{\tau > T\} = e^{-\eta^2 T/2} \, \mathbb{E}_{\mathbb{P}^*} \left( e^{\eta W_T^* - \alpha \int_0^T (\beta - W_u^*)^+ \, du} \right),$$

where $\alpha = \sigma \bar{\lambda} E_0 \hat{f}(E_0)$ and $\beta = (1 - \hat{F}(E_0))(\sigma E_0 \hat{f}(E_0))^{-1}$. In case of an exponential loss $\xi$ introduced in Example 8.6.4, Davydov et al. (2000) derive a closed-form solution for the Green function. In this way, they obtain a quasi-explicit expression for the risk-neutral probability of default in terms of the inverse Laplace transform.

### 8.6.4 Credit Spread Models

To effectively deal with some credit derivatives, such as credit-spread options, an alternative approach based on the direct modeling of the credit spreads seems to be more convenient. The modeling of credit spreads involves not only the *credit-spread curves*, but also *credit-spread volatilities*, and, if several distinct assets are modeled simultaneously, the *credit-spread correlations*. Since available market data are scarce, the estimation of the credit-spread curve is more problematic than the estimation of the risk-free yield curve.[6] Therefore, when dealing with the debt issued by a particular entity, one might use the rating-specific credit-spread curve as a proxy for the unobservable firm-specific credit-spread curve (see Fridson and Jónsson (1995)). Of course, there is a good chance that the difficulty in collecting of sufficient empirical data will lessen in the future, with the further development of the market. This also applies to the estimation of credit-spread volatilities that, in principle, can be statistically inferred from the observed variations of the credit-spread yield curve. Since we are not going to discuss these issues in detail here, the interested reader may consult the original papers by Fons (1987, 1994), Sarig and Warga (1989), Foss (1995), Schwartz (1998) or Helwege and Turner (1999). Let us only mention two examples of *credit spread models*.

**Nielsen and Ronn model.** The two-factor model proposed by Nielsen and Ronn (1997) assumes that the risk-free interest rate $r$ satisfies

$$dr_t = \mu_r r_t \, dt + \sigma_r r_t \, dW_t,$$

and the instantaneous yield spread $s$ for corporate bond is governed by

$$ds_t = \sigma_s s_t \, d\tilde{W}_t,$$

where $W$ and $\tilde{W}$ are correlated Brownian motions. After default the bond sells at a post-default price, which is a fraction of face value. They also postulate that the intensity of default $\lambda_t$ satisfies: $\lambda_t = s_t(1-\delta)^{-1}$, where $\delta$ is the recovery rate. They examine a discrete-time implementation of their model, based on the recombining (quadrinomial or trinomial) tree for $r$ and $s$. Of course, whenever default occurs, the branch of the tree terminates.

**Das model.** Das (1995) posits the following dynamics for the short-term credit spread $s_t$:

$$ds_t = (\tilde{a} - \tilde{b} s_t) \, dt + \tilde{\sigma} \sqrt{s_t} \, d\tilde{W}_t.$$

These dynamics are combined with the CIR model of the default-free short-term rate

$$dr_t = (a - br_t) \, dt + \sigma_r \sqrt{r_t} \, dW_t,$$

where the two Brownian motions $W$ and $\tilde{W}$ are correlated. Again, the lattice approach to the valuation of credit derivatives is examined.

---

[6] For an overview of methods of estimation of the yield curve, the interested reader is referred to Bliss (1997).

# 9. Conditionally Independent Defaults

The next two chapters are devoted to the study of mutually dependent default times within the framework of the intensity-based approach. In case of conditionally independent default times studied in this chapter, we are able to establish closed-form pricing results for the $i^{\text{th}}$-to-default contingent claim. In general, the issue becomes much more complicated, and we only provide partial results in the next chapter.

Sect. 9.1 is concerned with the arbitrage valuation, and thus it is natural to use there the risk-neutral probability $\mathbb{Q}^*$. In contrast, Sect. 9.2 is oriented mostly towards risk management applications, and thus it would be more appropriate to use the real-world probability $\mathbb{Q}$ in this section. Nevertheless, for the sake of notational uniformity, we shall always work with the spot martingale measure $\mathbb{Q}^*$. Whenever formulae under the real-world probability measure $\mathbb{Q}$ are of interest, they may be obtained in a fashion analogous to the derivation done under the risk-neutral probability $\mathbb{Q}^*$, after an appropriate modification of the assumptions, which are made relative to the probability $\mathbb{Q}^*$ throughout the entire chapter.

Let us describe the basic set-up adopted in this chapter. We shall consider a finite collection of random times $\tau_1, \ldots, \tau_n$, defined on a common probability space $(\Omega, \mathcal{G}, \mathbb{Q}^*)$. Unless explicitly stated otherwise, we shall postulate throughout that $\mathbb{Q}^*\{\tau_k = 0\} = 0$ and $\mathbb{Q}^*\{\tau_k > t\} > 0$ for every $t \in \mathbb{R}_+$ and $k = 1, \ldots, n$. The case of simultaneous defaults is not examined here; namely, we postulate that $\mathbb{Q}^*\{\tau_k = \tau_j\} = 0$ for arbitrary $k, j = 1, \ldots, n$ with $k \neq j$. Exposition in Chap. 9 and 10 is based on Bielecki and Rutkowski (2002), Kijima (2000) and Kijima and Muromachi (2000). For the study of the case of simultaneous defaults, we refer to Lando (2000b).

We find it convenient to introduce some auxiliary notation. We associate with the collection $\tau_1, \ldots, \tau_n$ of default times the ordered sequence of random times: $\tau_{(1)} \leq \tau_{(2)} \leq \cdots \leq \tau_{(n)}$. By definition, $\tau_{(1)} = \min(\tau_1, \tau_2, \ldots, \tau_n)$, and

$$\tau_{(i+1)} = \min\left(\tau_k : k = 1, \ldots, n, \ \tau_k > \tau_{(i)}\right)$$

for $i = 1, \ldots, n-1$. In particular, $\tau_{(n)} = \max(\tau_1, \tau_2, \ldots, \tau_n)$.

In addition to the family $\tau_1, \ldots, \tau_n$ of random times, we postulate that we are also given a reference filtration, $\mathbb{F}$ say, on the probability space $(\Omega, \mathcal{G}, \mathbb{Q}^*)$. We introduce the enlarged filtration $\mathbb{G}$ by setting $\mathbb{G} = \mathbb{F} \vee \mathbb{H}^1 \vee \mathbb{H}^2 \vee \ldots \vee \mathbb{H}^n$. It will be also convenient to denote $\mathbb{H} = \mathbb{H}^1 \vee \mathbb{H}^2 \vee \ldots \vee \mathbb{H}^n$.

## 9.1 Basket Credit Derivatives

Our goal is to derive valuation formulae for the $i^{\text{th}}$-to-default contingent claims. We shall consider a general $i^{\text{th}}$-to-default claim, $CCT^{(i)}$ say, which matures at time $T$ and is specified by the following covenants:

– if $\tau_{(i)} = \tau_k \leq T$ for some $k = 1, \ldots, n$, then the claim pays at time $\tau_{(i)}$ the amount $Z^k_{\tau_{(i)}}$, where $Z^k$ is a $\mathbb{G}$-predictable process, and it pays at time $T$ a $\mathcal{G}_T$-measurable amount $X_k$,

– if $\tau_{(i)} > T$, the claimholder receives at time $T$ a $\mathcal{G}_T$-measurable amount $X$.

According to the convention above, if the $i^{\text{th}}$ default occurs in the time interval $[0, T]$ – that is, if $\tau_{(i)} = \tau_k \leq T$ for some $k$ – an immediate recovery cash flow $Z^k_{\tau_{(i)}}$ is received at time $\tau_{(i)}$, and a delayed recovery cash flow $X_k$ is passed to the claimholder at the maturity date $T$. A more general convention (not examined here) for payoffs associated with the claim $CCT^{(i)}$ would also involve immediate recovery payoffs at each default time $\tau_j < \tau_{(i)}$.

*Example 9.1.1.* Duffie (1998a) considers an example of a first-to-default type claim $CCT^{(1)}$ that is defined by setting $X_k = 0$ for $k = 1, \ldots, n$. The corresponding last-to-default contract is the claim $CCT^{(n)}$ with $X_k = 0$ for $k = 1, \ldots, n$.

*Example 9.1.2.* Kijima and Muromachi (2000) examine a special case of a first-to-default type claim $CCT^{(1)}$, which is termed the *default swap of type F*. It is defined by setting $Z^k \equiv 0$ for $k = 1, \ldots, n$. Another contingent claim considered by Kijima and Muromachi (2000) – the so-called *default swap of type D* – may be seen as an example of the second-to-default contingent claim. Formally, they deal with the claim $CCT^{(2)}$ with the following specific features. First, they set $Z_k \equiv 0$ for $k = 1, \ldots, n$. Second, they postulate that, for each $k = 1, \ldots, n$, the recovery payoff on the set $\{\tau_{(2)} = \tau_k \leq T\}$ is

$$ X_k = \sum_{l \neq k} (\tilde{X}_k + \tilde{X}_l) \mathbb{1}_{\{\tau_{(1)} = \tau_l\}}, $$

where $\tilde{X}_j$ is a $\mathcal{G}_T$-measurable random variable for each $j = 1, \ldots, n$. Finally, the recovery payoff on the set $\{\tau_{(2)} > T\}$ equals

$$ X = \hat{X}_0 \mathbb{1}_{\{\tau_{(1)} > T\}} + \sum_{j=1}^{n} \hat{X}_j \mathbb{1}_{\{\tau_{(1)} = \tau_j \leq T\}}, $$

where $\hat{X}_j$ is a $\mathcal{G}_T$-measurable random variable for each $j = 0, \ldots, n$. In this general formulation, a default swap of type D protects its holder against the first two defaults, provided that they both have occurred before or at the maturity date of the contract.

*Example 9.1.3.* Li (1999b) examines still another example of the $i^{\text{th}}$-to-default claim. Namely, he sets $Z^k \equiv 1$, $X_k = 0$ for $k = 1, \ldots, n$ and $X = 0$. Such a contract is termed *digital default put of basket type*.

### 9.1.1 Mutually Independent Default Times

We shall first examine the most simple case of default times $\tau_1, \ldots, \tau_n$ that are mutually independent under the risk-neutral probability. In this case, it is not hard to derive the probability distribution of the time of the $i^{\text{th}}$ default in terms of the marginal distributions of individual defaults. Indeed, suppose that for every $k = 1, \ldots, n$ we know the cumulative distribution function $F_k(t) = \mathbb{Q}^*\{\tau_k \leq t\}, t \in \mathbb{R}_+$, of the default time of the $k^{\text{th}}$ reference entity. Then the cumulative distribution functions of $\tau_{(1)}$ and $\tau_{(n)}$ are:

$$F_{(1)}(t) := \mathbb{Q}^*\{\tau_{(1)} \leq t\} = 1 - \mathbb{Q}^*\{\tau_{(1)} > t\} = 1 - \prod_{k=1}^{n}(1 - F_k(t)),$$

and

$$F_{(n)}(t) := \mathbb{Q}^*\{\tau_{(n)} \leq t\} = \mathbb{Q}^*\{\tau_1 \leq t, \ldots, \tau_n \leq t\} = \prod_{k=1}^{n} F_k(t),$$

respectively. More generally, for any $i = 1, \ldots, n$ we have

$$F_{(i)}(t) := \mathbb{Q}^*\{\tau_{(i)} \leq t\} = \sum_{m=i} \sum_{\pi \in \Pi^m} \prod_{j \in \pi} F_{k_j}(t) \prod_{l \notin \pi}(1 - F_{k_l}(t)),$$

where $\Pi^m$ denote the family of subsets of $\{1, \ldots, n\}$ consisting of $m$ elements. Suppose also that the default times $\tau_1, \ldots, \tau_n$ admit intensity functions $\lambda_1(t), \ldots, \lambda_n(t)$. It is then clearly seen that the default time $\tau_{(1)}$ admits the intensity function $\lambda_{(1)}(t) = \lambda_1(t) + \cdots + \lambda_n(t)$ (cf. Lemma 7.1.1) and

$$\mathbb{Q}^*\{\tau_{(1)} > t\} = e^{-\int_0^u \lambda_{(1)}(v)dv}, \quad \forall t \in \mathbb{R}_+.$$

By direct calculations, it is also possible to find the intensity function of the $i^{\text{th}}$ default time. Let us notice that we do not necessarily need to assume that the reference filtration $\mathbb{F}$ is trivial, so that the case of random interest rates is not excluded a priori.

*Example 9.1.4. Digital default put of basket type.* As a simple example of a basket credit derivative, we consider a contract that pays a fixed amount (e.g., one unit of cash) at the $i^{\text{th}}$ default time, provided that it occurs prior to or at the contract's maturity date $T$. For simplicity, we assume that the interest rates are deterministic. Then the value at time 0 of the contract equals

$$S_0 = \mathbb{E}_{\mathbb{Q}^*}\left(B_{\tau}^{-1}\mathbb{1}_{\{\tau_{(i)} \leq T\}}\right) = \int_{]0,T]} B_u^{-1} \, dF_{(i)}(u).$$

In particular, if the default times $\tau_1, \ldots, \tau_n$ admit intensities, then the value of the $i^{\text{th}}$-to-default digital put equals

$$S_0 = \int_0^T B_u^{-1} \, dF_{(i)}(u) = \int_0^T B_u^{-1} \lambda_{(i)}(u) e^{-\int_0^u \lambda_{(i)}(v)dv} \, du.$$

### 9.1.2 Conditionally Independent Default Times

We shall now study the valuation of basket credit derivatives, under an additional assumption of conditional independence of default times with respect to the underlying filtration. This assumption underpins a vast majority of works devoted to the intensity-based valuation of basket derivatives (see, for example, Kijima (2000) and Kijima and Muromachi (2000)).

Before we proceed with a formal definition of conditional independence of default times, we provide the intuitive meaning of this assumption. Observe that all reference credit names are subject to common risk factors that may trigger credit (default) events. In addition, each credit name is also subject to the so-called *idiosyncratic risk* that is specific for this particular name, and may trigger credit (default) events associated with this credit name as well. At the intuitive level, the assumption of conditional independence of default times means that once the common risk factors are fixed, the idiosyncratic risk factors become independent of each other.

**Definition 9.1.1.** The random times $\tau_i$, $i = 1, \ldots, n$ are said to be *conditionally independent* with respect to the filtration $\mathbb{F}$ under $\mathbb{Q}^*$ if and only if the following condition is satisfied: for any $T > 0$, and arbitrary $t_1, \ldots, t_n \in [0, T]$,

$$\mathbb{Q}^*\{\tau_1 > t_1, \ldots, \tau_n > t_n \,|\, \mathcal{F}_T\} = \prod_{i=1}^{n} \mathbb{Q}^*\{\tau_i > t_i \,|\, \mathcal{F}_T\}.$$

*Remarks.* (i) Note that in general the conditional independence of random times does not imply their independence; the converse implication does not hold either. We find it convenient to also introduce a slightly more general formulation of the conditional independence property (see Definition 9.1.2). (ii) It should be emphasized that the property of conditional independence may not be invariant under an equivalent change of probability measure. Thus, if the random times $\tau_i$, $i = 1, \ldots, n$ are conditionally independent with respect to $\mathbb{F}$ under $\mathbb{Q}^*$, this does not imply that these random times are also conditionally independent with respect to $\mathbb{F}$ under an equivalent probability measure $\mathbb{Q}$.

Let us stress that the following equality does not necessarily hold for every $t_1, \ldots, t_n \in [0, u]$ and $u \in [0, T[$

$$\mathbb{Q}^*\{\tau_i > t_i \,|\, \mathcal{F}_T\} = \mathbb{Q}^*\{\tau_i > t_i \,|\, \mathcal{F}_u\}.$$

However, the family of random times constructed in Example 9.1.5 below enjoys the above property (this feature is reflected in equality (9.3)). Since this property will be frequently used in what follows, we now introduce the following assumption standing for the rest of this section.

**Condition (C.1)** For every $T > 0$, $u \in [0, T]$, and $i = 1, \ldots, n$ we have

$$\mathbb{Q}^*\{\tau_i > u \,|\, \mathcal{F}_T\} = \mathbb{Q}^*\{\tau_i > u \,|\, \mathcal{F}_u\}.$$

*Example 9.1.5. Canonical construction of conditionally independent default times.* We shall now provide an explicit construction of a conditionally independent family of random times with given $\mathbb{F}$-hazard processes.

Let $\Gamma^i$, $i = 1, \ldots, n$ be a given collection of $\mathbb{F}$-adapted increasing continuous stochastic processes, defined on a common filtered probability space $(\hat{\Omega}, \mathbb{F}, \mathbb{P}^*)$. We assume that $\Gamma_0^i = 0$ and $\Gamma_\infty^i = \infty$, for $i = 1, \ldots, n$ (clearly $\Gamma_t^i < \infty$ for every $t \in \mathbb{R}_+$). Let $(\tilde{\Omega}, \tilde{\mathcal{F}}, \tilde{\mathbb{P}})$ be an auxiliary probability space, endowed with a sequence $\xi_i$, $i = 1, \ldots, n$ of mutually independent random variables uniformly distributed on the interval $[0, 1]$. We consider the product space $(\Omega, \mathcal{G}, \mathbb{Q}^*) = (\hat{\Omega} \times \tilde{\Omega}, \mathcal{F}_\infty \otimes \tilde{\mathcal{F}}, \mathbb{P}^* \otimes \tilde{\mathbb{P}})$, and we set, for $i = 1, \ldots, n$,

$$\tau_i = \inf \{ t \in \mathbb{R}_+ : \Gamma_t^i \geq - \ln \xi_i \}. \tag{9.1}$$

It might be useful to observe that each random variable $\eta_i := - \ln \xi_i$ is exponentially distributed under $\mathbb{Q}^*$, with unit parameter. Therefore,

$$\tau_i = \inf \{ t \in \mathbb{R}_+ : \Gamma_t^i \geq \eta_i \},$$

where $\eta_i$, $i = 1, \ldots, n$ is a family of mutually independent random variables with unit exponential law. It is natural to endow the product space $(\Omega, \mathcal{G}, \mathbb{Q}^*)$ with the enlarged filtration $\mathbb{G} = \mathbb{F} \vee \mathbb{H}^1 \vee \ldots \vee \mathbb{H}^n$. For each $t$, the $\sigma$-field $\mathcal{G}_t$ represents all information available to an agent at time $t$, including the observations of all random times $\tau_i$, $i = 1, \ldots, n$. Formally,

$$\mathcal{G}_t = \mathcal{F}_t \vee \sigma \big( \{\tau_1 < t_1\}, \ldots, \{\tau_n < t_n\} : t_1 \leq t, \ldots, t_n \leq t \big).$$

Let us finally observe that the sequence of random times constructed above satisfies the desired property that the equality $\mathbb{Q}^* \{\tau_i = \tau_j\} = 0$ holds for every $i, j = 1, \ldots, n$ such that $i \neq j$.

**Lemma 9.1.1.** *For a family $\Gamma^i, \ldots, \Gamma^n$ of $\mathbb{F}$-adapted increasing continuous processes, let the random times $\tau_1, \ldots, \tau_n$ be defined as in Example 9.1.5.*
*(i) The joint conditional probability of survival satisfies, for $t_1, \ldots, t_n \in \mathbb{R}_+$,*

$$\mathbb{Q}^* \{\tau_1 > t_1, \ldots, \tau_n > t_n \,|\, \mathcal{F}_\infty\} = \prod_{i=1}^{n} e^{-\Gamma_{t_i}^i} = e^{-\sum_{i=1}^{n} \Gamma_{t_i}^i}. \tag{9.2}$$

*(ii) For arbitrary $t_1, \ldots, t_n \in \mathbb{R}_+$ and any $T \geq \max(t_1, \ldots, t_n)$ we have*

$$\mathbb{Q}^* \{\tau_1 > t_1, \ldots, \tau_n > t_n \,|\, \mathcal{F}_T\} = \prod_{i=1}^{n} e^{-\Gamma_{t_i}^i} = e^{-\sum_{i=1}^{n} \Gamma_{t_i}^i}. \tag{9.3}$$

*(iii) Random times $\tau_1, \ldots, \tau_n$ are conditionally independent with respect to the filtration $\mathbb{F}$ under $\mathbb{Q}^*$.*
*(iv) For each $i = 1, \ldots, n$, the process $\Gamma^i$ represents the $\mathbb{F}$-hazard process and the $(\mathbb{F}, \mathbb{G})$-martingale hazard process of the random time $\tau_i$. In other words, the equality $\Gamma^i = \Lambda^i$ is valid for every $i = 1, \ldots, n$.*

*Proof.* Observe first that $\{\tau_i > t\} = \{\Gamma_t^i < -\ln \xi_i\} = \{e^{-\Gamma_t^i} > \xi_i\}$. Let us take arbitrary numbers $t_1, \ldots, t_n \in \mathbb{R}_+$. Each random variable $\Gamma_{t_i}^i$ is obviously $\mathcal{F}_\infty$-measurable, and so

$$\mathbb{Q}^*\{\tau_1 > t_1, \ldots, \tau_n > t_n \mid \mathcal{F}_\infty\}$$
$$= \mathbb{Q}^*\{e^{-\Gamma_{t_1}^1} > \xi_1, \ldots, e^{-\Gamma_{t_n}^n} > \xi_n \mid \mathcal{F}_\infty\}$$
$$= \mathbb{Q}^*\{e^{-x_1} > \xi_1, \ldots, e^{-x_n} > \xi_n \mid \mathcal{F}_\infty\}_{x_1 = \Gamma_{t_1}^1, \ldots, x_n = \Gamma_{t_n}^n}$$
$$= \prod_{i=1}^n \mathbb{Q}^*\{e^{-x_i} > \xi_1\}_{x_i = \Gamma_{t_i}^i} = \prod_{i=1}^n \tilde{\mathbb{P}}\{e^{-x_i} > \xi_i\}_{x_i = \Gamma_{t_i}^i} = \prod_{i=1}^n e^{-\Gamma_{t_i}^i}.$$

This proves part (i). Equality (9.3) is a simple consequence of (9.2). Indeed, since for any $T \geq t_i$, the random variable $\Gamma_{t_i}^i$ is $\mathcal{F}_T$-measurable, it is apparent that the following chain of equalities holds

$$\mathbb{Q}^*\{\tau_1 > t_1, \ldots, \tau_n > t_n \mid \mathcal{F}_T\} = \mathbb{E}_{\mathbb{Q}^*}\left(\mathbb{Q}^*\{\tau_1 > t_1, \ldots, \tau_n > t_n \mid \mathcal{F}_\infty\} \mid \mathcal{F}_T\right)$$
$$= \mathbb{E}_{\mathbb{Q}^*}\left(e^{-\sum_{i=1}^n \Gamma_{t_i}^i} \mid \mathcal{F}_T\right) = e^{-\sum_{i=1}^n \Gamma_{t_i}^i}.$$

In particular, for any $i$ and every $t_i \leq T$ we have

$$\mathbb{Q}^*\{\tau_i > t_i \mid \mathcal{F}_T\} = \mathbb{Q}^*\{\tau_i > t_i \mid \mathcal{F}_\infty\} = e^{-\Gamma_{t_i}^i}.$$

To establish the conditional independence of random times $\tau_i$, $i = 1, \ldots, n$ with respect to the filtration $\mathbb{F}$, it is enough to observe that, by virtue of part (ii), for any fixed $T > 0$ and arbitrary $t_1, \ldots, t_n \leq T$ we have

$$\mathbb{Q}^*\{\tau_1 > t_1, \ldots, \tau_n > t_n \mid \mathcal{F}_T\} = \prod_{i=1}^n e^{-\Gamma_{t_i}^i} = \prod_{i=1}^n \mathbb{Q}^*\{\tau_i > t_i \mid \mathcal{F}_T\}.$$

For the last statement, notice that from Lemma 8.2.2 we know that $\Gamma^i$ represents the $\mathbb{F}$-hazard process of $\tau_i$ and the $(\mathbb{F}, \mathbb{G}^i)$-martingale hazard process of $\tau_i$, where $\mathbb{G}^i := \mathbb{F} \vee \mathbb{H}^i$. This means that the process $\tilde{M}_t^i = H_t^i - \Gamma_{t \wedge \tau_i}^i$ is a $\mathbb{G}^i$-martingale. We need to show that $\tilde{M}^i$ is also a $\mathbb{G}$-martingale. The process $\tilde{M}^i$ is manifestly $\mathbb{G}$-adapted. It suffices to check that for any $t \leq s$

$$\mathbb{E}_{\mathbb{Q}^*}(H_s^i - \Gamma_{s \wedge \tau_i}^i \mid \mathcal{G}_t) = \mathbb{E}_{\mathbb{Q}^*}(H_s^i - \Gamma_{s \wedge \tau_i}^i \mid \mathcal{G}_t^i).$$

Notice that the $\sigma$-fields $\mathcal{G}_s^i$ and $\tilde{\mathcal{H}}_t := \mathcal{H}_t^1 \vee \mathcal{H}_t^{i-1} \vee \ldots \vee \mathcal{H}_t^{i+1} \vee \mathcal{H}_t^n$ are conditionally independent given $\mathcal{G}_t^i$. Consequently,

$$\mathbb{E}_{\mathbb{Q}^*}(H_s^i - \Gamma_{s \wedge \tau_i}^i \mid \mathcal{G}_t) = \mathbb{E}_{\mathbb{Q}^*}(H_s^i - \Gamma_{s \wedge \tau_i}^i \mid \mathcal{G}_t^1 \vee \tilde{\mathcal{H}}_t) = \mathbb{E}_{\mathbb{Q}^*}(H_s^i - \Gamma_{s \wedge \tau_i}^i \mid \mathcal{G}_t^i).$$

We conclude that $\Gamma^i$ is the $(\mathbb{F}, \mathbb{G})$-martingale hazard process of $\tau_i$.    □

Let us introduce a property, which is apparently stronger than the conditional independence of random times with respect to a given filtration $\mathbb{F}$.

**Definition 9.1.2.** The random times $\tau_1, \ldots, \tau_n$ are *dynamically condition-ally independent* with respect to $\mathbb{F}$ under $\mathbb{Q}^*$ if and only if for any $0 \le t < T$ and arbitrary $t_1, \ldots, t_n \in [t, T]$ we have

$$\mathbb{Q}^*\{\tau_1 > t_1, \ldots, \tau_n > t_n \,|\, \mathcal{F}_T \vee \mathcal{H}_t\} = \prod_{i=1}^{n} \mathbb{Q}^*\{\tau_i > t_i \,|\, \mathcal{F}_T \vee \mathcal{H}_t\}.$$

Recall that we denote $\mathcal{H}_t = \mathcal{H}_t^1 \vee \ldots \vee \mathcal{H}_t^n$. Since Lemma 9.1.2 is merely a slight generalization of Lemma 5.1.4, its proof is left to the reader.

**Lemma 9.1.2.** *Let* $\tau_1, \ldots, \tau_n$ *be defined on the probability space* $(\Omega, \mathcal{G}, \mathbb{Q}^*)$. *For any sub-$\sigma$-field $\mathcal{F}$ of $\mathcal{G}$, we denote* $J = \mathbb{Q}^*\{\tau_1 > t_1, \ldots, \tau_n > t_n \,|\, \mathcal{F} \vee \mathcal{H}_t\}$. *If $t_i \ge t$ for $i = 1, \ldots, n$, then*

$$J = \mathbb{1}_{\{\tau_1 > t, \ldots, \tau_n > t\}} \frac{\mathbb{Q}^*\{\tau_1 > t_1, \ldots, \tau_n > t_n \,|\, \mathcal{F}\}}{\mathbb{Q}^*\{\tau_1 > t, \ldots, \tau_n > t \,|\, \mathcal{F}\}}.$$

**Proposition 9.1.1.** *The random times* $\tau_1, \ldots, \tau_n$ *are conditionally independent with respect to the filtration $\mathbb{F}$ under $\mathbb{Q}^*$ if and only if they are dynamically conditionally independent with respect to the filtration $\mathbb{F}$ under $\mathbb{Q}^*$.*

*Proof.* It is enough to show that the conditional independence implies the dynamical conditional independence. The conditional independence of $\tau_1, \ldots, \tau_n$ with respect to $\mathbb{F}$ is equivalent to the following property: for each $T > 0$ and for arbitrary Borel subsets $A_1, \ldots, A_n$ of the interval $[0, T]$ we have

$$\mathbb{Q}^*\{\tau_1 \in A_1, \ldots, \tau_n \in A_n \,|\, \mathcal{F}_T\} = \prod_{i=1}^{n} \mathbb{Q}^*\{\tau_i \in A_i \,|\, \mathcal{F}_T\}.$$

It is thus clear that it implies that for any $t \le T$, the $\sigma$-fields $\mathcal{H}_t^1, \ldots, \mathcal{H}_t^n$ are mutually conditionally independent given $\mathcal{F}_T$. For $t \le t_i \le T$ we have $\mathcal{H}_t^i \subseteq \mathcal{H}_{t_i}^i$ and the $\sigma$-fields $\mathcal{H}_{t_i}^i$ and $\tilde{\mathcal{H}}_{t_i} := \mathcal{H}_{t_i}^1 \vee \ldots \vee \mathcal{H}_{t_i}^{i-1} \vee \mathcal{H}_{t_i}^{i+1} \vee \ldots \vee \mathcal{H}_{t_i}^n$ are conditionally independent given $\mathcal{F}_T$. This yields

$$\mathbb{Q}^*\{\tau_i > t_i \,|\, \mathcal{F}_T \vee \mathcal{H}_t\} = \mathbb{Q}^*\{\tau_i > t_i \,|\, \mathcal{F}_T \vee \mathcal{H}_t^i\}.$$

Then, by virtue of Lemma 5.1.4 (or Lemma 9.1.2) we obtain

$$\mathbb{Q}^*\{\tau_i > t_i \,|\, \mathcal{F}_T \vee \mathcal{H}_t^i\} = \mathbb{1}_{\{\tau_i > t\}} \frac{\mathbb{Q}^*\{\tau_i > t_i \,|\, \mathcal{F}_T\}}{\mathbb{Q}^*\{\tau_i > t \,|\, \mathcal{F}_T\}}.$$

Let us denote $J = \mathbb{Q}^*\{\tau_1 > t_1, \ldots, \tau_n > t_n \,|\, \mathcal{F}_T \vee \mathcal{H}_t\}$. Applying Lemma 9.1.2 with $\mathcal{F} = \mathcal{F}_T$, we find that

$$J = \mathbb{1}_{\{\tau_1 > t, \ldots, \tau_n > t\}} \frac{\mathbb{Q}^*\{\tau_1 > t_1, \ldots, \tau_n > t_n \,|\, \mathcal{F}_T\}}{\mathbb{Q}^*\{\tau_1 > t, \ldots, \tau_n > t \,|\, \mathcal{F}_T\}}.$$

Consequently, using again the conditional independence of $\tau_1, \ldots, \tau_n$, we get

$$J = \prod_{i=1}^{n} \mathbb{1}_{\{\tau_i > t\}} \frac{\mathbb{Q}^*\{\tau_i > t_i \,|\, \mathcal{F}_T\}}{\mathbb{Q}^*\{\tau_i > t \,|\, \mathcal{F}_T\}} = \prod_{i=1}^{n} \mathbb{Q}^*\{\tau_i > t_i \,|\, \mathcal{F}_T \vee \mathcal{H}_t\}.$$

This ends the proof. $\qquad\qquad\square$

We have the following, rather obvious, modification of Lemma 9.1.1.

**Lemma 9.1.3.** *For a family $\Gamma^i, \ldots, \Gamma^n$ of $\mathbb{F}$-adapted increasing continuous processes, define the random times $\tau_1, \ldots, \tau_n$ as in Example 9.1.5. Then:*
*(i) The random times $\tau_i$, $i = 1, \ldots, n$ are dynamically conditionally independent with respect to the filtration $\mathbb{F}$ under $\mathbb{Q}^*$.*
*(ii) The joint conditional probability of survival satisfies, for every $t \geq 0$ and every $t_1, \ldots, t_n \in [t, \infty)$,*

$$\mathbb{Q}^*\{\tau_1 > t_1, \ldots, \tau_n > t_n \,|\, \mathcal{F}_\infty \vee \mathcal{G}_t\} = \mathbb{1}_{\{\tau_1 > t, \ldots, \tau_n > t\}} \, e^{\sum_{i=1}^n (\Gamma_t^i - \Gamma_{t_i}^i)}.$$

*(iii) For arbitrary $T > t \geq 0$ and any $t_1, \ldots, t_n \in [t, T]$ we have*

$$\mathbb{Q}^*\{\tau_1 > t_1, \ldots, \tau_n > t_n \,|\, \mathcal{F}_T \vee \mathcal{G}_t\} = \mathbb{1}_{\{\tau_1 > t, \ldots, \tau_n > t\}} \, e^{\sum_{i=1}^n (\Gamma_t^i - \Gamma_{t_i}^i)}.$$

*Proof.* The proof is left to the reader. □

Let us now examine the minimum of default times $\tau_1, \ldots, \tau_n$. If each hazard process $\Gamma^i$ admits the $\mathbb{F}$-intensity $\gamma^i$, equality (9.3) becomes

$$\mathbb{Q}^*\{\tau_1 > t_1, \ldots, \tau_n > t_n \,|\, \mathcal{F}_T\} = \prod_{i=1}^n \exp\left(-\int_0^{t_i} \gamma_u^i \, du\right). \qquad (9.4)$$

Let us now focus on the random time $\tau_{(1)} = \tau_1 \wedge \cdots \wedge \tau_n$. The $\mathbb{F}$-hazard process $\Gamma^{(1)}$ of this random time satisfies $\Gamma^{(1)} = \sum_{i=1}^n \Gamma^i$, since (9.3) implies that

$$e^{-\Gamma_t^{(1)}} = \mathbb{Q}^*\{\tau_{(1)} > t \,|\, \mathcal{F}_t\} = \mathbb{Q}^*\{\tau_1 > t, \ldots, \tau_n > t \,|\, \mathcal{F}_t\} = e^{-\sum_{i=1}^n \Gamma_t^i}.$$

Thus, in view of (5.11), for any $\mathcal{F}_s$-measurable random variable $Y$ and any $t \leq s$, the following equality holds

$$\mathbb{E}_{\mathbb{Q}^*}(\mathbb{1}_{\{\tau_{(1)} > s\}} Y \,|\, \mathcal{G}_t) = \mathbb{1}_{\{\tau_{(1)} > t\}} \mathbb{E}_{\mathbb{Q}^*}(Y e^{\Gamma_t^{(1)} - \Gamma_s^{(1)}} \,|\, \mathcal{F}_t). \qquad (9.5)$$

Notice that for any $\mathcal{G}_s$-measurable random variable $Y$ and any $t \leq s$ we have

$$\mathbb{E}_{\mathbb{Q}^*}(\mathbb{1}_{\{\tau_{(1)} > s\}} Y \,|\, \mathcal{G}_t) = \mathbb{E}_{\mathbb{Q}^*}(\mathbb{1}_{\{\tau_{(1)} > s\}} Y \,|\, \tilde{\mathcal{G}}_t),$$

where $\tilde{\mathbb{G}}$ stands for the filtration associated with $\tau_{(1)}$; that is, $\tilde{\mathbb{G}} = \mathbb{F} \vee \mathbb{H}^{(1)}$, where the filtration $\mathbb{H}^{(1)}$ is generated by the process $H_t^{(1)} = \mathbb{1}_{\{\tau_{(1)} \leq t\}}$.

*Remarks.* Equality (9.5) was derived through the direct approach, based on the concept of the $\mathbb{F}$-hazard process. Since we already know that $\Gamma^i = \Lambda^i$, it seems interesting to examine the martingale approach as well. To this end, we shall first show that the two filtrations, $\mathbb{F}$ and $\tilde{\mathbb{G}}$, satisfy the martingale invariance condition (M.1). Condition (M.1) is known to be equivalent to Condition (F1.a), which in the present context reads: for any $t > 0$

$$\mathbb{Q}^*\{\tau_{(1)} \leq u \,|\, \mathcal{F}_t\} = \mathbb{Q}^*\{\tau_{(1)} \leq u \,|\, \mathcal{F}_\infty\}, \quad \forall u \in [0, t].$$

The last condition is satisfied since, by virtue of parts (i)-(ii) of Lemma 9.1.1, we have

$$\mathbb{Q}^*\{\tau_{(1)} > u \,|\, \mathcal{F}_t\} = \mathbb{Q}^*\{\tau_{(1)} > u \,|\, \mathcal{F}_\infty\}, \quad \forall\, u \in [0, t].$$

We conclude that Condition (F1.a) – and thus also Condition (M.1) – are indeed satisfied in the present framework.

By virtue of Lemma 7.1.2, the $(\mathbb{F}, \tilde{\mathbb{G}})$-martingale hazard process of $\tau_{(1)}$ equals $\Lambda^{(1)} = \sum_{i=1}^n \Lambda^i$ (so that its $\mathbb{F}$-intensity $\lambda^{(1)} = \sum_{i=1}^n \lambda^i$, if it exists). Using Proposition 7.1.1, with the random variable $Y = 1$, we conclude that

$$\mathbb{Q}^*\{\tau_{(1)} > s \,|\, \tilde{\mathcal{G}}_t\} = \mathbb{1}_{\{\tau_{(1)} > t\}}\, \mathbb{E}_{\mathbb{Q}^*}\!\left(e^{\Lambda_t^{(1)} - \Lambda_s^{(1)}} \,|\, \mathcal{F}_t\right)$$

for $t \leq s$. However, to derive a counterpart of equality (9.5), we need to check a rather cumbersome continuity condition of Proposition 7.1.1. This shows that the direct approach, based on the notion of the $\mathbb{F}$-hazard process, is more convenient.

**Case of signed intensities.** Some authors (e.g., Kijima and Muromachi (2000)) examine credit risk models in which the negative values of intensities of random times involved are not precluded. They rightly indicate that negative values of the intensity process clearly contradict the interpretation of the intensity as the conditional probability of survival over an infinitesimal time interval. Nevertheless, the construction of conditionally independent random times also goes through in this case. Let us analyze this issue in some detail.

Assume that we are given a collection $\Gamma^i$, $i = 1, \ldots, n$ of $\mathbb{F}$-adapted continuous stochastic processes, with $\Gamma_0^i = 0$, defined on a filtered probability space $(\hat{\Omega}, \mathbb{F}, \hat{\mathbb{P}})$. We introduce a finite family $\tau_i$, $i = 1, \ldots, n$, of random times on the enlarged probability space $(\Omega, \mathcal{G}, \mathbb{Q}^*)$, through formula (9.1), i.e.,

$$\tau_i = \inf\{t \in \mathbb{R}_+ : \Gamma_t^i \geq -\ln \xi_i\}.$$

The random times $\tau_1, \ldots, \tau_n$ possess most of the required properties, but in general the hazard processes of these random times do not coincide with processes $\Gamma^i$ as the following result shows.

**Lemma 9.1.4.** *The random times $\tau_i$, $i = 1, \ldots, n$ are conditionally independent with respect to $\mathbb{F}$ under $\mathbb{Q}^*$. In particular, for every $t_1, \ldots, t_n \leq T$,*

$$\mathbb{Q}^*\{\tau_1 > t_1, \ldots, \tau_n > t_n \,|\, \mathcal{F}_T\} = \prod_{i=1}^n e^{-\tilde{\Gamma}_{t_i}^i} = e^{-\sum_{i=1}^n \tilde{\Gamma}_{t_i}^i}, \tag{9.6}$$

*where $\tilde{\Gamma}$ is the increasing process associated with $\Gamma^i$, i.e., $\tilde{\Gamma}_t^i := \sup_{u \leq t} \Gamma_u^i$.*

*Proof.* The proof goes along similar lines as the proof of Lemma 9.1.1. It is enough to observe that we have

$$\{\tau_i > t\} = \{\tilde{\Gamma}_t^i < -\ln \xi_i\} = \{e^{-\tilde{\Gamma}_t^i} > \xi_i\}.$$

Notice that the inclusion $\{\tilde{\Gamma}_t^i < -\ln \xi_i\} \subseteq \{\Gamma_t^i < -\ln \xi_i\}$ is always true, but in general $\{\tilde{\Gamma}_t^i < -\ln \xi_i\} \neq \{\Gamma_t^i < -\ln \xi_i\}$. $\qquad\square$

In view of the lemma, if default times are obtained through the construction described in this section, the intensity of each default time $\tau_i$ becomes automatically zero on the (random) set $\{\gamma^i < 0\}$. To conclude, the 'true' intensity of $\tau_i$ equals $\tilde{\gamma}^i = \max(\gamma^i, 0)$.

### 9.1.3 Valuation of the $i^{\text{th}}$-to-Default Contract

Our next goal is to compute the initial price $S_0^{(i)}$ for the $i^{\text{th}}$-to-default claim $CCT^{(i)}$ under the assumption of conditional independence of default times. We assume throughout that processes $Z^k$, $k = 1, \ldots, n$ are $\mathbb{F}$-predictable, and random payoffs $X_k$, $k = 1, \ldots, n$ and $X$ are $\mathcal{F}_T$-measurable. These assumptions make the subsequent results less universal than some results derived in the next chapter. For instance, the recovery payoffs that explicitly depend on the timing of previous defaults are formally excluded. However, if these restrictions were not imposed, we would not be able to benefit from the postulated conditional independence of default times with respect to the reference filtration $\mathbb{F}$.

To derive a representation for the value process of a general $i^{\text{th}}$-to-default claim $CCT^{(i)}$, we need to introduce some auxiliary notation. Let $i, j \in \{1, \ldots, n\}$ be fixed. By $\Pi^{(i,j)}$ we denote the collection of specific partitions of the set $\{1, \ldots, n\}$. Namely, if $\pi \in \Pi^{(i,j)}$ then $\pi = \{\pi_-, \{j\}, \pi_+\}$, where

$$\pi_- = \{k_1, k_2, \ldots, k_{i-1}\}, \quad \pi_+ = \{l_1, l_2, \ldots, l_{n-i}\},$$

and

$$j \notin \pi_-, \quad j \notin \pi_+, \quad \pi_- \cap \pi_+ = \emptyset, \quad \pi_- \cup \pi_+ \cup \{j\} = \{1, \ldots, n\}.$$

For a fixed $i \in \{1, \ldots, n\}$ and any $j \in \{1, \ldots, n\}$, the partition $\pi = \{\pi_-, \{j\}, \pi_+\}$ should be interpreted as follows: the index $j$ is the index of the $i^{\text{th}}$ defaulting entity. The set $\pi_-$ contains indices of all the names that default prior to the default of the $j^{\text{th}}$ entity. Finally, the set $\pi_+$ includes all indices corresponding to the entities whose defaults occur after the default of the $j^{\text{th}}$ entity.

*Example 9.1.6.* In this example, we consider $n = 2$ credit entities. For $i = 1$ (i.e., in the case of the first-to-default claim) and $j = 1, 2$ we have

$$\Pi^{(1,1)} = \big\{\{\emptyset, \{1\}, \{2\}\}\big\}, \quad \Pi^{(1,2)} = \big\{\{\emptyset, \{2\}, \{1\}\}\big\}.$$

Likewise, in the case of the second-to-default claim, we have

$$\Pi^{(2,1)} = \big\{\{\{2\}, \{1\}, \emptyset\}\big\}, \quad \Pi^{(2,2)} = \big\{\{\{1\}, \{2\}, \emptyset\}\big\}.$$

In this example, each set $\Pi^{(i,j)}$ contains only one partition; for example, the only element of $\Pi^{(1,1)}$ is the partition $\pi = \{\emptyset, \{1\}, \{2\}\}$.

*Example 9.1.7.* Let us now consider the case of $n = 4$. Let us take, for instance, $j = 3$. Then $\Pi^{(1,3)} = \left\{ \{\emptyset, \{3\}, \{1, 2, 4\}\} \right\}$,

$$\Pi^{(2,3)} = \left\{ \{\{1\}, \{3\}, \{2, 4\}\}, \{\{2\}, \{3\}, \{1, 4\}\}, \{\{4\}, \{3\}, \{1, 2\}\} \right\},$$

$$\Pi^{(3,3)} = \left\{ \{\{1, 2\}, \{3\}, \{4\}\}, \{\{1, 4\}, \{3\}, \{2\}\}, \{\{2, 4\}, \{3\}, \{1\}\} \right\},$$

and finally, $\Pi^{(4,3)} = \left\{ \{\{1, 2, 4\}, \{3\}, \emptyset\} \right\}$.

For any numbers $i, j \in \{1, \ldots, n\}$ and arbitrary $\pi \in \Pi^{(i,j)}$, we write $\tau(\pi_-) = \max \{ \tau_k : k \in \pi_- \}$ and $\tau(\pi_+) = \min \{ \tau_l : l \in \pi_+ \}$, where we set by convention: $\max \emptyset = -\infty$ and $\min \emptyset = \infty$. It is clear that $\tau(\pi_-)$ ($\tau(\pi_+)$, resp.) is the default time that immediately precedes (follows, resp.) the time of the $i^{\text{th}}$ default.

We shall first examine the general case, and subsequently the special case of non-random recovery payoffs and hazard processes. It is not difficult to check that the initial price of the $i^{\text{th}}$-to-default payoff satisfies (cf. Proposition 10.2.3)

$$S_0^{(i)} = \mathbb{E}_{\mathbb{Q}^*} \left( \sum_{j=1}^{n} B_{\tau_j}^{-1} Z_{\tau_j}^j \sum_{\pi \in \Pi^{(i,j)}} \mathbb{1}_{\{\tau(\pi_-) < \tau_j < \tau(\pi_+),\, \tau_j \leq T\}} \right)$$

$$+ \mathbb{E}_{\mathbb{Q}^*} \left( B_T^{-1} \sum_{j=1}^{n} X_j \sum_{\pi \in \Pi^{(i,j)}} \mathbb{1}_{\{\tau(\pi_-) < \tau_j < \tau(\pi_+),\, \tau_j \leq T\}} \right)$$

$$+ \mathbb{E}_{\mathbb{Q}^*} \left( B_T^{-1} X \sum_{j=1}^{n} \sum_{\pi \in \Pi^{(i,j)}} \mathbb{1}_{\{\tau(\pi_-) < \tau_j < \tau(\pi_+),\, \tau_j > T\}} \right)$$

$$=: J_1 + J_2 + J_3.$$

Since $B_0 = 1$, we shall frequently omit $B_0$ from the formulae. In view of the assumed conditional independence of random times $\tau_1, \ldots, \tau_n$, for the first term we obtain

$$J_1 = \mathbb{E}_{\mathbb{Q}^*} \left\{ \mathbb{E}_{\mathbb{Q}^*} \left( \sum_{j=1}^{n} Z_{\tau_j}^j B_{\tau_j}^{-1} \sum_{\pi \in \Pi^{(i,j)}} \mathbb{1}_{\{\tau(\pi_-) < \tau_j < \tau(\pi_+),\, \tau_j \leq T\}} \Big| \mathcal{F}_T \right) \right\}$$

$$= \mathbb{E}_{\mathbb{Q}^*} \left\{ \sum_{j=1}^{n} \int_0^T Z_u^j B_u^{-1} \left( \sum_{\pi \in \Pi^{(i,j)}} \left[ \prod_{k \in \pi_-} \mathbb{Q}^* \{\tau_k < u \,|\, \mathcal{F}_T\} \right] \right. \right.$$

$$\left. \left. \times \left[ \prod_{l \in \pi_+} \mathbb{Q}^* \{\tau_l > u \,|\, \mathcal{F}_T\} \right] \right) d\mathbb{Q}^* \{\tau_j \leq u \,|\, \mathcal{F}_T\} \right\}$$

$$= \mathbb{E}_{\mathbb{Q}^*} \left\{ \sum_{j=1}^{n} \int_0^T Z_u^j e^{-\int_0^u r_s ds} \left( \sum_{\pi \in \Pi^{(i,j)}} \left[ \prod_{k \in \pi_-} (1 - e^{-\Gamma_u^k}) \right] \right. \right. \tag{9.7}$$

$$\left. \left. \times \left[ e^{-\sum_{l \in \pi_+} \Gamma_u^l} \right] \right) \gamma_u^j e^{-\Gamma_u^j} \, du \right\}.$$

For $J_2$, we have

$$J_2 = \mathbb{E}_{\mathbb{Q}^*}\left\{\mathbb{E}_{\mathbb{Q}^*}\left(B_T^{-1}\sum_{j=1}^{n}X_j\sum_{\pi\in\Pi^{(i,j)}}\mathbb{1}_{\{\tau(\pi_-)<\tau_j<\tau(\pi_+),\,\tau_j\le T\}}\,\Big|\,\mathcal{F}_T\right)\right\}$$

$$= \mathbb{E}_{\mathbb{Q}^*}\left\{B_T^{-1}\sum_{j=1}^{n}X_j\int_0^T\left(\sum_{\pi\in\Pi^{(i,j)}}\left[\prod_{k\in\pi_-}\mathbb{Q}^*\{\tau_k<u\,|\,\mathcal{F}_T\}\right]\right.\right.$$
$$\left.\left.\times\left[\prod_{l\in\pi_+}\mathbb{Q}^*\{\tau_l>u\,|\,\mathcal{F}_T\}\right]\right)d\mathbb{Q}^*\{\tau_j\le u\,|\,\mathcal{F}_T\}\right\}$$

$$= \mathbb{E}_{\mathbb{Q}^*}\left\{e^{-\int_0^T r_s\,ds}\sum_{j=1}^{n}X_j\int_0^T\left(\sum_{\pi\in\Pi^{(i,j)}}\left[\prod_{k\in\pi_-}\left(1-e^{-\Gamma_u^k}\right)\right]\right.\right.$$
$$\left.\left.\times\left[e^{-\sum_{l\in\pi_+}\Gamma_u^l}\right]\right)\gamma_u^j e^{-\Gamma_u^j}\,du\right\},$$

and the last term satisfies

$$J_3 = \mathbb{E}_{\mathbb{Q}^*}\left\{\mathbb{E}_{\mathbb{Q}^*}\left(XB_T^{-1}\sum_{j=1}^{n}\sum_{\pi\in\Pi^{(i,j)}}\mathbb{1}_{\{\tau(\pi_-)<\tau_j<\tau(\pi_+),\,\tau_j>T\}}\,\Big|\,\mathcal{F}_\infty\right)\right\}$$

$$= \mathbb{E}_{\mathbb{Q}^*}\left\{XB_T^{-1}\sum_{j=1}^{n}\int_T^\infty\left(\sum_{\pi\in\Pi^{(i,j)}}\left[\prod_{k\in\pi_-}\mathbb{Q}^*\{\tau_k<u\,|\,\mathcal{F}_\infty\}\right]\right.\right.$$
$$\left.\left.\times\left[\prod_{l\in\pi_+}\mathbb{Q}^*\{\tau_l>u\,|\,\mathcal{F}_\infty\}\right]\right)d\mathbb{Q}^*\{\tau_j\le u\,|\,\mathcal{F}_\infty\}\right\}$$

$$= \mathbb{E}_{\mathbb{Q}^*}\left\{Xe^{-\int_0^T r_s\,ds}\sum_{j=1}^{n}\int_T^\infty\left(\sum_{\pi\in\Pi^{(i,j)}}\left[\prod_{k\in\pi_-}\left(1-e^{-\Gamma_u^k}\right)\right]\right.\right.$$
$$\left.\left.\times\left[e^{-\sum_{l\in\pi_+}\Gamma_u^l}\right]\right)\gamma_u^j e^{-\Gamma_u^j}\,du\right\}.$$

In case of $i = 1$, the above result agrees with the more abstract expression established in Proposition 10.2.4. Indeed, it suffices to check that, under the present assumptions, the process $\tilde{V}^{(1)}$ – defined in the statement of Proposition 10.2.4 – is continuous at $\tau_{(1)}$. Proposition 10.2.4 provides a stronger result, since the assumption of conditional independence of default times is not required for this more general result to hold.

Let us assume that $Z_t^j = z^j(t)$, where $z^j : [0,T] \to \mathbb{R}$, $i = j,\ldots,n$ are deterministic (integrable) functions of time. In addition, we assume that the terminal payoffs $X_j = x_j$ and $X = x$, where $x_j$, $j = 1,\ldots,n$ and $x$ are constants. Finally, the intensities of default times $\gamma_t^j = \gamma^j(t)$, $j = 1,\ldots,n$ are assumed to be non-random. Even though these assumptions are rather stringent, they are nevertheless widely common in literature, since they lead to a simple result for the values of various kinds of $i^{\text{th}}$-to-default claims. Let us stress that the interest rates are not assumed to be non-random here.

**Proposition 9.1.2.** *Let $B(0, T)$ stand for the price of a default-free zero-coupon bond maturing at time $T$. Assume that the default times $\tau_j$, $j = 1, \ldots, n$ are conditionally independent with respect to the filtration $\mathbb{F}$, with deterministic intensities $\gamma^j$. The price of the $i^{\mathrm{th}}$-to-default claim with deterministic recovery payoffs at time $t = 0$ equals*

$$
S_0^{(i)} = \sum_{j=1}^{n} \int_0^T B(0, u) z^j(u) g_{ij}(u) \gamma^j(u) e^{-\int_0^u \gamma^j(s) ds} \, du
$$

$$
+ B(0, T) \sum_{j=1}^{n} x_j \int_0^T g_{ij}(u) \gamma^j(u) e^{-\int_0^u \gamma^j(s) ds} \, du
$$

$$
+ B(0, T) x \sum_{j=1}^{n} \int_T^{\infty} g_{ij}(u) \gamma^j(u) e^{-\int_0^u \gamma^j(s) ds} \, du,
$$

*where, for every $u \in \mathbb{R}_+$,*

$$
g_{ij}(u) := \sum_{\pi \in \Pi^{(i,j)}} e^{-\sum_{l \in \pi_+} \int_0^u \gamma^l(s) ds} \prod_{k \in \pi_-} \left(1 - e^{-\int_0^u \gamma^k(s) ds}\right).
$$

*Proof.* It is enough to recall that $B(0, t) = \mathbb{E}_{\mathbb{Q}^*}(B_t^{-1})$ for $t \in [0, T]$.    $\square$

We shall now examine some particular cases of $i^{\mathrm{th}}$-to-default claims. We maintain the assumption that the default times are $\mathbb{F}$-conditionally independent, but we do not postulate that their $\mathbb{F}$-intensities are deterministic.

*Example 9.1.8.* Let us find the initial price, $S_0^{(F)}$ say, of a default swap of type F, which is an example of the first-to-default contract. To this end, we first consider a general first-to-default claim $CCT^{(1)}$. Using previously established formulae (or by direct calculations), we find that $S_0^{(1)} = J_1 + J_2 + J_3$, where the terms $J_1, J_2, J_3$ can be evaluated as follows. First, the term $J_1$ – associated with the recovery payoff $Z_{\tau_j}^j$ at default time $\tau_j \le T$ when the $j^{\mathrm{th}}$ reference entity is the first-to-default – is given by the formula

$$
J_1 = B_0 \, \mathbb{E}_{\mathbb{Q}^*} \left( \sum_{j=1}^{n} B_{\tau_j}^{-1} Z_{\tau_j}^j \, \mathbb{1}_{\{\tau_j = \tau_{(1)}, \, \tau_j \le T\}} \, \Big| \, \mathcal{G}_0 \right)
$$

$$
= \mathbb{E}_{\mathbb{Q}^*} \left( \sum_{j=1}^{n} \int_0^T Z_u^j e^{-\int_0^u r_s ds} e^{-\sum_{l \ne j} \Gamma_u^l} \gamma_u^j e^{-\Gamma_u^j} \, du \right).
$$

Then, the term $J_2$ – corresponding to the random payoff $X_j$ at maturity $T$ when the $j^{\mathrm{th}}$ reference entity is the first-to-default – satisfies

$$
J_2 = B_0 \, \mathbb{E}_{\mathbb{Q}^*} \left( B_T^{-1} \sum_{j=1}^{n} X_j \, \mathbb{1}_{\{\tau_j = \tau_{(1)}, \, \tau_j \le T\}} \, \Big| \, \mathcal{G}_0 \right)
$$

$$
= \mathbb{E}_{\mathbb{Q}^*} \left( e^{-\int_0^T r_s ds} \sum_{j=1}^{n} \int_0^T X_j e^{-\sum_{l \ne j} \Gamma_u^l} \gamma_u^j e^{-\Gamma_u^j} \, du \right).
$$

Finally, the last term – associated with the payoff $X$ at maturity $T$ in case there was no default prior to $T$ – is given by the formula

$$J_3 = B_0 \, \mathbb{E}_{\mathbb{Q}^*} \big( X B_T^{-1} \mathbb{1}_{\{\tau_{(1)} > T\}} \, | \, \mathcal{G}_0 \big) = \mathbb{E}_{\mathbb{Q}^*} \Big( X B_T^{-1} \sum_{j=1}^{n} \mathbb{1}_{\{\tau_j = \tau_{(1)}, \, \tau_j > T\}} \Big)$$

$$= \mathbb{E}_{\mathbb{Q}^*} \Big( X e^{-\int_0^T r_s ds} \sum_{j=1}^{n} \int_T^{\infty} e^{-\sum_{l \ne j} \Gamma_u^l} \gamma_u^j e^{-\Gamma_u^j} \, du \Big).$$

Under an additional assumption of constant recovery payoffs we have

$$J_2 = \hat{x} \, \mathbb{E}_{\mathbb{Q}^*} \big( B_T^{-1} \, \mathbb{1}_{\{\tau_{(1)} \le T\}} \big) = \hat{x} B(0,T) - \hat{x} B_0 \, \mathbb{E}_{\mathbb{Q}^*} \big( B_T^{-1} \mathbb{1}_{\{\tau_{(1)} > T\}} \big),$$

and

$$J_3 = x B_0 \, \mathbb{E}_{\mathbb{Q}^*} \big( B_T^{-1} \mathbb{1}_{\{\tau_{(1)} > T\}} \big),$$

provided that $X_j = \hat{x}$ for $j = 1, \dots, n$, and $X = x$, where $\hat{x}$ and $x$ are real numbers. In the case of the default swap of type F, we clearly have $Z^j \equiv 0$ for every $j$, and thus the following result – originally due to Kijima and Muromachi (2000) – is valid.

**Proposition 9.1.3.** *The value of a default swap of type F, with the constant payoffs $\hat{x}$ and $x$ at time $t = 0$ equals*

$$S_0^{(F)} = \hat{x} B(0,T) + (x - \hat{x}) B_0 \, \mathbb{E}_{\mathbb{Q}^*} \big( B_T^{-1} e^{-\sum_{j=1}^{n} \Gamma_T^j} \big). \tag{9.8}$$

*Proof.* It is enough to observe that

$$S_0^{(F)} = J_2 + J_3 = \hat{x} B(0,T) + (x - \hat{x}) B_0 \, \mathbb{E}_{\mathbb{Q}^*} \big( B_T^{-1} \mathbb{1}_{\{\tau_{(1)} > T\}} \big),$$

and to apply the conditioning with respect to the $\sigma$-field $\mathcal{F}_T$.    □

*Example 9.1.9.* We shall now calculate the initial price, $S_0^{(D)}$ say, of the default swap of type D. Recall that such a contract provides a protection against the first two defaults (see Example 9.1.2). Let us denote

$$I_j^{\pi} = \mathbb{1}_{\{\tau(\pi_-) < \tau_j < \tau(\pi_+), \, \tau_j \le T\}}, \quad \tilde{I}_j^{\pi} = \mathbb{1}_{\{\tau(\pi_-) < \tau_j < \tau(\pi_+), \, \tau_j > T\}}.$$

By virtue of Proposition 10.2.3 (or by direct considerations), it is clear that the price $S_0^{(D)}$ satisfies

$$S_0^{(D)} = \mathbb{E}_{\mathbb{Q}^*} \Big( B_T^{-1} \sum_{j=1}^{n} \Big( \sum_{m \ne j} (\tilde{X}_j + \tilde{X}_m) \mathbb{1}_{\{\tau_{(1)} = \tau_m\}} \Big) \sum_{\pi \in \Pi^{(2,j)}} I_j^{\pi} \Big)$$

$$+ \mathbb{E}_{\mathbb{Q}^*} \Big( B_T^{-1} \big( \hat{X}_0 \mathbb{1}_{\{\tau_{(1)} > T\}} + \sum_{m=1}^{n} \hat{X}_m \mathbb{1}_{\{\tau_{(1)} = \tau_m, \, \tau_{(1)} \le T\}} \big) \sum_{j=1}^{n} \sum_{\pi \in \Pi^{(2,j)}} \tilde{I}_j^{\pi} \Big)$$

$$=: I_1 + I_2.$$

Invoking the postulated conditional independence of default times, we obtain

$$I_1 = \mathbb{E}_{\mathbb{Q}^*}\left\{\mathbb{E}_{\mathbb{Q}^*}\left(B_T^{-1}\sum_{j=1}^{n}\left(\sum_{m\neq j}(\tilde{X}_j+\tilde{X}_m)\mathbb{1}_{\{\tau_m<\tau_j<\tau_l,\,l\neq j,\,l\neq m,\,\tau_j\leq T\}}\right)\Big|\mathcal{F}_T\right)\right\}$$

$$= \mathbb{E}_{\mathbb{Q}^*}\left\{B_T^{-1}\sum_{j=1}^{n}\int_0^T\left(\sum_{m\neq j}(\tilde{X}_j+\tilde{X}_m)\mathbb{Q}^*\{\tau_m<u\,|\,\mathcal{F}_T\}\right.\right.$$

$$\left.\left.\times\prod_{l\neq j,\,l\neq m}\mathbb{Q}^*\{\tau_l>u\,|\,\mathcal{F}_T\}\right)d\mathbb{Q}^*\{\tau_j\leq u\,|\,\mathcal{F}_T\}\right\}$$

$$= \mathbb{E}_{\mathbb{Q}^*}\left\{e^{-\int_0^T r_s\,ds}\sum_{j=1}^{n}\int_0^T\left(\sum_{m\neq j}(\tilde{X}_j+\tilde{X}_m)(1-e^{-\Gamma_u^m})e^{-\sum_{l\neq j,\,l\neq m}\Gamma_u^l}\right)\right.$$

$$\left.\times\gamma_u^j e^{-\Gamma_u^j}\,du\right\}$$

and

$$I_2 = \mathbb{E}_{\mathbb{Q}^*}\left\{\mathbb{E}_{\mathbb{Q}^*}\left(B_T^{-1}\Big(\hat{X}_0\mathbb{1}_{\{\tau_{(1)}>T\}}+\sum_{m=1}^{n}\hat{X}_m\mathbb{1}_{\{\tau_m\leq T,\,\tau_l>T,\,l\neq m\}}\Big)\Big|\mathcal{F}_T\right)\right\}$$

$$= \mathbb{E}_{\mathbb{Q}^*}\left\{B_T^{-1}\Big(\hat{X}_0\prod_{m=1}^{n}\mathbb{Q}^*\{\tau_m>T\,|\,\mathcal{F}_T\}+\sum_{m=1}^{n}\hat{X}_m\mathbb{Q}^*\{\tau_m<T\,|\,\mathcal{F}_T\}\right.$$

$$\left.\times\prod_{l\neq m}\mathbb{Q}^*\{\tau_l>T\,|\,\mathcal{F}_T\}\Big)\right\}$$

$$= \mathbb{E}_{\mathbb{Q}^*}\left\{e^{-\int_0^T r_s\,ds}\Big(\hat{X}_0 e^{-\sum_{m=1}^{n}\Gamma_T^m}+\sum_{m=1}^{n}\hat{X}_m(1-e^{-\Gamma_T^m})e^{-\sum_{l\neq m}\Gamma_T^l}\Big)\right\}.$$

Let us now consider a special case of a default swap of type D with constant recovery payoffs. Specifically, we postulate that $\hat{X}_0 = x$ and $\hat{X}_j = \tilde{X}_j = \hat{x}$, where $x$ and $\hat{x}$ are real numbers. Formally, we deal here with the following contingent claim $Y$, which settles at time $T$ and equals

$$Y = 2\hat{x}\mathbb{1}_{\{\tau_{(2)}\leq T\}}+\hat{x}\mathbb{1}_{\{\tau_{(1)}\leq T,\,\tau_{(2)}>T\}}+x\mathbb{1}_{\{\tau_{(1)}>T\}}.$$

The following result – due to Kijima and Muromachi (2000) – provides a simple representation for the price of such a claim at time 0. Recall that $S_0^{(F)}$ stands for the initial price of a default swap of type F with constant recovery payoffs $x$ and $\hat{x}$ (see formula (9.8)).

**Proposition 9.1.4.** *Consider the default swap of type D with constant recovery payoffs $x$ and $\hat{x}$. Its price at time 0 equals $S_0^{(D)} = S_0^{(F)} + \Delta$, where $\Delta$ is non-negative , specifically*

$$\Delta = \hat{x}B(0,T)-\hat{x}B_0\,\mathbb{E}_{\mathbb{Q}^*}\left\{B_T^{-1}\Big(e^{-\sum_{j=1}^{n}\Gamma_T^j}+\sum_{j=1}^{n}(1-e^{-\Gamma_T^j})e^{-\sum_{l\neq j}\Gamma_T^l}\Big)\right\}.$$

*Proof.* Since clearly $\{\tau_{(1)} > T, \tau_{(2)} > T\} = \{\tau_{(1)} > T\}$, the random variable $Y$ can be represented as follows

$$Y = \hat{x} + (x - \hat{x})\mathbb{1}_{\{\tau_{(1)}>T\}} + \hat{x}\big(1 - \mathbb{1}_{\{\tau_{(2)}>T\}}\big),$$

and so

$$S_0^{(D)} = S_0^{(F)} + \hat{x}\big(B(0,T) - B_0\,\mathbb{E}_{\mathbb{Q}^*}\big(B_T^{-1}\mathbb{1}_{\{\tau_{(2)}>T\}}\big)\big),$$

But the random variable $B_T^{-1}$ is $\mathcal{F}_T$-measurable, so that

$$\begin{aligned}
\mathbb{E}_{\mathbb{Q}^*}\big(B_T^{-1}\mathbb{1}_{\{\tau_{(2)}>T\}}\big) &= \mathbb{E}_{\mathbb{Q}^*}\big(B_T^{-1}\mathbb{Q}^*\{\tau_{(2)} > T \mid \mathcal{F}_T\}\big) \\
&= \mathbb{E}_{\mathbb{Q}^*}\big(B_T^{-1}\mathbb{Q}^*\{\tau_{(1)} > T \mid \mathcal{F}_T\}\big) \\
&\quad + \mathbb{E}_{\mathbb{Q}^*}\big(B_T^{-1}\mathbb{Q}^*\{\tau_{(1)} \leq T, \tau_{(2)} > T \mid \mathcal{F}_T\}\big).
\end{aligned}$$

To complete the proof, it suffices to make use of Lemma 9.1.1.    □

**Kijima and Muromachi approach.** Kijima and Muromachi (2000) and Kijima (2000) further specify the credit risk model, by postulating that the short-term rate $r$, as well as the intensity processes $\gamma^i$, $i = 1, \ldots, n$, of conditionally independent default times $\tau_i$, $i = 1, \ldots, n$, are modeled as mutually dependent Gaussian diffusions. To be more explicit, they assume that all processes involved are described by the extended Vasicek model:

$$dr_t = (\phi_t^0 - a_0 r_t)\,dt + \sigma_0\,dW_t^0,$$

and, for every $i = 1, \ldots, n$

$$d\gamma_t^i = (\phi_t^i - a_i\gamma_t^i)\,dt + \sigma_i\,dW_t^i,$$

where $W^0, \ldots, W^n$ are real-valued (possibly correlated) Brownian motions. The reference filtration $\mathbb{F}$ is thus generated by $(n+1)$-dimensional stochastic process $(W^0, \ldots, W^n)$.

Within this set-up – in which we manifestly encounter the signed intensities – they derive closed-form solutions for the values of certain default swaps. Let us observe that their calculations are based on formula (9.3), rather than on the correct equality (9.6). Therefore, their final expressions – that are too lengthy to be reported here – provide only approximate values for the 'true' prices of default swaps of types F and D. The accuracy of these approximations remains, to the best of our knowledge, an open problem.

This observation should be contrasted with the fact that in Vasicek's model the short-term interest rate also can take on negative values. Even though this feature of (nominal) interest rates is by no means appealing from the point of view of financial interpretation, the values of bond prices (bond options, etc.) evaluated through the risk-neutral valuation formula are nevertheless exact from the mathematical viewpoint.

### 9.1.4 Vanilla Default Swaps of Basket Type

In Example 9.1.2, we have described a default swap of type F and a default swap of type D. These two products are of a more abstract nature than the vanilla default swaps of basket type examined in this section. In Sect. 1.3.1, the concepts of vanilla default swaps and options were introduced relative to a given credit event – a default event of a particular corporate bond. Similar contracts may also be considered in the context of a basket of defaultable securities. Typically, the underlying credit event that triggers a recovery payment is defined as the event that occurs if the first $i$ among the $n$ reference credits default (the $i^{\text{th}}$-to-default contract).

We are going to illustrate, by means of an example, the method of calculating the credit swap premium for the $i^{\text{th}}$-to-default swap. Let us consider a basket of $n$ corporate bonds. The $k^{\text{th}}$ bond has the face value $L_k$, it matures at time $T_k$, and its price process is denoted by $D_k(t, T_k)$, $k = 1, \ldots, n$. By $\tau_k$ we denote the default time of the $k^{\text{th}}$ bond, and, as before, we denote by $\tau_{(i)}$ the time of the $i^{\text{th}}$ default (recall though, that the case of simultaneous defaults is not covered in this work). We then consider a default swap maturing at time $T < \min(T_1, \ldots, T_k)$, whose covenants are described as follows:
- if the $i^{\text{th}}$ default occurs prior to or at the maturity $T$ of the default swap, that is, if $\tau_{(i)} \leq T$, then the contract holder (i.e., the protection buyer) receives at time $\tau_{(i)}$ the recovery payment in the amount of:

$$\sum_{k=1}^{n} \left( L_k - D_k(\tau_{(i)}, T_k) \right) \mathbb{1}_{\{\tau_k = \tau_{(i)}\}};$$

  this means that if the $i^{\text{th}}$ defaulting bond corresponds to the $k^{\text{th}}$ reference entity, the recovery payment is based on the value of the $k^{\text{th}}$ bond only.
- a credit swap premium in the amount $\kappa$ is paid by the contract holder at each of pre-determined time instants $t_p \leq T$, $p = 1, 2, \ldots, J$ prior to the $i^{\text{th}}$ default time or to the maturity $T$, whichever comes first.

*Remarks.* (i) Notice that the face value $L_k$ represents the conventional face value of the $k^{\text{th}}$ bond in a default swap, rather than the actual face value of the issued corporate bonds.
(ii) Alternative covenants for the recovery payment may be considered. For instance, we may set

$$\sum_{k=1}^{n} \left( L_k B(\tau, T) - D_k(\tau_{(i)}, T) \right) \mathbb{1}_{\{\tau_k = \tau_{(i)}\}},$$

or

$$\sum_{k=1}^{n} \left( D_k(\tau_{(i)}-, T) - D_k(\tau_{(i)}, T) \right) \mathbb{1}_{\{\tau_k = \tau_{(i)}\}}.$$

Kijima (2000) studies a recovery payment analogous to the last specification. More general recovery covenants stipulating that the contract holder is compensated for all $i$ defaults are also of interest, and should be studied.

In formula (1.3), the cash flows corresponding to a generic default swap were represented, from the perspective of the protection buyer, in case when the underlying security is a single defaultable bond. An analogous formula in the present set-up takes the form (recall that $t_p \leq T$ for $p = 1, \ldots, J$)

$$\mathbb{1}_{\{t=\tau_{(i)}\}} \sum_{k=1}^{n} \left(L_k - D_k(\tau_{(i)}, T_k)\right) \mathbb{1}_{\{\tau_{(i)}=\tau_k \leq T\}} - \sum_{p=1}^{J} \kappa \mathbb{1}_{\{t_p < \tau_{(i)}\}} \mathbb{1}_{\{t=t_j\}}.$$

For the sake of concreteness and simplicity, we focus on the fractional recovery of par value. In other words, we postulate that the value of the $k^{\text{th}}$ bond immediately after its default equals $\delta_k L_k$, where $\delta_k \in [0,1)$ is a constant recovery rate specific to the $k^{\text{th}}$ bond. Thus, the last formula becomes

$$\mathbb{1}_{\{t=\tau_{(i)}\}} \sum_{k=1}^{n} L_k(1 - \delta_k) \mathbb{1}_{\{\tau_{(i)}=\tau_k \leq T\}} - \sum_{p=1}^{J} \kappa \mathbb{1}_{\{t_p < \tau_{(i)}\}} \mathbb{1}_{\{t=t_j\}}.$$

Let us take $t = 0$. The value of the above cash flows at time 0 equals

$$\mathbb{E}_{\mathbb{Q}^*} \left( B_{\tau_{(i)}}^{-1} \sum_{k=1}^{n} L_k(1 - \delta_k) \mathbb{1}_{\{\tau_{(i)}=\tau_k \leq T\}} - \sum_{p=1}^{J} \kappa B_{t_p}^{-1} \mathbb{1}_{\{t_p < \tau_{(i)}\}} \right). \qquad (9.9)$$

The constant *default swap premium* $\kappa$ is chosen in such a way that – as with every swap contract – the value of the swap at its inception date is zero. In the next result, we assume that the random times $\tau_k$ are conditionally independent with respect to $\mathbb{F}$, and they satisfy Condition (C.1).

**Proposition 9.1.5.** *Assume that each default time $\tau_i$ has the $\mathbb{F}$-hazard process $\Gamma^i$, which admits the intensity $\gamma^i$, $i = 1, \ldots, n$. Then $\kappa = J_1/J_2$, where*

$$J_1 = \sum_{j=1}^{n} \mathbb{E}_{\mathbb{Q}^*} \left\{ L_j(1 - \delta_j) \int_0^T e^{-\int_0^u r_s ds} \left( \sum_{\pi \in \Pi^{(i,j)}} \left[ \prod_{k \in \pi_-} (1 - e^{-\Gamma_u^k}) \right] \right. \right.$$
$$\left. \left. \times \left[ e^{-\sum_{l \in \pi_+} \Gamma_u^l} \right] \right) \gamma_u^j e^{-\Gamma_u^j} \, du \right\},$$

*and*

$$J_2 = \sum_{p=1}^{J} \mathbb{E}_{\mathbb{Q}^*} \left\{ e^{-\int_0^{t_p} r_u du} \sum_{j=1}^{n} \left( \int_{t_p}^T \left( \sum_{\pi \in \Pi^{(i,j)}} \left[ \prod_{k \in \pi_-} (1 - e^{-\Gamma_u^k}) \right] \right. \right. \right.$$
$$\left. \left. \left. \times \left[ e^{-\sum_{l \in \pi_+} \Gamma_u^l} \right] \right) \gamma_u^j e^{-\Gamma_u^j} \, du \right) \right\}$$
$$+ \sum_{p=1}^{J} \mathbb{E}_{\mathbb{Q}^*} \left\{ e^{-\int_0^{t_p} r_u du} \sum_{j=1}^{n} \left( \int_T^\infty \left( \sum_{\pi \in \Pi^{(i,j)}} \left[ \prod_{k \in \pi_-} (1 - e^{-\Gamma_u^k}) \right] \right. \right. \right.$$
$$\left. \left. \left. \times \left[ e^{-\sum_{l \in \pi_+} \Gamma_u^l} \right] \right) \gamma_u^j e^{-\Gamma_u^j} \, du \right) \right\}.$$

*Proof.* Since $\kappa$ is a constant, (9.9) yields[1]

$$\kappa = \frac{\mathbb{E}_{\mathbb{Q}^*}\left(B_{\tau_{(i)}}^{-1}\sum_{k=1}^{n}L_k(1-\delta_k)\mathbb{1}_{\{\tau_{(i)}=\tau_k\leq T\}}\right)}{\mathbb{E}_{\mathbb{Q}^*}\left(\sum_{p=1}^{J}B_{t_p}^{-1}\mathbb{1}_{\{t_p<\tau_{(i)}\}}\right)} =: \frac{J_1}{J_2}.$$

In order to verify the equality for $J_1$, we observe that $J_1$ may be interpreted as the value at time $t = 0$ of an $i^{\text{th}}$-to-default contract with $X = X^k = 0$ and $Z^k \equiv L_k(1-\delta_k)$ for $k = 1, \ldots, n$. The formula for $J_1$ thus easily follows from (9.7). To derive the expression for $J_2$, let us first note that

$$J_2 = \mathbb{E}_{\mathbb{Q}^*}\left(\sum_{p=1}^{J}B_{t_p}^{-1}\mathbb{1}_{\{t_p<\tau_{(i)}<T\}} + \sum_{p=1}^{J}B_{t_p}^{-1}\mathbb{1}_{\{\tau_{(i)}\geq T\}}\right) = K_1 + K_2,$$

where

$$K_1 = \mathbb{E}_{\mathbb{Q}^*}\left(\sum_{p=1}^{J}B_{t_p}^{-1}\mathbb{1}_{\{t_p<\tau_{(i)}<T\}}\right),$$

and

$$K_2 = \sum_{p=1}^{J}\mathbb{E}_{\mathbb{Q}^*}\left(B_{t_p}^{-1}\mathbb{1}_{\{\tau_{(i)}\geq T\}}\right).$$

Since the process $B$ is $\mathbb{F}$-adapted and $t_p \leq T$, $p = 1, \ldots, J$, we have

$$K_1 = \mathbb{E}_{\mathbb{Q}^*}\left(\sum_{p=1}^{J}B_{t_p}^{-1}\mathbb{E}_{\mathbb{Q}^*}\left(\mathbb{1}_{\{t_p<\tau_{(i)}<T\}}\,\big|\,\mathcal{F}_T\right)\right).$$

Consequently, using the conditional independence assumption as well as Condition (C.1), we conclude that

$$K_1 = \sum_{p=1}^{J}\mathbb{E}_{\mathbb{Q}^*}\left\{e^{-\int_0^{t_p}r_u\,du}\sum_{j=1}^{n}\left(\int_{t_p}^{T}\left(\sum_{\pi\in\Pi^{(i,j)}}\left[\prod_{k\in\pi_-}(1-e^{-\Gamma_u^k})\right]\right.\right.\right.$$
$$\left.\left.\left.\times\left[e^{-\sum_{l\in\pi_+}\Gamma_u^l}\right]\right)\gamma_u^j e^{-\Gamma_u^j}\,du\right)\right\}.$$

Finally, invoking once more the conditional independence assumption and Condition (C.1), we evaluate $K_2$ as

$$K_2 = \sum_{p=1}^{J}\mathbb{E}_{\mathbb{Q}^*}\left\{e^{-\int_0^{t_p}r_u\,du}\sum_{j=1}^{n}\left(\int_{T}^{\infty}\left(\sum_{\pi\in\Pi^{(i,j)}}\left[\prod_{k\in\pi_-}(1-e^{-\Gamma_u^k})\right]\right.\right.\right.$$
$$\left.\left.\left.\times\left[e^{-\sum_{l\in\pi_+}\Gamma_u^l}\right]\right)\gamma_u^j e^{-\Gamma_u^j}\,du\right)\right\}.$$

To conclude the proof, it suffices to recall that $J_2 = K_1 + K_2$.    $\square$

---

[1] According to our standing assumptions the probability $\mathbb{Q}^*\{\tau_{(i)}\geq t_1\}$ is positive, and so $\kappa$ is well defined.

## 9.2 Default Correlations and Conditional Probabilities

**Standing assumptions.** We postulate that the $\mathbb{F}$-hazard process $\Gamma^i$ exists for each of the default times $\tau_i$. For any subset $\mathcal{I} \subseteq \{1, \ldots, n\}$, we denote $\Gamma^{\mathcal{I}} = \sum_{i \in \mathcal{I}} \Gamma^i$. In a special case where $\mathcal{I} = \{i, j\}$, we shall write $\Gamma^{ij}$, rather than $\Gamma^{\mathcal{I}}$; that is, we set $\Gamma^{ij} = \Gamma^i + \Gamma^j$. In addition, we denote $\mathbb{G}^{\mathcal{I}} = \mathbb{F} \vee \mathbb{H}^{\mathcal{I}}$, where $\mathbb{H}^{\mathcal{I}} = \bigvee_{i \in \mathcal{I}} \mathbb{H}^i$. Finally, we shall use the notation $\mathbb{G}^i = \mathbb{F} \vee \mathbb{H}^i$ and $\mathbb{G}^{ij} = \mathbb{F} \vee \mathbb{H}^i \vee \mathbb{H}^j$. Unless explicitly stated otherwise, we make the following standing assumptions:

– the random times $\tau_i$, $i = 1, \ldots, n$ are dynamically conditionally independent with respect to the filtration $\mathbb{F}$,
– Condition (C.1) is satisfied.

The following technical result is an easy consequence of the assumed conditional independence of default times. Since the proof is based on a routine application of the properties of conditional expectations, it is omitted.

**Lemma 9.2.1.** *Let $0 \leq t < T$. For any subset $\mathcal{I} \subseteq \{1, \ldots, n\}$ and arbitrary numbers $t_k \in [t, T]$, $k \in \mathcal{I}$ we have*

$$\mathbb{Q}^* \{\tau_k > t_k, \, k \in \mathcal{I} \,|\, \mathcal{F}_T \vee \mathcal{G}_t^{\mathcal{I}}\} = \prod_{k \in \mathcal{I}} \mathbb{Q}^* \{\tau_k > t_k \,|\, \mathcal{F}_T \vee \mathcal{G}_t^{\mathcal{I}}\}.$$

### 9.2.1 Default Correlations

We shall first discuss the issue of computing linear correlation coefficients between individual default times $\tau_i$ and $\tau_j$. We write $D^i(t) = \{\tau_i \leq t\}$ to denote the event that the $i^{\text{th}}$ credit name defaults by time $t$. For each $0 \leq t < s$, we define the *conditional default correlation* $\rho_{ij}^D(t, s; \mathcal{I})$ by setting (by convention, $\frac{0}{0} := 0$)

$$\rho_{ij}^D(t, s; \mathcal{I}) = \frac{Q_{ij}^*(t, s; \mathcal{I}) - Q_i^*(t, s; \mathcal{I})Q_j^*(t, s; \mathcal{I})}{\sqrt{Q_i^*(t, s; \mathcal{I})\big(1 - Q_i^*(t, s; \mathcal{I})\big)}\sqrt{Q_j^*(t, s; \mathcal{I})\big(1 - Q_j^*(t, s; \mathcal{I})\big)}},$$

where

$$Q_i^*(t, s; \mathcal{I}) := \mathbb{Q}^* \{D^i(s) \,|\, \mathcal{G}_t^{\mathcal{I}}\} = \mathbb{Q}^* \{\tau_i \leq s \,|\, \mathcal{G}_t^{\mathcal{I}}\}$$

and

$$Q_{ij}^*(t, s; \mathcal{I}) := \mathbb{Q}^* \{D^i(s) \cap D^j(s) \,|\, \mathcal{G}_t^{\mathcal{I}}\} = \mathbb{Q}^* \{\tau_i \leq s, \, \tau_j \leq s \,|\, \mathcal{G}_t^{\mathcal{I}}\}.$$

It is clear that $\rho_{ij}^D(t, s; \mathcal{I})$ is the conditional Pearson's correlation coefficient between the random variables $\mathbb{1}_{D^i(s)}$ and $\mathbb{1}_{D^j(s)}$, given the $\sigma$-field $\mathcal{G}_t^{\mathcal{I}}$. Note that $\rho_{ij}^D(0, s; \mathcal{I})$ coincides with the unconditional correlation coefficient $\rho_{ij}^D(s)$ introduced in Sect. 3.6.

We shall focus on the case where $i, j \in \mathcal{I}$. The following proposition shows that the correlation coefficient $\rho_{ij}^D(t, s; \mathcal{I})$ can be expressed in terms of the hazard processes $\Gamma^i$ and $\Gamma^j$ (or in terms of the intensities $\gamma^i$ and $\gamma^j$ provided that they exist).

**Proposition 9.2.1.** *Let us fix a subset* $\mathcal{I} \subset \{1,\ldots,n\}$. *If* $i,j \in \mathcal{I}$, *then*

$$Q_k^*(t,s;\mathcal{I}) = 1 - 1\!\!1_{\{\tau_k > t\}} \mathbb{E}_{\mathbb{Q}^*}\left(e^{\Gamma_t^k - \Gamma_s^k} \,|\, \mathcal{F}_t\right)$$

*for* $k = i,j$, *and*

$$Q_{ij}^*(t,s;\mathcal{I}) = \left(1 - 1\!\!1_{\{\tau_i > t\}} \mathbb{E}_{\mathbb{Q}^*}\left(e^{\Gamma_t^i - \Gamma_s^i} \,|\, \mathcal{F}_t\right)\right)\left(1 - 1\!\!1_{\{\tau_j > t\}} \mathbb{E}_{\mathbb{Q}^*}\left(e^{\Gamma_t^j - \Gamma_s^j} \,|\, \mathcal{F}_t\right)\right).$$

*Proof.* Since $\mathbb{G}^k \subset \mathbb{G}^{\mathcal{I}}$ for $k = i,j$, the first equality immediately follows from formula (5.13). This also implies the second expression since, in view of Lemma 9.2.1, we have $Q_{ij}^*(t,s;\mathcal{I}) = Q_i^*(t,s;\mathcal{I})Q_j^*(t,s;\mathcal{I})$. □

*Remarks.* (i) Condition (C.1) is not required for Proposition 9.2.1 to be valid. The dynamic conditional independence is essential, but only for the second formula in Proposition 9.2.1.

(ii) Let us denote $\tau_{ij} = \tau_i \wedge \tau_j$. It might be useful to observe that

$$Q_{ij}^*(t,s;\mathcal{I}) = Q_i^*(t,s;\mathcal{I}) + Q_j^*(t,s;\mathcal{I}) - Q_*^{ij}(t,s;\mathcal{I}),$$

where

$$Q_*^{ij}(t,s;\mathcal{I}) := \mathbb{Q}^*\{D^i(s) \cup D^j(s) \,|\, \mathcal{G}_t^{\mathcal{I}}\} = \mathbb{Q}^*\{\tau_{ij} \le s \,|\, \mathcal{G}_t^{\mathcal{I}}\}.$$

(iii) A result analogous to Proposition 9.2.1 can be derived in terms of $(\mathbb{F},\mathbb{G})$-martingale hazard processes $\Lambda^i$ and $\Lambda^j$. In this case, we use the martingale approach, and we do not impose the dynamic conditional independence assumption. Let us set $\Lambda^{ij} = \Lambda^i + \Lambda^j$. Assume that Condition (M.1) holds, and the process $\mathbb{E}_{\mathbb{Q}^*}(\exp(\Lambda_t^{ij} - \Lambda_s^{ij}) \,|\, \mathcal{F}_t)$, $t \in [0,s]$, is continuous at $\tau_{ij}$. Then, as an immediate consequence of Propositions 6.4.1 and 7.1.1, we obtain the following equality

$$Q_*^{ij}(t,s;\mathcal{I}) = 1 - 1\!\!1_{\{\tau_{ij} > t\}} \mathbb{E}_{\mathbb{Q}^*}\left(e^{\Lambda_t^{ij} - \Lambda_s^{ij}} \,|\, \mathcal{F}_t\right).$$

One may be also interested in computing the correlation coefficient between the ordered random times $\tau_{(i)}$ and $\tau_{(j)}$, where, without loss of generality, $i < j$. For each $0 \le t < s$, we define the conditional correlation coefficient between $\tau_{(i)}$ and $\tau_{(j)}$ as

$$\rho_{(ij)}^D(t,s) = \frac{Q_{(ij)}^*(t,s) - Q_{(i)}^*(t,s)Q_{(j)}^*(t,s)}{\sqrt{Q_{(i)}^*(t,s)(1 - Q_{(i)}^*(t,s))}\sqrt{Q_{(j)}^*(t,s)(1 - Q_{(j)}^*(t,s))}},$$

where

$$Q_{(i)}^*(t,s) := \mathbb{Q}^*\{\tau_{(i)} \le s \,|\, \mathcal{G}_t\}$$

and

$$Q_{(ij)}^*(t,s) := \mathbb{Q}^*\{\tau_{(i)} \le s, \, \tau_{(j)} \le s \,|\, \mathcal{G}_t\} = \mathbb{Q}^*\{\tau_{(j)} \le s \,|\, \mathcal{G}_t\} = Q_{(j)}^*(t,s).$$

We thus have

$$\rho_{(ij)}^D(t,s) = \frac{(1 - Q_{(i)}^*(t,s))Q_{(j)}^*(t,s)}{\sqrt{Q_{(i)}^*(t,s)(1 - Q_{(i)}^*(t,s))}\sqrt{Q_{(j)}^*(t,s)(1 - Q_{(j)}^*(t,s))}}.$$

In order to explicitly compute $\rho^D_{(ij)}(t,s)$, one needs to evaluate the conditional probability $Q^*_{(i)}(t,s)$. We shall establish a formula for the unconditional probability $Q^*_{(i)}(t) := Q^*_{(i)}(0,t)$. Let $i,j \in \{1,\ldots,n\}$ be fixed. As in Sect. 10.2, we denote the collection of specific partitions of the set $\{1,\ldots,n\}$ by $\Pi^{i,j}$. For arbitrary $\pi \in \Pi^{i,j}$, we write $\tau(\pi_-) = \max\{\tau_k : k \in \pi_-\}$ and $\tau(\pi_+) = \min\{\tau_k : k \in \pi_+\}$. It is apparent that

$$Q^*_{(i)}(t) = \sum_{j=1}^{n} \sum_{\pi \in \Pi^{i,j}} \mathbb{Q}^*\{\tau(\pi_-) < \tau_j < \tau(\pi_+), \tau_j \le t\}. \qquad (9.10)$$

To make a practical use of formula (9.10), one needs to impose some sort of restrictions on the collection of default times $\tau_i$, $i = 1,\ldots,n$. Under the standing assumptions of this section, we have the following result, which provides a quasi-explicit formula for the probability $Q^*_{(i)}(t)$.

**Proposition 9.2.2.** *Suppose that each hazard process $\Gamma^i$, $i = 1,\ldots,n$ admits the intensity $\gamma^i$. Then*

$$Q^*_{(i)}(t) = \sum_{j=1}^{n} \sum_{\pi \in \Pi^{i,j}} \mathbb{E}_{\mathbb{Q}^*}\left\{ \int_0^t \gamma^j_u\, e^{-\left(\Gamma^j_u + \sum_{l \in \pi_+} \Gamma^l_u\right)} \prod_{k \in \pi_-} (1 - e^{-\Gamma^k_u})\, du \right\}.$$

*Proof.* We first notice that

$$\mathbb{Q}^*\{\tau(\pi_-) < \tau_j < \tau(\pi_+), \tau_j \le t\} = \mathbb{E}_{\mathbb{Q}^*}\big(\mathbb{Q}^*\{\tau(\pi_-) < \tau_j < \tau(\pi_+), \tau_j \le t \,|\, \mathcal{F}_t\}\big)$$

$$= \mathbb{E}_{\mathbb{Q}^*}\left(\int_0^t \gamma^j_u e^{-\Gamma^j_u}\, \mathbb{Q}^*\{\tau(\pi_-) < u < \tau(\pi_+) \,|\, \mathcal{F}_t\}\, du\right).$$

Using the conditional independence assumption, for every $u \in [0,t]$ we obtain

$$\mathbb{Q}^*\{\tau(\pi_-) < u < \tau(\pi_+) \,|\, \mathcal{F}_t\} = \prod_{l \in \pi_+} \mathbb{Q}^*\{\tau_l > u \,|\, \mathcal{F}_t\} \prod_{k \in \pi_-} (1 - \mathbb{Q}^*\{\tau_k \ge u \,|\, \mathcal{F}_t\}).$$

In view of Condition (C.1), we have $\mathbb{Q}^*\{\tau_k \ge u \,|\, \mathcal{F}_t\} = \mathbb{Q}^*\{\tau_k > u \,|\, \mathcal{F}_u\}$, and so

$$\mathbb{Q}^*\{\tau(\pi_-) < u < \tau(\pi_+) \,|\, \mathcal{F}_t\} = \prod_{l \in \pi_+} e^{-\Gamma^l_u} \prod_{k \in \pi_-} (1 - e^{-\Gamma^k_u}).$$

To complete the proof of the proposition, it suffices to make use of expression (9.10). □

If the hazard processes of default times do not admit intensities, the formula established in Proposition 9.2.2 becomes

$$Q^*_{(i)}(t) = \sum_{j=1}^{n} \sum_{\pi \in \Pi^{i,j}} \mathbb{E}_{\mathbb{Q}^*}\left\{ \int_0^t \Big(\prod_{k \in \pi_-} (1 - e^{-\Gamma^k_u})\Big)\Big(e^{-\sum_{l \in \pi_+} \Gamma^l_u}\Big) dF^j_u \right\},$$

where $F^j_u = \mathbb{Q}^*\{\tau_j \le u \,|\, \mathcal{F}_u\}$.

### 9.2.2 Conditional Probabilities

One may be interested in computing the conditional expectation of the form $\mathbb{E}_{\mathbb{Q}^*}(h(\tau_1,\ldots,\tau_n,\tau_{(1)},\ldots,\tau_{(n)}) \mid \mathcal{A})$, where $h : \mathbb{R}_+^{2n} \to \mathbb{R}$ is a certain integrable function and $\mathcal{A}$ is some $\sigma$-field. In particular, the following conditional probabilities are of interest:

$$\mathbb{Q}^*\{\tau_{(j)} = \tau_i, \ \tau_i \leq s \mid \mathcal{G}_t\}, \quad \mathbb{Q}^*\{\tau_{(k)} = \tau_i \mid \mathcal{G}_{\tau_{(j)}}\}, \quad \mathbb{Q}^*\{\tau_{(j)} = \tau_i \mid \mathcal{G}_{\tau_{(j)}-}\}.$$

Recall that the $\sigma$-field $\mathcal{G}_{\tau_{(j)}-}$ is defined as follows

$$\mathcal{G}_{\tau_{(j)}-} = \mathcal{G}_0 \vee \sigma\{A \cap \{\tau_{(j)} > t\} : A \in \mathcal{G}_t, \ t > 0\}.$$

The next result provides the probability of the event: the first default occurs by time $T$ and the first defaulting entity is the $i^{\text{th}}$ credit name.

**Proposition 9.2.3.** *If the hazard process $\Gamma^i$ admits an intensity $\gamma^i$, then*

$$\mathbb{Q}^*\{\tau_{(1)} = \tau_i, \ \tau_i \leq T\} = \mathbb{E}_{\mathbb{Q}^*}\left(\int_0^T \gamma_u^i e^{-\sum_{j=1}^n \Gamma_u^j} \, du\right).$$

*Proof.* The following chain of equalities:

$$\mathbb{Q}^*\{\tau_{(1)} = \tau_i, \ \tau_i \leq T\} = \mathbb{E}_{\mathbb{Q}^*}\left(\int_0^T \gamma_u^i e^{-\Gamma_u^i} \mathbb{Q}^*\{\tau_k > u, \ \forall\, k \neq i \mid \mathcal{F}_T\} \, du\right)$$

$$= \mathbb{E}_{\mathbb{Q}^*}\left(\int_0^T \gamma_u^i e^{-\Gamma_u^i} \prod_{k \neq i} \mathbb{Q}^*\{\tau_k > u \mid \mathcal{F}_T\} \, du\right)$$

$$= \mathbb{E}_{\mathbb{Q}^*}\left(\int_0^T \gamma_u^i e^{-\Gamma_u^i} \prod_{k \neq i} e^{-\Gamma_u^k} \, du\right)$$

proves the result.  $\square$

*Remarks.* (i) If the process $\Gamma^i$ does not admit an intensity, the result of the last proposition needs to be modified as follows:

$$\mathbb{Q}^*\{\tau_{(1)} = \tau_i, \ \tau_i \leq T\} = \mathbb{E}_{\mathbb{Q}^*}\left(\int_0^T \prod_{k \neq i} e^{-\Gamma_u^k} \, dF_u^i\right),$$

where $F_u^i = \mathbb{Q}^*\{\tau_i \leq u \mid \mathcal{F}_u\}$.
(ii) Assume that the intensities of all default times are constant – that is, $\gamma_t^i \equiv \gamma^i = \text{const.} > 0$. Then Proposition 9.2.3 yields

$$\mathbb{Q}^*\{\tau_{(1)} = \tau_i, \ \tau_i \leq T\} = \frac{\gamma^i}{\sum_{j=1}^n \gamma^j}\left(1 - e^{-T\sum_{j=1}^n \gamma^j}\right). \tag{9.11}$$

By letting $T$ tend to infinity, we obtain from the last formula:

$$\mathbb{Q}^*\{\tau_{(1)} = \tau_i\} = \frac{\gamma^i}{\sum_{j=1}^n \gamma^j}. \tag{9.12}$$

Let us observe that formula (9.12) is not unexpected for anyone familiar with the properties of (conditional) Markov chains.

**Proposition 9.2.4.** *Suppose that the hazard processes $\Gamma^i$ admit constant intensities $\gamma^i$. Then*

$$\mathbb{Q}^*\{\tau_{(1)} = \tau_i \mid \mathcal{G}_{\tau_{(1)}-}\} = \mathbb{Q}^*\{\tau_{(1)} = \tau_i\} = \frac{\gamma^i}{\sum_{j=1}^n \gamma^j}. \tag{9.13}$$

*Proof.* In view of (9.12), it suffices to verify that the first equality in (9.13) is valid. Let us fix $t > 0$ and $A \in \mathcal{G}_t$. We need to show that

$$\int_{A \cap \{\tau_{(1)} > t\}} \mathbb{1}_{\{\tau_{(1)} = \tau_i\}} \, d\mathbb{Q}^* = \int_{A \cap \{\tau_{(1)} > t\}} \mathbb{Q}^*\{\tau_{(1)} = \tau_i\} \, d\mathbb{Q}^*$$

or, equivalently,

$$\mathbb{Q}^*\{A \cap \{\tau_{(1)} > t\} \cap \{\tau_{(1)} = \tau_i\}\} = \mathbb{Q}^*\{A \cap \{\tau_{(1)} > t\}\} \mathbb{Q}^*\{\tau_{(1)} = \tau_i\}.$$

Put another way, we need to check that

$$\mathbb{Q}^*\{\tau_{(1)} = \tau_i\} = \mathbb{Q}^*\{\tau_{(1)} = \tau_i \mid A \cap \{\tau_{(1)} > t\}\}$$

Since the event $\{\tau_{(1)} > t\} = \{\tau_1 > t, \ldots, \tau_n > t\}$ belongs to the $\sigma$-field $\mathcal{G}_t$, and $\{\tau_{(1)} = \tau_i\} = \{\tau_i < \tau_j, \, j \neq i\}$, the last equality appears to be a rather straightforward consequence of the dynamical conditional independence of default times and Condition (C.1), combined with the assumption that their $\mathbb{F}$-intensities are constant. Details are left to the reader. $\square$

The knowledge of conditional probability $\mathbb{Q}^*\{\tau_{(j)} = \tau_i \mid \mathcal{G}_{\tau_{(j)}-}\}$ is essential for the purpose of simulating dependent defaults. Other probabilities of interest for simulation purposes are, among others, the probabilities of the form

$$\mathbb{Q}^*\{\tau_{(j+1)} = \tau_i, \, \tau_i \leq T \mid \mathcal{G}_{\tau_{(j)}}\}, \quad \mathbb{Q}^*\{\tau_{(j+1)} = \tau_i \mid \mathcal{G}_{\tau_{(j)}}\}. \tag{9.14}$$

It is clear that the above probabilities are both zero on the set $\{\tau_i \leq \tau_{(j)}\}$, which belongs, of course, the $\sigma$-field $\mathcal{G}_{\tau_{(j)}}$.

Thus, we only need to evaluate the probabilities on the set $\{\tau_i > \tau_{(j)}\}$. Given the information $\mathcal{G}_{\tau_{(j)}}$, we know whether the event $\{\tau_i > \tau_{(j)}\}$ has occurred or not by the time $\tau_{(j)}$. However, we can be more specific.

Let $n_{(j)}$ stand for the identity (the name) of the reference credit defaulting at time $\tau_{(j)}$. For any $m = 1, \ldots, n$, given the information $\mathcal{G}_{\tau_{(j)}}$, we know whether the event $\{n_{(j)} = m\}$ has occurred or not by the time $\tau_{(j)}$. Since we assume that $\mathbb{Q}^*\{\tau_i = \tau_j\} = 0$ for $i \neq j$, we see that the event $\{n_{(j)} = m\}$ coincides with the event $\{\tau_{(j)} = \tau_m\}$.

Likewise, given the information $\mathcal{G}_{\tau_{(j)}}$, we can say which of the reference credits have not yet defaulted by the time $\tau_{(j)}$. In other words, we know which of the random times $\tau_i$ satisfy the inequality $\tau_i > \tau_{(j)}$. Let $\pi^{(j)} = (\pi_-^{(j)}, n_{(j)}, \pi_+^{(j)}) \in \Pi^{(j,n_{(j)})}$ denote the corresponding (random) partition; this means that $\tau_l < \tau_{(j)}$ for every $l \in \pi_-^{(j)}$, and $\tau_l > \tau_{(j)}$ for every $l \in \pi_+^{(j)}$.

Let $\Pi^{n-j}$ denote the family of subsets of $\{1, \ldots, n\}$ consisting of $n-j$ elements. For any choice of the subset $\pi_+ \in \Pi^{n-j}$, we know by the time $\tau_{(j)}$ of the $j^{\text{th}}$ default, given the information $\mathcal{G}_{\tau_{(j)}}$ available at this time, whether the event $\{\pi_+^{(j)} = \pi_+\}$ has occurred or not. In conclusion, for any fixed numbers $i, j, m \in \{1, \ldots, n\}$ such that $i \neq m$, and any subset $\pi_+ \in \Pi^{n-j}$ such that $i$ is in $\pi_+$, but $m$ does not belong to $\pi_+$, by the time $\tau_{(j)}$ (i.e., given the $\sigma$-field $\mathcal{G}_{\tau_{(j)}}$) we know whether the event

$$A_{i,j,m,\pi_+} := \{\tau_i > \tau_{(j)}, \ \tau_{(j)} = \tau_m, \ \pi_+^{(j)} = \pi_+\} \tag{9.15}$$

has occurred or not. The proof of the following lemma is rather standard.

**Lemma 9.2.2.** *The set $A_{i,j,m,\pi_+}$ belongs to the $\sigma$-field $\mathcal{G}_{\tau_{(j)}}$ as well as to the $\sigma$-field $\mathcal{G}_{\tau_m}$. Moreover, $A_{i,j,m,\pi_+} \cap \mathcal{G}_{\tau_{(j)}} = A_{i,j,m,\pi_+} \cap \mathcal{G}_{\tau_m}$.*

We shall henceforth write $\hat{A} = A_{i,j,m,\pi_+}$. We find it convenient to introduce the following modifications of Definition 9.1.2 and Condition (C.1). Recall that for any subset $\mathcal{I} \subseteq \{1, \ldots, n\}$ we denote $\mathcal{H}_T^{\mathcal{I}} = \bigvee_{k \in \mathcal{I}} \mathcal{H}_T^k$.

**Definition 9.2.1.** The random times $\tau_1, \ldots, \tau_n$ are said to be *optionally conditionally independent* with respect to the filtration $\mathbb{F}$ under $\mathbb{Q}^*$ if and only if for any $T > 0$, $t_1, \ldots, t_n \in [0, T]$, an arbitrary subset $\mathcal{I} \subseteq \{1, \ldots, n\}$, and any $m \in \mathcal{I}$ we have

$$\mathbb{Q}^*\{\tau_i > t_i, \ i \notin \mathcal{I} \mid \tilde{\mathcal{F}}_T^m\} = \prod_{i \notin \mathcal{I}} \mathbb{Q}^*\{\tau_i > t_i \mid \tilde{\mathcal{F}}_T^m\},$$

where the $\sigma$-field $\tilde{\mathcal{F}}_T^m$ equals $\tilde{\mathcal{F}}_T^m = \sigma\big((\mathcal{F}_T \vee \mathcal{H}_T^{\mathcal{I}}) \cap \{\tau_m < \tau_i : i \notin \mathcal{I}\}\big)$.

Notice that the $\sigma$-field $\tilde{\mathcal{F}}_T^m$ includes the following information: all events from the $\sigma$-field $\mathcal{F}_T$, all events of the form $\{\tau_k \leq t\}$, $t \in [0, T]$, for any $k \in \mathcal{I}$, and finally, all events $\{\tau_m < \tau_i\}$ for $i \notin \mathcal{I}$.

**Condition (C.2)** For any $T > 0$, $u \in [0, T]$, any subset $\mathcal{I} \subseteq \{1, \ldots, n\}$, and arbitrary $i \notin \mathcal{I}$ and $m \in \mathcal{I}$ we have

$$\mathbb{Q}^*\{\tau_i > u \mid \tilde{\mathcal{F}}_T^m\} = \frac{1 - F_{u \wedge \tau_m}^i}{1 - F_{\tau_m}^i} = e^{\Gamma_{\tau_m}^i - \Gamma_{u \wedge \tau_m}^i},$$

where $F_t^i = \mathbb{Q}^*\{\tau_i \leq t \mid \mathcal{F}_t\} = e^{-\Gamma_t^i}$. In particular, $\mathbb{Q}^*\{\tau_i > u \mid \tilde{\mathcal{F}}_T^m\} = 1$ on the set $\{\tau_m > u\}$.

Although Definition 9.2.1 and Condition (C.2) may look complicated at first glance, the properties of default times stipulated by these two conditions are exactly what we shall need in the proof of Proposition 9.2.5.

Furthermore, it can be checked without difficulties that the random times $\tau_1, \ldots, \tau_n$ constructed through the canonical method, presented in Sect. 9.1.2, are optionally conditionally independent with respect to $\mathbb{F}$ under $\mathbb{Q}^*$. In addition, they also satisfy Condition (C.2). We are thus in a position to prove the following result.

**Proposition 9.2.5.** *Let $\tau_1, \ldots, \tau_n$ be constructed by the canonical method. Assume that the hazard processes $\Gamma^j$, $j = 1, \ldots, n$ are continuous, and the hazard process $\Gamma^i$ admits the $\mathbb{F}$-intensity process $\gamma^i$. Then we have*

$$\mathbb{Q}^* \{ \tau_{(j+1)} = \tau_i, \, \tau_i \le T \, | \, \mathcal{G}_{\tau_{(j)}} \} = \mathbb{E}_{\mathbb{Q}^*} \left( \int_{\tau_m}^T \psi_m(u) \, du \, \Big| \, \mathcal{G}_{\tau_m} \right),$$

*on the set $\hat{A}_T := \hat{A} \cap \{ \tau_m \le T \} \in \mathcal{G}_{\tau_m}$, and*

$$\mathbb{Q}^* \{ \tau_{(j+1)} = \tau_i \, | \, \mathcal{G}_{\tau_{(j)}} \} = \mathbb{E}_{\mathbb{Q}^*} \left( \int_{\tau_m}^\infty \psi_m(u) \, du \, \Big| \, \mathcal{G}_{\tau_m} \right),$$

*on the set $\hat{A}$, where*

$$\psi_m(u) = \gamma^i(u) e^{- \sum_{k \in \pi_+} (\Gamma_{\tau_m}^k - \Gamma_u^k)}.$$

Let us first state an auxiliary result, which gives some useful properties of conditional expectations.

**Lemma 9.2.3.** *Let $X$ and $Y$ be two integrable random variables on $(\Omega, \mathcal{G}, \mathbb{P})$. For any two sub-$\sigma$-fields $\mathcal{F}$ and $\tilde{\mathcal{F}}$ of $\mathcal{G}$ we have:*
*(i) if $A \in \mathcal{F}$ and $\mathbb{1}_A X = \mathbb{1}_A Y$, then $\mathbb{1}_A \mathbb{E}_{\mathbb{P}}(X \, | \, \mathcal{F}) = \mathbb{1}_A \mathbb{E}_{\mathbb{P}}(Y \, | \, \mathcal{F})$,*
*(ii) if $A \in \mathcal{F} \cap \tilde{\mathcal{F}}$ and $A \cap \mathcal{F} = A \cap \tilde{\mathcal{F}}$, then we have*

$$\mathbb{1}_A \mathbb{E}_{\mathbb{P}}(X \, | \, \mathcal{F}) = \mathbb{1}_A \mathbb{E}_{\mathbb{P}}(X \, | \, \tilde{\mathcal{F}}) = \mathbb{1}_A \mathbb{E}_{\mathbb{P}}(X \, | \, \tilde{\mathcal{G}}),$$

*where $\tilde{\mathcal{G}} = \sigma(A \cap \mathcal{F}) = \sigma(A \cap \tilde{\mathcal{F}})$.*

*Proof.* Part (i) is in fact trivial, since clearly

$$\mathbb{1}_A \, \mathbb{E}_{\mathbb{P}}(X \, | \, \mathcal{F}) = \mathbb{E}_{\mathbb{P}}(\mathbb{1}_A X \, | \, \mathcal{F}) = \mathbb{E}_{\mathbb{P}}(\mathbb{1}_A Y \, | \, \mathcal{F}) = \mathbb{1}_A \, \mathbb{E}_{\mathbb{P}}(Y \, | \, \mathcal{F}).$$

For the second statement, notice that since the random variables $\mathbb{1}_A \mathbb{E}_{\mathbb{P}}(X \, | \, \mathcal{F})$ and $\mathbb{1}_A \mathbb{E}_{\mathbb{P}}(X \, | \, \tilde{\mathcal{G}})$ are manifestly $\tilde{\mathcal{G}}$-measurable, it is enough to show that the equality

$$\int_C \mathbb{1}_A \, \mathbb{E}_{\mathbb{P}}(X \, | \, \mathcal{F}) \, d\mathbb{P} = \int_C \mathbb{1}_A \, \mathbb{E}_{\mathbb{P}}(X \, | \, \tilde{\mathcal{G}}) \, d\mathbb{P}$$

is satisfied for any event $C \in \tilde{\mathcal{G}}$ (the same property will then hold for $\tilde{\mathcal{F}}$). Let us take an arbitrary $C \in \tilde{\mathcal{G}}$. In view of the definition of $\tilde{\mathcal{G}}$, we may take an event $C$ of the form $C = A \cap B$, where $B \in \mathcal{F}$. Since $A \in \mathcal{F}$ and $B \in \mathcal{F}$, we obtain

$$\int_C \mathbb{1}_A \, \mathbb{E}_{\mathbb{P}}(X \, | \, \mathcal{F}) \, d\mathbb{P} = \int_B \mathbb{E}_{\mathbb{P}}(\mathbb{1}_A X \, | \, \mathcal{F}) \, d\mathbb{P} = \int_B \mathbb{1}_A X \, d\mathbb{P}.$$

On the other hand, since $A \in \tilde{\mathcal{G}}$, we also obtain, as expected:

$$\int_C \mathbb{1}_A \, \mathbb{E}_{\mathbb{P}}(X \, | \, \tilde{\mathcal{G}}) \, d\mathbb{P} = \int_C \mathbb{E}_{\mathbb{P}}(\mathbb{1}_A X \, | \, \tilde{\mathcal{G}}) \, d\mathbb{P} = \int_C \mathbb{1}_A X \, d\mathbb{P} = \int_B \mathbb{1}_A X \, d\mathbb{P}.$$

Since this reasoning is also true for $\tilde{\mathcal{F}}$, the second statement is verified. $\quad\square$

*Proof of Proposition 9.2.5.* Recall that $\mathbb{Q}^*\{\tau_{(j)} < \infty\} = 1$. Let us denote

$$I = \mathbb{Q}^*\{\tau_{(j+1)} = \tau_i, \, \tau_i \le T \,|\, \mathcal{G}_{\tau_{(j)}}\} \mathbb{1}_{\hat{A}_T}.$$

Part (i) of Lemma 9.2.3 and the definition of $\hat{A}_T$ yield (cf. (9.15))

$$I = \mathbb{Q}^*\{\tau_m < \tau_i \le T, \, \tau_i < \tau_l, \, l \in \pi_+, \, l \ne i \,|\, \mathcal{G}_{\tau_{(j)}}\} \mathbb{1}_{\hat{A}_T}.$$

Next, we have

$$\hat{A}_T := \hat{A} \cap \{\tau_m \le T\} = \hat{A} \cap \{\tau_{(j)} \le T\} \in \mathcal{G}_{\tau_m} \cap \mathcal{G}_{\tau_{(j)}}.$$

Thus, by combining Lemma 9.2.2 with part (ii) of Lemma 9.2.3, we obtain

$$I = \mathbb{Q}^*\{\tau_m < \tau_i \le T, \, \tau_i < \tau_l, \, l \in \pi_+, \, l \ne i \,|\, \tilde{\mathcal{G}}_{\tau_m}\} \mathbb{1}_{\hat{A}_T},$$

where $\tilde{\mathcal{G}}_{\tau_m} = \sigma(\mathcal{G}_{\tau_m} \cap \hat{A}_T)$. Let $\mathcal{I}$ stand for the set of these numbers $j = 1, \dots, n$, which are not in $\pi_+$ (recall that a subset $\pi_+ \in \Pi^{n-j}$ was fixed in the definition of $\hat{A}$). Then clearly $m$ is in $\mathcal{I}$, but $i$ does not belong to $\mathcal{I}$. Recall that $\mathcal{H}_t^{\mathcal{I}} = \bigvee_{k \in \mathcal{I}} \mathcal{H}_t^k$ for every $t \in \mathbb{R}_+$. For any $t \in [0, T]$ we denote (see also Definition 9.2.1)

$$\tilde{\mathcal{F}}_t^m = \sigma\big((\mathcal{F}_t \vee \mathcal{H}_t^{\mathcal{I}}) \cap \{\tau_m < \tau_k : k \in \pi_+\}\big) = \sigma\big((\mathcal{F}_t \vee \mathcal{H}_t^{\mathcal{I}}) \cap \{\tau_m < \tau_k : k \notin \mathcal{I}\}\big).$$

It is useful to observe that $\tilde{\mathcal{G}}_{\tau_m} \subset \tilde{\mathcal{F}}_T^m$. Using the optional conditional independence and Condition (C.2), we thus obtain:

$$I = \mathbb{E}_{\mathbb{Q}^*}\big(\mathbb{Q}^*\{\tau_m < \tau_i \le T, \, \tau_i < \tau_l, \, l \in \pi_+, \, l \ne i \,|\, \tilde{\mathcal{F}}_T^m\} \,\big|\, \tilde{\mathcal{G}}_{\tau_m}\big) \mathbb{1}_{\hat{A}_T}$$

$$= \mathbb{E}_{\mathbb{Q}^*}\bigg(\int_{\tau_m}^T \prod_{l \in \pi_+, \, l \ne i} \mathbb{Q}^*\{\tau_l > u \,|\, \tilde{\mathcal{F}}_T^m\} \, d\mathbb{Q}^*\{\tau_i \le u \,|\, \tilde{\mathcal{F}}_T^m\} \,\bigg|\, \tilde{\mathcal{G}}_{\tau_m}\bigg) \mathbb{1}_{\hat{A}_T}$$

$$= \mathbb{E}_{\mathbb{Q}^*}\bigg(\int_{\tau_m}^T \gamma_u^i e^{\Gamma_{\tau_m}^i - \Gamma_u^i} \prod_{l \in \pi_+, \, l \ne i} e^{\Gamma_{\tau_m}^l - \Gamma_u^l} \, du \,\bigg|\, \tilde{\mathcal{G}}_{\tau_m}\bigg) \mathbb{1}_{\hat{A}_T}$$

$$= \mathbb{E}_{\mathbb{Q}^*}\bigg(\int_{\tau_m}^T \gamma_u^i e^{-\sum_{k \in \pi_+}(\Gamma_{\tau_m}^k - \Gamma_u^k)} \, du \,\bigg|\, \mathcal{G}_{\tau_m}\bigg) \mathbb{1}_{\hat{A}_T}.$$

This ends the proof of the first part of the proposition. The proof of the second formula is left to the reader. $\qquad\square$

The following result is an immediate consequence of Proposition 9.2.5.

**Corollary 9.2.1.** *Under the assumptions of Proposition 9.2.5, on the set* $\{\tau_i > \tau_{(j)}\} \in \mathcal{G}_{\tau_{(j)}}$ *we have*

$$\mathbb{Q}^*\{\tau_{(j+1)} = \tau_i, \, \tau_i \le T \,|\, \mathcal{G}_{\tau_{(j)}}\} = \mathbb{E}_{\mathbb{Q}^*}\bigg(\int_{\tau_{(j)}}^T \psi_{(j)}(u) \, du \,\bigg|\, \mathcal{G}_{\tau_{(j)}}\bigg),$$

*and*

$$\mathbb{Q}^*\{\tau_{(j+1)} = \tau_i \,|\, \mathcal{G}_{\tau_{(j)}}\} = \mathbb{E}_{\mathbb{Q}^*}\left(\int_{\tau_{(j)}}^{\infty} \psi_{(j)}(u)\, du \,\Big|\, \mathcal{G}_{\tau_{(j)}}\right),$$

*where*

$$\psi_{(j)}(u) = \gamma^i(u) e^{\sum_{k \in \pi_+^{(j)}} (\Gamma_{\tau_{(j)}}^k - \Gamma_u^k)}.$$

Let us now consider the special case of default times $\tau_1, \ldots, \tau_n$ with constant intensities $\gamma^1, \ldots, \gamma^n$. Under these assumptions, on the set $\{\tau_i > \tau_{(j)}\}$, Corollary 9.2.1 yields,

$$\mathbb{Q}^*\{\tau_{(j+1)} = \tau_i,\ \tau_i \le T \,|\, \mathcal{G}_{\tau_{(j)}}\} = \frac{\gamma^i}{\sum_{k \in \pi_+^{(j)}} \gamma^k}\left(1 - e^{-(T - \tau_{(j)}) \sum_{k \in \pi_+^{(j)}} \gamma^k}\right)$$

and

$$\mathbb{Q}^*\{\tau_{(j+1)} = \tau_i \,|\, \mathcal{G}_{\tau_{(j)}}\} = \frac{\gamma^i}{\sum_{k \in \pi_+^{(j)}} \gamma^k}.$$

The last two formulae are straightforward generalizations of expressions (9.11) and (9.12), respectively.

*Remarks.* Let us conclude this chapter by mentioning that an econometric approach to the modeling of portfolio credit risk, with conditionally independent default times, was developed by Wilson (1997). He assumes that the probability density function of default for the $j^{\text{th}}$ segment of the market is given by the *logit* function:

$$f_j(t) = \frac{1}{1 + \exp(y_t^j)}, \quad \forall t \in \mathbb{R}_+,$$

for some segment-specific macroeconomic index $y_t^j$. An econometric hazard model for forecasting bankruptcy, which was put forward by Shumway (2001), is based on a similar assumption.

# 10. Dependent Defaults

In this chapter, we continue the study of the intensity-based approach to the modeling of dependent defaults. In Sect. 10.1, we shall analyze the ideas presented in a recent paper by Jarrow and Yu (2001). Then, in Sect. 10.2, we will analyze the martingale approach to the valuation of basket credit derivatives. For related results, the interested reader may consult Duffie (1998), Duffie and Singleton (1998b), Hull and White (2000, 2001), as well as to Lando (2000b), who examines the issue of modeling correlated defaults within the framework of a ratings-based model.

An important issue, which arise in the context of portfolio credit risk is the problem of deriving the joint probability distribution of a collection of default times $\tau_i, \ldots, \tau_n$ from the marginal probability distribution functions of these random variables. The problem is, of course, trivial, when default times are known to be mutually independent. If, on the contrary, the independence of default times is not taken for granted, the issue becomes fairly complex. To estimate the joint probability distribution of default times, one can start by estimating the marginal probability distributions of individual defaults, and then transform these individual estimates into the estimate of the joint distribution, using an appropriate methodology.

One methodology for performing such a transformation is based on the concept a *copula* function. Essentially, a *copula* is a function that links univariate marginals to the joint multivariate distribution. For the basic properties of copulas, statistical inference for copulas, and their applications to the risk modeling, the reader may consult, e.g., Nelsen (1998), Wang (1999b), Bouyé et al. (2000), Lindskog (2000), or Embrechts et al. (2002, 2003). Let us only mention that in the financial context, the copula function approach allows to produce efficient tools to measure the portfolio's risk when the convenient – but manifestly too restrictive in most practical situation – assumption of Gaussian character of assets returns is relaxed. This approach has also gained considerable popularity for credit risk related applications in recent years (see, for example, Frey and McNeil (2000) or Li (1999a, 2000)). Nevertheless, since the primary objective of this approach to the modeling of dependent defaults is the analysis of risk inherent in large (static) portfolios of loans, it is beyond the scope of the present text.

**Davis and Lo approach.** An alternative way of dealing with dependent defaults in bond portfolios was put forward by Davis and Lo (2001), who coined the term *infectious defaults*. At the intuitive level, they assume that a given bond in a portfolio of $n$ bonds may either default 'directly,' or as a result of 'infection' – that is, due to the default of some other bond. Specifically, the static version of their model postulates that

$$Z_i = X_i + (1 - X_i)\Big(1 - \prod_{j \neq i}(1 - X_j Y_{ji})\Big),$$

where the random variables $Z_i, X_i, Y_{ji}$ for $i, j = 1, \ldots, n$ are given the following interpretation:
- $Z_i = 1$ if the $i^{\text{th}}$ bond defaults, and $Z_i = 0$ otherwise,
- $X_i = 1$ if the $i^{\text{th}}$ bond defaults 'directly,' and $X_i = 0$ otherwise,
- if $X_j = 1$, then $Y_{ji}$ determines whether the 'infection' occurs (i.e., $Y_{ji} = 1$) or not (i.e., $Y_{ji} = 0$).

The total number of bonds defaulting in a given period of time thus equals $N = Z_1 + \cdots + Z_n$. In case when $Y_{ji} = 0$ for all $i, j$, and the random variables $X_i, i = 1, \ldots, n$ are mutually independent with $\mathbb{Q}^*\{X_i = 1\} = p$ for some $p \in [0, 1]$, $N$ has the binomial law:

$$\mathbb{Q}^*\{N = k\} = C_k^n p^k (1 - p)^{n-k}, \quad \forall k = 0, \ldots, n, \tag{10.1}$$

where $C_k^n = n!/k!(n - k)!$ is the binomial coefficient. In a more interesting case when $X_i, Y_{ji}, i, j = 1, \ldots, n$ are mutually independent random variables with $\mathbb{Q}^*\{X_i = 1\} = p$ and $\mathbb{Q}^*\{Y_{ji} = 1\} = q$ for $p, q \in [0, 1]$, Davis and Lo (2001) show that the distribution function $F_{p,q,n}(k) := \mathbb{Q}^*\{N = k\}$ satisfies $F_{p,q,n}(k) = C_k^n \alpha_{p,q,n}(k)$, where

$$\alpha_{p,q,n}(k) = p^k(1 - p)^{n-k}\tilde{q}^{k(n-k)} + \sum_{i=1}^{k-1} C_i^k p^i(1 - p)^{n-i}(1 - \tilde{q}^i)^{k-i}\tilde{q}^{i(n-k)},$$

with $\tilde{q} = 1 - q$. In particular, the expected value of $N$ equals

$$\mathbb{E}_{\mathbb{Q}^*}(N) = n\big(1 - (1 - p)(1 - pq)^{n-1}\big).$$

In the dynamic version of their approach, presented in Davis and Lo (2001), the authors apply a *piecewise deterministic Markov process* (for the definition, see Davis (1993)) to infer the default process $N_t$, where $N_t$ stands for the number of defaults in the interval $[0, t]$. At the intuitive level, the hazard rate of default of each bond is initially constant, and when a default happens, the hazard rate for all remaining bonds is increased by the factor $a > 1$. In addition, this factor disappears (or, strictly speaking, becomes 1) after an exponentially distributed random time with the parameter $\mu$.

*Remarks.* Expression (10.1) also underpins Moody's Binomial Expansion Technique (see Moody's Investment Service (1997)), which makes the assumption that bond issuers in the same industry sector are related, but issuers from different sectors can be treated as independent.

## 10.1 Dependent Intensities

In some circumstances, it is natural to assume that the default probability of a certain entity, firm A say, will increase if firm B defaults on its obligations to firm A. Therefore, we should be able to model a jump of the default intensity of one firm at the default time of another entity. This way of modeling dependent default was examined by, among others, Schmidt (1998), Kusuoka (1999), and Jarrow and Yu (2001). Although most of results presented in this section are valid for a finite number of reference entities, for the sake of expositional clarity, we shall mainly concentrate on the case of two entities.

### 10.1.1 Kusuoka's Approach

The example due to Kusuoka (1999) was discussed in detail in Sect. 7.3, and thus here we provide only a brief account of his results, and we shall slightly modify the notation. We begin by assuming that, under the original probability measure $\mathbb{Q}$, the default times $\tau_i$, $i = 1, 2$ are mutually independent, and have the exponential law with parameter $\lambda_1$ and $\lambda_2$, respectively. In other words, the joint probability law of $(\tau_1, \tau_2)$ under $\mathbb{Q}$ has the following probability density function:

$$f_{(\tau_1, \tau_2)}(x, y) = \lambda_1 \lambda_2 \, e^{-(\lambda_1 x + \lambda_2 y)}, \quad \forall \, (x, y) \in \mathbb{R}_+^2.$$

The crucial step is a judicious choice of an equivalent change of the underlying probability measure, from $\mathbb{Q}$ to $\mathbb{Q}^*$ say, to the effect that under a probability measure $\mathbb{Q}^*$ the 'intensity' of the default time $\tau_1$ will jump from $\lambda_1$ to a predetermined value $\alpha_1$ as soon as $\tau_2$ occurs (by symmetry, the behavior of the 'intensity' of $\tau_2$ will be analogous).

Let us introduce the joint filtration $\mathbb{G} = \mathbb{H}^1 \vee \mathbb{H}^2$. For a fixed $0 < T^* \leq \infty$, we define a probability measure $\mathbb{Q}^*$, equivalent to $\mathbb{Q}$ on $(\Omega, \mathcal{G}_{T^*})$, by setting

$$\frac{d\mathbb{Q}^*}{d\mathbb{Q}}\Big|_{\mathcal{G}_{T^*}} = \eta_{T^*}, \quad \mathbb{Q}\text{-a.s.,}$$

where the process $\eta_t$, $t \in [0, T^*]$ satisfies

$$\eta_t = 1 + \sum_{i=1}^{2} \int_{]0,t]} \eta_{u-} \kappa_u^i \, d\tilde{M}_u^i,$$

where the process $\tilde{M}_t^i := H_t^i - \int_0^{t \wedge \tau_i} \lambda_i \, du$ is known to follow a $\mathbb{G}$-martingale under $\mathbb{Q}$. The two processes $\kappa^1$ and $\kappa^2$ are given by the formulae

$$\kappa_t^1 = \mathbb{1}_{\{\tau_2 < t\}}\left(\frac{\alpha_1}{\lambda_1} - 1\right), \quad \kappa_t^2 = \mathbb{1}_{\{\tau_1 < t\}}\left(\frac{\alpha_2}{\lambda_2} - 1\right),$$

where in turn $\alpha_1$ and $\alpha_2$ are positive real numbers. From general results of Sect. 7.2 (or by the direct calculations), it follows that the martingale intensities $\lambda^{*1}$ and $\lambda^{*2}$ of default times $\tau_1$ and $\tau_2$ under $\mathbb{Q}^*$ satisfy

$$\lambda_t^{*1} = \lambda_1(1 - H_t^2) + \alpha_1 H_t^2 = \lambda_1 \mathbb{1}_{\{\tau_2 > t\}} + \alpha_1 \mathbb{1}_{\{\tau_2 \le t\}}$$

and

$$\lambda_t^{*2} = \lambda_2(1 - H_t^1) + \alpha_2 H_t^1 = \lambda_2 \mathbb{1}_{\{\tau_1 > t\}} + \alpha_2 \mathbb{1}_{\{\tau_1 \le t\}},$$

respectively. In other words, the two processes

$$H_t^1 - \int_0^{t \wedge \tau_1} \left( \lambda_1 \mathbb{1}_{\{\tau_2 > u\}} + \alpha_1 \mathbb{1}_{\{\tau_2 \le u\}} \right) du$$

and

$$H_t^2 - \int_0^{t \wedge \tau_2} \left( \lambda_2 \mathbb{1}_{\{\tau_1 > u\}} + \alpha_2 \mathbb{1}_{\{\tau_1 \le u\}} \right) du$$

are $\mathbb{Q}^*$-martingales with respect to the joint filtration $\mathbb{G} = \mathbb{H}^1 \vee \mathbb{H}^2$. It appears, however, that the intensities $\lambda^{*1}$ and $\lambda^{*2}$ are not suitable for the calculations of conditional expectations under $\mathbb{Q}^*$. We refer to Sect. 7.3 for details. Let us only observe that, by virtue of Proposition 7.3.1, for every $t \le s \le T^*$ we have:

$$\mathbb{Q}^* \{\tau_1 > s \,|\, \mathcal{H}_t^1 \vee \mathcal{H}_t^2\} = \mathbb{1}_{\{\tau_1 > t\}} \mathbb{E}_{\mathbb{Q}} \left( e^{-\int_t^s \lambda_u^{1*} \, du} \,\Big|\, \mathcal{H}_t^2 \right),$$

but

$$\mathbb{Q}^* \{\tau_1 > s \,|\, \mathcal{H}_t^1 \vee \mathcal{H}_t^2\} \ne \mathbb{1}_{\{\tau_1 > t\}} \mathbb{E}_{\mathbb{Q}^*} \left( e^{-\int_t^s \lambda_u^{1*} \, du} \,\Big|\, \mathcal{H}_t^2 \right).$$

The last inequality leads to an important observation that the process $\lambda^{*1}$ does not represent the $\mathbb{H}^2$-intensity of the default time $\tau_1$ under the probability measure $\mathbb{Q}^*$. Indeed, we have shown in Sect. 7.3 that the $\mathbb{H}^2$-hazard process of $\tau_1$ under $\mathbb{Q}^*$ has a discontinuity at $\tau_2$.

Put another way, the desirable interpretation of $\lambda^{*1}$ as the $\mathbb{H}^2$-conditional probability of default over a small time interval is not justified. This failure of $\lambda^{*1}$ to be the hazard rate is linked to the fact (also shown in Sect. 7.3) that the martingale invariance property (M.1) does not hold under $\mathbb{Q}^*$ in the present set-up, that is, when applied to the 'reference' filtration $\mathbb{H}^2$ and the 'enlarged' filtration $\mathbb{G} = \mathbb{H}^1 \vee \mathbb{H}^2$.

We conclude the brief presentation of Kusuoka's example by mentioning that the default times $\tau_1$ and $\tau_2$ are not conditionally independent with respect to $\mathbb{F}$. Indeed, in the present set-up, the reference filtration $\mathbb{F}$ is trivial, and the random variables $\tau_1$ and $\tau_2$ are not mutually independent under $\mathbb{Q}^*$.

### 10.1.2 Jarrow and Yu Approach

Kusuoka's counter-example shows that the modeling of the counterparty risk in default-risk sensitive contracts constitutes a rather delicate issue. Jarrow and Yu (2001) argue that some difficulties can be circumvented through a judicious choice of reference filtrations. To explain the ideas that underpin their approach, we start by assuming that there are $n$ firms in the economy; they are also informally referred to as 'counterparties' in the sequel.

Jarrow and Yu (2001) propose to make a distinction between the *primary firms* and the *secondary firms*. The former class encompasses these entities whose probabilities of default are influenced by macroeconomic conditions, but not by the credit risk of counterparties. The pricing of bonds issued by primary firms can be done through the standard intensity-based methodology described at some length in Chap. 8; in particular, it is natural to introduce in this context the state-variables process $Y$, representing the macroeconomic factors (see Example 8.2.2). Thus, it suffices to focus on securities issued by secondary firms, i.e., firms for which the intensity of default depends on the status of other firms.

*Remarks.* To resolve the problem of the mutual dependence of martingale intensities, Jarrow and Yu (2001) postulate that the information structure is asymmetric. Specifically, in their assessment of default probabilities investors take into account the observed defaults of primary firms, but they deliberately choose to disregard eventual defaults of secondary firms. Such an assumption is supported by real-life financial arguments of two kinds. First, a secondary firm may be seen as a financial institution that has a long or a short position in the debt of a primary firm (e.g., a large corporation), so that the likelihood of its default depends on the status of this corporation. It is natural to assume that the situation is not symmetric, and the default probability of a primary firm depends only on macroeconomic factors. In the second interpretation, a primary firm may be seen as a large corporation, and a secondary firm as one of many relatively small dependent manufacturers. For instance, a large firm can be a major supplier for many small manufacturers, or there may be a lot of small suppliers for a large corporation.

The following assumption underpins Jarrow and Yu (2001) approach.

**Assumption (J)** Let $\mathcal{I} = \{1, \ldots, n\}$ represent the set of all firms, and let $\mathbb{F}$ be the reference filtration. We postulate that:
- for any firm from the set $\{1, \ldots, k\}$ of primary firms, the 'default intensity' depends only on $\mathbb{F}$,
- the 'default intensity' of each firm from the set $\{k+1, \ldots, n\}$ of secondary firms may depend not only on the filtration $\mathbb{F}$, but also on the status (default or no-default) of the primary firms.

The construction of the collection of default times $\tau_1, \ldots, \tau_n$ with the desired properties runs as follows. In the first step, we assume that we are given a family of $\mathbb{F}$-adapted intensity processes $\lambda^1, \ldots, \lambda^k$, and we produce a collection $\tau_1, \ldots, \tau_k$ of $\mathbb{F}$-conditionally independent random times through the canonical method presented in Sect. 9.1.2. Specifically:

$$\tau_i = \inf \left\{ t \in \mathbb{R}_+ \ : \ \int_0^t \lambda_u^i \, du \geq \eta_i \right\}, \tag{10.2}$$

where $\eta_i$, $i = 1, \ldots, k$ are mutually independent, identically distributed, random variables with unit exponential law under $\mathbb{Q}^*$.

In the second step, we assume that the underlying probability space $(\Omega, \mathcal{G}, \mathbb{Q}^*)$ is large enough to accommodate a family $\eta_i$, $i = k + 1, \ldots, n$ of mutually independent random variables, with unit exponential law under $\mathbb{Q}^*$, and such that these random variables are independent not only of the filtration $\mathbb{F}$, but also of the already constructed in the previous step default times $\tau_1, \ldots, \tau_k$ of primary firms.

The default times $\tau_i$, $i = k+1, \ldots, n$ are also defined by means of equality (10.2). However, the 'intensity processes' $\lambda^i$, $i = k + 1, \ldots, n$ are now given by the following generic expression:

$$\lambda_t^i = \mu_t^i + \sum_{l=1}^{k} \nu_t^{i,l} \mathbb{1}_{\{\tau_l \leq t\}}, \tag{10.3}$$

where $\mu^i$ and $\nu^{i,l}$ are $\mathbb{F}$-adapted stochastic processes. If the default of the $j^{\text{th}}$ primary firm does not affect the default probability of the $i^{\text{th}}$ secondary firm, we set $\nu^{i,j} \equiv 0$ in (10.3).

*Remarks.* Let $\mathbb{G} = \mathbb{F} \vee \mathbb{H}^1 \vee \ldots \vee \mathbb{H}^n$ stand for the enlarged filtration, and let $\tilde{\mathbb{F}} = \mathbb{F} \vee \mathbb{H}^{k+1} \vee \ldots \vee \mathbb{H}^n$ be the filtration generated by the 'macroeconomic factors' and the observations of default of secondary firms. Then:
- the default times $\tau_1, \ldots, \tau_k$ of primary firms are no longer conditionally independent when we replace the filtration $\mathbb{F}$ by $\tilde{\mathbb{F}}$,
- in general, the intensity of default for a primary firm with respect to $\tilde{\mathbb{F}}$ differs from the corresponding intensity with respect to $\mathbb{F}$.

The last observation indicates that the processes $\lambda^1, \ldots, \lambda^k$ do not represent the conditional probabilities of survival, unless we disregard the information flow generated by default processes of secondary firms. Put another way, a one-way dependence in default intensities is not possible. If the intensity of default of firm A jumps at the time of default of firm B, a similar effect will appear in the default intensity of firm B at the time of default of firm A.

**Case of two firms.** To clarify the last statement, we shall examine in detail a special case of the Jarrow and Yu model. We consider only two firms, A and B say, and we postulate that the first one represents a primary firm, while the second is a secondary firm. Let the process $\lambda^1$ represent the $\mathbb{F}$-intensity of default for firm A. The default time $\tau_1$ is given by the standard formula

$$\tau_1 = \inf \left\{ t \in \mathbb{R}_+ : \int_0^t \lambda_u^1 \, du \geq \eta_1 \right\},$$

where $\eta_1$ is a random variable, independent of the filtration $\mathbb{F}$, and exponentially distributed under $\mathbb{Q}^*$. For the second firm, the 'intensity' of default is assumed to satisfy

$$\lambda_t^2 = \mu_t^2 + \nu_t^{1,2} \mathbb{1}_{\{\tau_1 \leq t\}},$$

where $\mu^2$ and $\nu^{1,2}$ are positive, $\mathbb{F}$-adapted processes.

We set

$$\tau_2 = \inf \{\, t \in \mathbb{R}_+ \,:\, \int_0^t \lambda_u^2 \, du \geq \eta_2 \,\},$$

where $\eta_2$ is a random variable with the unit exponential law under $\mathbb{Q}^*$, independent of $\mathbb{F}$, and such that $\eta_1$ and $\eta_2$ are mutually independent under $\mathbb{Q}^*$. From the construction above, it is apparent that the following properties are valid:

- the process $\lambda^1$ represents the intensity of $\tau_1$ with respect to $\mathbb{F}$,
- the process $\lambda^2$ represents the intensity of $\tau_2$ with respect to $\mathbb{F} \vee \mathbb{H}^1$.

However, as in Kusuoka's example, it is possible to check that the process $\lambda^1$ does not represent the intensity of $\tau_1$ with respect to the filtration $\mathbb{F} \vee \mathbb{H}^2$.

We shall now apply our model to the valuation of corporate bonds. To this end, we assume that we have already specified some model of the default-free term structure. In particular, we are given a filtered probability space $(\tilde{\Omega}, \mathbb{F}, \mathbb{P}^*)$, where $\mathbb{P}^*$ is the spot martingale measure for the Treasury bonds market. As usual, we denote by $B(t, T)$ the price at time $t$ of a zero-coupon Treasury bond with maturity $T$. To obtain closed-form solutions for the values of corporate bonds, in addition, we postulate that $\lambda_t^1 = \lambda_1$ for some positive constant $\lambda_1$, and $\lambda^2$ equals

$$\lambda_t^2 = \lambda_2 + (\alpha_2 - \lambda_2) \mathbb{1}_{\{\tau_1 \leq t\}},$$

for some positive constants $\lambda_2$ and $\alpha_2$. Notice the jump of $\lambda^2$ at time $\tau_1$ may be either positive or negative, depending on the financial interpretation.

To construct default times, we enlarge the probability space in a standard way, so that we end up with the enlarged probability space $(\Omega, \mathbb{G}, \mathbb{Q}^*)$, and the two mutually independent, exponentially distributed, random variables $\eta_1, \eta_2$ that are also independent of the filtration $\mathbb{F}$ under $\mathbb{Q}^*$. As usual, we write $\mathcal{G}_t = \mathcal{F}_t \vee \mathcal{H}_t^1 \vee \mathcal{H}_t^2$ for $t \in \mathbb{R}_+$.

For the sake of convenience, we shall also introduce the forward martingale measure $\mathbb{Q}_T$ for the date $T > 0$. To this end, recall that under $\mathbb{P}^*$ (and thus also under $\mathbb{Q}^*$) we have

$$dB(t, T) = B(t, T)\big(r_t \, dt + b(t, T) \, dW_t^*\big),$$

where the volatility $b(\cdot, T)$ is an $\mathbb{F}$-adapted stochastic process. The probability measure $\mathbb{Q}_T$ is given on $(\Omega, \mathcal{G}_T)$ through the Radon-Nikodým density

$$\frac{d\mathbb{Q}_T}{d\mathbb{Q}^*}\bigg|_{\mathcal{G}_T} = \exp\bigg(\int_0^T b(u, T) \, dW_u^* - \frac{1}{2} \int_0^T b^2(u, T) \, du\bigg), \quad \mathbb{Q}^*\text{-a.s.}$$

In view of the assumed independence, it is clear that the random variables $\eta_1, \eta_2$ have identical probabilistic properties under $\mathbb{Q}^*$ and $\mathbb{Q}_T$. We assume that the bonds issued by the firms A and B are subject to the fractional recovery of Treasury value scheme with constant recovery rates $\delta_1$ and $\delta_2$, respectively.

Jarrow and Yu (2001) show that at time $t \leq T$ the bond issued by the primary firm has the following value:

$$D_1(t,T) = B(t,T)\big(\delta_1 + (1-\delta_1)e^{-\lambda_1(T-t)}\mathbb{1}_{\{\tau_1>t\}}\big). \qquad (10.4)$$

This valuation formula is rather obvious when it is postulated that

$$D_1(t,T) = B(t,T)\,\mathbb{E}_{\mathbb{Q}_T}\big(\mathbb{1}_{\{\tau_1>T\}} + \delta_1\mathbb{1}_{\{\tau_1\leq T\}}\,\big|\,\mathcal{F}_t \vee \mathcal{H}_t^1\big).$$

It is noteworthy that the right-hand side of (10.4) also yields the correct value for $D_1(t,T)$ if it is defined through the standard formula:

$$D_1(t,T) = B(t,T)\,\mathbb{E}_{\mathbb{Q}_T}\big(\mathbb{1}_{\{\tau_1>T\}} + \delta_1\mathbb{1}_{\{\tau_1\leq T\}}\,\big|\,\mathcal{G}_t\big).$$

The intuitive difference between the last two formulae is clear: the former assumes a priori that the occurrence of default of the secondary firm is not relevant for the valuation of a bond issued by the primary firm. The latter formula relies on the complete information available at time $t$. For the secondary firm, we also adopt the usual formula – that is, we set

$$D_2(t,T) = B(t,T)\,\mathbb{E}_{\mathbb{Q}_T}\big(\mathbb{1}_{\{\tau_2>T\}} + \delta_2\mathbb{1}_{\{\tau_2\leq T\}}\,\big|\,\mathcal{G}_t\big).$$

Let us denote, for $\lambda_1 + \lambda_2 - \alpha_2 \neq 0$,

$$c_{\lambda_1,\lambda_2,\alpha_2}(u) = \frac{1}{\lambda_1+\lambda_2-\alpha_2}\big(\lambda_1 e^{-\alpha_2 u} + (\lambda_2-\alpha_2)e^{-(\lambda_1+\lambda_2)u}\big).$$

Otherwise, we set

$$c_{\lambda_1,\lambda_2,\alpha_2}(u) = (1+\lambda_1 u)e^{-(\lambda_1+\lambda_2)u}.$$

The following result is borrowed from Jarrow and Yu (2001). Since we are going to establish a general result that covers this special case, the proof of Proposition 10.1.1 is omitted.

**Proposition 10.1.1.** *The value of a zero-coupon bond issued by the secondary firm equals, on the event $\{\tau_1 > t\}$, that is, prior to default of the primary firm,*

$$D_2(t,T) = B(t,T)\big(\delta_2 + (1-\delta_2)c_{\lambda_1,\lambda_2,\alpha_2}(T-t)\mathbb{1}_{\{\tau_2>t\}}\big),$$

*and on the set $\{\tau_1 \leq t\}$, that is, after default of the primary firm,*

$$D_2(t,T) = B(t,T)\big(\delta_2 + (1-\delta_2)e^{-\alpha_2(T-t)}\mathbb{1}_{\{\tau_2>t\}}\big).$$

Assume, for the sake of simplicity that $\delta_2 = 0$. Then for every $t \leq T$ Proposition 10.1.1 yields(we denote hereafter $\lambda = \lambda_1 + \lambda_2$)

$$D_2(t,T) = \mathbb{1}_{\{\tau_1>t,\tau_2>t\}}\frac{1}{\lambda-\alpha_2}\big(\lambda_1 e^{-\alpha_2(T-t)} + (\lambda_2-\alpha_2)e^{-\lambda(T-t)}\big)$$
$$+ \mathbb{1}_{\{\tau_1\leq t<\tau_2\}}\,e^{-\alpha_2(T-t)}.$$

**Case of zero recovery.** We shall now argue that the assumption that some firms are primary while other firms are secondary is in fact not relevant, and thus it can be relaxed (see also Bielecki and Rutkowski (2003)). For the sake of simplicity, we maintain the assumption that $n = 2$; i.e., we consider two firms only, and we place ourselves in Kusuoka's set-up, with $T^* = \infty$. It is worth noticing that the calculations below are also valid in the extended Jarrow and Yu (2001) framework.

To start with, let us assume that both bonds are subject to the zero-recovery scheme, and the interest rate $r$ is constant, so that $B(t,T) = e^{-r(T-t)}$ for every $t \leq T$. Due to the last assumption we have $\mathbb{Q}_T = \mathbb{Q}^*$ and the filtration $\mathbb{F}$ is trivial. Since the situation is symmetric, it suffices to analyze one bond only, for instance, a bond issued by the first firm. By definition, the price of this bond equals

$$D_1(t,T) = B(t,T)\,\mathbb{Q}^*\{\tau_1 > T \,|\, \mathcal{G}_t\} = B(t,T)\,\mathbb{Q}^*\{\tau_1 > T \,|\, \mathcal{H}_t^1 \vee \mathcal{H}_t^2\}.$$

We shall also evaluate the following random variables:

$$\tilde{D}_1(t,T) := B(t,T)\,\mathbb{Q}^*\{\tau_1 > T \,|\, \mathcal{H}_t^2\}$$

and

$$\hat{D}_1(t,T) := B(t,T)\,\mathbb{Q}^*\{\tau_1 > T \,|\, \mathcal{H}_t^1\}.$$

For the sake of simplicity, we shall assume in the next result that $r = 0$, so that $B(t,T) = 1$ for every $t \in [0,T]$.

**Proposition 10.1.2.** *The price $D_1(t,T)$ equals*

$$D_1(t,T) = \mathbb{1}_{\{\tau_1 > t, \tau_2 > t\}} \frac{1}{\lambda - \alpha_1}\left(\lambda_2 e^{-\alpha_1(T-t)} + (\lambda_1 - \alpha_1)e^{-\lambda(T-t)}\right)$$
$$+ \mathbb{1}_{\{\tau_2 \leq t < \tau_1\}} e^{-\alpha_1(T-t)}.$$

*Processes $\tilde{D}_1(t,T)$ and $\hat{D}_1(t,T)$ satisfy*

$$\tilde{D}_1(t,T) = \mathbb{1}_{\{\tau_2 > t\}} \frac{\lambda - \alpha_2}{\lambda - \alpha_1}\, \frac{(\lambda_1 - \alpha_1)e^{-\lambda(T-t)} + \lambda_2 e^{-\alpha_1(T-t)}}{\lambda_1 e^{-(\lambda - \alpha_2)t} + \lambda_2 - \alpha_2}$$
$$+ \mathbb{1}_{\{\tau_2 \leq t\}} \frac{(\lambda - \alpha_2)\lambda_2 e^{-\alpha_1(T-\tau_2)}}{\lambda_1 \alpha_2 e^{(\lambda - \alpha_2)\tau_2} + \lambda(\lambda_2 - \alpha_2)}$$

*and*

$$\hat{D}_1(t,T) = \mathbb{1}_{\{\tau_1 > t\}} \frac{\lambda_2 e^{-\alpha_1 T} + (\lambda_1 - \alpha_1)e^{-\lambda T}}{\lambda_2 e^{-\alpha_1 t} + (\lambda_1 - \alpha_1)e^{-\lambda t}}.$$

*Proof.* The first equality is an immediate consequence of Lemma 7.3.1 combined with the definition of $D_1(t,T)$. Let us only mention that we use here the following representation for the price process $D_1(t,T)$ (cf. (7.19)):

$$D_1(t,T) = \mathbb{Q}^*\{\tau_1 > T \,|\, \mathcal{H}_t^1 \vee \mathcal{H}_t^2\} = \mathbb{1}_{\{\tau_1 > t\}} \frac{\mathbb{Q}^*\{\tau_1 > T \,|\, \mathcal{H}_t^2\}}{\mathbb{Q}^*\{\tau_1 > t \,|\, \mathcal{H}_t^2\}}. \qquad (10.5)$$

To derive an explicit valuation formula for $D_1(t,T)$, it suffices to combine (10.5) with the following equality, which holds for every $t \leq s \leq T$

$$Q^*\{\tau_1 > s \,|\, \mathcal{H}_t^2\} = 1\!\!1_{\{\tau_2 > t\}} \frac{Q^*\{\tau_1 > s, \tau_2 > t\}}{Q^*\{\tau_2 > t\}} + 1\!\!1_{\{\tau_2 \leq t\}} Q^*\{\tau_1 > s \,|\, \tau_2\}.$$

Using the last formula, we also obtain

$$\tilde{D}_1(t,T) = 1\!\!1_{\{\tau_2 > t\}} \frac{Q^*\{\tau_1 > T, \tau_2 > t\}}{Q^*\{\tau_2 > t\}} + 1\!\!1_{\{\tau_2 \leq t\}} Q^*\{\tau_1 > T \,|\, \tau_2\}.$$

where (see Lemma 7.3.1)

$$Q^*\{\tau_1 > T, \tau_2 > t\} = \frac{1}{\lambda - \alpha_1} \left( (\lambda_1 - \alpha_1)e^{-\lambda T} + \lambda_2 e^{-\alpha_1 T - (\lambda - \alpha_1)t} \right).$$

Furthermore (cf. (7.17))

$$Q^*\{\tau_2 > t\} = \frac{1}{\lambda - \alpha_2} \left( \lambda_1 e^{-\alpha_2 t} + (\lambda_2 - \alpha_2)e^{-\lambda t} \right)$$

and (cf. (7.20))

$$1\!\!1_{\{\tau_2 \leq t\}} Q^*\{\tau_1 > T \,|\, \tau_2\} = 1\!\!1_{\{\tau_2 \leq t\}} \frac{(\lambda - \alpha_2)\lambda_2 e^{-\alpha_1(T - \tau_2)}}{\lambda_1 \alpha_2 e^{(\lambda - \alpha_2)\tau_2} + \lambda(\lambda_2 - \alpha_2)}.$$

By combining the last three formulae, we obtain the expression for $\tilde{D}_1(t,T)$. Finally, the last conditional probability satisfies

$$\hat{D}_1(t,T) = 1\!\!1_{\{\tau_1 > t\}} Q^*\{\tau_1 > T \,|\, \tau_1 > t\} = 1\!\!1_{\{\tau_1 > t\}} \frac{Q^*\{\tau_1 > T\}}{Q^*\{\tau_1 > t\}},$$

which entails that

$$\hat{D}_1(t,T) = 1\!\!1_{\{\tau_1 > t\}} \frac{\lambda_2 e^{-\alpha_1 T} + (\lambda_1 - \alpha_1)e^{-\lambda T}}{\lambda_2 e^{-\alpha_1 t} + (\lambda_1 - \alpha_1)e^{-\lambda t}}.$$

This completes the proof.                                                  □

Few pertinent comments on the last result are in order. First, in case of a non-zero interest rate $r$, it suffices to multiply the right-hand sides in the valuation formulae of the last proposition by $B(t,T)$. Second, notice that $D_1(t,T)$ and $\hat{D}_1(t,T)$ represent the ex-dividend values of the bond, so that they vanish after default (this remark does not apply to $\tilde{D}_1(t,T)$, however).

Although derived in a different set-up, formula for $D_1(t,T)$ and an analogous formula for $D_2(t,T)$ coincide with the Jarrow and Yu result for the bond issued by the secondary firm. Also, it is essential to observe that the bond price $D_1(t,T)$ ($D_2(t,T)$, resp.) does not depend on the value of $\alpha_2$ ($\alpha_1$, resp.) This means, of course, that in calculations of $D_1(t,T)$ we may assume that the equality $\lambda_2 = \alpha_2$ holds. Similarly, when searching for the price $D_2(t,T)$, we may set $\alpha_1 = \lambda_1$.

The last observation is in fact crucial. It shows that when valuing a bond issued by a given firm, we may equally well postulate that the default of this firm affects the default intensity of the other firm, or, on the contrary, that it doesn't. This is also clear at the intuitive level: since we search for the pre-default value of the bond, the behavior of default probabilities of other firms after the default of the bond's issuer is irrelevant.

To conclude, for $D_1(t,T)$ we need to distinguish between only the two following cases: $\lambda_1 \neq \alpha_1$ and $\lambda_1 = \alpha_1$. In the former case, the value of $D_1(t,T)$ is given by the general formula of Proposition 10.1.2. In the latter, this formula reduces to the expected result:

$$D_1(t,T) = \mathbb{1}_{\{\tau_1>t\}} B(t,T) e^{-\lambda_1(T-t)} = \mathbb{1}_{\{\tau_1>t\}} e^{-(r+\lambda_1)(T-t)}.$$

The same arguments apply to $D_2(t,T)$, so that for $\lambda_2 = \alpha_2$ we have

$$D_2(t,T) = \mathbb{1}_{\{\tau_2>t\}} B(t,T) e^{-\lambda_2(T-t)} = \mathbb{1}_{\{\tau_2>t\}} e^{-(r+\lambda_2)(T-t)}.$$

The second valuation formula of Proposition 10.1.2 hinges on the assumption that we observe the default time $\tau_2$, but the default time of the first firm is not observed. For the special case when $\lambda_1 = \alpha_1$ and $\lambda_2 \neq \alpha_2$, the process $\tilde{D}_1(t,T)$ equals

$$\tilde{D}_1(t,T) = \mathbb{1}_{\{\tau_2>t\}} B(t,T) \frac{(\lambda - \alpha_2)e^{-\lambda_1(T-t)}}{\lambda_1 e^{(\lambda-\alpha_2)t} + \lambda_2 - \alpha_2}$$
$$+ \mathbb{1}_{\{\tau_2\leq t\}} B(t,T) \frac{(\lambda - \alpha_2)\lambda_2 e^{-\lambda_1(T-\tau_2)}}{\lambda_1 \alpha_2 e^{(\lambda-\alpha_2)\tau_2} + \lambda(\lambda_2 - \alpha_2)}.$$

If we assume instead that $\lambda_1 \neq \alpha_1$ but $\lambda_2 = \alpha_2$, we obtain

$$\tilde{D}_1(t,T) = \mathbb{1}_{\{\tau_2>t\}} B(t,T) \frac{(\lambda_1 - \alpha_1)e^{-\lambda(T-t)} + \lambda_2 e^{-\alpha_1(T-t)}}{(\lambda - \alpha_1)e^{\lambda_1 t}}$$
$$+ \mathbb{1}_{\{\tau_2\leq t\}} B(t,T) e^{-\lambda_1\tau_2-\alpha_1(T-\tau_2)}.$$

It is interesting to observe that in both cases considered above, the price process $D_1(t,T)$ admits a discontinuity at the default time $\tau_2$ of the second firm. This is a natural consequence of the fact that in both cases the default times $\tau_1$ and $\tau_2$ are mutually dependent under $\mathbb{Q}^*$. Finally, for $\lambda_1 = \alpha_1$ and $\lambda_2 = \alpha_2$, we get

$$\tilde{D}_1(t,T) = B(t,T) e^{-\lambda_1 T} = B(t,T) \mathbb{Q}^*\{\tau_1 > T\} = B(t,T) \mathbb{Q}\{\tau_1 > T\}.$$

This result is also clear, since in this case default times of both firms are mutually independent under $\mathbb{Q}^* = \mathbb{Q}$.

Finally, the last formula in Proposition 10.1.2 implicitly assumes that we observe only the default time of the first firm. It is apparent that this result is also independent of the value of $\alpha_2$. Again, in case when $\lambda_1 = \alpha_1$, we conclude that the equalities

$$\hat{D}_1(t,T) = \mathbb{1}_{\{\tau_1>t\}} B(t,T) e^{-\lambda_1(T-t)} = D_1(t,T)$$

are valid for every $t \leq T$.

**Case of non-zero recovery rates.** The case on non-zero recovery is not much different from the previous case. Indeed, the payoff $D_i(T,T)$ at maturity can be represented as follows:

$$D_i(T,T) = \mathbb{1}_{\{\tau_i > T\}} + \delta_i \mathbb{1}_{\{\tau_i \leq T\}} = \delta_i + (1 - \delta_i)\mathbb{1}_{\{\tau_i > T\}}$$

and so

$$D_i(t,T) = B(t,T)\big(\delta_i + (1 - \delta_i)\,\mathbb{Q}^*\{\tau_i > T \mid \mathcal{H}_t^1 \vee \mathcal{H}_t^2\}\big).$$

Explicit formulae for $D_i(t,T)$, as well as for $\tilde{D}_i(t,T)$ and $\hat{D}_i(t,T)$, can be thus obtained directly from Proposition 10.1.2. It is easy to see that the expressions for the values of $D_1(t,T)$ and $D_2(t,T)$ will coincide, up to a suitable change of notation, with the formula of Proposition 10.1.1. Moreover, when $\lambda_1 = \alpha_1$, we obtain (cf. formula (10.4))

$$D_1(t,T) = B(t,T)\big(\delta_1 + (1 - \delta_1)e^{-\lambda_1(T-t)}\mathbb{1}_{\{\tau_1 > t\}}\big),$$

while for $\lambda_2 = \alpha_2$, we get

$$D_2(t,T) = B(t,T)\big(\delta_2 + (1 - \delta_2)e^{-\lambda_2(T-t)}\mathbb{1}_{\{\tau_2 > t\}}\big).$$

**Interpretation of martingale intensities.** Let us provide an intuitive probabilistic interpretation for the martingale intensities $\lambda_1^*$ and $\lambda_2^*$. Recall that

$$\lambda_1^*(t) = \lambda_1 \mathbb{1}_{\{\tau_2 > t\}} + \alpha_1 \mathbb{1}_{\{\tau_2 \leq t\}}, \quad \lambda_2^*(t) = \lambda_2 \mathbb{1}_{\{\tau_1 > t\}} + \alpha_2 \mathbb{1}_{\{\tau_1 \leq t\}}.$$

The following result shows that the jump of the martingale intensity has the desired financial interpretation.

**Proposition 10.1.3.** *For $i = 1, 2$ and every $t \in \mathbb{R}_+$ we have*

$$\lambda_i = \lim_{h \downarrow 0} h^{-1}\,\mathbb{Q}^*\{t < \tau_i \leq t + h \mid \tau_1 > t,\, \tau_2 > t\} \qquad (10.6)$$

*and*

$$\alpha_i = \lim_{h \downarrow 0} h^{-1}\,\mathbb{Q}^*\{t < \tau_i \leq t + h \mid \tau_1 > t,\, \tau_2 \leq t\}. \qquad (10.7)$$

*Proof.* By symmetry, it suffices to examine only one intensity, e.g., $\lambda_1^*$. Using Lemma 7.3.1, we obtain

$$I(t,h) := \mathbb{Q}^*\{t < \tau_1 \leq t + h \mid \tau_1 > t,\, \tau_2 > t\} = 1 - \frac{\mathbb{Q}^*\{\tau_1 > t + h,\, \tau_2 > t\}}{\mathbb{Q}^*\{\tau_1 > t,\, \tau_2 > t\}}$$

$$= 1 - \frac{(\lambda_1 - \alpha_1)e^{-\lambda h} + \lambda_2 e^{-\alpha_1 h}}{\lambda - \alpha_1}.$$

Since clearly $\lim_{h \downarrow 0} h^{-1} I(t,h) = \lambda_1$ for every $t \in \mathbb{R}_+$, equality (10.6) is valid.

For the second convergence, notice that

$$J(t,h) := \mathbb{Q}^* \{t < \tau_1 \le t + h \mid \tau_1 > t, \tau_2 \le t\} = \frac{\mathbb{Q}^* \{t < \tau_1 \le t + h, \tau_2 \le t\}}{\mathbb{Q}^* \{\tau_1 > t, \tau_2 \le t\}}$$

$$= \frac{\mathbb{Q}^* \{\tau_1 > t\} - \mathbb{Q}^* \{\tau_1 > t + h\} - \mathbb{Q}^* \{t < \tau_1 \le t + h, \tau_2 > t\}}{\mathbb{Q}^* \{\tau_1 > t\} - \mathbb{Q}^* \{\tau_1 > t, \tau_2 > t\}}$$

and so

$$J(t,h) = 1 - \frac{\mathbb{Q}^* \{\tau_1 > t + h\} - \mathbb{Q}^* \{\tau_1 > t + h, \tau_2 > t\}}{\mathbb{Q}^* \{\tau_1 > t\} - \mathbb{Q}^* \{\tau_1 > t, \tau_2 > t\}}.$$

Equality (7.17) yields

$$\mathbb{Q}^* \{\tau_1 > t\} = \frac{1}{\lambda - \alpha_1} \left( \lambda_2 e^{-\alpha_1 t} + (\lambda_1 - \alpha_1) e^{-\lambda t} \right).$$

Combining the last formula with Lemma 7.3.1, we get, after simplifications,

$$J(t,h) = 1 - \frac{e^{-\alpha_1(t+h)} - e^{-\alpha_1 t - \lambda_1 t}}{e^{-\alpha_1 t} - e^{-\lambda_1 t}}.$$

The reader can readily verify that $\lim_{h \downarrow 0} h^{-1} J(t,h) = \alpha_1$ for $t \in \mathbb{R}_+$. □

**Equivalent constructions of default times.** Instead of using Kusuoka's approach, we could have defined $\tau_1$ and $\tau_2$ through the canonical method. Specifically, we could have postulated that the intensity processes satisfy

$$\lambda_t^1 = \lambda_1 \mathbb{1}_{\{\tau_2 > t\}} + \alpha_1 \mathbb{1}_{\{\tau_2 \le t\}},$$

$$\lambda_t^2 = \lambda_2 \mathbb{1}_{\{\tau_1 > t\}} + \alpha_2 \mathbb{1}_{\{\tau_1 \le t\}},$$

and we could have defined $\tau_1$ and $\tau_2$ through the standard formula

$$\tau_i = \inf \left\{ t \in \mathbb{R}_+ : \int_0^t \lambda_u^i \, du \ge \eta_i \right\}, \qquad (10.8)$$

where $\eta_i$, $i = 1, 2$ are mutually independent, identically distributed, random variables with unit exponential law under $\mathbb{Q}^*$. Of course, it is not quite obvious that the random times $\tau_1$ and $\tau_2$ satisfying (10.8) are well defined.

Another equivalent construction of $\tau_1$ and $\tau_2$ runs as follows: we take the two mutually independent, identically distributed, random variables $\eta_i$, $i = 1, 2$ with unit exponential law under $\mathbb{Q}^*$, and we set

$$\tau_1 = \begin{cases} \lambda_1^{-1} \eta_1, & \text{if } \lambda_1^{-1} \eta_1 \le \lambda_2^{-1} \eta_2, \\ \lambda_2^{-1} \eta_2 + \alpha_1^{-1} (\eta_1 - \lambda_1 \lambda_2^{-1} \eta_2), & \text{if } \lambda_1^{-1} \eta_1 > \lambda_2^{-1} \eta_2. \end{cases} \qquad (10.9)$$

and

$$\tau_2 = \begin{cases} \lambda_2^{-1} \eta_2, & \text{if } \lambda_2^{-1} \eta_2 \le \lambda_1^{-1} \eta_1, \\ \lambda_1^{-1} \eta_1 + \alpha_2^{-1} (\eta_2 - \lambda_2 \lambda_1^{-1} \eta_1), & \text{if } \lambda_2^{-1} \eta_2 > \lambda_1^{-1} \eta_1. \end{cases} \qquad (10.10)$$

It is worth noticing that the last two constructions are in fact identical; indeed, the random times given by expressions (10.9)–(10.10) are unique solutions to (10.8). The latter method of construction of several dependent random times was examined in detail by Shaked and Shanthikumar (1987).

## 10.2 Martingale Approach to Basket Credit Derivatives

In this section, in which we follow Bielecki and Rutkowski (2001b), we examine the martingale approach to the valuation of basket credit derivatives. To simplify the presentation, we assume that the $(\mathbb{F}, \mathbb{G})$-martingale hazard process $\Lambda^i$ of each random time $\tau_i$ admits an $\mathbb{F}$-intensity process $\lambda^i$. We shall use the following notation throughout this section: $H_t^i := \mathbb{1}_{\{\tau_i \le t\}}$, $\tilde{H}_t^i := 1 - H_t^i = \mathbb{1}_{\{\tau_i > t\}}$. Before formulating a general result, we proceed with some illustrative examples and comments.

**Case $n = 1$.** Although we deal here with only one default time, $\tau_1 = \tau_{(1)}$, some interesting features of the $i^{\text{th}}$-to-default valuation will already appear in this case. According to our convention, the dividend process for the claim $CCT^{(1)}$ equals

$$
D_t^{(1)} = \int_{]0,t]} Z_u^1 \, dH_u^1 + \mathbb{1}_T(t)\big(X_1 \mathbb{1}_{\{\tau_1 \le T\}} + X \mathbb{1}_{\{\tau_1 > T\}}\big)
$$

$$
= Z_{\tau_1}^1 \mathbb{1}_{\{\tau_1 \le t\}} + \mathbb{1}_T(t)\big((X - X_1)\mathbb{1}_{\{\tau_1 > T\}} + X_1\big).
$$

Thus, the (ex-dividend) pre-default value process $S^{(1)}$ satisfies

$$
S_t^{(1)} = B_t \, \mathbb{E}_{\mathbb{Q}^*}\Big(\int_{]t,T]} B_u^{-1} \, dD_u^{(1)} \,\Big|\, \mathcal{G}_t\Big) = \tilde{S}_t^{(1)} + B_t \, \mathbb{E}_{\mathbb{Q}^*}\big(B_T^{-1} X_1 \,|\, \mathcal{G}_t\big),
$$

where

$$
\tilde{S}_t^{(1)} = B_t \, \mathbb{E}_{\mathbb{Q}^*}\Big(B_{\tau_1}^{-1} Z_{\tau_1}^1 \mathbb{1}_{\{t < \tau_1 \le T\}} + B_T^{-1}(X - X_1)\mathbb{1}_{\{\tau_1 > T\}} \,\Big|\, \mathcal{G}_t\Big). \tag{10.11}
$$

From results in Sect. 8.3, we know that one may derive an equivalent representation of the price process $\tilde{S}^{(1)}$ in terms of the default-risk-adjusted interest rate $\tilde{r}_u = r_u + \lambda_u^1$. Specifically, we have

$$
\tilde{S}_t^{(1)} = \mathbb{1}_{\{\tau > t\}}\Big(\tilde{V}_t^{(1)} - \tilde{B}_t \, \mathbb{E}_{\mathbb{Q}^*}\big(\tilde{B}_{\tau_1}^{-1} \Delta \tilde{V}_{\tau_1}^{(1)} \,|\, \mathcal{G}_t\big)\Big), \tag{10.12}
$$

where

$$
\tilde{V}_t^{(1)} = \tilde{B}_t \, \mathbb{E}_{\mathbb{Q}^*}\Big(\int_t^T \tilde{B}_u^{-1} Z_u^1 \lambda_u^1 \, du + \tilde{B}_T^{-1}(X - X_1) \,\Big|\, \mathcal{G}_t\Big)
$$

and

$$
\tilde{B}_t = \exp\Big(\int_0^t \tilde{r}_u \, du\Big) = B_t \exp\Big(\int_0^t \lambda_u^1 \, du\Big).
$$

One may also establish a similar – but less convenient – representation:

$$
\tilde{S}_t^{(1)} = \Big(\hat{V}_t^{(1)} - \hat{B}_t \, \mathbb{E}_{\mathbb{Q}^*}\big(\hat{B}_{\tau_1}^{-1} \Delta \hat{V}_{\tau_1}^{(1)} \,|\, \mathcal{G}_t\big)\Big), \tag{10.13}
$$

where

$$
\hat{V}_t^{(1)} = \hat{B}_t \, \mathbb{E}_{\mathbb{Q}^*}\Big(\int_t^T \hat{B}_u^{-1} Z_u^1 \lambda_u^1 \tilde{H}_u^1 \, du + \hat{B}_T^{-1}(X - X_1) \,\Big|\, \mathcal{G}_t\Big)
$$

and

$$
\hat{B}_t = B_t \exp\Big(\int_0^{t \wedge \tau_1} \lambda_u^1 \, du\Big).
$$

As we shall see below, both kinds of representations – (10.12) and (10.13) – are valid for pre-default values of first-to-default claims. However, even in the simple setting considered here, it is apparent that representation (10.13) is not convenient from the computational point of view. In fact, the process $\hat{V}^{(1)}$ will typically have a discontinuity at $\tau_1$. Let us consider a trivial example: $Z^1 \equiv z^1 = \text{const.}$, $\lambda^1 = \text{const.}$, and $X_1 = X = 0$. In this case, the process $\hat{V}^{(1)}$ admits a discontinuity at $\tau_1$, whereas the process $\tilde{V}^{(1)}$ is continuous at $\tau_1$, and thus representation (10.12) simplifies to $\tilde{S}_t^{(1)} = \tilde{H}_t^1 \tilde{V}_t^{(1)}$.

*Remarks.* Unfortunately, only a representation similar to (10.13) can be derived for the price process of a general $i^{\text{th}}$-to-default claim, with the notable exception of the first-to-default claim for which the representation (10.12) can be obtained (see Proposition 10.2.4 below). The practical use of both kinds of the above representations is thus in general largely limited, and the basic formula (10.16) needs to be employed in calculations.

**Case $n = 2$.** In the case of two reference entities, the claim $CCT^{(1)}$ is associated with the random time $\tau_{(1)} = \min(\tau_1, \tau_2)$, whereas the claim $CCT^{(2)}$ is associated with the random time $\tau_{(2)} = \max(\tau_1, \tau_2)$. For the claim $CCT^{(1)}$, the dividend process equals, for every $t \in [0, T]$,

$$
\begin{aligned}
D_t^{(1)} &= Z_{\tau_1}^1 \mathbb{1}_{\{\tau_1 < \tau_2,\, \tau_1 \leq t\}} + Z_{\tau_2}^2 \mathbb{1}_{\{\tau_2 < \tau_1,\, \tau_2 \leq t\}} \\
&\quad + \mathbb{1}_T(t)\big(X_1 \mathbb{1}_{\{\tau_1 < \tau_2,\, \tau_1 \leq T\}} + X_2 \mathbb{1}_{\{\tau_2 < \tau_1,\, \tau_2 \leq T\}} + X \mathbb{1}_{\{\tau_{(1)} > T\}}\big) \\
&= Z_{\tau_1}^1 \mathbb{1}_{\{\tau_1 \leq t,\, \tau_1 < \tau_2\}} + Z_{\tau_2}^2 \mathbb{1}_{\{\tau_2 \leq t,\, \tau_2 < \tau_1\}} \\
&\quad + \mathbb{1}_T(t)\big(X_1 \mathbb{1}_{\{\tau_1 < \tau_2\}} + X_2 \mathbb{1}_{\{\tau_2 < \tau_1\}} + Y \mathbb{1}_{\{\tau_{(1)} > T\}}\big),
\end{aligned}
$$

where we denote

$$
Y = X - X_1 \mathbb{1}_{\{\tau_1 < \tau_2\}} - X_2 \mathbb{1}_{\{\tau_2 < \tau_1\}}.
$$

Thus, the (ex-dividend) pre-default value process $S^{(1)}$ satisfies

$$
\begin{aligned}
S_t^{(1)} &= B_t\, \mathbb{E}_{\mathbb{Q}^*}\left( \int_{]t,T]} B_u^{-1} dD_u^{(1)} \,\Big|\, \mathcal{G}_t \right) \\
&= B_t\, \mathbb{E}_{\mathbb{Q}^*}\left( B_{\tau_1}^{-1} Z_{\tau_1}^1 \mathbb{1}_{\{t < \tau_1 < \tau_2,\, \tau_1 \leq T\}} + B_{\tau_2}^{-1} Z_{\tau_2}^2 \mathbb{1}_{\{t < \tau_2 < \tau_1,\, \tau_2 \leq T\}} \,\Big|\, \mathcal{G}_t \right) \\
&\quad + B_t\, \mathbb{E}_{\mathbb{Q}^*}\left( B_T^{-1}(X_1 \mathbb{1}_{\{\tau_1 < \tau_2\}} + X_2 \mathbb{1}_{\{\tau_2 < \tau_1\}} + Y \mathbb{1}_{\{\tau_{(1)} > T\}}) \,\Big|\, \mathcal{G}_t \right) \\
&= \tilde{S}_t^{(1)} + B_t\, \mathbb{E}_{\mathbb{Q}^*}\left( B_T^{-1}(X_1 \mathbb{1}_{\{\tau_1 < \tau_2\}} + X_2 \mathbb{1}_{\{\tau_2 < \tau_1\}}) \,\Big|\, \mathcal{G}_t \right),
\end{aligned}
$$

where

$$
\begin{aligned}
\tilde{S}_t^{(1)} &= B_t\, \mathbb{E}_{\mathbb{Q}^*}\left( B_{\tau_1}^{-1} Z_{\tau_1}^1 \mathbb{1}_{\{t < \tau_1 < \tau_2,\, \tau_1 \leq T\}} + B_{\tau_2}^{-1} Z_{\tau_2}^2 \mathbb{1}_{\{t < \tau_2 < \tau_1,\, \tau_2 \leq T\}} \,\Big|\, \mathcal{G}_t \right) \\
&\quad + B_t\, \mathbb{E}_{\mathbb{Q}^*}\left( B_T^{-1} Y \mathbb{1}_{\{\tau_{(1)} > T\}} \,\Big|\, \mathcal{G}_t \right).
\end{aligned}
$$

Two other expressions for the price process $\tilde{S}^{(1)}$ – analogous to representations (10.12) and (10.13) – may also be derived here.

We shall provide detailed calculations leading to the representation analogous to (10.13) for the first-to-default claim. To this end, we define the process $\hat{B}$ by setting $\lambda_t^{(1)} = \lambda_t^1 + \lambda_t^2$, and

$$\hat{B}_t = B_t \exp\left( \int_0^{t \wedge \tau_{(1)}} \lambda_u^{(1)} \, du \right),$$

and we introduce an auxiliary process $\hat{V}^{(1)}$:

$$\hat{V}_t^{(1)} = \hat{B}_t \, \mathbb{E}_{\mathbb{Q}^*}\left( \int_t^T \hat{B}_u^{-1}(Z_u^1 \lambda_u^1 + Z_u^2 \lambda_u^2) \mathbb{1}_{\{\tau_{(1)} > u\}} \, du + \hat{B}_T^{-1} Y \,\Big|\, \mathcal{G}_t \right).$$

The next result furnishes a counterpart of representation (10.13).

**Proposition 10.2.1.** *For every $t \in [0, T]$ we have*

$$\tilde{S}_t^{(1)} = \mathbb{1}_{\{\tau_{(1)} > t\}} \left( \hat{V}_t^{(1)} - \hat{B}_t \, \mathbb{E}_{\mathbb{Q}^*}\left( \hat{B}_{\tau_{(1)}}^{-1} \Delta \hat{V}_{\tau_{(1)}}^{(1)} \,\big|\, \mathcal{G}_t \right) \right). \tag{10.14}$$

*Proof.* We begin by noting that

$$\hat{V}_t^{(1)} = \hat{B}_t \left( N_t^{(1)} - \int_0^t \hat{B}_u^{-1}(Z_u^1 \lambda_u^1 + Z_u^2 \lambda_u^2) \mathbb{1}_{\{\tau_{(1)} > u\}} \, du \right),$$

where the process $N^{(1)}$ is a $\mathbb{G}$-martingale (under mild technical assumptions), specifically,

$$N_t^{(1)} := \mathbb{E}_{\mathbb{Q}^*}\left( \int_0^T \hat{B}_u^{-1}(Z_u^1 \lambda_u^1 + Z_u^2 \lambda_u^2) \mathbb{1}_{\{\tau_{(1)} > u\}} \, du + \hat{B}_T^{-1} Y \,\Big|\, \mathcal{G}_t \right).$$

Next, we define $U_t^{(1)} = \mathbb{1}_{\{\tau_{(1)} > t\}} \hat{V}_t^{(1)} = \tilde{H}_t^1 \tilde{H}_t^2 \hat{V}_t^{(1)}$. Combining the product rule:

$$dU_t^{(1)} = \tilde{H}_{t-}^1 \tilde{H}_{t-}^2 \, d\hat{V}_t^{(1)} + \hat{V}_{t-}^{(1)} \, d(\tilde{H}_t^1 \tilde{H}_t^2) + \Delta(\tilde{H}_t^1 \tilde{H}_t^2) \Delta \hat{V}_t^{(1)},$$

with the fact that $\Delta H_t^1 \Delta H_t^2 = 0$, $\mathbb{Q}^*$-a.s., we obtain

$$\begin{aligned}
dU_t^{(1)} &= r_t U_t^{(1)} dt - \tilde{H}_t^1 \tilde{H}_t^2 \lambda_t^1 (Z_t^1 - \hat{V}_t^{(1)}) \, dt - \tilde{H}_t^1 \tilde{H}_t^2 \lambda_t^2 (Z_t^2 - \hat{V}_t^{(1)}) \, dt \\
&\quad + \tilde{H}_t^1 \tilde{H}_t^2 \hat{B}_t \, dN_t^{(1)} - \hat{V}_{t-}^{(1)}(\tilde{H}_{t-}^1 \, dH_t^2 + \tilde{H}_{t-}^2 \, dH_t^1) \\
&\quad - \Delta \hat{V}_t^{(1)}(\tilde{H}_{t-}^1 \, dH_t^2 + \tilde{H}_{t-}^2 \, dH_t^1) \\
&= r_t U_t^{(1)} dt - (Z_t^1 + \Delta \hat{V}_t^{(1)}) \tilde{H}_{t-}^2 \, dH_t^1 - (Z_t^2 + \Delta \hat{V}_t^{(1)}) \tilde{H}_{t-}^1 \, dH_t^2 \\
&\quad + (Z_t^1 - \hat{V}_t^{(1)}) \tilde{H}_{t-}^2 \, dM_t^1 - (Z_t^2 - \hat{V}_t^{(1)}) \tilde{H}_{t-}^1 \, dM_t^2 \\
&\quad + \tilde{H}_t^1 \tilde{H}_t^2 \hat{B}_t \, dN_t^{(1)},
\end{aligned}$$

where the processes $M^j$, $j = 1, 2$, are $\mathbb{G}$-local martingales:

$$M_t^j := H_t^j - \int_0^t \lambda_u^j \tilde{H}_u^j \, du = H_t^j - \int_0^{t \wedge \tau_j} \lambda_u^j \, du.$$

We shall assume from now on that the processes $M^j$, $j = 1, 2$ are 'true' martingales, rather than local martingales. Since $U_T^{(1)} = Y \mathbb{1}_{\{\tau_{(1)} > T\}}$, we conclude that

$$U_t^{(1)} = B_t \, \mathbb{E}_{\mathbb{Q}^*} \left( \int_{]t,T]} B_u^{-1}(Z_u^1 + \Delta \hat{V}_u^{(1)}) \tilde{H}_{u-}^2 \, dH_u^1 \, \middle| \, \mathcal{G}_t \right),$$

$$+ B_t \, \mathbb{E}_{\mathbb{Q}^*} \left( \int_{]t,T]} B_u^{-1}(Z_u^2 + \Delta \hat{V}_u^{(1)}) \tilde{H}_{u-}^1 \, dH_u^2 \, \middle| \, \mathcal{G}_t \right)$$

$$+ B_t \, \mathbb{E}_{\mathbb{Q}^*} \left( B_T^{-1} Y \mathbb{1}_{\{\tau_{(1)} > T\}} \, \middle| \, \mathcal{G}_t \right).$$

This shows that representation (10.14) is indeed correct.    □

Regrettably, representation (10.14) – although quite interesting on its own – is not a good basis for practical valuation calculations. It is thus important to notice that, using similar arguments, one can also derive a representation analogous to (10.12). More explicitly, we have

$$\tilde{S}_t^{(1)} = \mathbb{1}_{\{\tau_{(1)} > t\}} \left( \tilde{V}_t^{(1)} - B_t \, \mathbb{E}_{\mathbb{Q}^*} \left( \tilde{B}_{\tau_{(1)}}^{-1} \Delta \tilde{V}_{\tau_{(1)}}^{(1)} \, \middle| \, \mathcal{G}_t \right) \right), \tag{10.15}$$

where

$$\tilde{V}_t^{(1)} = \tilde{B}_t \, \mathbb{E}_{\mathbb{Q}^*} \left( \int_t^T \tilde{B}_u^{-1}(Z_u^1 \lambda_u^1 + Z_u^2 \lambda_u^2) \, du + \tilde{B}_T^{-1} Y \, \middle| \, \mathcal{G}_t \right)$$

and

$$\tilde{B}_t = \exp \left( \int_0^t (r_u + \lambda_u^{(1)}) \, du \right) = B_t \exp \left( \int_0^t \lambda_u^{(1)} \, du \right),$$

where $\lambda_u^{(1)} = \lambda_u^1 + \lambda_u^2$. Observe that, according to Lemma 7.1.2, the process $\lambda^{(1)}$ is the $\mathbb{F}$-martingale intensity process of the random time $\tau_{(1)}$. Representation (10.15) – previously established in Duffie (1998a) – is much more amenable for practical calculations and simulations than (10.14).

We proceed with the second part of our example, in which we shall consider the valuation problem for the last-to-default claim $CCT^{(2)}$. Our goal is to derive a representation analogous to (10.14); it seems to us that it is not possible to obtain a representation similar to (10.15) for the pre-default value process of the claim $CCT^{(2)}$. The dividend process of the last-to-default claim $CCT^{(2)}$ equals, for $t \in [0, T]$,

$$D_t^{(2)} = Z_{\tau_1}^1 \mathbb{1}_{\{\tau_2 < \tau_1, \, \tau_1 \le t\}} + Z_{\tau_2}^2 \mathbb{1}_{\{\tau_1 < \tau_2, \, \tau_2 \le t\}}$$
$$+ \mathbb{1}_T(t) \left( X_1 \mathbb{1}_{\{\tau_2 < \tau_1, \, \tau_1 \le T\}} + X_2 \mathbb{1}_{\{\tau_1 < \tau_2, \, \tau_2 \le T\}} + X \mathbb{1}_{\{\tau_{(2)} > T\}} \right)$$

or, equivalently,

$$D_t^{(2)} = Z_{\tau_1}^1 \mathbb{1}_{\{\tau_2 < \tau_1, \, \tau_1 \le t\}} + Z_{\tau_2}^2 \mathbb{1}_{\{\tau_1 < \tau_2, \, \tau_2 \le t\}}$$
$$+ \mathbb{1}_T(t) \left( X_1 \mathbb{1}_{\{\tau_2 < \tau_1\}} + X_2 \mathbb{1}_{\{\tau_1 < \tau_2\}} + \tilde{Y} \mathbb{1}_{\{\tau_{(2)} > T\}} \right),$$

where we have set

$$\tilde{Y} = X - X_1 \mathbb{1}_{\{\tau_2 < \tau_1\}} - X_2 \mathbb{1}_{\{\tau_1 < \tau_2\}}.$$

Consequently, the (ex-dividend) pre-default value process $S^{(2)}$ equals

$$S_t^{(2)} = B_t \, \mathbb{E}_{\mathbb{Q}^*} \left( \int_{]t,T]} B_u^{-1} dD_u^{(2)} \,\middle|\, \mathcal{G}_t \right)$$

$$= B_t \, \mathbb{E}_{\mathbb{Q}^*} \left( B_{\tau_1}^{-1} Z_{\tau_1}^1 \, \mathbb{1}_{\{\tau_2 < \tau_1, \, t < \tau_1 \leq T\}} + B_{\tau_2}^{-1} Z_{\tau_2}^2 \, \mathbb{1}_{\{\tau_1 < \tau_2, \, t < \tau_2 \leq T\}} \,\middle|\, \mathcal{G}_t \right)$$

$$+ B_t \, \mathbb{E}_{\mathbb{Q}^*} \left( B_T^{-1} \big( X_1 \mathbb{1}_{\{\tau_2 < \tau_1\}} + X_2 \mathbb{1}_{\{\tau_1 < \tau_2\}} + \tilde{Y} \mathbb{1}_{\{\tau_{(2)} > T\}} \big) \,\middle|\, \mathcal{G}_t \right).$$

Put another way,

$$S_t^{(2)} = \tilde{S}_t^{(2)} + B_t \, \mathbb{E}_{\mathbb{Q}^*} \left( B_T^{-1} X_1 \mathbb{1}_{\{\tau_2 < \tau_1\}} + B_T^{-1} X_2 \mathbb{1}_{\{\tau_1 < \tau_2\}} \,\middle|\, \mathcal{G}_t \right),$$

where

$$\tilde{S}_t^{(2)} = B_t \, \mathbb{E}_{\mathbb{Q}^*} \left( B_{\tau_1}^{-1} Z_{\tau_1}^1 \, \mathbb{1}_{\{\tau_2 < \tau_1, \, t < \tau_1 \leq T\}} + B_{\tau_2}^{-1} Z_{\tau_2}^2 \, \mathbb{1}_{\{\tau_1 < \tau_2, \, t < \tau_2 \leq T\}} \,\middle|\, \mathcal{G}_t \right)$$

$$+ B_t \, \mathbb{E}_{\mathbb{Q}^*} \left( B_T^{-1} \tilde{Y} \mathbb{1}_{\{\tau_{(2)} > T\}} \,\middle|\, \mathcal{G}_t \right).$$

We shall now formulate a representation for the process $\tilde{S}^{(2)}$ analogous to (10.14). Let us define processes $\lambda^{(2)}$ and $\bar{B}$ by setting

$$\lambda_t^{(2)} = \lambda_t^1 \tilde{H}_t^1 H_t^2 + \lambda_t^2 H_t^1 \tilde{H}_t^2,$$

and

$$\bar{B}_t = \exp \left( \int_0^t (r_u + \lambda_u^{(2)}) \, du \right) = B_t \exp \left( \int_0^t \lambda_u^{(2)} \, du \right).$$

The process $\hat{V}^{(2)}$ is now defined as follows:

$$\hat{V}_t^{(2)} = \bar{B}_t \, \mathbb{E}_{\mathbb{Q}^*} \left( \int_t^T \bar{B}_u^{-1} \big( Z_u^1 \lambda_u^1 H_u^2 \tilde{H}_u^1 + Z_u^2 \lambda_u^2 \tilde{H}_u^2 H_u^1 \big) \, du + \bar{B}_T^{-1} \tilde{Y} \,\middle|\, \mathcal{G}_t \right).$$

We then obtain the following representation (its derivation is left to the reader)

$$\tilde{S}_t^{(2)} = \mathbb{1}_{\{\tau_{(2)} > t\}} \left( \hat{V}_t^{(2)} - \bar{B}_t \, \mathbb{E}_{\mathbb{Q}^*} \big( \bar{B}_{\tau_{(2)}}^{-1} \Delta \hat{V}_{\tau_{(2)}}^{(2)} \,\middle|\, \mathcal{G}_t \big) \right).$$

Again, the last representation does not seem to facilitate the explicit calculations of the claim's value.

**Case $n = 3$.** The claims $CCT^{(1)}$, $CCT^{(2)}$ and $CCT^{(3)}$ are the first-to-default, the second-to-default and the last-to-default type claims, respectively. Here, we shall focus on the valuation problem for the first-to-default claim. Its dividend process equals, for $t \in [0, T]$,

$$D_t^{(1)} = Z_{\tau_1}^1 \mathbb{1}_{\{\tau_1 < \tau_2 \wedge \tau_3, \, \tau_1 \leq t\}} + Z_{\tau_2}^2 \mathbb{1}_{\{\tau_2 < \tau_1 \wedge \tau_3, \, \tau_2 \leq t\}} + Z_{\tau_3}^3 \mathbb{1}_{\{\tau_3 < \tau_1 \wedge \tau_2, \, \tau_3 \leq t\}}$$

$$+ \mathbb{1}_T(t) \big( X_1 \mathbb{1}_{\{\tau_1 < \tau_2 \wedge \tau_3, \, \tau_1 \leq T\}} + X_2 \mathbb{1}_{\{\tau_2 < \tau_1 \wedge \tau_3, \, \tau_2 \leq T\}} \big)$$

$$+ \mathbb{1}_T(t) \big( X_3 \mathbb{1}_{\{\tau_3 < \tau_1 \wedge \tau_2, \, \tau_3 \leq T\}} + X \mathbb{1}_{\{\tau_{(1)} > T\}} \big).$$

Consequently, the associated value process $S^{(1)}$ satisfies

$$
\begin{aligned}
S_t^{(1)} &= B_t \, \mathbb{E}_{\mathbb{Q}^*} \big( B_{\tau_1}^{-1} Z_{\tau_1}^1 \, \mathbb{1}_{\{t < \tau_1 < \tau_2 \wedge \tau_3, \, \tau_1 \le T\}} + B_{\tau_2}^{-1} Z_{\tau_2}^2 \, \mathbb{1}_{\{t < \tau_2 < \tau_1 \wedge \tau_3, \, \tau_2 \le T\}} \,\big|\, \mathcal{G}_t \big) \\
&\quad + B_t \, \mathbb{E}_{\mathbb{Q}^*} \big( B_{\tau_3}^{-1} Z_{\tau_3}^3 \, \mathbb{1}_{\{t < \tau_3 < \tau_1 \wedge \tau_2, \, \tau_3 \le T\}} + B_T^{-1} X_1 \mathbb{1}_{\{\tau_1 < \tau_2 \wedge \tau_3, \, \tau_1 \le T\}} \,\big|\, \mathcal{G}_t \big) \\
&\quad + B_t \, \mathbb{E}_{\mathbb{Q}^*} \big( B_T^{-1} X_2 \mathbb{1}_{\{\tau_2 < \tau_1 \wedge \tau_3, \, \tau_2 \le T\}} + B_T^{-1} X_3 \mathbb{1}_{\{\tau_3 < \tau_1 \wedge \tau_2, \, \tau_3 \le T\}} \,\big|\, \mathcal{G}_t \big) \\
&\quad + B_t \, \mathbb{E}_{\mathbb{Q}^*} \big( B_T^{-1} X \mathbb{1}_{\{\tau_{(1)} > T\}} \,\big|\, \mathcal{G}_t \big).
\end{aligned}
$$

Let us set $\lambda_t^{(1)} = \lambda_t^1 + \lambda_t^2 + \lambda_t^3$ (of course, $\lambda^{(1)}$ is thus the $\mathbb{F}$-martingale intensity of the random time $\tau_{(1)} = \tau_1 \wedge \tau_2 \wedge \tau_3$) and

$$
\tilde{B}_t = \exp \left( \int_0^t (r_u + \lambda_u^{(1)}) \, du \right) = B_t \exp \left( \int_0^t \lambda_u^{(1)} \, du \right).
$$

We find it convenient to denote

$$
\hat{Y} = X_1 \mathbb{1}_{\{\tau_1 < \tau_2 \wedge \tau_3\}} - X_2 \mathbb{1}_{\{\tau_2 < \tau_1 \wedge \tau_3\}} - X_3 \mathbb{1}_{\{\tau_3 < \tau_1 \wedge \tau_2\}}.
$$

Then we have the following result, whose proof is left to the reader.

**Proposition 10.2.2.** *For every $t \in [0, T]$ we have*

$$
S_t^{(1)} = \mathbb{1}_{\{\tau_{(1)} > t\}} \left( \tilde{V}_t^{(1)} - \tilde{B}_t \, \mathbb{E}_{\mathbb{Q}^*} \big( \tilde{B}_{\tau_{(1)}}^{-1} \Delta \tilde{V}_{\tau_{(1)}}^{(1)} \,\big|\, \mathcal{G}_t \big) \right) + B_t \, \mathbb{E}_{\mathbb{Q}^*} \big( B_T^{-1} \hat{Y} \,\big|\, \mathcal{G}_t \big),
$$

*where*

$$
\tilde{V}_t^{(1)} = \tilde{B}_t \, \mathbb{E}_{\mathbb{Q}^*} \left( \int_t^T \tilde{B}_u^{-1} (Z_u^1 \lambda_u^1 + Z_u^2 \lambda_u^2 + Z_u^3 \lambda_u^3) \, du + \tilde{B}_T^{-1} (X - \hat{Y}) \,\Big|\, \mathcal{G}_t \right).
$$

At this point, the following important warning is called for: the practical usefulness of formulae established in Propositions 10.2.1 and 10.2.2 hinges on one's ability to verify the following continuity conditions:

$$
\mathbb{E}_{\mathbb{Q}^*} \big( \hat{B}_{\tau_{(1)}}^{-1} \Delta \hat{V}_{\tau_{(1)}}^{(1)} \,\big|\, \mathcal{G}_t \big) = 0, \qquad \mathbb{E}_{\mathbb{Q}^*} \big( \tilde{B}_{\tau_{(1)}}^{-1} \Delta \tilde{V}_{\tau_{(1)}}^{(1)} \,\big|\, \mathcal{G}_t \big) = 0,
$$

respectively. These conditions seem to be rather hard to verify. In case of conditionally independent defaults, the valuation formulae of Propositions 10.2.1 and 10.2.2 necessarily reduce to the formula of Proposition 9.1.2, so that the continuity conditions are valid. To the best of our knowledge, this is the only (general) situation when these conditions are known to hold.

### 10.2.1 Valuation of the $i^{\text{th}}$-to-Default Claims

Let us recall that for any $i, j \in \{1, \ldots, n\}$, we denote by $\Pi^{(i,j)}$ the collection of specific partitions of the set $\{1, \ldots, n\}$. Namely, if $\pi \in \Pi^{(i,j)}$ then $\pi = \{\pi_-, \{j\}, \pi_+\}$, where $\pi_- = \{k_1, k_2, \ldots, k_{i-1}\}$, $\pi_+ = \{l_1, l_2, \ldots, l_{n-i}\}$, and: $j \notin \pi_-$, $j \notin \pi_+$, $\pi_- \cap \pi_+ = \emptyset$, $\pi_- \cup \pi_+ \cup \{j\} = \{1, \ldots, n\}$. For any numbers $i, j \in \{1, \ldots, n\}$ and an arbitrary partition $\pi \in \Pi^{(i,j)}$, we shall write $\tau(\pi_-) = \max \{\tau_k : k \in \pi_-\}$ and $\tau(\pi_+) = \min \{\tau_l : l \in \pi_+\}$.

The proof of the next result relies on rather straightforward, but tedious, algebraic considerations, and thus it is omitted.

**Proposition 10.2.3.** *The pre-default value process $S^{(i)}$ of the $i^{\text{th}}$-to-default claim $CCT^{(i)}$ equals*

$$S_t^{(i)} = B_t \, \mathbb{E}_{\mathbb{Q}^*} \left( \int_{]t,T]} B_u^{-1} dD_u^{(i)} \, \Big| \, \mathcal{G}_t \right) \tag{10.16}$$

*with the dividend process $D^{(i)}$ given as, for every $t \in [0, T]$,*

$$D_t^{(i)} = \sum_{j=1}^n \int_{]0,t]} Z_u^j H_{u-}^{(i,j)} dH_u^j + \mathbb{1}_{\{t=T\}} \sum_{j=1}^n X_j \mathbb{1}_{\{\tau_j \le T\}} \int_{]0,T]} H_{u-}^{(i,j)} dH_u^j$$
$$+ \, \mathbb{1}_{\{t=T, \, \tau_{(i)} > T\}} X,$$

*where*

$$H_t^{(i,j)} = \sum_{\pi \in \Pi^{(i,j)}} \prod_{k \in \pi_-} H_t^k \prod_{l \in \pi_+} \tilde{H}_t^l.$$

*More explicitly,*

$$S_t^{(i)} = B_t \, \mathbb{E}_{\mathbb{Q}^*} \left( \sum_{j=1}^n B_{\tau_j}^{-1} Z_{\tau_j}^j \sum_{\pi \in \Pi^{(i,j)}} \mathbb{1}_{\{\tau(\pi_-) < \tau_j < \tau(\pi_+), \, t < \tau_j \le T\}} \, \Big| \, \mathcal{G}_t \right)$$

$$+ \, B_t \, \mathbb{E}_{\mathbb{Q}^*} \left( B_T^{-1} \sum_{j=1}^n X_j \sum_{\pi \in \Pi^{(i,j)}} \mathbb{1}_{\{\tau(\pi_-) < \tau_j < \tau(\pi_+), \, \tau_j \le T\}} \, \Big| \, \mathcal{G}_t \right)$$

$$+ \, B_t \, \mathbb{E}_{\mathbb{Q}^*} \left( B_T^{-1} X \sum_{j=1}^n \sum_{\pi \in \Pi^{(i,j)}} \mathbb{1}_{\{\tau(\pi_-) < \tau_j < \tau(\pi_+), \, \tau_j > T\}} \, \Big| \, \mathcal{G}_t \right).$$

Let us set $\lambda_t^{(1)} = \sum_{j=1}^n \lambda_t^j$ for $t \in [0, T]$. Then $\lambda^{(1)}$ is the $\mathbb{F}$-martingale intensity of the random time $\tau_{(1)}$. For the first-to-default claim, it is possible to prove the following result extending expressions (10.12) and (10.15).

**Proposition 10.2.4.** *The pre-default value process $S^{(1)}$ of the first-to-default claim $CCT^{(1)}$ satisfies, for every $t \in [0, T]$,*

$$S_t^{(1)} = \mathbb{1}_{\{\tau_{(1)} > t\}} \left( \tilde{V}_t^{(1)} - \tilde{B}_t \, \mathbb{E}_{\mathbb{Q}^*} \left( \tilde{B}_{\tau_{(1)}}^{-1} \Delta \tilde{V}_{\tau_{(1)}}^{(1)} \, \big| \, \mathcal{G}_t \right) \right)$$

$$+ \, B_t \, \mathbb{E}_{\mathbb{Q}^*} \left( B_T^{-1} \sum_{j=1}^n X_j \sum_{\pi \in \Pi^{(1,j)}} \mathbb{1}_{\{\tau_j < \tau(\pi_+)\}} \, \Big| \, \mathcal{G}_t \right),$$

*where*

$$\tilde{V}_t^{(1)} = \tilde{B}_t \, \mathbb{E}_{\mathbb{Q}^*} \left( \int_t^T \tilde{B}_u^{-1} \sum_{j=1}^n Z_u^j \lambda_u^j \, du \, \Big| \, \mathcal{G}_t \right)$$

$$+ \, \tilde{B}_t \, \mathbb{E}_{\mathbb{Q}^*} \left( \tilde{B}_T^{-1} \left( X - \sum_{j=1}^n X_j \sum_{\pi \in \Pi^{(1,j)}} \mathbb{1}_{\{\tau_j < \tau(\pi_+)\}} \right) \, \Big| \, \mathcal{G}_t \right)$$

*and*

$$\tilde{B}_t = B_t \exp \left( \int_0^t \lambda_u^{(1)} \, du \right).$$

# 11. Markov Chains

Our next goal is to present the most basic notions and results from the theory of discrete- and continuous-time Markov chains. As expected, the emphasis is put here on these properties that are relevant from the viewpoint of credit risk modeling. Throughout this chapter, we fix an underlying probability space $(\Omega, \mathcal{G}, \mathbb{Q})$, as well as a finite set $\mathcal{K} = \{1, \ldots, K\}$, which plays the role of the *state space* for all considered Markov chains. Since the state space is finite, it is clear that any function $h : \mathcal{K} \to \mathbb{R}$ is bounded and measurable, provided that we endow the state space with the $\sigma$-field of all its subsets.

In Sect. 11.1, we examine rather succinctly the case of discrete-time Markov chains. Subsequently, continuous-time Markov chains are studied in some detail. We first examine the conditional expectations, as well as some 'fundamental' martingales with respect to certain relevant filtrations. Then, a few versions of martingale representation theorems based on these martingales are derived. We also deal with various examples of random times associated with a Markov chain, such as the jump times and the absorption time. Finally, we study the behavior of a time-homogeneous Markov chain under an equivalent change of a probability measure.

Another fundamental issue arising in the context of financial applications is the estimation of statistical characteristics of Markov chains: the one-step transition matrix in the discrete-time case, and the intensity matrix in the continuous-time case. We do not examine this question in this chapter; some relevant discussion is presented in Chap. 12. The reader is referred to Andersen et al. (1993), Küchler and Sørensen (1997), Gill (1999) and Lando and Skødeberg (2002) for a discussion of the estimation problem for continuous-time Markov chains within the context of point processes, survival analysis, and product integration. Israel et al. (2001) consider the same issue in the context of the so-called embedding problem (see Sect. 11.2.6).

Results provided in this chapter are – as a rule – well known, and they can be found in literature. Also, we do not pretend to cover in detail the vast area of the theory of Markov chains. For a more exhaustive treatment of discrete- and continuous-time Markov chains, we refer to any of a large variety of available monographs on the theory of stochastic processes, to mention a few: Bhattacharya and Waymire (1990), Syski (1992), Last and Brandt (1995), and Rogers and Williams (2000).

## 11.1 Discrete-Time Markov Chains

We shall write $\mathbb{N}^* = \{0, 1, \ldots\}$ to denote the set of all non-negative integers. Let $C_t$, $t \in \mathbb{N}^*$, be a sequence of random variables on $(\Omega, \mathcal{G}, \mathbb{Q})$ with values in $\mathcal{K}$, and let $\mathbb{F}^C$ be the natural filtration generated by the process $C$, i.e., $\mathcal{F}_t^C = \sigma(C_s : s = 0, \ldots, t)$. We write $\mathbb{G}$ to denote some filtration in $(\Omega, \mathcal{G}, \mathbb{Q})$, and we assume that $\mathbb{F}^C$ is a sub-filtration of $\mathbb{G}$: $\mathbb{F}^C \subseteq \mathbb{G}$.

**Definition 11.1.1.** A process $C$ is a *discrete-time Markov chain* under $\mathbb{Q}$ with respect to $\mathbb{G}$ (or briefly, a $\mathbb{G}$-*Markov chain*) if for any function $h : \mathcal{K} \to \mathbb{R}$ we have

$$\mathbb{E}_{\mathbb{Q}}(h(C_{t+s}) \,|\, \mathcal{G}_t) = \mathbb{E}_{\mathbb{Q}}(h(C_{t+s}) \,|\, C_t), \quad \forall s, t \in \mathbb{N}^*, \tag{11.1}$$

where $\mathbb{E}_{\mathbb{Q}}(h(C_{t+s}) \,|\, C_t)$ stands for $\mathbb{E}_{\mathbb{Q}}(h(C_{t+s}) \,|\, \sigma(C_t))$. If, in addition,

$$\mathbb{E}_{\mathbb{Q}}(h(C_{t+s}) \,|\, C_t) = \mathbb{E}_{\mathbb{Q}}(h(C_{u+s}) \,|\, C_u), \quad \forall s, t, u \in \mathbb{N}^*,$$

a Markov chain $C$ is said to be *time-homogeneous*.

If a process $C$ is a $\mathbb{G}$-Markov chain under $\mathbb{Q}$ then, of course, it is also an $\mathbb{F}^C$-Markov chain under $\mathbb{Q}$, but the converse implication is not true, in general. Since the state space $\mathcal{K}$ is finite, it is apparent that condition (11.1) is equivalent to the following one: for every $t \in \mathbb{N}^*$ and any $j \in \mathcal{K}$ we have

$$\mathbb{Q}\{C_{t+1} = j \,|\, \mathcal{G}_t\} = \mathbb{Q}\{C_{t+1} = j \,|\, C_t\}.$$

Likewise, a process $C$ is a *time-homogeneous* $\mathbb{G}$-Markov chain under $\mathbb{Q}$ if for every $s, t \in \mathbb{N}^*$ and any $j \in \mathcal{K}$ the following equalities are valid:

$$\mathbb{Q}\{C_{t+1} = j \,|\, \mathcal{G}_t\} = \mathbb{Q}\{C_{t+1} = j \,|\, C_t\} = \mathbb{Q}\{C_{s+1} = j \,|\, C_s\}.$$

It is well known that the Markov property (11.1) generalizes to the following condition: for any function $\bar{h} : \mathcal{K} \times \ldots \times \mathcal{K} \to \mathbb{R}$ and every $s_1, \ldots, s_n, t \in \mathbb{N}^*$

$$\mathbb{E}_{\mathbb{Q}}\big(\bar{h}(C_{t+s_1}, \ldots, C_{t+s_n}) \,|\, \mathcal{G}_t\big) = \mathbb{E}_{\mathbb{Q}}\big(\bar{h}(C_{t+s_1}, \ldots, C_{t+s_n}) \,|\, C_t\big). \tag{11.2}$$

From now on, we assume that $C$ is a time-homogeneous $\mathbb{G}$-Markov chain under the original probability measure $\mathbb{Q}$.

**Definition 11.1.2.** A matrix $P = [p_{ij}]_{1 \le i,j \le K}$ is called the *transition probability matrix* (or the *transition matrix*) for the $\mathbb{G}$-Markov chain $C$ if the equality $p_{ij} = \mathbb{Q}\{C_{t+1} = j \,|\, C_t = i\}$ is satisfied for every $t \in \mathbb{N}^*$ and arbitrary states $i, j \in \mathcal{K}$.

Any transition matrix $P$ is a *stochastic matrix*, meaning that $p_{ij} \ge 0$ for every $i, j \in \mathcal{K}$ and $\sum_{j=1}^{K} p_{ij} = 1$ for any fixed $i \in \mathcal{K}$. We denote by $P^{(s)} = [p_{ij}^{(s)}]_{1 \le i,j \le K}$ the transition matrix in $s$ steps, so that for any $s \in \mathbb{N}^*$ we have

$$p_{ij}^{(s)} = \mathbb{Q}\{C_s = j \,|\, C_0 = i\}, \quad \forall i, j \in \mathcal{K}.$$

The following result is standard, and thus its proof is omitted.

**Proposition 11.1.1.** *For every $s, t \in \mathbb{N}^*$ and any $i, j \in \mathcal{K}$ we have*

$$\mathbb{Q}\{C_{t+s} = j \mid C_t = i\} = p_{ij}^{(s)}.$$

*The Chapman-Kolmogorov equations are satisfied, specifically, the equalities $P^{(t+s)} = P^{(t)} P^{(s)} = P^{(s)} P^{(t)}$ are valid for every $s, t \in \mathbb{N}^*$. More explicitly, for every $s, t \in \mathbb{N}^*$ and $i, j \in \mathcal{K}$ we have*

$$p_{ij}^{(t+s)} = \sum_{k=1}^{K} p_{ik}^{(t)} p_{kj}^{(s)} = \sum_{k=1}^{K} p_{ik}^{(s)} p_{kj}^{(t)}.$$

*For any $s \in \mathbb{N}^*$ we have $P^{(s)} = P^s$, where $P^s$ stands for the $s^{\text{th}}$ power of the transition matrix $P$.*

The last statement in Proposition 11.1.1 provides the basis for the *first step analysis*:

$$P^{(t+1)} = P P^{(t)}, \quad \forall t \in \mathbb{N}^*,$$

and the *last step analysis*:

$$P^{(t+1)} = P^{(t)} P, \quad \forall t \in \mathbb{N}^*.$$

Let us write $\Delta P^{(t+1)} := P^{(t+1)} - P^{(t)}$. The last two equations can be rewritten as the *backward Kolmogorov* equation

$$\Delta P^{(t+1)} = \Lambda P^{(t)}, \quad P^{(0)} = \text{Id}, \quad \forall t \in \mathbb{N}^*,$$

and the *forward Kolmogorov* equation[1]

$$\Delta P^{(t+1)} = P^{(t)} \Lambda, \quad P^{(0)} = \text{Id}, \quad \forall t \in \mathbb{N}^*,$$

respectively, where $\Lambda = P - \text{Id}$ and Id is the $K$-dimensional identity matrix. Note that the entries on the diagonal of the matrix $\Lambda$ are nonpositive, and that the sum of all entries in each row equals 0. The matrix $\Lambda$ is called the *generator matrix* associated with the stochastic matrix $P$. The (row) vector

$$\mu_0 = [\mu_0(i)]_{1 \leq i \leq K} = [\mathbb{Q}\{C_0 = i\}]_{1 \leq i \leq K}$$

is called the *initial probability distribution* for $C$ under the probability measure $\mathbb{Q}$. More generally, for any $t \in \mathbb{N}^*$, the (row) vector

$$\mu_t = [\mu_t(i)]_{1 \leq i \leq K} = [\mathbb{Q}\{C_t = i\}]_{1 \leq i \leq K}$$

represents the *time-$t$ probability distribution* for $C$. The following property can be easily verified: for any $s, t \in \mathbb{N}^*$ we have

$$\mu_{t+s} = \mu_0 P^{(t+s)} = \mu_t P^{(s)} = \mu_s P^{(t)}.$$

Let us finally mention that standard constructions of an $\mathbb{F}^C$-Markov chain $C$, corresponding to a given transition matrix $P$ and a given initial distribution $\mu_0$, can be found in most of the several textbooks available.

---

[1] In the present setting, the backward and forward Kolmogorov equations coincide.

## 11.1.1 Change of a Probability Measure

In most financial applications, it is enough to study the behavior of a Markov chain only up to some horizon date $T^* < \infty$. For a fixed $T^*$, we introduce a probability measure $\mathbb{Q}^*$, which is equivalent to $\mathbb{Q}$ on $(\Omega, \mathcal{G}_{T^*})$, by setting:

$$\frac{d\mathbb{Q}^*}{d\mathbb{Q}}\Big|_{\mathcal{G}_{T^*}} = \eta_{T^*}, \quad \mathbb{Q}\text{-a.s.} \tag{11.3}$$

where the $\mathcal{G}_{T^*}$-measurable random variable $\eta_{T^*}$ is strictly positive $\mathbb{Q}$-a.s., and $\mathbb{E}_{\mathbb{Q}}(\eta_{T^*}) = 1$. Then the density process $\eta_t = \mathbb{E}_{\mathbb{Q}}(\eta_{T^*}|\mathcal{G}_t)$, $t = 0, \ldots, T^*$, follows a strictly positive martingale under $\mathbb{Q}$. Our next goal is to examine whether a time-homogeneous[2] $\mathbb{G}$-Markov chain $C$ under $\mathbb{Q}$ remains a (possibly time-inhomogeneous) $\mathbb{G}$-Markov chain under $\mathbb{Q}^*$. In case of a positive answer, that is, when the $\mathbb{G}$-Markov property is preserved under $\mathbb{Q}^*$, we would also like to relate the time-dependent transition probabilities $p_{ij}^*(t) := \mathbb{Q}^*\{C_{t+1} = j \mid C_t = i\}$ for $C$ under $\mathbb{Q}^*$ to the transition probabilities $p_{ij}$ for $C$ under the original probability $\mathbb{Q}$. The following condition will play an important role in this study.

**Condition (B.1)** The random variable $\eta_t^{-1}\eta_{t+1}$ is $\sigma(C_t, C_{t+1})$-measurable, for any $t = 0, \ldots, T^* - 1$. In other words, for every $t = 0, \ldots, T^* - 1$ we have $\eta_t^{-1}\eta_{t+1} = g_t(C_t, C_{t+1})$ for some function $g_t : \mathcal{K} \times \mathcal{K} \to \mathbb{R}$.

We start by the following general proposition, which shows that, under Condition (B.1), the Markov property of $C$ is preserved under an equivalent change of a probability measure.

**Proposition 11.1.2.** *Assume that Condition* (B.1) *is satisfied. If $C$ follows a time-homogeneous $\mathbb{G}$-Markov chain under $\mathbb{Q}$, then it follows a $\mathbb{G}$-Markov chain under $\mathbb{Q}^*$ and we have $p_{ij}^*(t) = p_{ij}g_t(i,j)$ for arbitrary states $i, j \in \mathcal{K}$ and every $t = 0, \ldots, T^* - 1$.*

*Proof.* Let us fix $t \in \mathbb{N}^*$. Using the abstract version of the Bayes theorem, for any state $j \in \mathcal{K}$ we obtain

$$\begin{aligned}
\mathbb{Q}^*\{C_{t+1} = j \mid \mathcal{G}_t\} &= \mathbb{E}_{\mathbb{Q}}\big(\eta_t^{-1}\eta_{T^*}\mathbb{1}_{\{C_{t+1}=j\}} \mid \mathcal{G}_t\big) \\
&= \mathbb{E}_{\mathbb{Q}}\big(\mathbb{E}_{\mathbb{Q}}(\eta_{T^*} \mid \mathcal{G}_{t+1})\eta_t^{-1}\mathbb{1}_{\{C_{t+1}=j\}} \mid \mathcal{G}_t\big) \\
&= \mathbb{E}_{\mathbb{Q}}\big(\eta_t^{-1}\eta_{t+1}\mathbb{1}_{\{C_{t+1}=j\}} \mid \mathcal{G}_t\big) \\
&= \mathbb{E}_{\mathbb{Q}}\big(g_t(C_t, C_{t+1})\mathbb{1}_{\{C_{t+1}=j\}} \mid \mathcal{G}_t\big) \\
&= \mathbb{E}_{\mathbb{Q}}\big(g_t(C_t, C_{t+1})\mathbb{1}_{\{C_{t+1}=j\}} \mid C_t\big),
\end{aligned}$$

where the last equality follows from formula (11.2) applied to the function $\bar{h}$ given by: $\bar{h}(k,l) = g_t(k,l)\mathbb{1}_{\{j\}}(l)$. This makes it clear that the conditional probability $\mathbb{Q}^*\{C_{t+1} = j \mid \mathcal{G}_t\}$ is in fact a $\sigma(C_t)$-measurable random variable.

---

[2] Results obtained in this section can be easily extended to the case when the process is a time-inhomogeneous Markov chain under the original measure.

Since $\sigma(C_t) \subseteq \mathcal{G}_t$, we conclude that

$$\mathbb{Q}^*\{C_{t+1} = j \mid \mathcal{G}_t\} = \mathbb{Q}^*\{C_{t+1} = j \mid C_t\},$$

so that $C$ manifestly follows a $\mathbb{G}$-Markov chain under $\mathbb{Q}^*$. Furthermore, we have

$$\begin{aligned}
p_{ij}^*(t) &= \mathbb{Q}^*\{C_{t+1} = j \mid C_t = i\} \\
&= \mathbb{E}_\mathbb{Q}\big(\eta_t^{-1}\eta_{t+1}\mathbb{1}_{\{C_{t+1}=j\}} \,\big|\, C_t = i\big) \\
&= \mathbb{E}_\mathbb{Q}\big(g_t(C_t, C_{t+1})\mathbb{1}_{\{C_{t+1}=j\}} \,\big|\, C_t = i\big) \\
&= p_{ij}g_t(i,j),
\end{aligned}$$

so that the proof of the proposition is completed. □

*Remarks.* Typically, a Markov chain $C$ is no longer a time-homogeneous process under an equivalent probability measure $\mathbb{Q}^*$. In some special cases, the convenient property of time-homogeneity is preserved, though.

The following Condition (B.2) is easily seen to be equivalent to (B.1).

**Condition (B.2)** There exists a finite set, denoted by $A$, such that for any $t = 0, \ldots, T^* - 1$ the product $\eta_t^{-1}\eta_{t+1}$ admits the following representation:

$$\eta_t^{-1}\eta_{t+1} = 1 + \sum_{\alpha \in A} \tilde{m}_t^\alpha \hat{m}_{t+1}^\alpha, \tag{11.4}$$

where $\tilde{m}_t^\alpha$ is $\sigma(C_t)$-measurable and $\hat{m}_{t+1}^\alpha$ is $\sigma(C_t, C_{t+1})$-measurable. In other words, $\tilde{m}_t = \tilde{g}_t^\alpha(C_t)$ for some function $\tilde{g}_t^\alpha : \mathcal{K} \to \mathbb{R}$, and $\hat{m}_{t+1} = \hat{g}_{t+1}^\alpha(C_t, C_{t+1})$ for some function $\hat{g}_{t+1}^\alpha : \mathcal{K} \times \mathcal{K} \to \mathbb{R}$.

The next result is an immediate consequence of Proposition 11.1.2.

**Corollary 11.1.1.** *Suppose that Condition (B.2) holds. If $C$ follows a time-homogeneous $\mathbb{G}$-Markov chain under $\mathbb{Q}$, then it follows a $\mathbb{G}$-Markov chain under $\mathbb{Q}^*$ and for every $i, j \in \mathcal{K}$ and $t = 0, \ldots, T^* - 1$ we have*

$$p_{ij}^*(t) = p_{ij}\Big(1 + \sum_{\alpha \in A} \tilde{g}_t^\alpha(i)\hat{g}_{t+1}^\alpha(i,j)\Big).$$

To get a more convenient representation for the transition probabilities $p_{ij}^*(t)$, we assume that the $\sigma$-field $\mathcal{F}_0^C$ is trivial (so that $\eta_0 = 1$), and we introduce an auxiliary sequence of random variables by setting: $m_0 = 0$ and

$$\Delta m_{t+1} := m_{t+1} - m_t = \eta_t^{-1}(\eta_{t+1} - \eta_t), \quad \forall t = 0, \ldots, T^* - 1.$$

It is easy to see that the process $m$ follows a $\mathbb{G}$-martingale under $\mathbb{Q}$, and $\Delta m_t > -1$ for every $t = 1, \ldots, T^*$. Moreover, for $t = 0, \ldots, T^*$ we have

$$\eta_t = 1 + \sum_{u=1}^t \eta_{u-1}\Delta m_u = \prod_{u=1}^t (1 + \Delta m_u). \tag{11.5}$$

Relationship (11.5) yields a one-to-one correspondence between equivalent probability measures $\mathbb{Q}^*$ on $(\Omega, \mathcal{G}_{T^*})$ and $\mathbb{G}$-martingales $m_t$, $t = 0, \ldots, T^*$ with $m_0 = 0$ and $\Delta m_t > -1$. Condition (B.2) may now be re-expressed in the following way: $\Delta m_{t+1} = \sum_{\alpha \in A} \tilde{m}_t^\alpha \hat{m}_{t+1}^\alpha$ for every $t = 0, \ldots, T^* - 1$.

We shall now make a specific choice for the sequence $m$. We need first to introduce some notation. For any $i, j \in \mathcal{K}$, we denote $H_t^i = \mathbb{1}_{\{C_t = i\}}$ and $H_t^{ij} = \sum_{u=1}^{t} H_{u-1}^i H_u^j$. Observe that $H_t^{ij}$ represents the number of one-step transitions of the process $C$ from the state $i$ to the state $j$ that occurred between time 0 and time $t$. Let $\delta_{ii} = 1$ and $\delta_{ij} = 0$ if $i \neq j$. We shall write

$$\tilde{H}_t^{ij} = H_t^{ij} - \delta_{ij} \sum_{u=1}^{t} H_{u-1}^i.$$

and

$$M_t^{ij} = \tilde{H}_t^{ij} - \sum_{u=1}^{t} (p_{ij} - \delta_{ij}) H_{u-1}^i. \tag{11.6}$$

It is easy to check that $M_t^{ij} = H_t^{ij} - \sum_{u=1}^{t} p_{ij} H_{u-1}^i$. We have chosen to define $M^{ij}$ through (11.6), because this formula directly corresponds to the continuous-time Lévy-type representation (11.33).

**Proposition 11.1.3.** *For any two states $i, j \in \mathcal{K}$, the process $M_t^{ij}$, $t \in \mathbb{N}^*$, follows a $\mathbb{G}$-martingale under $\mathbb{Q}$.*

*Proof.* For $i \neq j$ we have

$$\mathbb{E}_{\mathbb{Q}}(M_{t+1}^{ij} - M_t^{ij} \mid \mathcal{G}_t) = \mathbb{E}_{\mathbb{Q}}(H_t^i H_{t+1}^j - p_{ij} H_t^i \mid \mathcal{G}_t)$$
$$= \mathbb{E}_{\mathbb{Q}}((H_{t+1}^j - p_{ij}) H_t^i \mid C_t) = H_t^i (\mathbb{Q}\{C_{t+1} = j \mid C_t = i\} - p_{ij}) = 0.$$

For $i = j$, the martingale property of $M^{ij}$ is also clear.    □

We are ready to further specify the sequence $m$. Let $m_0 = 0$ and let

$$\Delta m_t = \sum_{k,l=1}^{K} \xi_{t-1}^{kl} \Delta M_t^{kl}, \quad \forall t = 1, \ldots, T^*, \tag{11.7}$$

where $\Delta M_t^{kl} = M_t^{kl} - M_{t-1}^{kl}$, and for any $k, l \in \mathcal{K}$ and $t \in \mathbb{N}^*$, the random variable $\xi_t^{kl} > -1$ is assumed to be $\sigma(C_t)$-measurable, so that $\xi_t^{kl} = \tilde{g}_t^{kl}(C_t)$ for some function $\tilde{g}^{kl} : \mathcal{K} \to \mathbb{R}$. Notice that

$$\Delta M_{t+1}^{kl} = M_{t+1}^{kl} - M_t^{kl} = H_t^k H_{t+1}^l - p_{kl} H_t^k = \mathbb{1}_{\{C_t = k, C_{t+1} = l\}} - p_{kl} \mathbb{1}_{\{C_t = k\}}$$

and so $\Delta M_{t+1}^{kl} = \hat{g}_{t+1}^{kl}(C_t, C_{t+1})$, where

$$\hat{g}_{t+1}^{kl}(i, j) = \mathbb{1}_{\{k\}}(i)(\mathbb{1}_{\{l\}}(j) - p_{kl}).$$

Finally, it is not difficult to check that the process $m$ is a $\mathbb{G}$-martingale under $\mathbb{Q}$ and $\Delta m_t > -1$ for every $t = 1, \ldots, T^*$.

Hence, the process $\eta$, given as (cf. (11.5))

$$\eta_t = \prod_{u=1}^{t} \left(1 + \Delta m_u\right) = 1 + \sum_{u=1}^{t} \eta_{u-1} \left( \sum_{k,l=1}^{K} \xi_{u-1}^{kl} \Delta M_u^{kl} \right), \tag{11.8}$$

follows a positive $\mathbb{G}$-martingale under $\mathbb{Q}$. We have $\Delta m_{t+1} = \sum_{k,l=1}^{K} \tilde{m}_t^{kl} \hat{m}_{t+1}^{kl}$, with $\tilde{m}_t^{kl} = \xi_t^{kl} = \tilde{g}_t^{kl}(C_t)$ and $\hat{m}_{t+1}^{kl} = \Delta M_{t+1}^{kl} = \hat{g}_{t+1}^{kl}(C_t, C_{t+1})$, so that Condition (B.2) holds. Using Corollary 11.1.1, we obtain the following result.

**Corollary 11.1.2.** *Let the probability $\mathbb{Q}^*$ be defined by (11.3), with the random variable $\eta_{T^*}$ given by (11.8). If $C$ is a $\mathbb{G}$-Markov chain under $\mathbb{Q}$, then $C$ is a $\mathbb{G}$-Markov chain under $\mathbb{Q}^*$ with the following transition probabilities, for every $i, j \in \mathcal{K}$ and $t \in \mathbb{N}^*$,*

$$p_{ij}^*(t) = p_{ij}\left(1 + \tilde{g}_t^{ij}(i) - \sum_{l=1}^{K} \tilde{g}_t^{il}(i)p_{il}\right). \tag{11.9}$$

*Example 11.1.1.* Let $\xi_t^{kl} = g_t^k(C_t)$ for every $k, l \in \mathcal{K}$ and $t = 0, \ldots, T^* - 1$. It is easy to see that $\Delta m_t = 0$ for every $t = 1, \ldots, T^* - 1$, and thus $\eta_t = 1$ for every $t = 0, \ldots, T$. Hence, we have $\mathbb{Q}^* = \mathbb{Q}$ and $p_{ij}^*(t) = p_{ij}$.

*Example 11.1.2.* Let us fix a function $k^* : \mathcal{K} \to \mathcal{K}$.[3] To simplify the presentation, we assume that $p_{ik^*(i)} \neq 0$ and $p_{ik^*(i)} \neq 1$ for $i = 1, \ldots, K - 1$. Suppose now that for every $k, l \in \mathcal{K}$ and any $t = 0, \ldots, T^* - 1$ we have

$$\xi_t^{kl} = \tilde{g}_t^{kl}(C_t) = \begin{cases} g_t^k(C_t), & l \neq k^*(k), \\ 0, & l = k^*(k), \end{cases}$$

for some functions $g_t^k : \mathcal{K} \to \mathbb{R}$. Then for every $i, j \in \mathcal{K}$ we obtain

$$p_{ij}^*(t) = p_{ij}\pi_{ij}(t), \tag{11.10}$$

where

$$\pi_{ij}(t) = \begin{cases} 1 + g_t^i(i)p_{ik^*(i)}, & j \neq k^*(i), \\ \left(1 + g_t^i(i)(p_{ik^*(i)} - 1)\right)p_{ik^*(i)}^{-1}, & j = k^*(i). \end{cases}$$

In the present case, we have

$$\Delta m_t = \sum_{i=1}^{K} g_{t-1}^i(C_{t-1})\left(p_{ik^*(i)} - H_t^{k^*(i)}\right) H_{t-1}^{k^*(i)}.$$

Thus, if we assume that for every $i \neq K - 1$ and $t = 0, \ldots, T^* - 1$:

$$-(p_{ik^*(i)})^{-1} < g_t^i(i) < (1 - p_{ik^*(i)})^{-1}, \tag{11.11}$$

then we have $\Delta m_t > -1$ and $p_{ij}^*(t) \in [0, 1]$ for any states $i, j = 1, 2, \ldots, K$.

---

[3] With $k^*(i) = i$ for every $i \in \mathcal{K}$, this example corresponds to the JLT (1997) model (see Sect. 12.1.1). If we set $k^*(i) = K$ for every $i \in \mathcal{K}$, it corresponds to the Kijima and Komoribayashi model (see Sect. 12.1.3).

Finally, if we denote $\pi_i(t) := 1 + g_t^i(i)p_{ik^*(i)}$, then we may rewrite (11.10) as follows: $p_{ij}^*(t) = p_{ij}\pi_i(t)$ for every $j \neq k^*(i)$, and

$$p_{ik^*(i)}^*(t) = p_{ik^*(i)}\big(\pi_i(t) - g_t^i(i)\big) = 1 - \pi_i(t)\big(1 - p_{ik^*(i)}\big).$$

Notice that condition (11.11) becomes: for every $i \neq K-1$ and $t = 0, \ldots, T^* - 1$

$$0 < \pi_i(t) < (1 - p_{ik^*(i)})^{-1}, \quad \forall i \neq K - 1, \; t \in \mathbb{N}^*. \tag{11.12}$$

### 11.1.2 The Law of the Absorption Time

We say that a state $k \in K$ is *absorbing* for a $\mathbb{G}$-Markov chain $C$ under $\mathbb{Q}$ if

$$\mathbb{Q}\{C_s = k \,|\, C_t = k\} = 1, \quad \forall t, s \in \mathbb{N}^*, \; t \leq s.$$

It is obvious that if a state $k$ is a absorbing for $C$ under $\mathbb{Q}$, then $p_{kj} = \delta_{kj}$. We have the following simple result; its proof is left to the reader.

**Proposition 11.1.4.** *Let a probability measure $\mathbb{Q}^*$ on $(\Omega, \mathcal{G})$ be equivalent to $\mathbb{Q}$, and let $C$ be a $\mathbb{G}$-Markov chain under both $\mathbb{Q}$ and $\mathbb{Q}^*$. Then a state $k$ is absorbing for $C$ under $\mathbb{Q}$ if and only if $k$ is absorbing for $C$ under $\mathbb{Q}^*$.*

Assume that the state $K$ is the only absorbing state for a $\mathbb{G}$-Markov chain $C$ under $\mathbb{Q}$. In terms of the transition matrix $P$, we have

$$P = \begin{pmatrix} p_{1,1} & \cdots & p_{1,K-1} & p_{1,K} \\ \cdot & \cdots & \cdot & \cdot \\ p_{K-1,1} & \cdots & p_{K-1,K-1} & p_{K-1,K} \\ 0 & \cdots & 0 & 1 \end{pmatrix}$$

where $p_{ii} < 1$ for every $i = 1, \ldots, K - 1$. The associated generator $\Lambda$ takes the following form

$$\Lambda = \begin{pmatrix} p_{1,1} - 1 & \cdots & p_{1,K-1} & p_{1,K} \\ \cdot & \cdots & \cdot & \cdot \\ p_{K-1,1} & \cdots & p_{K-1,K-1} - 1 & p_{K-1,K} \\ 0 & \cdots & 0 & 0 \end{pmatrix}.$$

Let us define the random time $\tau$ as the first moment when the process $C$ jumps to the absorbing state $K$, i.e., $\tau = \min\{t \geq 0 : C_t = K\}$. We shall find the probability distribution for $\tau$ under $\mathbb{Q}$, and under an equivalent probability measure $\mathbb{Q}^*$. To this end, we denote by $\tilde{P} = [\tilde{p}_{ij}]_{1 \leq i,j \leq K-1}$ the $(K-1)$-dimensional matrix obtained from $P$ by deleting the last column and the last row, and we write $\tilde{P}^{(s)} = [\tilde{p}_{ij}^{(s)}]_{1 \leq i,j \leq K-1}$ to denote the $s^{\text{th}}$ power of $\tilde{P}$. The following result is standard and rather easy to establish (see, e.g., Proposition 11.1 in Bhattacharya and Waymire (1990)).

**Proposition 11.1.5.** *For any* $t \in \mathbb{N}$ *and an arbitrary* $i \neq K$ *we have*

$$\mathbb{Q}\{\tau \leq t \,|\, C_0 = i\} = 1 - \sum_{j=1}^{K-1} \tilde{p}_{ij}^{(t)}.$$

**Corollary 11.1.3.** *For any* $t \in \mathbb{N}$ *and an arbitrary* $i \neq K$ *we have*

$$\mathbb{Q}\{\tau \leq t \,|\, C_0 = i\} = 1 - \sum_{j=1}^{K-1} p_{ij}^{(t)}. \tag{11.13}$$

*Proof.* Recall that $p_{Kj} = 0$ for every $j = 1, \ldots, K - 1$, so that the equality $\tilde{p}_{ij}^{(t)} = p_{ij}^{(t)}$ is satisfied for every $i, j = 1, \ldots, K - 1$ and $t \in \mathbb{N}$. $\square$

To directly derive equality (11.13), observe that for every $t \in \mathbb{N}$ and $i \neq K$ we have

$$\mathbb{Q}\{\tau \leq t \,|\, C_0 = i\} = 1 - \mathbb{Q}\{C_t \neq K \,|\, C_0 = i\} = 1 - \sum_{j=1}^{K-1} p_{ij}^{(t)}.$$

From the general theory of Markov chains with finite state space,[4] it is well known that the equality $\mathbb{Q}\{\tau = \infty \,|\, C_0 = i\} = 0$ is satisfied for any initial state $i = 1, \ldots, K - 1$ if and only if all the states $i = 1, \ldots, K - 1$ are *transient* for the chain $C$ under $\mathbb{Q}$ or, equivalently, if the state $K$ is the unique *recurrent* state for the chain $C$ under $\mathbb{Q}$. We shall be assuming from now on that the state $K$ is the only recurrent state for $C$ under $\mathbb{Q}$. Thus, $K$ is the only absorbing state for $C$ under $\mathbb{Q}$.

Let us introduce the matrix $\mathcal{P}^*(t) = [p_{ij}^*(t)]_{1 \leq i,j \leq K}$ of the time-$t$ one-step transition probabilities for the Markov chain $C$ under $\mathbb{Q}^*$, where $\mathbb{Q}^*$ stands for an arbitrary probability measure, which is equivalent to $\mathbb{Q}$ on $(\Omega, \mathcal{F}_{T^*}^C)$. We also define $\mathcal{P}^*(t, s) = [p_{ij}^*(t, s)]_{1 \leq i,j \leq K}$, where for every $i, j \in \mathcal{K}$

$$p_{ij}^*(t, s) := \mathbb{Q}^*\{C_{t+s} = j \,|\, C_t = i\}, \quad \forall \, t, s \geq 0.$$

Then we have $\mathcal{P}^*(t) = \mathcal{P}^*(t, 1)$ and

$$\mathcal{P}^*(t, s) = \prod_{u=0}^{s-1} \mathcal{P}^*(t + u). \tag{11.14}$$

If $C$ is a Markov chain under $\mathbb{Q}^*$, by virtue of Proposition 11.1.4, it is clear that the state $K$ is absorbing for the chain $C$ under $\mathbb{Q}^*$ as well. We define the *truncated absorption time* $\tau^*$ by setting[5]

$$\tau^* = \min\{t = 0, \ldots, T^* : C_t = K\}.$$

In the next result, we examine the probability distribution of $\tau^*$ under $\mathbb{Q}^*$.

---

[4] See, for instance, Bhattacharya and Waymire (1990).
[5] As usual, the infimum over an empty set is equal to $+\infty$.

**Corollary 11.1.4.** *For every* $t = 1, \ldots, T^*$ *and every* $i \neq K$ *we have*

$$Q^*\{\tau^* \le t \,|\, C_0 = i\} = 1 - \sum_{j=1}^{K-1} p_{ij}^*(0,t)$$

*and*

$$Q^*\{\tau^* = \infty \,|\, C_0 = i\} = 1 - Q^*\{\tau^* > T^* \,|\, C_0 = i\}.$$

*Proof.* For the first equality, it is enough to note that

$$Q^*\{\tau^* \le t \,|\, C_0 = i\} = 1 - Q^*\{C_t \neq K \,|\, C_0 = i\} = 1 - \sum_{j=1}^{K-1} p_{ij}^*(0,t).$$

The second formula is also clear.                                      □

### 11.1.3 Discrete-Time Conditionally Markov Chains

As we shall see in Chap. 12 and 13, an important role in the modeling of credit risk is played by the so-called *conditional Markov chains*. We present here a brief discussion of such processes in a discrete-time framework. The continuous-time case is treated in much more detail, including the canonical construction of an F-conditional Markov chain (see Sect. 11.3).

We consider a probability space $(\Omega, \mathcal{G}, \mathbb{Q})$ endowed with some filtrations $\mathbb{F} = (\mathcal{F}_t)_{t \in \mathbb{N}^*}$ and $\mathbb{G} = (\mathcal{G}_t)_{t \in \mathbb{N}^*}$, such that $\mathbb{F} \subseteq \mathbb{G}$. Let $C$ be a $\mathcal{K}$-valued stochastic process defined on this probability space, where $\mathcal{K} = \{1, \ldots, K\}$. As usual, $\mathbb{F}^C$ denotes the filtration generated by the process $C$. It is natural to assume that $C$ follows a $\mathbb{G}$-adapted process, so that $\mathbb{F}^C \subseteq \mathbb{G}$.

**Definition 11.1.3.** A discrete-time $\mathcal{K}$-valued process $C$ is said to be a *discrete-time conditionally $\mathbb{G}$-Markov chain relative to* $\mathbb{F}$ (under $\mathbb{Q}$) if for every $t, s \in \mathbb{N}^*$, $t \le s$, and any function $h : \mathcal{K} \to \mathbb{R}$ we have

$$\mathbb{E}_{\mathbb{Q}}(h(C_s) \,|\, \mathcal{G}_t) = \mathbb{E}_{\mathbb{Q}}(h(C_s) \,|\, \mathcal{F}_t \vee \sigma(C_t)). \tag{11.15}$$

For the sake of brevity, we shall say that $C$ is a *discrete-time $\mathbb{F}$-conditional $\mathbb{G}$-Markov chain* under $\mathbb{Q}$ if $C$ satisfies the last definition.

*Remarks.* Assume that the reference filtration $\mathbb{F}$ is trivial, i.e., $\mathbb{F} = \mathbb{F}^0$, where $\mathcal{F}_t^0 = \mathcal{F}_0^0 = \{\Omega, \emptyset\}$ for every $t \in \mathbb{N}^*$. In this case, equality (11.15) becomes

$$\mathbb{E}_{\mathbb{Q}}(h(C_s) \,|\, \mathcal{G}_t) = \mathbb{E}_{\mathbb{Q}}(h(C_s) \,|\, \sigma(C_t)),$$

and Definition 11.1.3 coincides with Definition 11.1.1 of a $\mathbb{G}$-Markov chain. Furthermore, since $\sigma(C_t) \subseteq \mathcal{F}_t \vee \sigma(C_t) \subseteq \mathcal{G}_t$, it is clear that if a given process $C$ follows a $\mathbb{G}$-Markov chain under $\mathbb{Q}$, then $C$ is also an $\mathbb{F}$-conditional Markov chain under $\mathbb{Q}$ for any choice of a sub-filtration $\mathbb{F}$ of $\mathbb{G}$. On the other hand, if a discrete-time process $C$ follows an $\mathbb{F}$-conditional $\mathbb{G}$-Markov chain under $\mathbb{Q}$, it does not necessarily satisfy the definition of an ordinary $\mathbb{G}$-Markov chain under $\mathbb{Q}$, so that the notion of an $\mathbb{F}$-conditional $\mathbb{G}$-Markov chain is weaker than the notion of a $\mathbb{G}$-Markov chain.

*Example 11.1.3.* Let us consider the case when $\mathbb{G} = \mathbb{F} \vee \mathbb{F}^C$. Then equality (11.15) becomes:

$$\mathbb{E}_{\mathbb{Q}}(h(C_s) \,|\, \mathcal{F}_t \vee \mathcal{F}_t^C) = \mathbb{E}_{\mathbb{Q}^*}\big(h(C_s) \,|\, \mathcal{F}_t \vee \sigma(C_t)\big). \qquad (11.16)$$

If $\mathbb{F}^C$ is a sub-filtration of the reference filtration $\mathbb{F}$, condition (11.16) is always trivially satisfied. This observation leads to a conclusion that in the case when $\mathbb{G} = \mathbb{F} \vee \mathbb{F}^C$, it is natural to require that the filtration $\mathbb{F}^C$ generated by $C$ is not a sub-filtration of the reference filtration $\mathbb{F}$.

We assume from now on that $C$ is a discrete-time $\mathbb{F}$-*conditional* $\mathbb{G}$-*Markov chain* under $\mathbb{Q}$, and that the state space of $C$ is the finite set $\mathcal{K} = \{1, \ldots, K\}$.

**Definition 11.1.4.** An $\mathbb{F}$-adapted, matrix-valued stochastic process

$$P(t) = [p_{ij}(t)]_{1 \leq i,j \leq K}, \quad \forall t \in \mathbb{N}^*,$$

is called the (one-step) $\mathbb{F}$-*conditional transition probability matrix process* (or, the $\mathbb{F}$-*conditional transition matrix process*) for the chain $C$ if for every $t \in \mathbb{N}^*$ and arbitrary $i, j \in \mathcal{K}$ we have, on the set $\{C_t = i\}$,

$$p_{ij}(t) = \mathbb{Q}\{C_{t+1} = j \,|\, \mathcal{F}_t \vee \sigma(C_t)\}, \quad \mathbb{Q}\text{-a.s.}$$

The last definition can be extended to the case of several steps.

**Definition 11.1.5.** Let us fix $t, s \in \mathbb{N}^*$. An $\mathcal{F}_t$-measurable, matrix-valued random variable $P(t, s) = [p_{ij}(t, s)]_{1 \leq i,j \leq K}$ is called the $s$-*step* $\mathbb{F}$-*conditional transition matrix at time* $t$ if for every $i, j \in \mathcal{K}$ we have, on the set $\{C_t = i\}$,

$$p_{ij}(t, s) = \mathbb{Q}\{C_{t+s} = j \,|\, \mathcal{F}_t \vee \sigma(C_t)\}, \quad \mathbb{Q}\text{-a.s.}$$

Finally, we shall say that the chain $C$ satisfies an extended $\mathbb{F}$-conditional $\mathbb{G}$-Markov property if $C$ satisfies a property analogous to (11.2). To be more specific, we introduce the following definition.

**Definition 11.1.6.** A process $C$ satisfies an *extended* $\mathbb{F}$-*conditional* $\mathbb{G}$-*Markov property* if for any function $\bar{h} : \mathcal{K} \times \ldots \times \mathcal{K} \to \mathbb{R}$ and every $s_1, \ldots, s_n, t \in \mathbb{N}^*$ we have

$$\mathbb{E}_{\mathbb{Q}}\big(\bar{h}(C_{t+s_1}, \ldots, C_{t+s_n}) \,|\, \mathcal{G}_t\big) = \mathbb{E}_{\mathbb{Q}}\big(\bar{h}(C_{t+s_1}, \ldots, C_{t+s_n}) \,|\, \mathcal{F}_t \vee \sigma(C_t)\big).$$

It is interesting to observe that the last property does not necessarily hold for a discrete-time $\mathbb{F}$-conditional $\mathbb{G}$-Markov chain under $\mathbb{Q}$. However, if a chain is obtained through the canonical construction, analogous to the continuous-time method presented in Sect. 11.3, this property is indeed valid.

*Remarks.* As we shall see in Sect. 12.2, some credit migration models developed in a discrete-time set-up may be categorized as conditionally Markov models, in the sense that they model the credit migration process as an $\mathbb{F}$-conditional $\mathbb{G}$-Markov chain. It should be stressed that a satisfactory theory of conditionally Markov credit migration models in discrete time, including the issues related to a model's calibration, is not yet available. This remark applies to the continuous-time conditionally Markov models as well.

## 11.2 Continuous-Time Markov Chains

Let $C_t$, $t \in \mathbb{R}_+$, be a right-continuous stochastic process on $(\Omega, \mathcal{G}, \mathbb{Q})$ with values in the finite set $\mathcal{K}$, and let $\mathbb{F}^C$ be the filtration generated by this process. Also, let $\mathbb{G}$ be some filtration such that $\mathbb{F}^C \subseteq \mathbb{G}$.

**Definition 11.2.1.** A process $C$ is a *continuous-time $\mathbb{G}$-Markov chain* if for an arbitrary function $h : \mathcal{K} \to \mathbb{R}$ and any $s, t \in \mathbb{N}^*$ we have

$$\mathbb{E}_{\mathbb{Q}}(h(C_{t+s}) \,|\, \mathcal{G}_t) = \mathbb{E}_{\mathbb{Q}}(h(C_{t+s}) \,|\, C_t).$$

A continuous-time $\mathbb{G}$-Markov chain $C$ is said to be *time-homogeneous* if, in addition, for any $s, t, u \in \mathbb{N}^*$ we have

$$\mathbb{E}_{\mathbb{Q}}(h(C_{t+s}) \,|\, C_t) = \mathbb{E}_{\mathbb{Q}}(h(C_{u+s}) \,|\, C_u).$$

**Definition 11.2.2.** A two-parameter family $\mathcal{P}(t, s)$, $t, s \in \mathbb{R}_+$, $t \le s$, of stochastic matrices is called the family of *transition probability matrices* for the $\mathbb{G}$-Markov chain $C$ under $\mathbb{Q}$ if, for every $t, s \in \mathbb{R}_+$, $t \le s$,

$$\mathbb{Q}\{C_s = j \,|\, C_t = i\} = p_{ij}(t, s), \quad \forall i, j \in \mathcal{K}.$$

In particular, the equality $\mathcal{P}(t, t) = \mathrm{Id}$ is satisfied for every $t \in \mathbb{R}_+$.

**Time-homogeneous chains.** In case of a time-homogeneous Markov chain $C$, we introduce the following definition.

**Definition 11.2.3.** The one-parameter family $\mathcal{P}(t)$, $t \in \mathbb{R}_+$, of stochastic matrices is called the family of *transition probability matrices* for the time-homogeneous $\mathbb{G}$-Markov chain $C$ under $\mathbb{Q}$ if, for every $t, s \in \mathbb{R}_+$,

$$\mathbb{Q}\{C_{s+t} = j \,|\, C_s = i\} = p_{ij}(t), \quad \forall i, j \in \mathcal{K}. \tag{11.17}$$

If $\mathcal{P}(t)$, $t \in \mathbb{R}_+$ is the family of transition matrices for $C$ then for any subset $A \subseteq \mathcal{K}$ we have

$$\mathbb{Q}\{C_{t+s} \in A \,|\, C_t\} = \sum_{j \in A} p_{C_t j}(s), \quad \forall s, t \in \mathbb{R}_+.$$

Moreover, the Chapman-Kolmogorov equation is satisfied, namely,

$$\mathcal{P}(t + s) = \mathcal{P}(t)\mathcal{P}(s) = \mathcal{P}(s)\mathcal{P}(t), \quad \forall s, t \in \mathbb{R}_+.$$

Equivalently, for every $s, t \in \mathbb{R}_+$ and $i, j \in \mathcal{K}$,

$$p_{ij}(t + s) = \sum_{k=1}^{K} p_{ik}(t)p_{kj}(s) = \sum_{k=1}^{K} p_{ik}(s)p_{kj}(t).$$

Let the $K$-dimensional (row) vector $\mu_0 = [\mu_0(i)]_{1 \le i \le K} = [\mathbb{Q}\{C_0 = i\}]_{1 \le i \le K}$ denote the initial probability distribution for the Markov chain $C$ under $\mathbb{Q}$.

Likewise, let the (row) vector $\mu_t = [\mu_t(i)]_{1 \leq i \leq K} = [\mathbb{Q}\{C_t = i\}]_{1 \leq i \leq K}$ stand for the probability distribution of $C$ at time $t \in \mathbb{R}_+$. It can be easily checked that

$$\mu_{t+s} = \mu_0 \mathcal{P}(t+s) = \mu_t \mathcal{P}(s) = \mu_s \mathcal{P}(t), \quad \forall s, t \in \mathbb{R}_+.$$

We now impose an important assumption on the family $\mathcal{P}(\cdot)$, specifically, that this family is right-continuous at time $t = 0$, that is, $\lim_{t \downarrow 0} \mathcal{P}(t) = \mathcal{P}(0)$. By virtue of the Chapman-Kolmogorov equation, this implies that

$$\lim_{s \to 0} \mathcal{P}(t+s) = \mathcal{P}(t), \quad \forall t > 0,$$

and thus

$$\lim_{s \to 0} \mathbb{Q}\{C_{t+s} = j \mid C_t = i\} = \delta_{ij}, \quad \forall i, j \in \mathcal{K}, \ t > 0.$$

It is a well-known fact (see, for instance, Theorem 8.1.2 in Rolski et al. (1998)) that the right-hand side continuity at time $t = 0$ of the family $\mathcal{P}(\cdot)$ implies the right-hand side differentiability at $t = 0$ of this family. More specifically, the following finite limits exist, for every $i, j \in \mathcal{K}$,

$$\lambda_{ij} := \lim_{t \downarrow 0} \frac{p_{ij}(t) - p_{ij}(0)}{t} = \lim_{t \downarrow 0} \frac{p_{ij}(t) - \delta_{ij}}{t}. \tag{11.18}$$

Observe that for every $i \neq j$ we have $\lambda_{ij} \geq 0$, and $\lambda_{ii} = -\sum_{j=1, j \neq i}^K \lambda_{ij}$. The matrix $\Lambda := [\lambda_{ij}]_{1 \leq i,j \leq K}$ is called the *infinitesimal generator matrix* for a Markov chain associated with the family $\mathcal{P}(\cdot)$ via (11.17). Since each entry $\lambda_{ij}$ of the matrix $\Lambda$ can be shown to represent the intensity of transition from the state $i$ to the state $j$, the infinitesimal generator matrix $\Lambda$ is also commonly known as the *intensity matrix*.

Invoking the Chapman-Kolmogorov equation and equality (11.18), one may derive the *backward Kolmogorov equation*

$$\frac{d\mathcal{P}(t)}{dt} = \Lambda \mathcal{P}(t), \quad \mathcal{P}(0) = \mathrm{Id}, \tag{11.19}$$

and the *forward Kolmogorov equation*

$$\frac{d\mathcal{P}(t)}{dt} = \mathcal{P}(t)\Lambda, \quad \mathcal{P}(0) = \mathrm{Id}, \tag{11.20}$$

where, at time $t = 0$, we take the right-hand side derivatives. It is well known that both these equations have the same unique solution:

$$\mathcal{P}(t) = e^{t\Lambda} := \sum_{n=0}^{\infty} \frac{\Lambda^n t^n}{n!}, \quad \forall t \in \mathbb{R}_+. \tag{11.21}$$

We conclude that the generator matrix $\Lambda$ uniquely determines all relevant probabilistic properties of a time-homogeneous Markov chain.

The following important result provides a martingale characterization of a time-homogeneous Markov chain $C$ in terms of its infinitesimal generator. For the proof of Proposition 11.2.1, we refer to Last and Brandt (1995) or Rogers and Williams (2000). In the quoted references, the corresponding result is stated for an $\mathbb{F}^C$-Markov chain, rather than for a $\mathbb{G}$-Markov chain. However, the proof of this more general version is analogous. For any state $i \in \mathcal{K}$ and any function $h : \mathcal{K} \to \mathbb{R}$, we denote $(\Lambda h)(i) = \sum_{j=1}^{K} \lambda_{ij} h(j)$.

**Proposition 11.2.1.** *A process $C$ is a time-homogeneous $\mathbb{G}$-Markov chain under $\mathbb{Q}$, with the initial distribution $\mu_0$ and with the infinitesimal generator matrix $\Lambda$, if and only if the following conditions are satisfied:*
(i) $\mathbb{Q}\{C_0 = i\} = \mu_0(i)$ *for every $i \in \mathcal{K}$,*
(ii) *for any function $h : \mathcal{K} \to \mathbb{R}$ the process $M^h$, defined by the formula*

$$M_t^h = h(C_t) - \int_0^t (\Lambda h)(C_u) \, du, \quad \forall t \in \mathbb{R}_+,$$

*follows a $\mathbb{G}$-martingale under $\mathbb{Q}$.*

*Example 11.2.1.* Let $C$ be a time-homogeneous $\mathbb{G}$-Markov chain with the infinitesimal generator matrix $\Lambda$. Applying Proposition 11.2.1 to the function $h(\cdot) = \mathbb{1}_{\{i\}}(\cdot)$, we conclude that the process

$$M_t^i = H_t^i - \int_0^t \lambda_{C_u i} \, du, \quad \forall t \in \mathbb{R}_+, \tag{11.22}$$

follows a $\mathbb{G}$-martingale (and an $\mathbb{F}^C$-martingale) under $\mathbb{Q}$. Conversely, if for every $i \in \mathcal{K}$ the process $M^i$ follows a $\mathbb{G}$-martingale, then for any function $h : \mathcal{K} \to \mathbb{R}$ the process $M^h$ is a $\mathbb{G}$-martingale under $\mathbb{Q}$.

**Time-inhomogeneous chains.** If a Markov chain is time-inhomogeneous, the time-dependent transition intensities are introduced through the formula[6]

$$\lambda_{ij}(t) = \lim_{h \downarrow 0} \frac{p_{ij}(t, t+h) - \delta_{ij}}{h}.$$

It is obvious that $\lambda_{ij}(t) \geq 0$ for arbitrary $i \neq j$, and

$$\lambda_{ii}(t) = \lim_{h \downarrow 0} \frac{p_{ii}(t, t+h) - 1}{h} = -\lim_{h \downarrow 0} \frac{\sum_{j=1, j \neq i}^{K} p_{ij}(t, t+h)}{h} = -\sum_{j=1, j \neq i}^{K} \lambda_{ij}(t),$$

where

$$p_{ij}(t, t+h) = \mathbb{Q}\{C_{t+h} = j \,|\, C_t = i\}, \quad \forall i, j \in \mathcal{K}.$$

We shall write $\Lambda(t) = [\lambda_{ij}(t)]_{1 \leq i, j \leq K}$ to denote the infinitesimal generator matrix function associated with a time-inhomogeneous Markov chain $C$.

---

[6] Let us mention that mild regularity conditions need to be satisfied by the probabilities $p_{ij}(s, t)$ for the results of this subsection to be valid.

The two parameter family $\mathcal{P}(t,s) = [p_{ij}(t,s)]_{1\leq i,j\leq K}$, $0 \leq t \leq s$, of transition matrices for $C$ satisfies the Chapman-Kolmogorov equation:

$$\mathcal{P}(t,s) = \mathcal{P}(t,u)\mathcal{P}(u,s), \quad \forall t \leq u \leq s,$$

the forward Kolmogorov equation:

$$\frac{d\mathcal{P}(t,s)}{ds} = \mathcal{P}(t,s)\Lambda(s), \quad \mathcal{P}(t,t) = \mathrm{Id}, \tag{11.23}$$

and the backward Kolmogorov equation:

$$\frac{d\mathcal{P}(t,s)}{dt} = -\Lambda(t)\mathcal{P}(t,s), \quad \mathcal{P}(s,s) = \mathrm{Id}. \tag{11.24}$$

Let us sketch the derivation of (11.23). For arbitrary two states $i,j \in \mathcal{K}$, the Chapman-Kolmogorov equation yields

$$p_{ij}(t,s+h) = \sum_{k=1}^{K} p_{ik}(t,s)p_{kj}(s,s+h), \tag{11.25}$$

where $0 \leq t \leq s \leq s + h$. Consequently, for any $0 \leq t \leq s$ we obtain

$$\lim_{h\downarrow 0} \frac{p_{ij}(t,s+h) - p_{ij}(t,s)}{h} = \lim_{h\downarrow 0} \frac{1}{h}\left( \sum_{k=1}^{K} p_{ik}(t,s)p_{kj}(s,s+h) - p_{ij}(t,s) \right)$$

$$= \lim_{h\downarrow 0} \left( p_{ij}(t,s)\frac{p_{jj}(s,s+h) - 1}{h} + \sum_{k=1, k\neq j}^{K} p_{ik}(t,s)\frac{p_{kj}(s,s+h)}{h} \right)$$

$$= \sum_{k=1}^{K} p_{ik}(t,s)\lambda_{kj}(s).$$

Using analogous arguments, one may verify that

$$\lim_{h\downarrow 0} \frac{p_{ij}(t,s-h) - p_{ij}(t,s)}{h} = \sum_{k=1}^{K} p_{ik}(t,s)\lambda_{kj}(s).$$

The next result is an immediate consequence of Kolmogorov's equations.

**Corollary 11.2.1.** *The family $\mathcal{P}(t,s)$, $0 \leq t \leq s$, satisfies the integral equations*

$$\mathcal{P}(t,s) = \mathrm{Id} + \int_{t}^{s} \mathcal{P}(t,u)\Lambda(u)\,du$$

*and*

$$\mathcal{P}(t,s) = \mathrm{Id} + \int_{t}^{s} \Lambda(u)\mathcal{P}(u,s)\,du.$$

The above equations can be used in order to derive some remarkable representations, which are important from the computational point of view, and which are counterparts of (11.21). For the proof of Corollary 11.2.2, the interested reader is referred to Rolski et al. (1998) (see Theorem 8.4.4 therein).

**Corollary 11.2.2.** *For every* $0 \leq t \leq s$ *we have*

$$\mathcal{P}(t,s) = \mathrm{Id} + \sum_{n=1}^{\infty} \int_t^s \int_{u_1}^s \cdots \int_{u_{n-1}}^s \Lambda(u_1)\ldots\Lambda(u_n)\,du_n\ldots du_1,$$

*and*

$$\mathcal{P}(t,s) = \mathrm{Id} + \sum_{n=1}^{\infty} \int_t^s \int_t^{u_1} \cdots \int_t^{u_{n-1}} \Lambda(u_1)\ldots\Lambda(u_n)\,du_n\ldots du_1.$$

Assume that the matrix function $\Lambda(t) = [\lambda_{ij}(t)]_{1 \leq i,j \leq K}$ satisfies the conditions, which characterize the infinitesimal generator of an inhomogeneous Markov chain, namely,

$$\lambda_{ij}(t) \geq 0, \quad i \neq j, \quad \lambda_{ii}(t) = -\sum_{j=1,\,j\neq i}^{K} \lambda_{ij}(t).$$

For any function $h : \mathcal{K} \to \mathbb{R}$, we introduce the mapping $\Lambda h : \mathcal{K} \times \mathbb{R}_+ \to \mathbb{R}$ by setting:

$$(\Lambda h)(i,t) = \sum_{j=1}^{K} \lambda_{ij}(t)h(j), \quad \forall i \in \mathcal{K}, t \in \mathbb{R}_+.$$

The following result, which is also taken for granted (see Proposition 11.3.1, though), is a natural extension of Proposition 11.2.1.

**Proposition 11.2.2.** *A process $C$ is a $\mathbb{G}$-Markov chain under $\mathbb{Q}$, with the initial distribution $\mu_0$ and with the infinitesimal generator matrix function $\Lambda(\cdot)$, if and only if:*
(i) $\mathbb{Q}\{C_0 = i\} = \mu_0(i)$ *for every $i \in \mathcal{K}$,*
(ii) *for any function $h : \mathcal{K} \to \mathbb{R}$ the process $M^h$, defined by the formula*

$$M_t^h = h(C_t) - \int_0^t (\Lambda h)(C_u, u)\,du, \quad t \in \mathbb{R}_+, \tag{11.26}$$

*follows a $\mathbb{G}$-martingale under $\mathbb{Q}$.*

*Example 11.2.2.* Let $C$ be a time-inhomogeneous $\mathbb{G}$-Markov chain with the infinitesimal generator matrix function $\Lambda(\cdot)$. By applying Proposition 11.2.2 to the function $h(\cdot) = \mathbb{1}_{\{i\}}(\cdot)$, we find that the process

$$M_t^i = H_t^i - \int_0^t \lambda_{C_u i}(u)\,du, \quad \forall t \in \mathbb{R}_+, \tag{11.27}$$

follows a $\mathbb{G}$-martingale (and an $\mathbb{F}^C$-martingale) under $\mathbb{Q}$. Similarly as in the time-homogeneous case, if for every $i \in \mathcal{K}$ the process $M^i$ is a $\mathbb{G}$-martingale, then for any function $h : \mathcal{K} \to \mathbb{R}$ the process $M^h$, given by formula (11.26), also follows a $\mathbb{G}$-martingale.

### 11.2.1 Embedded Discrete-Time Markov Chain

Let $C_t$, $t \in \mathbb{R}_+$, stand for a time-homogeneous $\mathbb{G}$-Markov chain (and thus an ordinary Markov chain as well) under $\mathbb{Q}$ with the infinitesimal generator $\Lambda$. Let $\tau_n$, $n \in \mathbb{N}$, denote the random sequence of successive jump times of $C$. More explicitly, for any $n \in \mathbb{N}$, the random variable $\tau_n$ defined as (by convention, $\tau_0 = 0$):

$$\tau_n = \inf\{t > \tau_{n-1} : C_t \neq C_{\tau_{n-1}}\},$$

represents the time of the $n^{\text{th}}$ jump (or transition) for $C$. Let us recall few classic results related to the behavior of a continuous-time Markov chain at its jump times.

First, it is well known that the following property holds, for any $n \in \mathbb{N}$ and every $t \in \mathbb{R}_+$,

$$\mathbb{Q}\{\tau_n - \tau_{n-1} > t \,|\, C_{\tau_{n-1}} = i\} = e^{\lambda_{ii}t}, \quad \forall i = 1, \ldots, K. \tag{11.28}$$

Equality (11.28) makes it clear that, conditionally on the position $C_{\tau_{n-1}} = i$ at the jump time $\tau_{n-1}$, the random time that elapses until the next jump occurs has an exponential probability law with the parameter $-\lambda_{ii} > 0$.

Second, the conditional probabilities of transitions are known to satisfy:

$$\mathbb{Q}\{C_{\tau_n} = j \,|\, C_{\tau_{n-1}} = i\} = p_{ij} := -\frac{\lambda_{ij}}{\lambda_{ii}}, \quad \forall i, j \in \mathcal{K}, \, i \neq j. \tag{11.29}$$

Formula (11.29) specifies the conditional probability law of a continuous-time Markov chain $C$ after its $n^{\text{th}}$ jump, given the position after the $(n-1)^{\text{th}}$ jump (it coincides, of course, with the position of $C$ just before the $n^{\text{th}}$ jump).

Let us emphasize that since $C$ is assumed to be a time-homogeneous Markov chain, both probability laws introduced above do not depend on the number of transitions in the past (that is, on $n$). They only depend on the value taken by $C$ after the previous jump.

Define a random sequence $\hat{C}_n = C_{\tau_n}$ for every $n \in \mathbb{N}^*$. It is well known that the sequence $\hat{C}$ is a time-homogeneous Markov chain under $\mathbb{Q}$ with the one-step transition probability matrix $P = [p_{ij}]_{1 \leq i,j \leq K}$. The discrete-time Markov chain $\hat{C}_n$, $n \in \mathbb{N}^*$, is called the *embedded Markov chain* corresponding to the continuous-time Markov chain $C$.

### 11.2.2 Conditional Expectations

In view of our further purposes, we are mainly interested in Markov chains, which admit absorbing states. We say that a state $k \in \mathcal{K}$ is *absorbing* for a time-homogeneous $\mathbb{G}$-Markov chain $C_t$, $t \in \mathbb{R}_+$, if the following holds:

$$\mathbb{Q}\{C_s = k \,|\, C_t = k\} = 1, \quad \forall t, s \in \mathbb{R}_+, \, t \leq s.$$

In view of (11.18), it is clear that if a state $k \in \mathcal{K}$ is absorbing, then we have $\lambda_{kj} = 0$ for every $j = 1, \ldots, K$.

From now on, we shall postulate that the state $K$ is absorbing. This implies that the infinitesimal generator of $C$ under $\mathbb{Q}$ is given by the intensity matrix $\Lambda$ of the following form:

$$\Lambda = \begin{pmatrix} \lambda_{1,1} & \cdots & \lambda_{1,K-1} & \lambda_{1,K} \\ \cdot & \cdots & \cdot & \cdot \\ \lambda_{K-1,1} & \cdots & \lambda_{K-1,K-1} & \lambda_{K-1,K} \\ 0 & \cdots & 0 & 0 \end{pmatrix}.$$

We assume that the initial state $C_0 = x \neq K$ is fixed, and we denote by $\tau$ the random time of absorption at $K$, i.e., $\tau = \inf \{t > 0 : C_t = K\}$. We assume that $\tau < \infty$, $\mathbb{Q}$-a.s.; this implies that the state $K$ is the only recurrent state for $C$. As usual, we write $H_t^i = \mathbb{1}_{\{C_t=i\}}$ and $H_t = \mathbb{1}_{\{\tau \leq t\}} = \mathbb{1}_{\{C_t=K\}} = H_t^K$.

In the next few auxiliary results, we shall deal with the conditional expectations with respect to the filtrations $\mathbb{G}$ and $\mathbb{F}^C$. The absorption time $\tau$ is, of course, an $\mathbb{F}^C$-stopping time and a $\mathbb{G}$-stopping time. In what follows, $Y$ will denote an integrable random variable, which is defined on the reference probability space $(\Omega, \mathcal{G}, \mathbb{Q})$.

**Lemma 11.2.1.** *We have*

$$\mathbb{1}_{\{\tau \leq t\}} \mathbb{E}_{\mathbb{Q}}(Y \mid \mathcal{G}_t) = \mathbb{E}_{\mathbb{Q}}(\mathbb{1}_{\{\tau \leq t\}} Y \mid \mathcal{G}_t \vee \sigma(\tau))$$

*and*

$$\mathbb{1}_{\{\tau \leq t\}} \mathbb{E}_{\mathbb{Q}}(Y \mid \mathcal{F}_t^C) = \mathbb{E}_{\mathbb{Q}}(\mathbb{1}_{\{\tau \leq t\}} Y \mid \mathcal{F}_t^C \vee \sigma(\tau)).$$

*Proof.* Consider an event $A \in \mathcal{G}_t \vee \sigma(\tau)$. Then $A \cap \mathbb{1}_{\{\tau \leq t\}} \in \mathcal{G}_t$, and so

$$\int_A \mathbb{E}_{\mathbb{Q}}(\mathbb{1}_{\{\tau \leq t\}} Y \mid \mathcal{G}_t \vee \sigma(\tau)) \, d\mathbb{Q} = \int_A \mathbb{1}_{\{\tau \leq t\}} Y \, d\mathbb{Q} = \int_{A \cap \{\tau \leq t\}} Y \, d\mathbb{Q}$$

$$= \int_{A \cap \{\tau \leq t\}} \mathbb{E}_{\mathbb{Q}}(Y \mid \mathcal{G}_t) \, d\mathbb{Q} = \int_A \mathbb{E}_{\mathbb{Q}}(\mathbb{1}_{\{\tau \leq t\}} Y \mid \mathcal{G}_t) \, d\mathbb{Q}$$

since the event $\{\tau \leq t\} = \{C_t = K\}$ manifestly belongs to $\mathcal{G}_t$. The proof of the second equality is analogous.  $\square$

In the next lemma, we examine the case when the random variable $Y$ has the form $h(C_{\tau-}, \tau)$ for some function $h : \mathcal{K} \times \mathbb{R}_+ \to \mathbb{R}$.

**Lemma 11.2.2.** *Let* $Y = h(C_{\tau-}, \tau)$ *for some function* $h : \mathcal{K} \times \mathbb{R}_+ \to \mathbb{R}$. *Then*

$$\mathbb{E}_{\mathbb{Q}}(\mathbb{1}_{\{\tau > t\}} Y \mid \mathcal{G}_t) = \sum_{i=1}^{K-1} H_t^i \, \mathbb{E}_{\mathbb{Q}}(\mathbb{1}_{\{\tau > t\}} Y \mid C_t = i).$$

*Proof.* Observe that the event $\{\tau > t\} = \{C_t \neq K\}$ belongs to $\sigma$-field $\mathcal{G}_t$, and thus the Markov property of $C$ yields

$$\mathbb{1}_{\{\tau > t\}} \mathbb{E}_{\mathbb{Q}}(Y \mid \mathcal{G}_t) = \mathbb{1}_{\{\tau > t\}} \mathbb{E}_{\mathbb{Q}}(\mathbb{1}_{\{\tau > t\}} Y \mid \mathcal{G}_t)$$
$$= \mathbb{1}_{\{\tau > t\}} \mathbb{E}_{\mathbb{Q}}(\mathbb{1}_{\{\tau > t\}} h(C_{\tau-}, \tau) \mid \mathcal{G}_t) = \mathbb{1}_{\{\tau > t\}} \mathbb{E}_{\mathbb{Q}}(\mathbb{1}_{\{\tau > t\}} h(C_{\tau-}, \tau) \mid C_t).$$

Since the event $\{\tau > t\}$ also belongs to the $\sigma$-field $\mathcal{F}_t^C$, a similar reasoning shows that the first equality holds. Furthermore,

$$\mathbb{1}_{\{\tau>t\}} \, \mathbb{E}_{\mathbb{Q}}(\mathbb{1}_{\{\tau>t\}} h(C_{\tau-}, \tau) \,|\, C_t) = \sum_{i=1}^{K-1} H_t^i \, \mathbb{E}_{\mathbb{Q}}(\mathbb{1}_{\{\tau>t\}} Y \,|\, C_t = i),$$

as expected.                                                                               □

The next two auxiliary results are simple corollaries to Lemma 11.2.2.

**Corollary 11.2.3.** *Let* $Y = h(C_{\tau-}, \tau)$ *for some function* $h : \mathcal{K} \times \mathbb{R}_+ \to \mathbb{R}$. *Then*

$$\mathbb{E}_{\mathbb{Q}}(Y \,|\, \mathcal{G}_t) = \mathbb{1}_{\{\tau \le t\}} Y + \sum_{i=1}^{K-1} H_t^i \, \mathbb{E}_{\mathbb{Q}}(\mathbb{1}_{\{\tau>t\}} Y \,|\, C_t = i).$$

**Corollary 11.2.4.** *For any* $s, t \in \mathbb{R}_+$, *the following equalities are valid*

$$\mathbb{Q}\{\tau > s \,|\, \mathcal{G}_t\} = \mathbb{1}_{\{s \le t\}} \mathbb{1}_{\{\tau > s\}} + \mathbb{1}_{\{s > t\}} \sum_{i=1}^{K-1} H_t^i \, \mathbb{Q}\{\tau > s \,|\, C_t = i\},$$

*and*

$$\mathbb{Q}\{\tau \ge s \,|\, \mathcal{G}_t\} = \mathbb{1}_{\{s \le t\}} \mathbb{1}_{\{\tau \ge s\}} + \mathbb{1}_{\{s > t\}} \sum_{i=1}^{K-1} H_t^i \, \mathbb{Q}\{\tau \ge s \,|\, C_t = i\}.$$

*Proof.* The first equality is obvious. To verify the second formula, we consider the random variable $Y = h(\tau)$, where, for a fixed $s$, we set $h(t) = \mathbb{1}_{\{t > s\}}$. Since $Y$ is a $\sigma(\tau)$-measurable random variable, making use of the second equality in Corollary 11.2.3, we obtain

$$\mathbb{Q}\{\tau > s \,|\, \mathcal{G}_t\} = \mathbb{1}_{\{\tau \le t\}} \mathbb{1}_{\{\tau > s\}} + \sum_{i=1}^{K-1} H_t^i \, \mathbb{E}_{\mathbb{Q}}(\mathbb{1}_{\{\tau > t, \, \tau > s\}} \,|\, C_t = i)$$

$$= \mathbb{1}_{\{s \le t\}} \mathbb{1}_{\{s < \tau \le t\}} + \mathbb{1}_{\{s \le t\}} \sum_{i=1}^{K-1} H_t^i \, \mathbb{Q}\{\tau > t \,|\, C_t = i\}$$

$$+ \mathbb{1}_{\{s > t\}} \sum_{i=1}^{K-1} H_t^i \, \mathbb{Q}\{\tau > s \,|\, C_t = i\}.$$

Invoking the identity

$$\sum_{i=1}^{K-1} H_t^i \, \mathbb{Q}\{\tau > t \,|\, C_t = i\} = \sum_{i=1}^{K-1} H_t^i \, \mathbb{1}_{\{\tau > t\}} = \mathbb{1}_{\{\tau > t\}},$$

we conclude that the second asserted formula is valid. The last equality follows by similar arguments.    □

### 11.2.3 Probability Distribution of the Absorption Time

We maintain the assumptions of Sect. 11.2.2. More explicit formulae for the conditional expectations with respect to the $\sigma$-field $\mathcal{G}_t$ can be obtained, if the knowledge of conditional laws of $C$ is used. Notice that for every $0 \le t \le s$ we have

$$\mathbb{Q}\{\tau > s \,|\, C_t = i\} = 1 - \mathbb{Q}\{C_s = K \,|\, C_t = i\} = 1 - p_{iK}(s - t),$$

hence, the first formula of Corollary 11.2.4 can be rewritten as follows:

$$\mathbb{Q}\{\tau > s \,|\, \mathcal{G}_t\} = \mathbb{1}_{\{s \le t\}}\mathbb{1}_{\{\tau > s\}} + \mathbb{1}_{\{s > t\}} \sum_{i=1}^{K-1} H_t^i\big(1 - p_{iK}(s - t)\big). \qquad (11.30)$$

To derive an alternative representation for the probability distribution of the absorption time, let us denote by $\tilde{\Lambda}$ the matrix obtained from $\Lambda$ by deleting the last row and the last column. Also, let $\tilde{\mathcal{P}}(t) = [\tilde{p}_{ij}(t)]_{i,j \in \tilde{\mathcal{K}}}$ stand for the associated transition matrix, where $\tilde{\mathcal{K}} = \{1, \ldots, K - 1\}$.

It is not difficult to check that the so-called *taboo probabilities* $\tilde{p}_{ij}(t)$, $i, j \in \tilde{\mathcal{K}}$ can be found by solving the following differential equation:

$$\frac{d}{dt}\tilde{\mathcal{P}}(t) = \tilde{\Lambda}\tilde{\mathcal{P}}(t), \quad t > 0, \qquad (11.31)$$

with the initial condition $\tilde{\mathcal{P}}(0) = \mathrm{Id}$.

It is also clearly seen that (recall that we have assumed that $C_0 = i \in \tilde{\mathcal{K}}$)

$$F(t) = 1 - \sum_{j=1}^{K-1} \tilde{p}_{ij}(t) = 1 - \sum_{j=1}^{K-1} p_{ij}(t). \qquad (11.32)$$

Since $F(t) < 1$ for every $t \in \mathbb{R}_+$, we may introduce the hazard function $\Gamma$ of $\tau$ by setting $\Gamma(t) = -\ln(1 - F(t))$. Denoting by $f(t)$ the density of $F(t)$ with respect to the Lebesgue measure, and setting $\gamma(t) = f(t)(1 - F(t))^{-1}$, we obtain $\Gamma(t) = \int_0^t \gamma(u)\,du$. In view of (11.32), we have

$$f(t) = -\sum_{j=1}^{K-1} \frac{d\tilde{p}_{ij}(t)}{dt} = -\sum_{j=1}^{K-1} \frac{dp_{ij}(t)}{dt}.$$

Let us finally mention, that in case of a time-inhomogeneous Markov chain, we have the following obvious counterpart of Corollary 11.1.4.

**Corollary 11.2.5.** *For every $t \in \mathbb{R}_+$ and any $i = 1, \ldots, K - 1$, the conditional law of the absorption time $\tau$ is given by the formula*

$$\mathbb{Q}\{\tau \le t \,|\, C_0 = i\} = 1 - \sum_{j=1}^{K-1} p_{ij}(0, t).$$

### 11.2.4 Martingales Associated with Transitions

We shall now introduce some important examples of martingales associated with the absorption time $\tau$ and with the number of transitions. For any fixed $i \neq j$, let $H_t^{ij}$ stand for the number of jumps of the process $C$ from $i$ to $j$ in the interval $(0, t]$. Formally, for any $i \neq j$ we set

$$H_t^{ij} := \sum_{0 < u \leq t} H_{u-}^i H_u^j, \quad \forall t \in \mathbb{R}_+.$$

The following result is classic (see Brémaud (1981), Last and Brandt (1995) or Rogers and Williams (2000)), and thus we omit the proof.

**Lemma 11.2.3.** *For every $i, j \in \mathcal{K}$, $i \neq j$, the processes*

$$M_t^{ij} = H_t^{ij} - \int_0^t \lambda_{ij} H_u^i \, du = H_t^{ij} - \int_0^t \lambda_{C_u j} H_u^i \, du \tag{11.33}$$

*and*

$$M_t^K = H_t - \int_0^t \sum_{i=1}^{K-1} \lambda_{iK} H_u^i \, du = H_t - \int_0^t \lambda_{C_u K}(1 - H_u) \, du \tag{11.34}$$

*follow $\mathbb{G}$-martingales (and $\mathbb{F}^C$-martingales).*

*Proof.* See, for instance, Theorem 7.5.5 in Last and Brandt (1995), or Lemma 21.12 in Rogers and Williams (2000). Let us only mention that in both cases the asserted martingale property is a rather straightforward consequence of the following general property: for any function $h : \mathcal{K} \times \mathcal{K} \to \mathbb{R}$ the compensated process $M^h$, given by the equality

$$M_t^h = \sum_{0 < u \leq t} h(C_{u-}, C_u) - \int_0^t \sum_{l=1, l \neq C_u}^K \lambda_{C_u l} h(C_u, l) \, du, \quad \forall t \in \mathbb{R}_+,$$

follows a $\mathbb{G}$-martingale under $\mathbb{Q}$. To check that the process $M^{ij}$ given by formula (11.33) is a $\mathbb{G}$-martingale, it suffices to examine the process $M^h$ associated with the following function:

$$h(k, l) = \mathbb{1}_{\{i\}}(k)\mathbb{1}_{\{j\}}(l), \quad \forall k, l \in \mathcal{K}.$$

Likewise, to establish the martingale property of the process $M^K$, we set

$$h(k, l) = \mathbb{1}_{\{K\}}(l) - \mathbb{1}_{\{K\}}(k), \quad \forall k, l \in \mathcal{K}.$$

Notice that the process $M^K$ given by (11.34) coincides with the process $M^K$ given by (11.22) for $i = K$. □

## 11.2.5 Change of a Probability Measure

We shall now examine how the Markov property and the generator $\Lambda$ of the time-homogeneous Markov chain $C$ are affected by a change of the reference probability measure $\mathbb{Q}$ to an equivalent probability measure $\mathbb{Q}^*$ on $(\Omega, \mathcal{G}_{T^*})$ for some fixed $T^* > 0$. Let us emphasize that we do not need to assume here that the state $K$ is absorbing.

Consider a family $\tilde{\kappa}^{kl}$, $k, l \in \mathcal{K}$, $k \neq l$, of bounded, $\mathbb{F}^C$-predictable, real-valued processes, such that $\tilde{\kappa}_t^{kl} > -1$. For the sake of notational convenience, we also introduce processes $\tilde{\kappa}^{kk} \equiv 0$ for $k = 1, \ldots, K$. Let us define an auxiliary $\mathbb{G}$-martingale $M$ (which is also an $\mathbb{F}^C$-martingale) by setting

$$M_t = \int_{]0,t]} \sum_{k,l=1}^{K} \tilde{\kappa}_u^{kl} \, dM_u^{kl} = \int_{]0,t]} \sum_{k,l=1}^{K} \tilde{\kappa}_u^{kl} \, dH_u^{kl} - M_t^c, \tag{11.35}$$

where $M_t^c$ is the path-by-path continuous component of the process $M$, i.e.,

$$M_t^c = \int_0^t \sum_{k,l=1}^{K} \tilde{\kappa}_u^{kl} \lambda_{kl} H_u^k \, du.$$

*Remarks.* From Theorem 21.15 in Rogers and Williams (2000), we know that an arbitrary $\mathbb{F}^C$-local martingale $M$ under $\mathbb{Q}$ admits the following representation

$$M_t = \sum_{0 < u \leq t} h_u(C_{u-}, C_u) - \int_0^t \sum_{j=1}^{K} \lambda_{C_{u-}j} \, h_u(C_{u-}, j) \, du,$$

where, for any states $i, j \in \mathcal{K}$, the process $h(i, j)$ is $\mathbb{F}^C$-predictable. In addition, we postulate that $h(j, j) \equiv 0$. Notice that the process $M$ as in (11.35) can be obtained by setting:

$$h_t(i, j) = \sum_{k,l=1}^{K} \tilde{\kappa}_t^{kl} \delta_{ik} \delta_{jl}.$$

Let us return to our problem. We fix a horizon date $T^* < \infty$, and we define an $\mathbb{G}$-martingale $\eta_t$, $t \in [0, T^*]$, by postulating that

$$\eta_t = 1 + \int_{]0,t]} \sum_{k,l=1}^{K} \eta_{u-} \tilde{\kappa}_u^{kl} \, dM_u^{kl}. \tag{11.36}$$

It is known that the unique solution to the SDE (11.36) equals, for every $t \in [0, T^*]$,

$$\eta_t = e^{-M_t^c} \prod_{0 < u \leq t} (1 + \Delta M_u).$$

More explicitly,

$$\eta_t = e^{-M_t^c} \prod_{0 < u \le t} \left( 1 + \sum_{k,l=1}^{K} \tilde{\kappa}_u^{kl} (M_u^{kl} - M_{u-}^{kl}) \right).$$

Observe that

$$1 + \sum_{k,l=1}^{K} \tilde{\kappa}_u^{kl} (M_u^{kl} - M_{u-}^{kl}) = 1 + \sum_{k,l=1}^{K} \tilde{\kappa}_u^{kl} (H_u^{kl} - H_{u-}^{kl}). \tag{11.37}$$

Since at most one of the differentials $H_u^{kl} - H_{u-}^{kl}$ is equal to one, and all those that are not equal to one are equal to zero, we see that the right-hand side of (11.37) is either equal to $1 + \tilde{\kappa}_u^{ij}$ for some $i \neq j \in \mathcal{K}$, or it is equal to 1. Thus, in view of our assumption that $\tilde{\kappa}_u^{kl} > -1$ for all $k \neq l$, we conclude that the product

$$\prod_{0 < u \le t} \left( 1 + \sum_{k,l=1}^{K} \tilde{\kappa}_u^{kl} (M_u^{kl} - M_{u-}^{kl}) \right)$$

is strictly positive. Consequently, the process $\eta$ is strictly positive. Since, in addition, $\mathbb{E}_{\mathbb{Q}}(\eta_{T^*}) = 1$, we may define a probability measure $\mathbb{Q}^*$, equivalent to $\mathbb{Q}$ on $(\Omega, \mathcal{G}_{T^*})$, by setting

$$\frac{d\mathbb{Q}^*}{d\mathbb{Q}} \bigg|_{\mathcal{G}_{T^*}} = \eta_{T^*}, \quad \mathbb{Q}\text{-a.s.} \tag{11.38}$$

It is clear that for any date $t \in [0, T^*]$ we have

$$\frac{d\mathbb{Q}^*}{d\mathbb{Q}} \bigg|_{\mathcal{G}_t} = \eta_t, \quad \mathbb{Q}\text{-a.s.}$$

Before proceeding further, we need to impose an additional measurability condition on processes $\tilde{\kappa}^{kl}$, namely, we postulate that, for any fixed $k, l \in \mathcal{K}$ and $t \in \mathbb{R}_+$, the random variable $\tilde{\kappa}_t^{kl}$ is measurable with respect to the $\sigma$-field $\sigma(C_t)$. This implies that, for any fixed $k, l \in \mathcal{K}$ and $t \in \mathbb{R}_+$, there exists a function $g_t^{kl} : \mathcal{K} \to \mathbb{R}$ such that $\tilde{\kappa}_t^{kl} = g_t^{kl}(C_t)$.

We assume that for any $i \in \mathcal{K}$ there exists a version of $g_t^{kl}(i)$, $t \in \mathbb{R}_+$ that is Borel measurable as a function of $t$, and we introduce a family of functions $\kappa_{kl} : \mathbb{R}_+ \to (-1, \infty)$ by setting $\kappa_{kl}(t) := g_t^{kl}(k)$ for every $k, l \in \mathcal{K}$ and $t \in \mathbb{R}_+$.

To further simplify the exposition, we shall only consider processes $\tilde{\kappa}_t^{kl}$ of the special form: $\tilde{\kappa}_t^{kl} = \kappa_{kl}(t)$, where for every $k, l \in \mathcal{K}$, $k \neq l$, the function $\kappa_{kl} : \mathbb{R}_+ \to (-1, \infty)$ is Borel measurable and bounded. We thus may and do assume that $\kappa_{kk} \equiv 0$ for every $k = 1, \ldots, K$. Under this assumption, we have the following result that provides sufficient conditions for a $\mathbb{G}$-Markov chain $C$ to remain a (time-inhomogeneous, in general) $\mathbb{G}$-Markov chain under $\mathbb{Q}^*$.

**Proposition 11.2.3.** *Let the probability measure* $\mathbb{Q}^*$ *by defined by* (11.38) *with the Radon-Nikodým density* $\eta_{T^*}$ *given by* (11.36). *Then*
(i) *the process* $C_t$, $t \in [0, T^*]$, *is a* $\mathbb{G}$*-Markov chain under* $\mathbb{Q}^*$,
(ii) *the infinitesimal generator matrix function* $\Lambda^*(t) = [\lambda^*_{ij}(t)]_{1 \leq i,j \leq K}$ *for* $C$ *under* $\mathbb{Q}^*$ *satisfies, for* $i \neq j$,

$$\lambda^*_{ij}(t) = (1 + \kappa_{ij}(t))\lambda_{ij}, \quad \forall t \in [0, T^*], \tag{11.39}$$

*and*

$$\lambda^*_{ii}(t) = - \sum_{j=1, j \neq i}^{K} \lambda^*_{ij}(t), \quad \forall t \in [0, T^*], \tag{11.40}$$

(iii) *the two parameter family* $\mathcal{P}^*(t, s)$, $0 \leq s \leq t \leq T^*$, *of transition matrices for* $C$ *relative to* $\mathbb{Q}^*$ *satisfies the forward Kolmogorov equation*

$$\frac{d\mathcal{P}^*(t, s)}{ds} = \mathcal{P}^*(t, s)\Lambda^*(s), \quad \mathcal{P}^*(t, t) = \mathrm{Id},$$

*and the backward Kolmogorov equation*

$$\frac{d\mathcal{P}^*(t, s)}{dt} = -\Lambda^*(t)\mathcal{P}^*(t, s), \quad \mathcal{P}^*(s, s) = \mathrm{Id}.$$

*Proof.* Since the matrix function $\Lambda^*(\cdot)$ specified by (11.39)–(11.40) satisfies the conditions of an infinitesimal generator of a time-inhomogeneous Markov chain, in view of Proposition 11.2.2, to establish the first two statements, it suffices to verify that for any $i \in \mathcal{K}$ the process

$$M^{*j}_t = H^j_t - \int_0^t \lambda_{C_u j}(1 + \kappa_{C_u j}(u)) \, du, \quad \forall t \in [0, T^*], \tag{11.41}$$

follows a $\mathbb{G}$-(local) martingale under the equivalent probability measure $\mathbb{Q}^*$ or, equivalently, that the product $\eta_t M^{*j}_t$, $t \in [0, T^*]$, is a $\mathbb{G}$-(local) martingale under the original probability measure $\mathbb{Q}$. Since $\eta$ and $M^{*j}$ follow processes of finite variation, Itô's product rule yields

$$\eta_t M^{*j}_t = H^j_0 + \int_{]0,t]} \eta_{u-} \, dM^{*j}_u + \int_{]0,t]} M^{*j}_{u-} \, d\eta_u + \sum_{0 < u \leq t} \Delta M^{*j}_u \Delta \eta_u$$

for every $t \in [0, T^*]$. Using (11.22) and (11.36), for every $t \in [0, T^*]$ we obtain

$$\eta_t M^{*j}_t = H^j_0 + \int_{]0,t]} \eta_{u-} \, dM^j_u + \int_{]0,t]} M^{*j}_{u-} \, d\eta_u - \int_0^t \eta_{u-} \lambda_{C_u j} \kappa_{C_u j}(u) \, du$$

$$+ \sum_{0 < u \leq t} \sum_{k,l=1}^{K} \eta_{u-} \kappa_{kl}(u) \Delta H^j_u \Delta H^{kl}_u.$$

The first three terms on the right-hand side of the last formula manifestly follow $\mathbb{G}$-(local) martingales; it is thus enough to examine the following process:

$$Z_t = \sum_{0<u\le t} \sum_{k,l=1}^{K} \kappa_{kl}(u)\Delta H_u^j \Delta H_u^{kl} - \int_0^t \lambda_{C_u j}\kappa_{C_u j}(u)\,du.$$

Simple considerations show that

$$\sum_{0<u\le t} \sum_{k,l=1}^{K} \kappa_{kl}(u)\Delta H_u^j \Delta H_u^{kl} = \sum_{i=1,\,i\ne j}^{K} \sum_{0<u\le t} \kappa_{ij}(u)\Delta H_u^{ij}$$

and (recall that $\kappa_{jj}\equiv 0$)

$$\int_0^t \lambda_{C_u j}\kappa_{C_u j}(u)\,du = \sum_{i=1,\,i\ne j}^{K} \int_0^t \lambda_{ij}\kappa_{ij}(u)H_u^i\,du.$$

Thus, we obtain

$$Z_t = \sum_{i=1,\,i\ne j}^{K} \int_{]0,t]} \kappa_{C_u j}(u)\,dM_u^{ij},$$

where the processes $M^{ij}$, $i\ne j$, are given by (11.33). By virtue of Lemma 11.2.3, we conclude that the process $Z$ follows a $\mathbb{G}$-martingale under $\mathbb{Q}$. Let us finally notice that

$$\eta_t M_t^{*j} = H_0^j + \int_{]0,t]} \eta_{u-}\,dM_u^j + \int_{]0,t]} M_{u-}^{*j}\,d\eta_u + \int_{]0,t]} \eta_{u-}\,dZ_u,$$

so that the product $\eta_t M_t^{*j}$, $t\in[0,T^*]$, follows a $\mathbb{G}$-(local) martingale under the original probability measure $\mathbb{Q}$. The validity of the forward and backward Kolmogorov equations is standard. □

It is clear that $\Lambda^*(t) = \Lambda$ for every $t\in[0,T^*]$ if and only if $\kappa_{kl}\equiv 0$ for all $k\ne l$. Letting $\phi_{ij}(t) = 1 + \kappa_{ij}(t)$, we obtain $\lambda_{ij}^*(t) = \phi_{ij}(t)\lambda_{ij}$. Analogous notation will be used later in this text (see Sect. 13.2.8).

*Example 11.2.3.* Suppose that $\kappa_{kl}(t) = v_k(t)$, for every $k,l\in\mathcal{K}$, $l\ne k$, and $t\in[0,T^*]$. Then we obtain

$$\Lambda^*(t) = U(t)\Lambda, \quad \forall\,t\in[0,T^*],$$

where $U(t)$ is a diagonal $K$-dimensional matrix, specifically,

$$U(t) = \mathrm{diag}\,[1 + v_1(t),\dots,1 + v_K(t)].$$

## 11.2.6 Identification of the Intensity Matrix

In this section, we follow Israel et al. (2001) and the references therein (we also refer to the original paper for the proofs). The classical *embedding problem* for Markov chains can be summarized as follows: Suppose that $C_t$, $t \in \mathbb{R}_+$, is a time-homogeneous, continuous-time Markov chain with finite state space $\mathcal{K} = \{1, \ldots, K\}$ under some probability measure $\mathbb{Q}$. Suppose that we are given the transition probability matrix for $C$ corresponding to time $t = 1$. Thus, denoting this matrix by $\mathcal{P}(1) = [p_{ij}(1)]_{1 \leq i,j \leq K}$, for every $i, j = 1, \ldots, K$ and every $t \in \mathbb{R}_+$ we have

$$p_{ij}(1) = \mathbb{Q}\{C_1 = j \mid C_0 = i\} = \mathbb{Q}\{C_{t+1} = j \mid C_t = i\}.$$

The embedding problem for $C$ relative to $\mathbb{Q}$ can be stated as follows.

**Embedding problem.** Find a $K \times K$ matrix $\hat{\Lambda}$ with non-negative off-diagonal entries and with all rows summing to 0, such that $e^{\hat{\Lambda}} = \mathcal{P}(1)$. More explicitly, $\hat{\Lambda} = [\hat{\lambda}_{ij}]_{1 \leq i,j \leq K}$, where $\hat{\lambda}_{ij} \geq 0$ for every $i, j = 1, \ldots, K$ with $i \neq j$, and

$$\hat{\lambda}_{ii} = -\sum_{j \neq i} \hat{\lambda}_{ij}, \quad \forall i = 1, \ldots, K.$$

From Sect. 11.1, we know that if the transition probability matrix function $\mathcal{P}(t)$, $t \in \mathbb{R}_+$, corresponding to a time-homogeneous, continuous-time Markov chain $C$, satisfies some mild regularity conditions, then there exists a (unique) infinitesimal generator matrix $\Lambda$ for $C$, and $\mathcal{P}(t) = e^{\Lambda t}$ for every $t \in \mathbb{R}_+$; in particular, the equality $\mathcal{P}(1) = e^{\Lambda}$ is valid. Consequently, one of the solutions of the embedding problem for $C$ relative to $\mathbb{Q}$ in this case is $\hat{\Lambda} = \Lambda$, so that the infinitesimal generator matrix is a solution to the embedding problem. It would appear then that the embedding problem is trivial. This is not true, in general, for the following reasons. First, in order to conclude that the infinitesimal generator matrix $\Lambda$ is a solution to the embedding problem we need to assume that the transition probability matrix function $\mathcal{P}(t)$, $t \in \mathbb{R}_+$, satisfies specific regularity conditions. Second, in order to find this solution – that is, to find $\Lambda$ – we need to be able to compute the derivative of $\mathcal{P}(t)$. Therefore, observing just one value of $\mathcal{P}(t)$, such as $\mathcal{P}(1)$, will not suffice, in general. From the practical point of view, trying to solve the embedding problem for $C$ relative to $\mathbb{Q}$ means attempting to determine the $\mathbb{Q}$-infinitesimal generator matrix for $C$ from just one observation of the transition matrix, the matrix $\mathcal{P}(1)$. It is clear that if the embedding problem does not admit any solutions, then the infinitesimal generator matrix $\Lambda$ does not exist. Israel et al. (2001) provide several criteria under which the embedding problem does not admit any solutions. We do not know, however, whether the existence of a solution $\hat{\Lambda}$ to the embedding problem implies existence of the infinitesimal generator matrix for the underlying Markov chain $C$. One can construct a Markov chain, say $\hat{C}$, with $\hat{\Lambda}$ as its infinitesimal generator matrix, but the chains $C$ and $\hat{C}$ may not be equivalent.

**Existence and uniqueness of solutions.** Let us consider the following series

$$\sum_{n=1}^{\infty} (-1)^{n+1} \frac{(\mathcal{P}(1) - \mathrm{Id})^n}{n!}. \tag{11.42}$$

If the above series is absolutely convergent, we may define

$$\ln \mathcal{P}(1) = \hat{\Lambda} := \sum_{n=1}^{\infty} (-1)^{n+1} \frac{(\mathcal{P}(1) - \mathrm{Id})^n}{n!}. \tag{11.43}$$

In this case we have $\mathcal{P}(1) = e^{\hat{\Lambda}}$. By virtue of Theorem 2.2 in Israel et al. (2001), a sufficient criterion for (11.42) to converge is that all the diagonal elements of the matrix $\mathcal{P}(1)$ are greater than 0.5. It should be stressed, however, that this is not a necessary condition.

The main drawback of the above logarithmic result is that the matrix $\hat{\Lambda}$ defined by (11.43) may fail to have all off-diagonal entries non-negative. We refer to Israel et al. (2001) for a discussion of this issue, as well as for a description of ways for correcting the problem. In Sect. 6, they furnish an algorithm for constructing a solution to the embedding problem in the case when series (11.42) fails to converge to a matrix with non-negative off-diagonal entries. Finally, it may also happen that the embedding problem admits more than one solution; we refer to Sect. 5 in Israel et al. (2001) for illustrative examples. Each of these solutions gives rise to a time-homogeneous Markov chain. Israel et al. (2001) provide some insight into the problem of selecting a particular solution as the most desirable candidate for the infinitesimal generator of the underlying Markov chain $C$.

The following result, borrowed from Israel et al. (2001) (see Theorem 5.2 therein), is also worth mentioning.

**Proposition 11.2.4.** *If the matrix $\mathcal{P}(1)$ has all real distinct and positive eigenvalues then $\ln \mathcal{P}(1)$ is the only real-valued matrix $\hat{\Lambda}$ such that $\mathcal{P}(1) = e^{\hat{\Lambda}}$.*

**Valid generator.** Any solution $\hat{\Lambda}$ of the embedding problem is called the *valid generator* for $\mathcal{P}(1)$. As we remarked above, we do not know whether a valid generator for $\mathcal{P}(1)$ is also an infinitesimal generator matrix for the underlying Markov chain $C$. On the other hand, the infinitesimal generator matrix for $C$ must coincide with one of the valid generators for $\mathcal{P}(1)$. The following proposition (see Theorem 5.1 in Israel et al. (2001)) seems to be of interest. We write here $\det(\mathcal{P}(1))$ to denote the determinant of $\mathcal{P}(1)$.

**Proposition 11.2.5.** (i) *If $\det(\mathcal{P}(1)) > 0.5$, then the matrix $\mathcal{P}(1)$ admits at most one valid generator.*
(ii) *If $\det(\mathcal{P}(1)) > 0.5$ and if $|\mathcal{P}(1) - \mathrm{Id}| < 0.5$ (using any matrix norm), then the only possible valid generator for $\mathcal{P}(1)$ is $\ln \mathcal{P}(1)$.*
(iii) *If $\mathcal{P}(1)$ has distinct eigenvalues and if $\det(\mathcal{P}(1)) > e^{-\pi}$, then the only possible generator for $\mathcal{P}(1)$ is $\ln \mathcal{P}(1)$.*

## 11.3 Continuous-Time Conditionally Markov Chains

In this section, we focus on the case of continuous-time processes, and thus the time parameter $t$ is assumed to take values in $\mathbb{R}_+$. For the sake of the reader's convenience, we shall work here under the risk-neutral probability. For this reason, the intensities of transitions are denoted by $\lambda_{ij}^*(t)$ rather than $\lambda_{ij}(t)$. Let us observe that the construction given below also provides a method for constructing an ordinary $\mathbb{G}$-Markov chain; in this case, the transition intensities are deterministic functions of time.

We consider a probability space $(\Omega, \mathcal{G}, \mathbb{Q}^*)$ endowed with some filtrations $\mathbb{F} = (\mathcal{F}_t)_{t \in \mathbb{R}_+}$ and $\mathbb{G} = (\mathcal{G}_t)_{t \in \mathbb{R}_+}$, such that $\mathbb{F} \subseteq \mathbb{G}$. Let $C$ be a $\mathcal{K}$-valued stochastic process defined on this probability space, where $\mathcal{K} = \{1, \ldots, K\}$. As usual, $\mathbb{F}^C$ denotes the filtration generated by the process $C$. It is natural to assume that $C$ is a $\mathbb{G}$-adapted process, so that $\mathbb{F}^C \subseteq \mathbb{G}$. The following definition is an obvious counterpart of Definition 11.1.3.

**Definition 11.3.1.** A process $C$ is called a *conditionally $\mathbb{G}$-Markov chain relative to $\mathbb{F}$ and under $\mathbb{Q}^*$* if for every $0 \leq t \leq s$ and any function $h : \mathcal{K} \to \mathbb{R}$ we have

$$\mathbb{E}_{\mathbb{Q}^*}(h(C_s) \,|\, \mathcal{G}_t) = \mathbb{E}_{\mathbb{Q}^*}(h(C_s) \,|\, \mathcal{F}_t \vee \sigma(C_t)). \tag{11.44}$$

For the sake of brevity, we say that $C$ is an *$\mathbb{F}$-conditional $\mathbb{G}$-Markov chain under $\mathbb{Q}^*$* if $C$ satisfies the above definition. If the reference filtration $\mathbb{F}$ is trivial, equality (11.44) reads:

$$\mathbb{E}_{\mathbb{Q}^*}(h(C_s) \,|\, \mathcal{G}_t) = \mathbb{E}_{\mathbb{Q}^*}(h(C_s) \,|\, \sigma(C_t)),$$

and thus Definition 11.3.1 coincides with Definition 11.2.1 of a $\mathbb{G}$-Markov chain. Moreover since $\sigma(C_t) \subseteq \mathcal{F}_t \vee \sigma(C_t) \subseteq \mathcal{G}_t$, it is easy to see that if a process $C$ is a $\mathbb{G}$-Markov chain under $\mathbb{Q}^*$, then $C$ is also an $\mathbb{F}$-conditional Markov chain under $\mathbb{Q}^*$, for any choice of a sub-filtration $\mathbb{F}$ of $\mathbb{G}$. On the other hand, when $C$ follows an $\mathbb{F}$-conditional $\mathbb{G}$-Markov chain under $\mathbb{Q}^*$, it is not necessarily a $\mathbb{G}$-Markov chain under $\mathbb{Q}^*$, in general.

*Example 11.3.1.* Let us first consider the most typical case when $\mathbb{G} = \mathbb{F} \vee \mathbb{F}^C$ for some reference filtration $\mathbb{F}$. Then (11.44) (and (11.46) below) becomes:

$$\mathbb{E}_{\mathbb{Q}^*}(h(C_s) \,|\, \mathcal{F}_t \vee \mathcal{F}_t^C) = \mathbb{E}_{\mathbb{Q}^*}(h(C_s) \,|\, \mathcal{F}_t \vee \sigma(C_t)). \tag{11.45}$$

If $\mathbb{F}^C$ is a sub-filtration of $\mathbb{F}$, condition (11.45) is trivially satisfied. Thus, in case when $\mathbb{G} = \mathbb{F} \vee \mathbb{F}^C$, it is natural to require that $\mathbb{F}^C$ is not a sub-filtration of a reference filtration $\mathbb{F}$.

*Remarks.* In case when $\tilde{\mathbb{G}}$ is an arbitrary filtration such that $\mathbb{F} \subseteq \tilde{\mathbb{G}}$, but the process $C$ is not $\tilde{\mathbb{G}}$-adapted, Definition 11.3.1 still makes sense, provided that we set $\mathbb{G} := \tilde{\mathbb{G}} \vee \mathbb{F}^C$. With this specification of the filtration $\mathbb{G}$, condition (11.44) takes the following form:

$$\mathbb{E}_{\mathbb{Q}^*}(h(C_s) \,|\, \tilde{\mathcal{G}}_t \vee \mathcal{F}_t^C) = \mathbb{E}_{\mathbb{Q}^*}(h(C_s) \,|\, \mathcal{F}_t \vee \sigma(C_t)). \tag{11.46}$$

We shall follow this convention in Sect. 11.3.1.

Let us also observe that the $\mathbb{F}$-conditional $\mathbb{G}$-Markov property (11.46) should not be confused with the following property: for every $0 \le t \le u \le s$ and any function $h : \mathcal{K} \to \mathbb{R}$,

$$\mathbb{E}_{\mathbb{Q}^*}\big(h(C_u)\,|\,\tilde{\mathcal{G}}_s \vee \mathcal{F}_t^C\big) = \mathbb{E}_{\mathbb{Q}^*}\big(h(C_u)\,|\,\tilde{\mathcal{G}}_s \vee \sigma(C_t)\big). \tag{11.47}$$

Intuitively speaking, the last property says that once a sample event within the filtration $\tilde{\mathcal{G}}_s$ is selected, then, conditionally on this event, the process $C_t$, $t \in [0, s]$, follows a (possibly time-inhomogeneous) Markov chain.

Finally, observe that property (11.47) yields the $\tilde{\mathbb{G}}$-conditional $\mathbb{G}$-Markov property, but not necessarily the $\mathbb{F}$-conditional $\mathbb{G}$-Markov property for some choice of the reference filtration $\mathbb{F}$. In other words, property (11.47) does not imply (11.46), in general. It is thus worthwhile to emphasize that a process $C$ obtained by means of the canonical construction presented in Sect. 11.3.1 has both property (11.46) and property (11.47).

Let $\Lambda_t^* = [\lambda_{ij}^*(t)]_{1 \le i,j \le K}$, $t \in \mathbb{R}_+$, denote an $\mathbb{F}$-progressively measurable, bounded, matrix-valued process (the boundedness is postulated for the sake of simplicity of presentation). The next definition follows in the footsteps of Propositions 11.2.1 and 11.2.2. For every $i \in \mathcal{K}$, $t \in \mathbb{R}_+$, and any function $h : \mathcal{K} \to \mathbb{R}$, we denote

$$\Lambda_t^* h(i) = \sum_{j=1}^{K} \lambda_{ij}^*(t)h(j).$$

**Definition 11.3.2.** An $\mathbb{F}$-progressively measurable, bounded, matrix-valued process $\Lambda^*$ is called an $\mathbb{F}$-*conditional infinitesimal generator* for a $\mathcal{K}$-valued $\mathbb{F}$-conditional $\mathbb{G}$-Markov chain $C$ under $\mathbb{Q}^*$ if for any function $h : \mathcal{K} \to \mathbb{R}$ the process $M^h$, given as

$$M_t^h = h(C_t) - h(C_0) - \int_0^t \Lambda_u^* h(C_u)\, du, \quad \forall\, t \in \mathbb{R}_+,$$

follows a $\mathbb{G}$-martingale under $\mathbb{Q}^*$.

In view of the natural interpretation of the process $\lambda_{ij}^*(t)$, $t \in \mathbb{R}_+$, as the $\mathbb{F}$-conditional intensity of transition from the state $i$ to the state $j$, the $\mathbb{F}$-conditional infinitesimal generator $\Lambda^*$ is also commonly referred to as the *matrix of stochastic intensities* for $C$ under $\mathbb{Q}^*$.

*Remarks.* It is important to stress the difference between the above definition and the martingale characterization stated in Proposition 11.2.2, which also deals with $\mathbb{G}$-Markov chains, but assumes a deterministic infinitesimal generator function $\Lambda^*(t)$, rather than a random infinitesimal generator process $\Lambda_t^*$. If a process $C$ is an $\mathbb{F}$-conditional $\mathbb{G}$-Markov chain under $\mathbb{Q}^*$ for some filtration $\mathbb{G}$, and if it admits a deterministic infinitesimal generator, in the sense of the last definition, then, by virtue of Proposition 11.2.2, $C$ is also a $\mathbb{G}$-Markov process under $\mathbb{Q}^*$.

*Example 11.3.2.* We return to the case of $\mathbb{G} = \mathbb{F} \vee \mathbb{F}^C$. Under an assumption that an $\mathbb{F}$-conditional infinitesimal generator process $\Lambda^*$ is given a priori, we make take the filtration $\mathbb{F}^{\Lambda^*}$ generated by $\Lambda^*$ as the reference filtration $\mathbb{F}$. Condition (11.45) then takes the following form

$$\mathbb{E}_{\mathbb{Q}^*}(h(C_s) \,|\, \mathcal{F}_t^{\Lambda^*} \vee \mathcal{F}_t^C) = \mathbb{E}_{\mathbb{Q}^*}(h(C_s) \,|\, \mathcal{F}_t^{\Lambda^*} \vee \sigma(C_t)). \qquad (11.48)$$

### 11.3.1 Construction of a Conditionally Markov Chain

We shall now provide a formal construction of an $\mathbb{F}$-conditional $\mathbb{G}$-Markov chain $C$ associated with a given infinitesimal generator. The construction given below is inspired, in particular, by Chap. 2 in Davis (1993), Sect. 7.3 in Last and Brandt (1995), and Sect. 2.3-2.4 in Yin and Zhang (1997).

We fix the underlying probability space $(\tilde{\Omega}, \mathcal{F}, \mathbb{P}^*)$, and we assume that it is endowed with the two filtrations, $\mathbb{F}$ and $\tilde{\mathbb{G}}$, satisfying the 'usual conditions' and such that $\mathbb{F} \subseteq \tilde{\mathbb{G}}$. We consider a $K \times K$ matrix $\Lambda^*$ of non-negative, bounded, $\mathbb{F}$-progressively measurable stochastic processes

$$\Lambda_t^* = \begin{pmatrix} \lambda_{1,1}^*(t) & \cdots & \lambda_{1,K-1}^*(t) & \lambda_{1,K}^*(t) \\ & \cdots & \cdot & \cdot \\ \lambda_{K-1,1}^*(t) & \cdots & \lambda_{K-1,K-1}^*(t) & \lambda_{K-1,K}^*(t) \\ 0 & \cdots & 0 & 0 \end{pmatrix}.$$

The matrix $\Lambda^*$ will play the role of the matrix of stochastic intensities. We assume that processes $\lambda_{ij}^*$, $i \neq j$, are non-negative and

$$\lambda_{ii}^*(t) = -\sum_{j \neq i} \lambda_{ij}^*(t), \quad \forall\, t \in \mathbb{R}_+.$$

Since the last row of the matrix $\Lambda_t^*$ is zero, the state $K$ will be an absorbing state for $C$ under $\mathbb{Q}^*$.

To construct an associated conditionally Markov chain, we need to enlarge the underlying probability space. To this end, we introduce two sequences, $U_{1,k}$, $U_{2,k}$, $k = 1, 2, \ldots$, of mutually independent random variables uniformly distributed on $[0, 1]$. We may and do assume that they are defined on a Hilbert cube $(\Omega^U, \mathcal{F}^U, \mathbb{P}^U)$ (see Sect. 23 in Davis (1993)). The generic elements of $\tilde{\Omega}$, $\Omega^U$ and of the set $\mathcal{K}$ are denoted by $\tilde{\omega}$, $\omega^U = (\omega_{1,1}^U, \omega_{2,1}^U, \omega_{1,2}^U, \omega_{2,2}^U, \omega_{1,3}^U, \ldots)$ and $i$, respectively. Assume that the initial law $\mu$ belongs to $\mu(\mathcal{K})$, where $\mu(\mathcal{K})$ stands for the set of all probability distributions on the space $\bar{\Omega} := \mathcal{K}$. Let $C_0 : \bar{\Omega} \to \mathcal{K}$ be a random variable distributed according to $\mu$. We may and do assume that $C_0(i) = i$ (since the generic element of $\bar{\Omega}$ is denoted by $\bar{\omega}$, we shall also write $C_0(\bar{\omega}) = \bar{\omega}$).

The following notation will be used throughout for the survival functions of the jump times of the process $C$ that we are going to construct,

$$G(t, i, \tilde{\omega}) := e^{\int_0^t \lambda_{ii}^*(v, \tilde{\omega})\, dv} = e^{-\int_0^t \lambda_i^*(v, \tilde{\omega})\, dv},$$

where we denote $\lambda_i^*(v, \tilde{\omega}) = -\lambda_{ii}^*(v, \tilde{\omega})$ for $i = 1, \ldots, K$.

We define an auxiliary mapping $\mathbb{T} : \mathcal{K} \times [0, \infty) \times [0, 1] \times \tilde{\Omega} \to [0, \infty]$ by setting (by convention, inf $\emptyset = \infty$)

$$\mathbb{T}(i, s, u, \tilde{\omega}) = \inf \left\{ t \geq 0 : \frac{G(t + s, i, \tilde{\omega})}{G(s, i, \tilde{\omega})} \leq u \right\}$$

or, equivalently,

$$\mathbb{T}(i, s, u, \tilde{\omega}) = \inf \left\{ t \geq 0 : e^{- \int_s^{t+s} \lambda_i^*(v, \tilde{\omega}) \, dv} \leq u \right\}.$$

Let $\mathbb{C} : [0, 1] \times \mathcal{K} \times [0, \infty) \times \tilde{\Omega} \to \mathcal{K}$ be any mapping such that, for every $i, j \in \mathcal{K}, j \neq i$,

$$\ell(\{u \in [0, 1] : \mathbb{C}(u, i, t, \tilde{\omega}) = j\}) = \begin{cases} \frac{\lambda_{ij}^*(t, \tilde{\omega})}{\lambda_i^*(t, \tilde{\omega})}, & \lambda_i^*(t, \tilde{\omega}) > 0, \\ 0, & \lambda_i^*(t, \tilde{\omega}) = 0, \end{cases}$$

where $\ell(A)$ stands for the Lebesgue measure of the set $A$. Finally, we define the enlarged probability space by setting:[7]

$$(\Omega, \mathcal{G}, \mathbb{Q}^*) = (\tilde{\Omega} \times \Omega^U \times \bar{\Omega}, \tilde{\mathcal{G}}_\infty \otimes \mathcal{F}^U \otimes 2^{\mathcal{K}}, \mathbb{P}^* \otimes \mathbb{P}^U \otimes \mu).$$

**Step 1: Construction of the 1ˢᵗ jump time.** Let $\tau_0 := 0$. We define (for brevity, we shall frequently write simply $\omega$ instead of $(\tilde{\omega}, \omega^U, \bar{\omega})$)

$$\eta_1(\omega) = \eta_1(\tilde{\omega}, \omega_{1,1}^U, \bar{\omega}) := \mathbb{T}(\bar{\omega}, 0, U_{1,1}(\omega_{1,1}^U), \tilde{\omega})$$

or, more explicitly,

$$\eta_1(\omega) = \inf \left\{ t \geq 0 : e^{- \int_0^t \lambda_{C_0}^*(v, \tilde{\omega}) \, dv} \leq U_{1,1} \right\}.$$

Put another way,

$$\eta_1(\omega) = \inf \left\{ t \geq 0 : \int_0^t \lambda_{C_0}^*(v, \tilde{\omega}) \, dv \geq \tilde{e}_{1,1} \right\},$$

where $\tilde{e}_{1,1} := - \ln U_{1,1}$ is a unit exponential random variable. We define the first jump time $\tau_1$ by setting $\tau_1 := \tau_0 + \eta_1$ so that $\tau_1 = \tau_1(\tilde{\omega}, \omega_{1,1}^U, \bar{\omega})$. It is thus clear that $\tau_1$ is a random variable on $(\Omega, \mathcal{G}, \mathbb{Q}^*)$.

In fact, $\tau_1$ only depends on the following variables: $\tilde{\omega}$, $\omega_{1,1}^U$ and $\bar{\omega}$. It is also clearly seen that, for every $t > 0$,

$$\mathbb{Q}^* \{\tau_1 > t \,|\, \mathcal{F}_t \vee \sigma(C_0)\}(\omega) = \frac{G(t, C_0, \tilde{\omega})}{G(0, C_0, \tilde{\omega})} = e^{- \int_0^t \lambda_{C_0}^*(v, \tilde{\omega}) \, dv}.$$

---

[7] Filtrations defined on the component subspaces are extended to the enlarged space in an obvious way and their denotation is preserved. So, for example, the filtration $\mathbb{F}$ defined on $(\tilde{\Omega}, \mathcal{F}, \mathbb{P}^*)$ is extended to $(\Omega, \mathcal{G}, \mathbb{Q}^*)$, and is still denoted as $\mathbb{F}$.

Consequently,

$$Q^*\{\tau_1 > t\} = \mathbb{E}_{Q^*}\big(G(t, C_0, \tilde{\omega})\big) = \mathbb{E}_{Q^*}\Big(e^{-\int_0^t \lambda_{C_0}^*(v,\tilde{\omega})\,dv}\Big).$$

Let us observe that the uniform boundedness of the processes $\lambda_{ij}^*$ implies that $Q^*\{\tau_1 = 0\} = 0$. Finally, since by assumption $\int_0^\infty \lambda_i^*(t)\,dt = \infty$ for any $i = 1, \ldots, K-1$ we have $Q^*\{\tau_1 < \infty\} = 1$.

We shall now check that $Q^*\{\lambda_{C_0}^*(\tau_1) = 0\} = 0$ or, equivalently, that the equality

$$Q^*\{\lambda_i^*(\tau_1) = 0, C_0 = i\} = 0$$

is valid for every $i = 1, \ldots, K-1$. From the construction of the jump time $\tau_1$, it can be easily deduced that for any bounded, $\mathbb{F}$-adapted stochastic process $Z$ we have

$$\mathbb{E}_{Q^*}\big(\mathbb{1}_{\{C_0=i\}} Z_{\tau_1(\omega)}(\tilde{\omega})\big) = \mathbb{E}_{Q^*}\Big(\mathbb{1}_{\{C_0=i\}} \int_0^\infty Z_t \lambda_i^*(t) e^{-\int_0^t \lambda_i^*(s)\,ds}\,dt\Big).$$

By applying the last formula to the bounded, $\mathbb{F}$-adapted process $Z_t = \mathbb{1}_{\tilde{B}}(t)$, where $\tilde{B} = \{(t, \tilde{\omega}) : \lambda_i^*(t, \tilde{\omega}) = 0\}$, we obtain

$$Q^*\{\lambda_i^*(\tau_1) = 0, C_0 = i\} = Q^*\{(\tau_1(\omega), \tilde{\omega}) \in \tilde{B}, C_0 = i\} = 0.$$

**Step 2: Construction of the 1$^{\text{st}}$ jump.** For any $\omega = (\tilde{\omega}, \omega^U, \bar{\omega})$, we define $\bar{C}_1(\omega)$ by setting

$$\bar{C}_1(\omega) = \bar{C}_1(\tilde{\omega}, \omega_{1,1}^U, \omega_{2,1}^U, \bar{\omega}) := \mathbb{C}(U_{2,1}(\omega_{2,1}^U), C_0(\bar{\omega}), \tau_1(\tilde{\omega}, \omega_{1,1}^U, \bar{\omega}), \tilde{\omega}).$$

It is clear that $\bar{C}_1$ is a random variable on $(\Omega, \mathcal{G}, Q^*)$. Also, it is apparent that $\bar{C}_1$ depends on $\tilde{\omega}$, $\omega_{1,1}^U$, $\omega_{2,1}^U$ and $\bar{\omega}$ only. Moreover, we have

$$Q^*\{\bar{C}_1 = j \,|\, \mathcal{G}_{\tau_1}^{1,0}\}(\tilde{\omega}, \omega_{1,1}^U, \bar{\omega}) = \frac{\lambda_{C_0,j}^*(\tau_1(\tilde{\omega}, \omega_{1,1}^U, \bar{\omega}), \tilde{\omega})}{\lambda_{C_0}^*(\tau_1(\tilde{\omega}, \omega_{1,1}^U, \bar{\omega}), \tilde{\omega})},$$

where we set $\mathcal{G}_t^{1,0} = \mathcal{F}_t \vee \mathcal{H}_t^1 \vee \sigma(C_0)$, where $\mathcal{H}_t^1 = \sigma\left(\mathbb{1}_{\{\tau_1 \leq s\}} : 0 \leq s \leq t\right)$ (notice that $\sigma(\tau_1) \subset \mathcal{H}_{\tau_1}^1$).

**Step 3: Construction of the 2$^{\text{nd}}$ jump time.** In order to define the second jump time, we first set

$$\eta_2(\omega) = \eta_2(\tilde{\omega}, \omega_{1,1}^U, \omega_{2,1}^U, \omega_{1,2}^U, \bar{\omega})$$
$$= \mathbb{T}(\bar{C}_1(\tilde{\omega}, \omega_{1,1}^U, \omega_{2,1}^U, \bar{\omega}), \tau_1(\tilde{\omega}, \omega_{1,1}^U, \bar{\omega}), U_{1,2}(\omega_{1,2}^U), \tilde{\omega}).$$

More explicitly,

$$\eta_2 = \inf\Big\{t \geq 0 : e^{-\int_{\tau_1}^{\tau_1+t} \lambda_{\bar{C}_1}^*(v)\,dv} \leq U_{1,2}\Big\}$$

or, equivalently,

$$\eta_2 = \inf\left\{ t \geq 0 : \int_0^t \lambda^*_{\bar{C}_1}(v)\,dv \geq \tilde{e}_{1,2} \right\},$$

where $\tilde{e}_{1,2} := -\ln U_{1,2}$. As expected, we define the time of the second jump by setting $\tau_2 := \tau_1 + \eta_2$. The random variable $\tau_2$, defined on the probability space $(\Omega, \mathcal{G}, \mathbb{Q}^*)$, depends only on the following variables: $\tilde{\omega}, \omega^U_{1,1}, \omega^U_{2,1}, \omega^U_{1,2}$, and $\bar{\omega}$. Again, it can be easily verified that

$$\mathbb{Q}^*\{\eta_2 > t \,|\, \mathcal{F}_{t+\tau_1} \vee \mathcal{H}^1_{\tau_1} \vee \sigma(\bar{C}_1)\}(\omega) = \frac{G(t + \tau_1(\tilde{\omega}, \omega^U_{1,1}, \bar{\omega}), \bar{C}_1, \tilde{\omega})}{G(\tau_1(\tilde{\omega}, \omega^U_{1,1}, \bar{\omega}), \bar{C}_1, \tilde{\omega})},$$

where $\bar{C}_1 = \bar{C}_1(\tilde{\omega}, \omega^U_{1,1}, \omega^U_{2,1}, \bar{\omega})$, and so

$$\mathbb{Q}^*\{\eta_2 > t\} = \mathbb{E}_{\mathbb{Q}^*}\left( \frac{G(t + \tau_1(\tilde{\omega}, \omega^U_{1,1}, \bar{\omega}), \bar{C}_1, \tilde{\omega})}{G(\tau_1(\tilde{\omega}, \omega^U_{1,1}, \bar{\omega}), \bar{C}_1, \tilde{\omega})} \right).$$

Arguing along similar lines as in Step 1, one may easily check that the following equalities hold true: $\mathbb{Q}^*\{\eta_2 = 0\} = 0$, $\mathbb{Q}^*\{\eta_2 < \infty\} = 1$, and $\mathbb{Q}^*\{\lambda^*_{\bar{C}_1}(\tau_2) = 0\} = 0$.

**Step 4: Construction of the 2$^{\text{nd}}$ jump.** The random variable $\bar{C}_2$ is defined through the following expression:

$$\bar{C}_2(\omega) := \mathbb{C}(U_{2,2}(\omega^U_{2,2}), \bar{C}_1(\tilde{\omega}, \omega^U_{1,1}, \omega^U_{2,1}, \bar{\omega}), \eta_2(\tilde{\omega}, \omega^U_{1,1}, \omega^U_{2,1}, \omega^U_{1,2}, \bar{\omega}), \tilde{\omega}).$$

As in Step 2, it can be checked that

$$\mathbb{Q}^*\{\bar{C}_2 = j \,|\, \mathcal{G}^{2,1}_{\tau_2}\} = \frac{\lambda^*_{\bar{C}_1, j}(\tau_2)}{\lambda^*_{\bar{C}_1}(\tau_2)},$$

where we set $\mathcal{G}^{2,1}_t = \mathcal{F}_t \vee \mathcal{H}^2_t \vee \sigma(\bar{C}_1)$, and $\mathcal{H}^2_t = \sigma\left(\mathbb{1}_{\{\tau_2 \leq s\}} : 0 \leq s \leq t\right)$ (notice that $\sigma(\tau_2) \subset \mathcal{H}^2_{\tau_1}$).

**Step 5: Construction of the $k^{\text{th}}$ jump time and the $k^{\text{th}}$ jump.** In a similar way as in previous steps, we may construct the $k^{\text{th}}$ jump time $\tau_k = \tau_{k-1} + \eta_k$ as well as the $k^{\text{th}}$ jump $\bar{C}_k$ for the process $C$. More specifically, for every $t > 0$ we have

$$\mathbb{Q}^*\{\eta_k > t \,|\, \mathcal{F}_{t+\tau_{k-1}} \vee \mathcal{H}^{k-1}_{\tau_{k-1}} \vee \sigma(\bar{C}_{k-1})\} = \frac{G(t + \tau_{k-1}, \bar{C}_{k-1})}{G(\tau_{k-1}, \bar{C}_{k-1})}$$

and

$$\mathbb{Q}^*\{\bar{C}_k = j \,|\, \mathcal{G}^{k,k-1}_{\tau_k}\} = \frac{\lambda^*_{\bar{C}_{k-1}, j}(\tau_k)}{\lambda^*_{\bar{C}_{k-1}}(\tau_k)},$$

where we write

$$\mathcal{G}^{k,k-1}_t = \mathcal{F}_t \vee \mathcal{H}^k_t \vee \sigma(\bar{C}_{k-1})$$

and

$$\mathcal{H}^k_t = \sigma\left(\mathbb{1}_{\{\tau_k \leq s\}} : 0 \leq s \leq t\right).$$

Let us finally observe that, in view of the assumed uniform boundedness of processes $\lambda^*_{ij}$, we have: $\tau_k \to \infty$ with probability 1 as $k$ tends to $\infty$.

**Step 6: Construction of $C$.** To obtain a conditionally Markov chain $C$ with values in the state space $\mathcal{K}$, it suffices to set $C_t := \bar{C}_{k-1}$ for $t \in [\tau_{k-1}, \tau_k)$ and any $k \geq 1$. This achieves the canonical construction of an $\mathbb{F}$-conditional $\mathbb{G}$-Markov chain associated with a given $\mathbb{F}$-adapted, matrix-valued, absolutely continuous stochastic process $\varLambda_t^*$, $t \in \mathbb{R}_+$.

*Remarks.* The one-dimensional migration process, which will be used in Chap. 12 to model credit ratings, can be constructed as the process $C$. On the other hand, in the framework of Chap. 13, we shall use a two-dimensional migration process, which will be specified as follows: the first component of the two-dimensional migration process is the process $C$ defined above. By definition, the second component of the migration process, denoted by $\hat{C}$, equals

$$\hat{C}_t = \begin{cases} C_0, & \text{if } t \in [0, \tau_2) \\ \bar{C}_{k-1}, & \text{if } t = [\tau_k, \tau_{k+1}), \ k \geq 2. \end{cases}$$

In this way, we obtain a two-dimensional migration process $\tilde{C}_t = (C_t, \hat{C}_t)$ with the finite state space $\mathcal{K} \times \mathcal{K}$.

### 11.3.2 Conditional Markov Property

Let $\tilde{\mathcal{F}}_t = \sigma(\tilde{C}_s : 0 \leq s \leq t)$, $t \in \mathbb{R}_+$, stand for the natural filtration generated by the process $\tilde{C}$. We shall now verify that the process $\tilde{C}$ has an $\mathbb{F}$-conditional $\mathbb{G}$-Markov property, where the enlarged filtration $\mathbb{G}$ is set to satisfy $\mathbb{G} = \tilde{\mathbb{G}} \vee \tilde{\mathbb{F}}$. It will be also clear that the process $C$ has the same property as well.

**Lemma 11.3.1.** *Let us denote $D = \{\tau_k \leq t < \tau_{k+1}\}$. For any $k = 0, 1, \ldots$ and for any $t \geq 0$ we have*

$$\mathbb{1}_D \, \mathbb{Q}^* \{ \tilde{C}_{\tau_{k+1}} = (j, i) \, | \, \mathcal{G}_t \} = \mathbb{1}_D \, \mathbb{Q}^* \{ \tilde{C}_{\tau_{k+1}} = (j, i) \, | \, \mathcal{F}_t \vee \sigma(\tilde{C}_t) \}.$$

*Proof.* From the construction the process $\tilde{C}$, it follows that jumps at random times $\tau_k$ only, so that

$$\begin{aligned} J &= \mathbb{1}_D \, \mathbb{Q}^* \{ \tilde{C}_{\tau_{k+1}} = (j, i) \, | \, \tilde{\mathcal{G}}_t \vee \tilde{\mathcal{F}}_t \} \\ &= \mathbb{1}_D \, \mathbb{Q}^* \{ \tilde{C}_{\tau_{k+1}} = (j, i) \, | \, \mathcal{F}_t \vee \tilde{\mathcal{F}}_t \} \\ &= \mathbb{1}_D \, \mathbb{Q}^* \{ \tilde{C}_{\tau_{k+1}} = (j, C_{\tau_k}) \, | \, \mathcal{F}_t \vee \tilde{\mathcal{F}}_{\tau_k} \}. \end{aligned}$$

Consequently,

$$\begin{aligned} J &= \mathbb{1}_D \, \mathbb{Q}^* \{ \tilde{C}_{\tau_{k+1}} = (j, C_{\tau_k}) \, | \, \mathcal{F}_t \vee \tilde{\mathcal{F}}_{\tau_k} \} \\ &= \mathbb{1}_D \, \mathbb{E}_{\mathbb{Q}^*} \left( \mathbb{Q}^* \{ \tilde{C}_{\tau_{k+1}} = (j, C_{\tau_k}) \, | \, \mathcal{G}_{\tau_{k+1}}^{k+1,k} \vee \tilde{\mathcal{F}}_{\tau_k} \} \, \Big| \, \mathcal{F}_t \vee \tilde{\mathcal{F}}_{\tau_k} \right) \\ &= \mathbb{1}_D \, \mathbb{E}_{\mathbb{Q}^*} \left( \frac{\lambda_{\bar{C}_k, j}^*(\tau_{k+1})}{\lambda_{\bar{C}_k}^*(\tau_{k+1})} \, \Big| \, \mathcal{F}_t \vee \tilde{\mathcal{F}}_{\tau_k} \right) \\ &= \mathbb{1}_D \, \mathbb{E}_{\mathbb{Q}^*} \left( \frac{\lambda_{\bar{C}_k, j}^*(\tau_{k+1})}{\lambda_{\bar{C}_k}^*(\tau_{k+1})} \, \Big| \, \mathcal{F}_t \vee \sigma(\tilde{C}_{\tau_k}) \right) \end{aligned}$$

since $\mathcal{G}^{k+1,k}_{\tau_{k+1}} := \mathcal{F}_{\tau_{k+1}} \vee \mathcal{H}^{k+1}_{\tau_{k+1}} \vee \sigma(\bar{C}_k)$ and $C_t = C_{\tau_k} = \bar{C}_k$ on the random interval $[\tau_k, \tau_{k+1})$. On the other hand, by reasoning similarly as above, we obtain

$$\mathbb{1}_D \, \mathbb{Q}^* \big\{ \tilde{C}_{\tau_{k+1}} = (j,i) \,\big|\, \mathcal{F}_t \vee \sigma(\tilde{C}_t) \big\} = \mathbb{1}_D \, \mathbb{E}_{\mathbb{Q}^*} \Big( \frac{\lambda^*_{\bar{C}_k, j}(\tau_{k+1})}{\lambda^*_{\bar{C}_k}(\tau_{k+1})} \,\Big|\, \mathcal{F}_t \vee \sigma(\tilde{C}_{\tau_k}) \Big).$$

This proves the lemma.    □

For any $t \geq 0$, we set $\tilde{\tau}(t) := \inf \{ u \geq t : \tilde{C}_u \neq \tilde{C}_{u-} \}$. By virtue of Lemma 11.3.1, for every $t \in \mathbb{R}_+$ we have

$$\begin{aligned} \mathbb{Q}^* \big\{ \tilde{C}_{\tilde{\tau}(t)} = (j,i) \,\big|\, \mathcal{G}_t \big\} &= \mathbb{Q}^* \big\{ \tilde{C}_{\tilde{\tau}(t)} = (j,i) \,\big|\, \mathcal{F}_t \vee \tilde{\mathcal{F}}_t \big\} \\ &= \mathbb{Q}^* \big\{ \tilde{C}_{\tilde{\tau}(t)} = (j,i) \,\big|\, \mathcal{F}_t \vee \sigma(\tilde{C}_t) \big\}. \end{aligned}$$

The last equality makes it clear that the conditional Markov property of $\tilde{C}$ with respect to the filtration $\mathbb{F}$ is indeed satisfied (cf. Definition 11.3.1).

### 11.3.3 Associated Local Martingales

We need to introduce some notation. Let $Z$ be marked point process:

$$Z := \big\{ (\tau_k, \bar{C}_k), \, k = 0, 1, \dots \big\}.$$

For every $j \in \mathcal{K}$ and $t \geq 0$, we define

$$\Phi(t,j) = \sum_{k=1}^{\infty} \mathbb{1}_{\{ \tau_k \leq t \,:\, \bar{C}_k = j \}}$$

and

$$\nu(t,j) = \int_0^t \lambda^*_{C_{u-}, j}(u) \, du.$$

Thus, the process $\Phi(\cdot, j)$ is the counting process associated with the marked point process $Z$. Finally, we shall write $q(t,j) = \Phi(t,j) - \nu(t,j)$. Then have the following auxiliary result.

**Lemma 11.3.2.** *For any $j \in \mathcal{K}$, the process $\nu(\cdot, j)$ is the compensator of the increasing process $\Phi(\cdot, j)$. In other words, the process $q(t \wedge \tau_k, j)$, $t \in \mathbb{R}_+$, is a $\mathbb{G}$-martingale for every $k$ and every $j \in \mathcal{K}$.*

*Proof.* We shall first verify that $q(t \wedge \tau_1, j)$ is a $\mathbb{G}$-martingale. Fix $0 \leq s < t$. Let us denote

$$J_s(\Phi) = \mathbb{E}_{\mathbb{Q}^*} \Big( \Phi(t \wedge \tau_1, j) - \Phi(s \wedge \tau_1, j) \,\Big|\, \mathcal{G}_s \Big)$$

and

$$J_s(\nu) = \mathbb{E}_{\mathbb{Q}^*} \Big( \nu(t \wedge \tau_1, j) - \nu(s \wedge \tau_1, j) \,\Big|\, \mathcal{G}_s \Big).$$

For $J_s(\Phi)$, we obtain

$$
\begin{aligned}
J_s(\Phi) &= \mathbb{1}_{\{s<\tau_1\}} \mathbb{E}_{\mathbb{Q}^*}\left(\mathbb{1}_{\{t\geq\tau_1\}}\mathbb{1}_{\{\bar{C}_1=j\}} - \mathbb{1}_{\{s\geq\tau_1\}}\mathbb{1}_{\{\bar{C}_1=j\}}\,\Big|\,\mathcal{G}_s\right) \\
&= \mathbb{1}_{\{s<\tau_1\}} \mathbb{E}_{\mathbb{Q}^*}\left(\mathbb{1}_{\{t\geq\tau_1\}}\mathbb{1}_{\{\bar{C}_1=j\}}\,\Big|\,\mathcal{G}_s\right) \\
&= \mathbb{1}_{\{s<\tau_1\}} \mathbb{E}_{\mathbb{Q}^*}\left(\frac{\lambda_{ij}^*(\tau_1)}{\lambda_i^*(\tau_1)}\,\Big|\,\mathcal{G}_s\right) - \mathbb{1}_{\{s<\tau_1\}}\mathbb{E}_{\mathbb{Q}^*}\left(\mathbb{1}_{\{t<\tau_1\}}\frac{\lambda_{ij}^*(\tau_1)}{\lambda_i^*(\tau_1)}\,\Big|\,\mathcal{G}_s\right) \\
&= \mathbb{1}_{\{s<\tau_1\}}\left\{\mathbb{E}_{\mathbb{Q}^*}\left(\int_s^\infty \lambda_{ij}^*(r)e^{\int_s^r \lambda_{ij}^*(u)\,du}\,dr\,\Big|\,\mathcal{G}_s\right)\right. \\
&\qquad \left. - \mathbb{E}_{\mathbb{Q}^*}\left(\int_t^\infty \lambda_{ij}^*(r)e^{\int_s^r \lambda_{ij}^*(u)\,du}\,dr\,\Big|\,\mathcal{G}_s\right)\right\} \\
&= \mathbb{1}_{\{s<\tau_1\}}\mathbb{E}_{\mathbb{Q}^*}\left(\int_s^t \lambda_{ij}^*(r)e^{\int_s^r \lambda_{ij}^*(u)\,du}\,dr\,\Big|\,\mathcal{G}_s\right).
\end{aligned}
$$

On the other hand, $J_s(\nu)$ equals

$$
\begin{aligned}
J_s(\nu) &= \mathbb{1}_{\{s<\tau_1\}}\mathbb{E}_{\mathbb{Q}^*}\left(\int_s^{t\wedge\tau_1} \lambda_{ij}^*(r)\,dr\,\Big|\,\mathcal{G}_s\right) \\
&= \mathbb{1}_{\{s<\tau_1\}}\mathbb{E}_{\mathbb{Q}^*}\left(\int_s^\infty \left(\int_s^{t\wedge u}\lambda_{ij}^*(r)\,dr\right)\lambda_i^*(u)e^{\int_s^u \lambda_{ij}^*(v)\,dv}\,du\,\Big|\,\mathcal{G}_s\right) \\
&= \mathbb{1}_{\{s<\tau_1\}}\mathbb{E}_{\mathbb{Q}^*}\left(\int_s^t \left(\int_s^u\lambda_{ij}^*(r)\,dr\right)\lambda_i^*(u)e^{\int_s^u \lambda_{ij}^*(v)\,dv}\,du\,\Big|\,\mathcal{G}_s\right) \\
&\quad + \mathbb{1}_{\{s<\tau_1\}}\mathbb{E}_{\mathbb{Q}^*}\left(\int_t^\infty \left(\int_s^t\lambda_{ij}^*(r)\,dr\right)\lambda_i^*(u)e^{\int_s^u \lambda_{ij}^*(v)\,dv}\,du\,\Big|\,\mathcal{G}_s\right) \\
&= \mathbb{1}_{\{s<\tau_1\}}\mathbb{E}_{\mathbb{Q}^*}\left(\int_s^t \lambda_{ij}^*(r)e^{\int_s^r \lambda_{ij}^*(u)\,du}\,dr\,\Big|\,\mathcal{G}_s\right).
\end{aligned}
$$

We conclude that the process $q(t\wedge\tau_1,j)$ follows a $\mathbb{G}$-martingale under $\mathbb{Q}^*$. Using an analogous reasoning, combined with the conditional Markov property and the concatenation argument, one may prove the martingale property of the process $q(t\wedge\tau_k,j)$ for any $k>1$. Details are left to the reader.   □

Consider a bounded, measurable mapping $g:\mathcal{K}\times[0,\infty)\times\Omega\to\mathbb{R}$. In addition, assume that for any $i\in\mathcal{K}$ the process $g(i,t)$ is $\mathbb{G}$-predictable. Let us define the associated adapted process $M^g$ by setting

$$
M_t^g = \int_0^t \sum_{i=1}^K g(i,u)q(du,i), \quad \forall t\in\mathbb{R}_+. \tag{11.49}
$$

The following useful corollary is a straightforward consequence of Lemma 11.3.2, and thus its proof is omitted.

**Corollary 11.3.1.** *The process $M^g$ given by formula (11.49) follows a $\mathbb{G}$-martingale under the probability measure $\mathbb{Q}^*$.*

We are in a position to show that the process $\Lambda^*$ represents the conditional infinitesimal generator process or, equivalently, the matrix-valued process of stochastic intensities for the first component of the process $\tilde{C}$, i.e., for the $\mathcal{K}$-valued process $C$.

To this end, let us first recall that for any function $h : \mathcal{K} \to \mathbb{R}$ we denote, for every $i \in \mathcal{K}$ and $t \in \mathbb{R}_+$,

$$\Lambda_t^* h(i) = \sum_{j=1}^K \lambda_{ij}^*(t) h(j).$$

The next result is a counterpart of Proposition 11.2.2.

**Proposition 11.3.1.** *For any function $h : \mathcal{K} \to \mathbb{R}$, the process $M^h$, given by the formula*

$$M_t^h = h(C_t) - \int_0^t \Lambda_u^* h(C_u) \, du, \quad \forall \, t \in \mathbb{R}_+, \tag{11.50}$$

*is a $\mathbb{G}$-martingale under $\mathbb{Q}^*$.*

*Proof.* To prove the proposition, it suffices to apply Corollary 11.3.1 to the function $g(i, t, \omega) = h(i) - h(C_{t-}(\omega))$. $\qquad\square$

In view of Definition 11.3.2, the last proposition demonstrates that the process $\Lambda^*$ is indeed a matrix-valued process of stochastic intensities associated with the $\mathbb{F}$-conditional Markov chain $C$.

We now verify the martingale property for some auxiliary processes that were used earlier in this chapter. Let us define (cf. (11.33))

$$M_t^{ij} := H_t^{ij} - \int_0^t \lambda_{ij}^*(u) H_u^i \, du, \tag{11.51}$$

where $H_t^i = \mathbb{1}_{\{C_t = i\}}$ and $H_t^{ij}$ represents the number of one-step transitions from the class $i$ to $j$ by the migration process $C$ over the time interval $(0, t]$. The following result is another important consequence of Lemma 11.3.2 (cf. Corollary 7.5.3 in Last and Brandt (1995)).

**Corollary 11.3.2.** (i) *Let $h$ be a real valued function on $\mathcal{K} \times \mathcal{K}$. Then the process $N^h$ defined as*

$$N_t^h = \sum_{0 < u \leq t} h(C_{u-}, C_u) - \int_0^t \sum_{k \neq C_u} \lambda_{C_u, k}^*(u) h(C_u, k) \, du$$

*is a $\mathbb{G}$-martingale.*
(ii) *For any $i, j \in \mathcal{K}$, $i \neq j$, the process $M^{ij}$ given by formula (11.51) follows a $\mathbb{G}$-martingale.*

*Proof.* To establish part (i), it is enough to apply Corollary 11.3.1 to the function $g(k, t, \omega) = h(C_{t-}(\omega), k)$. The second statement follows from the first, upon setting $h(c, c') = \delta_{ic}\delta_{ic'}$. $\qquad\square$

### 11.3.4 Forward Kolmogorov Equation

Let $\mathcal{P}^*(t, s)$ be the $\mathbb{F}$-conditional transition probability matrix for the process $C$ under the probability measure $\mathbb{Q}^*$, specifically, for $t \leq s$,

$$\mathcal{P}^*(t, s) := \left[p_{ij}^*(t, s)\right]_{1 \leq i, j \leq K},$$

where, for every $i, j = 1, \ldots, K$,

$$p_{ij}^*(t, s) := \mathbb{Q}^*\left\{C_s = j \,\middle|\, \mathcal{F}_t \vee \{C_t = i\}\right\}.$$

Let us now define another matrix-valued process: for $t \leq s$, we set

$$\mathbf{P}^*(t, s) := \left[\mathbf{p}_{ij}^*(t, s)\right]_{1 \leq i, j \leq K},$$

where, for every $i, j = 1, \ldots, K$,

$$\mathbf{p}_{ij}^*(t, s) := \mathbb{Q}^*\left\{C_s = j \,\middle|\, \mathcal{F}_{T^*} \vee \{C_t = i\}\right\}.$$

Given our construction of the process $C$ it is clear that for a fixed $\tilde{\omega}$ the process $\tilde{C}$, defined as $\tilde{C}_t(\omega^U, \bar{\omega}) = C_t(\tilde{\omega}, \omega^U, \bar{\omega})$ is an ordinary, time-inhomogeneous Markov chain, and that $\Lambda_t^*(\tilde{\omega})$ is the associated infinitesimal generator function, so that

$$\frac{d\mathbf{P}^*(t, s)(\tilde{\omega})}{ds} = \mathbf{P}^*(t, s)(\tilde{\omega})\Lambda_s^*(\tilde{\omega}), \quad t \leq s \leq T^*.$$

Taking into account the set of regularity conditions that were imposed on the matrix process $\Lambda^*$ of transition intensities, it is also possible to establish the following result, which is a counterpart of Corollary 11.2.2.

**Corollary 11.3.3.** *For every $0 \leq t \leq s \leq T^*$ we have*

$$\mathbf{P}^*(t, s)(\tilde{\omega}) = \mathrm{Id} + \sum_{n=1}^{\infty} \int_t^s \int_{u_1}^s \cdots \int_{u_{n-1}}^s \Lambda_{u_1}^*(\tilde{\omega}) \ldots \Lambda_{u_n}^*(\tilde{\omega}) \, du_n \ldots du_1.$$

Now, observe that we have the following equality:

$$\mathcal{P}^*(t, s) = \mathbb{E}_{\mathbb{Q}^*}\left\{\mathbf{P}^*(t, s) \,\middle|\, \mathcal{F}_t\right\}, \quad \forall t \leq s \leq T^*,$$

which, together with Corollary 11.3.3 leads to

**Corollary 11.3.4.** *For every $0 \leq t \leq s \leq T^*$ we have*

$$\mathcal{P}^*(t, s) = \mathrm{Id} + \mathbb{E}_{\mathbb{Q}^*}\left\{\sum_{n=1}^{\infty} \int_t^s \int_{u_1}^s \cdots \int_{u_{n-1}}^s \Lambda_{u_1}^* \ldots \Lambda_{u_n}^* \, du_n \ldots du_1 \,\middle|\, \mathcal{F}_t\right\}.$$

# 12. Markovian Models of Credit Migrations

In Chap. 8, the reduced-form approach was applied to the study of a particular kind of a credit event, namely, the default. In this chapter, we shall consider several possible credit events within the framework of the intensity-based methodology. More specifically, we are going to examine the issue of dynamical modeling of credit migrations of a corporate bond between several possible *rating grades* (or *credit ratings*). In other words, we shall focus on the modeling of changes over time in the credit quality of reference names; such changes are henceforth referred to as *credit migrations*.

The most popular way of modeling credit migrations is in terms of either discrete- or continuous-time Markov chains (or conditionally Markov chains). The aim of this chapter is thus to discuss several Markov-type models of credit migrations that have been proposed by various authors in recent years. For the sake of expositional clarity, we restrict ourselves to the issue of credit migration of corporate zero-coupon bonds. The exposition in this chapter is based on papers by Jarrow and Turnbull (1995), Jarrow et al. (1997), Das and Tufano (1996), Kijima (1998), Kijima and Komoribayashi (1998), Lando (1998, 2000a, 2000b), Thomas et al. (1998), Arvanitis et al. (1999) and Duffie and Singleton (1999). Of course, due to the limited space, it is not possible to present in detail all issues and results related to the modeling of credit migrations. In fact, we provide a relatively detailed presentation of the Jarrow, Lando and Turnbull approach, but we comment rather succinctly on alternative approaches.

A brief account of mathematical results underlying the models presented in this chapter can be found in Chap. 11; as a rule, we preserve here the notation introduced in this chapter. Let us only mention that the symbol $\mathbb{G}$ will be used as a generic symbol to denote the underlying filtration. Thus, depending on a particular model, the symbol $\mathbb{G}$ will be given a specific interpretation. Since an equivalent change of a probability measure will play an important role in the sequel, in Chap. 12 and 13 we adopt the following notational convention: the intensities of migrations under the real-world probability $\mathbb{Q}$ and under the spot martingale measure $\mathbb{Q}^*$ will be denoted by $\lambda_{ij}$ and $\lambda_{ij}^*$, respectively. Finally, let us add that the important issue of statistical verification of the validity of various types of Markovian modeling postulates is not discussed in detail in the present text.

## 12.1 JLT Markovian Model and its Extensions

A firm's credit rating is a measure of the firm's propensity to default. Credit ratings are typically identified with elements of a certain set, referred to as the set of *credit classes* (or *credit grades*). In financial literature, the credit classes are typically understood as the credit ratings attributed by a commercial rating agency. A particular credit rating assigned to a corporate bond at a given time thus reflects the credit quality of a corporation as perceived by specialized financial analysts at this time. This does not mean, however, that in the theoretical approach the credit ratings should necessarily be understood as being attributed by a commercial rating agency. Indeed, most of the major financial institutions maintain their own credit rating systems, based on internally developed methodologies (the so-called *internal ratings*). In addition, the official credit ratings primarily reflect the likelihood of default, as opposed to a more general concept of debt's quality. Finally, if changes in the firm's credit quality occur, the upgrade or downgrade in the 'official' rating occurs with a considerable delay. To sum up, in the present text, the generic term *credit rating* (or *credit quality*) refers to any kind of a debt classification that can be justified for specific purposes.

Formally, we postulate that the credit quality of corporate debt is quantified and categorized into a finite number of (mutually disjoint) *credit rating classes* (*credit classes,* for short). Each credit class is represented by an element of a finite set, say $\mathcal{K} = \{1, \ldots, K\}$. By convention, the element $K$ is always assumed to correspond to the default event.

As manifested in practice, the credit quality of a given corporate debt undergoes variations when time passes. We shall refer to this feature by saying that the credit quality migrates between various credit classes; the corresponding credit risk models are given a generic name of *multiple credit ratings models*. The traditional intensity-based approach, discussed at some length in Chap. 8 and 9, focuses on the pre-default value of a corporate bond in a single rating class model. However, several recent studies have extended this methodology to the case of multiple ratings. Specifically, they engaged in the credit migration analysis – that is, the analysis of the dynamics of the credit quality of the underlying credit instrument, such as: a corporate bond or a commercial loan. Credit migrations of both of these types of instruments are, of course, tied to credit migrations of the associated debtor.

Credit migrations have been typically modeled in terms of a Markov chain $C$ with finite state space and either discrete- or continuous-time parameter. The Markov chain $C$ is referred to as the *credit migration process*. Typically, the multiple defaults are excluded; in other words, the default class is assumed to represent the absorbing state for the Markov chain $C$. The main issue in the Markov based approach is thus the specification of the matrix of transition probabilities (discrete time) or transition intensities (continuous time) for $C$, both under the risk-neutral and the real-world probabilities.

It is important to realize that, typically, the overall filtration involved in a credit risk model that involves credit migrations in terms of a Markov chain $C$, will be larger than the natural filtration $\mathbb{F}^C$ of the migration process. Therefore, in general, one will need to deal with either the $\mathbb{G}$-Markov property (see Chap. 11) or with some kind of the conditional Markov property (see Sect. 11.3), rather than with the ordinary Markov property of the migration process $C$. On the intuitive level, we may say that assuming either the $\mathbb{G}$-Markov property or the conditional Markov property of the migration process addresses the presence of several sources of uncertainty, involving market risk (typically, the interest rate risk), credit risk, as well as the uncertainty involved in other economic factors. An important problem arising in this context is thus the issue of invariance (or preservation) of the $\mathbb{G}$-Markov property under an equivalent change of probability measure, such as the change from the real-world probability to a risk-neutral probability.

The question of preservation of the $\mathbb{G}$-Markov property of the migration process in a credit risk model admitting the presence of the market/factor risk is still an open problem, when the equivalent change of probability measure affects all sources of uncertainty involved in the model. From our discussion in Chap. 11, one may conclude that the conditions imposed therein for the respective Radon-Nikodým densities so that the $\mathbb{G}$-Markov property is preserved, are manifestly too restrictive for these credit risk models, in which the change from the real-world to the risk-neutral probability involves all sources of uncertainty present in a particular model. This is because the Radon-Nikodým densities are assumed to be only adapted with respect to the natural filtration of the Markov chain, rather than adapted to the filtration $\mathbb{G}$. However, if the given credit risk model admits structural properties such as some sort of decomposition of the overall risk into the market risk and credit risk, mathematically $\mathbb{G} = \mathbb{F} \otimes \mathbb{F}^C$, then it seems natural to expect that the $\mathbb{G}$-Markov property of the migration process will still be preserved, provided that the component of the Radon-Nikodým density corresponding to the filtration $\mathbb{F}^C$ will satisfy sufficient conditions analogous to those specified in Chap. 11. Such a decomposition would correspond to an underlying product probability space, which may also support the assumption of independence between market risk and default risk (imposed, for instance, in the Jarrow et al. (1997) approach). In the rest of this section, we shall be pointing out to the above technical issues when discussing specific models.

The next two subsections are devoted to the study of discrete- and continuous-time versions of the JLT model (the abbreviation JLT refers to R. Jarrow, D. Lando and S. Turnbull, who co-authored the Jarrow et al. (1997) paper). Subsequently, we shall examine briefly some extensions of this approach, due to Das and Tufano (1996), Arvanitis et al. (1998), and Kijima and Komoribayashi (1998). The study of models based on conditionally Markov credit migration process with *state variables* (or *factors*) is postponed to Sect. 12.2.

### 12.1.1 JLT Model: Discrete-Time Case

We are given a filtered probability space $(\Omega, \mathbb{G}, \mathbb{Q})$, where $\mathbb{Q}$ is interpreted as the real-world probability measure, and the filtration $\mathbb{G}$ models the flow of all the observations available to traders. Let $T^* > 0$ be a fixed horizon date. To create a tractable model that accounts for the migration of a corporate bond between rating grades, Jarrow et al. (1997) make a set of (simplifying) assumptions. We somewhat modify these assumptions, and we refer to them as Conditions (JLT.1)–(JLT.7) in the discrete-time case.

**Condition (JLT.1)** There exists a (unique) equivalent martingale measure $\mathbb{Q}^*$, equivalent to $\mathbb{Q}$ on $(\Omega, \mathcal{G}_{T^*})$, such that all default-free and default-risky zero-coupon bond prices follow $\mathbb{G}$-martingales, after discounting by the savings account.

Condition (JLT.1) is standard in the intensity-based approach to the modeling of defaultable term structure, presented in some detail in Chap. 8. The next two assumptions are related to the short-term interest rate.

**Condition (JLT.2)** The interest rate risk is modeled by means of an $\mathbb{F}$-adapted stochastic process $r$ of the default-free short-term interest rate, where $\mathbb{F}$ is some sub-filtration of $\mathbb{G}$.

**Condition (JLT.3)** The default time $\tau$ is a random variable independent of the default-free interest rate process $r$, conditionally upon the filtration $\mathbb{G}$ under the martingale measure $\mathbb{Q}^*$. More specifically, for any integrable functional $\phi$ of the interest rate process $r$, and any integrable function $f$ of the random time $\tau$ we have, for every $t \in \mathbb{R}_+$,

$$\mathbb{E}_{\mathbb{Q}^*}(\phi(r_.)f(\tau) \mid \mathcal{G}_t) = \mathbb{E}_{\mathbb{Q}^*}(\phi(r_.) \mid \mathcal{G}_t)\mathbb{E}_{\mathbb{Q}^*}(f(\tau) \mid \mathcal{G}_t).$$

Condition (JLT.3) is quite convenient from the perspective of computations; although from the practical viewpoint it is too restrictive. In further developments of the JLT methodology this rather unrealistic assumption was essentially relaxed. The next condition makes a specific choice of the recovery scheme; the fractional recovery of Treasury value can be replaced here by any alternative recovery rule.

**Condition (JLT.4)** A corporate bond is subject to the fractional recovery of Treasury value scheme, with the constant recovery coefficient $\delta$.

In the first step, Jarrow et al. (1997) produce a discrete-time model of defaultable term structure that makes account for the migrations of a defaultable bond in the finite set of credit rating classes. Subsequently, they provide a continuous-time version of their model; we shall present it in Sect. 12.1.2. Let us finally mention that the methodology developed in Jarrow et al. (1997) is based on the approach put forward by Jarrow and Turnbull (1995). For this reason, we shall first briefly describe the latter approach.

**Jarrow and Turnbull approach.** Jarrow and Turnbull (1995) assume that a defaulted bond pays at maturity a fixed fraction of its par value – that is, they assume the fractional recovery of Treasury value scheme. Consequently, the price $\tilde{D}^\delta(t,T)$ at time $t \le T \le T^*$ of a $T$-maturity corporate bond equals[1]

$$\tilde{D}^\delta(t,T) = B_t \, \mathbb{E}_{\mathbb{Q}^*} \left( B_T^{-1} (\delta \mathbb{1}_{\{T \ge \tau\}} + \mathbb{1}_{\{T < \tau\}}) \, \big| \, \mathcal{G}_t \right), \qquad (12.1)$$

where $\delta$ is the constant recovery rate. Suppose that we have chosen some model for the short-term rate $r$. As apparent from expression (12.1), we only need to model a random time $\tau$. If, in addition, Condition (JLT.3) is satisfied – so that the law of $\tau$ under $\mathbb{Q}^*$ is independent of the interest rate risk – formula (12.1) can be substantially simplified, as the following result shows.

**Proposition 12.1.1.** *Under assumptions (JLT.1)–(JLT.4), for every $t \le T$ the price of a corporate bond equals*

$$\tilde{D}^\delta(t,T) = B(t,T) \left( \delta + (1-\delta)\mathbb{Q}^*\{T < \tau \,|\, \mathcal{G}_t\} \right). \qquad (12.2)$$

*Proof.* In view of Condition (JLT.3), expression (12.1) yields

$$\begin{aligned}
\tilde{D}^\delta(t,T) &= B_t \, \mathbb{E}_{\mathbb{Q}^*} \left( B_T^{-1} (\delta \mathbb{1}_{\{T \ge \tau\}} + \mathbb{1}_{\{T < \tau\}}) \, \big| \, \mathcal{G}_t \right) \\
&= B_t \, \mathbb{E}_{\mathbb{Q}^*} (B_T^{-1} \,|\, \mathcal{G}_t) \, \mathbb{E}_{\mathbb{Q}^*} (\delta \mathbb{1}_{\{T \ge \tau\}} + \mathbb{1}_{\{T < \tau\}} \,|\, \mathcal{G}_t) \\
&= B(t,T) (\delta \mathbb{Q}^*\{\tau \le T \,|\, \mathcal{G}_t\} + \mathbb{Q}^*\{T < \tau \,|\, \mathcal{G}_t\}) \\
&= B(t,T) (\delta(1 - \mathbb{Q}^*\{T < \tau \,|\, \mathcal{G}_t\}) + \mathbb{Q}^*\{T < \tau \,|\, \mathcal{G}_t\}).
\end{aligned}$$

After simplifications, this proves formula (12.2). $\qquad\qquad\square$

*Remarks.* If the conditional independence assumption (JLT.3) is relaxed, we obtain

$$\tilde{D}^\delta(t,T) = B(t,T) \left( \delta + (1-\delta)\mathbb{Q}_T\{\tau > T \,|\, \mathcal{G}_t\} \right), \qquad (12.3)$$

where $\mathbb{Q}_T$ is the forward martingale measure for the date $T \le T^*$; that is, the probability measure on $(\Omega, \mathcal{G}_T)$ given by the equality

$$\frac{d\mathbb{Q}_T}{d\mathbb{Q}^*} = \frac{1}{B(0,T)B_T}, \qquad \mathbb{Q}^*\text{-a.s.} \qquad (12.4)$$

Indeed, by applying the Bayes rule to (12.1), we obtain

$$\tilde{D}^\delta(t,T) = B(t,T) \, \mathbb{E}_{\mathbb{Q}_T} (\delta \mathbb{1}_{\{T \ge \tau\}} + \mathbb{1}_{\{T < \tau\}} \,|\, \mathcal{G}_t),$$

and this in turn yields (12.3). It is interesting to observe that under (JLT.3) we have

$$\mathbb{Q}^*\{\tau > T \,|\, \mathcal{G}_t\} = \mathbb{Q}_T\{\tau > T \,|\, \mathcal{G}_t\}$$

almost surely with respect to the spot martingale measure $\mathbb{Q}^*$ or, equivalently, almost surely with respect to the forward martingale measure $\mathbb{Q}_T$.

---

[1] In a discrete-time model, the savings account process $B$ is defined in accordance with the simple compounding convention – that is, $B_t = \prod_{u=0}^{t-1}(1 + r_u)$.

**Credit migrations.** We are now ready to present the discrete-time version of the JLT approach. We consider the dates $t = 0, \ldots, T^*$, where the horizon date $T^*$ is assumed to be a positive integer. Formally, given an initial rating $C_0$ of a defaultable bond, the future changes in its ratings are described by a stochastic process $C$, referred to as the *migration process*. Again, formally the value of the associated migration process $C$ at time $t$ coincides with the current rating of a defaultable bond at time $t$.

As already mentioned, there is no loss of generality if we assume that the set of rating classes is $\{1, \ldots, K\}$, where the state $K$ is assumed to correspond to the default event. In addition, according to Jarrow et al. (1997) convention the states are ordered so that the state $i = 1$ represents the highest ranking, whereas the state $i = K - 1$ represents the lowest ranking. Recall that, by definition $p_{ij} := \mathbb{Q}\{C_{t+1} = j \mid C_t = i\}$.

**Condition (JLT.5)** The migration process $C$ follows a $\mathbb{G}$-Markov chain under the real-world probability $\mathbb{Q}$. The transition matrix of $C$ under $\mathbb{Q}$ equals

$$P = [p_{ij}]_{1 \leq i,j \leq K}, \quad p_{ij} \geq 0, \quad \sum_{j=1}^{K} p_{ij} = 1,$$

where $p_{Kj} = 0$ for every $j = 1, \ldots, K$, so that $p_{KK} = 1$. In other words, the state $K$ is absorbing.

As mentioned in Chap. 11, a time-homogeneous $\mathbb{G}$-Markov chain also follows an ordinary time-homogeneous Markov chain. At the intuitive level, the last assumption means that the future probabilistic evolution of credit ratings of a particular bond does not depend on the history of the market and on the past ratings of this bond. On the contrary, it is assumed to depend exclusively on the current rating of a bond. In view of (JLT.5), the transition matrix for $C$ under $\mathbb{Q}$ equals

$$P = \begin{pmatrix} p_{1,1} & \cdots & p_{1,K-1} & p_{1,K} \\ \cdot & \cdots & \cdot & \cdot \\ p_{K-1,1} & \cdots & p_{K-1,K-1} & p_{K-1,K} \\ 0 & \cdots & 0 & 1 \end{pmatrix}$$

so that the associated discrete-time generator matrix $\Lambda$ takes the following form

$$\Lambda = \begin{pmatrix} p_{1,1} - 1 & \cdots & p_{1,K-1} & p_{1,K} \\ \cdot & \cdots & \cdot & \cdot \\ p_{K-1,1} & \cdots & p_{K-1,K-1} - 1 & p_{K-1,K} \\ 0 & \cdots & 0 & 0 \end{pmatrix}.$$

*Remarks.* In accordance with the ordering of states, the following inequalities for one-step default probabilities should be obeyed

$$p_{iK} \leq p_{jK} \quad \text{for} \quad 1 \leq i \leq j \leq K. \tag{12.5}$$

Kijima (1998) examines the issue of the stochastic monotonicity of absorbing discrete-time Markov chains.

**Condition (JLT.6)** The migration process $C$ follows a (time-inhomogeneous) $\mathbb{G}$-Markov chain under the spot martingale measure $\mathbb{Q}^*$, with the time-dependent transition matrix[2]

$$\mathcal{P}^*(t) = [p_{ij}^*(t)]_{1 \leq i,j \leq K},$$

where

$$p_{ij}^*(t) \geq 0, \quad \sum_{j=1}^{K} p_{ij}^*(t) = 1,$$

and finally $p_{Kj}^*(t) = 0$ for every $j < K$ and $t = 0, \dots, T^* - 1$, so that once more the state $K$ is absorbing.

Recall that $p_{ij}^*(t) := \mathbb{Q}^* \{C_{t+1} = j \mid C_t = i\}$ for every $t = 0, \dots, T^* - 1$. In view of Condition (JLT.6), we have

$$\mathcal{P}^*(t) = \begin{pmatrix} p_{1,1}^*(t) & \cdots & p_{1,K-1}^*(t) & p_{1,K}^*(t) \\ \cdot & \cdots & \cdot & \cdot \\ p_{K-1,1}^*(t) & \cdots & p_{K-1,K-1}^*(t) & p_{K-1,K}^*(t) \\ 0 & \cdots & 0 & 1 \end{pmatrix}.$$

By definition, the default time $\tau$ is the first moment the rating process hits the state $K$:

$$\tau := \inf \{ t \in \{0, \dots, T^*\} : C_t = K \},$$

where, by convention, $\inf \emptyset = +\infty$. In Sect. 11.1, we have already discussed the issue of preservation of the Markov property relative to the reference filtration $\mathbb{G}$ under the equivalent change of probability measure, and we have studied there the effect of an equivalent change of probability measure on one-step transition probabilities. To ensure analytical tractability of the model, Jarrow et al. (1997) postulate one more technical condition (note an analogy between Condition (JLT.7) and Example 11.1.2).

**Condition (JLT.7)** The following relationship holds

$$p_{ij}^*(t) = \pi_i(t) p_{ij}, \quad \forall j \neq i, \tag{12.6}$$

where time-dependent, deterministic coefficients $\pi_i(t)$ are interpreted as discrete-time *risk premiums*. Since clearly $\pi_K(t) = 1$ for any $t$, we shall refer to the vector $(\pi_1(t), \dots, \pi_{K-1}(t))$ as the risk premia at time $t$.

The last condition implies, in particular, that

$$p_{ii}^*(t) = 1 + \pi_i(t)(p_{ii} - 1).$$

In other words, for any state $i$, the probability under the martingale measure $\mathbb{Q}^*$ of jumping to the state $j \neq i$ is assumed to be proportional to the corresponding probability under the real-world probability $\mathbb{Q}$, with the proportionality factor, which may depend on $i$ and $t$, but not on $j$.

---

[2] To be consistent with the notation used in Jarrow et al. (1997), we should have written $p_{ij}^*(t, t+1)$, rather than $p_{ij}^*(t)$.

**Model construction.** A mathematical model supporting the above assumptions might be constructed as follows. We set $\Omega = \tilde{\Omega} \otimes \hat{\Omega}$, $\mathbb{Q} = \mathbb{P} \otimes \hat{\mathbb{Q}}$ and $\mathbb{G} = \mathbb{F} \otimes \mathbb{F}^C$, where $\mathbb{F}$ ($\mathbb{F}^C$, resp.) is a filtration of events in $\tilde{\Omega}$ (in $\hat{\Omega}$, resp.) We introduce the associated sub-filtrations $\mathbb{F}$ and $\mathbb{F}^C$ of $\mathbb{G}$ by formally identifying $\mathbb{F}$ with $\mathbb{F} \otimes \{\emptyset, \hat{\Omega}\}$ and $\mathbb{F}^C$ with $\{\emptyset, \tilde{\Omega}\} \otimes \mathbb{F}^C$. This is, of course, a rather standard notational convention. In such a model, the migration process $C$ is essentially supported by $\hat{\Omega}$. An appropriate change of the product probability measure $\mathbb{Q}$ to an equivalent product probability measure $\mathbb{Q}^* = \mathbb{P}^* \otimes \hat{\mathbb{Q}}^*$ would preserve the $\mathbb{G}$-Markov property of $C$, as well as the independence assumption (JLT.3).

A sufficient condition for the preservation of the $\mathbb{G}$-Markov property would be a condition such as (B.1) (or (B.2)) of Sect. 11.1, applied to the component of the Radon-Nikodým density that corresponds to the pair $\hat{\mathbb{Q}}, \hat{\mathbb{Q}}^*$. In this kind of a model, the $\mathbb{G}$-Markov property of $C$ on the product space is rather trivial, as it is essentially equivalent to the Markov property of $C$ on the component space. Observe, though, that in such a model the Markov property of the migration process is in fact detached from the market fundamentals (it is sort of superimposed on the market of defaultable claims).

It is worth noticing that in the model suggested above, we also have $\mathbb{G} = \mathbb{F} \vee \mathbb{F}^C$. Thus, $C$ is also an $\mathbb{F}$-conditionally $\mathbb{G}$-Markov chain[3] under $\mathbb{Q}$ (and under $\mathbb{Q}^*$). We see that in such a model the Markov property, the $\mathbb{G}$-Markov property and the $\mathbb{F}$-conditional $\mathbb{G}$-Markov property of the migration process are essentially equivalent, which is not surprising as the filtrations $\mathbb{F}$ and $\mathbb{F}^C$ are essentially independent. More realistic – in our opinion – are those models that postulate a non-trivial $\mathbb{F}$-conditional $\mathbb{G}$-Markov property, rather than a simple $\mathbb{G}$-Markov property. Some examples of such models will be discussed later in the text. Let us finally point out that similar observations will apply to the continuous-time version of the JLT approach.

**Valuation of a defaultable bond.** In view of (JLT.1)–(JLT.7) and Corollary 11.1.4, it is easy to derive an expression for the risk-neutral conditional probability of solvency, namely (cf. Lemma 1 in Jarrow et al. (1997)),

$$\mathbb{Q}^*\{\tau > T \,|\, \mathcal{G}_t\} = \mathbb{Q}^*\{\tau > T \,|\, C_t\} = \sum_{j \neq K} p^*_{C_t j}(t, T), \quad t = 0, \dots, T,$$

where, for every $0 \le t \le s \le T$,

$$p^*_{ij}(t, s) := \mathbb{Q}^*\{C_s = j \,|\, C_t = i\}, \quad \forall i, j \in \mathcal{K}.$$

Observe that the probabilities $p^*_{ij}(t, s)$ can be obtained from the one step probabilities $p^*_{ij}(t)$ by multiplication of the transition matrices $\mathcal{P}^*(t)$, just as in (11.14). Combining the last equality with (12.2), we conclude that under conditions (JLT.1)–(JLT.7) the following result holds.

---

[3] For the definition and properties of an $\mathbb{F}$-conditionally $\mathbb{G}$-Markov chain, we refer to Sect. 11.3.

**Proposition 12.1.2.** *For any $i = 1, \ldots, K - 1$, let the conditional value of defaultable bond be denoted by $D_i(t, T)$, more specifically,*

$$D_i(t, T) := \mathbb{E}_{\mathbb{Q}^*}\left(\tilde{D}^\delta(t, T) \,\middle|\, C_t = i\right).$$

*Then*

$$D_i(t, T) = B(t, T)\left(\delta + (1 - \delta) \sum_{j \neq K} p_{ij}^*(t, T)\right) \tag{12.7}$$

*and*

$$\tilde{D}^\delta(t, T) = D_{C_t}(t, T) = B(t, T)\left(\delta + (1 - \delta) \sum_{j \neq K} p_{C_t j}^*(t, T)\right). \tag{12.8}$$

For the ease of further reference, let us consider a special case of only two credit classes: the non-default class $i = 1$ and the default class $i = 2$. Then formulae (12.7)–(12.8) become

$$D_1(t, T) = B(t, T)\left(\delta + (1 - \delta)\mathbb{Q}^*\{\tau > T \mid C_t = 1\}\right) \tag{12.9}$$

and

$$\tilde{D}^\delta(t, T) = D_{C_t}(t, T) = B(t, T)\left(\delta + (1 - \delta)\mathbb{Q}^*\{\tau > T \mid C_t\}\right),$$

respectively.

**Credit spreads.** By definition, the one-step forward rate on the defaultable bond for the future date $T$, as seen from time $t \leq T$, is given as

$$g_{C_t}(t, T) := -\ln\left(\frac{D_{C_t}(t, T + 1)}{D_{C_t}(t, T)}\right).$$

Likewise, for the default-free bond we define the one-step forward rate by the standard formula

$$f(t, T) := -\ln\left(\frac{B(t, T + 1)}{B(t, T)}\right).$$

We thus obtain the following formula for the credit spread process

$$s_{C_t}(t, T) := g_{C_t}(t, T) - f(t, T) = \ln\left(\frac{\delta + (1 - \delta)\sum_{j \neq K} p_{C_t j}^*(t, T)}{\delta + (1 - \delta)\sum_{j \neq K} p_{C_t j}^*(t, T + 1)}\right).$$

In particular, on the set $\{C_t = i\}$ we have (see formula (9) in Jarrow et al. (1997))

$$s_i(t, T) := g_i(t, T) - f(t, T) = \ln\left(\frac{\delta + (1 - \delta)\sum_{j \neq K} p_{ij}^*(t, T)}{\delta + (1 - \delta)\sum_{j \neq K} p_{ij}^*(t, T + 1)}\right),$$

where $g_i(t, T)$ represents the one-step forward return on defaultable bonds that are at time $t$ in the $i^{\text{th}}$ credit class (of course, $g_K(t, T) = f(t, T)$).

**Model calibration.** We are in a position to discuss the important issue of calibration of the discrete-time version of the JLT model. For any $i \leq K - 1$, formula (12.7) gives the theoretical value of a $T$-maturity defaultable discount bond, which is in the $i^{\text{th}}$ credit class at time 0:

$$D_i(0, T) = B(0, T) \left( \delta + (1 - \delta) \sum_{j \neq K} p_{ij}^*(0, T) \right). \qquad (12.10)$$

Assume that we are given the following inputs:
- the initial term structure of default-free bonds – that is, the market values $B(0, T)$ for $T = 1, \ldots, T^*$,
- the observed initial term structures of defaultable bonds from various credit classes: $D_i(0, T)$ for $T = 1, \ldots, T^*$ and $i = 1, \ldots, K - 1$.

Our aim is to identify the risk-neutral transition matrices $\mathcal{P}^*(t)$ that make the observed market prices $D_i(0, T)$ coincide with the theoretical values predicted by the model through equations (12.10). Towards this end, we first use the fact that the probabilities $p_{ij}^*(0, T)$ can be obtained from one-step transition probabilities $p_{ij}^*(t)$. Indeed, defining the matrix $\mathcal{P}^*(0, T) = [p_{ij}^*(0, T)]_{1 \leq i,j \leq K}$ for $T = 1, \ldots, T^*$ we have (cf. (11.14))

$$\mathcal{P}^*(0, T) = \prod_{t=0}^{T-1} \mathcal{P}^*(t). \qquad (12.11)$$

In view of Condition (JLT.7), we may also write, for $t = 0, \ldots, T^* - 1$,

$$\mathcal{P}^*(t) = \Pi(t)\Lambda + \text{Id}, \qquad (12.12)$$

where $\Lambda$ is the discrete-time generator matrix, $\Pi(t)$ is the diagonal matrix

$$\Pi(t) = \text{diag}\,[\pi_1(t), \ldots, \pi_{K-1}(t), 1],$$

and Id is the $K \times K$ identity matrix. It is reasonable to make an additional assumption that the real-world transition matrix $P$ is known. For instance, the matrix $P$ can be estimated from the historical data on credit migrations. As apparent from (12.10)–(12.12), the calibration procedure is thus reduced to determining the sequence $\Pi(t)$, $t = 0, \ldots, T^* - 1$. A straightforward algebra suffices to conclude that for every $t = 0, \ldots, T^* - 1$ the risk premia $(\pi_1(t), \ldots, \pi_{K-1}(t))$ satisfy the following equation[4]

$$\tilde{\mathcal{P}}^*(0, t) \begin{pmatrix} \pi_1(t)p_{1K} \\ \vdots \\ \pi_{K-1}(t)p_{K-1,K} \\ 1 \end{pmatrix} = \begin{pmatrix} \frac{B(0,t+1)-D_1(0,t+1)}{(1-\delta)B(0,t+1)} \\ \vdots \\ \frac{B(0,t+1)-D_{K-1}(0,t+1)}{(1-\delta)B(0,t+1)} \\ 1 \end{pmatrix}, \qquad (12.13)$$

where $\tilde{\mathcal{P}}^*(0, 0) = \text{Id}$ and $\tilde{\mathcal{P}}^*(0, t) = \mathcal{P}^*(0, t)$ for $t = 1, \ldots, T^* - 1$.

---

[4] Note that this equation slightly differs from the respective equation on Page 493 in Jarrow et al. (1997).

Equations (12.13) lead to a simple recursive procedure, which can be summarized as follows:
- first, for $t = 0$, compute from (12.13) the initial values of credit risk premiums, i.e., the vector $(\pi_1(0), \ldots, \pi_{K-1}(0))$,
- subsequently, by combining (12.6) with (12.11), find the one-step risk-neutral probability matrix $\mathcal{P}^*(0, 1)$,
- use the matrix $\mathcal{P}^*(0, 1)$ in order to compute from (12.13) the vector $(\pi_1(t), \ldots, \pi_{K-1}(t))$ of credit risk premia at time $t = 1$,
- again, use (12.6) and (12.11) in order to find the two-step risk-neutral probability matrix $\mathcal{P}^*(0, 2)$, and so on.

The steps described above are repeated as many times as required – that is, until all vectors of risk premia $(\pi_1(t), \ldots, \pi_{K-1}(t))$, $t = 0, \ldots, T^* - 1$, and, simultaneously, all values of $\mathcal{P}^*(0, T)$, $T = 1, \ldots, T^*$, are found. Observe that as soon as the risk premia are calibrated to the market data, the risk-neutral one-step transition matrices $\mathcal{P}^*(t)$, $t = 0, \ldots, T^* - 1$ can be found from Condition (JLT.7). Since we also have that, for every $0 \le t \le T - 1$ (cf. (11.14)),

$$\mathcal{P}^*(t, T) := [p_{ij}^*(t, T)]_{1 \le i,j \le k} = \prod_{u=t}^{T-1} \mathcal{P}^*(u),$$

formula (12.8) can be applied for every $t$.

*Remarks.* (i) Given a set of initial data, there is no guarantee that equation (12.13) admits at least one solution for every $t = 0, \ldots, T^* - 1$. Furthermore, even if a solution exists for every $t = 0, \ldots, T^* - 1$, it is still required that at least one of them is non-negative. Finally, a solution must be such that the matrices $\mathcal{P}^*(t)$ resulting from (12.6) are indeed the one-step transition matrices. We refer to Jarrow et al. (1997) for a discussion of the issue of 'data inconsistency' – that is, the situation when the calibration procedure described above fails to produce (in a unique way) transition matrices.

(ii) Recall that Condition (JLT.1) postulates existence of a (unique) spot martingale measure $\mathbb{Q}^*$. In principle, one can use the theory developed in Sect. 11.1 to construct a measure $\mathbb{Q}^*$ given the credit risk premia $\pi_i(t)$, $i \ne K$, $0 \le t \le T^* - 1$, as well as the market (interest rate) risk premiums, which are not discussed in this section. The open problem remains what kind of additional conditions on the risk premiums, if any, will suffice to guarantee that the constructed probability is indeed the risk-neutral probability, in the sense of Condition (JLT.1). Similar remark will also apply to the continuous-time version of the JLT model, which will be examined later on.

(iii) It is important to emphasize that the model calibration procedure described above indeed consists of two major steps. The first step involves the estimation of recovery rates $\delta$ of corporate bonds, as well as the estimation of the statistical transition matrix $P$. In practical implementations, the value of the recovery rates and the entries of the real-world transition matrix $P$ are estimated from historical data. The second step is the recursive procedure for computing the credit risk premia described above.

(iv) Lando (2000a) proposes an alternative calibration procedure, which is capable of matching only the risk-neutral default probabilities $p^*_{iK}(0, T)$ to the observed market data $D_i(0, T)$, $i = 1, \ldots, K - 1$ and $B(0, T)$ for every $T = 1, \ldots, T^*$.

(v) The JLT model considered in this section is built from a migration process, which is assumed to be a time-homogeneous Markov chain under the real-world probability $\mathbb{Q}$. As observed in Jarrow et al. (1997), the time-homogeneity of the migration process $C$ is not a restriction of a fundamental nature, and it was postulated mainly for the purpose of simplifying the estimation of the transition matrix of $C$ under the real-world probability. In order to account for dependence of these probabilities on such factors as business and/or credit cycles, some authors (e.g., Wei (2000)) model the statistical migration probabilities as functions of time-dependent *state variables*. Models of this kind – which will typically fall into the category of conditionally Markov models – allow for keeping the risk premia almost constant over time. This seems to be an important modification of the original JLT approach since, according to Wei (2000), there is no good theoretical justification why credit risk premia should dramatically change over time.

(vi) For an interesting re-interpretation of the risk premia $\pi_i(t)$, we refer to Thomas et al. (1998). They argue that the risk premia may be seen as the belief that the market puts on the extreme risky future scenarios – i.e., scenarios that allow for the default event to occur for bonds in all rating classes within the same unit of time.

### 12.1.2 JLT Model: Continuous-Time Case

In the continuous-time setting, Jarrow et al. (1997) propose to substitute conditions (JLT.5)–(JLT.6) with the following natural counterparts.

**Condition (JLT.5c)** Under the real-world probability measure $\mathbb{Q}$, the migration process $C$ follows a time-homogeneous $\mathbb{G}$-Markov chain, with the intensity matrix $\Lambda$ of the following form:

$$\Lambda = \begin{pmatrix} \lambda_{1,1} & \cdots & \lambda_{1,K-1} & \lambda_{1,K} \\ \cdot & \cdots & \cdot & \cdot \\ \lambda_{K-1,1} & \cdots & \lambda_{K-1,K-1} & \lambda_{K-1,K} \\ 0 & \cdots & 0 & 0 \end{pmatrix}.$$

**Condition (JLT.6c)** Under the spot martingale measure $\mathbb{Q}^*$, the credit migration process $C$ follows a (time-inhomogeneous) $\mathbb{G}$-Markov chain, with a time-dependent intensity matrix $\Lambda^*(t)$, where

$$\Lambda^*(t) = \begin{pmatrix} \lambda^*_{1,1}(t) & \cdots & \lambda^*_{1,K-1}(t) & \lambda^*_{1,K}(t) \\ \cdot & \cdots & \cdot & \cdot \\ \lambda^*_{K-1,1}(t) & \cdots & \lambda^*_{K-1,K-1}(t) & \lambda^*_{K-1,K}(t) \\ 0 & \cdots & 0 & 0 \end{pmatrix}$$

and the entries of the matrix $\Lambda^*(t)$ are functions $\lambda^*_{ij} : [0, T^*] \to \mathbb{R}_+$.

The matrix $\Lambda$ is also assumed to satisfy mild technical conditions, which are aimed to guarantee that a suitable monotonicity condition for the probability of default, similar to (12.5), is valid. As in the discrete-time setting, the default time $\tau$ is defined as the first time the credit migration process jumps to the absorbing state $K$, specifically:

$$\tau := \inf \{ t \in [0, T^*] : C_t = K \}.$$

The model's tractability condition becomes (cf. Example 11.2.3).

**Condition (JLT.7c)** There exists a matrix function $U(t)$ of the form:

$$U(t) = \begin{pmatrix} u_{1,1}(t) & \cdots & 0 & 0 \\ \cdot & \cdots & \cdot & \cdot \\ 0 & \cdots & u_{K-1,K-1}(t) & 0 \\ 0 & \cdots & 0 & 1 \end{pmatrix},$$

where the entries, $u_{ii}(t)$, $i = 1, \ldots, K - 1$, are strictly positive, integrable functions, such that the risk-neutral and real-world intensity matrices satisfy $\Lambda^*(t) = U(t)\Lambda$ for every $t \in [0, T^*]$.

**Bond valuation formulae.** For every $0 \le t \le s \le T^*$, we denote

$$p_{ij}^*(t, s) = \mathbb{Q}^*\{C_s = j \,|\, C_t = i\}, \quad \forall 1 \le i, j \le K,$$

and we introduce the transition matrix $\mathcal{P}^*(t, s) = [p_{ij}^*(t, s)]_{1 \le i,j \le K}$. It follows from Corollary 11.2.2 that the matrix $\mathcal{P}^*(t, s)$ can be represented as follows:

$$\mathcal{P}^*(t, s) = \mathrm{Id} + \sum_{n=1}^{\infty} \int_t^s \int_{u_1}^s \cdots \int_{u_{n-1}}^s \Lambda^*(u_1) \ldots \Lambda^*(u_n) \, du_n \ldots du_1, \quad (12.14)$$

and

$$\mathcal{P}^*(t, s) = \mathrm{Id} + \sum_{n=1}^{\infty} \int_t^s \int_t^{u_1} \cdots \int_t^{u_{n-1}} \Lambda^*(u_1) \ldots \Lambda^*(u_n) \, du_n \ldots du_1. \quad (12.15)$$

*Remarks.* Let us consider a special case when the risk premium function $U(t)$ is assumed to be a constant function – i.e., when

$$U(t) = U := \mathrm{diag}\,[u_1, \ldots, u_{K-1}, 1], \quad \forall 0 \le t \le T^*.$$

Then, in view of Condition (JLT.6c), the above formulae reduce to the exponential formula $\mathcal{P}^*(t, s) = e^{U\Lambda(s-t)}$ for every $0 \le t \le s \le T^*$.

Similarly as in the discrete-time case (see also (11.30)), we obtain

$$\mathbb{Q}^*\{\tau > T \,|\, \mathcal{G}_t\} = \mathbb{Q}^*\{\tau > T \,|\, C_t\} = 1 - p_{C_t K}^*(t, T) = \sum_{j \ne K} p_{C_t j}^*(t, T)$$

for every $t \in [0, T]$.

Consequently, as soon as the transition matrices $\mathcal{P}^*(t,T)$, $t \in [0,T]$, are found, we can apply in the continuous-time case exactly the same valuation formulae as those derived for the discrete-time case in Proposition 12.1.2. More specifically, we have

$$D_i(t,T) = B(t,T) \left( \delta + (1-\delta)(1 - p^*_{iK}(t,T)) \right) \qquad (12.16)$$

and

$$\tilde{D}^\delta(t,T) = D_{C_t}(t,T) = B(t,T) \left( \delta + (1-\delta)(1 - p^*_{C_tK}(t,T)) \right). \qquad (12.17)$$

**Credit spreads.** By definition, the instantaneous forward rate on the defaultable bond for the future date $T$, as seen from time $t$, is given as:

$$g_{C_t}(t,T) = -\frac{\partial \ln D_{C_t}(t,T)}{\partial T}.$$

Likewise, for the default-free bond we define the instantaneous forward rate as

$$f(t,T) = -\frac{\partial \ln B(t,T)}{\partial T}.$$

Of course, it is assumed that the above derivatives are well defined. In view of (12.17) and (12.15), we see that for the first derivative to be well defined, it suffices that the second exists.

In view of relationship (12.17), the *instantaneous forward credit spread* process $s_{C_t}(t,T) = g_{C_t}(t,T) - f(t,T)$ satisfies

$$s_{C_t}(t,T) = \frac{(1-\delta)}{\delta + (1-\delta)(1 - p^*_{C_tK}(t,T))} \frac{\partial p^*_{C_tK}(t,T)}{\partial T}.$$

The last formula makes it clear that within the Jarrow et al. (1997) framework, the randomness enters the forward credit spreads only through the uncertain credit ranking $C_t$. The $i^{\text{th}}$ instantaneous forward credit spread $s_i(t,T) = g_i(t,T) - f(t,T)$ equals (see formula (23) in Jarrow et al. (1997))

$$s_i(t,T) = \frac{(1-\delta)}{\delta + (1-\delta)(1 - p^*_{iK}(t,T))} \frac{\partial p^*_{iK}(t,T)}{\partial T}, \qquad (12.18)$$

where $g_i(t,T)$ represents the instantaneous forward rate on a defaultable bond, which belongs at time $t$ to the $i^{\text{th}}$ credit class (as before, $g_K(t,T) = f(t,T)$).

**Short-term credit spreads.** As explained in Sect. 11.2, the matrix-valued function $\mathcal{P}^*(t,T)$, $0 \le t \le T \le T^*$ satisfies the forward Kolmogorov equation:

$$\frac{d\mathcal{P}^*(t,T)}{dT} = \mathcal{P}^*(t,T)\Lambda^*(t), \qquad \mathcal{P}^*(t,t) = \text{Id}. \qquad (12.19)$$

Letting $t$ tend to $T$ in (12.19), we get, for every $T \le T^*$,

$$\frac{d\mathcal{P}^*(t,T)}{dT}\bigg|_{t=T} = \Lambda^*(T).$$

In view of (JLT.7c), the last equality implies that, for every $T \leq T^*$,

$$\frac{d\mathcal{P}^*(t,T)}{dT}\bigg|_{t=T} = U(T)\Lambda.$$

Consequently, for every $T \leq T^*$ we also obtain

$$\frac{dp^*_{iK}(t,T)}{dT}\bigg|_{t=T} = u_i(T)\lambda_{iK}. \tag{12.20}$$

The short-term interest rate process implied by corporate bonds belonging to the $i^{\text{th}}$ credit class is denoted by $r^i$. By definition, we have $r^i_t = g_i(t,t)$ for $t \in \mathbb{R}_+$. Likewise, for the default-free bond we set $r_t = f(t,t)$. The process $r$ represents, of course, the short-term interest rate process corresponding to default-free term structure. By virtue of (12.20) and (12.18), in the continuous-time version of the JLT model, for every $t \in [0, T^*]$ we have[5]

$$s^i_t := r^i_t - r_t = (1 - \delta)\lambda_{iK} u_i(t). \tag{12.21}$$

The $i^{\text{th}}$ *short-term credit spread* $s^i$ thus follows a positive process, proportional to the complement $(1 - \delta)$ of the recovery rate $\delta$, to the intensity of default $\lambda_{iK}$, and to the $i^{\text{th}}$ credit risk premium $u_i(T)$.

**Model calibration.** Assume that we are given the following inputs:
- the initial term structure of default-free bonds – that is, the prices $B(0,T)$ for $T \in [0, T^*]$,
- the initial defaultable term structures for various credit classes: $D_i(0,T)$ for every $T \in [0, T^*]$ and any $i = 1, \ldots, K - 1$.

The aim is to identify the risk-neutral intensity matrices $\Lambda^*(t)$ that make the observed market prices $D_i(0,T)$ coincide with the theoretical values predicted by the model through equations (12.16) in conjunction with representation (12.14) or (12.15). In view of (JLT.7c), similarly as in the discrete-time case, the calibration procedure splits into two major steps: (i) estimation of the recovery rate $\delta$ and the statistical generator matrix $\Lambda$, and (ii) computation of the credit risk premium $U(t)$. Once $\delta$ and $\Lambda$ are estimated, Jarrow et al. (1997) propose to compute the risk premia in an approximate way by a time-discretization combined with a recursive procedure, as explained in the discrete-time setting. We refer to the original paper for more details.

**Estimation of the intensity matrix $\Lambda$.** In general, estimation of the intensity matrix for a continuous-time Markov is a complex issue. Let us briefly discuss three alternative approaches to this problem. For a more detailed study of estimation of transition intensities, we refer, for instance, to Israel et al. (2001) or Kavvathas (2000).

**First method.** We begin by presenting an approach based on the concept of an embedded Markov chain (cf. Sect. 11.2.1). From (11.28), it follows that

---

[5] Compare (12.21) with equation (24) in Jarrow et al. (1997); the latter equation contains an extra term $\mathbb{1}_{\{\tau > t\}}$, which is spurious.

for any $n \in \mathbb{N}^*$ and for any $i = 1, \ldots, K - 1$, the conditional probability distribution of $\tau_n - \tau_{n-1}$ given that $C_{\tau_{n-1}} = i$ is exponential with parameter $-\lambda_{ii}$. Consequently, for every $n \in \mathbb{N}^*$ and $i = 1, \ldots, K - 1$,

$$\mu_i := \mathbb{E}_{\mathbb{Q}}(\tau_n - \tau_{n-1} \mid C_{\tau_{n-1}} = i) = -\lambda_{ii}^{-1}. \tag{12.22}$$

In principle, the conditional probabilities of transitions given by (11.29) and the *mean sojourn times* $\mu_i$, $i = 1, \ldots, K - 1$ specified by (12.22) can be estimated by observing the past behavior of the migration process $C$. In this way, we obtain $(K - 1)^2$ estimates $\hat{p}_{ij}$, $i = 1, \ldots, K - 1$, $j = 1, \ldots, K$, $i \neq j$, of transition probabilities, as well as $K - 1$ estimates $\hat{\mu}_i$, $i = 1, \ldots, K - 1$ of sojourn times. We may thus attempt to produce simple estimates of the $K(K - 1)$ intensities $\lambda_{ij}$, $i = 1, \ldots, K - 1$, $j = 1, \ldots, K$ by setting:

$$\hat{\lambda}_{ii} = -\hat{\mu}_i^{-1} \text{ for } i = 1, \ldots K - 1,$$

and

$$\hat{\lambda}_{ij} = \hat{p}_{ij} \hat{\mu}_i^{-1} \text{ for } i \neq j.$$

Unfortunately, in view of the scarcity of migration data, the approach described above is not likely to yield satisfactory results. Thus, despite it's simplicity, it does not seem to be of practical interest.

**Second method.** An alternative approach is based on solving the *embedding problem,* discussed in Sect. 11.2.6. In this method, one uses the historical data provided by credit rating agencies regarding migration probabilities during a certain period (typically, in one year) to obtain an estimate, $\hat{P}(1)$ say, of the transition probability matrix $P(1)$ for the migration process $C$. A valid estimate of the infinitesimal generator matrix $\Lambda$ will be obtained if the matrix $\hat{P}(1)$ admits a unique valid generator (cf. Sect. 11.2.6). We refer to Israel et al. (2001) for a more complete discussion of this estimation method. Let us only mention that, typically, the estimation of the transition probability $P(1)$ relies on a 'cohort' method, which is rather problematic. A thorough discussion of this issue is given in Lando and Skødeberg (2002).

**Third method.** Another way of estimation is discussed in Lando and Skødeberg (2002) (see also Jarrow et al. (1997)). It is based on the maximum likelihood estimation methods for stochastic processes, studied previously by Küchler and Sørensen (1997).

Recall that $H^{ij}(T)$ stands for the number of transitions from class $i$ to class $j$, $i \neq j$, $i, j < K$, over the time period $[0, T]$. In addition, let $Y_i(t)$ denote the number of firms that are in the $i^{\text{th}}$ rating class at time $t \in [0, T]$. The maximum likelihood estimate of the intensity $\lambda_{ij}$ for $i \neq j$, $i, j < K$, based on the data collected over the time period $[0, T]$, is computed as the following ratio:

$$\hat{\lambda}_{ij} = \frac{H^{ij}(T)}{\int_0^T Y_i(t)dt}.$$

For a numerical study of this procedure, the interested reader may consult Lando and Skødeberg (2002).

### 12.1.3 Kijima and Komoribayashi Model

Kijima and Komoribayashi (1998) argue that Condition (JLT.7) may fail to hold in the practical implementation of the JLT methodology. First, they observe that in view of (11.12) applied to the function $k^*(i) = i$, $i = 1, \ldots, K$, corresponding to Condition (JLT.7), the risk premiums in the JLT model must satisfy[6]

$$0 < \pi_i(t) < (1 - p_{ii})^{-1}, \quad \forall i \neq K - 1, \ t \in \mathbb{N}^*. \tag{12.23}$$

Next, they observe that, in view of (12.13), for $i = 1, \ldots, K - 1$ we have

$$\pi_i(0) = \frac{B(0, 1) - D_i(0, 1)}{(1 - \delta)B(0, 1)p_{iK}}. \tag{12.24}$$

Finally, they argue that if the default probabilities $p_{iK}$ are sufficiently small, as compared to the differences $B(0, 1) - D_i(0, 1)$, then – as it is seen from (12.24) – conditions (12.23) may be violated (at least for $t = 0$).

*Remarks.* Both theoretical and statistical grounds for this critique of Condition (JLT.7) need some careful analysis. This is because if $p_{iK}$ is close to zero, so that default probability for the bond in class $i$ is small (meaning that the bond is a high grade bond), the value of the credit spread $B(0, 1) - D_i(0, 1)$ should also be small. Consequently, it may happen that the default probabilities $p_{iK}$ will not be sufficiently small as compared to the differences $B(0, 1) - D_i(0, 1)$. Thus, the assumed possibility that underpins the critique of Condition (JLT.7), namely, the possibility that 'the default probabilities $p_{iK}$ will not be sufficiently small as compared to the differences $B(0, 1) - D_i(0, 1)$' may not occur.

To construct a more 'practical' model, Kijima and Komoribayashi (1998) postulate a slight modification of the JLT model, in which assumptions (JLT.1)–(JLT.6) are preserved, and (JLT.7) is substituted with the following postulate.

**Condition (K)** The following relationship holds

$$p^*_{ij}(t) = \pi_i(t)p_{ij}, \quad \forall j \neq K,$$

where time-dependent, deterministic coefficients $\pi_i(t)$ are interpreted as discrete-time *risk premiums*.

Note that the above assumption corresponds to setting $k^*(i) = K$ for every $i = 1, \ldots, K$ in Example 11.1.2. Hence, the counterpart of condition (12.23) now reads

$$0 < \pi_i(t) < (1 - p_{iK})^{-1}, \tag{12.25}$$

for every $i \neq K - 1$, and every $t \in \mathbb{N}^*$.

---

[6] Kijima and Komoribayashi (1998) consider $K + 1$ rating states: $1, \ldots, K + 1$, where $K + 1$ represents the default state. We preserve our convention of $K$ possible rating states: $1, \ldots, K$, where $K$ is the default state.

**Model calibration.** Bond valuation formulae in Kijima and Komoribayashi (1998) coincide with formulae (12.16)–(12.17); however, the risk-neutral default probabilities $p^*_{iK}(t, T)$ are now specified by Condition (K). To find the calibration equations for the risk premiums, let us first observe that the transition matrix $P$ admits the following representation:

$$P = \begin{pmatrix} A & R \\ 0' & 1 \end{pmatrix}, \quad A = \begin{pmatrix} p_{1,1} & \cdots & p_{1,K-1} \\ \cdot & \cdots & \cdot \\ p_{K-1,1} & \cdots & p_{K-1,K-1} \end{pmatrix}, \quad R = \begin{pmatrix} p_{1,K} \\ \vdots \\ p_{K-1,K} \end{pmatrix},$$

where $0$ is the $K-1$ dimensional column vector of zeros. Consequently, under Condition (K), the transition matrix $\mathcal{P}^*(t)$ can be represented as follows:

$$\mathcal{P}^*(t) = \begin{pmatrix} \mathcal{A}(t) & \mathcal{R}(t) \\ 0' & 1 \end{pmatrix},$$

where $\mathcal{A}(t) = \Pi(t)A$, $\mathcal{R}(t) = 1 - \Pi(t)A1$, $\Pi(t) = \mathrm{diag}\,[\pi_1(t), \ldots, \pi_{K-1}(t)]$, and where $1$ denotes the $(K-1)$-dimensional column vector with all components equal to 1. Finally, for every $T \le T^*$ we define (cf. (12.11))

$$\mathcal{A}^*(0, T) = \prod_{t=0}^{T-1} \mathcal{A}^*(t).$$

Now, a straightforward algebra yields a system of the following calibration equations, for every $t = 0, \ldots, T^* - 1$,

$$\tilde{\mathcal{A}}^*(0, t) \begin{pmatrix} \pi_1(t)(1 - p_{1K}) \\ \vdots \\ \pi_{K-1}(t)(1 - p_{K-1,K}) \end{pmatrix} = \begin{pmatrix} \frac{D_1(0,t+1) - \delta B(0,t+1)}{(1-\delta)B(0,t+1)} \\ \vdots \\ \frac{D_{K-1}(0,t+1) - \delta B(0,t+1)}{(1-\delta)B(0,t+1)} \end{pmatrix},$$

where $\tilde{\mathcal{A}}^*(0, 0) = \mathrm{Id}$ and $\tilde{\mathcal{A}}^*(0, t) = \mathcal{A}^*(0, t)$ for $t = 1, \ldots, T^* - 1$. In particular, for every $i = 1, \ldots, K - 1$ we obtain

$$\pi_i(0) = \frac{D_i(0, 1) - \delta B(0, 1)}{(1 - \delta)B(0, 1)(1 - p_{iK})}. \tag{12.26}$$

As already mentioned, the right-hand side inequality in (12.23) for $t = 0$ may fail to hold if the one-step probabilities of default $p_{iK}$ are close to zero. From (12.26), we see that even if the one-step probabilities of default $p_{iK}$ are close to zero, the quantities $\pi_i(0)$ will not explode, and so the right-hand side inequality in (12.25) may still be valid for $t = 0$.

Observe, however, that the left-hand side inequality in formula (12.25) may fail to hold for $t = 0$, unless the inequality $D_i(0, 1) - \delta B(0, 1) > 0$ holds for $i = 1, \ldots, K - 1$. More generally, in the present setting, for the positivity of $\pi_i(t)$, $i = 1, \ldots, K - 1$, it is required that $D_i(0, t) - \delta B(0, t) > 0$ for $i = 1, \ldots, K - 1$. An analogous condition is discussed in Sect. 13.1.5.

### 12.1.4 Das and Tufano Model

Das and Tufano (1996) extend the discrete-time version of the Jarrow et al. (1997) model to the case of random recovery rates. They assume discrete-time HJM framework for the modeling of the default-free term structure and the Markov chain dynamics for credit migrations. However, the recovery rates are no longer assumed to be constant. Let us briefly describe their approach, focusing on modifications with respect to the original Jarrow et al. (1997) model.

We consider a discrete-time framework with the time step of length $h$. For a fixed $T^* > 0$, we denote $T^* = Mh$ and $t_i = ih$ for $i = 0, \ldots, M$. Assume – as in Heath et al. (1990) or Amin and Bodurtha (1995) – that for every $T = mh \leq T^*$, the dynamics of the instantaneous forward interest rate $f(t, T)$ are governed by the following expression

$$f(t_i + h, T) = f(t_i, T) + \alpha(t_i, T)h + \sigma(t_i, T)\eta_i \sqrt{h}$$

with $f(0, T)$ given, where $\eta_i$, $i = 0, \ldots, M - 1$ are mutually independent, identically distributed, random variables with the standard Gaussian law under the martingale measure $\mathbb{Q}^*$. In particular, the spot rate $r(t) = f(t, t)$ satisfies

$$r(t_i + h) = f(0, t_i) + \sum_{j=0}^{i} \left( \alpha(t_j, t_i)h + \sigma(t_j, t_i)\eta_i \sqrt{h} \right).$$

It is well known that the no-arbitrage condition takes the following form

$$\sum_{j=i+1}^{M-1} \alpha(t_i, jh)h = h^{-1} \ln \mathbb{E}_{\mathbb{Q}^*} \left( \sum_{j=i+1}^{M-1} \sigma(t_i, jh)\eta_i \sqrt{h} \right).$$

The above discrete-time model of the default-free term structure is the first component of the Das and Tufano model. The second component of the model – the default model in their terminology – consists of two parts: a model of the default process (or, more appropriately, a model of the credit migration process), and a model of random recovery rates. There is nothing new regarding the model of the credit migration process in the Das and Tufano approach, as compared with the Jarrow et al. (1997) formulation.[7] Thus, it appears that the only novelty with respect to the Jarrow et al. (1997) approach is that the recovery rates are now modeled as random processes, rather than deterministic quantities. Specifically, Das and Tufano (1996) postulate that the recovery rate process $\delta(t)$ obeys the following recurrence relationship

$$\delta(t_i + h) = \left( 1 + \frac{1 - \delta(t_i)}{\delta(t_i)} \exp\left(\sigma_\delta \zeta_i \sqrt{h}\right) \right)^{-1}$$

with the initial condition $\delta(0) \in [0, 1]$ (so that $\delta(t_i) \in [0, 1]$ for any $t_i$).

---

[7] Our discussion from Sect. 12.1.1, regarding the $\mathbb{G}$-Markov nature of the migration process, applies here as well.

The random variables $\zeta_i$, $i = 0, \ldots, M - 1$ are assumed to be mutually independent and identically distributed, with the standard Gaussian law under $\mathbb{Q}^*$. Finally, the joint law of $(\eta_i, \zeta_i)$ is Gaussian, with the correlation coefficient $\rho$.

Let us assume, following Das and Tufano, that $\sigma(t, T) = \sigma_r \exp(-\lambda(T - t))$ for some constant $\sigma_r$. In other words, we deal here with the HJM version of Vasicek's model. Then we have, for some real number $K(t_i)$,

$$R(t_i) := r(t_i + h) - r(t_i) = K(t_i) + \sigma_r \eta_i \sqrt{h}$$

and

$$A(t_i) := \ln\left(\frac{1 - \delta(t_i + h)}{\delta(t_i + h)} \frac{1}{1 - \delta(t_i)}\right) = \sigma_\delta \zeta_i \sqrt{h}.$$

Consequently, the corresponding variance-covariance matrices satisfy

$$\begin{pmatrix} \sigma_r^2 & \rho\sigma_r\sigma_\delta \\ \rho\sigma_r\sigma_\delta & \sigma_\delta^2 \end{pmatrix} = h^{-2} \begin{pmatrix} \sigma_R^2 & \rho\sigma_R\sigma_A \\ \rho\sigma_R\sigma_A & \sigma_A^2 \end{pmatrix}$$

and thus the parameters of the two-dimensional process $(r(t_i), \delta(t_i))$ can be easily estimated.

Since the recovery rate is stochastic, the credit spreads may change even when the credit rating of the firm is unchanged (at the intuitive level, this corresponds to the two-factor model of credit spreads). Let us recall that Jarrow et al. (1997) assume that the debtholder receives the recovery amount at the maturity of the debt. When the recovery rate is stochastic, the need for an analysis of intermediate cash flows, at any date before the debt's maturity, arises in a natural way.

By allowing for randomness of recovery rates, Das and Tufano (1996) achieve the following main goals. First, the credit spreads generated by the model are closer to those observed in the real-world debt market. Second, the credit spreads are tied not only to ratings, but also to other factors. In particular, they move even when the rating remains unchanged. In other words, the variability of credit spreads is firm-specific, rather than rating-class-specific. Finally, the recovery rates, and thus also the credit spreads, are correlated with the default-free interest rates (this feature is supported by empirical studies).

Das and Tufano (1996) argue that due to the flexibility of their model, it allows for a wide range of behavior within the same rating class, so that it is more suitable to price a wide range of spread-based exotic debt and option contracts than simpler models with constant recovery rates.

*Remarks.* Let us stress that the main focus in Das and Tufano (1996) is put on a tree implementation of their model of defaultable term structure. Due to the absence of explicitly stated assumptions, some formulae appearing in this paper (for instance, the representation of the bond price in Sect. 3.2.2) are educated guesses, rather than strict mathematical results.

### 12.1.5 Thomas, Allen and Morkel-Kingsbury Model

Thomas et al. (1998) propose a discrete-time model that builds upon the discrete-time JLT model. We shall summarize the main interesting features of their approach, as we see them, without engaging in a detailed probabilistic construction. The presentation below does not exactly follow the exposition of the model in Thomas et al. (1998); we adapt their exposition in a way that we deem necessary for the formal correctness of the model.

**State variables.** Thomas et al. (1998) postulate the existence of three sequences of random variables, defined on the underlying probability space $(\Omega, \mathcal{G}, \mathbb{Q}^*)$, where $\mathbb{Q}^*$ is a risk-neutral probability measure:

- a sequence $Y$ of state variables: $Y_t$, $t \in \mathcal{T}^* := \{0, \dots, T^*\}$; each random variable $Y_t$ takes values in the set $\{g, b\}$; the states $g$ and $b$ correspond to the two states of the economy: $g = \text{good}$ and $b = \text{bad}$,
- a sequence $I$ of state variables:[8] $I_t$, $t \in \mathcal{T}^*$, which take values in the set $\mathcal{T}^*$; the level of spot interest rate at time $t$ is assumed to be $r_t = r_t(I_t, Y_t)$ where $r_t(\cdot, \cdot)$ are some functions; by convention $I_0 = 0$,
- a migration process $C = C_t$, $t \in \mathcal{T}^*$, with values in the set $\mathcal{K} = \{1, \dots, K\}$; as usual, the state $K$ corresponds to bankruptcy; it is natural to assume that the initial rating $C_0 \neq K$.

It is worth noting here that Thomas et al. (1998) write $E_t$ rather than $Y_t$, and $R_t$ rather than $C_t$. We endow the probability space $(\Omega, \mathcal{G}, \mathbb{Q}^*)$ with the filtration $\mathbb{G} = \mathbb{F}^C \vee \mathbb{F}^I \vee \mathbb{F}^I = \mathbb{F}^{(C, I, Y)}$. It is assumed that the information flow available to market participants is carried by the processes $Y$, $I$ and $C$. In other words, the market information is modeled by the filtration $\mathbb{G}$.

**Markovian features.** It is assumed that the three-dimensional stochastic process $(C, I, Y)$ follows a Markov chain. Since here $\mathbb{G}$ is the natural filtration of $(C, I, Y)$, this process follows also a $\mathbb{G}$-Markov chain. In addition, we impose some restrictions on one-step transition probabilities of this process.

**Condition (D.1)** On the set $\{I_t = u\}$ we have

$$\mathbb{Q}^*\{I_{t+1} = u \text{ or } I_{t+1} = u + 1 \mid Y_t, C_t, I_t\} = 1$$

for every $u \in \mathcal{T}^*$ and $t = 0, \dots, T^* - 1$.

Notice that (D.1), combined with the equality $I_0 = 0$, imply that $I_t \leq t$.

**Condition (D.2)** The state $K$ is absorbing for the migration process $C$ under $\mathbb{Q}^*$, so that we have

$$\mathbb{Q}^*\{C_{t+1} = K \mid Y_t, C_t = k, I_t\} = 1$$

for every $t = 0, \dots, T^* - 1$.

---

[8] This is motivated by the so called *lattice-type Markov chain* interest rate model of Pliska (1997), Sect. 6.2. In the present set-up, it suffices to take the finite set $\{0, \dots, T^*\}$, as opposed to the infinite set considered in Sect. 6.2 of Pliska (1997).

*Remarks.* (i) As in the JLT model, the default time is given in terms of the migration process, namely,

$$\tau = \min \{t = 0, \dots, T^* : C_t = K\}.$$

Observe that the validity of Condition (JLT.3) now depends on the statistical properties of the sequences $Y$, $I$ and $C$.

(ii) Markovian postulates with regard to the processes $C$, $I$ and $Y$ formulated above are sufficient for the results presented below (see Corollary 12.1.1 and Proposition 12.1.3) to hold. Unfortunately, we were not able to justify the valuation results obtained in Thomas et al. (1998) based on the respective Markovian postulates formulated therein.

**Bond valuation.** Similarly as in the JLT model, we assume that a fixed fraction $\delta$ of a corporate zero-coupon bond is received by the bondholder at the bond's maturity date $T$ if the bond defaults before or on this date. Hence, for every $t = 0, \dots, T$ we have

$$D^\delta(t, T) = B_t \, \mathbb{E}_{\mathbb{Q}^*} \big( B_T^{-1}(\mathbb{1}_{\{C_T \neq K\}} + \delta \mathbb{1}_{\{C_T = K\}}) \,\big|\, \mathcal{G}_t \big).$$

From the assumed Markov property of the process $(C, I, Y)$ it follows that the value of $D^\delta(t, T)$ depends on $\mathcal{G}_t$ through the values of the random variable $(C_t, I_t, Y_t)$ only. The following result is thus rather obvious.

**Lemma 12.1.1.** *Let us fix the dates $t \leq T$. Then there exists a function $D^\delta(t, T; \cdot, \cdot, \cdot) : \mathcal{K} \times \mathcal{T}^* \times \{g, b\} \to \mathbb{R}$ such that the equality*

$$D^\delta(t, T) = D^\delta(t, T; i, u, y)$$

*is satisfied $\mathbb{Q}^*$-a.s. on the set $\{C_t = i, I_t = u, Y_t = y\} \in \mathcal{G}_t$.*

In fact, it is also possible to provide a recursive formula for the pricing function $D^\delta(t, T; i, u, y)$.

**Proposition 12.1.3.** *Let us fix the bond's maturity date $T$. Then for every $i \leq K - 1$, $u \in \{0, \dots, T^*\}$, $y \in \{g, b\}$ and $t = 0, \dots, T - 1$ we have*

$$D^\delta(t, T; i, u, y) = \frac{1}{1 + r_t(u, y)} \sum_{j, u', y'} \big( D^\delta(t+1, T; j, u', y') p(i, i', u, u', y, y') \big),$$

*where we sum over $j \in K$, $u' \in \{u, u+1\}$, $y' \in \{g, b\}$, and where we denote*

$$p(i, i', u, u', y, y') := \mathbb{Q}^* \{ C_{t+1} = j, I_{t=1} = u', Y_{t+1} = y' \,|\, C_t = i, I_t = u, Y_t = y \}.$$

*In addition, for $t = T$ and for every $u \in \{0, \dots, T^*\}$ and $y \in \{g, b\}$ we have*

$$D^\delta(T, T; i, u, y) = \begin{cases} 1, & \text{for } i \neq K, \\ \delta, & \text{for } i = K. \end{cases}$$

*Proof.* We fix $i \leq K - 1$, $u \in \{0, \ldots, T^*\}$, $y \in \{g, b\}$ and $t = 0, \ldots, T - 1$. Recalling that $B_t = \prod_{u=0}^{t-1}(1 + r_u)$, we see that

$$D^\delta(t, T) = B_t \, \mathbb{E}_{\mathbb{Q}^*}\big(B_{t+1}^{-1} D^\delta(t+1, T) \,\big|\, \mathcal{G}_t\big) = (1 + r_t)^{-1} \mathbb{E}_{\mathbb{Q}^*}\big(D^\delta(t+1, T) \,\big|\, \mathcal{G}_t\big),$$

and so

$$D^\delta(t, T) = (1 + r_t(I_t, Y_t))^{-1} \mathbb{E}_{\mathbb{Q}^*}\big(D^\delta(t+1, T) \,\big|\, \mathcal{G}_t\big).$$

Consequently, taking into account the Markovian features of the model, as well as Condition (D.1), on the set $\{C_t = i, I_t = u, Y_t = y\}$ we obtain[9]

$$D^\delta(t, T) = (1 + r_t(I_t, Y_t))^{-1} \sum_{j, u', y'} \Big( D^d(t+1, T; j, u', y')$$

$$\times \, \mathbb{Q}^*\{C_{t+1} = j, I_{t=1} = u', Y_{t+1} = y' \,|\, C_t, I_t, Y_t\}\Big),$$

where the summation is over $j \in K$, $u' \in \{u, u + 1\}$ and $y' \in \{g, b\}$. To complete the proof of the proposition, it suffices to combine the last formula with Lemma 12.1.1. $\qquad \square$

In addition, Thomas et al. (1998) postulate a certain decomposition of one-step transition probabilities $p(i, i', u, u', y, y')$. It is not clear to us, however, why such a decomposition should be satisfied for a non-trivial model. For this reason, we have decided to omit this step.

## 12.2 Conditionally Markov Models

As we said earlier, we believe that modeling credit migrations in terms of an $\mathbb{F}$-conditional $\mathbb{G}$-Markov chain[10] is more appropriate than modeling credit migrations in terms of a $\mathbb{G}$-Markov chain. In this section, we present an example of an approach towards the former methodology, due to Lando (1998); another model of this type is discussed in Sect. 13.2.

In both cases, we shall see that modeling intensities of migrations as functionals of underlying state variables is quite natural in the present context. Let us recall that in Sect. 8.6, we have already discussed the modeling of the default event in terms of state variables. The present section may thus be seen as a natural continuation of Sect. 8.6.

For other examples of models that may be categorized as conditionally Markov models of defaultable term structures, the interested reader is referred to papers by Arvanitis et al. (1998), for a continuous-time setting, and to Wei (2000) and McNulty and Levin (2000), for a discrete-time framework.

---

[9] In view of our assumptions about the process $I$, this set is empty if $u > t$. Thus, we only consider the case when $u \leq t$.

[10] We refer to Sect. 11.1.3 and to Sect. 11.3 for a discussion of $\mathbb{F}$-conditional $\mathbb{G}$-Markov chains.

### 12.2.1 Lando's Approach

We shall describe here an extension – due to Lando (1998) – of the continuous-time credit ratings model elaborated by Jarrow et al. (1997). We preserve the probabilistic set-up of Sect. 11.3.1; that is, we consider the underlying enlarged probability space

$$(\Omega, \mathcal{G}, \mathbb{Q}^*) := (\tilde{\Omega} \times \Omega^U \times \bar{\Omega}, \tilde{\mathcal{G}}_\infty \otimes \mathcal{F}^U \otimes 2^K, \mathbb{P}^* \otimes \mathbb{P}^U \otimes \mu)$$

endowed with the filtration $\mathbb{G}$. Recall, that the filtration $\mathbb{F}$, originally given on the component space $(\tilde{\Omega}, \mathcal{F}, \mathbb{P}^*)$, is then extended to the enlarged space $(\Omega, \mathcal{G}, \mathbb{Q}^*)$, and it is still denoted as $\mathbb{F}$. Thus, it becomes a sub-filtration of $\mathbb{G}$. The filtration $\mathbb{F}$ reflects the information carried by the *state variables* (also known as *factors*). We shall now describe the assumption postulated in Lando (1998); as usual, we are going to use our notation, rather than that of the original paper.

**Condition (L.1)** We are given an $\mathbb{F}$-adapted process $Y_t$, $t \in [0, T^*]$, which takes values in some state space, e.g., $\mathbb{R}^k$. The process $Y$ is meant to represent the evolution of the vector of state variables $(Y^1, \ldots, Y^k)$.

Lando (1998) maintains Condition (JLT.1) with regard to the spot martingale measure $\mathbb{Q}^*$ and the filtration $\mathbb{G}$; for the sake of convenience, we shall henceforth refer to this condition as Condition (L.2). He extends the original JLT approach by introducing a *conditionally Markov* migration process, which makes account for both the presence of different rating classes and for the postulated existence of the underlying state variables. This is achieved by a suitable modification of the migration process $C$ introduced in Sect. 12.1.2. In fact, since Lando (1998) does not consider the real-world probability measure $\mathbb{Q}$, no counterparts of conditions (JLT.5c) and (JLT.7c) are given in his paper, and Condition (JLT.6c) is generalized as follows.

**Condition (L.3)** The migration process $C$ is assumed to be an $\mathbb{F}$-conditional $\mathbb{G}$-Markov chain under the spot martingale measure $\mathbb{Q}^*$, with the stochastic intensity matrix $\Lambda(Y_t) = [\lambda_{ij}(Y_t)]_{1 \leq i,j \leq K}$. The matrix $\Lambda(Y_t)$ is assumed to satisfy, for every $t \in [0, T^*]$ and $i = 1, \ldots, K$,

$$\lambda_{ii}(Y_t) = - \sum_{j=1, j \neq i}^{K} \lambda_{ij}(Y_t), \quad \lambda_{K,i}(Y_t) = 0, \tag{12.27}$$

where $\lambda_{ij} : \mathbb{R}^k \to \mathbb{R}_+$ are non-negative functions.

For any such matrix, given the state-variables process $Y$ and the initial rating $i$, it is possible to construct a migration process $C$ associated with the stochastic intensity matrix $\Lambda(Y_t)$. More specifically, the migration process $C$ is determined in such a way that, conditionally on a particular sample path $Y_t(\omega)$, $t \in [0, T^*]$ of the state-variables process $Y$, the migration process $C$ is a time-inhomogeneous Markov chain with finite state space $\{1, \ldots, K\}$ and time-dependent (but deterministic) intensity matrix $\Lambda(Y_t(\omega))$.

As apparent from (12.27), the $K^{\text{th}}$ row of the matrix $\Lambda(Y_t)$ is assumed to vanish identically, and thus, as usual, the state $K$ is absorbing for $C$. The construction of the migration process $C$ with the required properties can be done in a direct analogy to the construction presented in Sect. 11.3.1. To complete the model's specification, we need to describe the default-free interest rate process $r$ and the default time $\tau$. To this end, we replace conditions (JLT.2) and (JLT.3) by the following postulates (L.4) and (L.5), respectively.

**Condition (L.4)** The short-term interest rate $r$ satisfies $r_t = R(Y_t)$, $t \in [0, T^*]$, for some function $R : \mathbb{R}^k \to \mathbb{R}$.

**Condition (L.5)** The default time $\tau$ is the first time that the migration process $C$ jumps to the absorbing state: $\tau = \inf\{t \in [0, T^*] : C_t = K\}$.

Due to the nature of the short-term interest rate process $r$ and the default time $\tau$, the valuation of defaultable claims becomes more cumbersome than in the original JLT set-up. In fact, the default time $\tau$ and the short-term rate $r$ are no longer mutually independent, as was postulated in Condition (JLT.3) of Jarrow et al. (1997). For this reason, no explicit general pricing results, such as formula (12.2), are available in the present setting. Lando (1998) discusses the valuation of defaultable claims under additional assumptions on the statistical properties of the promised payment $X$ and the recovery process $Z$. We shall limit our discussion to the case of a $T$-maturity zero-coupon corporate bond with zero recovery – that is, we take $X \equiv 1$ and $Z \equiv 0$. By definition, the price of such a bond at time $t \leq T$ equals

$$D_i^0(t, T) := B_t \, \mathbb{E}_{\mathbb{Q}^*} \big( B_T^{-1} \mathbb{1}_{\{T < \tau\}} \,\big|\, \mathcal{F}_t \vee \{C_t = i\} \big),$$

where we assume that at time $t$ the bond belongs to the $i^{\text{th}}$ rating class, for some $i < K$. Using a similar reasoning as in the proof of Proposition 8.6.1 (i.e., conditioning on the future evolution of the process $Y$), we obtain the following result.

**Proposition 12.2.1.** *We have*

$$D_i^0(t, T) = B_t \, \mathbb{E}_{\mathbb{Q}^*} \big( B_T^{-1}(1 - p_{iK}^Y(t, T)) \,\big|\, \mathcal{F}_t \big),$$

*where*

$$p_{iK}^Y(t, T) = \mathbb{Q}^*\big\{ C_T = K \,\big|\, \sigma(Y_u : u \in [t, T]) \vee \{C_t = i\} \big\}.$$

Notice that $p_{iK}^Y(t, T)$ is nothing else than the conditional transition probability of the migration process $C$ over the time interval $[t, T]$, with conditioning on the future behavior of the state-variables process $Y$ over the same time period. To simplify the calculations further, Lando (1998) also postulates that $\Lambda(Y_t) = L_t \, \Gamma(Y_t) L_t^{-1}$, where $\Gamma(Y_t)$ is a diagonal matrix and $L_t$ is a $K \times K$ matrix whose columns are the eigenvectors of $\Lambda(Y_t)$. Under this rather restrictive condition, he derives a quasi-explicit valuation formula for a defaultable bond and, indeed, for any defaultable European contingent claim with the promised payoff $X = g(Y_T, C_T)$.

## 12.3 Correlated Migrations

We shall now discuss a few examples of practical problems pertaining to a finite family of migration processes, associated with a collection of reference credit names. It seems natural to assume that these migration processes are 'correlated' – that is, statistically mutually dependent. We do not provide an in-depth study of the issue of modeling dependent migrations, though. Such a study will be done elsewhere. In this section, we just want to illustrate, by means of selected examples, the necessity of mathematically sound modeling of mutually dependent migrations. Some examples presented below were previously discussed in other chapters. In this section, these examples are analyzed from the perspective of dependent migrations. In addition, we shall also summarize recent results of Lando (2000b), where dependent defaults are modeled within the context of credit migrations.

It is important to keep in mind that either the real-world (statistical) probability measure, or the risk-neutral probability measure, should be used when assessing the dependence between credit migrations, depending on the targeted application. In the context of credit-risk management applications – for example, in applications related to the Value-at-Risk (VaR) type calculations for credit portfolios – the dependence between credit migrations should be specified under the real-world probability. By contrast, when dealing with the issues of dependence between credit migrations in the context of valuation and hedging of credit risk – for example, in the context of basket credit derivatives – it is natural to consider the dependence between credit migrations under the risk-neutral probability.

In the rest of this section, we will fix a probability space $(\Omega, \mathcal{G}, \mathbb{Q})$, where $\mathbb{Q}$ may play the role of either the real-world probability, or the risk-neutral probability. We are given $n \geq 2$ credit entities or obligors. Let $\mathcal{K}_i := \{k_1^i, k_2^i, \ldots, k_{m_i}^i\}$ denote the set of possible credit ratings for the $i^{\text{th}}$ entity. For each $i = 1, \ldots, n$, the state $k_{m_i}^i$ is absorbing, and it represents the default state for the $i^{\text{th}}$ entity.

We adhere to the continuous-time set up throughout this section, and we postulate that the credit ratings of the $i^{\text{th}}$ obligor vary over time according to a continuous-time random process $C^i$ with finite state space $\mathcal{K}_i$, defined on the underlying probability space $(\Omega, \mathcal{G}, \mathbb{Q})$. Hence, the credit ratings of all the obligors evolve according to the process $C = (C^1, C^2, \ldots, C^n)$, defined on $(\Omega, \mathcal{G}, \mathbb{Q})$, and with the state space $\mathbf{K} := \mathcal{K}_1 \times \mathcal{K}_2 \times \ldots \mathcal{K}_n$.[11] We denote by $\mathbb{F}^C$ the natural filtration of the credit migration process $C$ – that is, $\mathcal{F}_t^C = \sigma(C_s : 0 \leq s \leq t)$ for $t \in \mathbb{R}_+$. We suppose that the space $(\Omega, \mathcal{G}, \mathbb{Q})$ is endowed with some underlying filtration $\mathbb{F}$, and that the information available to an economic agent is carried by some filtration $\mathbb{G} = (\mathcal{G}_t)_{t \geq 0}$, where for every $t \in \mathbb{R}_+$: $\mathcal{F}_t^C \subseteq \mathcal{G}_t \subseteq \mathcal{F}_t^C \vee \mathcal{F}_t$.

---

[11] In general, the process $C$ is thus a marked point process. We refer to Last and Brandt (1995) for an exhaustive treatment of the marked point processes.

We are interested in various conditional probability distributions associated with the migration process $C$ and the filtration $\mathbb{G}$. Knowledge of such distributions allows for the computation of quantities that may be of interest to economic agents. We now proceed with presenting a few pertinent examples of computational problems related to the modeling of dependent credit migrations.

*Example 12.3.1. Probability of no-default in a unit of time.* Let us fix $t \geq 0$. Consider the event $A_t^0$ defined as follows: $A_t^0$ occurs if and only if none of the $n$ credit entities defaults in the interval $[t, t+1]$ (of course, here 1 represents any time unit that is convenient for practical purposes). Formally,

$$A_t^0 = \bigcap_{i=1}^{n} \{C_s^i \neq k_{m_i}^i : \forall s \in [t, t+1]\}.$$

The quantity of interest is the conditional probability $\mathbb{Q}\{A_t^0 \mid \mathcal{G}_t\}$ – that is, the conditional probability that there will be no defaults in the time interval $[t, t+1]$, given the information $\mathcal{G}_t$ available at time $t$. We shall be assuming that the state of default is the absorbing state for the migration process; put another way, no debt renegotiations are allowed. Formally, as soon as the process $C^i$ jumps to the default state $k_{m_i}^i$, it stays there forever. In an unlikely situation of mutually independent credit migrations, we have

$$\mathbb{Q}\{A_t^0 \mid \mathcal{G}_t\} = \prod_{i=1}^{n} \mathbb{Q}\{C_s^i \neq k_{m_i}^i : \forall s \in [t, t+1]\}. \tag{12.28}$$

As we saw in Chap. 9, it is not uncommon to postulate the conditional independence, with respect to the reference filtration $\mathbb{F}$, between credit migrations of various credit entities. Under this assumption, a representation similar to (12.28) can be easily derived.

*Example 12.3.2. Probability of no changes in credit ratings occurring in a unit of time.* As before, we fix $t \geq 0$. Then, we consider the event $A_t^0$ defined as follows: $A_t^0$ occurs if and only if none of the $n$ credit entities changes its credit rating in the time interval $[t, t+1]$. Formally, the event $A_t^0$ satisfies:

$$A_t^0 = \bigcap_{i=1}^{n} \{C_s^i = C_t^i : \forall s \in [t, t+1]\}.$$

The quantity we are now interested in is the conditional probability $\mathbb{Q}\{A_t^0 \mid \mathcal{G}_t\}$, i.e., the conditional probability that there will be no credit changes at all during the time period $[t, t+1]$, given the information $\mathcal{G}_t$ available at time $t$.

*Example 12.3.3. Probability of $m \leq n$ defaults occurring in a unit of time.* We fix $t \geq 0$, and we choose an integer $m$ such that $0 \leq m \leq n$. We define the event $A_t^m$ as follows: $A_t^m$ occurs if and only if exactly $m$ of $n$ reference credit entities default in the time interval $[t, t+1]$. Again, one is interested in computing the conditional probability $\mathbb{Q}\{A_t^m \mid \mathcal{G}_t\}$. We shall illustrate such computations in the most simplified model in which, in particular, it is assumed that defaults are independent. Manifestly, computations in case of dependent defaults will be much more involved and will depend on the type of dependence structure (see Davis and Lo (1999), for example).

Assume that the probabilities of default in a unit of time are equal for all entities – that is, for $i = 1, \ldots, n$ we have

$$\mathbb{Q}\{\, B_t^i \mid \mathcal{G}_t\} = \mathbb{Q}\{\, \exists\, s \in [t, t+1] : C_s^i = k_{m_i}^i \mid \mathcal{G}_t\} =: p_t,$$

where $B_t^i$ represents the event: entity $i$ defaults in the time interval $[t, t+1]$. We assume that none of the credit entities has yet defaulted by time $t$, and we assume that the defaults occur independently for various obligors. Let $K_t$ be the random variable representing the number of defaulting entities in the time interval $[t, t+1]$. Under the present assumptions, the conditional distribution of $K_t$ is binomial, and so

$$\mathbb{Q}\{K_t = m \mid \mathcal{G}_t\} = \mathbb{Q}\{A_t^m \mid \mathcal{G}_t\} = \frac{n!}{m!(m-n)!}\, p_t^m (1 - p_t)^{n-m}.$$

It is noteworthy that Moody's Binomial Expansion Technique (see Moody's Investment Service (1997)), which is in fact a methodology for computing distribution of a number of bonds defaulting within a given portfolio of bonds, hinges on the simplified model described above. More realistic models – such as, for example, the approach of Davis and Lo (1999) briefly described in Chap. 10 – dispense with the assumption of independence between the default events for various reference credit entities.

*Example 12.3.4. Variablity of the value of a loan portfolio.* Suppose that each of the $n$ credit entities represents a commercial loan. The value of the $i^{\text{th}}$ loan at time $t$ is denoted by $V_t^i$. Besides the loan's notional amount, the loans's contractual interest rate, and the time to maturity, the current value of the loan may depend on the current credit rating of that loan and the current credit spread corresponding to this rating. For some simple methodologies for calculating present values of commercial loans, the interested reader is referred to Saunders (1999).

The value of the entire portfolio at time $t$ is clearly $V_t = \sum_{i=1}^n V_t^i$. We are interested in the conditional distribution of the value of the loan portfolio at some future date $s > t$ ($t$ represents today's date), given the information $\mathcal{G}_t$. To illustrate one method of deriving such a distribution, let us first denote by $V_s^i(k^i)$ the value of the $i^{\text{th}}$ loan at time $s$, if the current credit rating of this loan at time $s$ is $k^i \in \mathcal{K}_i$.[12] It is clear that

---

[12] We implicitly assume that once the credit rating of the given loan is known, the value of this loan is known as well; such a simplifying assumption is frequently made within credit risk management methodologies (see, e.g., Saunders (1999)).

$$\mathbb{Q}\{V_s = \sum_{i=1}^{n} V_s^i(k^i) \mid \mathcal{G}_t\} = \mathbb{Q}\{C_s = (k^1, k^2, \dots, k^n) \mid \mathcal{G}_t\}.$$

In particular, one may easily compute the conditional expected value for the loan portfolio at time $s$:

$$\mathbb{E}_{\mathbb{Q}}(V_s \mid \mathcal{G}_t) = \sum_{(k^1, k^2, \dots, k^n) \in \mathbf{K}} \left[ \mathbb{Q}\{C_s = (k^1, k^2, \dots, k^n) \mid \mathcal{G}_t\} \sum_{i=1}^{n} V_s^i(k^i) \right].$$

Similar representations can be derived for higher conditional moments of the random variable $V_s$. Since we do not go into details here, the interested reader is referred to Saunders (1999) (see, e.g., Page 127 therein).

Once the conditional distribution of the loan portfolio's value is known, then one can use it to perform the Value-at-Risk (VaR) type calculations, quite popular in risk management. The importance of modeling of dependent credit migrations (and not just defaults) is particularly pronounced in the context of applications of the type described in this example.

*Example 12.3.5. Valuation of basket credit derivatives.* Recall that a basket credit derivative (BCD) is a credit-risk sensitive product that derives its value from credit risks associated with a portfolio of credit entities (such as a loan portfolio or such as a bond portfolio). As we saw in Chap. 9, a typical structure of a BCD product involves a payment contingent upon the timing and the identity of several consecutive defaults (typically the first and the second) within a portfolio of credit entities (credit risks). A related structure involves termination of coupon and/or principal payments to various tranches of a Collateralized Bond Obligation (CBO) or to various tranches of a Collateralized Loan Obligation (CLO) contingent on consecutive defaults of the reference credits.[13] Yet another structure of a basket credit derivative is a derivative contract in which the payment is contingent on the event that the value of the loss in a credit portfolio exceeds a predetermined level.

As in Chap. 9, we denote by $\tau_j$ the default time of the $j^{\text{th}}$ credit entity. In the present setting, the default times satisfy, for every $j = 1, \dots, n$,

$$\tau_j = \inf\{t \geq 0 : C_t^j = k_{m_j}\}. \tag{12.29}$$

Formal definitions of the $i^{\text{th}}$-to-default contingent claim and the vanilla default swaps of basket type described in Sect. 9.1 can now be expressed in terms of default times defined in (12.29). Analogous products may be defined relative to random times associated with any other rating class. Specifically, fixing some class $k_j \in \mathcal{K}_j$ for any $j = 1, \dots, n$, we may first define a family $\tau_1(k_1), \dots, \tau_n(k_n)$ of random times by setting

$$\tau_j(k_j) = \inf\{t \geq 0 : C_t^j = k_j\}, \tag{12.30}$$

and subsequently, we may introduce various kinds of basket products relative to the random times defined in (12.30).

---

[13] See, e.g., Tavakoli (1998) or Nelken (1999) for a discussion of CBOs and CLOs.

*Example 12.3.6. Default swaps with counterparty credit risk and/or multiple reference securities.* In Sect. 1.3.1, we summarized the main features of default swaps and default options. However, in our discussion there, we were assuming that the counterparty credit risk was negligible and we have only considered the case of a single reference entity. If the credit risk of a protection seller in a default swap cannot be neglected, one needs to take into account possible dependencies between the credit rankings of a protection seller and of the reference instrument. A rating-based approach to this issue, proposed by Lando (2000b), is discussed in Sect. 12.3.1 below.

Likewise, if one attempts to value a credit swap with respect to multiple reference credits, the need for the modeling of dependent migrations of these reference credits becomes obvious.

### 12.3.1 Huge and Lando Approach

Lando (2000b) is concerned with the issue of valuation of default swaps in the situation when there are possible dependencies between the credit rankings of the protection seller and the reference instrument. Let us recall that in Sect. 1.3.1, we have summarized the most relevant covenants of plain-vanilla default swaps and options, which do not admit counterparty risk, such as a seller's credit risk.

Lando (2000b) draws on the methodology laid out in Huge and Lando (1999). We thus start by analyzing a general mathematical model, which is largely motivated by the latter work. As usual, $\mathbb{Q}^*$ will stand for the spot martingale measure.

**Mathematical model.** Let $W^*$ be a standard Brownian motion process given on some probability space $(\Omega, \mathcal{G}, \mathbb{Q}^*)$, endowed with the filtration $\mathbb{F}$ (in particular, $W^*$ is $\mathbb{F}$-adapted). The short-term interest rate process $r$ is postulated to follow a diffusion process specified in terms of an SDE, namely,

$$dr_t = \mu(r_t)\,dt + \sigma(r_t)\,dW_t^*, \quad t \in \mathbb{R}_+, \tag{12.31}$$

for some non-random coefficients $\mu : \mathbb{R} \to \mathbb{R}$ and $\sigma : \mathbb{R} \to \mathbb{R}$. A priori, the state space for the process $r$ is thus the entire real line $\mathbb{R}$, so that, for example, the Vasicek model of the short-term interest rate is not excluded. In Lando (2000b), the Cox-Ingersoll-Ross model is selected for the short-term process $r$, and thus the state space for $r$ is the positive half-line $(0, \infty)$ in this paper.

A two-dimensional migration process $C = (C^1, C^2)$ is also defined on the probability space $(\Omega, \mathcal{G}, \mathbb{Q}^*)$. The first component, $C^1$, lives on the state space $\mathcal{K}_1 = \{1, 2, \dots, K_1\}$; it accounts for the evolution of the credit rating of the protection seller. The second component, $C^2$, takes values in the state space $\mathcal{K}_2 = \{1, 2, \dots, K_2\}$; it is meant to describe the evolution of the credit rating of the reference corporate bond. Thus, the state space of the two-dimensional migration process $C$ is defined to be $\mathbf{K} = \{1, 2, \dots, K_1\} \times \{1, 2, \dots, K_2\}$. The states $K_1$ and $K_2$ are assumed to be absorbing states for the respective components of the process $C$.

Recall that, according to our convention adopted earlier, the information available to each market participant is carried by the filtration $\mathbb{G}$, such that $\mathbb{G} \subseteq \mathbb{F}^C \vee \mathbb{F}$. In particular, each market participant is assumed to be able to observe the short-term rate process $r$, as well as the migration process $C$.

To further specify the model, we postulate that the process $(r, C)$ is jointly a $\mathbb{G}$-Markov process. A $\mathbb{G}$-Markov process can be formally defined in an exact analogy with the definition of a $\mathbb{G}$-Markov chain. However, since we did not study general $\mathbb{G}$-Markov processes in the present text, we prefer to assume from now on that $\mathbb{G} = \mathbb{F}^r \vee \mathbb{F}^C = \mathbb{F}^{(r,C)}$, so that the process $(r, C)$ becomes an ordinary Markov process. Consequently, the process $X_t = (t, r_t, C_t)$, where $t$ is the (continuous) running time variable, is also a Markov process. As commonly known, it is convenient to characterize a Markov process in terms of its infinitesimal generator. Because our presentation in this section is rather informal, we shall not engage in discussing various possible notions of infinitesimal generators for the Markov process $X$ relative to $\mathbb{Q}^*$, such as: the strong generator, the weak generator, the generalized generator, etc. We shall be satisfied with recalling the so-called Dynkin's formula.

We henceforth denote by $\mathcal{A}$ an infinitesimal generator of the process $X$ with state space $\mathcal{X}$, relative to the probability measure $\mathbb{Q}^*$. The detailed specification of the state space $\mathcal{X}$ corresponding to our particular case will be given later.

**Dynkin's formula.** For an arbitrary function $f : \mathcal{X} \to \mathbb{R}$, which belongs to the domain of the infinitesimal generator $\mathcal{A}$, the real-valued stochastic process $M^f$, given as:

$$M_t^f = f(X_t) - f(X_0) - \int_0^t \mathcal{A}f(X_s)\,ds,$$

follows a (local) $\mathbb{F}^X$-martingale under $\mathbb{Q}^*$.

It is convenient to distinguish the three components of the infinitesimal generator $\mathcal{A}$, namely:
- the first-order differential operator corresponding to the (continuous) running time $t$; we shall denote this part of $\mathcal{A}$ by $\mathcal{D}_1$ in what follows,
- the component of the operator $\mathcal{A}$ associated with the diffusion part of $X$, that is, with the short-term rate process $r$; this part is a second-order differential operator, $\mathcal{D}_2$ say, which is consistent with dynamics (12.31),
- the part of $\mathcal{A}$ determined by the migration process $C$; it can be represented as an intensity matrix operator of a suitable dimension, $\mathcal{L}$ say.

Possible interactions (dependencies) between the two processes, $r$ and $C$, may be accounted for through the coefficients of the above operators. However, for our further discussion, it is enough to postulate that the coefficients of the operator $\mathcal{L}$ may depend on the short-term rate process $r$ in a Markovian way. This assumption will allow the intensities of rating transitions to depend on the short-term interest rate. In other words, the process $r$ may be now interpreted as a state variable.

To summarize, we postulate that for any sufficiently regular function $f : \mathcal{X} \to \mathbb{R}$, the infinitesimal operator $\mathcal{A}$ is given as:

$$\mathcal{A}f(x) = (\mathcal{D} + \mathcal{L})f(x),$$

where $\mathcal{D} = \mathcal{D}_1 + \mathcal{D}_2$ with (the subscripts denote here partial derivatives)

$$\mathcal{D}_1 f(x) = f_t(x), \quad \mathcal{D}_2 f(x) = \mu(r)f_r(x) + \tfrac{1}{2}\sigma^2(r)f_{rr}(x),$$

and

$$\mathcal{L}f(x) = \sum_{c' \in \mathbf{K}} \lambda_{cc'}(r)f(x).$$

A function $f$ is defined on the state space $\mathcal{X} = \mathbb{R}_+ \times \mathbb{R} \times \mathbf{K}$ and takes values in $\mathbb{R}$. Thus, $x = (t, r, c)$ for some $t \in \mathbb{R}_+$, $r \in \mathbb{R}$, and $c \in \mathbf{K}$.

*Remarks.* We hope that the reader will not be confused by our convention of using the symbol $r$ to denote in the same time the short-term rate process and the variable representing a particular value of this process.

The intensity functions $\lambda_{cc'} : \mathbb{R} \to \mathbb{R}$ are assumed to be non-negative and such that

$$\lambda_{cc}(r) = -\sum_{c' \neq c} \lambda_{cc'}(r), \quad \forall\, r \in \mathbb{R}.$$

We see that, according to the above specification of the operator $\mathcal{A}$, the short-term interest rate $r$ follows a time-homogeneous Markov diffusion process with the differential infinitesimal operator $\mathcal{D}_1$, whereas the process $C$ is a $\mathbb{F}^r$-conditional $\mathbb{G}$-Markov chain with the conditional infinitesimal generator matrix process $\Lambda_t = \Lambda(r_t) = [\lambda_{cc'}(r_t)]_{c,c' \in \mathbf{K}}$.[14]

*Remarks.* There are, of course, several different orderings of the $K_1 K_2$ states of $C$ possible, each of them leading to a different layout of the generator matrix $\Lambda_t$. Observe, however, that the rows of $\Lambda_t$ corresponding to the states $(K_1, c^2)$, $c^2 \neq K_2$, and $(c^1, K_2)$, $c^1 \neq K_1$, need not to be rows of zeros. Only the row corresponding to the state of the 'complete' absorption – i.e., the row corresponding to the state $K := (K_1, K_2)$ – must be a row of zeros.

The correlation (dependence) structure between migrations of credit rankings (default events, in particular) of the protection seller and the reference credit is encoded in the operator $\mathcal{L}$ – i.e., in the transition intensities $\lambda_{cc'}(r)$. For example, let us observe that simultaneous default of the protection seller and the reference bond may occur if the process $C$ jumps from some state $c = (c^1, c^2)$, where $c^1 \neq K_1$ and $c^2 \neq K_2$, to the absorbing state $K = (K_1, K_2)$. For any state $c \in \mathbf{K}$ such that $c \neq K$, the conditional infinitesimal probability of the simultaneous default of both entities is given as the ratio $-\lambda_{cK}(r)/\lambda_{cc}(r)$. Lando (2000b) proposes to analyze three alternative specifications of the correlation structure; the interested reader is advised to consult the original paper for details.

---

[14] Recall that, for the sake of expositional simplicity, the processes $r$ and $C$ were assumed to be jointly Markov. Therefore, in the present framework, the notion of $\mathbb{F}^r$-conditional $\mathbb{G}$-Markov chain is not fully exploited.

**Evaluation of relevant functionals.** Our next goal is to present few results and comments regarding the evaluation of functionals associated with the Markov process $X$. We shall give a rather exhaustive discussion concerning the evaluation of the functional given by expression (12.32) below, to indicate the type of techniques that may be used. It should be acknowledged, however, that this is not exactly the type of functional that occurs in the valuation of a default swap with counterparty default risk. For the latter purpose, one needs to consider functionals of a more general form, as given by (12.33), rather than (12.32).

We fix a horizon date $T > 0$, and for any $t \leq T$ we consider the following functional of the sample paths of the process $X$:

$$\Phi_t := \mathbb{E}_{\mathbb{Q}^*}\left(e^{-\int_t^T R(X_u)\,du} h(X_T) \,\Big|\, \mathcal{G}_t\right), \tag{12.32}$$

for some functions $R : \mathcal{X} \to \mathbb{R}$ and $h : \mathcal{X} \to \mathbb{R}$. Needless to say, the functions $R$ and $h$ are implicitly assumed to also satisfy suitable integrability conditions, so that the conditional expectation $\Phi_t$ is well defined. The next auxiliary result is a straightforward consequence of the Markovian feature of the process $X$.

**Lemma 12.3.1.** *There exists a (measurable) function $\phi : \mathcal{X} \to \mathbb{R}$ such that $\Phi_t = \phi(X_t)$ for every $t \in \mathbb{R}_+$.*

Assuming that the function $\phi$ is sufficiently regular,[15] as a consequence of Dynkin's formula, we obtain the following result.

**Proposition 12.3.1.** *The process $M^\Phi$, given as*

$$M_t^\Phi = \phi(X_t) - \phi(X_0) - \int_0^t \mathcal{A}\phi(X_s)\,ds, \quad \forall\, t \in [0, T],$$

*follows an $\mathbb{F}^X$-martingale under $\mathbb{Q}^*$.*

It is important to notice that the process $\Phi$ may be represented as the product of the two processes, specifically $\Phi_t = \tilde{B}_t \tilde{M}_t$ for $t \in [0, T]$, where we set $\tilde{B}_t = e^{-\int_0^t R(X_u)\,du}$, and where the martingale $\tilde{M}$ is given by the formula

$$\tilde{M}_t = \mathbb{E}_{\mathbb{Q}^*}\left(e^{-\int_0^T R(X_u)\,du} h(X_T) \,\Big|\, \mathcal{G}_t\right), \quad \forall\, t \in [0, T].$$

A simple application of the Itô product rule leads to the following proposition.

**Proposition 12.3.2.** *For every $t \leq s \leq T$, the process $\Phi$ satisfies*

$$\Phi_s - \Phi_t = \int_t^s \tilde{B}_u \, d\tilde{M}_u + \int_t^s R(X_u)\Phi_u \, du.$$

Finally, combining Lemma 12.3.1 with Propositions 12.3.1 and 12.3.2, we obtain the following corollary.

---

[15] Continuously differentiable with respect to $t$ and twice continuously differentiable with respect to $r$, e.g., with bounded partial derivatives.

**Corollary 12.3.1.** *For every $t \in [0, T]$ we have*

$$\mathbb{E}_{\mathbb{Q}^*}\left( \int_0^t \left( \mathcal{A}\phi(X_s) - R(X_s)\phi(X_s) \right) ds \right) = 0.$$

We are in a position to state the following result, which corresponds to equation (31) in Huge and Lando (1999). Recall that we denote $x = (t, r, c)$. Also, we denote by $\mathcal{R}$ a multiplication operator defined as $\mathcal{R}\phi(x) = R(x)\phi(x)$.

**Proposition 12.3.3.** *The function $\phi$ solves the following integro-differential equation on $[0, T] \times \mathbb{R} \times \mathbf{K}$*

$$(\mathcal{A} - \mathcal{R})\phi(x) = 0$$

*or, more explicitly,*

$$\phi_t(x) + \mu(r)\phi_r(x) + \tfrac{1}{2}\sigma^2(r)\phi_{rr}(x) + \sum_{c' \in \mathbf{K}} \lambda_{cc'}(r)\phi(x) - R(x)\phi(x) = 0$$

*subject to the terminal condition $\phi(T, r, c) = h(T, r, c)$ for $(r, C) \in \mathbb{R} \times \mathbf{K}$.*

A rather standard proof of Proposition 12.3.3 is omitted. The reader might have noticed that this result could have been derived through a direct application of a general version of the well-known Feynman-Kac formula.

**Valuation of a default swap.** Recall that we are interested in studying a default swap in the presence of a counterparty credit risk and credit migrations. More specifically, we allow for the possibility of credit migrations, also including the possibility of default, with regard to both of the protection seller and the underlying reference credit. The credit risk of the protection buyer is disregarded in this model. Such a swap is called the *protection-default-sensitive default swap*. Note that there are in fact three entities directly or indirectly involved in such a contract: the party that issues the reference defaultable bond, i.e., the reference party, the holder of the reference instrument, i.e., the protection buyer, and the party that sells the protection against adverse credit migrations (default in the present case) of the reference instrument, i.e., the protection seller. It is thus natural to expect that the payoff structure of a protection-default-sensitive default swap will amalgamate the cash flows of a vanilla default swap (as explained in Sect. 1.3.1), with the payoff structure of a defaultable swap with bilateral default risk (see Sect. 14.5 in this regard).

In order to develop an effective method for the valuation of a protection-default-sensitive default swap, we need to consider a functional of the form

$$\Phi_t = \mathbb{E}_{\mathbb{Q}^*}\left( \int_t^T e^{-\int_t^s R(X_u)du} dD_s \,\Big|\, \mathcal{G}_t \right), \tag{12.33}$$

where $D$ is the dividend process (i.e., a certain $\mathbb{G}$-adapted stochastic process of finite variation). It is expected that techniques similar to these presented above will also work for the evaluation of the functional above.

# 13. Heath-Jarrow-Morton Type Models

In the context of the modeling of the defaultable term structure, the HJM methodology was first examined by Jarrow and Turnbull (1995) and Duffie and Singleton (1999). Their studies were undertaken by Schönbucher (1996, 1998a), who has studied in a systematic way various forms of the no-arbitrage condition between the default-free and defaultable term structures. More recently, some of these results were re-discovered by Maksymiuk and Gątarek (1999) and Pugachevsky (1999), who focused on the arbitrage-free dynamics under the spot martingale measure of the instantaneous forward credit spreads. Subsequently, the HJM methodology was extended by Bielecki and Rutkowski (1999, 2000a, 2000b) and Schönbucher (2000a) to cover the cases of term structure models with multiple ratings for corporate bonds. Eberlein and Özkan (2001) generalize this approach by considering models driven by Lévy motions (for related results, also see Eberlein and Raible (1999) and Eberlein (2001)). In contrast with models presented in the previous chapter, the credit migration process is not exogenously specified, but it is endogenous in a model. It follows a conditionally Markov process with respect to a reference filtration under the spot (or forward) martingale measure.

This chapter is organized as follows. In Sect. 13.1, we provide a detailed analysis of the HJM-type approach to the modeling of term structures of interest rates with two rating classes: default-free bonds and defaultable bonds. The main results concerning the HJM-type modeling with credit migrations are presented in Sect. 13.2. In both cases, as inputs in the construction of a model of credit risk we use dynamics corresponding to instantaneous forward rates for bonds in various rating classes.

Let us stress that the underlying standard Brownian motion $W$ that governs the dynamics of forward rates is $d$-dimensional, and so are some other related processes. However, we have chosen not to introduce an explicit notation for the Euclidean inner product in $\mathbb{R}^d$. Thus, for example,

$$\int_0^t \beta_u \, dW_u = \sum_{i=1}^d \int_0^t \beta_u^i \, dW_u^i, \quad \sigma(t,T)b(t,T) = \sum_{i=1}^d \sigma^i(t,T)b^i(t,T).$$

For the sake of notational convenience, we assume throughout the chapter, without loss of generality, that the face value of a (default-free or defaultable) bond equals $L = 1$.

## 13.1 HJM Model with Default

In this section, we focus on a defaultable bond from a given rating class and we assume that it cannot migrate to another class before default. We assume that the dynamics of default-free and defaultable instantaneous forward rates are specified through the HJM approach. In other words, the coefficients of the real-world dynamics of instantaneous, continuously compounded, forward rates are taken as model's inputs. We assume, of course, that the model of default-free bond market is arbitrage-free. Our goal is to explain the dynamics of the defaultable instantaneous forward rate by introducing a judiciously chosen random time with a stochastic intensity, interpreted as the bond's default time. This random time will be defined as a totally inaccessible stopping time on an enlarged probability space. Unless explicitly stated otherwise, we assume here the fractional recovery of Treasury value scheme, with a constant recovery rate. However, in Sect. 13.1.9, we shall argue that other alternative recovery schemes can also be fitted in this approach.

Main results in this section are Propositions 13.1.3 and 13.1.4, as well as results of Sect. 13.1.9. Nevertheless, we believe that Proposition 13.1.1 also deserves attention, as it establishes the main contribution of the HJM theory in the context of defaultable term structure with only two credit grades: default-free (Treasury) bonds and defaultable (corporate) bonds.

### 13.1.1 Model's Assumptions

In this section, we shall work under the following standing assumptions (HJM.1)–(HJM.3). For mild technical conditions under which expressions (13.1)–(13.3) are well defined, the reader is referred to Heath et al. (1992) or Chap. 13 in Musiela and Rutkowski (1997a).

**Condition (HJM.1)** We are given a $d$-dimensional standard Brownian motion $W$, defined on the filtered probability space $(\tilde{\Omega}, \mathbb{F}, \mathbb{P})$, with $\mathbb{P}$ playing the role of the real-world probability measure. The reference filtration $\mathbb{F}$ is generated by the process $W$.

The dynamics of default-free instantaneous forward rates are postulated to be given by the standard HJM expression.

**Condition (HJM.2)** For any fixed maturity $T \leq T^*$, the *default-free instantaneous forward rate* $f(t, T)$ satisfies

$$df(t, T) = \alpha(t, T) \, dt + \sigma(t, T) \, dW_t, \tag{13.1}$$

where $\alpha(\cdot, T)$ and $\sigma(\cdot, T)$ are $\mathbb{F}$-adapted processes with values in $\mathbb{R}$ and $\mathbb{R}^d$, respectively. In an integrated form, for every $t \in [0, T]$ we have

$$f(t, T) = f(0, T) + \int_0^t \alpha(u, T) \, du + \int_0^t \sigma(u, T) \, dW_u \tag{13.2}$$

for some function $f(0, \cdot) : [0, T^*] \to \mathbb{R}$.

In the next assumption, we specify the dynamics for defaultable instantaneous forward rates. The meaning of the concept of such a rate will become
clear soon (in particular, see formula (13.6) and the interpretation of the
pre-default value $\tilde{D}(t,T)$ in Sect. 13.1.3). It will appear that these rates are
indeed pre-default instantaneous forward rates.

**Condition (HJM.3)** For any fixed maturity $T \leq T^*$, the *defaultable instantaneous forward rate* $g(t,T)$ satisfies

$$dg(t,T) = \tilde{\alpha}(t,T)\,dt + \tilde{\sigma}(t,T)\,dW_t, \tag{13.3}$$

for some $\mathbb{F}$-adapted stochastic processes $\tilde{\alpha}(\cdot,T)$ and $\tilde{\sigma}(\cdot,T)$ with values in $\mathbb{R}$
and $\mathbb{R}^d$, respectively. Equivalently, for every $t \in [0,T]$ we have

$$g(t,T) = g(0,T) + \int_0^t \tilde{\alpha}(u,T)\,du + \int_0^t \tilde{\sigma}(u,T)\,dW_u \tag{13.4}$$

for some function $g(0,\cdot) : [0,T^*] \to \mathbb{R}$.

Assumptions (HJM.1)–(HJM.2) are standard hypotheses of the Heath-
Jarrow-Morton methodology of term structure modeling. By definition, at
time $t \leq T$, the price of a unit default-free zero-coupon bond with the maturity date $T$ equals

$$B(t,T) := \exp\left(-\int_t^T f(t,u)\,du\right). \tag{13.5}$$

For any $t \leq T$, we also set

$$\tilde{D}(t,T) := \exp\left(-\int_t^T g(t,u)\,du\right). \tag{13.6}$$

**Definition 13.1.1.** For any date $t \leq T \leq T^*$, the *instantaneous forward
credit spread* $s(t,T)$ equals $s(t,T) = g(t,T) - f(t,T)$. The *short-term credit
spread* $s$ equals $s_t := s(t,t)$ for every $t \in [0,T^*]$.

It is clear that

$$\tilde{D}(t,T) = B(t,T)\exp\left(-\int_t^T s(t,u)\,du\right). \tag{13.7}$$

It is natural to assume that the instantaneous forward credit spread $s(t,T)$
is strictly positive, so that $\tilde{D}(t,T) < B(t,T)$. We shall interpret $\tilde{D}(t,T)$
as the *pre-default value* of a $T$-maturity zero-coupon corporate bond with
fractional recovery of Treasury value. Formally, $\tilde{D}(t,T)$ represents the value
of a zero-coupon corporate bond conditioned on the event that the bond
has not yet defaulted by time $t$. To justify this heuristic interpretation, we
need to develop an arbitrage-free model of default-free and defaultable term
structures, in which the process $\tilde{D}(t,T)$ does indeed represent the arbitrage
price of a $T$-maturity defaultable bond before default.

## 13.1.2 Default-Free Term Structure

In the first step, we focus on a default-free term structure of interest rates. For the reader's convenience, we quote the following standard result, due to Heath et al. (1992).

**Lemma 13.1.1.** *The dynamics of the default-free bond price* $B(t,T)$ *are*

$$dB(t,T) = B(t,T)\big(a(t,T)\,dt + b(t,T)\,dW_t\big), \tag{13.8}$$

*where*

$$a(t,T) = f(t,t) - \alpha^*(t,T) + \tfrac{1}{2}|\sigma^*(t,T)|^2, \quad b(t,T) = -\sigma^*(t,T), \tag{13.9}$$

*with* $\alpha^*(t,T) = \int_t^T \alpha(t,u)\,du$ *and* $\sigma^*(t,T) = \int_t^T \sigma(t,u)\,du$.

*Proof.* Let us denote $I_t = \ln B(t,T)$. In view of (13.2) and (13.5), we have

$$I_t = -\int_t^T f(0,u)\,du - \int_t^T \int_0^t \alpha(v,u)\,dv\,du - \int_t^T \int_0^t \sigma(v,u)\,dW_v\,du.$$

Applying Fubini's standard and stochastic theorems (for the latter, see Theorem IV.45 in Protter (1990)), we find that

$$I_t = -\int_t^T f(0,u)\,du - \int_0^t \int_t^T \alpha(v,u)\,du\,dv - \int_0^t \int_t^T \sigma(v,u)\,du\,dW_v$$

or, equivalently,

$$\begin{aligned}
I_t &= -\int_0^T f(0,u)\,du - \int_0^t \int_v^T \alpha(v,u)\,du\,dv - \int_0^t \int_v^T \sigma(v,u)\,du\,dW_v \\
&\quad + \int_0^t f(0,u)\,du + \int_0^t \int_v^t \alpha(v,u)\,du\,dv + \int_0^t \int_v^t \sigma(v,u)\,du\,dW_v.
\end{aligned}$$

Consequently,

$$I_t = I_0 + \int_0^t r_u\,du - \int_0^t \int_u^T \alpha(u,v)\,dv\,du - \int_0^t \int_u^T \sigma(u,v)\,dv\,dW_u,$$

where we have used the representation

$$r_u = f(u,u) = f(0,u) + \int_0^u \alpha(v,u)\,dv + \int_0^u \sigma(v,u)\,dW_v.$$

Taking into account (13.9), we obtain

$$I_t = I_0 + \int_0^t r_u\,du - \int_0^t \alpha^*(u,T)\,du - \int_0^t \sigma^*(u,T)\,dW_u.$$

To check that (13.8) holds, it suffices to apply Itô's formula.    □

We assume from now on that it is also possible to invest in the savings account $B_t = \exp(\int_0^t r_u \, du)$, corresponding to the short-term rate $r_t = f(t,t)$. In view of (13.8), the relative price $Z(t,T) = B_t^{-1} B(t,T)$ satisfies under $\mathbb{P}$

$$dZ(t,T) = Z(t,T)\Big(\big(\tfrac{1}{2}|b(t,T)|^2 - \alpha^*(t,T)\big)\,dt + b(t,T)\,dW_t\Big).$$

The following condition is known to exclude arbitrage across default-free bonds for all maturities $T \le T^*$ and the savings account (see Heath et al. (1992) or Chap. 13 in Musiela and Rutkowski (1997a)).

**Condition (HJM.4)** There exists an adapted $\mathbb{R}^d$-valued process $\beta$ such that

$$\mathbb{E}_{\mathbb{P}}\Big\{ \exp\Big(\int_0^{T^*} \beta_u \, dW_u - \frac{1}{2} \int_0^{T^*} |\beta_u|^2 \, du\Big)\Big\} = 1$$

and for any maturity $T \le T^*$ and any $t \in [0,T]$ we have:

$$\tfrac{1}{2}|\sigma^*(t,T)|^2 - \alpha^*(t,T) = \sigma^*(t,T)\beta_t$$

or, equivalently,

$$\alpha(t,T) + \sigma(t,T)(\beta_t - \sigma^*(t,T)) = 0. \tag{13.10}$$

Let $\beta$ be some process satisfying the last condition. Then the probability measure $\mathbb{P}^*$, given by the formula

$$\frac{d\mathbb{P}^*}{d\mathbb{P}} = \exp\Big(\int_0^{T^*} \beta_u \, dW_u - \frac{1}{2} \int_0^{T^*} |\beta_u|^2 \, du\Big), \quad \mathbb{P}\text{-a.s.}, \tag{13.11}$$

is a spot martingale measure for the default-free term structure. Define a Brownian motion $W^*$ under $\mathbb{P}^*$ by setting $W_t^* = W_t - \int_0^t \beta_u \, du$ for $t \in [0,T^*]$. Then, for any fixed maturity $T \le T^*$, the discounted price of default-free bond satisfies under $\mathbb{P}^*$

$$dZ(t,T) = Z(t,T)b(t,T)\,dW_t^*. \tag{13.12}$$

We shall assume from now on that the process $\beta$ is uniquely determined; in other words, the default-free bonds market is complete (strictly speaking, this assumption is not required for our further development). This means that any default-free contingent claim can be priced through the risk-neutral valuation formula. Let $\mathbb{P}_T$ be the forward martingale measure for the date $T$:

$$\frac{d\mathbb{P}_T}{d\mathbb{P}^*} = \frac{1}{B(0,T)B_T}, \quad \mathbb{P}^*\text{-a.s.}$$

Then the process $W_t^T := W_t^* - \int_0^t b(u,T) \, du$, $t \in [0,T]$, follows a standard Brownian motion under $\mathbb{P}_T$. Moreover, the price $B(t,U)$ of any $U$-maturity zero-coupon bond follows a martingale under $\mathbb{P}_T$ after discounting by the price $B(t,T)$ of the $T$-maturity bond. Specifically,

$$d\left(\frac{B(t,U)}{B(t,T)}\right) = \frac{B(t,U)}{B(t,T)}\big(b(t,U) - b(t,T)\big)\,dW_t^T. \tag{13.13}$$

### 13.1.3 Pre-Default Value of a Corporate Bond

We shall now focus on dynamics of the relative pre-default value of a corporate bond. The proof of the next result is analogous to the proof of Lemma 13.1.1.

**Lemma 13.1.2.** *We have*

$$d\tilde{D}(t,T) = \tilde{D}(t,T)\big(\tilde{a}(t,T)\,dt + \tilde{b}(t,T)\,dW_t\big)$$

*with*

$$\tilde{a}(t,T) = g(t,t) - \tilde{\alpha}^*(t,T) + \tfrac{1}{2}\,|\tilde{\sigma}^*(t,T)|^2, \quad \tilde{b}(t,T) = -\tilde{\sigma}^*(t,T).$$

In view of Lemma 13.1.2, under $\mathbb{P}$ the process $\tilde{Z}(t,T) = B_t^{-1}\tilde{D}(t,T)$ satisfies

$$d\tilde{Z}(t,T) = \tilde{Z}(t,T)\big((\tilde{a}(t,T) - r_t)\,dt + \tilde{b}(t,T)\,dW_t\big). \tag{13.14}$$

Consequently, under the (unique) spot martingale measure $\mathbb{P}^*$, we have

$$d\tilde{Z}(t,T) = \tilde{Z}(t,T)\big(\lambda^*(t,T)\,dt + \tilde{b}(t,T)\,dW_t^*\big), \tag{13.15}$$

where for every $t \in [0,T]$ we set

$$\lambda^*(t,T) := \tilde{a}(t,T) - r_t + \tilde{b}(t,T)\beta_t.$$

It is useful to notice that

$$\lambda^*(t,T) = s_t - \tilde{\alpha}^*(t,T) + \tfrac{1}{2}\,|\tilde{\sigma}^*(t,T)|^2 - \tilde{\sigma}^*(t,T)\beta_t, \tag{13.16}$$

where $s_t = g(t,t) - f(t,t)$ is the short-term credit spread. As apparent from (13.16), the process $\lambda^*(t,T)$ may depend on the maturity date $T$, in general. However, we shall assume that this is not the case, so that the following assumption is satisfied.

**Condition (HJM.5)** Processes $\lambda^*(t,T)$ given by (13.16) do not depend on $T$, namely, $\lambda^*(t,U) = \lambda^*(t,T)$ for every $t \in [0, U \wedge T]$ and every $U,T \le T^*$.

To emphasize that we work under assumption (HJM.5), we shall henceforth write $\lambda_t^*$, rather than $\lambda^*(t,T)$. From the property that the credit spread $s(t,u) = g(t,u) - f(t,u)$ is strictly positive, it is also possible to deduce that $\lambda^*$ follows a strictly positive process. To this end, we observe that the process

$$\tilde{Z}(t,T)\exp\Big(-\int_t^T \lambda_u^*\,du\Big)$$

is a $\mathbb{P}^*$-martingale. Put another way, for every $t \in [0,T]$ we have

$$\tilde{D}(t,T) = \mathbb{E}_{\mathbb{P}^*}\Big\{\exp\Big(-\int_t^T (r_u + \lambda_u^*)\,du\Big)\,\Big|\,\mathcal{F}_t\Big\}.$$

Since we assume that $\tilde{D}(t,T) < B(t,T)$ for every $t \in [0,T)$ and for any maturity $T > 0$, it must hold that $\int_s^t \lambda_u^*\,du > 0$ for every $s < t$, thereby implying that $\lambda_t^* > 0$ almost surely, for almost all $t$.

The following simple lemma appears to be useful.

**Lemma 13.1.3.** *Under Assumptions* (HJM.1)–(HJM.5), *we have*

$$d\left(\frac{\tilde{D}(t,U)}{\tilde{D}(t,T)}\right) = \frac{\tilde{D}(t,U)}{\tilde{D}(t,T)}\left(b(t,U) - b(t,T)\right)\left(dW_t^* - \tilde{b}(t,U)\,dt\right).$$

*Proof.* It is enough to make use of (13.15) and apply Itô's formula.     □

It is convenient to introduce the following concept (for more details on defaultable forward martingale measures, see Sect. 15.2.2).

**Definition 13.1.2.** The probability measure $\tilde{\mathbb{P}}_T$ on $(\Omega, \mathcal{F}_T)$ given by

$$\frac{d\tilde{\mathbb{P}}_T}{d\mathbb{P}^*} = \exp\left(\int_0^T \tilde{b}(u,T)\,dW_u^* - \frac{1}{2}\int_0^T |\tilde{b}(u,T)|^2\,du\right), \quad \mathbb{P}^*\text{-a.s.}, \quad (13.17)$$

is called the *restricted defaultable forward martingale measure* for the date $T$.

By virtue of Lemma 13.1.3, we have

$$d\left(\frac{\tilde{D}(t,U)}{\tilde{D}(t,T)}\right) = \frac{\tilde{D}(t,U)}{\tilde{D}(t,T)}\left(b(t,U) - b(t,T)\right)d\tilde{W}_t^T,$$

where the process

$$\tilde{W}_t^T = W_t^* - \int_0^t \tilde{b}(u,T)\,du, \quad \forall t \in [0,T], \quad (13.18)$$

is a standard Brownian motion under $\tilde{\mathbb{P}}_T$. It is clear that

$$\tilde{W}_t^T = W_t^T - \int_0^t \left(\tilde{b}(u,T) - b(u,T)\right)du. \quad (13.19)$$

In some instances, it will be convenient to assume that the coefficients in (13.16) are chosen in such a way that the following property is valid.[1]

**Condition (HJM.6)** We have $\lambda_t^* = s_t$ for every $t \in [0, T^*]$.

In view of (13.16), Condition (HJM.6) is satisfied whenever for every $0 \le t \le T \le T^*$ we have

$$\tfrac{1}{2}|\tilde{\sigma}^*(t,T)|^2 - \tilde{\alpha}^*(t,T) = \tilde{\sigma}^*(t,T)\beta_t$$

or, equivalently (cf. (13.10)),

$$\tilde{\alpha}(t,T) + \tilde{\sigma}(t,T)(\beta_t - \tilde{\sigma}^*(t,T)) = 0. \quad (13.20)$$

If the coefficients $\alpha(t,T)$, $\sigma(t,T)$ and $\tilde{\sigma}(t,T)$ are given, the last equality uniquely specifies the drift coefficient $\tilde{\alpha}(t,T)$ in (13.3).

---

[1] In the special case of zero recovery, Condition (HJM.6) implies that the intensity of default equals the short-term credit spread.

### 13.1.4 Dynamics of Forward Credit Spreads

In this section, we shall work under assumptions (HJM.1)–(HJM.6). Let us recall that for any two dates $t < T$, the instantaneous forward credit spread $s(t, T)$ equals $g(t, T) - f(t, T)$. To prove the next lemma, it suffices to combine (13.1) and (13.3) with the definition of the Brownian motion $W^*$.

**Lemma 13.1.4.** *For each fixed $T$, the instantaneous forward credit spread satisfies*

$$ds(t, T) = \alpha_s(t, T)\, dt + \sigma_s(t, T)\, dW_t^*,$$

*where $\sigma_s(t, T) = \tilde{\sigma}(t, T) - \sigma(t, T)$ and*

$$\alpha_s(t, T) = \tilde{\alpha}(t, T) - \alpha(t, T) + (\tilde{\sigma}(t, T) - \sigma(t, T))\beta_t. \qquad (13.21)$$

Schönbucher (1998a) and Pugachevsky (1999) analyze relationships between $f(t, T), g(t, T)$ and $s(t, T)$ under the spot martingale measure $\mathbb{P}^*$. For the sake of convenience, we shall first assume that $\beta \equiv 0$. Then conditions (13.10) and (13.20) become (see equations (6) and (23) in Pugachevsky (1999))

$$\alpha(t, T) = \sigma(t, T) \int_t^T \sigma(t, u)\, du = \sigma(t, T)\sigma^*(t, T),$$

and

$$\tilde{\alpha}(t, T) = \tilde{\sigma}(t, T) \int_t^T \tilde{\sigma}(t, u)\, du = \tilde{\sigma}(t, T)\tilde{\sigma}^*(t, T),$$

respectively. By applying Lemma 13.1.4 with $\beta \equiv 0$, we obtain

$$\alpha_s(t, T) = \tilde{\alpha}(t, T) - \alpha(t, T) = \tilde{\sigma}(t, T)\tilde{\sigma}^*(t, T) - \sigma(t, T)\sigma^*(t, T),$$

so that, after simple manipulations, we arrive at the following equality:

$$\alpha_s(t, T) = \sigma_s(t, T)\sigma^*(t, T) + (\sigma_s(t, T) + \sigma(t, T))\sigma_s^*(t, T), \qquad (13.22)$$

where $\sigma_s^*(t, T) = \int_t^T \sigma_s(t, u)\, du$. The last condition coincides with equation (54) in Schönbucher (1998a) and equation (33) in Pugachevsky (1999).[2] It specifies the relationship under the spot martingale measure $\mathbb{P}^*$ between the drift term $\alpha_s$ in the dynamics of $s(t, T)$ and the diffusion terms $\sigma_s$ and $\sigma$ in the dynamics of processes $s(t, T)$ and $f(t, T)$, respectively.

In the case when $\sigma(t, T)\sigma_s(t, T) = 0$ for every $t \in [0, T]$, the no-arbitrage condition simplifies to the following equality:

$$\alpha_s(t, T) = \sigma_s(t, T)\sigma_s^*(t, T). \qquad (13.23)$$

In case of deterministic coefficients $\sigma$ and $\tilde{\sigma}$, the condition: $\sigma(t, T)\sigma_s(t, T) = 0$ for every $t \in [0, T]$, corresponds to the mutual independence under $\mathbb{P}^*$ of the default-free rate process $f(t, T)$ and the forward credit spread process $s(t, T)$.

---

[2] Pugachevsky (1999) considers correlated Brownian motions; his convention does not lead to a more general model, though.

For a general case of non-zero process $\beta$, it suffices to replace the real-world drifts $\alpha(t, T)$ and $\tilde{\alpha}(t, T)$ by the risk-neutral drifts:

$$\zeta(t, T) = \alpha(t, T) + \sigma(t, T)\beta_t, \quad \tilde{\zeta}(t, T) = \tilde{\alpha}(t, T) + \tilde{\sigma}(t, T)\beta_t.$$

We have

$$df(t, T) = \zeta(t, T)\, dt + \sigma(t, T)\, dW_t^*,$$

and

$$dg(t, T) = \tilde{\zeta}(t, T)\, dt + \tilde{\sigma}(t, T)\, dW_t^*.$$

In terms of risk-neutral drifts, (13.21) becomes: $\alpha_s(t, T) = \tilde{\zeta}(t, T) - \zeta(t, T)$. Likewise, conditions (13.10) and (13.20) become $\zeta(t, T) = \sigma(t, T)\sigma^*(t, T)$ and $\tilde{\zeta}(t, T) = \tilde{\sigma}(t, T)\tilde{\sigma}^*(t, T)$, respectively. All considerations above remain valid, subject to an obvious change of notation. Thus, we are in a position to state the following result.

**Proposition 13.1.1.** (i) *Under* (HJM.1)–(HJM.6), *the drift term in the dynamics of* $s(t, T)$ *under the spot martingale measure* $\mathbb{P}^*$ *satisfies* (13.22), *i.e.,*

$$\alpha_s(t, T) = \sigma_s(t, T)\sigma^*(t, T) + (\sigma_s(t, T) + \sigma(t, T))\sigma_s^*(t, T).$$

*If, in addition,* $\sigma(t, T)\sigma_s(t, T) = 0$ *for every* $t \in [0, T]$, *then*

$$\alpha_s(t, T) = \sigma_s(t, T)\sigma_s^*(t, T).$$

(ii) *Conversely, if assumptions* (HJM.1)–(HJM.5) *and condition* (13.22) *hold, then* (HJM.6) *is satisfied, so that* $\lambda_t^* = s(t, t)$ *for every* $t \in [0, T^*]$, *and so*

$$d\tilde{Z}(t, T) = \tilde{Z}(t, T)\big(s(t, t)\, dt + \tilde{b}(t, T)\, dW_t^*\big).$$

*Remarks.* (i) Condition (13.23) was previously derived by Maksymiuk and Gątarek (1999). They focused on the case of a zero recovery rate and postulated the mutual independence of interest rate and default risk. They also noticed that, under their assumptions, condition (13.23) is equivalent to the property that the $\mathbb{F}$-intensity of the default time is equal to the credit spread $s(t, t)$. We shall examine the latter issue in Sect. 13.1.6, in which we shall show that the last feature is valid provided that (HJM.6) holds and a corporate bond is subject to the zero recovery rule.

(ii) Recall that

$$\frac{\tilde{D}(t, T)}{B(t, T)} = \exp\left(-\int_t^T s(t, u)\, du\right).$$

Combining the last equality with Proposition 13.1.1, we deduce that the relative price of a $T$-maturity defaultable bond is fully specified by the diffusion coefficient $\sigma(t, T)$ in the dynamics of the default-free term structure and the volatility coefficient $\sigma_s(t, T)$ of the credit spread $s(t, T)$. Of course, both $\sigma(t, T)$ and $\sigma_s(t, T)$ are invariant with respect to an equivalent change of a probability measure (in other words, if $\sigma(t, T)$ and $\sigma_s(t, T)$ are deterministic, they can be estimated under the real-life probability).

### 13.1.5 Default Time of a Corporate Bond

We henceforth assume that conditions (HJM.1)–(HJM.5) are satisfied. Let $\delta \in [0, 1)$ be a fixed number. By virtue (13.15), we have

$$d\tilde{Z}(t, T) = \tilde{Z}(t, T)\big(\lambda_t^* \, dt + \tilde{b}(t, T) \, dW_t^*\big).$$

We introduce an auxiliary process $\lambda_{1,2}^*$, which satisfies, for every $t \in [0, T^*]$,

$$(\tilde{Z}(t, T) - \delta Z(t, T))\lambda_{1,2}^*(t) = \tilde{Z}(t, T)\lambda_t^*. \tag{13.24}$$

Notice that for $\delta = 0$ we simply have $\lambda_{1,2}^*(t) = \lambda_t^*$ for every $t \in [0, T]$. On the other hand, if we take $\delta > 0$ then the process $\lambda_{1,2}^*$ is strictly positive provided that $\tilde{D}(t, T) > \delta B(t, T)$ (recall that we have assumed that $\tilde{D}(t, T) < B(t, T)$).

*Remarks.* If the assumption $\tilde{D}(t, T) > \delta B(t, T)$ is relaxed, the process $\lambda_{1,2}^*$ is strictly positive provided that $\lambda_t^*(\tilde{Z}(t, T) - \delta Z(t, T)) > 0$ for every $t \in [0, T]$. Notice also that in general $\lambda_{1,2}^*$ depends both on the recovery rate $\delta$ and on maturity date $T$. We assume that the process $\lambda_{1,2}^*$ is strictly positive and that it is integrable on $[0, T^*]$.

We shall show that there exists a random time $\tau$, such that the process (as before, $H_t = \mathbb{1}_{\{\tau \le t\}}$)

$$M_t = H_t - \int_0^t \lambda_{1,2}^*(u)\mathbb{1}_{\{\tau > u\}} \, du, \quad \forall t \in [0, T], \tag{13.25}$$

follows a martingale under a suitable extension $\mathbb{Q}^*$ of the spot martingale measure $\mathbb{P}^*$. The existence of $\tau$ easily follows from the results of Chap. 5 and 8, if we allow for a suitable enlargement of the underlying probability space. In general, we cannot expect a stopping time $\tau$ with the desired properties to exist on the original probability space $(\tilde{\Omega}, \mathbb{F}, \mathbb{P}^*)$. For instance, if the underlying filtration is generated by a standard Brownian motion, which is the usual assumption imposed to ensure the uniqueness of the spot martingale measure $\mathbb{P}^*$, no stopping time with desired properties exists on the original space. The necessity of enlarging the underlying probability space is also closely related to the fact that it is not possible to replicate a defaultable bond using risk-free bonds. More exactly, the process $D^\delta(t, T)$ does not correspond to the wealth process of a self-financing portfolio of risk-free bonds. This means that it is not a redundant security in the default-free bond market.

Let us denote by $(\Omega, \mathcal{G}, \mathbb{Q}^*)$ the enlarged probability space. Our additional requirement is that $W^*$ remains a standard Brownian motion when we move from $\mathbb{P}^*$ to $\mathbb{Q}^*$. To satisfy all these requirements, it suffices to take a product space $(\tilde{\Omega} \times \hat{\Omega}, (\mathcal{F}_t \otimes \hat{\mathcal{F}})_{t \in [0, T^*]}, \mathbb{P}^* \otimes \hat{\mathbb{Q}})$, where the probability space $(\hat{\Omega}, \hat{\mathcal{F}}, \hat{\mathbb{Q}})$ is large enough to support a unit exponential random variable, denoted by $\eta$ in what follows. Then we may put (cf. (8.58))

$$\tau = \inf \Big\{ t \in \mathbb{R}_+ : \int_0^t \lambda_{1,2}^*(u) \, du \ge \eta \Big\}.$$

We extend $W^*$ (and all other processes) to the enlarged space by setting $W_t^*(\tilde{\omega}, \hat{\omega}) = W_t^*(\tilde{\omega})$, etc. We preserve the notation $\mathbb{F}$ for the trivial extension of $\mathbb{F}$ to the enlarged probability space $(\Omega, \mathcal{G}, \mathbb{Q}^*)$, and we introduce the filtration $\mathbb{H} = (\mathcal{H}_t)_{t \in [0,T^*]}$ generated by the random time $\tau$: $\mathcal{H}_t = \sigma(H_u : u \leq t)$, where $H_u = \mathbb{1}_{\{\tau \leq u\}}$ is the jump process associated with $\tau$. Finally, we set $\mathcal{G}_t = \mathcal{F}_t \vee \mathcal{H}_t = \sigma(\mathcal{F}_t, \mathcal{H}_t)$ for every $t$. Then, it is clearly seen that the desired properties hold under $\mathbb{Q}^* = \mathbb{P}^* \otimes \hat{\mathbb{Q}}$. In particular, the process $M$ given by (13.25) is a $\mathbb{G}$-local martingale under $\mathbb{Q}^*$ and $W^*$ is a $\mathbb{G}$-Brownian motion under $\mathbb{Q}^*$. Notice that for obvious reasons the independence of $\tau$ and $W^*$ does not hold. The proof of the following auxiliary result is left to the reader.

**Lemma 13.1.5.** *For $t \in [0, T^*]$, let $\xi, \eta$ be the two $\mathcal{F}_t$-measurable random variables such that $\xi = \eta$ on $\{\tau > t\}$. Then $\xi = \eta$, $\mathbb{Q}^*$-a.s.*

We are in a position to specify the price process of a $T$-maturity defaultable bond with fractional recovery of Treasury value. We first introduce an auxiliary process $\hat{Z}(t, T)$ by postulating that $\hat{Z}(t, T)$ solves the following SDE

$$d\hat{Z}(t, T) = \hat{Z}(t, T)\big(\tilde{b}(t, T)\mathbb{1}_{\{\tau > t\}} + b(t, T)\mathbb{1}_{\{\tau \leq t\}}\big)\, dW_t^*$$
$$+ (\delta Z(t, T) - \hat{Z}(t-, T))\, dM_t \qquad (13.26)$$

with the initial condition $\hat{Z}(0, T) = \tilde{Z}(0, T)$. For obvious reasons, the process $\hat{Z}(t, T)$, if well defined, follows a local martingale under $\mathbb{Q}^*$. Combining (13.26) with (13.25), we obtain

$$d\hat{Z}(t, T) = \hat{Z}(t, T)\big(\tilde{b}(t, T)\mathbb{1}_{\{\tau > t\}} + b(t, T)\mathbb{1}_{\{\tau \leq t\}}\big)\, dW_t^*$$
$$+ (\hat{Z}(t, T) - \delta Z(t, T))\lambda_{1,2}^*(t)\mathbb{1}_{\{\tau > t\}}\, dt + (\delta Z(t, T) - \hat{Z}(t-, T))\, dH_t.$$

On the other hand, inserting (13.15) into (13.24), we find that $\tilde{Z}(t, T)$ obeys

$$d\tilde{Z}(t, T) = (\tilde{Z}(t, T) - \delta Z(t, T))\lambda_{1,2}^*(t)\, dt + \tilde{Z}(t, T)\tilde{b}(t, T)\, dW_t^*. \qquad (13.27)$$

It is thus easy to see that $\hat{Z}(t, T) = \tilde{Z}(t, T)$ on $[0, \tau[$, and thus $\hat{Z}(t, T)$ also satisfies the following SDE:

$$d\hat{Z}(t, T) = \hat{Z}(t, T)\big(\tilde{b}(t, T)\mathbb{1}_{\{\tau > t\}} + b(t, T)\mathbb{1}_{\{\tau \leq t\}}\big)\, dW_t^*$$
$$+ \hat{Z}(t, T)\lambda_t^*\mathbb{1}_{\{\tau > t\}}\, dt + (\delta Z(t, T) - \hat{Z}(t-, T))\, dH_t.$$

Next, from (13.12), for any $t \in [0, T]$ we obtain

$$\hat{Z}(t, T) = \mathbb{1}_{\{\tau > t\}}\tilde{Z}(t, T) + \delta\mathbb{1}_{\{\tau \leq t\}}Z(t, T). \qquad (13.28)$$

To check (13.28), it is enough to solve the SDE in question first on the random interval $[0, \tau[$ and subsequently on $[\tau, T]$. In view of the last equality, we may represent the Itô differential of $\hat{Z}(t, T)$ in another way, namely,

$$d\hat{Z}(t, T) = \big(\tilde{Z}(t, T)\tilde{b}(t, T)\mathbb{1}_{\{\tau > t\}} + \delta Z(t, T)b(t, T)\mathbb{1}_{\{\tau \leq t\}}\big)\, dW_t^*$$
$$+ \tilde{Z}(t, T)\lambda_t^*\mathbb{1}_{\{\tau > t\}}\, dt + (\delta Z(t, T) - \tilde{Z}(t-, T))\, dH_t.$$

In the next step, we introduce the price process $D^\delta(t, T)$ of a $T$-maturity defaultable bond.

For any $t \in [0, T]$, the process $D^\delta(t, T)$ is defined through the formula

$$D^\delta(t, T) := B_t \hat{Z}(t, T) = \mathbb{1}_{\{\tau > t\}} \tilde{D}(t, T) + \delta \mathbb{1}_{\{\tau \le t\}} B(t, T), \qquad (13.29)$$

where the second equality is an immediate consequence of (13.28). In the case when $\delta = 0$, the process $\hat{Z}(t, T)$ vanishes on the stochastic interval $[\tau, T]$, and we have

$$d\hat{Z}(t, T) = \hat{Z}(t, T)(\lambda_t^* dt + \tilde{b}(t, T) dW_t^*) - \hat{Z}(t-, T) dH_t. \qquad (13.30)$$

It is interesting to notice that $\hat{Z}(t, T)$ also satisfies

$$\begin{aligned} d\hat{Z}(t, T) = {}& \left( \tilde{Z}(t, T)\tilde{b}(t, T)\mathbb{1}_{\{\tau > t\}} + \delta Z(t, T)b(t, T)\mathbb{1}_{\{\tau \le t\}} \right) dW_t^* \\ & + (\tilde{Z}(t, T) - \delta Z(t, T))\lambda_{1,2}^*(t)\mathbb{1}_{\{\tau > t\}} dt \\ & + (\delta Z(t, T) - \tilde{Z}(t, T)) dH_t. \end{aligned}$$

This means that the process $\hat{Z}(t, T)$ can alternatively be introduced through the expression

$$\begin{aligned} d\hat{Z}(t, T) = {}& \left( \tilde{Z}(t, T)\tilde{b}(t, T)\mathbb{1}_{\{\tau > t\}} + \delta Z(t, T)b(t, T)\mathbb{1}_{\{\tau \le t\}} \right) dW_t^* \\ & + (\delta Z(t, T) - \tilde{Z}(t, T)) dM_t \end{aligned} \qquad (13.31)$$

with $\hat{Z}(0, T) = \tilde{Z}(0, T)$. We shall use an analogous approach in the next section.

To simplify the exposition, we shall make throughout the following technical assumption that will also be in force in Sect. 13.2.1 (although the process $\hat{Z}(t, T)$ is defined differently in the next section).

**Condition (HJM.7)** The process $\hat{Z}(t, T)$, given by the stochastic differential equation (13.26) (or, equivalently, by expression (13.31)), follows a $\mathbb{G}$-martingale (as opposed to a local martingale) under $\mathbb{Q}^*$.

Let us now focus on the migration process. In the present setting, the two-dimensional migration process $\tilde{C} = (C, \hat{C})$, cf. Sect. 11.3.1, lives on four states, since we have $K = 2$. We may and do assume that $\tilde{C}_0 = (C_0, \hat{C}_0) = (1, 1)$. We also assume that $\hat{C}_t = 1$ for every $t$. Thus, the only relevant states for the process $\tilde{C}$ are $(1, 1)$ and $(2, 1)$. The state $(1, 1)$ is the *pre-default state*, and the state $(2, 1)$ is the absorbing *default state*. Since the component $\hat{C}$ is uniquely determined by the past of the first component, $C$, it is clear that we only need to specify the dynamics for $C$. We postulate that the $\mathbb{F}$-conditional intensity matrix for $C$ equals

$$\Lambda_t^* = \begin{pmatrix} -\lambda_{1,2}^*(t) & \lambda_{1,2}^*(t) \\ 0 & 0 \end{pmatrix}.$$

The default time $\tau$ is given by the formula

$$\tau = \inf\{t \in \mathbb{R}_+ : C_t = 2\} = \inf\{t \in \mathbb{R}_+ : (C_t, \hat{C}_t) = (2, 1)\}. \qquad (13.32)$$

Using (13.29), for every $t \in [0, T]$ we obtain

$$
\begin{aligned}
D_C(t, T) &:= 1\!\!1_{\{C_t=1\}} \, \tilde{D}(t, T) + \delta 1\!\!1_{\{C_t=2\}} \, B(t, T) \\
&= 1\!\!1_{\{\tau>t\}} \, \tilde{D}(t, T) + \delta 1\!\!1_{\{\tau \le t\}} \, B(t, T) = D^\delta(t, T)
\end{aligned}
$$

as expected. The component $\hat{C}$ plays no essential role in the present setting. Its relevance will show up in case of multiple credit ratings, though.

In the remaining part of this section, we shall frequently use the notation $1\!\!1_{\{\tau>t\}}$ and $1\!\!1_{\{\tau \le t\}}$, rather than $1\!\!1_{\{C_t=1\}}$ and $1\!\!1_{\{C_t=2\}}$, respectively.

### 13.1.6 Case of Zero Recovery

Conditions (HJM.1)–(HJM.5) and (HJM.7) are assumed below. We shall now examine in detail the case of zero recovery rate. We already know that for $\delta = 0$, the matrix $\Lambda^*$ takes the following form (cf. (13.24)):

$$
\Lambda_t^* = \begin{pmatrix} -\lambda_t^* & \lambda_t^* \\ 0 & 0 \end{pmatrix}.
$$

This means that the $\mathbb{F}$-intensity of the default time equals $\lambda^*$. In particular, it coincides with the short-term credit spread $s(t, t)$ if (HJM.6) is valid. Let $D^0(t, T)$ be given by (13.29) with $\delta = 0$, i.e., $D^0(t, T) = 1\!\!1_{\{\tau>t\}} \tilde{D}(t, T)$, and let $\mathbb{Q}_T$ be the forward martingale measure associated with $\mathbb{Q}^*$ through the formula

$$
\frac{d\mathbb{Q}_T}{d\mathbb{Q}^*} = \frac{1}{B(0, T) B_T}, \qquad \mathbb{Q}^*\text{-a.s.} \tag{13.33}
$$

It is apparent that we have $\mathbb{Q}_T = \mathbb{P}_T$ on $(\Omega, \mathcal{F}_T)$.

**Proposition 13.1.2.** (i) *Under the spot martingale measure* $\mathbb{Q}^*$, *we have*

$$
dD^0(t, T) = D^0(t, T)\Big( \big(\tilde{a}(t, T) + \tilde{b}(t, T)\beta_t\big) \, dt + \tilde{b}(t, T) \, dW_t^* \Big) - D^0(t-, T) \, dH_t.
$$

(ii) *The following risk-neutral valuation formulae are valid*

$$
D^0(t, T) = B_t \, \mathbb{E}_{\mathbb{Q}^*}\big(B_T^{-1} 1\!\!1_{\{\tau>T\}} \,|\, \mathcal{G}_t\big) = B(t, T) \, \mathbb{Q}_T\{\tau > T \,|\, \mathcal{G}_t\}. \tag{13.34}
$$

(iii) *The pre-default value process satisfies*

$$
\tilde{D}(t, T) = B(t, T) \, \frac{\mathbb{Q}_T\{\tau > T \,|\, \mathcal{F}_t\}}{\mathbb{Q}_T\{\tau > t \,|\, \mathcal{F}_t\}}.
$$

(iv) *For any fixed* $T$, *the process* $K(t, T)$, *given by the expression*

$$
K(t, T) = \frac{\tilde{D}(t, T)}{B(t, T)} \, e^{-\int_0^t \lambda_u^* \, du}, \qquad \forall t \in [0, T], \tag{13.35}
$$

*follows a martingale under the forward martingale measure* $\mathbb{P}_T$, *and we have* (*see* (13.17))

$$
\frac{d\tilde{\mathbb{P}}_T}{d\mathbb{P}_T}\bigg|_{\mathcal{F}_t} = K(t, T), \qquad \mathbb{P}_T\text{-a.s.} \tag{13.36}
$$

*Proof.* The first statement is an immediate consequence of (13.29), combined with (13.14), (13.28) and (13.30). From (13.15), we obtain

$$d\tilde{D}(t,T) = \tilde{D}(t,T)\big((r_t + \lambda_t^*)\,dt + \tilde{b}(t,T)\,dW_t^*\big),$$

and thus (recall that $\tilde{D}(T,T) = 1$)

$$\tilde{D}(t,T) = \tilde{B}_t\,\mathbb{E}_{\mathbb{P}^*}(\tilde{B}_T^{-1}\,|\,\mathcal{F}_t) = \tilde{B}_t\,\mathbb{E}_{\mathbb{Q}^*}(\tilde{B}_T^{-1}\,|\,\mathcal{G}_t), \tag{13.37}$$

where we denote (cf. (8.28)): $\tilde{B}_t = \exp\big(\int_0^t (r_u + \lambda_u^*)\,du\big)$. If we now define the process $V_t = \tilde{D}(t,T)$, then this process is just like the process $V$ introduced in Proposition 8.3.2, with $Z = 0$ and $X = 1$. Since $\Delta V_\tau = 0$ (this holds since we know that the process $\tilde{D}(t,T)$ is continuous), using Corollary 8.3.1 we obtain the first equality in (13.34):

$$D^0(t,T) = \mathbb{1}_{\{\tau>t\}}\tilde{D}(t,T) = B_t\,\mathbb{E}_{\mathbb{Q}^*}(B_T^{-1}\mathbb{1}_{\{\tau>T\}}\,|\,\mathcal{G}_t).$$

The second equality in (13.34) follows from the Bayes rule and (13.33). Part (iii) follows from part (ii) and Lemma 13.1.5. For the last statement, recall that

$$\eta_t := \frac{d\mathbb{P}_T}{d\mathbb{P}^*}\Big|_{\mathcal{F}_t} = \mathbb{E}_{\mathbb{P}^*}\Big(\frac{1}{B(0,T)B_T}\,\Big|\,\mathcal{F}_t\Big) = \frac{B(t,T)}{B(0,T)B_t}.$$

It is enough to check that $\eta_t K(t,T)$ follows a $\mathbb{P}^*$-martingale. But

$$\eta_t K(t,T) = \frac{\tilde{D}(t,T)}{B(0,T)\tilde{B}_t},$$

hence, the martingale property is a consequence of the first equality in (13.37), since $\tilde{D}(T,T) = 1$. The last statement follows from (13.19). □

### 13.1.7 Default-Free and Defaultable LIBOR Rates

Our goal is to derive few auxiliary results concerning the LIBOR rates, which will prove useful in Chap 15. For a financial background and a discussion of the concept of default-free and defaultable LIBOR rates, the interested reader is referred to Sect. 14.1 (in particular, see formulae (14.3) and (14.11)). The modeling of forward LIBOR rates is examined in Sect. 15.1 and 15.2.

For a fixed $\Delta > 0$ and any dates $t < T$, the default-free forward LIBOR rate $L(t,T)$ over the accrual period $[T, T+\Delta]$ equals

$$L(t,T) = \frac{1}{\Delta}\Big(\frac{B(t,T)}{B(t,T+\Delta)} - 1\Big). \tag{13.38}$$

Likewise, the defaultable forward LIBOR rate $\tilde{L}(t,T)$ is defined by the formula

$$\tilde{L}(t,T) = \frac{1}{\Delta}\Big(\frac{\tilde{D}(t,T)}{\tilde{D}(t,T+\Delta)} - 1\Big), \tag{13.39}$$

where $\tilde{D}(t,T)$ represents the pre-default value of a corporate bond with zero recovery, so that $D^0(t,T) = \mathbb{1}_{\{\tau>t\}}\tilde{D}(t,T)$.

As a simple consequence of (13.13) and the definition of the forward LI-BOR rate $L(t, T)$, we obtain the following result.

**Lemma 13.1.6.** *The dynamics of $L(t, T)$ under the forward martingale measure $\mathbb{P}_{T+\Delta}$ are*

$$dL(t, T) = L(t, T)\nu(t, T)\, dW_t^{T+\Delta},$$

*where*

$$\nu(t, T) = \frac{1 + \Delta L(t, T)}{\Delta L(t, T)}\left(b(t, T) - b(t, T + \Delta)\right)$$

*and the process $W^{T+\Delta}$ follows a standard Brownian motion under $\mathbb{P}_{T+\Delta}$.*

By virtue of Lemma 13.1.3, the dynamics of $\tilde{L}(t, T)$ under the spot martingale measure $\mathbb{P}^*$ are

$$d\tilde{L}(t, T) = \Delta^{-1}\left(1 + \Delta\tilde{L}(t, T)\right)\left(\tilde{b}(t, T) - \tilde{b}(t, T + \Delta)\right)\left(dW_t^* - \tilde{b}(t, T + \Delta)\, dt\right).$$

The following auxiliary result is thus obvious. Let us recall that the probability measure $\tilde{\mathbb{P}}_{T+\Delta}$ and the associated standard Brownian motion $\tilde{W}^{T+\Delta}$ are given by Definition 13.1.2 and formula (13.18), respectively.

**Lemma 13.1.7.** *The defaultable forward LIBOR rate $\tilde{L}(t, T)$ satisfies under $\tilde{\mathbb{P}}_{T+\Delta}$*

$$d\tilde{L}(t, T) = \tilde{L}(t, T)\tilde{\nu}(t, T)\, d\tilde{W}_t^{T+\Delta},$$

*where*

$$\tilde{\nu}(t, T) = \frac{1 + \Delta\tilde{L}(t, T)}{\Delta\tilde{L}(t, T)}\left(\tilde{b}(t, T) - \tilde{b}(t, T + \Delta)\right)$$

*and the process $\tilde{W}^{T+\Delta}$ follows a standard Brownian motion under $\tilde{\mathbb{P}}_{T+\Delta}$.*

We are in a position to introduce the *forward survival process* $G(t, T)$. For any dates $t < T$, we set (for the second equality below, see part (iii) of Proposition 13.1.2; the last follows from (13.7))

$$G(t, T) := \frac{\tilde{D}(t, T)}{B(t, T)} = \frac{\mathbb{Q}_T\{\tau > T \mid \mathcal{F}_t\}}{\mathbb{Q}_T\{\tau > t \mid \mathcal{F}_t\}} = \exp\left(-\int_t^T s(t, u)\, du\right).$$

It appears that the dynamics of the forward survival process under the forward martingale measure have a neat form, as the following result shows.

**Lemma 13.1.8.** *For any fixed $T > 0$, the dynamics of the forward survival process $G(t, T)$ under $\mathbb{P}_T$ are*

$$dG(t, T) = G(t, T)\left(\lambda_{12}^*(t)\, dt + \left(\tilde{b}(t, T) - b(t, T)\right)dW_t^T\right). \tag{13.40}$$

*Proof.* To establish (13.40), it suffices to combine (13.12) with (13.15), and to make use of Itô's formula. $\qquad\square$

Recall that in case of zero recovery we have $\lambda_t^* = \lambda_{12}^*(t)$, where $\lambda^*$ is the $\mathbb{F}$-intensity of the default time $\tau$. Formula (13.40) thus confirms that the process $K(t, T)$, given by formula (13.35), follows a $\mathbb{P}_T$-martingale.

**13.1.8 Case of a Non-Zero Recovery Rate**

We shall work under assumptions (HJM.1)–(HJM.5) and (HJM.7). The next result deals with the fractional recovery of Treasury value scheme with an arbitrary recovery rate $\delta \in [0, 1]$.[3] Since Proposition 13.1.3 below covers the case of zero recovery, equality (13.34) can also be seen as a special case of formula (13.41).

**Proposition 13.1.3.** *The price process $D^\delta(t, T)$ of a corporate bond equals*

$$D^\delta(t, T) = \mathbb{1}_{\{C_t = 1\}} \exp\left(-\int_t^T g(t, u)\, du\right) + \delta \mathbb{1}_{\{C_t = 2\}} \exp\left(-\int_t^T f(t, u)\, du\right)$$

*or, equivalently,*

$$D^\delta(t, T) = \mathbb{1}_{\{C_t = 1\}} \tilde{D}(t, T) + \delta \mathbb{1}_{\{C_t = 2\}} B(t, T).$$

*Moreover, the risk-neutral valuation formula holds*

$$D^\delta(t, T) = B_t \, \mathbb{E}_{\mathbb{Q}^*}\left(\delta B_T^{-1} \mathbb{1}_{\{\tau \le T\}} + B_T^{-1} \mathbb{1}_{\{\tau > T\}} \,\big|\, \mathcal{G}_t\right). \tag{13.41}$$

*Furthermore,*

$$D^\delta(t, T) = B(t, T) \, \mathbb{E}_{\mathbb{Q}_T}\left(\delta \mathbb{1}_{\{\tau \le T\}} + \mathbb{1}_{\{\tau > T\}} \,\big|\, \mathcal{G}_t\right),$$

*where $\mathbb{Q}_T$ is the forward martingale measure for the date $T$, associated with the spot martingale measure $\mathbb{Q}^*$ through (13.33).*

*Proof.* The first two formulae follow from (13.5)–(13.6), combined with (13.29) and (13.32). In view of representation (13.29), it is also clear that $D^\delta(T, T) = \delta \mathbb{1}_{\{\tau \le T\}} + \mathbb{1}_{\{\tau > T\}}$. To establish (13.41), it is thus enough to show that the discounted process $B_t^{-1} D^\delta(t, T)$ follows a martingale under $\mathbb{Q}^*$. This is obvious, however, since by virtue of (13.29) we have $B_t^{-1} D^\delta(t, T) = \hat{Z}(t, T)$. In view of (13.41), the last equality is an immediate consequence of the Bayes rule and the definition of $\mathbb{Q}_T$.    $\square$

Combining (13.37) with (13.29), we obtain

$$D^\delta(t, T) = \mathbb{1}_{\{\tau > t\}} \tilde{B}_t \, \mathbb{E}_{\mathbb{P}^*}\left(\tilde{B}_T^{-1} \,\big|\, \mathcal{F}_t\right) + \delta \mathbb{1}_{\{\tau \le t\}} B_t \, \mathbb{E}_{\mathbb{P}^*}\left(B_T^{-1} \,\big|\, \mathcal{F}_t\right).$$

In view of the last equality and (13.41), it is tempting to conjecture that

$$I_1(t) := B_t \, \mathbb{E}_{\mathbb{Q}^*}\left(B_T^{-1} \mathbb{1}_{\{\tau \le T\}} \,\big|\, \mathcal{G}_t\right) = \mathbb{1}_{\{\tau \le t\}} B_t \, \mathbb{E}_{\mathbb{P}^*}\left(B_T^{-1} \,\big|\, \mathcal{F}_t\right)$$

and

$$I_2(t) := B_t \, \mathbb{E}_{\mathbb{Q}^*}\left(B_T^{-1} \mathbb{1}_{\{\tau > T\}} \,\big|\, \mathcal{G}_t\right) = \mathbb{1}_{\{\tau > t\}} \tilde{B}_t \, \mathbb{E}_{\mathbb{P}^*}\left(\tilde{B}_T^{-1} \,\big|\, \mathcal{F}_t\right).$$

This conjecture is false, however, as the following proposition shows.

---

[3] The case of full recovery, $\delta = 1$, is added here for the sake of completeness. As expected, in this case the valuation formula leads to an obvious result: $D^1(t, T) = B(t, T)$.

**Proposition 13.1.4.** *The following equalities are valid*

$$I_1(t) = B(t, T) - \mathbb{1}_{\{\tau > t\}} \bar{B}_t \, \mathbb{E}_{\mathbb{P}^*}(\bar{B}_T^{-1} \,|\, \mathcal{F}_t), \qquad (13.42)$$

*and*

$$I_2(t) = \mathbb{1}_{\{\tau > t\}} \bar{B}_t \, \mathbb{E}_{\mathbb{P}^*}(\bar{B}_T^{-1} \,|\, \mathcal{F}_t), \qquad (13.43)$$

*where*

$$\bar{B}_t = \exp\left(\int_0^t \left(r_u + \lambda_{1,2}^*(u)\right) du\right).$$

*Furthermore,*

$$D^\delta(t, T) = \delta B(t, T) + (1 - \delta) \mathbb{1}_{\{\tau > t\}} \bar{B}_t \, \mathbb{E}_{\mathbb{P}^*}(\bar{B}_T^{-1} \,|\, \mathcal{F}_t) \qquad (13.44)$$

*or, equivalently,*

$$D^\delta(t, T) = B(t, T) - (1 - \delta)\left(B(t, T) - \mathbb{1}_{\{\tau > t\}} \bar{B}_t \, \mathbb{E}_{\mathbb{P}^*}(\bar{B}_T^{-1} \,|\, \mathcal{F}_t)\right). \qquad (13.45)$$

*Finally, we have*

$$D_C(t, T) = B(t, T)\left(\delta + (1 - \delta)\mathbb{1}_{\{\tau > t\}} \mathbb{E}_{\mathbb{P}_T}\left(e^{-\int_t^T \lambda_{1,2}^*(u)\, du} \,\Big|\, \mathcal{F}_t\right)\right),$$

*where $\mathbb{P}_T$ is the forward martingale measure for the date $T$.*

*Proof.* Let us rewrite $I_1(t)$ as follows:

$$I_1(t) = B_t \, \mathbb{E}_{\mathbb{Q}^*}(B_T^{-1} H_T \,|\, \mathcal{G}_t) = B_t \, \mathbb{E}_{\mathbb{Q}^*}(B_T^{-1} \,|\, \mathcal{G}_t) - B_t \, \mathbb{E}_{\mathbb{Q}^*}(B_T^{-1}(1 - H_T)\,|\,\mathcal{G}_t).$$

It is clear that

$$\mathbb{E}_{\mathbb{Q}^*}(B_T^{-1}(1 - H_T)\,|\,\mathcal{G}_t) = \mathbb{E}_{\mathbb{Q}^*}\left(B_T^{-1}\mathbb{E}_{\mathbb{Q}^*}(1 - H_T \,|\, \mathcal{F}_T \vee \mathcal{H}_t)\,|\,\mathcal{G}_t\right).$$

Reasoning as in the proof of Proposition 8.6.1, we obtain (as usual, we set $\mathcal{H}_t = \sigma(H_u : u \le t)$)

$$\mathbb{E}_{\mathbb{Q}^*}(1 - H_T \,|\, \mathcal{F}_T \vee \mathcal{H}_t) = \mathbb{Q}^*\{\tau > T \,|\, \mathcal{F}_T \vee \mathcal{H}_t\} = (1 - H_t)\, e^{-\int_t^T \lambda_{1,2}^*(u)\, du}.$$

Combining the formulae above, we obtain

$$
\begin{aligned}
I_1(t) &= B_t \, \mathbb{E}_{\mathbb{Q}^*}(B_T^{-1} \,|\, \mathcal{G}_t) - B_t \, \mathbb{E}_{\mathbb{Q}^*}\left(B_T^{-1}(1 - H_t)\, e^{-\int_t^T \lambda_{1,2}^*(u)\, du} \,\Big|\, \mathcal{G}_t\right) \\
&= B_t \, \mathbb{E}_{\mathbb{P}^*}(B_T^{-1} \,|\, \mathcal{F}_t) - (1 - H_t)\bar{B}_t \, \mathbb{E}_{\mathbb{Q}^*}(\bar{B}_T^{-1} \,|\, \mathcal{G}_t) \\
&= B(t, T) - (1 - H_t)\bar{B}_t \, \mathbb{E}_{\mathbb{P}^*}(\bar{B}_T^{-1} \,|\, \mathcal{F}_t).
\end{aligned}
$$

Since for $I_2(t)$ we have

$$I_2(t) = B_t \, \mathbb{E}_{\mathbb{Q}^*}(B_T^{-1}(1 - H_T)\,|\,\mathcal{G}_t),$$

using the same arguments as for $I_1(t)$, we arrive at

$$I_2(t) = (1 - H_t)\bar{B}_t \, \mathbb{E}_{\mathbb{Q}^*}(\bar{B}_T^{-1} \,|\, \mathcal{G}_t).$$

Finally, $D^\delta(t,T) = \delta I_1(t) + I_2(t)$, and thus (13.44)–(13.45) are trivial consequences of (13.42)–(13.43). The last equality follows from (13.44) and the properties of the forward measure $\mathbb{P}_T$. □

Notice that for $\delta = 0$ we have $\bar{B} = \tilde{B}$, and so formula (13.44) reduces to $D^0(t,T) = \mathbb{1}_{\{\tau > t\}} \tilde{D}(t,T)$. On the other hand, for $\delta = 1$ we have, as expected, $D^1(t,T) = B(t,T)$. Finally, when $0 < \delta < 1$ expression (13.44) yields a decomposition of the price $D^\delta(t,T)$ of a defaultable bond into its predicted *post-default value* $\delta B(t,T)$ and the *pre-default premium* $D^\delta(t,T) - \delta B(t,T)$. Similarly, (13.45) represents $D^\delta(t,T)$ as the difference between its *default-free value* $B(t,T)$ and the *expected loss in value* due to the credit risk. One might also look at (13.45) from the perspective of the buyer of a defaultable bond: the price $D^\delta(t,T)$ equals the price of the default-free bond minus a compensation for credit risk.

*Remarks.* Let us denote

$$J(t) = \mathbb{1}_{\{\tau > t\}} \bar{B}_t \, \mathbb{E}_{\mathbb{Q}^*}(\bar{B}_T^{-1} \,|\, \mathcal{G}_t) = B_t \, \mathbb{E}_{\mathbb{Q}^*}\left(B_T^{-1}(1 - H_t)e^{-\int_t^T \lambda_{1,2}^*(u)\,du} \,\Big|\, \mathcal{G}_t\right).$$

From the proof of Proposition 13.1.4, we know that

$$(1 - H_t)\,e^{-\int_t^T \lambda_{1,2}^*(u)\,du} = \mathbb{Q}^*\{\tau > T \,|\, \mathcal{F}_T \vee \mathcal{H}_t\}.$$

Consequently,

$$J(t) = B_t \, \mathbb{E}_{\mathbb{Q}^*}\left(B_T^{-1}\mathbb{Q}^*\{\tau > T \,|\, \mathcal{F}_T \vee \mathcal{H}_t\} \,\big|\, \mathcal{F}_t\right).$$

As already mentioned, in the present setting the stopping time $\tau$ and the underlying Brownian motion $W^*$ (and consequently $\tau$ and $B$) are not usually mutually independent. Assume, that $\tau$ and $B$ are mutually independent.[4] Under this rather restrictive assumption, $J(t)$ becomes

$$J(t) = B(t,T)\mathbb{Q}^*\{\tau > T \,|\, \mathcal{H}_t\}.$$

Consequently, we are able to rewrite the valuation formula (13.44) on the set $\{\tau > t\} = \{C_t = 1\}$ in the following way:

$$D^\delta(t,T) = \tilde{D}(t,T) = B(t,T)\big(\delta + (1 - \delta)\mathbb{Q}^*\{\tau > T \,|\, C_t = 1\}\big). \qquad (13.46)$$

The last formula corresponds to expression (12.9), obtained in a different set-up by Jarrow et al. (1997). Let us recall that in the Jarrow et al. (1997) approach, it is explicitly assumed that the migration process is independent of the underlying short-term interest rate $r$. Needless to say that representation (13.44) is more general than (13.46), since it does not exclude the mutual dependence between the migration process of a defaultable bond and the risk-free term structure.

---

[4] More precisely, we assume that the default time $\tau$ is independent of $\mathcal{F}_T$ and that the process $B$ is independent of the filtration $\mathbb{H}$.

### 13.1.9 Alternative Recovery Rules

So far, we have been assuming that the recovery payment is fixed and takes place at the maturity date $T$ of a corporate bond. In this section, we shall assume instead that the constant (or random) payment is done at the default time, rather than at the bond's maturity date, and we shall focus on two important special cases: *fractional recovery of par value* and *fractional recovery of market value*. In case of the fractional recovery of par value scheme, the bondholder receives a fixed fraction of the bond's face value upon default. The constant payoff $\delta$ at some date $t < T$ is equivalent to the payoff $\delta B^{-1}(t, T)$ at the terminal date $T$. Likewise, the payoff $\delta \tilde{D}(t, T)$, which corresponds to the *fractional recovery of market value*, can be represented by the payoff $\delta \tilde{D}(t, T) B^{-1}(t, T)$ at the bond's maturity. We conclude that to cover typical cases when the recovery payment is made at time of default, it is enough to extend the construction provided in the previous section to the case of an $\mathbb{F}$-adapted stochastic process $\delta_t$. We thus introduce the general recovery process $\delta_t$ through the following condition.

**Condition (HJM.8)** The *terminal recovery process* $\delta_t$ is an $\mathbb{F}$-adapted and (locally) bounded process defined on the original probability space $(\tilde{\Omega}, \mathbb{F}, \mathbb{P})$. In financial interpretation, if the default occurs at time $\tau = t \leq T$, then the recovery payment in the amount of $\delta_t$ is made at the maturity date $T$.

*Remarks.* (i) The recovery process $Z_t$ of Sect. 8.1 and the terminal recovery process $\delta_t$ introduced above are tied to each other through the relationship $\delta_t = Z_t B^{-1}(t, T)$.
(ii) Observe that the fractional recovery of Treasury scheme considered earlier corresponds to $\delta_t \equiv \delta$. In general, the payment $\delta_t$ represents the time-$T$ equivalent of a random payment $\delta_t B(t, T)$ made at time $t$.

Condition (13.24), which serves to specify the intensity of default time $\tau$, now takes the following form

$$(\tilde{Z}(t, T) - \delta_t Z(t, T)) \lambda^*_{1,2}(t) = \tilde{Z}(t, T) \lambda^*_t, \quad \forall t \in [0, T].$$

As before, we assume that the last condition defines a strictly positive $\mathbb{F}$-adapted process $\lambda^*_{1,2}(t)$. We shall now show how to modify the basic equations (13.26)–(13.29). To this end, we introduce an auxiliary process $\hat{Z}(t, T)$ that obeys the following SDE:

$$d\hat{Z}(t, T) = \hat{Z}(t, T) \big( \tilde{b}(t, T) \mathbb{1}_{\{\tau > t\}} + b(t, T) \mathbb{1}_{\{\tau \leq t\}} \big) dW^*_t$$
$$+ (\delta_t Z(t, T) - \hat{Z}(t-, T)) dM_t, \tag{13.47}$$

with the initial condition $\hat{Z}(0, T) = \tilde{Z}(0, T)$. Notice that, as before, the process $\hat{Z}(t, T)$ follows a local martingale under $\mathbb{Q}^*$. Reasoning along the same lines as in the previous section, we conclude that $\hat{Z}(t, T)$ satisfies:

$$d\hat{Z}(t,T) = \hat{Z}(t,T)\big(\tilde{b}(t,T)\mathbb{1}_{\{\tau>t\}} + b(t,T)\mathbb{1}_{\{\tau\leq t\}}\big)\,dW_t^*$$
$$+ \hat{Z}(t,T)\lambda_t^*\mathbb{1}_{\{\tau>t\}}\,dt + (\delta_t Z(t,T) - \hat{Z}(t-,T))\,dH_t,$$

so that, for any $t \in [0,T]$ we have:

$$\hat{Z}(t,T) = \mathbb{1}_{\{\tau>t\}}\tilde{Z}(t,T) + \delta_\tau\mathbb{1}_{\{\tau\leq t\}}Z(t,T). \tag{13.48}$$

In the present setting, Condition (HJM.7) is substituted with the following assumption.

**Condition (HJM.9)** The process $\hat{Z}(t,T)$, given by the stochastic differential equation (13.47) (or, equivalently, by expression (13.48)), follows a $\mathbb{G}$-martingale (as opposed to a local martingale) under $\mathbb{Q}^*$.

The price process $\hat{D}^\delta(t,T)$ of a $T$-maturity defaultable bond is given by the following expression

$$\hat{D}^\delta(t,T) := B_t\hat{Z}(t,T) = \mathbb{1}_{\{\tau>t\}}\tilde{D}(t,T) + \delta_\tau\mathbb{1}_{\{\tau\leq t\}}B(t,T).$$

Arguing similarly as in the proof of Proposition 13.1.3, we may then show that the following result is true.

**Proposition 13.1.5.** *Let us assume that Conditions* (HJM.1)–(HJM.4) *and* (HJM.8)-(HJM.9) *are valid. Then*

$$\hat{D}^\delta(t,T) = B_t\,\mathbb{E}_{\mathbb{Q}^*}\big(\delta^* B_T^{-1}\mathbb{1}_{\{\tau\leq T\}} + B_T^{-1}\mathbb{1}_{\{\tau>T\}}\,\big|\,\mathcal{G}_t\big),$$

*where* $\delta^* = \delta_\tau$.

We shall apply the foregoing result to the two examples of recovery schemes.

**Fractional recovery of par value.** By setting $\delta_t = \delta B^{-1}(t,T)$, we obtain

$$\hat{D}^\delta(t,T) = \mathbb{1}_{\{\tau>t\}}\tilde{D}(t,T) + \delta B^{-1}(\tau,T)\mathbb{1}_{\{\tau\leq t\}}B(t,T).$$

This corresponds to the random payoff $\delta^* = \delta B^{-1}(\tau,T)$ at time $T$. Consequently, we obtain the following expression for the price process of a $T$-maturity defaultable bond

$$\hat{D}^\delta(t,T) = \mathbb{1}_{\{\tau>t\}}\tilde{D}(t,T) + \delta^*\mathbb{1}_{\{\tau\leq t\}}B(t,T),$$

and thus also

$$\hat{D}^\delta(t,T) = B_t\,\mathbb{E}_{\mathbb{Q}^*}\big(\delta B^{-1}(\tau,T)B_T^{-1}\mathbb{1}_{\{\tau\leq T\}} + B_T^{-1}\mathbb{1}_{\{\tau>T\}}\,\big|\,\mathcal{G}_t\big).$$

**Fractional recovery of market value.** In this scheme, the recovery process equals $\delta_t = \delta\tilde{D}(t,T)B^{-1}(t,T)$, and so

$$\hat{D}^\delta(t,T) = \mathbb{1}_{\{\tau>t\}}\tilde{D}(t,T) + \delta\tilde{D}(\tau,T)B^{-1}(\tau,T)\mathbb{1}_{\{\tau\leq t\}}B(t,T)$$

or, equivalently,

$$\hat{D}^\delta(t,T) = \mathbb{1}_{\{\tau>t\}}\tilde{D}(t,T) + \delta^*\mathbb{1}_{\{\tau\leq t\}}B(t,T),$$

where $\delta^* = \delta\tilde{D}(\tau,T)B^{-1}(\tau,T)$. Of course, we also have

$$\hat{D}^\delta(t,T) = B_t\,\mathbb{E}_{\mathbb{Q}^*}\big(\delta\tilde{D}(\tau,T)B^{-1}(\tau,T)B_T^{-1}\mathbb{1}_{\{\tau\leq T\}} + B_T^{-1}\mathbb{1}_{\{\tau>T\}}\,\big|\,\mathcal{G}_t\big).$$

## 13.2 HJM Model with Credit Migrations

We shall now construct an arbitrage-free model of a defaultable term structure of interest rates with migrations of the credit rankings of a corporate bond between several possible rating classes. In this section, we start with:
- a pre-specified default-free term structure, given in terms of the corresponding instantaneous forward rates,
- pre-specified term structures corresponding to a given finite collection of credit classes, represented by a finite family of instantaneous forward rates.

Our goal is to create a model that supports an exogenously given defaultable term structure through a judiciously chosen migration process, defined on a suitable enlargement of the underlying probability space. Put another way, we assume that the pre-default dynamics of defaultable bonds are given a priori, and we search for an arbitrage-free set-up that would support these values. The term structure model developed in this section relies on a presumption that the credit risk inherent in the credit-sensitive securities is fully explained by the credit-spread curve and its volatility. This statement should not be misunderstood; it does not mean that several relevant quantities, which are typically present in credit-risk considerations, should be totally neglected in this setting. On the contrary, all other quantities commonly used in most models of credit risk (such as: default probabilities, recovery rates, and default correlations) are also present in the approach proposed here. Results presented in this section are mainly drawn from papers by Bielecki and Rutkowski (1999, 2000a) (for related results, see Schönbucher (2000a)).

We shall maintain here Conditions (HJM.1), (HJM.2) and (HJM.4) of Sect. 13.1; however, we shall rename them as (BR.1), (BR.2) and (BR.4), respectively. The main results in this section are Proposition 13.2.1 and Theorem 13.2.1. The discussion of the market price of interest rate risk and the market prices of credit risk in Sect. 13.2.8 should also be acknowledged. Similarly as in Sect. 13.1, we first assume the fractional recovery of Treasury value scheme. Alternative recovery schemes are studied in Sect. 13.2.5.

### 13.2.1 Model's Assumption

We now assume that the set of credit rating classes is $\mathcal{K} = \{1, \ldots, K\}$, where the class $K$ corresponds to the default event. For any $i = 1, \ldots, K-1$, we write $\delta_i \in [0, 1)$ to denote the corresponding (deterministic) recovery rate. We shall first focus on the fractional recovery of Treasury value scheme, so that $\delta_i$ can be seen as the fraction of par paid at bond's maturity, if the bond, which is currently in the $i^{\text{th}}$ rating class, defaults. For the sake of brevity, we shall denote $\delta = (\delta_1, \ldots, \delta_{K-1})$.

We shall combine the risk-free term structure of Sect. 13.1.2 with $K-1$ different term structures corresponding to the $K-1$ pre-default credit rating classes (the discussion in the previous section regarded the case of $K = 2$). Within the present framework, Condition (HJM.3) takes the following form.

**Condition (BR.3)** For any $T \leq T^*$, the instantaneous forward rate $g_i(t, T)$, corresponding to the rating class $i = 1, \ldots, K$ satisfies under $\mathbb{P}$

$$dg_i(t, T) = \alpha_i(t, T)\, dt + \sigma_i(t, T)\, dW_t,$$

where $\alpha_i(t, T)$, $\sigma_i(t, T)$, $t \in [0, T]$, are adapted stochastic processes with values in $\mathbb{R}$ and $\mathbb{R}^d$, respectively.

It is tempting, but not necessary, to assume in addition that

$$g_{K-1}(t, T) > g_{K-2}(t, T) > \cdots > g_1(t, T) > f(t, T). \tag{13.49}$$

As before, the price of a $T$-maturity default-free discount bond is denoted by $B(t, T)$, so that

$$B(t, T) = \exp\left(-\int_t^T f(t, u)\, du\right), \tag{13.50}$$

and we denote $Z(t, T) = B_t^{-1} B(t, T)$. For any $i = 1, \ldots, K - 1$ we set

$$D_i(t, T) := \exp\left(-\int_t^T g_i(t, u)\, du\right). \tag{13.51}$$

Formulae analogous to (13.9) hold for processes $B(t, T)$ and $D_i(t, T)$, $i = 1, \ldots, K-1$, after a suitable change of notation. In particular, we now denote

$$a_i(t, T) = g_i(t, t) - \alpha_i^*(t, T) + \tfrac{1}{2}|\sigma_i^*(t, T)|^2, \quad b_i(t, T) = -\sigma_i^*(t, T),$$

where

$$\alpha_i^*(t, T) = \int_t^T \alpha_i(t, u)\, du, \quad \sigma_i^*(t, T) = \int_t^T \sigma_i(t, u)\, du.$$

Recall that Condition (HJM.4), which is assumed throughout this section and is now called Condition (BR.4), defines the process $\beta$. Given the process $\beta$, Condition (HJM.5) now takes the following form.

**Condition (BR.5)** For $i = 1, \ldots, K - 1$, the process $\lambda_i^*$, which is given by the formula

$$\lambda_i^*(t, T) := a_i(t, T) - f(t, t) + b_i(t, T)\beta_t, \quad \forall t \in [0, T],$$

does not depend on the maturity $T$.

*Remarks.* If we also assume that

$$a_i(t, T) + b_i(t, T)\beta_t = g_i(t, T),$$

then $\lambda_i^*(t) = g_i(t, t) - f(t, t)$, so that obviously $\lambda_i^*(t) > 0$ for $i = 1, \ldots, K$. More generally, arguing along the same lines as in the preceding section, one can show that processes $\lambda_i^*$ are strictly positive (this is a consequence of (13.49)). It is worth stressing, however, that neither the strict positivity of $\lambda_i^*$s, nor their independence of maturity $T$, are necessary requirements for our further developments.

We make the standing assumptions (BR.1)–(BR.5). Proceeding as in Sect. 13.1, we construct the spot martingale measure $\mathbb{P}^*$ for the risk-free term structure of interest rates. Under $\mathbb{P}^*$, the process $Z(t,T) = B_t^{-1} B(t,T)$ satisfies

$$dZ(t,T) = Z(t,T)b(t,T)\,dW_t^*. \qquad (13.52)$$

Likewise, if we define processes $Z_i(t,T) = B_t^{-1} D_i(t,T)$ for $i = 1,\ldots,K-1$, we obtain the following dynamics for $Z_i(t,T)$ under $\mathbb{P}^*$ (cf. (13.15))

$$dZ_i(t,T) = Z_i(t,T)\big(\lambda_i^*(t)\,dt + b_i(t,T)\,dW_t^*\big). \qquad (13.53)$$

### 13.2.2 Migration Process

The next step is to introduce a conditionally Markov chain $C$ on the state space $\mathcal{K} = \{1,\ldots,K\}$. To construct $C$, one needs to enlarge the underlying probability space. Suitable extensions of $\mathcal{F}_t$ and $\mathbb{P}^*$ will be denoted by $\mathcal{G}_t$ and $\mathbb{Q}^*$, respectively (the detailed construction of $C$ was given in Sect. 11.3). The $\mathbb{F}$-conditional infinitesimal generator of $C$ equals

$$\Lambda_t^* = \begin{pmatrix} \lambda_{1,1}^*(t) & \cdots & \lambda_{1,K}^*(t) \\ \cdot & \cdots & \cdot \\ \lambda_{K-1,1}^*(t) & \cdots & \lambda_{K-1,K}^*(t) \\ 0 & \cdots & 0 \end{pmatrix},$$

where $\lambda_{ii}^*(t) = -\sum_{j\neq i} \lambda_{ij}^*(t)$ for $i = 1,\ldots,K-1$ and $\lambda_{ij}^*$ are $\mathbb{F}$-adapted, strictly positive processes. To ensure that our pricing model is arbitrage free, the processes $\lambda_{ij}^*$ will be additionally assumed to satisfy the consistency condition (BR.6) (or (13.58) if $K = 3$).

It will be apparent from Condition (BR.6) that the intensities $\lambda_{ij}^*$ of credit migrations may depend on the maturity $T$ and on the vector of recovery rates $\delta$. Thus, to each maturity $T$ and to every recovery vector $\delta$ there may correspond a different migration process. Nevertheless, this feature of the model will not interfere with the property of absence of arbitrage between defaultable bonds of various maturities (and possibly various recovery profiles). This is because the enlarged probability space $(\Omega, (\mathcal{G}_t)_{t\in[0,T^*]}, \mathbb{Q}^*)$ does not depend on neither $T$ nor $\delta$, and the processes $D^\delta(\cdot,T)$, introduced later in this section, are martingales on $(\Omega, (\mathcal{G}_t)_{t\in[0,T^*]}, \mathbb{Q}^*)$, regardless of the particular values of $T \leq T^*$ and $\delta \in [0,1)^{K-1}$.

As usual, we shall write $H_t^i = \mathbb{1}_{\{C_t=i\}}$ for $i = 1,\ldots,K$. Let us define

$$M_t^{ij} := H_t^{ij} - \int_0^t \lambda_{ij}^*(s)H_s^i\,ds, \quad \forall\,t \in [0,T], \qquad (13.54)$$

for $i = 1,\ldots,K-1$, $j = 1,\ldots,K$, and $j \neq i$, where, as before, $H_t^{ij}$ represents the number of transitions from $i$ to $j$ by $C$ over the time interval $(0,t]$. It can be shown (see Sect. 11.3) that $M_t^{ij}$ is a local martingale on the enlarged probability space $(\Omega, (\mathcal{G}_t)_{t\in[0,T^*]}, \mathbb{Q}^*)$.

Recall that in Sect. 11.3.1 we also constructed the 'pre-jump' component $\hat{C}$ of the two-dimensional conditional Markov chain $\tilde{C} = (C, \hat{C})$. This component will play an important role with regard to our present model. Observe that $\hat{C}_t = C_{u(t)-}$, where $u(t) = \sup\{u \leq t : C_u \neq C_t\}$. By convention, $\sup \emptyset = 0$, therefore $\hat{C}_t = C_t$ if $C_u = C_0$ for every $u \in [0, t]$. In other words, $u(t)$ is the time of the last jump of $C$ before (and including) time $t$, and $\hat{C}_t$ represents the last state of $C$ before the jump to the current state.

### 13.2.3 Special Case

We will first examine the case when $K = 3$. We assume that $(C_0, \hat{C}_0) \in \{(1, 1), (2, 2)\}$ so that $H_0^1 + H_0^2 = \mathbb{1}_{\{C_0 = 1\}} + \mathbb{1}_{\{C_0 = 2\}} = 1$. We also observe that for $i, j = 1, 2$, $i \neq j$ and for every $t \in [0, T]$ we have

$$H_t^i = H_0^i + H_t^{ji} - H_t^{ij} - H_t^{i3} \tag{13.55}$$

and

$$H_t^{i3} = \mathbb{1}_{\{C_t = 3, \hat{C}_t = i\}}. \tag{13.56}$$

Next, we define an auxiliary process $\hat{Z}(t, T)$, which also follows a $\mathbb{G}$-local martingale under $\mathbb{Q}^*$, by setting (the formula below is a straightforward generalization of (13.31))

$$
\begin{aligned}
d\hat{Z}(t, T) :=\ & \left(Z_2(t, T) - Z_1(t, T)\right) dM_t^{1,2} + \left(Z_1(t, T) - Z_2(t, T)\right) dM_t^{2,1} \\
& + \left(\delta_1 Z(t, T) - Z_1(t, T)\right) dM_t^{1,3} + \left(\delta_2 Z(t, T) - Z_2(t, T)\right) dM_t^{2,3} \\
& + \left(H_t^1 Z_1(t, T) b_1(t, T) + H_t^2 Z_2(t, T) b_2(t, T)\right) dW_t^* \\
& + \left(\delta_1 H_t^{1,3} + \delta_2 H_t^{2,3}\right) Z(t, T) b(t, T) dW_t^*
\end{aligned}
$$

with the initial condition:

$$\hat{Z}(0, T) = H_0^1 Z_1(0, T) + H_0^2 Z_2(0, T). \tag{13.57}$$

Using (13.54), we arrive at the following representation for the dynamics of $\hat{Z}(t, T)$

$$
\begin{aligned}
d\hat{Z}(t, T) =\ & Z_1(t)\left(dH_t^{2,1} - dH_t^{1,2} - dH_t^{1,3}\right) + H_t^1 dZ_1(t) \\
& + Z_2(t)\left(dH_t^{1,2} - dH_t^{2,1} - dH_t^{2,3}\right) + H_t^2 dZ_2(t) \\
& + Z(t)\left(\delta_1 dH_t^{1,3} + \delta_2 dH_t^{2,3}\right) + \left(\delta_1 H_t^{1,3} + \delta_2 H_t^{2,3}\right) dZ(t) \\
& + \left[\lambda_{1,2}^*(t)\left(Z_1(t) - Z_2(t)\right) + \lambda_{1,3}^*(t)\left(Z_1(t) - \delta_1 Z(t)\right)\right] H_t^1 dt \\
& + \left[\lambda_{2,1}^*(t)\left(Z_2(t) - Z_1(t)\right) + \lambda_{2,3}^*(t)\left(Z_2(t) - \delta_2 Z(t)\right)\right] H_t^2 dt \\
& - \left(\lambda_1^*(t) Z_1(t) H_t^1 + \lambda_2^*(t) Z_2(t) H_t^2\right) dt,
\end{aligned}
$$

where we write $Z_i(t) = Z_i(t, T)$ and $Z(t) = Z(t, T)$. To construct a consistent model of the term structure of interest rates, we need to specify the intensity matrix $\Lambda^*$ in a judicious way.

We postulate that the entries of $\Lambda^*$ are chosen in such a way that the equalities

$$\begin{cases} \lambda_{1,2}^*(t)(Z_1(t) - Z_2(t)) + \lambda_{1,3}^*(t)(Z_1(t) - \delta_1 Z(t)) = \lambda_1^*(t)Z_1(t), \\ \lambda_{2,1}^*(t)(Z_2(t) - Z_1(t)) + \lambda_{2,3}^*(t)(Z_2(t) - \delta_2 Z(t)) = \lambda_2^*(t)Z_2(t), \end{cases} \quad (13.58)$$

are satisfied for all $t \in [0, T]$.

*Remarks.* First suppose that $\delta_1 = \delta_2 = 0$. In this case, we postulate that the entries of $\Lambda^*$ satisfy

$$\begin{cases} \lambda_{1,2}^*(t)(1 - D_{21}(t)) + \lambda_{1,3}^*(t) = \lambda_1^*(t), \\ \lambda_{2,1}^*(t)(1 - D_{12}(t)) + \lambda_{2,3}^*(t) = \lambda_2^*(t), \end{cases}$$

where we set $D_{ij}(t) = Z_i(t,T)/Z_j(t,T) = D_i(t,T)/D_j(t,T)$. Notice that the coefficients $\lambda_{ij}^*(t)$ are not uniquely determined. We may take, for instance, $\lambda_{1,2}^*(t) = \lambda_{2,1}^*(t) = 0$ (no migration between classes 1 and 2) to obtain $\lambda_{1,3}^*(t) = \lambda_1^*(t)$ and $\lambda_{2,3}^*(t) = \lambda_2^*(t)$, but other choices are not excluded a priori. Notice also that we cannot set $\lambda_{1,3}^*(t) = \lambda_{2,3}^*(t) = 0$ (i.e., no default is possible), since we would then have either $\lambda_{1,2}^*(t) < 0$ or $\lambda_{2,1}^*(t) < 0$. Suppose now that $\delta_1 + \delta_2 > 0$. In this case, we have

$$\begin{cases} \lambda_{1,2}^*(t)(1 - D_{21}(t)) + \lambda_{1,3}^*(t)(1 - \delta_1 d_{31}(t)) = \lambda_1^*(t), \\ \lambda_{2,1}^*(t)(1 - D_{12}(t)) + \lambda_{2,3}^*(t)(1 - \delta_2 d_{32}(t)) = \lambda_2^*(t), \end{cases}$$

where $d_{ij}(t) = Z(t,T)/Z_j(t,T) = B(t,T)/D_j(t,T)$.

Let us return to the analysis of the process $\hat{Z}(t,T)$. Under (13.58), $\hat{Z}(t,T)$ satisfies

$$\begin{aligned} d\hat{Z}(t,T) := {}& \left(Z_2(t,T) - Z_1(t,T)\right) dH_t^{1,2} + \left(Z_1(t,T) - Z_2(t,T)\right) dH_t^{2,1} \\ & + \left(\delta_1 Z(t,T) - Z_1(t,T)\right) dH_t^{1,3} + \left(\delta_2 Z(t,T) - Z_2(t,T)\right) dH_t^{2,3} \\ & + H_t^1 dZ_1(t,T) + H_t^2 dZ_2(t,T) + \left(\delta_1 H_t^{1,3} + \delta_2 H_t^{2,3}\right) dZ(t,T) \end{aligned}$$

with the initial condition (13.57). The foregoing representation of the process $\hat{Z}(t,T)$, combined with (13.55) and (13.56), results in the following important formula:

$$\hat{Z}(t,T) = \mathbb{1}_{\{C_t=1\}} Z_1(t,T) + \mathbb{1}_{\{C_t=2\}} Z_2(t,T) + \left(\delta_1 H_t^{1,3} + \delta_2 H_t^{2,3}\right)Z(t,T).$$

Put another way

$$\hat{Z}(t,T) = \mathbb{1}_{\{C_t\neq 3\}} Z_{C_t}(t,T) + \delta_{\hat{C}_t} \mathbb{1}_{\{C_t=3\}} Z(t,T). \quad (13.59)$$

Finally, we introduce the price process of a $T$-maturity defaultable bond by setting (also cf. (13.29))

$$D^\delta(t,T) := B_t \hat{Z}(t,T) = \mathbb{1}_{\{C_t\neq 3\}} D_{C_t}(t,T) + \delta_{\hat{C}_t} \mathbb{1}_{\{C_t=3\}} B(t,T).$$

*Remarks.* Under the present assumptions, the process $\hat{Z}(t) := \hat{Z}(t, T)$, given by (13.59), can be also defined as the unique solution of the following SDE (cf. (13.26))

$$
\begin{aligned}
d\hat{Z}(t) = {} & \left(Z_2(t) - H_t^1 \hat{Z}(t-)\right) dM_t^{1,2} + \left(Z_1(t) - H_t^2 \hat{Z}(t-)\right) dM_t^{2,1} \\
& + \left(\delta_1 Z(t) - H_t^1 \hat{Z}(t-)\right) dM_t^{1,3} + \left(\delta_2 Z(t) - H_t^2 \hat{Z}(t-)\right) dM_t^{2,3} \\
& + \left(H_t^1 \hat{Z}(t) b_1(t, T) + H_t^2 \hat{Z}(t) b_2(t, T) + H_t^3 \hat{Z}(t) b(t, T)\right) dW_t^*
\end{aligned}
$$

with the initial condition (13.57). Indeed, since

$$
H_t^3 = 1 - H_t^1 - H_t^2 = H_t^{13} + H_t^{23}
$$

we may rewrite this SDE as follows:

$$
\begin{aligned}
d\hat{Z}(t) = {} & \left(Z_2(t) - H_t^1 \hat{Z}(t-)\right) dH_t^{1,2} + H_t^1 \hat{Z}(t)\left(\lambda_1^*(t)\, dt + b_1(t, T)\right) dW_t^* \\
& + \left(Z_1(t) - H_t^2 \hat{Z}(t-)\right) dH_t^{2,1} + H_t^2 \hat{Z}(t)\left(\lambda_2^*(t)\, dt + b_2(t, T)\right) dW_t^* \\
& + \left(\delta_1 Z(t) - H_t^1 \hat{Z}(t-)\right) dH_t^{1,3} + \left(\delta_2 Z(t) - H_t^2 \hat{Z}(t-)\right) dH_t^{2,3} \\
& + \left(H_t^{1,3} + H_t^{2,3}\right) \hat{Z}(t) b(t, T)\, dW_t^* \\
& - H_t^1 \left[\lambda_{1,2}^*(t)\left(Z_2(t) - \hat{Z}(t)\right) + \lambda_{1,3}^*(t)\left(\delta_1 Z(t) - \hat{Z}(t)\right) + \lambda_1^*(t)\hat{Z}(t)\right] dt \\
& - H_t^2 \left[\lambda_{2,1}^*(t)\left(Z_1(t) - \hat{Z}(t)\right) + \lambda_{2,3}^*(t)\left(\delta_2 Z(t) - \hat{Z}(t)\right) + \lambda_2^*(t)\hat{Z}(t)\right] dt.
\end{aligned}
$$

In view of (13.52)–(13.53) and (13.58), it is not difficult to check that the unique solution $\hat{Z}(t, T)$ to the SDE above coincides with the process given by the right-hand side of (13.59).

### 13.2.4 General Case

We are in a position to examine the general case. For any $K \geq 3$, we define an auxiliary process $\hat{Z}(t, T)$ by postulating that it satisfies:

$$
\begin{aligned}
d\hat{Z}(t, T) = {} & \sum_{i=1}^{K-1} H_t^i Z_i(t, T) b_i(t, T)\, dW_t^* + \sum_{i=1}^{K-1} \delta_i H_t^{iK} Z(t, T) b(t, T)\, dW_t^* \\
& + \sum_{i,j=1, i \neq j}^{K-1} \left(Z_j(t, T) - Z_i(t, T)\right) dM_t^{ij} + \sum_{i=1}^{K-1} \left(\delta_i Z(t, T) - Z_i(t, T)\right) dM_t^{iK}
\end{aligned}
$$

with the initial condition:

$$
\hat{Z}(0, T) = \sum_{i=1}^{K-1} H_0^i Z_i(0, T).
$$

We shall now generalize the consistency condition (13.58). For the sake of brevity, we shall write $Z_i(t)$ rather than $Z_i(t, T)$.

**Condition (BR.6)** For any $i = 1, \ldots, K-1$ and every $t \in [0, T]$ the following equalities are satisfied:

$$\sum_{j=1}^{K-1} \lambda_{ij}^*(t)\big(Z_j(t) - Z_i(t)\big) + \lambda_{iK}^*(t)\big(\delta_i Z(t) - Z_i(t)\big) + \lambda_i^*(t) Z_i(t) = 0.$$

Under the assumption above, the process $\hat{Z}(t, T)$ is governed by the following expression:

$$d\hat{Z}(t, T) = \sum_{i,j=1, i \neq j}^{K-1} \big(Z_j(t, T) - Z_i(t, T)\big) dH_t^{ij}$$

$$+ \sum_{i=1}^{K-1} \big(\delta_i Z(t, T) - Z_i(t, T)\big) dH_t^{iK}$$

$$+ \sum_{i=1}^{K-1} H_t^i \, dZ_i(t, T) + \sum_{i=1}^{K-1} \delta_i H_t^{iK} \, dZ(t, T).$$

The following result furnishes more convenient representations for the auxiliary process $\hat{Z}(t, T)$.

**Lemma 13.2.1.** *Under assumption* (BR.6), *the process* $\hat{Z}(t, T)$ *satisfies*

$$\hat{Z}(t, T) = \sum_{i=1}^{K-1} \big(H_t^i Z_i(t, T) + \delta_i H_t^{iK} Z(t, T)\big)$$

*or, equivalently,*

$$\hat{Z}(t, T) = \mathbb{1}_{\{C_t \neq K\}} Z_{C_t}(t, T) + \delta_{\hat{C}_t} \mathbb{1}_{\{C_t = K\}} Z(t, T). \tag{13.60}$$

*Moreover,* $\hat{Z}(t, T)$ *is the unique solution to the SDE*

$$d\hat{Z}(t, T) = \sum_{i,j=1, i \neq j}^{K-1} \big(Z_j(t, T) - H_t^i \hat{Z}(t-, T)\big) dM_t^{ij}$$

$$+ \sum_{i=1}^{K-1} \big(\delta_i Z(t, T) - H_t^i \hat{Z}(t-, T)\big) dM_t^{iK}$$

$$+ \sum_{i=1}^{K-1} H_t^i \hat{Z}(t, T) b_i(t, T) \, dW_t^* + H_t^K \hat{Z}(t, T) b(t, T) \, dW_t^*$$

*with the initial condition* $\hat{Z}(0, T) = \sum_{i=1}^{K-1} H_0^i Z_i(0, T)$.

*Proof.* The lemma can be proved by reasoning as in the case of $K = 3$. For this reason, its proof is left to the reader. $\qquad\square$

As expected, we define the value process of a $T$-maturity zero-coupon corporate bond by setting

$$D^\delta(t,T) := B_t \hat{Z}(t,T) = \mathbb{1}_{\{C_t \neq K\}} D_{C_t}(t,T) + \delta_{\hat{C}_t} \mathbb{1}_{\{C_t=K\}} B(t,T). \quad (13.61)$$

The following result is an immediate consequence of the properties of the auxiliary process $\hat{Z}(t,T)$.

**Proposition 13.2.1.** *Under the risk-neutral probability $\mathbb{Q}^*$, the dynamics of the price process $D^\delta(t,T)$ are*

$$dD^\delta(t,T) = \sum_{i,j=1,\, i \neq j}^{K-1} \left( D_j(t,T) - D_i(t,T) \right) dH_t^{ij}$$

$$+ \sum_{i=1}^{K-1} \left( \delta_i B(t,T) - D_i(t,T) \right) dH_t^{iK} + \sum_{i=1}^{K-1} H_t^i \, dD_i(t,T)$$

$$+ \sum_{i=1}^{K-1} \delta_i H_t^{iK} \, dB(t,T) + r_t D^\delta(t,T) \, dt,$$

*where the Itô differentials $dB(t,T)$ and $dD_i(t,T)$ are given by the formulae*

$$dB(t,T) = B(t,T)\left(r_t \, dt + b(t,T) \, dW_t^*\right)$$

*and*

$$dD_i(t,T) = D_i(t,T)\left((r_t + \lambda_i^*(t)) \, dt + b_i(t,T) \, dW_t^*\right).$$

The next theorem shows that the process $D^\delta(t,T)$, formally introduced through (13.61), can be given an intuitive interpretation in terms of the default time and recovery rates. To this end, we make the following technical assumption (cf. Condition (HJM.7) of Sect. 13.1.5).

**Condition (BR.7)** The process $\hat{Z}(t,T)$, given by (13.60), follows a $\mathbb{G}$-martingale (as opposed to a local martingale) under $\mathbb{Q}^*$.

The main result of this section holds under assumptions (BR.1)–(BR.7).

**Theorem 13.2.1.** *For any $i = 1,\ldots,K-1$, let $\delta_i \in [0,1)$ be the recovery rate for a defaultable bond from the $i^{\text{th}}$ rating class. The price process $D^\delta(t,T)$ of a $T$-maturity defaultable bond equals*

$$D^\delta(t,T) = \mathbb{1}_{\{C_t \neq K\}} e^{-\int_t^T g_{C_t}(t,u)\, du} + \delta_{\hat{C}_t} \mathbb{1}_{\{C_t=K\}} e^{-\int_t^T f(t,u)\, du},$$

*that is,*

$$D^\delta(t,T) = \mathbb{1}_{\{C_t \neq K\}} D_{C_t}(t,T) + \delta_{\hat{C}_t} \mathbb{1}_{\{C_t=K\}} B(t,T),$$

*Equivalently,*

$$D^\delta(t,T) = B(t,T)\left(\mathbb{1}_{\{C_t \neq K\}} e^{-\int_t^T s_{C_t}(t,u)\, du} + \delta_{\hat{C}_t} \mathbb{1}_{\{C_t=K\}}\right),$$

*where $s_i(t,u) = g_i(t,u) - f(t,u)$ represents the $i^{\text{th}}$ credit spread.*

*Moreover, $D^\delta(t,T)$ is given by the following version of the risk-neutral valuation formula*

$$D^\delta(t,T) = B_t \, \mathbb{E}_{\mathbb{Q}^*} \big( \delta_{\hat{C}_T} B_T^{-1} \mathbb{1}_{\{\tau \leq T\}} + B_T^{-1} \mathbb{1}_{\{\tau > T\}} \,|\, \mathcal{G}_t \big), \qquad (13.62)$$

*where $\tau$ is the default time, i.e., $\tau = \inf\{t \in \mathbb{R}_+ : C_t = K\}$. The last formula can also be rewritten as follows:*

$$D^\delta(t,T) = B(t,T) \, \mathbb{E}_{\mathbb{Q}_T} \big( \delta_{\hat{C}_T} \mathbb{1}_{\{\tau \leq T\}} + \mathbb{1}_{\{\tau > T\}} \,|\, \mathcal{G}_t \big),$$

*where $\mathbb{Q}_T$ is the forward martingale measure for the date $T$, associated with $\mathbb{Q}^*$ through (13.33).*

*Proof.* The first formula is an immediate consequence of (13.61) combined with (13.50)–(13.51). For the second, notice first that in view of the second equality in (13.61) and the definition of $\tau$, the process $D^\delta(t,T)$ satisfies the terminal condition

$$D^\delta(T,T) = \delta_{\hat{C}_T} \mathbb{1}_{\{\tau \leq T\}} + \mathbb{1}_{\{\tau > T\}}.$$

Furthermore, using the first equality in (13.61), we deduce the discounted process $B_t^{-1} D^\delta(t,T)$ coincides with $\hat{Z}(t,T)$, and thus it follows a $\mathbb{Q}^*$-martingale. Equality (13.62) is thus obvious. $\qquad \square$

### 13.2.5 Alternative Recovery Schemes

In Sect. 13.1.9, we dealt with alternative recovery rules in case of only one pre-default rating class, i.e., for $K = 2$. Valuation results for $K > 2$ in case of a general recovery scheme can be obtained in a straightforward manner by combining the ideas presented in Sect. 13.1.9 and 13.2.4. Condition (HJM.8) of Sect. 13.1.9 is now replaced by the following one.

**Condition (BR.8)** The (terminal) *recovery profile* is given in terms of a $(K-1)$-dimensional process $\delta(t) = \big(\delta_1(t), \ldots, \delta_{K-1}(t)\big)$. Each process $\delta_i(t)$, $i = 1, \ldots, K-1$ is an $\mathbb{F}$-adapted and (locally) bounded process on the original probability space $(\tilde{\Omega}, \mathbb{F}, \mathbb{P})$.

Before we complete the description of the general recovery structure, we define an auxiliary process $\hat{Z}(t,T)$ by setting

$$\hat{Z}(t,T) = \sum_{i=1}^{K-1} \big( H_t^i Z_i(t,T) + \delta_i(t) H_t^{iK} Z(t,T) \big)$$

or, equivalently,

$$\hat{Z}(t,T) = \mathbb{1}_{\{C_t \neq K\}} Z_{C_t}(t,T) + \delta_{\hat{C}_t}(t) \mathbb{1}_{\{C_t = K\}} Z(t,T).$$

Condition (BR.6) of Sect. 13.2.4, which serves to specify the migration intensities of the migration process $C$, now takes the following form.

**Condition (BR.9)** For any $i = 1, \ldots, K - 1$ and every $t \in [0, T]$, the following equalities are satisfied:

$$\sum_{j=1}^{K-1} \lambda_{ij}^*(t)\big(Z_j(t) - Z_i(t)\big) + \lambda_{iK}^*(t)\big(\delta_i(t)Z(t) - Z_i(t)\big) + \lambda_i^*(t)Z_i(t) = 0.$$

We may now complete the description of the financial interpretation of the general recovery structure. Given the migration process $C$ specified via Condition (BR.9), it is postulated that the recovery payment of $\delta_{\hat{C}_\tau}(\tau)$ occurs at the maturity date $T$ (provided that $\tau \le T$). Specifically, if $\tau \le T$ and if the rating class immediately preceding the default is the $i^{\text{th}}$ class (i.e., $\hat{C}_T(\tau) = i$), then the recovery payment received by the bondholders at time $T$ equals $\delta_i(\tau)$. Observe now that under (BR.9) the auxiliary process $\hat{Z}(t, T)$ is a $\mathbb{G}$-local martingale under $\mathbb{Q}^*$. As before, we postulate that this process is a martingale, and we define the price process $\hat{D}^\delta(t, T)$ of a $T$-maturity defaultable bond by the following expression

$$\hat{D}^\delta(t, T) := B_t \hat{Z}(t, T) = \mathbb{1}_{\{\tau > t\}} D_{C_t}(t, T) + \mathbb{1}_{\{\tau \le t\}} \delta_{\hat{C}_\tau}(\tau) B(t, T). \quad (13.63)$$

Consequently, we obtain a result analogous to Proposition 13.1.5, i.e.,

$$\hat{D}^\delta(t, T) = B_t \mathbb{E}_{\mathbb{Q}^*}\big(\delta^* B_T^{-1} \mathbb{1}_{\{\tau \le T\}} + B_T^{-1} \mathbb{1}_{\{\tau > T\}} \,\big|\, \mathcal{G}_t\big), \quad (13.64)$$

where $\delta^* = \delta_{\hat{C}_\tau}(\tau)$. We shall specify the above results to the two particular recovery schemes that were also considered in the previous section.

**Fractional recovery of par value.** First suppose that $\delta_i(t) = \delta_i B^{-1}(t, T)$ for $i = 1, \ldots, K - 1$. Then we obtain

$$\hat{D}^\delta(t, T) = \mathbb{1}_{\{\tau > t\}} D_{C_t}(t, T) + \mathbb{1}_{\{\tau \le t\}} \delta_{\hat{C}_\tau} B^{-1}(\tau, T) B(t, T).$$

This corresponds to the random payoff $\delta^* = \delta_{\hat{C}_\tau} B^{-1}(\tau, T)$ at time $T$. Consequently, we obtain the following expression for the price process of a $T$-maturity defaultable bond

$$\hat{D}^\delta(t, T) = \mathbb{1}_{\{\tau > t\}} D_{C_t}(t, T) + \mathbb{1}_{\{\tau \le t\}} \delta^* B(t, T),$$

and so

$$\hat{D}^\delta(t, T) = B_t \mathbb{E}_{\mathbb{Q}^*}\big(\delta_{\hat{C}_\tau} B^{-1}(\tau, T) B_T^{-1} \mathbb{1}_{\{\tau \le T\}} + B_T^{-1} \mathbb{1}_{\{\tau > T\}} \,\big|\, \mathcal{G}_t\big).$$

**Fractional recovery of market value.** Assume now that the recovery processes are given as

$$\delta_i(t) = \delta_i D_i(t, T) B^{-1}(t, T), \quad \forall\, i = 1, \ldots, K - 1.$$

Then

$$\hat{D}^\delta(t, T) = \mathbb{1}_{\{\tau > t\}} D_{C_t}(t, T) + \mathbb{1}_{\{\tau \le t\}} \delta_{\hat{C}_\tau} D_{\hat{C}_\tau}(\tau, T) B^{-1}(\tau, T) B(t, T).$$

Consequently,

$$\hat{D}^\delta(t, T) = B_t \mathbb{E}_{\mathbb{Q}^*}\big(\delta_{\hat{C}_\tau} D_{\hat{C}_\tau}(\tau, T) B^{-1}(\tau, T) B_T^{-1} \mathbb{1}_{\{\tau \le T\}} + B_T^{-1} \mathbb{1}_{\{\tau > T\}} \,\big|\, \mathcal{G}_t\big).$$

### 13.2.6 Defaultable Coupon Bonds

Consider a corporate coupon bond with the face value $L$ and maturity date $T$, which promises to pay (non-random) coupons $c_k$ at times $T_k < T$, $k = 1, \ldots, n$. The coupon payments are only made prior to default and there is no recovery payment due in conjunction with the unpaid coupons whose payment dates are past the default time. The only recovery payment is due relative to the bond's face value. We postulate that this recovery payment is made at maturity $T$, in case the bond defaults before or at the maturity. The bond's recovery profile is given in terms of a vector of recovery rates $\delta = (\delta_1, \ldots, \delta_{K-1})$, where $\delta_i \in [0, 1)$ are deterministic quantities.

We consider a migration process $C$, whose transition intensities satisfy Condition (BR.9), with the recovery processes $\delta_i(t)$ given by:

$$\delta_i(t) = \sum_{k=1}^{n} c_k B^{-1}(T_k, T) \mathbb{1}_{\{T_k < t\}} + \delta_i. \tag{13.65}$$

The process $C$ is called the *migration process of the corporate bond*. It is clear that this migration process may depend on the maturity date $T$, the coupon payment schedule $(c_k, T_k)$, $k = 1, \ldots, n$ and the recovery profile $\delta$.

Formally, a defaultable coupon bond is represented by the following cash flows:

$$\sum_{k=1}^{n} c_k \mathbb{1}_{\{\tau > T_k\}} \mathbb{1}_{T_k}(t) + \left( L \mathbb{1}_{\{\tau > T\}} + \delta_{\hat{C}_\tau} L \mathbb{1}_{\{\tau \leq T\}} \right) \mathbb{1}_T(t),$$

where $\tau$ stands for the bond's default time: $\tau = \inf \{t \in [0, T^*] : C_t = K\}$. Observe that we deal here with the case of a *mixed recovery* of a stream of promised payments. The promised payoff at the bond's maturity equals

$$X = \sum_{k=1}^{n} c_k B^{-1}(T_k, T) + L,$$

and the recovery processes are given by (13.65). Therefore, a defaultable coupon bond is formally equivalent to a single random payoff at the maturity date $T$ in the amount of

$$\sum_{k=1}^{n} c_k \mathbb{1}_{\{\tau > T_k\}} B^{-1}(T_k, T) + \left( L \mathbb{1}_{\{\tau > T\}} + \delta_{\hat{C}_\tau} L \mathbb{1}_{\{\tau \leq T\}} \right).$$

In view of (13.64), we conclude that the arbitrage price $D_c(t, T) = \hat{D}^\delta(t, T)$ of the corporate coupon bond, with the above cash flows, is given by

$$D_c(t, T) = B_t \, \mathbb{E}_{\mathbb{Q}^*} \left( B_T^{-1} U \mathbb{1}_{\{\tau \leq T\}} + B_T^{-1} L \mathbb{1}_{\{\tau > T\}} \,\big|\, \mathcal{G}_t \right),$$

where we denote

$$U = \sum_{k=1}^{n} c_k B^{-1}(T_k, T) \mathbb{1}_{\{t \leq T_k < \tau\}} + \delta_{\hat{C}_\tau} L.$$

## 13.2.7 Default Correlations

So far, we have considered the migration process governing the behavior of
a particular defaultable bond with a given recovery profile (see Condition
(BR.8)) and maturity. Suppose now that we are given a family of recovery
profiles:

$$\delta^m(t) = \big(\delta_1^m(t), \ldots, \delta_{K-1}^m(t)\big), \quad \forall\, t \in [0, T^*],$$

for $m = 1, \ldots, M$. Let us consider a securities market model consisting of
the following primary assets:

- zero-coupon default-free bonds with maturities $T \in [0, T^*]$,
- $M$ classes of defaultable zero-coupon bonds with maturities $T \in [0, T^*]$;
  for each $m = 1, \ldots, M$, we deal with a class-specific recovery profile $\delta^m(t)$.

For each class $m$ and any maturity $T \in [0, T^*]$, we consider a migration
process with intensities $\lambda_{ij}^*(t)$ satisfying the consistency condition (BR.9).
As apparent from Condition (BR.9), the intensities $\lambda_{ij}^*(t)$ may depend both
on the class $m$ (more specifically, on the recovery profile $\delta^m(t)$) and on the
maturity date $T$. We shall emphasize this possible dependence by writing
$\lambda_{ij}^{*m}(t, T)$. By convention, the migration process corresponding to the maturity $T < T^*$ is extended by constancy to the interval $[T, T^*]$.

**Statistical dependence between corporate bonds.** It is essential to
observe that in general even if all four pieces of data – namely: the maturity
date, the transition intensities, the recovery scheme and the initial rating –
are identical for the two zero-coupon bonds, the bonds themselves may not
be identical. In fact, if they are issued by two different entities, the associated
migration processes $C$ and $C'$ are also generally different. The point we would
like to make here is that if all four pieces of data listed above are the same, the
finite dimensional distributions of the two processes $C$ and $C'$ are necessarily
identical, but the two processes themselves do not necessarily coincide.

In other words, if we consider the joint migration process $(C, C')$, then
the marginal finite-dimensional distributions for $C$ and $C'$ are identical, but
in general $C \not\equiv C'$. If $C \not\equiv C'$, the credit migration processes $C$ and $C'$ may
be either (conditionally) independent or dependent. In case of independent
migration processes $C$ and $C'$, no statistical dependence between credit migrations of the two bonds appears. In case of mutually dependent migration
processes, one needs to calibrate the dependence structure (or, more crudely,
the correlation structure) between $C$ and $C'$.

The foregoing remarks are valid if one considers an application of the general methodology presented in this chapter to the valuation and hedging of
individual defaultable bonds – that is, corporate bonds issued by particular
institutions, as well as to the valuation and hedging of related credit derivatives. As an alternative, let us mention that the methodology presented in
this chapter may be applied to a totality of alike defaultable bonds – that is,
to the totality of bonds for which all four features listed above coincide.

In the latter approach, we identify all such bonds and we substitute them with a *representative bond* with an associated representative migration process. This application of our methodology aims at valuation and hedging of credit derivatives that are tied to the average market value of corporate bonds of a given credit quality. Thus, the correlation structure between individual bonds is deliberately disregarded. All that really matters in this interpretation are the marginal statistical properties of individual corporate bonds, and they are identical for all bonds in a given class.

**Correlation coefficients.** Let us consider two different defaultable bonds, and let us denote the associated migration processes as $C$ and $C'$. The respective default events are:

$$A = \bigcup_{t \leq T^*} \{C_{t-} \neq K, C_t = K\}, \quad A' = \bigcup_{t \leq T^*} \{C'_{t-} \neq K, C'_t = K\},$$

and the respective default times are:

$$\tau = \inf\{t \in [0, T^*] : C_t = K\}, \quad \tau' = \inf\{t \in [0, T^*] : C'_t = K\}.$$

We may study two types of default correlations: the correlation between random variables $\mathbb{1}_A$ and $\mathbb{1}_{A'}$ and the correlation between random variables $\tau$ and $\tau'$. Various correlation coefficients, such as Pearson's (or linear) correlation coefficient, may be used to measure the strength of these correlations. Likewise, we may analyze the correlations between the *survival events* of the form: $S(t) = \{\tau > t\}$ and $S'(t) = \{\tau' > t\}$. Of course, the correlation structure will typically vary depending on whether one uses the risk-neutral probability $\mathbb{Q}^*$ or the real-world probability $\mathbb{Q}$.

### 13.2.8 Market Prices of Interest Rate and Credit Risk

As made clear in the preceding section, the migration process and the jump martingale associated with a defaultable bond may depend on $T$ and $m$, because of such a dependence of migration intensities. The migration process for the recovery class $m$ and the maturity $T$ should thus be denoted by $C^{m,T}$. Accordingly, we should use the notation $H^{i,m,T}$, $H^{ij,m,T}$ and $M^{ij,m,T}$ for the associated jump processes and jump martingales. However, in order to simplify further considerations we shall impose the following condition.

**Condition (BR.10)** The migration intensities $\lambda_{ij}^{*m}(t, T)$ do not depend on the maturity date $T$.

In view of the last condition, we may (and will) skip the superscript $T$ in what follows. Thus, we shall only write $\lambda_{ij}^{*m}$, $C^m$, etc.

*Remarks.* (i) Since our goal is to simultaneously construct all processes $C^m$, $m = 1, \ldots, M$, we shall typically need an obvious modification of the enlarged probability space $(\Omega, \mathbb{G}, \mathbb{Q}^*)$. We shall, however, preserve the notation, so that from now on $(\Omega, \mathbb{G}, \mathbb{Q}^*)$ will stand for this modified enlargement.

(ii) For each $m$ and every $T \in [0, T^*]$, the process $\hat{D}^{\delta^m}(\cdot, T)$ (see (13.63)) follows a $\mathbb{G}$-martingale on the enlarged probability space $(\Omega, \mathbb{G}, \mathbb{Q}^*)$ (under all technical assumptions introduced above). Thus, the securities market model considered in this section is arbitrage-free.

We shall now change, using a suitable version of Girsanov's theorem, the probability measure $\mathbb{Q}^*$ to an equivalent probability measure $\mathbb{Q}$ on $(\Omega, \mathcal{G}_{T^*})$. The probability measure $\mathbb{Q}$ introduced below is postulated to play the role of the real-world probability in our model. For this reason, we require that the restriction of $\mathbb{Q}$ to the original probability space $\tilde{\Omega}$ necessarily coincides with the real-world probability $\mathbb{P}$ for the default-free term structure model. Recall that in the present set-up we have (cf. (13.11))

$$\frac{d\mathbb{P}^*}{d\mathbb{P}}\Big|_{\mathcal{F}_t} = \tilde{L}_t, \quad \mathbb{P}\text{-a.s.}$$

with the process $\tilde{L}$ satisfying the SDE

$$d\tilde{L}_t = \tilde{L}_t \beta_t \, dW_t, \quad \tilde{L}_0 = 1.$$

Thus, it is natural to set

$$\frac{d\mathbb{Q}}{d\mathbb{Q}^*}\Big|_{\mathcal{G}_t} = L_t, \quad \mathbb{Q}^*\text{-a.s.}, \tag{13.66}$$

where the process $L$ is governed by the following SDE

$$dL_t = -L_t \beta_t \, dW_t^* + L_{t-} \, dM_t, \quad L_0 = 1. \tag{13.67}$$

The process $M$ in the last formula is defined as follows:

$$dM_t = \sum_{m=1}^{M} \sum_{i \neq j} \kappa_{ij}^m(t) \, dM_t^{ij,m},$$

where in turn, for any $m = 1, \ldots, M$ and $i \neq j$, we denote by $\kappa_{ij}^m$ a nonnegative, $\mathbb{F}$-predictable process such that

$$\int_0^{T^*} (1 + \kappa_{ij}^m(t)) \lambda_{ij}^{*m}(t) \, dt < \infty, \quad \mathbb{Q}^*\text{-a.s.}$$

More explicitly,

$$dM_t = \sum_{m=1}^{M} \sum_{i \neq j} \kappa_{ij}^m(t) \left( dH_t^{ij,m} - \lambda_{ij}^{*m}(t) H_t^{i,m} \, dt \right).$$

We know already that the process $M$ follows a $\mathbb{G}$-local martingale under $\mathbb{Q}^*$. Typically, this process will depend on recovery profiles $\delta^m$, $m = 1, \ldots, M$.

**Definition 13.2.1.** The processes $\beta$ and $-\kappa_{ij}^m$, where $m = 1, \ldots, M$ and $i, j = 1, \ldots, K$, $i \neq j$ are referred to as the *market price of interest rate risk* and the *market price of credit risk*, respectively.

*Remarks.* (i) It is quite natural that the market price of credit risk should depend on the recovery profile, so that credit risks of corporate bonds with different recovery profiles are priced differently by the market.

(ii) There is no direct analogy between the set-up analyzed in this section and the results of Proposition 11.2.3. The main difference is that we assume here that the processes $\kappa_{ij}^m$, corresponding to processes $\tilde{\kappa}^{ij}$ in Sect. 11.2.5, are $\mathbb{F}$-adapted. In Sect. 11.2.5, the reference filtration $\mathbb{F}$ was assumed to be trivial, so that any $\mathbb{F}$-adapted process followed a deterministic function.

In view of the discussion above, we see that the process $L$ is a $\mathbb{G}$-local martingale under $\mathbb{Q}^*$; it is defined on the interval $[0, T^*]$ and may depend on recovery profiles $\delta^m$, $m = 1, \ldots, M$. From now on, we assume that $\mathbb{E}_{\mathbb{Q}^*}(L_{T^*}) = 1$, so that the probability measure $\mathbb{Q}$ is well defined on $(\Omega, \mathcal{G}_{T^*})$ and, by virtue of Condition (BR.10), it does not depend on $T$. However, the probability $\mathbb{Q}$ may depend on the recovery profiles $\delta^m$, $m = 1, \ldots, M$, that is, it may depend on all the defaultable market instruments through their recovery profiles. To simplify the following discussion, we shall also assume that there is only one recovery class in our securities market. However, we maintain Condition (BR.10). Thus, we deal with only one migration process, denoted by $C$. Accordingly, the superscript $m$ will not show up in the rest of this section.

To analyze the behavior of the migration process $C$ under $\mathbb{Q}$, we introduce the point process $\bar{Z} := (\tau_k, (\tilde{C}_k, \tilde{C}_{k-1}))$, $k \in \mathbb{N}$, and the associated 'transition' process (for the definition of the sequence $\tilde{C}_k$, see Sect. 11.3.1)

$$\bar{\Phi}(t, i, l) := \sum_{k=1}^{\infty} \mathbb{1}_{\{\tau_k \le t, \, \tilde{C}_k = i, \, \tilde{C}_{k-1} = l\}}.$$

We also introduce a family of processes $\bar{\lambda}_{(i,k),(j,l)}(t)$, where $i, j = 1, \ldots, K$ and $k, l = 1, \ldots, K - 1$, by setting:

$$\bar{\lambda}_{(i,k),(j,l)}(t) = \begin{cases} \lambda_{ij}^*(t), & \text{if } l = i \ne K, \\ 0, & \text{if } l = i \text{ or } i = K. \end{cases}$$

It is clear that the processes $\bar{\lambda}_{(i,k),(j,l)}$ are the migration intensities under $\mathbb{Q}^*$ for the two-dimensional process $(C, \hat{C})$, defined on the finite state space $\{1, \ldots, K\} \times \{1, \ldots, K - 1\}$. Next, we define

$$\bar{\nu}_{\mathbb{Q}^*}(t, i, l) := \int_0^t \bar{\lambda}_{(C_{u-}, \hat{C}_{u-}),(i,l)}(u) \, du.$$

Observe that the following identities are valid: $\bar{\Phi}(t, i, l) = H_t^{li}$ and

$$\bar{\nu}_{\mathbb{Q}^*}(t, i, l) = \int_0^t \lambda_{C_{u-}, i}^*(u) H_{u-}^l \, du = \int_0^t \lambda_{li}^*(u) H_u^l \, du.$$

In view of Corollary 11.3.2, the process $\bar{\nu}_{\mathbb{Q}^*}(t,i,l)$ represents the compensator for the process $\bar{\Phi}(t,i,l)$ under $\mathbb{Q}^*$. Consequently, from the Girsanov theorem for local martingales,[5] it results that the process

$$\bar{\nu}_{\mathbb{Q}}(t,i,l) := \int_0^t (1 + \kappa_{li}(u))\bar{\lambda}_{(C_{u-},\hat{C}_{u-}),(i,l)}(u)\,du$$

is the compensator for $\bar{\Phi}(t,i,l)$ under $\mathbb{Q}$. Now observe that

$$\Phi(t,i) := \sum_{k=1}^{\infty} \mathbb{1}_{\{\tau_k \le t,\, \tilde{C}_k = i\}} = \sum_{l=1}^{K} \bar{\Phi}(t,i,l).$$

We conclude that the compensator for $\Phi(t,i)$ under $\mathbb{Q}$ equals

$$\nu_{\mathbb{Q}}(t,i) = \sum_{l=1}^{K} \bar{\nu}_{\mathbb{Q}}(t,i,l) = \int_0^t \sum_{l=1}^{K} (1 + \kappa_{li}(u))\bar{\lambda}_{(C_{u-},\hat{C}_{u-}),(i,l)}(u)\,du$$

$$= \int_0^t \sum_{l=1}^{K} (1 + \kappa_{C_{u-},i}(u))\lambda^*_{C_{u-},i}(u)H_{u-}^l\,du = \int_0^t \lambda_{C_{u-},i}(u)\,du,$$

where the processes $\lambda_{ij}(t)$, $t \in [0, T^*]$, satisfy

$$\lambda_{ij}(t) = (1 + \kappa_{ij}(t))\lambda^*_{ij}(t), \quad \forall\, i, j = 1, \ldots, K,\ i \ne j,$$

and, as usual, $\lambda_{ii}(t) = -\sum_{j \ne i} \lambda_{ij}(t)$ for $i = 1, \ldots, K$. Repeating the procedure presented in Sect. 11.3.1, one may construct a continuous-time $\mathbb{F}$-conditionally Markov chain $\bar{C}$ on the state space $\mathcal{K}$ with the following intensity matrix under $\mathbb{Q}$:

$$\Lambda_t = \begin{pmatrix} \lambda_{1,1}(t) & \cdots & \lambda_{1,K}(t) \\ \cdot & \cdots & \\ \lambda_{K-1,1}(t) & \cdots & \lambda_{K-1,K}(t) \\ 0 & \cdots & 0 \end{pmatrix}.$$

All results established in Sect. 11.3.2 and 11.3.3 are valid for the process $\bar{C}$ defined in this way. We thus arrive at the following proposition.

**Proposition 13.2.2.** *The finite-dimensional laws under the probability measure $\mathbb{Q}$ of the two processes $C$ and $\bar{C}$ are identical.*

Consequently, under the probability measure $\mathbb{Q}$ given by (13.66)–(13.67), the migration process is still an $\mathbb{F}$-conditionally Markov process and it has under $\mathbb{Q}$ the conditional infinitesimal generator $\Lambda_t$ for every $t \in [0, T^*]$.

*Remarks.* Of course, the equality $\Lambda \equiv \Lambda^*$ holds if and only if $\kappa_{ij} \equiv 0$ for every $i$ and $j$. If the market price for credit risk depends only on the current rating $i$ (and not on the rating $j$ after jump) so that $\kappa_{ij} = \kappa_{ii} =: v_i$ for every $j$, then $\Lambda_t = V\Lambda_t^*$, where $V = \text{diag}\,[v_i]$ is the diagonal matrix. A similar relationship was postulated, for example, in Jarrow et al. (1997) (cf. Condition (JLT.7c) in Sect. 12.1.1).

---

[5] See, for instance, Theorem III.3.11 in Jacod and Shiryaev (1987).

## 13.3 Applications to Credit Derivatives

In this section, we examine the valuation problem for two examples of credit derivatives, and we comment briefly on hedging of credit derivatives.

### 13.3.1 Valuation of Credit Derivatives

**Total rate of return swap.** We first consider the valuation problem for a total rate of return swap (TROR). As a reference asset we take a coupon bond described in Sect. 13.2.6, with the promised cash flows $c_i$ at times $T_i$. Assume the maturity of the total rate of return swap is $\tilde{T} \leq T$. In addition, suppose that the *reference rate* payments (the annuity payments) are made by the investor at fixed scheduled times $t_i \leq \tilde{T}$, $i = 1, \ldots, m$.

As explained in Sect. 1.3.2, the owner of a TROR is entitled not only to all coupon payments during the life of the contract, but also to the change in the value of the reference bond, paid as a lump sum at the contract's termination. We denote by $\kappa$ the fixed rate to be paid by the investor. From the receiver's perspective the cash flows are:

$$\sum_{i=1}^{n} c_i \mathbb{1}_{\{T_i < \tilde{\tau}\}} \mathbb{1}_{\{T_i\}}(t) + \left(D_c(\tilde{\tau}, T) - D_c(0, T)\right) \mathbb{1}_{\{\tilde{\tau}\}}(t) - \kappa \sum_{i=1}^{m} \mathbb{1}_{\{t_i < \tilde{\tau}\}} \mathbb{1}_{\{t_i\}}(t),$$

where $\tilde{\tau} = \tau \wedge \tilde{T}$ and $\tau = \inf \{t \geq 0 : C_t = K\}$. Consequently, the rate $\kappa$ should be computed from the equality:

$$\sum_{k=1}^{n} c_i \mathbb{Q}^*\{T_k < \tilde{\tau}\} \, \mathbb{1}_{[0,\tilde{T}]}(T_i) + \mathbb{E}_{\mathbb{Q}^*} \left(B_{\tilde{\tau}}^{-1} \left(D_c(\tilde{\tau}, T) - D_c(0, T)\right)\right)$$

$$= \kappa \sum_{i=1}^{m} \mathbb{Q}^*\{t_i < \tilde{\tau}\} \, \mathbb{1}_{[0,\tilde{T}]}(t_i).$$

**Default swaps and option.** We shall consider a plain-vanilla default swap and option related to a corporate zero-coupon bond with the fractional recovery of Treasury value. We assume that the derivative contract matures at time $U \leq T$, where $T$ is the maturity of a reference bond. The contingent payment is triggered by the default event $\{C_t = K\}$, where the migration process $C$ is specified as in Sect. 13.2.4. The claim is settled at time $\tau$, and equals (we assume that the notional principal $L = 1$)

$$Y = \left(1 - \delta_{\hat{C}_U} B(\tau, T)\right) \mathbb{1}_{\{\tau \leq U\}}.$$

Notice the dependence of $Y$ on the initial rating $C_0$ through the default time $\tau$ and the recovery rate $\delta_{\hat{C}_U}$. We distinguish the two alternative conventions regarding the payment of the credit insurance premium by the buyer:
- the buyer pays a lump sum at the contract's inception date (*default option*),
- the buyer pays an annuity at fixed time instants $t_i < U$, $i = 1, \ldots, m$ (*default swap*).

The value of a default option at time $t = 0$ equals

$$\pi_0(Y) = \mathbb{E}_{\mathbb{Q}^*}\left(B_\tau^{-1}\left(1 - \delta_{\hat{C}_U} B(\tau, T)\right) \mathbb{1}_{\{\tau \leq U\}}\right).$$

In the case of a default swap, the payment of the credit insurance premium is distributed in time. The annuity $\kappa$ can be found from the equality

$$\pi_0(Y) = \kappa\, \mathbb{E}_{\mathbb{Q}^*}\left(\sum_{i=1}^m B_{t_i}^{-1} \mathbb{1}_{\{\tau > t_i\}}\right) = \kappa \sum_{i=1}^m D^0(0, t_i).$$

Note that both the price $\pi_0(Y)$ and the annuity $\kappa$ depend on the initial rating $C_0$ of the reference corporate bond.

### 13.3.2 Hedging of Credit Derivatives

By a *perfect hedging* of a credit-risk sensitive instrument, such as a credit derivative, we mean the ability to completely eliminate the risk associated with the instrument. This means that perfect hedging involves complete risk elimination through all the credit migration times, including the default time, that may occur during the life-time of an instrument. A perfect hedging of a credit risk sensitive instrument is typically possible if liquid instruments, which are sensitive to the same market and credit risk as the reference instruments, are traded. For example, a plain-vanilla default swap can be, in principle, fully hedged by rolling a portfolio of shorter maturity default swaps that are sensitive to the same default risk as the hedged swap. For an outline of this practical approach, we refer to papers by Arvanitis and Laurent (1999) and Arvanitis (2000), as well as to the recent monograph by Arvanitis and Gregory (2001).

If the market for credit-risk sensitive instruments is illiquid, the perfect hedging of credit derivatives is not possible; in other words, we deal with the case of an incomplete model of the securities market. Several alternative techniques of hedging contingent claims in incomplete financial models were developed in recent years, to mention a few: the risk-minimizing hedging (for a survey of these approaches, see Schweizer (2001)), the quantile hedging (see Föllmer and Leukert (1999)), as well as the shortfall hedging (see Föllmer and Leukert (2000)). In the framework of a simple model with a constant default intensity, hedging of credit risk via the local risk minimization was examined by Lotz (1998), and the optimal shortfall hedging of credit risk was studied in Lotz (1999). Let us finally mention that the utility-based pricing of defaultable claims was analyzed by Collin-Dufresne and Hugonnier (1999). It should be acknowledged, however, that the issue of hedging credit derivatives (in complete and incomplete models) requires further intensive studies.

# 14. Defaultable Market Rates

In this chapter, we formally introduce several possible interest rate contract structures in the presence of the counterparty risk. Such contracts in turn give rise to several concepts of credit-risk related LIBOR and swap rates, referred to as the *defaultable market rates,* or more specifically, *defaultable LIBOR rates* and *defaultable swap rates.* We examine spot and forward rates associated with single- and multi-period contracts that are subject to either unilateral or bilateral counterparty risk, and we derive several formulae for various kinds of defaultable market rates. The classification and the terminology introduced in what follows is merely tentative, though. It is interesting to note that certain interest rate contracts that are subject to the counterparty credit risk may be equivalently restated as interest rate contracts involving a reference credit risk. However, when we analyze defaultable interest rate swaps, we always assume that the underlying reference floating interest rate is the default-free LIBOR rate.

It should also be made clear that we restrict our attention to formal definitions of interest rate agreements with default risk. A more practical issue of whether the default risk of counterparties has indeed a non-negligible impact on observed swap rates in contracts with counterparties of different credit quality is not treated here (see: Hull and White (1995), Duffee (1996) or Collin-Dufresne and Solnik (2001) in this regard).

As already mentioned, defaultable swaps may be defined in a large variety of ways. In addition, the number of potential models that can be used to value these contracts is practically unlimited. For this reason, we have decided not to make any attempts to establish a specific valuation result for a defaultable swap, and we restrict our attention to the derivation of generic expressions. The reader interested in analytic valuation results for swap contracts that are subject to default risk is referred to papers by Cooper and Mello (1991), Hull and White (1995), Li (1998a), Hübner (2001) and Yu and Kwok (2002), who all worked within the structural approach. Duffie and Huang (1996) extend the results of Duffie and Singleton (1994) and Duffie et al. (1996) by applying the intensity-based approach to the valuation of swaps, in which the two counterparties have asymmetric default risk. In particular, they provide a quasi-explicit formula for the marginal impact of an increase in a credit-risk asymmetry on the market value of a defaultable swap.

## 14.1 Interest Rate Contracts with Default Risk

The goal of this section is to briefly describe simple examples of defaultable interest rate contracts and to derive expressions for the associated default-able forward LIBOR and swap rates. We shall focus here on single-period contracts; the multi-period contracts are studied later in the chapter. We assume throughout that we are given a finite family of reset/settlement dates $T_0 < T_1 < \cdots < T_m$. By convention, and without loss of generality, the nominal principal of each contract analyzed below is set to be equal to 1.

### 14.1.1 Default-Free LIBOR and Swap Rates

The LIBOR (London Interbank Offered Rate) is the interest rate offered by banks on deposits from other banks in Eurocurrency markets. Since LIBOR represents the interest rate at which banks lend money to each other, it is not a default-free rate. However, in the mathematical approach, it is a common practice to formally identify the LIBOR rate with the default-free nominal rate implied by the market prices of Treasury bonds. We shall follow this convention in this chapter. Consequently, we shall refer to the real-life LIBOR rate as the *defaultable LIBOR rate*.

Let us fix the date $T_i$. A default-free interest rate agreement (IRA) with the fixed rate $\kappa$, initiated at time $T_i$ for the accrual period $[T_i, T_{i+1}]$, is specified by the following covenants:

– at the contract's inception date $T_i$, the payer of the fixed rate receives one unit of cash (i.e., the nominal value of the contract),
– the payer pays at the settlement date $T_{i+1}$ the amount $1 + \kappa(T_{i+1} - T_i)$.

From the payer's perspective, the value of such a contract at time $T_i$ equals

$$\mathbf{IRA}(T_i; \kappa) := 1 - (1 + \kappa \Delta_i) B(T_i, T_{i+1}),$$

where $\Delta_i = T_{i+1} - T_i$. The one-period IRA implicitly defines the *spot LIBOR rate* $L(T_i)$; that is, the level of $\kappa$ that makes the contract valueless at the inception date $T_i$. It is easy to see that $L(T_i)$ satisfies

$$1 + L(T_i)\Delta_i = \frac{1}{B(T_i, T_{i+1})}. \tag{14.1}$$

Likewise, the *forward LIBOR rate* $L(t, T_i)$ for the accrual period $[T_i, T_{i+1}]$ is specified through a forward rate agreement (FRA) entered at time $t$, in which one party deposits 1 unit of cash at time $T_i$ and receives at time $T_{i+1}$ from the other party the cash amount $1 + \kappa(T_{i+1} - T_i) = 1 + \kappa \Delta_i$. The level of $\kappa = L(t, T_i)$ is determined at the contract's inception date $t$ in such a way that the contract is valueless at this date. Put more explicitly, $\kappa$ solves the equation $\mathbf{FRA}(t, T_i; \kappa) = 0$, where

$$\mathbf{FRA}(t, T_i; \kappa) := B(t, T_i) - (1 + \kappa \Delta_i) B(t, T_{i+1}). \tag{14.2}$$

It is clear that

$$L(t, T_i) = \frac{1}{\Delta_i} \left( \frac{B(t, T_i)}{B(t, T_{i+1})} - 1 \right). \tag{14.3}$$

As defined above, the spot and forward LIBOR rates are linked to the term deposits that start either at a current date (spot rate) or at some date in the future (forward rate). It is not difficult to show, however, that in a default-free environment the spot LIBOR rate $L(T_i)$ also coincides with the *swap rate*; that is, the rate implied by a single-period fixed-for-floating interest rate swap initiated at time $T_i$ and settled at time $T_{i+1}$. Similarly, the forward LIBOR rate $L(t, T_i)$ can be seen as the *forward swap rate* associated with a single-period fixed-for-floating forward swap settled *in arrears,* i.e., with the reset date $T_i$ and the settlement date $T_{i+1}$.

To justify the last statement, we fix the rate $\kappa$ and we consider the forward swap (the spot swap rate is merely a special case of the forward swap rate). By the contractual features of the fixed-for-floating forward swap settled in arrears, the long party, commonly referred as the *payer,* pays the fixed amount $\kappa(T_{i+1} - T_i) = \kappa \Delta_i$ at time $T_{i+1}$, and at the same date he collects the floating amount

$$L(T_i)\Delta_i = B^{-1}(T_i, T_{i+1}) - 1.$$

The values of these payoffs at time $t \le T_i$ are $B(t, T_{i+1})\kappa \Delta_i$ and $B(t, T_i) - B(t, T_{i+1})$, respectively. Consequently, for any fixed $t \le T_j$, the level of the *forward swap rate* that makes the forward swap valueless at time $t$ can be found by solving for $\kappa = \kappa(t, T_i)$ the equation $\mathbf{FS}(t, T_i; \kappa) = 0$, where $\mathbf{FS}(t, T_i; \kappa)$ represents the value of the forward swap to the payer. Specifically,

$$\mathbf{FS}(t, T_i; \kappa) := B(t, T_i) - B(t, T_{i+1}) - \kappa \Delta_i B(t, T_{i+1}).$$

The value at the inception date $T_i$ of a single-period interest rate swap equals $\mathbf{FS}(T_i, T_i; \kappa) =: \mathbf{IRS}(T_i; \kappa)$, where

$$\mathbf{IRS}(T_i; \kappa) := 1 + B(T_i, T_{i+1}) - \kappa \Delta_i B(T_i, T_{i+1}). \tag{14.4}$$

It is easy to see that $\mathbf{FS}(t, T_i; \kappa) = \mathbf{FRA}(t, T_i; \kappa)$ and $\mathbf{IRS}(T_i; \kappa) = \mathbf{IRA}(T_i; \kappa)$. It is also apparent that the forward swap rate $\kappa(t, T_i)$ satisfies

$$\kappa(t, T_i) = \frac{B(t, T_i) - B(t, T_{i+1})}{\Delta_i B(t, T_{i+1})}. \tag{14.5}$$

In view of (14.3) and (14.5), we conclude that in a default-free setting the forward swap rate $\kappa(t, T_j)$ and the forward LIBOR rate $L(t, T_j)$ coincide. Let us finally mention, that the forward swap rate $\kappa(t, T_0; m)$, associated with the $m$-period fixed-for-floating swap with the first reset date $T_0$, is known to satisfy

$$\kappa(t, T_0; m) = \frac{B(t, T_0) - B(t, T_m)}{\sum_{j=1}^{m} \Delta_{j-1} B(t, T_j)}. \tag{14.6}$$

For a more detailed analysis of these issues, we refer to Chap. 14 and 16 in Musiela and Rutkowski (1997a).

### 14.1.2 Defaultable Spot LIBOR Rates

We focus here on contracts with one-sided default risk. We make the following standing assumption: there are two parties to the contract: the *payer* (of the fixed rate $\kappa$) and the *receiver* (of the fixed rate).

**Payer's default risk.** We assume that only the payer is prone to default. Let us first describe a basic example of a defaultable spot interest rate agreement with a constant recovery rate. A defaultable IRA initiated at time $T_i$ for the accrual period $[T_i, T_{i+1}]$, with the fixed rate $\kappa$ and the constant recovery rate $\delta$, is specified by the following covenants:

- at time $T_i$, the receiver of the fixed rate pays 1 (the nominal value of the contract),
- if the payer does not default in the time period $(T_i, T_{i+1}]$, the receiver is paid at the settlement date $T_{i+1}$ the full due amount: $1 + \kappa \Delta_i$,
- if the payer defaults in the time period $(T_i, T_{i+1}]$, the receiver collects at time $T_{i+1}$ the reduced amount: $\delta(1 + \kappa \Delta_i)$.

If the payer has already defaulted prior to or at the contract's inception date $T_i$, the contract is not initiated at all (that is, all cash flows are zero by definition). We call the above contract an IRA with unilateral credit risk since only the default risk of the payer appears to be relevant. Let $\tau$ be the default time (of the payer party in the present context). From the payer's perspective the value at the inception date $T_i$ of the contract equals, on the set $\{\tau > T_i\}$,

$$\mathbf{IRA}^{\delta}(T_i; \kappa) := 1 - (1 + \kappa \Delta_i) D^{\delta}(T_i, T_{i+1}), \tag{14.7}$$

where $D^{\delta}(T_i, T_{i+1})$ is the price of a discount bond issued by the payer, with fractional recovery of Treasury value and constant recovery rate $\delta$.

By convention, the value of the contract is zero after default – that is, on the set $\{\tau \leq T_i\}$. The covenants of the IRA can be also restated as follows. At time $T_i$:

- the receiver pays to the payer the amount $(1 + \kappa \Delta_i)^{-1}$,
- he receives a discount bond with unit face value and maturity $T_{i+1}$ with fractional recovery of Treasury value and constant recovery rate $\delta$.

We thus formally deal here with an outright purchase of a defaultable bond. In this interpretation, the counterparty risk involved is thus the default risk of the bond's issuer. The determination of the right level of the fixed rate $\kappa$ is thus essentially equivalent to the valuation of the bond issued by the payer party. The *defaultable spot LIBOR rate* $L^{\delta}(T_i)$ (or, more specifically, the *payer-risk-adjusted spot LIBOR rate* $L_p^{\delta}(T_i)$) can be defined as that level of the fixed rate $\kappa$ that makes the contract valueless at the inception date $T_i$. Using (14.7), we obtain

$$L_p^{\delta}(T_i) = L^{\delta}(T_i) = \frac{1}{\Delta_i}\left(\frac{1}{D^{\delta}(T_i, T_{i+1})} - 1\right). \tag{14.8}$$

The rate $L^\delta(T_i)$ is directly linked to the yield on a discount bond issued by the counterparty; it is well defined provided that the issuer has not yet defaulted at time $T_i$. It can also be seen as the spot swap rate for a single-period fixed-for-floating *default-free* swap settled in arrears, provided that the floating rate in the swap is computed off a given defaultable bond maturing at $T_{i+1}$. To see this, consider a default-free contract in which, at time $T_{i+1}$, the long party pays $\kappa \Delta_i$ and receives

$$\left(D^\delta(T_i, T_{i+1})\right)^{-1} - 1 = L^\delta(T_i)\Delta_i.$$

It is easy to see that the values of these payoffs at time $T_i$ are: $B(T_i, T_{i+1})\kappa\Delta_i$ and $B(T_i, T_{i+1})L^\delta(T_i)\Delta_i$, respectively. Consequently, the *default related swap rate*, which makes the swap described above valueless at time $T_i$, can be found by solving for $\kappa = \hat{\kappa}^\delta(T_i)$ the following equation:

$$L^\delta(T_i)\Delta_i B(T_i, T_{i+1}) - \kappa\Delta_i B(T_i, T_{i+1}) = 0.$$

It is apparent that $\hat{\kappa}^\delta(T_i) = L^\delta(T_i)$; i.e., the default related swap rate $\hat{\kappa}^\delta(T_i)$ coincides with the spot LIBOR rate $L^\delta(T_i)$. To summarize, the defaultable spot IRA can be formally interpreted as a default-free swap with the reference credit risk – the default risk of the reference bond.

**Receiver's default risk.** Assume now that the default-prone party is the receiver of the fixed rate and that the payer of the fixed rate is default-free. The cash flows at the settlement date $T_{i+1}$ of the corresponding defaultable spot IRA can be summarized as follows:
– the receiver of the fixed rate collects the full due amount: $1 + \kappa\Delta_i$,
– if he does not default in $(T_i, T_{i+1}]$, he pays the full amount: $B^{-1}(T_i, T_{i+1})$,
– if default occurs in $(T_i, T_{i+1}]$, he pays the reduced amount: $\delta B^{-1}(T_i, T_{i+1})$.
Let $\tau$ be the default time of the receiver. From the payer's perspective, the value at time $T_i$ of the contract equals, on the set $\{\tau > T_i\}$,

$$\textbf{IRA}_r^\delta(T_i; \kappa) := D^\delta(T_i, T_{i+1})B^{-1}(T_i, T_{i+1}) - (1 + \kappa\Delta_i)B(T_i, T_{i+1}),$$

where $D^\delta(T_i, T_{i+1})$ is the price of a bond issued by the receiver, with fractional recovery of Treasury value and constant rate $\delta$. The *receiver-risk-adjusted spot LIBOR rate* $L_r^\delta(T_i)$ makes the contract valueless at $T_i$. It is clear that

$$L_r^\delta(T_i) = \frac{1}{\Delta_i}\left(\frac{D^\delta(T_i, T_{i+1})}{B^2(T_i, T_{i+1})} - 1\right).$$

### 14.1.3 Defaultable Spot Swap Rates

The main difference between an interest rate agreement and an interest rate swap is that in the latter the notional principals are not exchanged, and thus only the net payments are relevant. This feature can be disregarded in a default-free set-up, but it appears to be essential when the default-risk is taken into account. We shall examine fixed-for-floating defaultable swaps, in which the floating rate is the LIBOR rate.

**Payer's default risk.** Consider first a defaultable swap, in which only the payer of the fixed rate is default prone. For a fixed reset date $T_i$, the contract initiated at time $T_i$, for the accrual period $[T_i, T_{i+1}]$, with the fixed rate $\kappa$ and the constant recovery rate $\delta$, is specified by the following covenants:
- the receiver of the fixed rate pays $L(T_i)\Delta_i$ at time $T_{i+1}$,
- if the payer does not default in the time period $(T_i, T_{i+1}]$, the receiver is paid at the settlement date $T_{i+1}$ the full due amount: $\kappa\Delta_i$,
- if the payer defaults in the time period $(T_i, T_{i+1}]$, the receiver collects at time $T_{i+1}$ the reduced amount: $\delta\kappa\Delta_i$.

Let us denote by $\kappa^\delta(T_i)$ the value of the fixed rate $\kappa$ that makes the defaultable swap described above valueless at time $T_i$. We shall refer to $\kappa^\delta(T_i)$ as the *defaultable spot swap rate* with the unilateral default risk from the payer's side (i.e., the *payer-risk-adjusted spot swap rate*). Since the value at time $T_i$ of the defaultable interest rate swap equals

$$\mathbf{IRS}^\delta(T_i; \kappa) = 1 - B(T_i, T_{i+1}) - \kappa\Delta_i D^\delta(T_i, T_{i+1}),$$

we conclude that on the set $\{\tau > T_i\}$ the defaultable spot swap rate equals

$$\kappa^\delta(T_i) = \frac{1 - B(T_i, T_{i+1})}{\Delta_i D^\delta(T_i, T_{i+1})}. \tag{14.9}$$

**Receiver's default risk.** Assume now that the default-prone party is the receiver of the fixed rate and the payer of the fixed rate $\kappa$ is default-free. The corresponding defaultable swap rate at time $T_i$, denoted by $\kappa_r^\delta(T_i)$, is referred to as the *receiver-risk-adjusted spot swap rate*. One easily checks that

$$\kappa_r^\delta(T_i) = L(T_i)D^\delta(T_i, T_{i+1})B^{-1}(T_i, T_{i+1}),$$

where $D^\delta(T_i, T_{i+1})$ is the price of a defaultable bond issued by the receiver.

### 14.1.4 FRAs with Unilateral Default Risk

To define defaultable counterparts of forward LIBOR and swap rates, we need to examine interest rate agreements with the inception date $t < T_i$. We postulate that if the default has not occurred prior to or at the reset date $T_i$, then the contract becomes a defaultable spot interest rate agreement. To complete the contract's specification we only need to describe the way in which the contract is settled if the default occurs prior to $T_i$. We shall examine the following four schemes: if the default occurs at some date $s$ before the date $T_i$, the contract is unwound as follows:
- 1st kind: at time $s$ the receiver gets the value of defaulted bond multiplied by $1 + \kappa\Delta_i$, and makes the promised payoff 1 to the payer at time $T_i$ (or, equivalently, the payoff $B(t, T_i)$ at time $t$),
- 2nd kind: the contract is immediately unwound, with no cash flows,
- 3rd kind: the contract becomes a default-free FRA with the reduced nominal principal $\delta$,
- 4th kind: the contract is terminated and settled using a pre-specified marking-to-market procedure.

At the intuitive level, the above settlement schemes differ between themselves in the way the default risk is understood. In some of them, it is interpreted as the counterparty risk, while in some others it plays the role of the reference risk. The first scheme refers to the previously mentioned interpretation of a defaultable forward agreement as the outright purchase of a defaultable bond. The payoffs at default time correspond to the values at this date of the defaultable bond with the face value equal to $1 + \kappa\Delta_i$ and the default-free bond with the face value 1, respectively. In this scheme we deal with the one-sided counterparty risk from the payer's side. The implied forward rate is referred to as the *payer-risk-adjusted forward nominal rate*.

The settlement of the second kind stipulates that the contract becomes void if default occurs prior to $T_i$. Such a convention seems to be more natural in case of an FRA, rather than in case of a forward swap. Indeed, if the prospective lender goes bankrupt before $T_i$, it is natural to expect that the contract would be canceled, with no compensation to any party involved. Alternatively, the default risk may here be treated as the reference risk (both counterparties being default-free). The resulting forward rate is termed the *reference-risk-adjusted forward nominal rate*.

The third scheme describes a specific procedure of reducing the nominal principal of a forward rate agreement to accommodate for the loss of value of the reference entity, if default occurs before the reset date $T_i$. The reduced-size contract then becomes default-free; this supports the intuition that we are dealing here with the reference risk, rather than the counterparty risk. The implied forward rate is called the *defaultable forward LIBOR rate*.

The settlement scheme of the fourth kind in fact encompasses numerous differing procedures. It may describe, for instance, the commonly standard market practice of termination of a defaultable swap, which stipulates that:
- if default occurs prior to or at[1] $T_i$, the defaulted party receives the market value of a comparable non-defaultable forward swap, provided that this value is positive to the defaulted party,
- the defaulted party pays nothing (or a certain recovery value) to the other party if the market value mentioned above is negative to him.

The implied forward interest rates are given the generic name of *mark-to-market defaultable forward nominal rates*.

A still another scheme – proposed by Lotz and Schlögl (2000) – postulates that in case of default prior to $T_i$, the underlying FRA is revalued on a no-default-prior-to-$T_i$ basis. One first needs to find the value at time $t$ of a non-defaultable before $T_i$ security, which at time $T_i$ pays the amount

$$(1 + \kappa\Delta_i)D^\delta(T_i, T_{i+1}) - 1.$$

If this value is positive to the defaulting party, he receives it from the counterparty. If, on the contrary, it is negative, the defaulted party either pays nothing or a certain recovery value to the solvent counterparty.

---

[1] In what follows, we shall write briefly 'prior to $T_i$' instead of 'prior to or at $T_i$'.

**Payer-risk-adjusted forward nominal rate.** According to the first settlement scheme, the value of the contract to the payer at time $t < T_i$ is given by the following expression:

$$\widehat{\mathbf{FRA}}^\delta(t, T_i; \kappa) := B(t, T_i) - (1 + \kappa \Delta_i) D^\delta(t, T_{i+1}).$$

The corresponding rate $\widehat{L}^\delta(t, T_i)$, termed the *payer-risk-adjusted forward nominal rate*, is defined as that value of $\kappa$ that makes the contract valueless at time $t$. It is apparent that on the set $\{\tau > t\}$

$$\widehat{L}^\delta(t, T_i) = \frac{1}{\Delta_i} \left( \frac{B(t, T_i)}{D^\delta(t, T_{i+1})} - 1 \right).$$

At time $T_i$ we have $\widehat{L}^\delta(T_i, T_i) = L^\delta(T_i)$, where $L^\delta(T_i)$ is given by (14.8).

**Reference-risk-adjusted forward nominal rate.** Let us examine the payoff at time $T_i$ associated with the second scheme. It is clear that on the set $\{\tau > T_i\}$ the payoff coincides with the value at time $T_i$ of a defaultable IRA. To be more specific, it equals $\mathbf{IRA}^\delta(T_i; \kappa)$, where $\mathbf{IRA}^\delta(T_i; \kappa)$ is given by (14.7). Otherwise, i.e., on the set $\{\tau \le T_i\}$, the payoff at time $T_i$ equals zero. We are thus dealing here with the claim $Y$, which settles at time $T_i$, and equals

$$Y = \mathbb{1}_{\{\tau > T_i\}} \left( 1 - (1 + \kappa \Delta_i) D^\delta(T_i, T_{i+1}) \right). \tag{14.10}$$

As apparent from (14.10), to evaluate the *reference-risk-adjusted forward nominal rate* $\widetilde{L}^\delta(t, T_i)$ (i.e., the value of $\kappa$ that makes $Y$ valueless at time $t$) one needs to rely on a particular stochastic model of the default time. However, let us notice that the equality $\widetilde{L}^\delta(T_i, T_i) = L^\delta(T_i)$ will always hold, regardless of which particular model is chosen. In the case of zero recovery, we have

$$D^0(T_i, T_{i+1}) = \mathbb{1}_{\{\tau > T_i\}} D^0(T_i, T_{i+1}), \quad D^0(T_i, T_i) = \mathbb{1}_{\{\tau > T_i\}},$$

and so (14.10) may be rewritten as follows:

$$Y = D^0(T_i, T_i) - (1 + \kappa \Delta_i) D^0(T_i, T_{i+1}).$$

The arbitrage price of $Y$ at time $t \le T_i$ equals

$$\pi_t(Y) = D^0(t, T_i) - (1 + \kappa \Delta_i) D^0(t, T_{i+1}).$$

We conclude that in the case of zero recovery we have, on the set $\{\tau > t\}$,

$$\widetilde{L}^0(t, T_i) = \frac{1}{\Delta_i} \left( \frac{D^0(t, T_i)}{D^0(t, T_{i+1})} - 1 \right) = \frac{1}{\Delta_i} \left( \frac{\widetilde{D}(t, T_i)}{\widetilde{D}(t, T_{i+1})} - 1 \right). \tag{14.11}$$

We shall henceforth write briefly $\widetilde{L}(t, T_i)$, rather than $\widetilde{L}^0(t, T_i)$. It appears that the last expression can be extended to the case of positive recovery rate $\delta$ (see formula (14.12)). However, to justify this generalization, we need to impose a different settlement scheme.

**Defaultable forward LIBOR rate.** According to the third settlement scheme, a defaultable FRA is specified as follows:
- if the payer defaults in the time interval $(t, T_i]$, the receiver pays $\delta$ at time $T_i$ and he receives $\delta(1 + \kappa\Delta_i)$ at time $T_{i+1}$,
- otherwise, the receiver pays 1 at time $T_i$, and he collects at time $T_{i+1}$ either the reduced amount $\delta(1 + \kappa\Delta_i)$, or the full amount $1 + \kappa\Delta_i$, depending on whether there has been default in $(T_i, T_{i+1}]$ or not.

The value to the payer of the contract specified above equals, on $\{\tau > t\}$,

$$\mathbf{FRA}^\delta(t, T_i; \kappa) = D^\delta(t, T_i) - (1 + \kappa\Delta_i)D^\delta(t, T_{i+1}).$$

We define the *defaultable forward LIBOR rate* $L^\delta(t, T_i)$ as that level of the fixed rate $\kappa$ for which $\mathbf{FRA}^\delta(t, T_i; \kappa) = 0$. Using the formula above, we get

$$L^\delta(t, T_i) = \frac{1}{\Delta_i}\left(\frac{D^\delta(t, T_i)}{D^\delta(t, T_{i+1})} - 1\right). \tag{14.12}$$

Letting $t = T_i$ in the last formula we obtain $L^\delta(T_i, T_i) = L^\delta(T_i)$, on the set $\{\tau > T_i\}$, where $L^\delta(T_i)$ is given by (14.8). Formula (14.12) can be extended in a natural way to the case of an abstract defaultable bond. Namely, we set

$$L^d(t, T_i) := \frac{1}{\Delta_i}\left(\frac{D(t, T_i)}{D(t, T_{i+1})} - 1\right). \tag{14.13}$$

The contract specification that leads to equality (14.12) is rather artifical, as the default prior to the reset date $T_i$ results only in the reduction of the contract's nominal principal, and the contract then becomes default-free. The default risk is thus here treated as the reference risk, rather than the counterparty risk. Nevertheless, since (14.13) mimics (14.3), it may serve as a convenient abstract definition of the defaultable forward LIBOR rate.

**Mark-to-market forward nominal rate.** Assume that the contract is marked to market if default occurs prior to or at $T_i$. We postulate that at the time of default the value of a default-free FRA with otherwise identical covenants is found, and:
- if the value $\mathbf{FRA}(\tau, T_i; \kappa)$ of the equivalent default-free FRA is positive to the payer, he gets this value from the receiver,
- if the value $\mathbf{FRA}(\tau, T_i; \kappa)$ is positive to the receiver, he gets from the payer only a fixed fraction $\delta$ of this value.

From the payer's perspective, the contract can be represented as the following contingent claim $Y$, which settles at time $T_i$,

$$Y = \mathbb{1}_{\{t < \tau \leq T_i\}} B^{-1}(\tau, T_i)\left(\left(\mathbf{FRA}(\tau, T_i; \kappa)\right)^+ - \delta\left(-\mathbf{FRA}(\tau, T_i; \kappa)\right)^+\right)$$
$$+ \mathbb{1}_{\{\tau > T_i\}} \mathbf{IRA}^\delta(T_i; \kappa),$$

where $\mathbf{FRA}(\tau, T_i; \kappa)$ and $\mathbf{IRA}^\delta(T_i; \kappa)$ are given by (14.2) and (14.7), respectively.

Put another way,

$$Y = \mathbb{1}_{\{t < \tau \le T_i\}} B^{-1}(\tau, T_i) Z(\tau, T_i; \kappa) + \mathbb{1}_{\{\tau > T_i\}} \mathbf{IRA}^\delta(T_i; \kappa),$$

where for $t \le T_i$ the recovery process $Z$ satisfies

$$Z(t, T_i; \kappa) = \mathbf{FRA}(t, T_i; \kappa) + (1 - \delta)\big(-\mathbf{FRA}(t, T_i; \kappa)\big)^+. \qquad (14.14)$$

The determination of the implied mark-to-market forward nominal rate $\bar{L}(t, T_i)$, which is defined as the value of $\kappa$ that makes the claim $Y$ valueless at time $t$, requires a joint modeling of default and interest rate risks.

### 14.1.5 Forward Swaps with Unilateral Default Risk

Our next goal is to analyze various rates implicit in defaultable forward swaps.
**Payer-risk-adjusted forward swap rate.** According to the first scheme, the value at time $t < T_i$ of the forward swap to the payer equals:

$$\widehat{\mathbf{FS}}^\delta(t, T_i; \kappa) := B(t, T_i) - B(t, T_{i+1}) - \kappa \Delta_i D^\delta(t, T_{i+1}).$$

Hence, the associated forward swap rate $\widehat{\kappa}^\delta(t, T_i)$ satisfies, on the set $\{\tau > t\}$,

$$\widehat{\kappa}^\delta(t, T_i) = \frac{B(t, T_i) - B(t, T_{i+1})}{\Delta_i D^\delta(t, T_{i+1})}.$$

In particular, we have $\widehat{\kappa}^\delta(T_i, T_i) = \kappa^\delta(T_i)$ with $\kappa^\delta(T_i)$ given by (14.9).
**Reference-risk-adjusted forward swap rate.** Assume now that the forward swap is subject to the second settlement scheme. In this case, it can be formally represented as the following contingent claim, which settles at $T_i$,

$$Y = \mathbb{1}_{\{\tau > T_i\}} \big(1 - B(T_i, T_{i+1}) - \kappa \Delta_i D^\delta(T_i, T_{i+1})\big).$$

It is clear that the *reference-risk-adjusted forward swap rate* $\widetilde{\kappa}^\delta(t, T_i)$ cannot be found, unless we introduce some model for default and interest rate risks. In the case of $\delta = 0$, we obtain

$$\widetilde{\kappa}^\delta(t, T_i) = \frac{D^0(t, T_{i+1}) - \pi_t(\widetilde{Y})}{\Delta_i D^0(t, T_{i+1})},$$

where $\pi_t(\widetilde{Y})$ denotes the value at time $t$ of the payoff $\widetilde{Y} = \mathbb{1}_{\{\tau > T_i\}} B(T_i, T_{i+1})$ which settles at time $T_i$.
**Defaultable forward swap rate.** Under the third scheme, the forward swap corresponds to the claim

$$Y = D^\delta(T_i, T_i)\big(1 - B(T_i, T_{i+1})\big) - \kappa \Delta_i D^\delta(T_i, T_{i+1}),$$

hence, the *defaultable forward swap rate* $\kappa^\delta(t, T_i)$ equals

$$\kappa^\delta(t, T_i) = \frac{D^\delta(t, T_{i+1}) - \pi_t(Y)}{\Delta_i D^\delta(t, T_{i+1})},$$

where $\pi_t(Y)$ stands for the arbitrage price at time $t \le T_i$ of the payoff $Y = D^\delta(T_i, T_{i+1}) B(T_i, T_{i+1})$ which settles at time $T_i$.

**Mark-to-market forward swap rate.** Let us now focus on a particular case of the last settlement scheme, in which the position is unwound at time $\tau < T_i$, according to a marking-to-market procedure. A defaultable forward swap initiated at time $t$ for the accrual period $[T_i, T_{i+1}]$, with the fixed rate $\kappa$ and the recovery rate $\delta$, is now specified as follows:

- if there is no default prior to $T_i$, the contract becomes a defaultable swap with unilateral credit risk,
- if default occurs prior to $T_i$, and if the value $\mathbf{FS}(\tau, T_i; \kappa)$ of the equivalent default-free forward swap is negative to the receiver, he must meet his obligation in full,
- if default occurs prior to $T_i$, and the value $\mathbf{FS}(\tau, T_i; \kappa)$ is positive to the receiver, he only gets a fixed fraction $\delta$ of this value from the payer.

By convention, if the payer has defaulted prior to the contract's inception date $t$, the contract is not initiated at all, and thus the cash flows are zero by definition. Thus, the contract can be represented as a single contingent claim $Y$, which settles at time $T_i$, and equals

$$Y = \mathbb{1}_{\{t<\tau\leq T_i\}} B^{-1}(\tau, T_i) \Big( \big(\mathbf{FS}(\tau, T_i; \kappa)\big)^+ - \delta \big( -\mathbf{FS}(\tau, T_i; \kappa)\big)^+ \Big)$$
$$+ \mathbb{1}_{\{\tau > T_i\}} \mathbf{IRS}^\delta(T_i; \kappa)$$

or, equivalently,

$$Y = \mathbb{1}_{\{t<\tau\leq T_i\}} B^{-1}(\tau, T_i) Z(\tau, T_i; \kappa) + \mathbb{1}_{\{\tau > T_i\}} \mathbf{IRS}^\delta(T_i; \kappa),$$

where the *recovery process* $Z$ is defined by setting (notice that the process $Z$ given by (14.15) coincides with the process $Z$ given by (14.14))

$$Z(t, T_i; \kappa) = \mathbf{FS}(t, T_i; \kappa) + (1 - \delta)\big( -\mathbf{FS}(t, T_i; \kappa)\big)^+. \tag{14.15}$$

Let us denote by $\mathbf{FS}^\delta(t, T_i; \kappa)$ the value at time $t$ of the claim $Y$. The *mark-to-market forward swap rate* $\bar{\kappa}^\delta(t, T_i)$ is defined as that level of $\kappa$ for which the price at time $t$ of the defaultable forward swap described above is zero. It is implicitly defined through the equation $\mathbf{FS}^\delta\big(t, T_i; \bar{\kappa}^\delta(t, T_i)\big) = 0$.

Before we conclude this section, let us make clear that the various forward rates introduced above are by no means standard. In fact, our goal was merely to show that the concept of a spot or a forward interest rate derived from a defaultable interest rate agreement is rather ambiguous; it was shown to rely heavily on specific, sometimes rather artificial, covenants. Market practice is largely based on the concept of the credit spread – that is, the spread over the rate implied by a default-free counterpart of a given contract, rather than on a defaultable rate. In other words, the practical approach hinges on the assumption that the market risk and the credit risk can be disentangled, and that the contract-specific credit spread reflects the level of the credit risk involved. Let us observe that, for the sake of simplicity, we have only examined contracts that were subject to the unilateral default risk. The study of the case of bilateral risk is postponed to Sect. 14.5.

## 14.2 Multi-Period IRAs with Unilateral Default Risk

The goal of this section is to discuss the multi-period IRAs in the presence of unilateral default risk. There are many ways of defining the covenants for such contracts. It would be gratuitous to consider all possibilities; we only discuss a few examples of settlement schemes. We shall focus on the basic two cases: a fixed-rate defaultable IRA and a floating-rate defaultable IRA.

To simplify the description of the contracts, we shall follow here the covenant of *advance payments* by the solvent party (the lender). Let us observe that if the agreement is modified so that the non-defaulting party is supposed to make payments at the settlement dates, the contracts described below become equivalent to defaultable swaps.

Assume that we are given a pre-specified finite family of reset/settlement dates $T_0 < T_1 < \cdots < T_m$. By convention, and without loss of generality, the nominal principal of all contracts analyzed below is set to be equal to 1. If any of the counterparties defaults prior to or at the contract's inception date the contract is not initiated, so that all cash flows are zero.

**Fixed-rate defaultable IRA.** We first assume that the only party prone to default is the payer of the fixed rate; the IRA can thus be seen here as the fixed-rate loan. A defaultable multi-period IRA initiated at time $T_i$ for the accrual periods $[T_i, T_{i+1}], \ldots, [T_{m-1}, T_m]$, with the fixed rate $\kappa$ and the recovery rate $\delta$, is specified by the following covenants (as before, we write $\Delta_j = T_{j+1} - T_j$):

− at time $T_i$, the receiver (of the fixed rate) pays 1,
− if $i \leq j < m-1$ and if the payer has not defaulted in $(T_j, T_{j+1}]$, the receiver collects at time $T_{j+1}$ the full due amount: $1 + \kappa\Delta_j$, and he pays 1 at time $T_{j+1}$; the net payment to the receiver at $T_{j+1}$ is thus $\kappa\Delta_j$,
− if $i \leq j < m-1$ and if the payer has defaulted in the time period $(T_j, T_{j+1}]$, the contract settles according to the following rules: the receiver collects at time $T_{j+1}$ the reduced present value of the promised future payments:

$$\delta \sum_{l=j+1}^{m} B(T_{j+1}, T_l)(1 + \kappa\Delta_{l-1}),$$

and he pays at time $T_{j+1}$ the reduced amount:

$$\delta \sum_{l=j+1}^{m-1} B(T_{j+1}, T_l),$$

− if $j = m - 1$ and if the payer has not defaulted in the time period $(T_{m-1}, T_m]$, the receiver is paid at the settlement date $T_m$ the full due amount: $1 + \kappa\Delta_{m-1}$,
− if $j = m - 1$ and if the payer has defaulted in the time period $(T_{m-1}, T_m]$, then the receiver gets at time $T_m$ the amount $\delta(1 + \kappa\Delta_{m-1})$.

For the reader's convenience, we shall first examine the case when $m = i+3$. Let us denote $\mathbb{1}_j = \mathbb{1}_{\{T_j < \tau \le T_{j+1}\}}$. From the payer's perspective, the cash flows are:

$$\mathbb{1}_{\{t=T_i\}} - \mathbb{1}_{\{\tau > T_{i+3}\}}\left[\mathbb{1}_{\{t=T_{i+1}\}}\kappa\Delta_i + \mathbb{1}_{\{t=T_{i+2}\}}\kappa\Delta_i\mathbb{1}_{\{t=T_{i+3}\}}(1+\kappa\Delta_i)\right]$$

$$- \mathbb{1}_{i+2}\left[\mathbb{1}_{\{t=T_{i+1}\}}\kappa\Delta_i + \mathbb{1}_{\{t=T_{i+2}\}}\kappa\Delta_i + \delta\mathbb{1}_{\{t=T_{i+3}\}}(1+\kappa\Delta_i)\right]$$

$$- \mathbb{1}_{i+1}\left[\mathbb{1}_{\{t=T_{i+1}\}}\kappa\Delta_i + \mathbb{1}_{\{t=T_{i+2}\}}\delta\Big(\sum_{l=i+2}^{i+3} B(T_{i+2},T_l)(1+\kappa\Delta_{l-1})-1\Big)\right]$$

$$- \mathbb{1}_i\left[\mathbb{1}_{\{t=T_{i+1}\}}\delta\Big(\sum_{l=i+1}^{i+3} B(T_{i+1},T_l)(1+\kappa\Delta_{l-1}) - \sum_{l=i+1}^{i+2} B(T_{i+1},T_l)\Big)\right].$$

Since the payment of 1 at time $t$ is equivalent to payment of $B(s,t)$ at time $s \le t$ and manifestly $\mathbb{1}_{\{s<\tau\le t\}} = \mathbb{1}_{\{\tau\le t\}} - \mathbb{1}_{\{\tau\le s\}}$, the above cash flows can be represented as follows:

$$\mathbb{1}_{\{t=T_i\}} - \mathbb{1}_{\{t=T_{i+1}\}}\kappa\Delta_i\Big(\mathbb{1}_{\{\tau>T_{i+1}\}} + \delta\mathbb{1}_{\{T_i<\tau\le T_{i+1}\}}\Big)$$

$$- \mathbb{1}_{\{t=T_{i+2}\}}\kappa\Delta_{i+1}\Big(\mathbb{1}_{\{\tau>T_{i+2}\}} + \delta\mathbb{1}_{\{T_i<\tau\le T_{i+2}\}}\Big)$$

$$- \mathbb{1}_{\{t=T_{i+3}\}}(1+\kappa\Delta_{i+2})\Big(\mathbb{1}_{\{\tau>T_{i+3}\}} + \delta\mathbb{1}_{\{T_i<\tau\le T_{i+3}\}}\Big).$$

The value at time $T_i$ of the defaultable three-period IRA thus equals, from the payer's perspective,

$$\mathbf{IRA}_p^\delta(T_i;\kappa,3) = 1 - \kappa\Delta_i D^\delta(T_i,T_{i+1}) - \kappa\Delta_{i+1}D^\delta(T_i,T_{i+2})$$
$$- (1+\kappa\Delta_{i+2})D^\delta(T_i,T_{i+3}).$$

Let us denote $c_j = \kappa\Delta_{j-1}$, $j = i+1,\ldots,m-1$ and $c_m = 1+\kappa\Delta_{m-1}$. For any $m \ge i+1$, the value of the defaultable multi-period IRA at its inception date $T_i$ is given as, on the set $\{\tau > T_i\}$,

$$\mathbf{IRA}_p^\delta(T_i;\kappa,m-i) = 1 - \sum_{j=i+1}^{m-1} c_j D^\delta(T_i,T_j) - c_m D^\delta(T_i,T_m), \qquad (14.16)$$

where $D^\delta(t,T)$ denotes the price of a bond issued by the payer, with fractional recovery of Treasury value and constant recovery rate $\delta$. The *payer-risk-adjusted nominal rate* $L_p^\delta(T_i;m-i)$ can now be formally defined off the contract described above, as that level of the fixed rate $\kappa$ that makes the contract valueless at inception, i.e., as the value of $\kappa$ which solves the equation $\mathbf{IRA}^\delta(T_i;\kappa,m-i) = 0$. It is easy to see that the unique solution to his equation is given by the formula

$$L_p^\delta(T_i;m-i) = \frac{1 - D^\delta(T_i,T_m)}{\sum_{j=i+1}^m \Delta_{j-1}D^\delta(T_i,T_j)}. \qquad (14.17)$$

*Remarks.* (i) The covenants of the defaultable multi-period spot IRA described above formally correspond to an outright purchase of a defaultable coupon bond. At the initiation date $T_i$, the receiver pays 1 and he receives a coupon bond issued by the payer, with the payment dates $T_{i+1}, \ldots, T_m$ and with the coupons: $c_j = \kappa \Delta_{j-1}$, $j = i+1, \ldots, m-1$ and $c_m = 1 + \kappa \Delta_{m-1}$.

(ii) Suppose that, in a defaultable multi-period spot IRA described above, the fractions of the payments after default are different for the two parties: the payer party pays a fraction $\delta_p$, and the receiver party pays a fraction $\delta_r$, of what would be paid in the absence of default. It is easy to show that the corresponding defaultable multi-period nominal rate is (we denote $\hat{\delta} = \delta_r - \delta_p$)

$$
L^{\delta_p, \delta_r}(T_i; m - i) = \frac{1 - D^{\delta_p}(T_i, T_m) - \hat{\delta} \sum_{j=i+1}^{m-1} \left( B(T_i, T_j) - D^0(T_i, T_j) \right)}{\sum_{j=i+1}^{m} \Delta_{j-1} D^{\delta_p}(T_i, T_j)},
$$

where $D^0(T_i, T_j)$ ($D^{\delta_p}(T_i, T_j)$, resp.) is the price of a discount bond with zero recovery (with the recovery rate $\delta_p$, resp.) issued by the payer.

**Floating-rate defaultable IRA.** Assume now that the default-prone party is the payer of the floating rate, and the payer of the fixed rate is default-free. As already mentioned, we assume that the non-defaulting party makes payments is advance, i.e., at the reset dates $T_i, \ldots, T_{m-1}$, and that the default prone party (the lender) makes payments in arrears, i.e., at the settlement dates $T_{i+1}, \ldots, T_m$. By convention, we refer to such a floating-rate loan as the IRA with the receiver's default risk. The cash flows of the corresponding multi-period defaultable spot IRA can be summarized as follows:

– at time $T_i$, the receiver (of the fixed rate) is paid $B(T_i, T_{i+1})(1 + \kappa \Delta_i)$,

– if $i+1 \leq j \leq m-1$ and if the receiver has not defaulted in $(T_{j-1}, T_j]$, the receiver is paid at time $T_j$ the amount: $B(T_j, T_{j+1})(1 + \kappa \Delta_j)$, and he pays the full due amount: $B^{-1}(T_{j-1}, T_j)$,

– if $i+1 \leq j \leq m-1$ and if the receiver has defaulted in $(T_{j-1}, T_j]$, the contract is unwound as follows: the receiver gets at time $T_j$ the amount:

$$
R_j := \delta \sum_{l=j}^{m-1} B(T_j, T_{l+1})(1 + \kappa \Delta_l),
$$

and he pays at time $T_j$ the amount:

$$
P_j := \delta B^{-1}(T_{j-1}, T_j) + \delta \sum_{l=j}^{m-1} B(T_j, T_l),
$$

– if $j = m$ and if the receiver has not defaulted in $(T_{m-1}, T_m]$, he pays at the settlement date $T_m$ the full due amount: $B^{-1}(T_{m-1}, T_m)$,

– if $j = m$ and if the receiver has defaulted in $(T_{m-1}, T_m]$, he pays at time $T_m$ the reduced amount: $\delta B^{-1}(T_{m-1}, T_m)$.

It is clear that if default occurs in $(T_{j-1}, T_j]$, the net cash flow to the payer at time $T_j$ equals $N_j = P_j - R_j$ or, more explicitly,

$$N_j = \delta\Big(1 + B^{-1}(T_{j-1}, T_j) - \sum_{l=j+1}^{m-1} \kappa\Delta_{l-1}B(T_j, T_l) - (1 + \kappa\Delta_{m-1})B(T_j, T_m)\Big).$$

We find it convenient to introduce the following definition.

**Definition 14.2.1.** For any random time $\tau$, the random time $\tau_*$ satisfies: $\tau_* = j$ on $\{T_j < \tau \le T_{j+1}\}$ for $j = 0, \ldots, m-1$, and $\tau_* = m$ on $\{\tau > T_m\}$.

Let $\tau$ denote the default time of the receiver. From the payer's perspective, at time $T_i$ the value of the contract equals

$$\mathbf{IRA}_r^\delta(T_i; \kappa, m-i) = B_{T_i} \mathbb{E}_{\mathbb{P}^*}\Big(\sum_{j=i+1}^{\tau_*} B_{T_j}^{-1}B^{-1}(T_{j-1}, T_j)\Big| \mathcal{F}_{T_i}\Big)$$

$$- B_{T_i} \mathbb{E}_{\mathbb{P}^*}\Big(\sum_{j=i}^{\tau_* \wedge (m-1)} B_{T_j}^{-1}B(T_j, T_{j+1})(1 + \kappa\Delta_j)\Big| \mathcal{F}_{T_i}\Big)$$

$$- B_{T_i} \mathbb{E}_{\mathbb{P}^*}\Big(\sum_{j=i}^{m-1} \mathbb{1}_j B_{T_{j+1}}^{-1}N_{j+1}\Big| \mathcal{F}_{T_i}\Big),$$

where $\mathbb{1}_j = \mathbb{1}_{\{\tau_*=j\}}$. Observe that $I_j$ and $K_j$ can be represented as follows:

$$I_j = \delta \sum_{l=j}^{m-1} B_{T_j} \mathbb{E}_{\mathbb{P}^*}\Big(B_{T_l}^{-1}B(T_l, T_{l+1})(1 + \kappa\Delta_l)\Big| \mathcal{F}_{T_j}\Big),$$

and

$$K_j = \delta \sum_{l=j}^{m} B_{T_j} \mathbb{E}_{\mathbb{P}^*}\Big(B_{T_l}^{-1}B^{-1}(T_{l-1}, T_l)\Big| \mathcal{F}_{T_j}\Big).$$

We conclude that

$$\mathbf{IRA}_r^\delta(T_i; \kappa, m-i) = B_{T_i} \mathbb{E}_{\mathbb{P}^*}\Big(\sum_{j=i+1}^{\tau_*} B_{T_j}^{-1}B^{-1}(T_{j-1}, T_j)\Big| \mathcal{F}_{T_i}\Big)$$

$$- B_{T_i} \mathbb{E}_{\mathbb{P}^*}\Big(\sum_{j=i}^{\tau_* \wedge (m-1)} B_{T_j}^{-1}B(T_j, T_{j+1})(1 + \kappa\Delta_j)\Big| \mathcal{F}_{T_i}\Big)$$

$$- \delta B_{T_i} \mathbb{E}_{\mathbb{P}^*}\Big(\sum_{j=i}^{m-1} \mathbb{1}_j \sum_{l=j+1}^{m-1} \mathbb{E}_{\mathbb{P}^*}\big(B_{T_l}^{-1}B(T_l, T_{l+1})(1 + \kappa\Delta_l)\big| \mathcal{F}_{T_j}\big)\Big| \mathcal{F}_{T_i}\Big)$$

$$+ \delta B_{T_i} \mathbb{E}_{\mathbb{P}^*}\Big(\sum_{j=i}^{m-1} \mathbb{1}_j \sum_{l=j+1}^{m} \mathbb{E}_{\mathbb{P}^*}\big(B_{T_l}^{-1}B^{-1}(T_{l-1}, T_l)\big| \mathcal{F}_{T_j}\big)\Big| \mathcal{F}_{T_i}\Big).$$

The *receiver-risk-adjusted nominal rate* $L_r^\delta(T_i; m-i)$ can now be formally defined as that level of $\kappa$ for which $\mathbf{IRA}_r^\delta(T_i; \kappa, m-i) = 0$.

## 14.3 Multi-Period Defaultable Forward Nominal Rates

In Sect. 14.1.4–14.1.5, we have discussed various settlement schemes underlying single-period defaultable FRAs and swaps. We shall now study analogous schemes in case of multi-period interest rate agreements. Similarly as in the single-period case, we examine contracts with the inception date $T_i$, and we postulate that if both counterparties remain solvent at the first reset date $T_i$, then the contract becomes a corresponding defaultable multi-period spot contract, as described in the previous section.

**Fixed-rate defaultable FRA.** Assume that the only default prone party is the payer of the fixed rate. Similarly as in Sect. 14.1.4, we shall examine the following four possible settlement schemes: if the default occurs at some date $s < T_i$ then:

- 1$^{st}$ kind: at time $s$ the receiver gets the value of defaulted coupon bond issued by the payer, and he makes the promised payoff 1 to the payer at time $T_i$ (or, equivalently, the payoff $B(t, T_i)$ at time $t$); this scheme refers to the interpretation of the defaultable, multi-period spot IRA with the payer's default risk as the outright purchase by the receiver of a coupon bond issued by the payer (cf. Sect. 14.2),
- 2$^{nd}$ kind: the contract is immediately unwound, with no cash flows,
- 3$^{rd}$ kind: the contract becomes a default-free multi-period FRA with the reduced nominal principal $\delta$,
- 4$^{th}$ kind: the contract unfolds according to a pre-specified marking-to-market procedure.

We refer to Sect. 14.1.4 for an intuitive discussion of such analogous settlement rules in the case of single-period defaultable FRAs. The four settlement schemes listed above lead to the four distinct definitions of the associated multi-period defaultable forward rates.

**Payer-risk-adjusted forward nominal rate.** According to the first settlement scheme, at time $t < T_i$ the value of the contract to the payer is given by the following expression:

$$\widehat{\mathbf{FRA}}_p^\delta(t, T_i; \kappa, m - i) = B(t, T_i) - \sum_{j=i+1}^{m-1} c_j D^\delta(t, T_j) - c_m D^\delta(t, T_m),$$

where $c_j = \kappa \Delta_{j-1}$, $j = i + 1, \ldots, m - 1$ and $c_m = 1 + \kappa \Delta_{m-1}$. The rate $\widehat{L}_p^\delta(t, T_i; m - i)$, termed the *payer-risk-adjusted forward nominal rate*, is defined as the level of $\kappa$ that makes the contract valueless at time $t$. It is apparent that on the set $\{\tau > t\}$

$$\widehat{L}_p^\delta(t, T_i; m - i) = \frac{B(t, T_i) - D^\delta(t, T_m)}{\sum_{j=i+1}^m \Delta_{j-1} D^\delta(t, T_j)}. \tag{14.18}$$

Notice that at the reset date $T_i$ we have $\widehat{L}^\delta(T_i, T_i; m - i) = L^\delta(T_i; m - i)$, where $L^\delta(T_i; m - i)$ is given by (14.17).

**Reference-risk-adjusted forward nominal rate.** Let us examine the payoff at time $T_i$ associated with the second scheme. On the set $\{\tau > T_i\}$, the payoff coincides with the value of a defaultable multi-period spot IRA at time $T_i$. More specifically, it equals $\mathbf{IRA}^\delta(T_i; \kappa, m - i)$. Otherwise, i.e., on the set $\{\tau \leq T_i\}$, the payoff at time $T_i$ equals zero. We are thus dealing here with the contingent claim $Y$, which settles at time $T_i$, and equals

$$Y = \mathbb{1}_{\{\tau > T_i\}} \mathbf{IRA}^\delta(T_i; \kappa, m - i). \tag{14.19}$$

Clearly, in order to evaluate the *reference-risk-adjusted forward nominal rate* $\widetilde{L}_p^\delta(t, T_i; m - i)$ (i.e., the value of $\kappa$ that makes $Y$ valueless at time $t$), it is indispensable to introduce some stochastic model of the default time; the equality $\widetilde{L}^\delta(T_i, T_i; m - i) = L_p^\delta(T_i; m - i)$ is always satisfied, though.

**Defaultable forward multi-period LIBOR rate.** According to the third settlement scheme, a defaultable multi-period FRA is specified as follows:
- if the payer defaults in the time interval $(t, T_i]$, the net cash flow to the payer at time $T_i$ is

$$\delta \sum_{j=i}^{m-1} B(T_i, T_j) - \delta \sum_{j=i+1}^{m} (1 + \kappa\Delta_{j-1})B(T_i, T_j),$$

- otherwise, the net cash flow to the payer at the inception date $T_i$ is equal to $\mathbf{IRA}^\delta(T_i; \kappa, m - i)$.

The value of the contract specified above to the payer equals, on $\{\tau > t\}$,

$$\mathbf{FRA}^\delta(t, T_i; \kappa, m - i) = D^\delta(t, T_i) - \sum_{j=i+1}^{m-1} c_j D^\delta(t, T_j) - c_m D^\delta(t, T_m),$$

where $D^\delta(t, T_j)$ is the price of the bond issued by the payer, $c_j = \kappa\Delta_{j-1}$ for $j = i+1, \ldots, m-1$ and $c_m = 1 + \kappa\Delta_{m-1}$. We define the *defaultable forward multi-period LIBOR rate* $L^\delta(t, T_i; m - i)$ as that level of the fixed rate $\kappa$ for which $\mathbf{FRA}^\delta(t, T_i; \kappa) = 0$. Using the formula above, we obtain

$$L^\delta(t, T_i; m - i) = \frac{D^\delta(t, T_i) - D^\delta(t, T_m)}{\sum_{j=i+1}^{m} \Delta_{j-1} D^\delta(t, T_j)}. \tag{14.20}$$

As expected, if the recovery rate $\delta = 1$ (i.e., in case of full recovery), the rate $L^1(t, T_i; m-i)$ coincides with the default-free forward swap rate $\kappa(t, T_i; m-i)$.

*Remarks.* Similar observations can be made here as in the single-period case (cf. Sect. 14.1.4). In particular, the contract specification that leads to equality (14.20) is rather artifical. Indeed, if the default occurs before the reset date $T_i$, then the only consequence is the (symmetric) reduction of the contract's nominal principal, and the contract then becomes default-free. This means that here the default risk is treated as the reference risk, rather than as the counterparty risk.

**Mark-to-market forward multi-period nominal rate.** Now assume that the contract is marked to market if default occurs prior to or at the inception date $T_i$. It is natural to postulate that at the time of default the value $\mathbf{FS}(\tau, T_i; \kappa, m - i)$ of a default-free fixed-for-floating forward swap settled in arrears, with otherwise identical covenants, is found from the default-free term structure,[2] and:

– if the value $\mathbf{FS}(\tau, T_i; \kappa, m - i)$ of the equivalent default-free swap is positive to the payer, then he gets this value from the receiver,

– if the value $\mathbf{FS}(\tau, T_i; \kappa, m - i)$ is positive to the receiver, then he gets from the payer only a fixed fraction $\delta$ of this value.

From the payer's perspective, the contract can be represented as the following contingent claim, which settles at time $T_i$:

$$Y = \mathbb{1}_{\{t < \tau \leq T_i\}} B^{-1}(\tau, T_i)\left(\left(\mathbf{FS}(\tau, T_i)\right)^+ - \delta\left(-\mathbf{FS}(\tau, T_i)\right)^+\right)$$
$$+ \mathbb{1}_{\{\tau > T_i\}} \mathbf{IRA}_p^\delta(T_i; \kappa, m - i),$$

where $\mathbf{FS}(t, T_i) = \mathbf{FS}(t, T_i; \kappa, m - i)$, and $\tau$ stands for the payer's default time. It is known (see, e.g., formula (16.3) in Musiela and Rutkowski (1997a)) that for any $t \leq T_i$

$$\mathbf{FS}(t, T_i; \kappa, m - i) = B(t, T_i) - \sum_{j=i+1}^{m-1} c_j B(t, T_j) - c_m B(t, T_m),$$

where $c_j = \kappa \Delta_{j-1}$, $j = i + 1, \ldots, m - 1$ and $c_m = \kappa \Delta_{m-1} + 1$. On the other hand, $\mathbf{IRA}_p^\delta(T_i; \kappa, m - i)$ is given by (14.16), namely,

$$\mathbf{IRA}_p^\delta(T_i; \kappa, m - i) = 1 - \sum_{j=i+1}^{m-1} c_j D^\delta(T_i, T_j) - c_m D^\delta(T_i, T_m).$$

To summarize, we are dealing here with a contingent claim of the form:

$$Y = \mathbb{1}_{\{t < \tau \leq T_i\}} B^{-1}(\tau, T_i) Z(\tau, T_i; \kappa, m - i) + \mathbb{1}_{\{\tau > T_i\}} \mathbf{IRA}_p^\delta(T_i; \kappa, m - i),$$

in which the recovery process $Z$ satisfies, for $t \leq T_i$,

$$Z(t, T_i; \kappa, m - i) = \mathbf{FS}(t, T_i; \kappa, m - i) + (1 - \delta)\left(-\mathbf{FS}(t, T_i; \kappa, m - i)\right)^+.$$

The determination of the implied mark-to-market forward nominal rate $\bar{L}(t, T_i; m - i)$ – which is defined as the value of $\kappa$ that makes the claim $Y$ valueless at time $t$ – requires joint modeling of the default time $\tau$ and the default-free term structure of interest rates.

Let us again stress that we have examined only the case of a defaultable fixed-rate FRA; the case of a defaultable floating-rate FRA can be analyzed along similar lines as in Sect. 14.2.

---

[2] As already mentioned, a default-free forward interest rate swap settled in arrears may as well be interpreted as a default-free multi-period forward rate agreement.

## 14.4 Defaultable Swaps with Unilateral Default Risk

We shall focus on the most widely used settlement rules for defaultable swaps, namely, on the case of *mark-to-market* swaps. The settlement payment will be related to the market value of an equivalent non-defaultable swap. This convention is not unanimously accepted neither in the real-life contracts, nor in financial literature. An alternative settlement rule stipulates that the recovery payment is a fraction the pre-default market value of the contract. The latter convention results in the backward SDE for the value of a defaultable swap (Duffie et al. (1996), Duffie and Huang (1996), Huge and Lando (1999)). In the sequel, we examine only spot contracts and rates.

For any $t \in \mathbb{R}_+$, we write $t_*$ to denote the integer $j$ such that $t \in (T_j, T_{j+1}]$. We shall make frequent use of the following notation:

$$\alpha_t = \frac{B(t, T_{t_*+1})}{B(T_{t_*}, T_{t_*+1})} - B(t, T_m) - \sum_{l=t_*+1}^{m} \kappa \Delta_{l-1} B(t, T_l). \tag{14.21}$$

It is not difficult to check that in any arbitrage-free model of the term structure, $\alpha_t$ can be given the following probabilistic interpretation

$$\alpha_t = B_t \, \mathbb{E}_{\mathbb{P}^*} \left( \sum_{l=t_*+1}^{m} B_{T_l}^{-1} \big( B^{-1}(T_{l-1}, T_l) - 1 - \kappa \Delta_{l-1} \big) \, \Big| \, \mathcal{F}_t \right)$$

$$= B_t \, \mathbb{E}_{\mathbb{P}^*} \left( \sum_{l=t_*+1}^{m} B_{T_l}^{-1} \big( L(T_{l-1}) - \kappa \big) \Delta_{l-1} \, \Big| \, \mathcal{F}_t \right),$$

where the second equality follows from the obvious relationship:

$$B^{-1}(T_{l-1}, T_l) - 1 - \kappa \Delta_{l-1} = \big( L(T_{l-1}) - \kappa \big) \Delta_{l-1}.$$

However, expression (14.21) makes it clear that $\alpha_t$ does not depend on the choice of a model. The fact that $\alpha_t$ can be inferred from the observed market prices of default-free bonds makes the settlement rules described below easy to implement in practice. In financial interpretation, $\alpha_t$ represents the arbitrage value at time $t$ of the remaining cash flows of the swap, under the assumption that all future cash flows after time $t$ up to the maturity $T_m$ are default-free, and that both counterparties make their respective payments in arrears. Put in a different way, $\alpha_t$ represents the value at time $t$ of a default-free fixed-for-floating interest rate swap with the first reset date $T_{t_*} < t$ and the first settlement date $T_{t_*+1} > t$. We shall examine the following three alternative marking-to-market rules:

- 1st kind: the contract is marked to market and settled immediately at the default time $\tau$,
- 2nd kind: the contract is marked to market at time $\tau$; however, the settlement payment is postponed to the first reset date after the default time,
- 3rd kind: the marking-to-market procedure and the contract's settlement are postponed to the first reset date after the default time.

### 14.4.1 Settlement of the 1st Kind

**Payer's default risk.** Assume that only the payer of the fixed rate $\kappa$ is prone to default, and denote by $\tau$ the default time of the payer. The swap is then specified as follows:

- for $i \leq j \leq m - 1$, if the payer does not default in $(T_j, T_{j+1}]$, at time $T_{j+1}$ he pays the full fixed amount $\kappa \Delta_j$, and he collects the full floating amount $L(T_j)\Delta_j$,

- for $i \leq j \leq m - 1$, if the payer defaults at time $\tau \in (T_j, T_{j+1}]$, the contract unfolds in the following way:

(a) if the payer is in-the-money at default time, i.e., if $\alpha_\tau > 0$, the net cash flow at time $\tau$ to the payer is equal to $\alpha_\tau$,

(b) if the receiver is in-the-money at default time, i.e., if $\alpha_\tau < 0$, the net cash flow to the receiver at time $\tau$ is $-\delta\alpha_\tau$.

From the payer's perspective, the value $\mathbf{IRS}_p^\delta(T_i; \kappa, m - i)$ of this contract at time $T_i$ equals (recall that $\tau_*$ was introduced in Definition 14.2.1)

$$
B_{T_i} \, \mathbb{E}_{\mathbb{P}^*} \left( \sum_{l=i+1}^{\tau_*} B_{T_l}^{-1} \big( L(T_{l-1}) - \kappa \big) \Delta_{l-1} + \mathbb{1}_{\{T_i < \tau \leq T_m\}} B_\tau^{-1} \big( \alpha_\tau^+ - \delta\alpha_\tau^- \big) \Big| \mathcal{F}_{T_i} \right)
$$

or, equivalently,

$$
\mathbf{IRS}_p^\delta(T_i; \kappa, m - i) = B_{T_i} \, \mathbb{E}_{\mathbb{P}^*} \left( \sum_{l=i+1}^{\tau_*} B_{T_l}^{-1} \big( L(T_{l-1}) - \kappa \big) \Delta_{l-1} \Big| \mathcal{F}_{T_i} \right)
$$

$$
+ B_{T_i} \, \mathbb{E}_{\mathbb{P}^*} \left( \mathbb{1}_{\{T_i < \tau \leq T_m\}} B_{T_{\tau_*+1}}^{-1} B^{-1}(\tau, T_{\tau_*+1}) \big( \alpha_\tau^+ - \delta\alpha_\tau^- \big) \Big| \mathcal{F}_{T_i} \right).
$$

The *payer-risk-adjusted swap rate* of the first kind is defined as the level of the fixed-rate $\kappa$, which makes the contract valueless at inception, i.e., it is the solution to the equation $\mathbf{IRS}_p^\delta(T_i; \kappa, m - i) = 0$.

*Remarks.* It is interesting to observe that the following representation for $\mathbf{IRS}_p^\delta(T_i; \kappa, m - i)$ is valid:

$$
\mathbf{IRS}_p^\delta(T_i; \kappa, m - i) = B_{T_i} \, \mathbb{E}_{\mathbb{P}^*} \left( \sum_{l=i+1}^{m} B_{T_l}^{-1} \big( L(T_{l-1}) - \kappa \big) \Delta_{l-1} \Big| \mathcal{F}_{T_i} \right)
$$

$$
+ (1 - \delta) B_{T_i} \, \mathbb{E}_{\mathbb{P}^*} \left( \mathbb{1}_{\{T_i < \tau \leq T_m\}} B_\tau^{-1} \alpha_\tau^- \Big| \mathcal{F}_{T_i} \right). \tag{14.22}
$$

It is easy to see that the first term in the last formula represents the value of a standard default-free swap settled in arrears. The second term can be seen as the reduced value at time $T_i$ of a forward receiver swaption that is to be exercised at the random time $\tau$ (recall that all valuation formulae discussed in this section are valid on the set $\{\tau > T_i\}$; otherwise, i.e., on the set $\{\tau \leq T_i\}$, the value of each examined contract is zero). A similar observation was already made by Lotz and Schlögl (2000).

*Example 14.4.1.* We shall show how to compute the payer-risk-adjusted swap rate, under special assumptions imposed on the default time $\tau$. We postulate that $\tau$ and the short-term interest rate are conditionally independent given the $\sigma$-field $\mathcal{F}_{T_i}$. Let us denote by $F(t; i)$ a version of the conditional cumulative probability distribution function of $\tau$ given $\mathcal{F}_{T_i}$. We obtain

$$\mathbf{IRS}_p^\delta(T_i; \kappa, m - i) = \mathbf{FS}(T_i; \kappa, m - i) + \mathbf{PS}(T_i, T_m; \kappa), \qquad (14.23)$$

where $\mathbf{FS}(T_i; \kappa, m - i)$ stands for the value of an equivalent default-free fixed-for-floating swap, specifically,

$$\mathbf{FS}(T_i; \kappa, m - i) = B_{T_i}\, \mathbb{E}_{\mathbb{P}^*}\left( \sum_{l=i+1}^{m} B_{T_l}^{-1}\big(L(T_{l-1}) - \kappa\big)\Delta_{l-1} \,\Big|\, \mathcal{F}_{T_i} \right)$$

$$= 1 - B(T_i, T_m) - \sum_{l=i+1}^{m} \kappa\Delta_{l-1} B(T_i, T_l)$$

and $\mathbf{PS}(T_i, T_m; \kappa)$ equals

$$\mathbf{PS}(T_i, T_m; \kappa) = (1 - \delta) \int_{T_i}^{T_m} B_{T_i}\, \mathbb{E}_{\mathbb{P}^*}(B_t^{-1}\alpha_t^- \,|\, \mathcal{F}_{T_i})\, dF(t; i). \qquad (14.24)$$

For each $t \in (T_i, T_m]$, the term $B_{T_i}\mathbb{E}_{\mathbb{P}^*}(B_t^{-1}\alpha_t^- \,|\, \mathcal{F}_{T_i})$ represents the value at time $T_i$ of a default-free forward receiver swaption with the exercise date $t$. Under the assumption of the lognormal distribution of default-free forward swap rates, this value can be found explicitly, using the so-called Black swaptions formula (for more details, see, e.g., Jamshidian (1997), Chap. 16 in Musiela and Rutkowski (1997a) or Sect. 15.1.2 below).

If we also specify the conditional distribution function $F(t; i)$ of the default time $\tau$, we may use formulae (14.23)–(14.24) in order to solve for $\kappa$ the equation $\mathbf{IRS}_p^\delta(T_i; \kappa, m - i) = 0$. In view of the non-linearity of this equation, we will only be able to find an approximate value of the payer-risk-adjusted swap rate predicted by the model (cf. Lotz and Schlögl (2000)).

**Receiver's default risk.** When only the receiver is prone to default, then the contract can be described as follows ($\tau$ is the receiver's default time):
- for $i \leq j \leq m - 1$, if the receiver does not default in $(T_j, T_{j+1}]$, then at time $T_{j+1}$ he pays the full floating amount: $L(T_j)\Delta_j$, and he collects the full fixed amount: $\kappa\Delta_j$,
- for $i \leq j \leq m - 1$, if the receiver defaults at time $\tau \in (T_j, T_{j+1}]$, then the contract unwinds as follows:
  (a) if the payer is in-the-money at default time, i.e., if $\alpha_\tau > 0$, then the net cash flow to the payer at the reset date $T_{j+1}$ equals $\delta B^{-1}(\tau, T_{j+1})\alpha_\tau$,
  (b) if the receiver is in-the-money at default time $\tau$, i.e., if $\alpha_\tau < 0$, then he receives at the reset date $T_{j+1}$ the amount: $-B^{-1}(\tau, T_{j+1})\alpha_\tau$.

From the payer's perspective, the value $\mathbf{IRS}_r^\delta(T_i; \kappa, m - i)$ of this contract at its inception date $T_i$ is equal to

$$B_{T_i} \mathbb{E}_{\mathbb{P}*} \left( \sum_{l=i+1}^{\tau_*} B_{T_l}^{-1} \big(L(T_{l-1}) - \kappa\big) \Delta_{l-1} + \mathbb{1}_{\{T_i < \tau \leq T_m\}} B_\tau^{-1} \big(\delta \alpha_\tau^+ - \alpha_\tau^-\big) \,\Big|\, \mathcal{F}_{T_i} \right)$$

or, equivalently,

$$\mathbf{IRS}_r^\delta(T_i; \kappa, m - i) = B_{T_i} \mathbb{E}_{\mathbb{P}*} \left( \sum_{l=i+1}^{\tau_*} B_{T_l}^{-1} \big(L(T_{l-1}) - \kappa\big) \Delta_{l-1} \,\Big|\, \mathcal{F}_{T_i} \right)$$

$$+ B_{T_i} \mathbb{E}_{\mathbb{P}*} \left( \mathbb{1}_{\{T_i < \tau \leq T_m\}} B_{T_{\tau_*+1}}^{-1} B^{-1}(\tau, T_{\tau_*+1}) \big(\delta \alpha_\tau^+ - \alpha_\tau^-\big) \,\Big|\, \mathcal{F}_{T_i} \right).$$

The *receiver-risk-adjusted swap rate* of the first kind is defined as the solution to the equation $\mathbf{IRS}_r^\delta(T_i; \kappa, m - i) = 0$.

### 14.4.2 Settlement of the 2$^{\text{nd}}$ Kind

Let us recall that the settlement scheme of the second kind stipulates that the contract is marked to market as soon as the default occurs, but the settlement payment (if any) is postponed to the first settlement date after the time of default. In our notation, the first settlement date after the random time $\tau$ is denoted by $T_{\tau_*+1}$.

**Payer's default risk.** In case of the payer's default risk, the contract is governed by the following rules:
- for $i \leq j \leq m - 1$, if the payer does not default in $(T_j, T_{j+1}]$, then he pays at time $T_{j+1}$ the full fixed amount: $\kappa \Delta_j$, and he collects the full floating amount: $L(T_j) \Delta_j$,
- for $i \leq j \leq m - 1$, if the payer defaults at time $\tau \in (T_j, T_{j+1}]$, then the contract unfolds in the following way:
  (a) if the payer is in-the-money at default time, i.e., if $\alpha_\tau > 0$, then the net cash flow at $T_{j+1}$ to the payer is equal to $\alpha_\tau$,
  (b) if the receiver is in-the-money at default time, i.e., if $\alpha_\tau < 0$, then the net cash flow at $T_{j+1}$ to the receiver is $-\delta \alpha_\tau$.

From the payer's perspective, the value of this contract at time $T_i$ equals

$$\mathbf{IRS}_p^\delta(T_i; \kappa, m - i) = B_{T_i} \mathbb{E}_{\mathbb{P}*} \left( \sum_{l=i+1}^{\tau_*} B_{T_l}^{-1} \big(L(T_{l-1}) - \kappa\big) \Delta_{l-1} \,\Big|\, \mathcal{F}_{T_i} \right)$$

$$+ B_{T_i} \mathbb{E}_{\mathbb{P}*} \left( \mathbb{1}_{\{T_i < \tau \leq T_m\}} B_{T_{\tau_*+1}}^{-1} \big(\alpha_\tau^+ - \delta \alpha_\tau^-\big) \,\Big|\, \mathcal{F}_{T_i} \right).$$

The *payer-risk-adjusted swap rate* of the second kind is the level of the fixed rate $\kappa$ that makes the contract valueless at inception, i.e., it is the solution to the equation $\mathbf{IRS}^\delta(T_i; \kappa, m - i) = 0$.

**Receiver's default risk.** When only the receiver is prone to default, the contract can be described as follows ($\tau$ stands for the receiver's default time):
- for $i \leq j \leq m - 1$, if the receiver does not default in $(T_j, T_{j+1}]$, then at time $T_{j+1}$ he pays the full floating amount due: $L(T_j)\Delta_j$, and is paid the full fixed amount due: $\kappa\Delta_j$,
- for $i \leq j \leq m - 1$, if the receiver defaults at time $\tau \in (T_j, T_{j+1}]$, then
  (a) if the payer is in-the-money at time of the default, i.e., if $\alpha_\tau > 0$, then at the reset date $T_{j+1}$ the net cash flow to the payer is equal to $\delta\alpha_\tau$,
  (b) if the receiver is in-the-money at time $\tau$, i.e., if $\alpha_\tau < 0$, then at the reset date $T_{j+1}$ the net cash flow to the receiver is equal to $-\alpha_\tau$, and the contract terminates

From the payer's perspective, the value of this contract at its inception date $T_i$ is equal to

$$\mathbf{IRS}_r^\delta(T_i; \kappa, m - i) = B_{T_i} \, \mathbb{E}_{\mathbb{P}^*}\Big( \sum_{l=i+1}^{\tau_*} B_{T_l}^{-1}\big(L(T_{l-1}) - \kappa\big)\Delta_{l-1} \,\Big|\, \mathcal{F}_{T_i} \Big)$$

$$+ B_{T_i} \, \mathbb{E}_{\mathbb{P}^*}\Big( \mathbb{1}_{\{T_i < \tau \leq T_m\}} B_{T_{\tau_*+1}}^{-1} \big(\delta\alpha_\tau^+ - \alpha_\tau^-\big) \,\Big|\, \mathcal{F}_{T_i} \Big).$$

The *receiver-risk-adjusted swap rate* of the second kind is defined as the solution to the equation $\mathbf{IRS}_r^\delta(T_i; \kappa, m - i) = 0$.

### 14.4.3 Settlement of the 3$^{\text{rd}}$ Kind

We now postulate that the contract is marked to market at the first settlement date after default and the contract unwinds at this date.

**Payer's default risk.** We first assume that the only party prone to default is the payer of the fixed rate. In this case:
- for $i \leq j \leq m - 1$, if the payer does not default in $(T_j, T_{j+1}]$, then he pays at time $T_{j+1}$ the full fixed amount: $\kappa\Delta_j$, and he is paid the full floating amount: $L(T_j)\Delta_j$,
- for $i \leq j \leq m - 1$, if the payer defaults at time $\tau \in (T_j, T_{j+1}]$, then the contract unfolds in the following way:
  (a) if the payer is in-the-money at the first reset date after default, i.e., if $\alpha_{T_{\tau_*+1}} > 0$, then the net cash flow at time $T_{\tau_*+1}$ to the payer is equal to $\alpha_{T_{\tau_*+1}}$,
  (b) if the receiver is in-the-money at default time, i.e., if $\alpha_{T_{\tau_*+1}} < 0$, then the net cash flow at time $\tau$ to the receiver is $-\delta\alpha_{T_{\tau_*+1}}$.

The value to the payer of such a contract at its inception date $T_i$ is given as:

$$\mathbf{IRS}_p^\delta(T_i; \kappa, m - i) = B_{T_i} \, \mathbb{E}_{\mathbb{P}^*}\Big( \sum_{l=i+1}^{\tau_*} B_{T_l}^{-1}\big(L(T_{l-1}) - \kappa\big)\Delta_{l-1} \,\Big|\, \mathcal{F}_{T_i} \Big)$$

$$+ B_{T_i} \, \mathbb{E}_{\mathbb{P}^*}\Big( \mathbb{1}_{\{T_i < \tau \leq T_m\}} B_{T_{\tau_*+1}}^{-1} \big(\alpha_{T_{\tau_*+1}}^+ - \delta\alpha_{T_{\tau_*+1}}^-\big) \,\Big|\, \mathcal{F}_{T_i} \Big).$$

The value of $\kappa$ that makes the contract valueless at $T_i$ is referred as the *payer-risk-adjusted swap rate* of the third kind.

**Receiver's default risk.** A straightforward modification of the above contract to the case when only the receiver is prone to default reads as follows:
- for $i \leq j \leq m-1$, if the payer does not default in $(T_j, T_{j+1}]$, then he pays at time $T_{j+1}$ the full fixed amount: $\kappa \Delta_j$, and he is paid the full floating amount: $L(T_j)\Delta_j$,
- for $i \leq j \leq m-1$, if the payer defaults at time $\tau \in (T_j, T_{j+1}]$, then the contract unfolds in the following way:
   (a) if the payer is in-the-money at the first reset date after default, i.e., if $\alpha_{T_{\tau_*+1}} > 0$, then the net cash flow at time $T_{\tau_*+1}$ to the payer is equal to $\delta \alpha_{T_{\tau_*+1}}$,
   (b) if the receiver is in-the-money at default time, i.e., if $\alpha_{T_{\tau_*+1}} < 0$, then the net cash flow at time $\tau$ to the receiver is $-\alpha_{T_{\tau_*+1}}$.

The value to the payer of such a contract at its inception date $T_i$ is given as

$$\mathbf{IRS}_r^\delta(T_i; \kappa, m-i) = B_{T_i} \, \mathbb{E}_{\mathbb{P}^*}\left( \sum_{l=i+1}^{\tau_*} B_{T_l}^{-1}\left(L(T_{l-1}) - \kappa\right)\Delta_{l-1} \,\Big|\, \mathcal{F}_{T_i} \right)$$

$$+ \, B_{T_i} \, \mathbb{E}_{\mathbb{P}^*}\left( \mathbb{1}_{\{T_i < \tau \leq T_m\}} B_{T_{\tau_*+1}}^{-1}\left(\delta \alpha_{T_{\tau_*+1}}^+ - \alpha_{T_{\tau_*+1}}^-\right) \,\Big|\, \mathcal{F}_{T_i} \right).$$

The value of $\kappa$ which makes the contract valueless at $T_i$ is referred as the *receiver-risk-adjusted swap rate* of the third kind.

### 14.4.4 Market Conventions

All settlement schemes introduced in this section can be seen as special cases of the market convention known as the full two-way payment rule. The generic name *full two-way payment rule* refers to any specific settlement scheme, which stipulates that in case of default the defaulting party receives the full market (or pre-default) value of the swap if this value is positive to the defaulting party, and it pays a fraction of this value to the non-defaulting party otherwise. An alternative market convention – known under the generic name of the *limited two-way payment rule* – postulates that both payments are fractional amounts of the value of the swap at time of default. In other words, in case of default, if the value of the swap is positive to the defaulting party, he is entitled only to a fractional payment, as opposed to the full termination payment the defaulting party is entitled to according to the full two-way payment rule. A particular case of the limited two-way payment rule stipulates that the two recovery rates are identical, so that the settlement rule in case of default is symmetric.

It is worth to stressing that in practice the notion of the value of the swap at time of default is by far more ambiguous than in the academic literature. Since swaps are traded over-the-counter, the 'pre-default value' of the swap may be specified in a large variety of ways, and thus the generic term *full/limited two-way payment rule* does not completely determine the complex real-life covenants of the marking-to-market swap settlement.

## 14.5 Defaultable Swaps with Bilateral Default Risk

We now turn to the case when the default risk of each parties of an interest rate contract is non-negligible. We shall only examine a few particular cases of a defaultable multi-period spot swap agreement with mark-to-market settlement covenants in case of default of any of the parties.

Let $\tau_p$ and $\tau_r$ be the default times of the payer and of the receiver, respectively. We make a technical assumption that $\tau_p \neq \tau_r$ almost surely, under the real-world probability, and thus also under any equivalent martingale probability measure. In contrast to the previous sections, we shall now write $\tau$ to denote the minimum of the two default times, i.e., we set:

$$\tau = \tau_p \wedge \tau_r := \begin{cases} \tau_p & \text{if } \tau_p \leq \tau_r, \\ \tau_r & \text{if } \tau_p \geq \tau_r. \end{cases}$$

We shall focus on the scheme of the first kind with the full two-way settlement (other cases are not much different). A mark-to-market defaultable multi-period spot swap is now specified as follows:

- for $i \leq j \leq m - 1$, if there is no default in $(T_j, T_{j+1}]$, then at the reset time $T_{j+1}$ the payer pays the full fixed amount: $\kappa \Delta_j$, and he collects the full floating amount: $L(T_j)\Delta_j$,
- for $i \leq j \leq m - 1$, if there is default in $(T_j, T_{j+1}]$ and the (first) defaulting party is the payer, i.e., if $\tau \in (T_j, T_{j+1}]$ and $\tau = \tau_p$, then the contract unwinds according to the marking-to-market procedure, namely,
  (a) if the payer is in-the-money at default time, i.e., if $\alpha_\tau > 0$, then the net cash flow to the payer at $T_{j+1}$ is equal to $B^{-1}(\tau, T_{j+1})\alpha_\tau$,
  (b) if the receiver is in-the-money at default time, i.e., if $\alpha_\tau > 0$, then the net cash flow to the receiver at $T_{j+1}$ is $-\delta_p B^{-1}(\tau, T_{j+1})\alpha_\tau$,
- for $i \leq j \leq m-1$, if the default occurs in $(T_j, T_{j+1}]$ and the (first) defaulting party is the receiver, i.e., if $\tau \in (T_j, T_{j+1}]$ and $\tau = \tau_r$, then the contract unwinds in the following way:
  (a) if the payer is in-the-money at time $\tau$, i.e., if $\alpha_\tau > 0$, then the net cash flow to the payer on the reset date $T_{j+1}$ equals $\delta_r B^{-1}(\tau, T_{j+1})\alpha_\tau$,
  (b) if the receiver is in-the-money at time $\tau$, i.e., if $\alpha_\tau < 0$, then the net cash flow to the receiver at $T_{j+1}$ equals $-B^{-1}(\tau_r, T_{j+1})\alpha_\tau$.

From the payer's perspective, the value of this contract at its inception date $T_i$ is given as

$$\mathbf{IRS}_f^\delta(T_i; \kappa, m - i) = B_{T_i} \, \mathbb{E}_{\mathbb{P}^*} \left( \sum_{j=i+1}^{\tau_*} B_{T_j}^{-1} \left(L(T_{j-1}) - \kappa\right)\Delta_{j-1} \,\Big|\, \mathcal{F}_{T_i} \right)$$

$$+ B_{T_i} \, \mathbb{E}_{\mathbb{P}^*} \left( \mathbb{1}_{\{T_i < \tau = \tau_p \leq T_m\}} B_{T_{\tau_*+1}}^{-1} B^{-1}(\tau, T_{\tau_*+1})\left(\alpha_\tau^+ - \delta_p \alpha_\tau^-\right) \Big|\, \mathcal{F}_{T_i} \right)$$

$$+ B_{T_i} \, \mathbb{E}_{\mathbb{P}^*} \left( \mathbb{1}_{\{T_i < \tau = \tau_r \leq T_m\}} B_{T_{\tau_*+1}}^{-1} B^{-1}(\tau, T_{\tau_*+1})\left(\delta_r \alpha_\tau^+ - \alpha_\tau^-\right) \Big|\, \mathcal{F}_{T_i} \right),$$

where $\delta_p \in [0, 1)$ ($\delta_r \in [0, 1)$, resp.) is the recovery rate of the payer (of the receiver, resp.)

It is interesting to observe that

$$\mathbf{IRS}_f^\delta(T_i; \kappa, m - i) = B_{T_i}\, \mathbb{E}_{\mathbb{P}^*}\Big( \sum_{j=i+1}^{m} B_{T_j}^{-1}\big(L(T_{j-1}) - \kappa\big)\Delta_{j-1}\,\Big|\, \mathcal{F}_{T_i} \Big)$$
$$+ (1 - \delta_p)B_{T_i}\, \mathbb{E}_{\mathbb{P}^*}\Big( \mathbb{1}_{\{T_i < \tau = \tau_p \leq T_m\}} B_{\tau_p}^{-1}\alpha_{\tau_p}^-\,\Big|\, \mathcal{F}_{T_i} \Big) \quad (14.25)$$
$$- (1 - \delta_r)B_{T_i}\, \mathbb{E}_{\mathbb{P}^*}\Big( \mathbb{1}_{\{T_i < \tau = \tau_r \leq T_m\}} B_{\tau_r}^{-1}\alpha_{\tau_r}^+\,\Big|\, \mathcal{F}_{T_i} \Big).$$

The first term in (14.25) is the value of a default-free swap settled in arrears. The second term is the reduced value at time $T_i$ of a forward receiver swaption exercised at the time of default of the payer, if the payer's default comes earlier than that of the receiver. Finally, the third term is the negative of the reduced value at time $T_i$ of a forward payer swaption, exercised at receiver's default, if the receiver defaults prior to the default of the payer. In both cases, the swaption expires valueless at the maturity date $T_m$ if none of the parties defaults before this date. Lotz and Schlögl (2000) provide an analysis of expression (14.25) under some specific model assumptions.

*Remarks.* Observe that if either $\tau_r = \infty$ or $\tau_p = \infty$, then the contract reduces to the defaultable swap with the payer (receiver) as the only party prone to default. The case introduced above corresponds to the full two-way settlement rule. In case of the limited two-way settlement rule, we have

$$\mathbf{IRS}_l^\delta(T_i; \kappa, m - i) = B_{T_i}\, \mathbb{E}_{\mathbb{P}^*}\Big( \sum_{j=i+1}^{\tau_*} B_{T_j}^{-1}\big(L(T_{j-1}) - \kappa\big)\Delta_{j-1}\,\Big|\, \mathcal{F}_{T_i} \Big)$$
$$+ B_{T_i}\, \mathbb{E}_{\mathbb{P}^*}\Big( \mathbb{1}_{\{T_i < \tau = \tau_p \leq T_m\}} B_{T_{\tau_*+1}}^{-1} B^{-1}(\tau, T_{\tau_*+1})\big(\hat{\delta}_r \alpha_\tau^+ - \delta_p \alpha_\tau^-\big)\,\Big|\, \mathcal{F}_{T_i} \Big)$$
$$+ B_{T_i}\, \mathbb{E}_{\mathbb{P}^*}\Big( \mathbb{1}_{\{T_i < \tau = \tau_r \leq T_m\}} B_{T_{\tau_*+1}}^{-1} B^{-1}(\tau, T_{\tau_*+1})\big(\delta_r \alpha_\tau^+ - \hat{\delta}_p \alpha_\tau^-\big)\,\Big|\, \mathcal{F}_{T_i} \Big),$$

where $\hat{\delta}_p, \hat{\delta}_r \in [0, 1)$ are constants. In particular, under the additional assumption that $\delta_p = \hat{\delta}_p$ and $\delta_r = \hat{\delta}_r$, we obtain

$$\mathbf{IRS}_l^\delta(T_i; \kappa, m - i) = B_{T_i}\, \mathbb{E}_{\mathbb{P}^*}\Big( \sum_{j=i+1}^{\tau_*} B_{T_j}^{-1}\big(L(T_{j-1}) - \kappa\big)\Delta_{j-1}\,\Big|\, \mathcal{F}_{T_i} \Big)$$
$$+ B_{T_i}\, \mathbb{E}_{\mathbb{P}^*}\Big( \mathbb{1}_{\{T_i < \tau \leq T_m\}} B_{T_{\tau_*+1}}^{-1} B^{-1}(\tau, T_{\tau_*+1})\big(\delta_r \alpha_\tau^+ - \delta_p \alpha_\tau^-\big)\,\Big|\, \mathcal{F}_{T_i} \Big).$$

Finally, in the special case of the limited two-way settlement rule with a common recovery rate $\delta$, i.e., when $\delta_p = \hat{\delta}_p = \delta_r = \hat{\delta}_r = \delta$, the value of the defaultable swap equals

$$\mathbf{IRS}_l^\delta(T_i; \kappa, m - i) = B_{T_i}\, \mathbb{E}_{\mathbb{P}^*}\Big( \sum_{j=i+1}^{\tau_*} B_{T_j}^{-1}\big(L(T_{j-1}) - \kappa\big)\Delta_{j-1}\,\Big|\, \mathcal{F}_{T_i} \Big)$$
$$+ B_{T_i}\, \mathbb{E}_{\mathbb{P}^*}\Big( \mathbb{1}_{\{T_i < \tau \leq T_m\}} B_{T_{\tau_*+1}}^{-1} B^{-1}(\tau, T_{\tau_*+1})\delta\alpha_\tau\,\Big|\, \mathcal{F}_{T_i} \Big).$$

## 14.6 Defaultable Forward Swap Rates

In Sect. 14.1.5, we have discussed various settlement rules for single-period defaultable forward swaps. We shall now briefly study two examples of such schemes in the case of multi-period forward swaps. We shall focus on swaps that are marked to market if default occurs prior to the inception date. As usual, we examine contracts with the inception date $T_i$, and we postulate that if both counterparties are solvent at time $T_i$, then the contract becomes a corresponding defaultable multi-period spot swap.

### 14.6.1 Forward Swaps with Unilateral Default Risk

We only examine the case when the payer alone is the default prone party, and we tie the forward contract considered here to the mark-to-market defaultable spot swap analyzed in Sect. 14.4. It is clear that we do not need to be more specific about the settlement scheme for the underlying spot swap. The value $\mathbf{IRS}_p^\delta(T_i; \kappa, m - i)$ below is thus given by any of the formulae of Sect. 14.4.1–14.4.3.

**Mark-to-market forward swap rate.** If the payer becomes insolvent at time $\tau$ prior to or at the inception date $T_i$, the defaultable forward swap considered here is marked to market in the following way: at time $\tau$ the value $\mathbf{FS}(\tau, T_i; \kappa, m-i)$ of an equivalent default-free fixed-for-floating forward swap settled in arrears is found from the default-free term structure, and:

– if the value $\mathbf{FS}(\tau, T_i; \kappa, m-i)$ of the equivalent default-free swap is positive to the payer, then he collects this value from the receiver,
– if $\mathbf{FS}(\tau, T_i; \kappa, m - i) < 0$; that is, the value of the equivalent default-free swap is positive to the receiver, then he gets from the payer only a fixed fraction $\delta$ of $-\mathbf{FS}(\tau, T_i; \kappa, m - i)$.

From the payer's perspective, such a contract corresponds to the contingent claim, which settles at time $T_i$, and equals

$$
Y = 1\!\!1_{\{t < \tau \le T_i\}} B^{-1}(\tau, T_i) \Big( \big( \mathbf{FS}(\tau, T_i) \big)^+ - \delta \big( - \mathbf{FS}(\tau, T_i) \big)^+ \Big)
$$
$$
+ 1\!\!1_{\{\tau > T_i\}} \mathbf{IRS}_p^\delta(T_i; \kappa, m - i),
$$

where we write $\mathbf{FS}(\tau, T_i) = \mathbf{FS}(\tau, T_i; \kappa, m - i)$. Equivalently,

$$
Y = 1\!\!1_{\{t < \tau \le T_i\}} B^{-1}(\tau, T_i) Z(\tau, T_i; \kappa, m - i) + 1\!\!1_{\{\tau > T_i\}} \mathbf{IRS}_p^\delta(T_i; \kappa, m - i),
$$

where the recovery process $Z$ satisfies, for $t \le T_i$,

$$
Z(t, T_i; \kappa, m - i) = \mathbf{FS}(t, T_i; \kappa, m - i) + (1 - \delta) \big( - \mathbf{FS}(t, T_i; \kappa, m - i) \big)^+.
$$

To find the implied forward swap rate $\bar\kappa_p(t, T_i; m - i)$ with unilateral risk (i.e., the level of $\kappa$ that makes the claim $Y$ valueless at time $t$) we need to jointly specify the default time $\tau$ and the default-free term structure.

### 14.6.2 Forward Swaps with Bilateral Default Risk

We continue the analysis of Sect. 14.5, and we focus on the case of full two-way settlement rule. As before, $\tau_p$ and $\tau_r$ denote the time of the payer's default and the time of the receiver's default, respectively; also $\tau$ is the moment of the first default, i.e., $\tau = \tau_p \wedge \tau_r$. Again, we assume that the two default times are different with probability one.

**Mark-to-market forward swap rate.** If either of the counterparties becomes insolvent prior to the inception date $T_i$, then the defaultable forward swap considered here is marked to market. More specifically, at the default time $\tau$ the value $\mathbf{FS}(\tau, T_i; \kappa, m - i)$ of an equivalent default-free fixed-for-floating forward swap settled in arrears is determined and:
- if the first defaulting party is the payer, the contract unwinds as follows:
    (a) if the payer is in-the-money at default time, i.e., if the value of the equivalent default-free swap is positive to the payer: $\mathbf{FS}(\tau, T_i; \kappa, m-i) > 0$, then he is paid this value by the solvent counterparty,
    (b) if $\mathbf{FS}(\tau, T_i; \kappa, m - i) < 0$, then the receiver gets from the payer only a fixed fraction $\delta_p$ of $-\mathbf{FS}(\tau, T_i; \kappa, m - i)$,
- if the first defaulting party is the receiver, the contract unwinds as follows:
    (a) if the receiver is in-the-money at default time, i.e., if the value of the equivalent default-free swap satisfies: $\mathbf{FS}(\tau, T_i; \kappa, m - i) < 0$, then he gets from the payer the full amount $-\mathbf{FS}(\tau, T_i; \kappa, m - i)$,
    (b) if $\mathbf{FS}(\tau, T_i; \kappa, m - i) > 0$, then the payer gets from the receiver only a fixed fraction $\delta_r$ of this value.

To find the value of this contract to the payer at time $t < T_i$, it is enough to consider a contingent claim $Y$, which settles at time $T_i$, and is given by the formula

$$Y = \mathbb{1}_{\{t < \tau = \tau_p \leq T_i\}} B^{-1}(\tau, T_i)\left(\left(\mathbf{FS}(\tau, T_i)\right)^+ - \delta_p\left(-\mathbf{FS}(\tau, T_i)\right)^+\right)$$

$$- \mathbb{1}_{\{t < \tau = \tau_r \leq T_i\}} B^{-1}(\tau, T_i)\left(\left(\mathbf{FS}(\tau, T_i)\right)^+ - \delta_r\left(-\mathbf{FS}(\tau, T_i)\right)^+\right)$$

$$+ \mathbb{1}_{\{\tau > T_i\}} \mathbf{IRS}_f^\delta(T_i; \kappa, m - i),$$

where $\mathbf{FS}(\tau, T_i) = \mathbf{FS}(\tau, T_i; \kappa, m - i)$. Once again, we conclude that the determination of the implied forward swap rate $\bar{\kappa}_f(t, T_i; m - i)$ with bilateral risk requires joint modeling of the default time $\tau$ and the default-free term structure of interest rates. It is apparent that the limited two-way settlement rule corresponds to the following claim:

$$Y = \mathbb{1}_{\{t < \tau = \tau_p \leq T_i\}} B^{-1}(\tau, T_i)\left(\hat{\delta}_r\left(\mathbf{FS}(\tau, T_i)\right)^+ - \delta_p\left(-\mathbf{FS}(\tau, T_i)\right)^+\right)$$

$$- \mathbb{1}_{\{t < \tau = \tau_r \leq T_i\}} B^{-1}(\tau, T_i)\left(\hat{\delta}_p\left(\mathbf{FS}(\tau, T_i)\right)^+ - \delta_r\left(-\mathbf{FS}(\tau, T_i)\right)^+\right)$$

$$+ \mathbb{1}_{\{\tau > T_i\}} \mathbf{IRS}_f^\delta(T_i; \kappa, m - i),$$

where $\hat{\delta}_p$ ($\hat{\delta}_r$, resp.) is the recovery rate of the payer (receiver, resp.) if the defaulted counterparty is in-the-money.

# 15. Modeling of Market Rates

In Chap. 13, we presented an approach to the modeling of defaultable term structure based on the Heath-Jarrow-Morton modeling methodology. As the underlying building blocks that served to produce a model of default-free and defaultable term structures, we have used there the dynamics of instantaneous, continuously compounded, forward interest rates.

Starting with papers by Musiela and Sondermann (1993) and Sandmann and Sondermann (1994), the so-called *market models* of interest rates were developed for fixed-income products in absence of the counterparty credit risk. A major breakthrough in this area was achieved in Miltersen et al. (1997), who postulated the existence of a lognormal model of forward LIBOR rates, and who have found the closed-form solution for the value of a cap, corresponding to the commonly standard market formula. Their ideas were subsequently undertaken by Brace et al. (1997), Jamshidian (1997), and Musiela and Rutkowski (1997b), in which various alternative mathematically rigorous approaches to the modeling of default-free forward LIBOR and swap rates were developed. It is noteworthy that in all these papers, the focus is put on the modeling of easily observed market rates, rather than on the abstract concept of the instantaneous, continuously compounded, forward rate.

We shall present a brief overview of some of these developments in Sect. 15.1. For an introduction to the modeling of forward LIBOR and swap rates, the interested reader is referred to Chap. 14 in Musiela and Rutkowski (1997a), Chap. 18 in Hunt and Kennedy (2000) or Chap 6 in Brigo and Mercurio (2001). A more exhaustive analysis of various market models and their implementations can be found in original papers by, among others, Glasserman and Zhao (1999, 2000), Rebonato (1999, 2000), Rutkowski (1999, 2001), Schlögl (1999), Schoenmakers and Coffey (1999), Andersen and Andreasen (2000), Brace and Womersley (2000), Hunt et al. (2000), and Sidenius (2000). Practical issues related to the calibration of market models are thoroughly examined in Chap. 7 and 8 of Brigo and Mercurio (2001).

More recently, some attempts to extend the modeling of market rates in order to also cover the case of defaultable interest rate contracts were done. Some preliminary results concerning defaultable market rates, established by Lotz and Schlögl (2000) and Schönbucher (2000b), are analyzed in Sect. 15.2.

## 15.1 Models of Default-Free Market Rates

We assume that we are given a pre-determined collection of reset/settlement dates $0 < T_0 < T_1 < \cdots < T_m = T^*$, referred to as the *tenor structure*, and we write $\Delta_i = T_{i+1} - T_i$ for $i = 0, \ldots, m-1$, where, by convention, $T_{-1} = 0$. We find it convenient to denote $T_n^* = T_{m-n}$ for every $n = 0, \ldots, m$. As before, $B(t, T_i)$ stands for the price at time $t$ of a default-free zero-coupon bond which matures at $T_i$. Finally, we assume that we are given a filtered probability space $(\Omega, \mathbb{F}, \mathbb{P})$, endowed with a $d$-dimensional standard Brownian motion $W$ that models the uncertainty in the economy.

### 15.1.1 Modeling of Forward LIBOR Rates

For any $i = 0, \ldots, m-1$, the forward LIBOR rate $L(\cdot, T_i)$ satisfies (cf. (14.3))

$$L(t, T_i) = \frac{B(t, T_i) - B(t, T_{i+1})}{\Delta_i B(t, T_{i+1})}, \quad \forall\, t \in [0, T_i].$$

**Definition 15.1.1.** For any $i = 0, \ldots, m$, a probability measure $\mathbb{P}_{T_i}$ on $(\Omega, \mathcal{F}_{T_i})$ is said to be the *forward LIBOR measure* for the date $T_i$ if for every $k = 0, \ldots, m$ the relative bond price

$$U_{m-i+1}(t, T_k) := \frac{B(t, T_k)}{B(t, T_i)}, \quad \forall\, t \in [0, T_k \wedge T_i],$$

follows an $\mathbb{F}$-(local) martingale under $\mathbb{P}_{T_i}$.

As apparent from Definition 15.1.1, the notion of the forward LIBOR measure is very closely related to the classic concept of a forward martingale measure for a given date (hence the notation). It is also easy to observe that the forward LIBOR rate $L(\cdot, T_i)$ follows a local martingale under the forward LIBOR measure for the date $T_{i+1}$. If, in addition, it is a strictly positive process, the existence of the associated volatility process can be justified by standard arguments. In our further development, we shall go the other way around; that is, we will assume that for any date $T_i$, the volatility $\nu(\cdot, T_i)$ of the forward LIBOR rate $L(\cdot, T_i)$ is pre-specified. In principle, it can be a deterministic function of time, a function of the underlying forward LIBOR rates, or it can follow an adapted stochastic process. For simplicity, we assume that the volatilities of forward LIBOR rates are bounded processes. We make the following standing assumptions (cf. Brace et al. (1997), Miltersen et al. (1997), and Musiela and Rutkowski (1997b)).

**Assumption (LR).** We are given a family of bounded $\mathbb{F}$-adapted processes $\nu(\cdot, T_i)$, $i = 0, \ldots, m-1$, which represent the volatilities of forward LIBOR rates $L(\cdot, T_i)$. In addition, we are given an initial term structure of interest rates, specified by a family $B(0, T_i)$, $i = 0, \ldots, m$ of bond prices. We assume here that $B(0, T_i) > B(0, T_{i+1})$ for $i = 0, \ldots, m-1$.

Our aim is to construct forward LIBOR rates $L(\cdot, T_i)$, $i = 0, \ldots, m-1$, a collection of mutually equivalent probability measures $\mathbb{P}_{T_i}$, $i = 1, \ldots, m$, and a family $W^{T_i}$, $i = 1, \ldots, m$ of processes in such a way that:

- for any $i = 1, \ldots, m$ the process $W^{T_i}$ follows a standard Brownian motion under the probability measure $\mathbb{P}_{T_i}$,
- for any $i = 0, \ldots, m-1$, the forward LIBOR rate $L(\cdot, T_i)$ satisfies the SDE

$$dL(t, T_i) = L(t, T_i)\nu(t, T_i)\, dW_t^{T_{i+1}}, \quad \forall t \in [0, T_i], \tag{15.1}$$

with the initial condition

$$L(0, T_i) = \frac{B(0, T_i) - B(0, T_{i+1})}{\Delta_i B(0, T_{i+1})}.$$

The construction provided below relies on the backward induction.

**First step.** We begin by defining the forward LIBOR rate with the longest maturity: $T_{m-1}$. We postulate that the process $L(\cdot, T_{m-1}) = L(\cdot, T_1^*)$ is governed, under the underlying probability measure $\mathbb{P}$, by the following SDE

$$dL(t, T_1^*) = L(t, T_1^*)\,\nu(t, T_1^*)\, dW_t, \quad \forall t \in [0, T_1^*], \tag{15.2}$$

with the initial condition

$$L(0, T_1^*) = \frac{B(0, T_1^*) - B(0, T^*)}{\Delta_{m-1} B(0, T^*)}. \tag{15.3}$$

Notice that, for simplicity, we have chosen the underlying probability measure $\mathbb{P}$ as the forward LIBOR measure for the date $T^*$, so that $\mathbb{P}_{T^*} = \mathbb{P}$; this choice is not essential, though. From (15.2)–(15.3), we obtain

$$L(t, T_1^*) = \frac{B(0, T_1^*) - B(0, T^*)}{\Delta_{m-1} B(0, T^*)} \, \mathcal{E}_t\left(\int_0^{\cdot} \nu(u, T_1^*)\, dW_u\right),$$

where $\mathcal{E}_t$ is the Doléans exponential. Since $B(0, T_1^*) > B(0, T^*)$, it is clear that the forward LIBOR rate process $L(\cdot, T_1^*)$ follows a strictly positive $\mathbb{F}$-martingale under $\mathbb{P}_{T^*} = \mathbb{P}$.

**Second step.** In the next step, we are going to specify the forward LIBOR rate for the date $T_2^*$. For this purpose, we first need to construct the forward martingale measure for the date $T_1^*$. By virtue of Definition 15.1.1, it is a probability measure $\mathbb{Q}$ defined on $(\Omega, \mathcal{F}_{T_1^*})$, which is equivalent to $\mathbb{P}$, and such that each process

$$U_2(t, T_k^*) = \frac{B(t, T_k^*)}{B(t, T_1^*)}, \quad \forall t \in [0, T_k^* \wedge T_1^*],$$

is a $\mathbb{Q}$-(local) martingale. It is important to observe that the process $U_2(\cdot, T_k^*)$ admits the following representation

$$U_2(t, T_k^*) = \frac{U_1(t, T_k^*)}{\Delta_{m-1} L(t, T_1^*) + 1}.$$

Let us state an auxiliary result, which is a straightforward consequence of Itô's formula.

**Lemma 15.1.1.** *Let $K$ and $H$ be the two real-valued processes, such that $dK_t = \alpha_t\, dW_t$ and $dH_t = \beta_t\, dW_t$ for some $\mathbb{F}$-adapted processes $\alpha$ and $\beta$. Assume, in addition, that $H_t > -1$ for every $t$ and denote $Y_t = (1 + H_t)^{-1}$. Then*

$$d(Y_t K_t) = Y_t\big(\alpha_t - Y_t K_t \beta_t\big)\big(dW_t - Y_t \beta_t\, dt\big).$$

It immediately follows from Lemma 15.1.1 that for an arbitrary $k = 1,\dots,m$ we have

$$dU_2(t, T_k^*) = \psi_t^k\left(dW_t - \frac{\Delta_{m-1} L(t, T_1^*)}{1 + \Delta_{m-1} L(t, T_1^*)}\, \nu(t, T_1^*)\, dt\right)$$

for some process $\psi^k$ (the exact knowledge of $\psi^k$ is not essential for our purposes). Therefore, it suffices to find a probability measure under which the process $W_t^{T_1^*}$, $t \in [0, T_1^*]$, given as

$$W_t^{T_1^*} = W_t - \int_0^t \frac{\Delta_{m-1} L(u, T_1^*)}{1 + \Delta_{m-1} L(u, T_1^*)}\, \nu(u, T_1^*)\, du = W_t - \int_0^t \gamma(u, T_1^*)\, du,$$

follows a standard Brownian motion (the definition of $\gamma(\cdot, T_1^*)$ is clear from the context). This goal can be easily achieved using Girsanov's theorem, as we may put, on $(\Omega, \mathcal{F}_{T_1^*})$,

$$\frac{d\mathbb{P}_{T_1^*}}{d\mathbb{P}} = \mathcal{E}_{T_1^*}\left(\int_0^{\cdot} \gamma(u, T_1^*)\, dW_u\right), \quad \mathbb{P}\text{-a.s.}$$

We are in a position to specify the dynamics of the forward LIBOR rate for the date $T_2^*$ under $\mathbb{P}_{T_1^*}$, namely,

$$dL(t, T_2^*) = L(t, T_2^*)\nu(t, T_2^*)\, dW_t^{T_1^*}, \quad \forall\, t \in [0, T_2^*],$$

with the initial condition

$$L(0, T_2^*) = \frac{B(0, T_2^*) - B(0, T_1^*)}{\Delta_{m-2} B(0, T_1^*)}.$$

**General induction step.** Let us now assume that we have found processes $L(\cdot, T_1^*), \dots, L(\cdot, T_n^*)$. This means, in particular, that the forward LIBOR measure $\mathbb{P}_{T_{n-1}^*}$ and the associated Brownian motion $W^{T_{n-1}^*}$ are already specified. Our aim is to find the forward LIBOR measure $\mathbb{P}_{T_n^*}$. It is easy to check that

$$U_{n+1}(t, T_k^*) = \frac{U_n(t, T_k^*)}{\Delta_{m-n} L(t, T_n^*) + 1}.$$

Using again Lemma 15.1.1, we obtain the following relationship, for every $t \in [0, T_n^*]$,

$$W_t^{T_n^*} = W_t^{T_{n-1}^*} - \int_0^t \frac{\Delta_{m-n} L(u, T_n^*)}{1 + \Delta_{m-n} L(u, T_n^*)}\, \nu(u, T_n^*)\, du.$$

The forward LIBOR measure $\mathbb{P}_{T_n^*}$ can thus be easily found using Girsanov's theorem. Finally, we define the process $L(\cdot, T_{n+1}^*)$ as the solution to the SDE

$$dL(t, T_{n+1}^*) = L(t, T_{n+1}^*)\nu(t, T_{n+1}^*)\, dW_t^{T_n^*}$$

with the initial condition

$$L(0, T_{n+1}^*) = \frac{B(0, T_{n+1}^*) - B(0, T_n^*)}{\Delta_{m-n+1} B(0, T_n^*)}.$$

Suppose now that the volatility coefficients $\nu(\cdot, T_i) : [0, T_i] \to \mathbb{R}^d$, where $d \geq 1$ is the dimension of the reference Brownian motion $W$, are deterministic functions. Then for any $i = 1, \ldots, m$ and any date $t \in [0, T_i]$, the random variable $L(t, T_i)$ has a lognormal probability law under the forward martingale measure $\mathbb{P}_{T_{i+1}}$. In this case, the model is referred to as the *lognormal model of forward LIBOR rates*; such a model was first put forward by Miltersen et al. (1997) and Brace et al. (1997).

**Caps and floors.** An *interest rate cap* (or a *ceiling rate agreement*) is a contractual arrangement where the grantor (seller) has an obligation to pay cash to the holder (buyer) if a particular interest rate exceeds a mutually agreed level at some future date or dates. Likewise, in an *interest rate floor*, the grantor has an obligation to pay cash to the holder if the interest rate is below a pre-assigned level. When cash is paid to the holder, the holder's net position is equivalent to borrowing (or depositing) at a rate fixed at that agreed level. This assumes that the holder of a cap (or floor) agreement also holds an underlying asset (such as a deposit) or an underlying liability (such as a loan). Finally, the holder is not affected by the agreement if the interest rate is ultimately more favorable to him than the agreed level. This feature of a cap (or floor) agreement makes it similar to an option. Specifically, a *forward start cap* (a *forward start floor*, resp.) is a strip of *caplets* (*floorlets*, resp.), each of which is a call (put, resp.) option on a forward rate, respectively.

Let us denote by $\kappa$ the cap strike rate. In a forward cap or floor, which starts at time $T_0$, and is settled in arrears at dates $T_{i+1}$, $i = 0, \ldots, m-1$, the cash flows at times $T_{i+1}$ are, per dollar of notional principal, $(L(T_i) - \kappa)^+ \Delta_i$ and $(\kappa - L(T_i))^+ \Delta_i$, respectively. As usual, the rate $L(T_i) = L(T_i, T_i)$ is determined at the reset date $T_i$, and it satisfies

$$1 + \Delta_i L(T_i) = B(T_i, T_{i+1})^{-1}. \tag{15.4}$$

Let the nominal value of a cap be 1. The price $\mathbf{FC}_t$ of a *forward cap* at time $t \leq T_0$ can be found from the standard risk-neutral valuation formula:

$$\mathbf{FC}_t = \sum_{i=0}^{m-1} \mathbb{E}_{\mathbb{P}^*}\left( \frac{B_t}{B_{T_{i+1}}} (L(T_i) - \kappa)^+ \Delta_i \,\Big|\, \mathcal{F}_t \right),$$

where $\mathbb{P}^*$ represents the *spot martingale measure*. The last formula is thus useful only if this probability is readily available in a given arbitrage-free model of the term structure of interest rates.

In our case, it is more convenient to use the following equivalent representation for the price $\mathbf{FC}_t$:

$$\mathbf{FC}_t = \sum_{i=0}^{m-1} B(t, T_{i+1}) \, \mathbb{E}_{\mathbb{P}_{T_{i+1}}} \left( (L(T_i) - \kappa)^+ \Delta_i \, \middle| \, \mathcal{F}_t \right). \tag{15.5}$$

Using the last formula, we shall now check that a caplet (that is, one leg of a cap) may also be seen as a put option with strike price 1 (per dollar of notional principal), which expires at the caplet start day and is written on a zero-coupon bond with face value $1 + \kappa \Delta_i$, which matures at the caplet end date. Indeed, since the cash flow of the $i^{\text{th}}$ caplet at time $T_i$ manifestly represents an $\mathcal{F}_{T_{i-1}}$-measurable random variable, we also have

$$\mathbf{FC}_t = \sum_{i=0}^{m-1} B(t, T_i) \, \mathbb{E}_{\mathbb{P}_{T_i}} \left( B(T_i, T_{i+1})(L(T_i) - \kappa)^+ \Delta_i \, \middle| \, \mathcal{F}_t \right).$$

Consequently, using (15.4), we obtain the following equality, for $t \in [0, T]$,

$$\mathbf{FC}_t = \sum_{i=0}^{m-1} B(t, T_i) \, \mathbb{E}_{\mathbb{P}_{T_i}} \left( \left( 1 - \tilde{\Delta}_i B(T_i, T_{i+1}) \right)^+ \, \middle| \, \mathcal{F}_t \right),$$

where $\tilde{\Delta}_i = 1 + \kappa \Delta_i$. It is thus apparent that an interest rate cap is essentially equivalent to a portfolio of European put options on zero-coupon bonds. The equivalence of a caplet and a European put option on a zero-coupon bond can also be explained in a more intuitive way: a caplet is exercised at time $T_i$ if and only if $L(T_i) - \kappa > 0$ or, equivalently, if

$$B(T_i, T_{i+1})^{-1} = 1 + L(T_i)(T_{i+1} - T_i) > 1 + \kappa \Delta_i = \tilde{\Delta}_i.$$

The last inequality holds whenever $\tilde{\Delta}_i B(T_i, T_{i+1}) < 1$. This shows that both of the considered options are exercised in the same circumstances. If exercised, the caplet pays $(L(T_i) - \kappa)\Delta_i$ at time $T_{i+1}$ or, equivalently,

$$\Delta_i B(T_i, T_{i+1})(L(T_i) - \kappa) = 1 - \tilde{\Delta}_i B(T_i, T_{i+1}) = \tilde{\Delta}_i \left( \tilde{\Delta}_i^{-1} - B(T_i, T_{i+1}) \right)$$

at time $T_i$. This reasoning confirms that the caplet, with strike level $\kappa$ and nominal value 1, is equivalent to a put option with strike price $\tilde{\Delta}_i^{-1}$ and nominal value $\tilde{\Delta}_i$ written on the unit zero-coupon bond with maturity $T_{i+1}$.

The analysis of a floor contract can be done along similar lines. By definition, the $i^{\text{th}}$ floorlet pays $(\kappa - L(T_i))^+ \Delta_i$ at time $T_{i+1}$ for any $i = 0, \ldots, m-1$. Thus, the price $\mathbf{FF}_t$ of a floor equals

$$\mathbf{FF}_t = \sum_{i=0}^{m-1} B(t, T_{i+1}) \, \mathbb{E}_{\mathbb{P}_{T_{i+1}}} \left( (\kappa - L(T_i))^+ \Delta_i \, \middle| \, \mathcal{F}_t \right), \tag{15.6}$$

but also

$$\mathbf{FF}_t = \sum_{i=0}^{m-1} B(t,T_i)\, \mathbb{E}_{\mathbb{P}_{T_i}}\Big(\big(\tilde{\Delta}_i B(T_i,T_{i+1}) - 1\big)^+ \,\big|\, \mathcal{F}_t\Big),$$

where, as before, $\tilde{\Delta}_i = 1 + \kappa\Delta_i$. Combining (15.5) with (15.6), we obtain the following *cap-floor parity* relationship:

$$\mathbf{FC}_t - \mathbf{FF}_t = \sum_{i=0}^{m-1} \big(B(t,T_i) - \tilde{\Delta}_i B(t,T_{i+1})\big).$$

**Valuation of caps.** As expected, the valuation of caps within the lognormal model of forward LIBOR rates is extremely simple. The dynamics of the forward LIBOR rate $L(t,T_i)$ under the forward martingale measure $\mathbb{P}_{T_{i+1}}$ are

$$dL(t,T_i) = L(t,T_i)\nu(t,T_i)\, dW_t^{T_{i+1}},$$

where $W^{T_{i+1}}$ follows a standard Brownian motion under the forward measure $\mathbb{P}_{T_{i+1}}$, and $\nu(\cdot,T_i) : [0,T_i] \to \mathbb{R}^d$ is a deterministic function. Consequently, for every $t \in [0,T_i]$ we have

$$L(t,T_i) = L(0,T_i)\mathcal{E}_t\Big(\int_0^{\cdot} \nu(u,T_i)\, dW_u^{T_{i+1}}\Big). \tag{15.7}$$

The following proposition is a straightforward consequence of formula (15.5), combined with expression (15.7). As usual, $N$ is the standard Gaussian probability distribution function, and $|\cdot|$ stands for the Euclidean norm in $\mathbb{R}^d$. The proof of Proposition 15.1.1 is left to the reader (see also Sect. 16.3 in Musiela and Rutkowski (1997a) or Rutkowski (1999, 2001)).

**Proposition 15.1.1.** *Consider an interest rate cap with strike level $\kappa$, settled in arrears at times $T_i$, $i = 1,\ldots,m$. Assuming the lognormal model of LIBOR rates, the price of a cap at time $t \in [0,T]$ equals*

$$\mathbf{FC}_t = \sum_{i=0}^{m-1} \Delta_i B(t,T_{i+1})\Big(L(t,T_i)N\big(d_1(t,T_i)\big) - \kappa N\big(d_2(t,T_i)\big)\Big),$$

*where, for every $i = 0,\ldots,m-1$,*

$$d_{1,2}(t,T_i) = \frac{\ln(L(t,T_i)/\kappa) \pm \frac{1}{2}v^2(t,T_i)}{v(t,T_i)},$$

*and*

$$v^2(t,T_i) = \int_t^{T_i} |\nu(u,T_i)|^2\, du.$$

Since a cap is basically a portfolio of caplets, a similar valuation formula can be derived for other caps with settlement dates in the set $\{T_1,\ldots,T_m\}$.

*Remarks.* In the framework presented in this section, the cap valuation formula given above was first established through the PDE approach by Miltersen et al. (1997). It was subsequently rederived through a probabilistic approach by Goldys (1997) and Rady (1997). Finally, the same result was established by means of the forward measure approach in Brace et al. (1997).

### 15.1.2 Modeling of Forward Swap Rates

The arbitrage-free methodology for the direct modeling of forward swap rates presented in this section, was put forward by Jamshidian (1996, 1997); we follow here the approach developed in Rutkowski (1999). Assume that the tenor structure $0 < T_0 < T_1 < \cdots < T_m = T^*$ is given. For any fixed $i$, we consider a fixed-for-floating forward (payer) swap which starts at time $T_i$ and has $m - i$ accrual periods, whose consecutive lengths are $\Delta_i, \ldots, \Delta_{m-1}$, where $\Delta_j = T_{j+1} - T_j$, $j = i, \ldots, m - 1$. The fixed rate paid at each of the reset dates $T_l$ for $l = i + 1, \ldots, m$ equals $\kappa$, and the corresponding floating rate is the LIBOR rate $L(T_{l-1})$; the net payment to the long party at time $T_l$ thus equals $(L(T_{l-1}) - \kappa)\Delta_{l-1}$ per unit of the notional principal. We shall assume that the notional principal equals 1. It is not difficult to check, using no-arbitrage arguments, that the value $\mathbf{FS}_t(\kappa)$ of the $i^{\text{th}}$ swap equals, for $t \in [0, T_i]$

$$\mathbf{FS}_t^i(\kappa) = B(t, T_i) - \sum_{l=i+1}^m c_l B(t, T_l),$$

where $c_l = \kappa \Delta_{l-1}$ for $l = i+1, \ldots, m-1$, and $c_m = 1 + \kappa \Delta_{m-1}$. The *forward swap rate*, $\kappa(t, T_i; m - i)$, that is, that value of a fixed rate $\kappa$ for which the $i^{\text{th}}$ swap is valueless at time $t$, equals, for $t \in [0, T_i]$ and $i = 0, \ldots, m - 1$,

$$\kappa(t, T_i; m - i) = \frac{B(t, T_i) - B(t, T_m)}{\Delta_i B(t, T_{i+1}) + \cdots + \Delta_{m-1} B(t, T_m)}. \tag{15.8}$$

We shall consider a family of forward swap rates $\hat{\kappa}(t, T_i) := \kappa(t, T_i; m - i)$ for $i = 0, \ldots, m - 1$. Let us stress that the underlying swaps differ in length; however, they all have a common expiration date $T^* = T_m$.

Suppose first that we are given a family of bond prices $B(t, T_i)$, $i = 1, \ldots, m$, defined on a filtered probability space $(\Omega, \mathbb{F}, \mathbb{P})$ equipped with a standard Brownian motion $W$. For any $n = 1, \ldots, m - 1$, we introduce the *coupon process* $\hat{G}(n)$ by setting, for $t \in [0, T_{m-n+1}]$,

$$\hat{G}_t(n) = \sum_{i=m-n+1}^m \Delta_{i-1} B(t, T_i) = \sum_{k=1}^n \Delta_{m-k} B(t, T_{k-1}^*). \tag{15.9}$$

A *forward swap measure* is that probability measure equivalent to $\mathbb{P}$, which corresponds to the choice of the coupon process as a numéraire asset. Formally, we have the following definition.

**Definition 15.1.2.** For $i = 1, \ldots, m$, a probability measure $\hat{\mathbb{P}}_{T_i}$ on $(\Omega, \mathcal{F}_{T_i})$, equivalent to $\mathbb{P}$, is said to be the *forward swap measure* for the date $T_i$ if, for every $k = 0, \ldots, m$, the relative bond price $Z_{m-i+1}(t, T_k)$, $t \in [0, T_k \wedge T_i]$, given as:

$$Z_{m-i+1}(t, T_k) := \frac{B(t, T_k)}{\hat{G}_t(m - i + 1)} = \frac{B(t, T_k)}{\Delta_{i-1} B(t, T_i) + \cdots + \Delta_{m-1} B(t, T_m)}$$

follows an $\mathbb{F}$-(local) martingale under $\hat{\mathbb{P}}_{T_i}$.

Put another way, for a fixed $n = 1, \ldots, m$, and any $k = 0, \ldots, m$, the relative bond price $Z_n(t, T_k^*)$, which equals, for $t \in [0, T_k^* \wedge T_{n-1}^*]$,

$$Z_n(t, T_k^*) = \frac{B(t, T_k^*)}{\hat{G}_t(n)} = \frac{B(t, T_k^*)}{\Delta_{m-n} B(t, T_{n-1}^*) + \cdots + \Delta_{m-1} B(t, T^*)}$$

follows an $\mathbb{F}$-(local) martingale under the forward swap measure $\hat{\mathbb{P}}_{T_{n-1}^*}$. The crucial observation is that the forward swap rate for the date $T_n^*$, which equals, for $t \in [0, T_n^*]$,

$$\hat{\kappa}(t, T_n^*) = \frac{B(t, T_n^*) - B(t, T^*)}{\Delta_{m-n} B(t, T_{n-1}^*) + \cdots + \Delta_{m-1} B(t, T^*)},$$

can also be represented as follows: $\hat{\kappa}(t, T_n^*) = Z_n(t, T_n^*) - Z_n(t, T^*)$. Therefore, the process $\hat{\kappa}(\cdot, T_n^*)$ also follows an $\mathbb{F}$-(local) martingale under the forward swap measure $\hat{\mathbb{P}}_{T_{n-1}^*}$. Moreover, since obviously $\hat{G}_t(1) = \Delta_{m-1} B(t, T^*)$, we conclude that the probability measure $\hat{\mathbb{P}}_{T^*}$ can be chosen to coincide with the forward martingale measure $\mathbb{P}_{T^*}$.

*Remarks.* Since obviously $\hat{G}_t(1) = \Delta_{m-1} B(t, T^*)$, the forward swap measure $\hat{\mathbb{P}}_{T^*}$ can also be interpreted as the forward martingale measure $\mathbb{P}_{T^*}$. This remark does not apply to probabilities $\hat{\mathbb{P}}_{T_n^*}$ for $n \geq 1$, though.

We wish to construct a family of forward swap rates in such a way that

$$d\hat{\kappa}(t, T_i) = \hat{\kappa}(t, T_i) \hat{\nu}(t, T_i) \, d\hat{W}_t^{T_{i+1}} \tag{15.10}$$

for any $i = 0, \ldots, m-1$, where each process $\hat{W}^{T_{i+1}}$ follows a standard Brownian motion under the corresponding forward swap measure $\hat{\mathbb{P}}_{T_{i+1}}$. The model should also be consistent with the initial term structure of interest rates, meaning that

$$\hat{\kappa}(0, T_i) = \frac{B(0, T_i) - B(0, T^*)}{\Delta_i B(0, T_{i+1}) + \cdots + \Delta_{m-1} B(0, T_m)}. \tag{15.11}$$

Alternatively, one may assume that the initial values of forward swap rates $\hat{\kappa}(0, T_i)$, $i = 0, \ldots, m - 1$ are known (e.g., they are taken from the real-life swap market). We make the following standing assumptions, which describe the model's inputs.

**Assumptions (SR).** We are given a family of bounded, $\mathbb{F}$-adapted processes $\hat{\nu}(\cdot, T_i)$, $i = 0, \ldots, m - 1$; each process $\hat{\nu}(t, T_i)$, $t \in [0, T_i]$, is assumed to represent the volatility of the forward swap rate $\hat{\kappa}(\cdot, T_i)$. In addition, we are given an initial term structure of interest rates, specified by a family $B(0, T_i)$, $i = 0, \ldots, m$, of bond prices. We assume that $B(0, T_i) > B(0, T_{i+1})$ for every $i = 0, \ldots, m - 1$.

Similarly as in the case of a model of forward LIBOR rates, we shall proceed by backward induction.

**First step.** We find it convenient to postulate that the reference probability measure $\mathbb{P}$ is the forward swap measure $\hat{\mathbb{P}}_{T^*}$ for the date $T^*$, and that the process $W = W^{T^*}$ is the corresponding Brownian motion. In the first step, we specify the forward swap rate for the date $T_1^*$ by postulating that the process $\hat{\kappa}(\cdot, T_1^*)$ solves the SDE

$$d\hat{\kappa}(t, T_1^*) = \hat{\kappa}(t, T_1^*)\hat{\nu}(t, T_1^*)\, d\hat{W}_t^{T^*}, \quad \forall t \in [0, T_1^*],$$

where $\hat{W}^{T^*} = W$, subject to the initial condition:

$$\hat{\kappa}(0, T_1^*) = \frac{B(0, T_1^*) - B(0, T^*)}{\Delta_{m-1}B(0, T^*)}.$$

**Second step.** To be able to define the process $\hat{\kappa}(\cdot, T_2^*)$, we first need to introduce a forward swap measure $\hat{\mathbb{P}}_{T_1^*}$ and an associated Brownian motion $\hat{W}^{T_1^*}$. To this end, it is useful to observe that each process $Z_1(\cdot, T_k^*) = B(\cdot, T_k^*)/\Delta_{m-1}B(\cdot, T^*)$ follows a strictly positive local martingale under $\hat{\mathbb{P}}_{T^*} = \mathbb{P}_{T^*}$. Specifically, we have

$$dZ_1(t, T_k^*) = Z_1(t, T_k^*)\gamma_1(t, T_k^*)\, d\hat{W}_t^{T^*}$$

for some adapted process $\gamma_1(\cdot, T_k^*)$. According to the definition of a forward swap measure, we postulate that for every $k$ the process

$$Z_2(t, T_k^*) = \frac{B(t, T_k^*)}{\Delta_{m-2}B(t, T_1^*) + \Delta_{m-1}B(t, T^*)} = \frac{Z_1(t, T_k^*)}{1 + \Delta_{m-2}Z_1(t, T_1^*)}$$

follows a local martingale under $\hat{\mathbb{P}}_{T_1^*}$. By applying Lemma 15.1.1 to processes $K_t = Z_1(t, T_k^*)$ and $H_t = \Delta_{m-2}Z_1(t, T_1^*)$, we conclude that for this property to hold, it suffices to assume that the process $\hat{W}^{T_1^*}$, which equals, for $t \in [0, T_1^*]$,

$$\hat{W}_t^{T_1^*} = \hat{W}_t^{T^*} - \int_0^t \frac{\Delta_{m-2}Z_1(u, T_1^*)}{1 + \Delta_{m-2}Z_1(u, T_1^*)}\,\gamma_1(u, T_1^*)\, du,$$

follows a Brownian motion under $\hat{\mathbb{P}}_{T_1^*}$ (the probability measure $\hat{\mathbb{P}}_{T_1^*}$ is yet unspecified, but will be soon obtained through Girsanov's theorem). Note that

$$Z_1(t, T_1^*) = \frac{B(t, T_1^*)}{\Delta_{m-1}B(t, T^*)} = \hat{\kappa}(t, T_1^*) + Z_1(t, T^*) = \hat{\kappa}(t, T_1^*) + \Delta_{m-1}^{-1}.$$

Differentiating both sides of the last equality, we get

$$Z_1(t, T_1^*)\gamma_1(t, T_1^*) = \hat{\kappa}(t, T_1^*)\hat{\nu}(t, T_1^*).$$

Consequently, $\hat{W}^{T_1^*}$ is explicitly given by the formula, for $t \in [0, T_1^*]$,

$$\hat{W}_t^{T_1^*} = \hat{W}_t^{T^*} - \int_0^t \frac{\Delta_{m-2}\hat{\kappa}(u, T_1^*)}{1 + \Delta_{m-2}\Delta_{m-1}^{-1} + \Delta_{m-2}\hat{\kappa}(u, T_1^*)}\,\hat{\nu}(u, T_1^*)\, du.$$

We are in a position to define, using the classic version of Girsanov's theorem, the associated forward swap measure $\hat{\mathbb{P}}_{T_1^*}$. Subsequently, we introduce the process $\hat{\kappa}(\cdot, T_2^*)$, by postulating that it is governed by the following SDE:

$$d\hat{\kappa}(t, T_2^*) = \hat{\kappa}(t, T_2^*)\hat{\nu}(t, T_2^*)\, d\hat{W}_t^{T_1^*}, \quad \forall t \in [0, T_2^*],$$

with the initial condition

$$\hat{\kappa}(0, T_2^*) = \frac{B(0, T_2^*) - B(0, T^*)}{\Delta_{m-2}B(0, T_1^*) + \Delta_{m-1}B(0, T^*)}.$$

**Third step.** For the reader's convenience, let us briefly sketch one more inductive step, in which we search for $\hat{\kappa}(t, T_3^*)$. We now consider processes

$$Z_3(t, T_k^*) = \frac{B(t, T_k^*)}{\Delta_{m-3}B(t, T_2^*) + \Delta_{m-2}B(t, T_1^*) + \Delta_{m-1}B(t, T^*)}.$$

Observe that

$$Z_3(t, T_k^*) = \frac{Z_2(t, T_k^*)}{1 + \Delta_{m-3}Z_2(t, T_2^*)},$$

and thus for every $t \in [0, T_2^*]$

$$\hat{W}_t^{T_2^*} = \hat{W}_t^{T_1^*} - \int_0^t \frac{\Delta_{m-3}Z_2(u, T_2^*)}{1 + \Delta_{m-3}Z_2(u, T_2^*)}\, \gamma_2(u, T_2^*)\, du.$$

It is useful to note that

$$Z_2(t, T_2^*) = \frac{B(t, T_2^*)}{\Delta_{m-2}B(t, T_1^*) + \Delta_{m-1}B(t, T^*)} = \hat{\kappa}(t, T_2^*) + Z_2(t, T^*),$$

where in turn

$$Z_2(t, T^*) = \frac{Z_1(t, T^*)}{1 + \Delta_{m-2}Z_1(t, T^*) + \Delta_{m-2}\hat{\kappa}(t, T_1^*)}$$

and the process $Z_1(\cdot, T^*)$ is already known from the previous step (clearly, $Z_1(\cdot, T^*) = 1/\Delta_{m-1}$). Differentiating the last equality, we may thus find the volatility of the process $Z_2(\cdot, T^*)$ and, consequently, we may define $\hat{\mathbb{P}}_{T_2^*}$.

**General induction step.** Suppose that we have found forward swap rates $\hat{\kappa}(\cdot, T_1^*), \ldots, \hat{\kappa}(\cdot, T_n^*)$, the forward swap measure $\hat{\mathbb{P}}_{T_{n-1}^*}$ and the associated Brownian motion $\hat{W}^{T_{n-1}^*}$. Our aim is to determine the forward swap measure $\hat{\mathbb{P}}_{T_n^*}$, the associated Brownian motion $\hat{W}^{T_n^*}$, and the forward swap rate $\hat{\kappa}(\cdot, T_{n+1}^*)$. To this end, we postulate that the process $Z_{n+1}(t, T_k^*)$, defined as

$$Z_{n+1}(t, T_k^*) = \frac{B(t, T_k^*)}{\hat{G}_t(n+1)} = \frac{B(t, T_k^*)}{\Delta_{m-n-1}B(t, T_n^*) + \cdots + \Delta_{m-1}B(t, T^*)}$$

$$= \frac{Z_n(t, T_k^*)}{1 + \Delta_{m-n-1}Z_n(t, T_n^*)},$$

follows an $\mathbb{F}$-(local) martingale under $\hat{\mathbb{P}}_{T_n^*}$ for every $k = 0, \ldots, m$.

In view of Lemma 15.1.1, applied to processes $K_t = Z_n(t, T_k^*)$ and $H_t = Z_n(t, T_n^*)$, it is clear that we may set

$$\hat{W}_t^{T_n^*} = \hat{W}_t^{T^*} - \int_0^t \frac{\Delta_{m-n-1} Z_n(u, T_n^*)}{1 + \Delta_{m-n-1} Z_n(u, T_n^*)} \gamma_n(u, T_n^*)\, du, \qquad (15.12)$$

for $t \in [0, T_n^*]$. Therefore it is sufficient to analyze the process

$$Z_n(t, T_n^*) = \frac{B(t, T_n^*)}{\Delta_{m-n} B(t, T_{n-1}^*) + \cdots + \Delta_{m-1} B(t, T^*)} = \hat{\kappa}(t, T_n^*) + Z_n(t, T^*).$$

To conclude, it is enough to notice that

$$Z_n(t, T^*) = \frac{Z_{n-1}(t, T^*)}{1 + \Delta_{m-n} Z_{n-1}(t, T^*) + \Delta_{m-n} \hat{\kappa}(t, T_{n-1}^*)}.$$

Indeed, from the preceding step, we know that the process $Z_{n-1}(\cdot, T^*)$ is a (rational) function of forward swap rates $\hat{\kappa}(\cdot, T_1^*), \ldots, \hat{\kappa}(\cdot, T_{n-1}^*)$. Consequently, the process under the integral sign on the right-hand side of (15.12) can be expressed using the terms $\hat{\kappa}(\cdot, T_1^*), \ldots, \hat{\kappa}(\cdot, T_{n-1}^*)$ and their volatilities (since the explicit formula is rather lengthy, it is not reported here). Having found the process $\hat{W}^{T_n^*}$ and probability measure $\hat{\mathbb{P}}_{T_n^*}$, we introduce the forward swap rate $\hat{\kappa}(\cdot, T_{n+1}^*)$ through (15.10)–(15.11), and so forth.

If all volatilities of forward swap rate are deterministic, the model is termed the *lognormal model of forward swap rates*. The main reason for developing this model lies in the fact that the valuation result for swaptions coincides in this framework with the conventional market formula – that is, the so-called Black swaptions formula.

**Payer and receiver swaptions.** The owner of a *payer* (*receiver*, respectively) *swaption* with strike rate $\kappa$, maturing at time $T = T_0$, has the right to enter at time $T$ the underlying forward payer (receiver, respectively) swap settled in arrears. Let us focus on a particular example of an $m$-period payer swaption; namely, we assume that the underlying fixed-for-floating interest rate swap starts at $T_0$, has $m$ accrual periods and the fixed interest rate $\kappa$. By convention, the notional principal of the underlying swap, and thus also the notional principal of the swaption, equals 1.

Let $\mathbf{FS}_T(\kappa)$ stand for the value of the underlying swap at time $T$. The price $\mathbf{PS}_t$ of the payer swaption at time $t \in [0, T]$ thus equals

$$\mathbf{PS}_t = B(t, T)\, \mathbb{E}_{\mathbb{P}_T}\left( (\mathbf{FS}_T(\kappa))^+ \,\big|\, \mathcal{F}_t \right).$$

Let us examine few equivalent representation of the value $\mathbf{FS}_T(\kappa)$, which give rise to alternative interpretations of a payer swaption. Let us first observe that

$$\mathbf{FS}_T(\kappa) = \sum_{i=1}^n B(T, T_i)\, \mathbb{E}_{\mathbb{P}_{T_i}}\left( (L(T_{j-1}) - \kappa)\Delta_{i-1} \,\big|\, \mathcal{F}_T \right).$$

On the other hand, since obviously $\mathbf{FS}_t(\kappa(t,T,m)) = 0$, we also have

$$\mathbf{FS}_t(\kappa) = \mathbf{FS}_t(\kappa) - \mathbf{FS}_t(\kappa(t,T,m)) = \sum_{i=1}^{m} \Delta_{i-1} B(t,T_i)(\kappa(t,T,m) - \kappa),$$

so that

$$\mathbf{FS}_T(\kappa) = \sum_{i=1}^{m} \Delta_{i-1} B(T,T_i)(\kappa(T,T,m) - \kappa),$$

and, finally,

$$\left(\mathbf{FS}_T(\kappa)\right)^+ = \sum_{i=1}^{m} \Delta_{i-1} B(T,T_i)(\kappa(T,T,m) - \kappa)^+,$$

An $m$-period payer swaption is thus formally represented by a finite sequence of cash flows $(\kappa(T,T,m) - \kappa)^+ \Delta_{i-1}$ which are received at settlement dates $T_1, \ldots, T_m$, but whose value is already known at the expiry date $T$ of the swaption. Put another way, a payer swaption can be seen as a specific call option on a forward swap rate, with a fixed strike level $\kappa$. The exercise date of the option is $T$, but the payoff takes place at each date $T_1, \ldots, T_m$. It is also not difficult to check that

$$\left(\mathbf{FS}_T(\kappa)\right)^+ = \left(1 - \sum_{j=1}^{m} c_j B(T,T_j)\right)^+,$$

so that the payer swaption may also be seen as a standard put option, with an exercise date $T$ and a strike price 1, written on a coupon-bearing bond with the coupon rate $\kappa$. To summarize, a payer swaption can be seen as:
- a European call option with an exercise date $T$ and a strike level 0, written on the value of the underlying swap,
- a specific call option on a forward swap rate, with a fixed strike $\kappa$, an exercise date $T$, and a sequence of payoffs at the dates $T_1, \ldots, T_m$,
- a European put option with an exercise date $T$ and a strike price 1, written on a coupon-bearing bond with the coupon rate $\kappa$.

Similar observations can be done for a receiver swaption. In particular, an $m$-period receiver swaption can also be viewed as a sequence of put options on a swap rate which are not allowed to be exercised separately. At time $T$ the long party receives the value of a sequence of cash flows, discounted from time $T_i$, $j = i, \ldots, m$, to the date $T$, defined by $\Delta_{i-1}(\kappa - \kappa(T,T,m))^+$. A receiver swaption may also be seen as a call option, with strike price 1 and expiry date $T$, written on a coupon bond with coupon rate equal to the strike rate $\kappa$ of the underlying forward swap. Let us finally mention the put-call parity relationship for swaptions:

$$\textit{Payer Swaption}(t) \quad - \quad \textit{Receiver Swaption}(t) \quad = \quad \textit{Forward Swap}(t)$$

provided that both swaptions expire at the same date $T$ (and, of course, have the same contractual features).

**Valuation of swaptions.** For a fixed, but otherwise arbitrary, initial date $T_i$, $i = 0, \ldots, m-1$, we consider a payer swaption with expiry date $T_i$, written on a forward payer swap settled in arrears. The underlying swap starts at date $T_i$, has the fixed rate $\kappa$ and $m - i$ accrual periods. For the sake of brevity, we shall henceforth refer to such a contract as the $i^{\text{th}}$ swaption.

The $i^{\text{th}}$ swaption can be seen as a security that pays to its owner the amount $(\kappa(T_i, T_i, m - i) - \kappa)^+ \Delta_{k-1}$ at each settlement date $T_k$, where $k = i + 1, \ldots, m$ (let us recall here that, by convention, the notional principal of the swaption equals 1). Equivalently, the $i^{\text{th}}$ swaption can be represented as a single cash flow $\hat{Y}_i$, where

$$\hat{Y}_i := \mathbf{PS}\,_{T_i}^{i} = \big(\mathbf{FS}\,_{T_i}^{i}(\kappa)\big)^+ = \sum_{k=i+1}^{m} \Delta_{k-1} B(T_i, T_k)\big(\hat{\kappa}(T_i, T_i) - \kappa\big)^+,$$

which settles at the maturity date $T_i$ of the $i^{\text{th}}$ swaption. It is essential to observe that the payoff $\hat{Y}_i$ admits a multiplicative decomposition in terms of the coupon process $\hat{G}(m - i)$ given by (15.9), namely,

$$\hat{Y}_i = \hat{G}_{T_i}(m - i)\big(\hat{\kappa}(T_i, T_i) - \kappa\big)^+.$$

The model of forward swap rates specifies the dynamics of the process $\hat{\kappa}(\cdot, T_i)$ by means of the following SDE:

$$d\hat{\kappa}(t, T_i) = \hat{\kappa}(t, T_i)\hat{\nu}(t, T_i)\, d\hat{W}_t^{T_{i+1}},$$

where $\hat{W}^{T_{i+1}}$ follows a standard Brownian motion under the corresponding forward swap measure $\hat{\mathbb{P}}_{T_{i+1}}$. The definition of the forward swap measure $\hat{\mathbb{P}}_{T_{i+1}}$ implies that the relative price $B(t, T_k)/\hat{G}_t(m - i)$ follows a local martingale under $\hat{\mathbb{P}}_{T_{i+1}}$ for any $k = 0, \ldots, m$. Furthermore, from the general considerations concerning the role of the choice of a numéraire asset in arbitrage pricing,[1] it is easy to see that the arbitrage price $\pi_t(X)$ of any attainable contingent claim $X$ of the form $X = g(B(T_i, T_{i+1}), \ldots, B(T_i, T_m))$ equals, for $t \in [0, T_j]$,

$$\pi_t(X) = \hat{G}_t(m - i)\, \mathbb{E}_{\hat{\mathbb{P}}_{T_{i+1}}}\big(\hat{G}_{T_i}^{-1}(m - i)Y \mid \mathcal{F}_t\big),$$

provided that $X$ settles at time $T_i$. Applying the last formula to the payoff $\hat{Y}_i$, we obtain the following representation for the arbitrage price $\mathbf{PS}\,_t^{i}$ of the $i^{\text{th}}$ swaption at time $t \in [0, T_i]$

$$\mathbf{PS}\,_t^{i} = \pi_t(\hat{Y}_i) = \hat{G}_t(m - i)\, \mathbb{E}_{\hat{\mathbb{P}}_{T_{i+1}}}\big((\hat{\kappa}(T_i, T_i) - \kappa)^+ \mid \mathcal{F}_t\big).$$

We assume from now on that $\hat{\nu}(\cdot, T_i)$ is a bounded deterministic function on $[0, T_i]$. In other words, we place ourselves within the framework of the lognormal model of forward swap rates. The proof of the next result, first established by Jamshidian (1996, 1997), presents no difficulties.

---

[1] See, for instance, Geman et al. (1995) or Musiela and Rutkowski (1997a).

**Proposition 15.1.2.** *For any $i = 1, \ldots, m - 1$, the arbitrage price of the $i^{\text{th}}$ swaption equals, for every $t \in [0, T_i]$,*

$$\mathbf{PS}_t^i = \sum_{k=i+1}^{m} \Delta_{k-1} B(t, T_k) \Big( \hat{\kappa}(t, T_i) N\big(\hat{d}_1(t, T_i)\big) - \kappa N\big(\hat{d}_2(t, T_i)\big) \Big),$$

*where $N$ denotes the standard Gaussian cumulative distribution function and*

$$\hat{d}_{1,2}(t, T_i) = \frac{\ln(\hat{\kappa}(t, T_i)/\kappa) \pm \frac{1}{2} \hat{v}^2(t, T_i)}{\hat{v}(t, T_i)},$$

*where in turn $\hat{v}^2(t, T_i) = \int_t^{T_i} |\hat{\nu}(u, T_i)|^2 \, du$.*

For the valuation results established in this section to be fully satisfactory, it is necessary to provide suitable replication-based arguments that would support the closed-form expressions derived in Propositions 15.1.1 and 15.1.2. It appears that the dynamic hedging of caps (swaptions, resp.) in the lognormal model of forward LIBOR (swap, resp.) rates is rather straightforward. For more details in this regard, see, for example, Musiela and Rutkowski (1997a) or Rutkowski (2001).

## 15.2 Modeling of Defaultable Forward LIBOR Rates

In this section, we shall present some ideas from recent papers by Lotz and Schlögl (2000) and Schönbucher (2000b). In both cases, we postulate a lognormal model of default-free forward LIBOR rates. As usual, the reference (Brownian) filtration for these models is denoted by $\mathbb{F}$.

### 15.2.1 Lotz and Schlögl Approach

To the best of our knowledge, the first attempt to extend the market model of LIBOR rates to the case of defaultable contracts was done by Lotz and Schlögl (2000). The convention adopted in this work was already mentioned in Sect. 14.1.4; we shall now describe it in detail. To this end, let us recall that in Sect. 14.1.2 we have introduced a defaultable IRA with one-sided default risk. A contract initiated at time $T$ for the accrual period $[T, T + \Delta]$, with the fixed rate $\kappa$, and the constant recovery rate $\delta$, was specified as follows:
- at time $T$, the receiver of the fixed rate pays 1 (the nominal value of the contract),
- if the payer does not default in the time period $(T, T + \Delta]$, the receiver is paid at the settlement date $T + \Delta$ the full due amount: $1 + \kappa\Delta$,
- if the payer's default occurs in $(T, T + \Delta]$, the receiver collects at time $T + \Delta$ the reduced amount: $\delta(1 + \kappa\Delta)$.

From the payer's perspective, the value of the contract at the inception date $T$ equals, on the set $\{\tau > T\}$,

$$\mathbf{IRA}^\delta(T; \kappa) := 1 - (1 + \kappa\Delta)D^\delta(T, T + \Delta), \tag{15.13}$$

where $\tau$ is the default time of the payer, and $D^\delta(T, T + \Delta)$ is the price of a zero-coupon bond issued by the payer, with fractional recovery of Treasury value and constant recovery rate $\delta$.

Now consider a non-defaultable contract initiated at time $t < T$, with the payoff $Y := \mathbf{IRA}^\delta(T; \kappa)$ at time $T$. For the sake of analytical tractability, Lotz and Schlögl (2000) also postulate that the $\mathbb{F}$-intensity $\gamma$ of the default time $\tau$ is deterministic, so that the price of a defaultable bond at time $T$ equals

$$D^\delta(T, T + \Delta) = B(T, T + \Delta)\Big(\delta + (1 - \delta)e^{\Gamma(T) - \Gamma(T + \Delta)}\Big),$$

where the hazard function $\Gamma$ equals $\Gamma(t) = \int_0^t \gamma(u)\, du$. Under this assumption, the arbitrage price $U_t(\kappa)$ of the payoff $Y = \mathbf{IRA}^\delta(T; \kappa)$ at time $t$ is easy to find. Indeed, we have

$$U_t(\kappa) = B(t, T) - B(t, T + \Delta)(1 + \kappa\Delta)\Big(\delta + (1 - \delta)e^{\Gamma(T) - \Gamma(T + \Delta)}\Big). \tag{15.14}$$

Let us stress that the last equality, as well as the subsequent results, are valid only under the restrictive assumption that the $\mathbb{F}$-intensity $\gamma$ of the time of default is deterministic.

A defaultable forward rate agreement considered by Lotz and Schlögl (2000) is specified as follows:
- if no default occurs up to time $T$, the contract then becomes a defaultable IRA, as described above,
- if default occurs at time $t < T$, the contract is marked to market; specifically, if $U_t(\kappa) \geq 0$, then the payer receives $U_t(\kappa)$, if $U_t(\kappa) < 0$, then the receiver is paid the reduced amount $-\delta U_t(\kappa)$.

Using techniques of Chap. 8, combined with the forward measure approach, it is easy to establish the following lemma.

**Lemma 15.2.1.** *For any date $t < T$, the value $V_t(\kappa)$ for the payer of the defaultable forward agreement described above equals, on the set $\{\tau > t\}$,*

$$V_t(\kappa) = B(t, T + \Delta) \int_t^T \gamma(u)e^{\Gamma(t) - \Gamma(u)}\, \mathbb{E}_{\mathbb{P}_{T+\Delta}}\Big(\frac{Z_u}{B(u, T + \Delta)}\, \Big|\, \mathcal{F}_t\Big)\, du$$

$$+ B(t, T + \Delta)\, \mathbb{E}_{\mathbb{P}_{T+\Delta}}\Big(e^{\Gamma(t) - \Gamma(T + \Delta)} B^{-1}(T, T + \delta)\mathbf{IRA}^\delta(T; \kappa)\, \Big|\, \mathcal{F}_t\Big),$$

*where $Z_u = U_t(\kappa) + (1 - \delta)U_t^-(\kappa)$ and $U_t^-(\kappa) = \max(-U_t(\kappa), 0)$.*

*Proof.* The formula can be seen as a variant of the results of Sect. 8.2.2 with the bond price of maturity $T + \Delta$ chosen to be the discount factor.  $\square$

**Corollary 15.2.1.** *We have*

$$V_t(\kappa) = U_t(\kappa) + (1 - \delta)B(t, T + \Delta) \int_t^T \gamma(u)e^{\Gamma(t)-\Gamma(u)} J(t, u, \kappa) \, du,$$

*where*

$$J(t, u, \kappa) := \mathbb{E}_{\mathbb{P}_{T+\Delta}} \left( \frac{U_u^-(\kappa)}{B(u, T + \Delta)} \, \Big| \, \mathcal{F}_t \right).$$

*Proof.* In view of Lemma 15.2.1, it suffices to show that, on the set $\{\tau > t\}$,

$$U_t(\kappa) = B(t, T + \Delta) \int_t^T \gamma(u)e^{\Gamma(t)-\Gamma(u)} \mathbb{E}_{\mathbb{P}_{T+\Delta}} \left( \frac{U_u(\kappa)}{B(u, T + \Delta)} \, \Big| \, \mathcal{F}_t \right) du$$

$$+ B(t, T + \Delta) \mathbb{E}_{\mathbb{P}_{T+\Delta}} \left( e^{\Gamma(t)-\Gamma(T+\Delta)} B^{-1}(T, T + \delta)\mathbf{IRA}^\delta(T; \kappa) \, \Big| \, \mathcal{F}_t \right).$$

But this is clear since $U_T(\kappa) = \mathbf{IRA}^\delta(T; \kappa)$, so that the right-hand side of the last formula represents the price at time $t$ of a contract, in which in all circumstances its holder receives the value $U_s(\kappa)$ at some (possibly random) time $s$. Formally, the right-hand side equals the value at time $t < \tau$ of a defaultable claim $DCT = (X, 0, 0, Z, \tau)$ with the recovery process $Z_u = U_u(\kappa)$ and the promised payoff $X = U_T(\kappa)$. Since the process $U_u(\kappa)$, $u \in [t, T]$, is known to represent the arbitrage price of a non-defaultable contingent claim $Y = \mathbf{IRA}^\delta(T; \kappa)$, the asserted equality is obvious. $\qquad\square$

For a fixed $t < T$, Lotz and Schlögl (2000) propose to associate with the defaultable forward rate agreement described above the *defaultable forward rate* $\kappa_t^\delta$ by postulating that the random variable $\kappa_t^\delta$ satisfies $V_t(\kappa_t^\delta) = 0$. To find an integral equation satisfied by $\kappa_t^\delta$, observe that by combining (15.14) with the definition of the default-free forward LIBOR rate:

$$1 + \Delta L(u, T) = \frac{B(u, T)}{B(u, T + \Delta)},$$

we obtain

$$\frac{U_u^-(\kappa)}{B(u, T + \Delta)} = \left( K - \Delta L(u, T) \right)^+,$$

where the constant $K$ equals

$$K = (1 + \kappa\Delta)\left( \delta + (1 - \delta)e^{\Gamma(T)-\Gamma(T+\Delta)} \right) - 1.$$

It is thus obvious that

$$J(t, u, \kappa) = \mathbb{E}_{\mathbb{P}_{T+\Delta}} \left( \left( K - \Delta L(u, T) \right)^+ \, \Big| \, \mathcal{F}_t \right).$$

Under the present assumptions, the default-free forward rate $L(u, T)$ has a lognormal probability law under the forward martingale measure $\mathbb{P}_{T+\Delta}$, specifically,

$$dL(t, T) = L(t, T)\nu(t, T) \, dW_t^{T+\Delta}$$

for some function $\nu(\cdot, T) : [0, T] \to \mathbb{R}^d$, where $W^{T+\Delta}$ follows a $d$-dimensional standard Brownian motion under $\mathbb{P}_{T+\Delta}$.

Consequently,

$$J(t, u, \kappa) = KN\big( - h_2(t, u, T)\big) - \Delta L(t, T)N\big( - h_1(t, u, T)\big), \tag{15.15}$$

where

$$h_{1,2}(t, u, T) = \frac{\ln(\Delta L(t, T)/K) \pm \frac{1}{2}v^2(t, u, T)}{v(t, u, T)}, \tag{15.16}$$

and

$$v^2(t, u, T) = \int_t^u |\nu(s, T)|^2 \, ds. \tag{15.17}$$

In the case of unilateral payer's default risk, the main result established in Lotz and Schlögl (2000) reads as follows.

**Proposition 15.2.1.** *For any fixed $t < T$, the defaultable forward rate $\kappa_t^\delta$ solves the following equation*

$$U_t(\kappa_t^\delta) + (1 - \delta)B(t, T + \Delta) \int_t^T \gamma(u)e^{\Gamma(t) - \Gamma(u)} J(t, u, \kappa_t^\delta) \, du = 0,$$

*where $U_t(\kappa)$ is given by (15.14) and $J(t, u, \kappa)$ satisfies (15.15)–(15.17).*

Lotz and Schlögl (2000) also examine a case of defaultable forward rate agreements and defaultable swaps with bilateral default risk. As above, they postulate that for each counterparty the $\mathbb{F}$-intensity of default is a deterministic function: $\gamma_1$ and $\gamma_2$, respectively. In case of mutually independent defaults with identical intensity $\gamma = \gamma_1 = \gamma_2$, they show that the associated defaultable forward rate $\tilde{\kappa}_t^\delta$ equals

$$\tilde{\kappa}_t^\delta = \frac{1}{\Delta}\Big(\big(1 + \Delta L(t, T)\big) \frac{B(T, T + \Delta)}{D^\delta(T, T + \Delta)} - 1\Big)$$

or, equivalently,

$$\tilde{\kappa}_t^\delta = \frac{1}{\Delta}\Big(\frac{B(t, T)B(T, T + \Delta)}{B(t, T + \Delta)D^\delta(T, T + \Delta)} - 1\Big).$$

Recall that the ratio

$$\frac{B(T, T + \Delta)}{D^\delta(T, T + \Delta)} = \Big(\delta + (1 - \delta)e^{\Gamma(T) - \Gamma(T + \Delta)}\Big)^{-1} \geq 1$$

is non-random under the present assumptions. It also follows from the last inequality that $\tilde{\kappa}_t^\delta \geq L(t, T)$ (equality holds here when $\delta = 1$ and/or $\gamma = 0$).

*Remarks.* The rates $\kappa_t^\delta$ and $\tilde{\kappa}_t^\delta$ introduced above are termed *defaultable forward LIBOR rates* in Lotz and Schlögl (2000). As shown in Sect. 14.1.4, the concept of such a rate is rather ambiguous. In this text, the name of a defaultable forward LIBOR rate (with unilateral default risk) is attributed to the process $L^\delta(t, T)$, given by expression (14.12), corresponding to the fractional recovery of Treasury value scheme or, more generally, to the process $L^d(t, T)$ given by (14.13). Following Schönbucher (2000b), we shall examine in the next section the modeling of forward LIBOR rates corresponding to the zero recovery scheme.

### 15.2.2 Schönbucher's Approach

Our last goal is to describe an attempt, due to Schönbucher (2000b), to extend the market model of LIBOR rates to the case of specific defaultable market rates. For a pre-determined collection of dates $0 < T_0 < \cdots < T_m = T^*$, he focuses on the defaultable forward LIBOR rate $\tilde{L}(t, T_i)$, $t \in [0, T_i]$,

$$\tilde{L}(t, T_i) := \frac{1}{\Delta_i} \left( \frac{\tilde{D}(t, T_i)}{\tilde{D}(t, T_{i+1})} - 1 \right), \tag{15.18}$$

where $\tilde{D}(t, T_i)$ represents the pre-default value of a corporate bond with zero recovery, so that $D^0(t, T_i) = \mathbb{1}_{\{\tau > t\}} \tilde{D}(t, T_i)$. The price of a unit zero-coupon bond with zero recovery at default satisfies (cf. (8.50))

$$D^0(t, T_i) = B(t, T_i)\, \mathbb{Q}_{T_i}\{\tau > T_i \,|\, \mathcal{G}_t\} = \mathbb{1}_{\{\tau > t\}} B(t, T_i)\, \frac{\mathbb{Q}_{T_i}\{\tau > T_i \,|\, \mathcal{F}_t\}}{\mathbb{Q}_{T_i}\{\tau > t \,|\, \mathcal{F}_t\}}.$$

If $\tau$ is constructed through the canonical approach, Lemma 14.2.1 yields

$$\tilde{D}(t, T_i) = B(t, T_i)\, \frac{\mathbb{Q}_{T_i}\{\tau > T_i \,|\, \mathcal{F}_t\}}{\mathbb{Q}_{T_i}\{\tau > t \,|\, \mathcal{F}_t\}}.$$

For any $t \in [0, T_i]$, we define the *forward LIBOR credit spread*

$$\tilde{S}(t, T_i) := \tilde{L}(t, T_i) - L(t, T_i), \tag{15.19}$$

and the *forward LIBOR hazard rate process*

$$H(t, T_i) := \frac{1}{\Delta_i} \left( \frac{G(t, T_i)}{G(t, T_{i+1})} - 1 \right), \tag{15.20}$$

where in turn the *forward survival process* $G(t, T_i)$ equals

$$G(t, T_i) := \frac{\tilde{D}(t, T_i)}{B(t, T_i)} = \frac{\mathbb{Q}_{T_i}\{\tau > T_i \,|\, \mathcal{F}_t\}}{\mathbb{Q}_{T_i}\{\tau > t \,|\, \mathcal{F}_t\}} = \mathbb{Q}_{T_i}\{\tau > T_i \,|\, \mathcal{G}_t\},$$

where the last equality holds on the set $\{\tau > t\}$. Let us notice that

$$\tilde{S}(t, T_i) = H(t, T_i)\big(1 + \Delta_i L(t, T_i)\big). \tag{15.21}$$

*Remarks.* Assume that the spot martingale measure $\mathbb{Q}^*$ exists, the equality $\mathbb{Q}_{T_i}\{\tau > t \,|\, \mathcal{F}_t\} = \mathbb{Q}^*\{\tau > t \,|\, \mathcal{F}_t\}$ holds for every $t \in [0, T_i]$ and we have, for every $i = 0, \ldots, m$,

$$\mathbb{Q}_{T_i}\{\tau > T_i \,|\, \mathcal{F}_t\} = \mathbb{Q}^*\{\tau > T_i \,|\, \mathcal{F}_t\}.$$

Then (15.20) becomes

$$H(t, T_i) := \frac{1}{\Delta_i} \left( \frac{\mathbb{Q}^*\{\tau > T_{i+1} \,|\, \mathcal{F}_t\}}{\mathbb{Q}^*\{\tau > T_i \,|\, \mathcal{F}_t\}} - 1 \right).$$

The abovementioned assumptions concerning the properties of $\tau$ under the martingale measures $\mathbb{Q}_{T_i}$ and $\mathbb{Q}^*$ are by no means restrictive. In effect, they are not difficult to establish, if the default time is obtained through the canonical construction.

**Basic assumptions.** The first assumption specifies the dynamics of the family of forward LIBOR rates $L(\cdot, T_i)$, $i = 0, \ldots, m - 1$ under the associated forward LIBOR measures $\mathbb{P}_{T_{i+1}}$. Recall that these probability measures coincide with forward martingale measures if the bond prices $B(t, T_i)$ are specified in a model.

**Assumption (S.1).** For any $i = 0, \ldots, m - 1$, the forward LIBOR rate $L(\cdot, T_i)$ satisfies the following SDE, for every $t \in [0, T_i]$,

$$dL(t, T_i) = L(t, T_i)\nu(t, T_i)\, dW_t^{T_{i+1}} \tag{15.22}$$

with the initial condition $L(0, T_i) > 0$.

It is common to postulate that the volatilities $\nu(\cdot, T_i) : [0, T_i] \to \mathbb{R}^d$ are deterministic, so that each forward LIBOR rate follows a lognormal process under the corresponding 'risk-neutral probability' $\mathbb{P}_{T_{i+1}}$. Following, in particular, Brace et al. (1997) and Miltersen et al. (1997), we have established in Sect. 15.1.1 the existence of this variant of a market model, termed the *lognormal model of LIBOR rates*. We place ourselves in this set-up.

In Sect. 13.1, we have shown how to extend the classic default-free HJM term structure methodology in order to also cover a defaultable term structure. The main focus was put on the construction of a default time $\tau$, associated with pre-specified pre-default dynamics of a corporate bond. In the present context, one also needs to deal with the issue of a judicious construction of $\tau$. However, instead of specifying the dynamics of a corporate bond, we shall now make either Assumption (S.2) or Assumption (S.3) below.

**Assumption (S.2).** For any $i = 0, \ldots, m - 1$, the forward LIBOR credit spread $\tilde{S}(\cdot, T_i)$ satisfies, for every $t \in [0, T_i]$,

$$d\tilde{S}(t, T_i) = \tilde{S}(t, T_i)\left(\mu_{\tilde{S}}(t, T_i)\, dt + \nu_{\tilde{S}}(t, T_i)\, dW_t^{T_{i+1}}\right) \tag{15.23}$$

with the initial condition $\tilde{S}(0, T_i) = \tilde{L}(0, T_i) - L(0, T_i)$, where the volatility $\nu_{\tilde{S}}(\cdot, T_i) : [0, T_i] \to \mathbb{R}^d$ follows a (bounded) measurable function and $\mu_{\tilde{S}}(\cdot, T_i)$ is an $\mathbb{F}$-adapted, real-valued stochastic process.

**Assumption (S.3).** For any $i = 0, \ldots, m - 1$, the forward LIBOR hazard rate process $H(\cdot, T_i)$ satisfies, for every $t \in [0, T_i]$,

$$dH(t, T_i) = H(t, T_i)\left(\mu_H(t, T_i)\, dt + \nu_H(t, T_i)\, dW_t^{T_{i+1}}\right) \tag{15.24}$$

with the initial condition (15.20), where $\nu_H(\cdot, T_i) : [0, T_i] \to \mathbb{R}^d$ is a (bounded) measurable function and $\mu_H(\cdot, T_i)$ is an $\mathbb{F}$-adapted, real-valued process.

In view of (15.21), it can be easily checked that for every $t \in [0, T_i]$ we have

$$\nu_{\tilde{S}}(t, T_i) = \nu_H(t, T_i) + \frac{\Delta_i L(t, T_i)\nu(t, T_i)}{1 + \Delta_i L(t, T_i)}. \tag{15.25}$$

The knowledge of $\nu_{\tilde{S}}(t, T_i)$ is thus equivalent to the knowledge of $\nu_H(t, T_i)$. Of course, if $\nu_{\tilde{S}}(t, T_i)$ is deterministic, then $\nu_H(t, T_i)$ is random and vice versa.

We shall henceforth work under an implicit assumption that the default time $\tau$ has been already specified through the canonical construction. Our main goal is to examine the most relevant features defaultable forward LIBOR rates under Assumptions (S.1)–(S.2) (or (S.1) and (S.3)), rather than to provide a detailed construction of the corresponding credit risk model. We make the standard assumption that the default at time 0 is excluded, so that $\mathbb{Q}^*\{\tau > 0\} = \mathbb{Q}_{T_i}\{\tau > 0\} = 1$.

**Defaultable forward measure.** We are in a position to introduce a probability measure that will play the role of the *pricing probability* for defaultable claims with zero recovery and settlement date $T_i$.

**Definition 15.2.1.** The *defaultable forward martingale measure* $\tilde{\mathbb{Q}}_{T_i}$ for the date $T_i$ is given on $(\Omega, \mathcal{G}_{T_i})$ through the equality

$$\frac{d\tilde{\mathbb{Q}}_{T_i}}{d\mathbb{Q}_{T_i}} = \frac{B(0, T_i)}{D^0(0, T_i)} D^0(T_i, T_i), \quad \mathbb{Q}_{T_i}\text{-a.s.}$$

We shall now provide few equivalent representations for the Radon-Nikodým density of $\tilde{\mathbb{Q}}_{T_i}$ with respect to $\mathbb{Q}_{T_i}$. On one hand, since $D^0(T_i, T_i) = \mathbb{1}_{\{\tau > T_i\}}$ and $D^0(0, T_i) = \tilde{D}(0, T_i)$, we obtain

$$\frac{d\tilde{\mathbb{Q}}_{T_i}}{d\mathbb{Q}_{T_i}} = \frac{B(0, T_i)}{\tilde{D}(0, T_i)} \mathbb{1}_{\{\tau > T_i\}}, \quad \mathbb{Q}_{T_i}\text{-a.s.}$$

On the other hand, using the equalities $\tilde{D}(T_i, T_i) = B(T_i, T_i) = 1$, we also find that

$$\frac{d\tilde{\mathbb{Q}}_{T_i}}{d\mathbb{Q}_{T_i}} = \mathbb{1}_{\{\tau > T_i\}} \frac{B(0, T_i)\tilde{D}(T_i, T_i)}{\tilde{D}(0, T_i)B(T_i, T_i)} = \mathbb{1}_{\{\tau > T_i\}} \frac{G(T_i, T_i)}{G(0, T_i)} = \frac{\mathbb{1}_{\{\tau > T_i\}}}{\mathbb{Q}_{T_i}\{\tau > T_i\}}.$$

It is apparent that $\tilde{\mathbb{Q}}_{T_i}$ is absolutely continuous with respect to $\mathbb{Q}_{T_i}$, i.e., for any event $A \in \mathcal{G}_{T_i}$ such that $\mathbb{Q}_{T_i}\{A\} = 0$, we also have $\tilde{\mathbb{Q}}_{T_i}\{A\} = 0$. It should be noted that the two probability measures, $\tilde{\mathbb{Q}}_{T_i}$ and $\mathbb{Q}_{T_i}$, are not mutually equivalent. In effect, for any $t \leq T_i$, the event $A = \{\tau \leq t\}$ has null probability under $\tilde{\mathbb{Q}}_{T_i}$, but it usually has a strictly positive probability under $\mathbb{Q}_{T_i}$. This also shows that the intensity of default under $\tilde{\mathbb{Q}}_{T_i}$ vanishes on the interval $[0, T_i]$. In this regard, it is interesting to recall that the intensity of default remains the same under the spot martingale measure $\mathbb{Q}^*$ and under the forward martingale measure $\mathbb{Q}_T$.

From another perspective, the probability measure $\tilde{\mathbb{Q}}_{T_i}$ is easily seen to correspond to the choice of the price process $D^0(\cdot, T_i)$ of a defaultable bond as a discount factor (i.e., as a numéraire). Notice, however, that the value process $D^0(\cdot, T_i)$ is not strictly positive with probability one under $\mathbb{Q}_{T_i}$. The latter property makes it clear why the two martingale measures, $\mathbb{Q}_{T_i}$ and $\tilde{\mathbb{Q}}_{T_i}$, are not mutually equivalent in most cases.

In Schönbucher (2000b), the probability measure $\tilde{Q}_{T_i}$ is termed the $T_i$-*survival measure*. To justify this terminology, it suffices to notice that the last representation of the Radon-Nikodým density yields, for any event $A \in \mathcal{G}_{T_i}$,

$$\tilde{Q}_{T_i}\{A\} = \frac{Q_{T_i}(A \cap \{\tau > T_i\})}{Q_{T_i}\{\tau > T_i\}} = Q_{T_i}(A \mid \{\tau > T_i\}),$$

where the last term is to be understood in the elementary sense, i.e., as a real number, rather than a random variable. In this sense, the defaultable forward martingale measure $\tilde{Q}_{T_i}$ can be seen as the forward martingale measure $Q_{T_i}$ conditioned on survival until time $T_i$.

In the next step, we shall examine the restriction of the above Radon-Nikodým density to the $\sigma$-field $\mathcal{G}_t$. Since $D^0(t, T_i)$ represents the price process of a tradeable security, the discounted process

$$\frac{D^0(t, T_i)}{B(t, T_i)} = Q_{T_i}\{\tau > T_i \mid \mathcal{G}_t\}$$

follows a $\mathbb{G}$-martingale under the forward martingale measure $Q_{T_i}$. Consequently, we obtain (recall that the $\sigma$-fields $\mathcal{F}_0$ and $\mathcal{G}_0$ are trivial)

$$\frac{d\tilde{Q}_{T_i}}{dQ_{T_i}}\Big|_{\mathcal{G}_t} = \frac{B(0, T_i)D^0(t, T_i)}{D^0(0, T_i)B(t, T_i)} = \mathbb{1}_{\{\tau > t\}} \frac{B(0, T_i)\tilde{D}(t, T_i)}{\tilde{D}(0, T_i)B(t, T_i)}$$
$$= \mathbb{1}_{\{\tau > t\}} \frac{G(t, T_i)}{G(0, T_i)} = \frac{\mathbb{1}_{\{\tau > t\}}}{Q_{T_i}\{\tau > T_i\}} \frac{Q_{T_i}\{\tau > T_i \mid \mathcal{F}_t\}}{Q_{T_i}\{\tau > t \mid \mathcal{F}_t\}}.$$

**Restricted defaultable forward measure.** In some circumstances, it will be convenient to consider the restriction $\tilde{\mathbb{P}}_{T_i}$ of the probability measure $\tilde{Q}_{T_i}$ to the sub-$\sigma$-field $\mathcal{F}_{T_i}$ of $\mathcal{G}_{T_i}$. We have

$$\mathbb{E}_{Q_{T_i}}\left(\frac{d\tilde{Q}_{T_i}}{dQ_{T_i}} \,\Big|\, \mathcal{F}_{T_i}\right) = \frac{Q_{T_i}\{\tau > T_i \mid \mathcal{F}_{T_i}\}}{Q_{T_i}\{\tau > T_i\}} = \frac{B(0, T_i)}{\tilde{D}(0, T_i)} Q_{T_i}\{\tau > T_i \mid \mathcal{F}_{T_i}\}.$$

Let us notice that the probability measure $\mathbb{P}_{T_i}$ is the restriction of $Q_{T_i}$ to the $\sigma$-field $\mathcal{F}_{T_i}$. This observation leads to the following definition of the restricted probability measure $\tilde{\mathbb{P}}_{T_i}$. Let us recall that a counterpart of this concept was already introduced within the HJM framework (see Definition 13.1.2).

**Definition 15.2.2.** The *restricted defaultable forward martingale measure* $\tilde{\mathbb{P}}_{T_i}$ for the date $T_i$ is given on $(\Omega, \mathcal{F}_{T_i})$ by the formula

$$\frac{d\tilde{\mathbb{P}}_{T_i}}{d\mathbb{P}_{T_i}} = \frac{B(0, T_i)}{\tilde{D}(0, T_i)} \mathbb{P}_{T_i}\{\tau > T_i \mid \mathcal{F}_{T_i}\}, \quad \mathbb{P}_{T_i}\text{-a.s.}$$

It is apparent that

$$\frac{d\tilde{\mathbb{P}}_{T_i}}{d\mathbb{P}_{T_i}}\Big|_{\mathcal{F}_t} = \mathbb{E}_{\mathbb{P}_{T_i}}\left(\frac{d\tilde{\mathbb{P}}_{T_i}}{d\mathbb{P}_{T_i}} \,\Big|\, \mathcal{F}_t\right) = \frac{B(0, T_i)}{\tilde{D}(0, T_i)} \mathbb{P}_{T_i}\{\tau > T_i \mid \mathcal{F}_t\}.$$

The forward survival process $G(t, T_i)$ follows under $\mathbb{P}_{T_i}$

$$dG(t, T_i) = G(t, T_i)\big(\mu_G(t, T_i)\, dt + \nu_G(t, T_i)\, dW_t^{T_i}\big) \tag{15.26}$$

for some $\mathbb{F}$-adapted process $\mu_G(\cdot, T_i)$ and $\nu_G(\cdot, T_i) = \tilde{b}(t, T_i) - b(t, T_i)$. Recall that for every $t \in [0, T_i]$

$$\frac{d\tilde{\mathbb{Q}}_{T_i}}{d\mathbb{Q}_{T_i}}\bigg|_{\mathcal{G}_t} = \mathbb{1}_{\{\tau > t\}} \frac{B(0, T_i)\tilde{D}(t, T_i)}{\tilde{D}(0, T_i)B(t, T_i)} = \mathbb{1}_{\{\tau > t\}} \frac{G(t, T_i)}{G(0, T_i)}.$$

We conclude that

$$\frac{d\tilde{\mathbb{Q}}_{T_i}}{d\mathbb{Q}_{T_i}}\bigg|_{\mathcal{G}_t} = \mathbb{1}_{\{\tau > t\}} \exp\left(\int_0^t \mu_G(u, T_i)\, du\right)\mathcal{E}_t\left(\int_0^\cdot \nu_G(u, T_i)\, dW_u^{T_i}\right), \tag{15.27}$$

where, as usual, the symbol $\mathcal{E}$ denotes the Doléans exponential. In view of the results of Chap. 5 (in particular, Lemma 5.1.7), it is natural to expect that

$$\frac{d\tilde{\mathbb{Q}}_{T_i}}{d\mathbb{Q}_{T_i}}\bigg|_{\mathcal{G}_t} = \mathbb{1}_{\{\tau > t\}} \exp\left(\int_0^t \gamma_u\, du\right)\mathcal{E}_t\left(\int_0^\cdot \nu_G(u, T_i)\, dW_u^{T_i}\right),$$

where $\gamma$ represents the $\mathbb{F}$-intensity process of $\tau$ under $\mathbb{Q}^*$ (and under $\mathbb{Q}_{T_i}$), and that the drift term $\mu_G(t, T_i)$ is independent of $i$. Let us stress that, in contrast with the HJM approach, the default time $\tau$ is not explicitly modeled in the present set-up. However, we have the following useful result.

**Lemma 15.2.2.** *We have*

$$\frac{d\tilde{\mathbb{P}}_{T_i}}{d\mathbb{P}_{T_i}}\bigg|_{\mathcal{F}_t} = \mathcal{E}_t\left(\int_0^\cdot \nu_G(u, T_i)\, dW_u^{T_i}\right) = \exp\left(-\int_0^t \mu_G(u, T_i)\, du\right)G(t, T_i).$$

*The process $\tilde{W}_t^{T_i}$, $t \in [0, T_i]$, given by*

$$\tilde{W}_t^{T_i} = W_t^{T_i} - \int_0^t \nu_G(u, T_i)\, du, \tag{15.28}$$

*follows a standard Brownian motion with respect to the filtration $\mathbb{F}$ under $\tilde{\mathbb{P}}_{T_i}$.*

*Proof.* We shall merely sketch the proof. Formula (15.27) provides a multiplicative representation for the $\mathbb{G}$-martingale as a product of a $\mathbb{G}$-jump martingale and a continuous $\mathbb{F}$-martingale (which also follows a $\mathbb{G}$-martingale). However, the reference filtration $\mathbb{F}$ is known to support only continuous martingales, and thus the assertion is valid.  □

*Remarks.* Let $\tilde{b}(t, T_i)$ and $b(t, T_i)$ stand for the volatility of $B(t, T_i)$ and $\tilde{D}(t, T_i)$, respectively. Then $\nu_G(t, T_i) = \tilde{b}(t, T_i) - b(t, T_i)$ and (cf. Sect. 13.1)

$$W_t^{T_i} = W_t^* - \int_0^t b(u, T_i)\, du, \quad \tilde{W}_t^{T_i} = W_t^* - \int_0^t \tilde{b}(u, T_i)\, du.$$

In the present setting, the volatilities $b(t, T_i)$ and $\tilde{b}(t, T_i)$ are not specified, so that expression (15.28) is more suitable.

**Dynamics of $\tilde{L}(t, T_i)$ and $\tilde{S}(t, T_i)$ under $\tilde{\mathbb{P}}_{T_{i+1}}$.** Under Assumptions (S.1) and (S.2) (or (S.1) and (S.3)) we may find explicit representations for the dynamics of the processes $\tilde{L}(t, T_i)$, $\tilde{S}(t, T_i)$ and $H(t, T_i)$ under $\tilde{\mathbb{P}}_{T_{i+1}}$. The drift restrictions derived below provide a useful hint on the method of construction of the market model for default-free and defaultable term structures of forward LIBOR rates.

**Proposition 15.2.2.** *For any $i = 0, \ldots, m - 1$, the defaultable forward LI-BOR rate $\tilde{L}(\cdot, T_i)$ satisfies the following SDE under $\tilde{\mathbb{P}}_{T_{i+1}}$*

$$d\tilde{L}(t, T_i) = \tilde{L}(t, T_i)\tilde{\nu}(t, T_i)\, d\tilde{W}_t^{T_{i+1}}, \tag{15.29}$$

*where*

$$\tilde{\nu}(t, T_i) = \tilde{L}^{-1}(t, T_i)\big(\tilde{S}(t, T_i)\nu_{\tilde{S}}(t, T_i) - L(t, T_i)\nu(t, T_i)\big). \tag{15.30}$$

*Furthermore,*

$$d\tilde{S}(t, T_i) = L(t, T_i)\nu(t, T_i)\nu_G(t, T_{i+1})\, dt + \tilde{S}(t, T_i)\nu_{\tilde{S}}(t, T_i)\, d\tilde{W}_t^{T_{i+1}}.$$

*Proof.* Formula (15.29) was already derived in the HJM framework (see Lemma 13.1.7). In the present set-up, it follows from the fact that the ratio $\tilde{D}(t, T_i)/\tilde{D}(t, T_{i+1})$ follows a martingale under $\tilde{\mathbb{P}}_{T_{i+1}}$. Indeed, we have

$$\frac{\tilde{D}(t, T_i)}{\tilde{D}(t, T_{i+1})} = \big(1 + \Delta_i L(t, T_i)\big)\frac{G(t, T_i)}{G(t, T_{i+1})}.$$

The first process on the right-hand side of the last formula follows a $\mathbb{P}_{T_{i+1}}$-martingale, and the second is proportional to the Radon-Nikodým density of $\tilde{\mathbb{P}}_{T_i}$ with respect to $\tilde{\mathbb{P}}_{T_{i+1}}$. Representation (15.30) is a consequence of the equality $\tilde{L}(t, T_i) = L(t, T_i) + \tilde{S}(t, T_i)$ and assumptions (15.22)–(15.23). To establish the last formula, it suffices to combine the first with the equality $\tilde{S}(t, T_i) = \tilde{L}(t, T_i) - L(t, T_i)$. The details are left to the reader.    □

Equality (15.30) makes it clear that if the volatilities $\nu(t, T_i)$ and $\nu_{\tilde{S}}(t, T_i)$ are deterministic then the volatility $\tilde{\nu}(t, T_i)$ is necessarily random. It does not seem convenient to directly specify the dynamics of defaultable LIBOR rates through a counterpart of expression (15.22), with deterministic volatilities $\tilde{\nu}(t, T_i)$, $i = 0, \ldots, m - 1$. Indeed, it is essential to guarantee the non-negativity of the forward LIBOR credit spreads $\tilde{S}(\cdot, T_i)$, $i = 0, \ldots, m - 1$.

The next result is an easy consequence of Proposition 15.2.2 and relationship (15.21).

**Corollary 15.2.2.** *The dynamics of $H(t, T_i)$ under $\tilde{\mathbb{P}}_{T_{i+1}}$ are*

$$dH(t, T_i) = \tilde{\mu}_H(t, T_i)\, dt + H(t, T_i)\nu_H(t, T_i)\, d\tilde{W}_t^{T_{i+1}},$$

*where*

$$\tilde{\mu}_H(t, T_i) = \frac{L(t, T_i)\nu(t, T_i)}{1 + \Delta_i L(t, T_i)}\left(\big(1 + \Delta_i H(t, T_i)\big)\nu_G(t, T_i) - \Delta_i H(t, T_i)\nu_H(t, T_i)\right).$$

**Valuation of defaultable claims.** We are in a position to proceed to the valuation of defaultable claims through defaultable forward martingale measures $\tilde{\mathbb{Q}}_{T_i}$ and $\tilde{\mathbb{P}}_{T_i}$. Consider a defaultable claim $(X, 0, 0, 0, \tau)$, with the settlement date $T_i$. According to the convention introduced in Sect. 8.1, we are dealing here with a defaultable claim with the promised payoff $X$, the promised dividends $A \equiv 0$ and the zero recovery upon default. We are going to study two particular cases: the general case of a $\mathcal{G}_{T_i}$-measurable payoff $X$ and the standard case of an $\mathcal{F}_{T_i}$-measurable payoff $X$. Assume that a defaultable claim $(X, 0, 0, 0, \tau)$ is attainable. Then its pre-default equals

$$S_t^0 = B(t, T_i)\, \mathbb{E}_{\mathbb{Q}_{T_i}}\big(X \mathbb{1}_{\{\tau > T_i\}} \,\big|\, \mathcal{G}_t\big). \tag{15.31}$$

**Proposition 15.2.3.** *Assume that the promised payoff $X$ is $\mathcal{G}_{T_i}$-measurable and integrable with respect to $\mathbb{Q}_{T_i}$. Then for every $t \in [0, T_i]$ we have*

$$S_t^0 = \mathbb{1}_{\{\tau > t\}} \tilde{D}(t, T_i)\, \mathbb{E}_{\tilde{\mathbb{Q}}_{T_i}}(X \,|\, \mathcal{G}_t) = D^0(t, T_i)\, \mathbb{E}_{\tilde{\mathbb{Q}}_{T_i}}(X \,|\, \mathcal{G}_t).$$

*If $X$ is $\mathcal{F}_{T_i}$-measurable, then*

$$S_t^0 = \mathbb{1}_{\{\tau > t\}} B(t, T_i)\, \mathbb{E}_{\tilde{\mathbb{P}}_{T_i}}(X \,|\, \mathcal{F}_t). \tag{15.32}$$

*Proof.* Let us denote

$$\bar{\eta}_{T_i} = \frac{d\tilde{\mathbb{Q}}_{T_i}}{d\mathbb{Q}_{T_i}}, \quad \tilde{\eta}_t = \frac{d\tilde{\mathbb{Q}}_{T_i}}{d\mathbb{Q}_{T_i}}\bigg|_{\mathcal{G}_t} = \mathbb{E}_{\mathbb{Q}_{T_i}}(\eta_{T_i} \,|\, \mathcal{G}_t).$$

Then for any $\mathcal{G}_{T_i}$-measurable and $\tilde{\mathbb{Q}}_{T_i}$-integrable random variable $X$ we have

$$\eta_t\, \mathbb{E}_{\tilde{\mathbb{Q}}_{T_i}}(X \,|\, \mathcal{G}_t) = \mathbb{E}_{\mathbb{Q}_{T_i}}(\eta_{T_i} X \,|\, \mathcal{G}_t), \quad \forall t \in [0, T_i],$$

or, equivalently,

$$\mathbb{1}_{\{\tau > t\}} \frac{B(0, T_i)\tilde{D}(t, T_i)}{\tilde{D}(0, T_i)B(t, T_i)}\, \mathbb{E}_{\tilde{\mathbb{Q}}_{T_i}}(X \,|\, \mathcal{G}_t) = \frac{B(0, T_i)}{\tilde{D}(0, T_i)}\, \mathbb{E}_{\mathbb{Q}_{T_i}}\big(X \mathbb{1}_{\{\tau > T_i\}} \,\big|\, \mathcal{G}_t\big).$$

After simple manipulations, we thus obtain

$$\mathbb{1}_{\{\tau > t\}} \tilde{D}(t, T_i)\, \mathbb{E}_{\tilde{\mathbb{Q}}_{T_i}}(X \,|\, \mathcal{G}_t) = B(t, T_i)\, \mathbb{E}_{\mathbb{Q}_{T_i}}\big(X \mathbb{1}_{\{\tau > T_i\}} \,\big|\, \mathcal{G}_t\big) = S_t^0,$$

which is the desired formula. Let us now consider the case of $\mathcal{F}_{T_i}$-measurable random variable $X$. In this case, we have (see, for example, Lemma 5.1.2)

$$\mathbb{E}_{\mathbb{Q}_{T_i}}\big(X \mathbb{1}_{\{\tau > T_i\}} \,\big|\, \mathcal{G}_t\big) = \mathbb{1}_{\{\tau > t\}} \frac{\mathbb{E}_{\mathbb{Q}_{T_i}}\big(X \mathbb{1}_{\{\tau > T_i\}} \,\big|\, \mathcal{F}_t\big)}{\mathbb{Q}_{T_i}\{\tau > T_i \,|\, \mathcal{F}_t\}}$$

$$= \mathbb{1}_{\{\tau > t\}} \frac{\mathbb{E}_{\mathbb{P}_{T_i}}\big(X \mathbb{Q}_{T_i}\{\tau > T_i \,|\, \mathcal{F}_{T_i}\} \,\big|\, \mathcal{F}_t\big)}{\mathbb{Q}_{T_i}\{\tau > T_i \,|\, \mathcal{F}_t\}} = \mathbb{1}_{\{\tau > t\}} \mathbb{E}_{\tilde{\mathbb{P}}_{T_i}}(X \,|\, \mathcal{F}_t),$$

where the last equality easily follows from the abstract Bayes rule. Combining just established equality with (15.31), we obtain the second formula of the proposition. $\square$

**Modeling of forward default swap rates.** Let us recall (see Sect. 13.3) that in a standard default swap:
- the buyer pays annuities $\kappa$ at pre-determined dates $T_i$, $i = 1, \ldots, m$ prior to default or the contract's maturity date $U$, whichever comes first,
- if default occurs prior to or at the maturity date $U$, the buyer receives at time of default the difference between the par value of the reference asset and its post-default market value.

Schönbucher (2000) examines a default swap in which the role of the underlying asset is played by a corporate coupon bond with fixed coupon rate $c$. By assumption, the reference bond is subject to the fractional recovery of par. Specifically, if the bond defaults in the time interval $]T_i, T_{i+1}]$, its holder receives at time of default the amount $\delta(1 + c)$. The value at time 0 of the fee stream equals

$$\kappa \, \mathbb{E}_{\mathbb{Q}^*} \left( \sum_{i=1}^{m} B_{T_i}^{-1} \mathbb{1}_{\{\tau > T_i\}} \right) = \kappa \sum_{i=1}^{m} D^0(0, T_i).$$

Since the value of the reference bond in default is $\delta(1 + c)$, the payoff in default equals $\tilde{X} = 1 - \delta(1 + c)$. For practical purposes, it is reasonable to assume that the recovery payoff is postponed to the next coupon date. Under this convention, the value of the recovery payoff at time 0 equals

$$\pi_0(\tilde{X}) = \sum_{i=1}^{m} \mathbb{E}_{\mathbb{Q}^*} \left( B_{T_i}^{-1} \tilde{X} \, \mathbb{1}_{\{T_{i-1} < \tau \leq T_i\}} \right) = \left( 1 - \delta(1 + c) \right) \sum_{i=1}^{m} D_i,$$

where for simplicity we have assumed that $U = T_m$, and where we denote

$$D_i = \mathbb{E}_{\mathbb{Q}^*} \left( B_{T_i}^{-1} \mathbb{1}_{\{T_{i-1} < \tau \leq T_i\}} \right).$$

The default swap rate at time 0, i.e., the value of $\kappa$, which makes the contract valueless at time 0, thus equals

$$\kappa_0 = \left( 1 - \delta(1 + c) \right) \frac{\sum_{i=1}^{m} D^0(0, T_i)}{\sum_{i=1}^{m} D_i}.$$

A call option on a default swap gives its buyer the right to enter a default swap at pre-determined spread at exercise date $T$. Let $\bar{\kappa}(t, T_i)$ denote the forward default swap rate at time $t$ for the date $T_i$. Schönbucher (2000) proposes to directly specify the dynamics of this process under the *defaultable forward swap measure* associated with the choice of the *defaultable coupon process* $\bar{G}_t(m - i)$ as a numeraire, where

$$\bar{G}_t(m - i) = \sum_{j=i+1}^{m} \Delta_{j-1} D^0(t, T_j).$$

Mimicking Jamshidian's approach presented in Sect. 15.1.2, he then postulates that

$$d\bar{\kappa}(t, T_i) = \bar{\kappa}(t, T_i) \bar{\nu}(t, T_i) \, d\bar{W}_t^{T_{i+1}}$$

with a deterministic volatility coefficient $\bar{\nu}(t, T_i)$. A result analogous to Proposition 15.1.2 is then valid. For details, we refer to Schönbucher (2000).

# A Guide to References

**Mathematical background:** Itô and McKean (1965), Dellacherie (1970, 1972), Chou and Meyer (1975), Dellacherie and Meyer (1978a, 1978b), Davis (1976), Elliott (1977), Jeulin and Yor (1978), Jacod (1979), Mazziotto and Szpirglas (1979), Çinlar, Jacod, Protter and Sharpe (1980), Jeulin (1980), Aven (1985), Brémaud (1981), Jennen and Lerche (1981), Jacod and Shiryaev (1987), Shaked and Shanthikumar (1987), Ikeda and Watanabe (1989), Bhattacharya and Waymire (1990), Pardoux and Peng (1990), Protter (1990), Karatzas and Shreve (1991), Revuz and Yor (1999), Williams (1991), Durbin (1992), He, Wang and Yan (1992), Syski (1992), Yor (1997, 2001), Davis (1993), Peng (1993), Artzner and Delbaen (1995), Krylov (1995), Last and Brandt (1995), Borodin and Salminen (1996), Yin and Zhang (1997), Øksendal (1998), Gill (1999), Rolski, Schmidli, Schmidt and Teuggels (1998), Ma and Yong (1999), Elliott, Jeanblanc and Yor (2000), Rogers and Williams (2000), Steele (2000).

**Arbitrage pricing theory:** Black and Scholes (1973), Merton (1973), Harrison and Pliska (1981), Merton (1990), Duffie and Stanton (1992), Cvitanić and Karatzas (1993), Geman, El Karoui and Rochet (1995), Baxter and Rennie (1996), Duffie (1996), Lamberton and Lapeyre (1996), Neftci (1996), El Karoui and Quenez (1997a, 1997b), Hull (1997), Pliska (1997), Musiela and Rutkowski (1997a), Bingham and Kiesel (1998), Björk (1998), Karatzas and Shreve (1998), Shiryaev (1998), Elliott and Kopp (1999), Föllmer and Leukert (1999, 2000), Mel'nikov (1999), Hunt and Kennedy (2000), Jarrow and Turnbull (2000a), Schweizer (2001).

**Term structure modeling:** Vasicek (1977), Cox, Ingersoll and Ross (1985a, 1985b), Jamshidian (1989, 1997), Heath, Jarrow and Morton (1990, 1992), Brace, Gątarek and Musiela (1997), Miltersen, Sandmann and Sondermann (1997), Musiela and Rutkowski (1997a, 1997b), Rutkowski (1999, 2001), Schlögl (1999), Hunt and Kennedy (2000), Hunt, Kennedy and Pelsser (2000), Jarrow and Turnbull (2000a), Pelsser (2000), Brigo and Mercurio (2001), Martellini and Priaulet (2001).

**Credit risk:** Litzenberger (1992), Das (1998a, 1998b), Caouette, Altman and Narayanan (1998), Tavakoli (1998), Francis et al. (1999), Nelken (1999), Saunders (1999), Arvanitis and Laurent (1999), Arvanitis (2000), Cossin and Pirotte (2000), Duffie and Singleton (2003).

**Structural approach:** Merton (1974), Black and Cox (1976), Geske (1977), Brennan and Schwartz (1977, 1978, 1980), Pitts and Selby (1983), Vasicek (1984), Chance (1990), Rendleman (1992), Kim, Ramaswamy and Sundaresan (1993), Nielsen, Saá-Requejo and Santa-Clara (1993), Hull and White (1995), Longstaff and Schwartz (1995), Klein (1996), Zhou (1996), Briys and de Varenne (1997), Pierides (1997), Rich and Leipus (1997), Cathcart and El-Jahel (1998), Crouhy, Galai and Mark (1998), Delianedis and Geske (1998), Ericsson and Reneby (1998, 1999), Li (1998), Barone-Adesi and Colwell (1999), Shirakawa (1999), Wang (1999a), Buffet (2000), Ericsson and Renault (2000), Collin-Dufresne and Goldstein (2001), Klein and Inglis (2001), Yu and Kwok (2002), Eom, Helwege and Huang (2003).

**Structural approach with strategic behavior:** Leland (1994), Anderson and Sundaresan (1996), Anderson, Sundaresan and Tychon (1996), Leland and Toft (1996), Fan and Sundaresan, (1997), Mella-Barral and Perraudin (1997), Ericsson (1999), Mella-Barral and Tychon (1999), Anderson, Pan and Sundaresan (2000), Anderson and Sundaresan (2000), Sarkar (2001).

**Reduced-form approach:** Pye (1974), Ramaswamy and Sundaresan (1986), Litterman and Iben (1991), Artzner and Delbaen (1995), Hull and White (1995), Jarrow and Turnbull (1995), Das and Tufano (1996), Duffie, Schroder and Skiadas (1996), Hughston (1996, 1997, 2000), Schönbucher (1996, 1998a, 1998b), Duffie and Singleton (1994, 1997, 1999), Lando (1997), Monkkonen (1997), Lando (1998), Lotz (1998, 1999), Schlögl (1998), Wong (1998), Collin-Dufresne and Hugonnier (1999), Kusuoka (1999), Maksymiuk and Gątarek (1999), Pugachevsky (1999), Blanchet-Scalliet and Jeanblanc (2000), Elliott, Jeanblanc and Yor (2000), Greenfield (2000), Jarrow, Lando and Yu (2000), Jeanblanc and Rutkowski (2000a, 2000b, 2002), Kijima (2000), Kijima and Muromachi (2000), Laurent (2000), Lotz and Schlögl (2000), Bélanger, Shreve and Wong (2001), Collin-Dufresne and Solnik (2001), Duffie and Lando (2001), Hübner (2001), Jarrow and Yu (2001).

**Ratings-based approach:** Das and Tufano (1996), Jarrow, Lando and Turnbull (1997), Nakazato (1997), Duffie and Singleton (1998a), Arvanitis, Gregory and Laurent (1998), Kijima (1998), Kijima and Komoribayashi (1998), Thomas, Allen and Morkel-Kingsbury (1998), Bielecki and Rutkowski (1999, 2000a, 2000b), Lando (2000a), Schönbucher (2000a), Wei (2000), Crouhy, Galai and Mark (2001), Eberlein and Özkan (2001), Israel, Rosenthal and Wei (2001), Krahnen and Weber (2001), Lando and Skødeberg (2002).

**Hybrid approach:** Crouhy, Galai and Mark (1998), Madan and Unal (1998, 2000), Wong (1998), Davydov, Linetsky and Lotz (1999), Bélanger, Shreve and Wong (2001), Duffie and Lando (2001).

**Modeling of credit spreads:** Das and Tufano (1996), Nielsen and Ronn (1997), Zheng (2000), Brunel (2001).

**Dependent defaults:** Lucas (1995), Gersbach and Lipponer (1997a, 1997b), Zhou (2001), Duffie (1998a), Duffie and Singleton (1998b), Davis and Lo (1999, 2001), Jarrow and Yu (1999), Jarrow, Lando and Yu (1999), Kusuoka (1999), Wang (1999b), Li (1999a, 2000), Erlenmaier and Gersbach (2000), Frey and McNeal (2000), Kijima (2000), Kijima and Muromachi (2000), Lando (2000b), Lindskog (2000), Nyfeler (2000), Bielecki and Rutkowski (2001b, 2002, 2003), Embrechts, McNeal and Straumann (2002), Embrechts, Lindskog and McNeal (2003).

**Liquidity risk:** Amihud and Mendelson (1991), Boudoukh and Whitelaw (1991), Longstaff (1995, 2001), Bangia et al. (1999), Ericsson and Renault (2000).

**Econometric studies and implementations:** Johnson (1967), Fons (1987), Altman (1989), Sarig and Warga (1989), Sun, Sundaresan and Wang (1993), Duffie and Singleton (1994, 1997, 1999), Fons (1994), Altman and Bencivenga (1995), Duffee (1996), Fridson and Jónsson (1995), Foss (1995), Altman and Kishore (1996), Uhrig (1996), Carty (1997), Carty and Lieberman (1997), Lehrbass (1997), Monkkonen (1997), Wei and Guo (1997), Wilson (1997a, 1999b), Altman and Saunders (1998), Delianedis and Geske (1998), Duffee (1998, 1999), Schwartz (1998), Kiesel, Perraudin and Taylor (1999, 2002), Taurén (1999), Altman and Suggit (2000), Christiansen (2000), Dai and Singleton (2000), Diaz and Skinner (2000), Finger (2000), Kavvathas (2000), Liu, Longstaff and Mandell (2000), Rachev, Schwartz and Khindanova (2000), Bakshi, Madan and Zhang (2001), Carey and Hrycay (2001), Collin-Dufresne and Solnik (2001), Collin-Dufresne, Goldstein and Martin (2001), Shumway (2001), Lando and Skødeberg (2002).

# References

Abken, P. (1993) Valuation of default-risky interest-rate swaps. *Adv. Futures Options Res.* 6, 93–116.

Altman, E.I. (1989) Measuring corporate bond mortality and performance. *J. Finance* 44, 909–922.

Altman, E.I. (1997) The importance and subtlety of credit rating migration. Working paper, Stern School of Business, New York University.

Altman, E.I. (1998) Market dynamics and investment performance of distressed and defaulted debt securities. Working paper, Stern School of Business, New York University.

Altman, E.I., Bencivenga, J.C. (1995) A yield premium model for the high-yield debt market. *Finan. Analysts J.* 51(5), 49–56.

Altman, E.I., Kishore, V.M. (1996) Almost everything you wanted to know about recoveries on defaulted bonds. *Finan. Analysts J.* 52(6), 57–64.

Altman, E.I, Saunders, A. (1998) Credit risk measurements: Developments over the last 20 years. *J. Bank. Finance* 21, 1721–1742.

Altman, E.I., Suggit, H. (2000) Default rates in the syndicated bank loan market: A mortality analysis. *J. Bank. Finance* 24, 229–253.

Amihud, Y., Mendelson, H. (1991) Liquidity, maturity, and yields on U.S. Treasury securities. *J. Finance* 46, 1411–1425.

Amin, K., Bodurtha, J. (1995) Discrete time valuation of American options with stochastic interest rates. *Rev. Finan. Stud.* 8, 193–234.

Ammann, M. (1999) *Pricing Derivative Credit Risk. Lecture Notes in Econ. Math. Systems* 470. Springer-Verlag, Berlin Heidelberg New York.

Ammann, M. (2001) *Credit Risk Valuation: Methods, Models, and Applications.* Springer-Verlag, Berlin Heidelberg New York.

Andersen, P.K., Borgan, Ø., Gill, R.D., Keiding, N. (1993) *Statistical Models Based on Counting Processes.* Springer-Verlag, Berlin Heidelberg New York.

Andersen, L., Andreasen, J. (2000) Volatility skews and extensions of the Libor market model. *Appl. Math. Finance* 7, 1–32.

Anderson, R., Sundaresan, S. (1996) Design and valuation of debt contracts. *Rev. Finan. Stud.* 9, 37–68.

Anderson, R., Sundaresan, S. (2000) A comparative study of structural models of corporate bond yields: An exploratory investigation. *J. Bank. Finance* 24, 255–269.

Anderson, R., Sundaresan, S., Tychon, P. (1996) Strategic analysis of contingent claims. *European Econ. Rev.* 40, 871–881.

Anderson, R., Pan, Y., Sundaresan, S. (2000) Corporate bond yield spreads and the term structure. *Finance* 21(2), 15–37.

Anderson, W.J. (1991) *Continuous-Time Markov Chains. An Applications-Oriented Approach.* Springer-Verlag, Berlin Heidelberg New York.

Artzner, P., Delbaen, F. (1992) Credit risk and prepayment option. *ASTIN Bulletin* 22, 81–96.

Artzner, P., Delbaen, F. (1995) Default risk insurance and incomplete markets. *Math. Finance* 5, 187–195.

Arvanitis, A. (2000) Getting the pricing right. *Risk* 13(9), 115–119.

Arvanitis, A., Gregory, J. (2001) *Credit: The Complete Guide to Pricing, Hedging and Risk Management.* Risk Books, London.

Arvanitis, A., Laurent, J.-P. (1999) On the edge of completeness. *Risk,* October.

Arvanitis, A., Gregory, J., Laurent, J.-P. (1999) Building models for credit spreads. *J. Derivatives* 6(3), 27–43.

Aven, T. (1985) A theorem for determining the compensator of a counting process. *Scand. J. Statist.* 12, 69–72.

Bakshi, G., Madan, D.B., Zhang, F. (2001) Investigating the sources of default risk: Lessons from empirically evaluating credit risk models. Working paper.

Bangia, A., Diebold, F.X., Schuermann, T., Stroughair, J.D. (1999) Modeling liquidity risk, with implications for traditional market risk measurement and management. Working paper, University of Pennsylvania.

Barone-Adesi, G., Colwell, D.B. (1999) Valuing risky debt with constant elasticity of variance effects. Working paper, University of Alberta and UNSW.

Baxter, M., Rennie A. (1996) *Financial Calculus. An Introduction to Derivative Pricing.* Cambridge University Press, Cambridge.

Baz, J., Pascutti, M.J. (1996) Alternative swap contracts: Analysis and pricing. *J. Derivatives* 4(2), 7–21.

Bélanger, A., Shreve, S.E., Wong, D. (2001) A unified model for credit derivatives. Forthcoming in *Math. Finance.*

Bhattacharya, R.N., Waymire, E.C. (1990) *Stochastic Processes with Applications.* J. Wiley, Chichester.

Bicksler, J., Chen, A.H. (1986) An economic analysis of interest rate swaps. *J. Finance* 41, 645–655.

Bielecki, T.R., Rutkowski, M. (1999) Defaultable term structure: Conditionally Markov approach. Forthcoming in *IEEE Trans. Automatic Control.*

Bielecki, T.R., Rutkowski, M. (2000a) HJM with multiples. *Risk* 13(4), 95–97.

Bielecki, T.R., Rutkowski, M. (2000b) Multiple ratings model of defaultable term structure. *Math. Finance* 10, 125–139.

Bielecki, T.R., Rutkowski, M. (2001a) Credit risk modelling: Intensity-based approach. In: *Option Pricing, Interest Rates and Risk Management,* E. Jouini, J. Cvitanić, M. Musiela, eds., Cambridge University Press, Cambridge, pp. 399–457.

Bielecki, T.R., Rutkowski, M. (2001b) Martingale approach to basket credit derivatives. Working paper, Northeastern Illinois University and UNSW.

Bielecki, T.R., Rutkowski, M. (2002) Intensity-based valuation of basket credit derivatives. In: *Mathematical Finance,* J. Yong, ed., World Scientific, Singapore, pp. 12–27.

Bielecki, T.R., Rutkowski, M. (2003) Dependent defaults and credit migrations. *Appl. Math.* 30, 121-145.

Bingham, N.H., Kiesel, R. (1998) *Risk-Neutral Valuation.* Springer-Verlag, Berlin Heidelberg New York.

Björk, T. (1998) *Arbitrage Theory in Continuous Time.* Oxford University Press, Oxford.

Black, F., Cox, J.C. (1976) Valuing corporate securities: Some effects of bond indenture provisions. *J. Finance* 31, 351–367.

Black, F., Scholes M. (1973) The pricing of options and corporate liabilities. *J. Political Econom.* 81, 637–654.

Blanchet-Scalliet, C., Jeanblanc, M. (2001) Hazard rate for credit risk and hedging defaultable contingent claims. Forthcoming in *Finance Stochast.*

Bliss, R. (1997) Testing term structure estimation methods. *Adv. Futures Options Res.* 9, 197–231.

Bollier, T.F., Sorensen, E.H. (1994) Pricing swap default risk. *Finan. Analysts J.* 50(3), 23–33.

Borodin, A., Salminen, P. (1996) *Handbook of Brownian Motion. Facts and Formulae.* Birkhäuser, Basel Boston Berlin.

Boudoukh, J., Whitelaw, R.F. (2001) The benchmark effect in the Japanese government bond market. *J. Fixed Income* 1(3), 52–59.

Bouyé, E., Durrleman, V., Nikeghbali, A., Riboulet, G., Roncalli, T. (2000) Copulas for finance: A reading guide and some applications. Working paper.

Brace, A., Womersley, R.S. (2000) Exact fit to the swaption volatility matrix using semidefinite programming. Working paper, National Australia Bank and UNSW.

Brace, A., Gątarek, D., Musiela, M. (1997) The market model of interest rate dynamics. *Math. Finance* 7, 127–154.

Brace, A., Dun, T., Barton, G. (2001) Towards a central interest rate model. In: *Option Pricing, Interest Rates and Risk Management,* E. Jouini, J. Cvitanić, M. Musiela, eds., Cambridge University Press, Cambridge, pp. 287–313.

Brémaud, P. (1981) *Point Processes and Queues. Martingale Dynamics.* Springer-Verlag, Berlin Heidelberg New York.

Brémaud, P., Yor, M. (1978) Changes of filtrations and of probability measures. *Z. Wahrsch. Verw. Gebiete* 45, 269–295.

Brennan, M.J., Schwartz, E.S. (1977) Convertible bonds: Valuation and optimal strategies for call and conversion. *J. Finance* 32, 1699–1715.

Brennan, M.J., Schwartz, E.S. (1978) Corporate income taxes, valuation and the problem of optimal capital structure. *J. Business* 51, 103–114.

Brennan, M.J., Schwartz, E.S. (1980) Analyzing convertible bonds. *J. Finan. Quant. Anal.* 15, 907–929.

Brigo, D., Mercurio, F. (2001) *Interest Rate Models: Theory and Practice.* Springer-Verlag, Berlin Heidelberg New York.

Briys, E., de Varenne, F. (1997) Valuing risky fixed rate debt: An extension. *J. Finan. Quant. Anal.* 32, 239–248.

Brooks, R., Yan, D.Y. (1999) London Inter-Bank Offer Rate (LIBOR) versus Treasury rate: Evidence from the parsimonious term structure model. *J. Fixed Income* 9(1), 71–83.

Brunel, V. (2001) Pricing credit derivatives with uncertain default probabilities. Working paper, HSBC CCF.

Buffet, E. (2000) Credit risk: The structural approach revisited. *Proceedings of the Third Seminar on Stochastic Analysis, Random Fields and Applications, Ascona 1999,* R. Dalang, M. Dozzi, M. Russo, eds., Birkhäuser, Basel Boston Berlin.

Cao, M., Wei, J. (2001) Vulnerable options, risky corporate bond and credit spread. *J. Futures Markets* 21.

Caouette, J.B., Altman, E.I., Narayanan, P. (1998) *Managing Credit Risk: The Next Great Financial Challenge.* J. Wiley, Chichester.

Carey, M. (2001) Dimensions of credit risk and their relationship to economic capital requirements. In: *Prudential Supervision: Why Is It Important and What Are the Issues,* F.S. Mishkin, ed., University of Chicago Press and NBER.

Carey, M., Hrycay, M. (2001) Parametrizing credit risk models with rating data. *J. Bank. Finance* 25, 197–270.

Carty, L.V. (1997) Moody's ratings migration and credit quality correlations, 1920-1996. Moody's Investors Service.

Carty, L.V., Lieberman, D. (1997) Historical default rates of corporate bond issuers, 1920-1996. Moody's Investors Service.

Cathcart, L., El-Jahel, L. (1998) Valuation of defaultable bonds. *J. Fixed Income* 8(1), 65–78.

Chance, D. (1990) Default risk and the duration of zero-coupon bonds. *J. Finance* 45, 265–274.

Chance, D., Rich, D. (1998) The pricing of equity swaps and swaptions. *J. Derivatives* 6(4), 19–31.

Chou, C.-S., Meyer, P.-A. (1975) Sur la représentation des martingales commes intégrales stochastiques dans les processus ponctuels. In: *Lecture Notes in Math.* 465, Springer-Verlag, Berlin Heidelberg New York, pp. 226–236.

Christiansen, C. (2002) Credit spreads and the term structure of interest rates. *Intern. Rev. Finan. Anal.* 11, 279–295.

Çinlar, E., Jacod, J., Protter, P., Sharpe, M.J. (1980) Semimartingales and Markov processes. *Z. Wahrsch. Verw. Gebiete* 54 (1980), 161–219.

Collin-Dufresne, P., Goldstein, R.S. (2001) Do credit spread reflect stationary leverage ratios? *J. Finance* 56, 1929–1957.

Collin-Dufresne, P., Hugonnier, J.-N. (1999) On the pricing and hedging of contingent claims in the presence of extraneous risks. Working paper, Carnegie Mellon University.

Collin-Dufresne, P., Solnik, B. (2001) On the term structure of default premia in the swap and LIBOR markets. *J. Finance* 56, 1095–1115.

Collin-Dufresne, P., Goldstein, R.S., Martin, J.S. (2001) The determinants of credit spread changes. *J. Finance* 56, 2177-2207.

Cooper, I.A., Martin, M. (1996) Default risk and derivative products. *Appl. Math. Finance* 3, 53–74.

Cooper, I.A., Mello, A.S. (1988) Default spreads in the fixed and in the floating interest rate markets: A contingent claims approach. *Adv. Futures Options Res.* 3, 269–289.

Cooper, I.A., Mello, A.S. (1991) The default risk of swaps. *J. Finance* 46, 597–620.

Cooper, I.A., Mello, A.S. (1992) Pricing and optimal use of forward contracts with default risk. Working paper, London School of Economics.

Cossin, D., Pirotte, H. (1998) Swap credit risk: An empirical investigation on transaction data. *J. Bank. Finance* 21, 1351–1373.

Cossin, D., Pirotte, H. (2000) *Advanced Credit Risk Analysis.* J. Wiley, Chichester.

Cox, D. (1955) Some statistical methods connected with series of events. *J. Roy. Stat. Soc.* B17, 129–164.

Cox, J.C., Ingersoll, J.E., Ross, S.A. (1980) An analysis of variable rate loan contracts. *J. Finance* 35, 389–403.

Cox, J.C., Ingersoll, J.E., Ross, S.A. (1985a) An intertemporal general equilibrium model of asset prices. *Econometrica* 53, 363–384.

Cox, J.C., Ingersoll, J.E., Ross, S.A. (1985b) A theory of the term structure of interest rates. *Econometrica* 53, 385–407.

CreditRisk$^+$ (1997) *CreditRisk$^+$: A Credit Risk Management Framework.* Credit Suisse Financial Products, London [www.csfb.com/creditrisk].

Crouhy, M., Galai, D., Mark, R. (1998) Credit risk revisited. *Risk – Credit Risk Supplement,* March, 40–44.

Crouhy, M., Galai, D., Mark, R. (2000) A comparative analysis of current credit risk models. *J. Bank. Finance* 24, 59–117.

Crouhy, M., Galai, D., Mark, R. (2001) Prototype risk rating system. *J. Bank. Finance* 25, 47–95.

Crosbie, P.J. (1997) Modeling default risk. KMV Corporation, San Francisco [www.kmv.com].

Cvitanić, J., Karatzas, I. (1993) Hedging contingent claims with constrained port-folios. *Ann. Appl. Probab.* 6, 652–681.

Dai, Q., Singleton, K. (2000) Specification analysis of affine term structure models. *J. Finance* 55, 1943–1978.

Das, S. (1998a) Credit derivatives – instruments. In: *Credit Derivatives: Trading and Management of Credit and Default Risk*, S.Das, ed., J. Wiley, Singapore, pp. 7–77.

Das, S. (1998b) Valuation and pricing of credit derivatives. In: *Credit Derivatives: Trading and Management of Credit and Default Risk*, S.Das, ed., J. Wiley, Singapore, pp. 173–231.

Das, S.R. (1995) Credit risk derivatives. *J. Derivatives* 2(3), 7–23.

Das, S.R. (1997) Pricing credit derivatives. Working paper, Harvard Business School and National Bureau of Economic Research.

Das, S.R., Tufano, P. (1996) Pricing credit-sensitive debt when interest rates, credit ratings, and credit spreads are stochastic. *J. Finan. Engrg* 5(2), 161–198.

Das, S.R., Sundaram, R.K. (2000) A discrete-time approach to arbitrage-free pricing of credit derivatives. *Management Science* 46(1), 46–62.

Davis, M.H.A. (1976) The representation of martingales of jump processes. *SIAM J. Control* 14, 623–638.

Davis, M.H.A. (1993) *Markov Models and Optimization.* Chapman & Hall, London.

Davis, M.H.A., Lischka, F. (1999) Convertible bonds with market risk and credit risk. Working paper, Tokyo-Mitsubishi International.

Davis, M.H.A., Lo, V. (1999) Modelling default correlation in bond portfolios. Working paper, Tokyo-Mitsubishi International.

Davis, M.H.A., Lo, V. (2001) Infectious defaults. *Quantitative Finance* 1, 382–386.

Davydov, D., Linetsky, V., Lotz, C. (1999) The hazard-rate approach to pricing risky debt: Two analytically tractable examples. Working paper.

Delianedis, G., Geske, R. (2001) Credit risk and risk neutral default probabilities: Information about rating migrations and defaults. Forthcoming.

Dellacherie, C. (1970) Un exemple de la théorie générale des processus. In: *Lecture Notes in Math.* 124, Springer-Verlag, Berlin Heidelberg New York, pp. 60–70.

Dellacherie, C. (1972) *Capacités et processus stochastiques.* Springer-Verlag, Berlin Heidelberg New York.

Dellacherie, C., Meyer, P.-A. (1978a) *Probabilities and potential.* Hermann, Paris.

Dellacherie, C., Meyer, P.-A. (1978b) A propos du travail de Yor sur les grossisse-ments des tribus. In: *Lecture Notes in Math.* 649, Springer-Verlag, Berlin Heidel-berg New York, pp. 69–78.

Diaz, A., Skinner, F.S. (2000) Term structure misspecification and arbitrage free models of credit risk. Working paper.

Duffee, G. (1996) On measuring credit risks of derivative instruments. *J. Bank. Finance* 20, 805–833.

Duffee, G. (1998) The relation between Treasury yields and corporate bond yield spreads. *J. Finance* 53, 2225–2242.

Duffee, G. (1999) Estimating the price of default. *Rev. Finan. Stud.* 12, 197–226.

Duffee, G., Zhou, C. (1996) Credit derivatives in banking: useful tools for loan risk management? Working paper, Federal Reserve Board.

Duffie, D. (1994) Forward rate curves with default risk. Working paper, Stanford University.

Duffie, D. (1996) *Dynamic Asset Pricing Theory.* 2$^{nd}$ edition. Princeton University Press, Princeton.

Duffie, D. (1998a) First-to-default valuation. Working paper, Stanford University.

Duffie, D. (1998b) Defaultable term structure models with fractional recovery of par. Working paper, Stanford University.

Duffie, D. (1999) Credit swap valuation. *Finan. Analysts J.* 55(1), 73–87.

Duffie, D., Epstein, L. (1992) Stochastic differential utility. *Econometrica* 60, 353–394.

Duffie, D., Gârleanu, N. (2001) Risk and the valuation of collateralized debt obligations. *Finan. Analysts J.* 57(1), 41–59.

Duffie, D., Huang, M. (1996) Swap rates and credit quality. *J. Finance* 51, 921–949.

Duffie, D., Kan, R. (1996) A yield-factor model of interest rates. *Math. Finance* 6, 379–406.

Duffie, D., Lando, D. (2001) The term structure of credit spreads with incomplete accounting information. *Econometrica* 69, 633–664.

Duffie, D., Liu, J. (2001) Floating-fixed credit spreads. *Finan. Analysts J.* 57(3), 76–87.

Duffie, D., Pan, J. (2001) Analytical value-at-risk with jumps and credit risk. *Finance Stochast.* 5, 155–180.

Duffie, D., Singleton, K.J. (1994) Econometric modeling of term structures of defaultable bonds. Working paper, Stanford University.

Duffie, D., Singleton, K.J. (1997) An econometric model of the term structure of interest-rate swap yields. *J. Finance* 52, 1287–1321.

Duffie, D., Singleton, K.J. (1998a) Ratings-based term structures of credit spreads. Working paper, Stanford University.

Duffie, D., Singleton, K.J. (1998b) Simulating correlated defaults. Working paper, Stanford University.

Duffie, D., Singleton, K.J. (1999) Modeling term structures of defaultable bonds. *Rev. Finan. Stud.* 12, 687–720.

Duffie, D., Singleton, K.J. (2003) *Credit Risk. Pricing, Measurement and Management.* Princeton University Press, Princeton.

Duffie, D., Stanton, R. (1992) Pricing continuously resettled contingent claims. *J. Econom. Dynamics Control* 16, 561–573.

Duffie, D., Schroder, M., Skiadas, C. (1996) Recursive valuation of defaultable securities and the timing of resolution of uncertainty. *Ann. Appl. Probab.* 6, 1075–1090.

Duffie, D., Schroder, M., Skiadas, C. (1997) A term structure model with preferences for the timing of resolution of uncertainty. *Economic Theory* 9, 3–22.

Durbin, J. (1992) The first passage density of the Brownian motion process to a curved boundary. *J. Appl. Probab.* 29, 291–304.

Eberhart, A.C., Moore, W.T., Roenfeldt, R.L. (1990) Security pricing and deviations from the absolute priority rule in bankruptcy proceedings. *J. Finance* 45, 1457–1469.

Eberlein, E. (2001) Application of generalized hyperbolic Lévy motions to finance. Working paper.

Eberlein, E., Özkan, F. (2001) The defaultable Lévy term structure: Ratings and restructuring. Working paper.

Eberlein, E., Raible, S. (1999) Term structure models driven by general Lévy processes. *Math. Finance* 9, 31–53.

El Karoui, N., Quenez, M.C. (1997a) Nonlinear pricing theory and backward stochastic differential equations. In: *Financial Mathematics, Bressanone, 1996*, W.Runggaldier, ed., Springer-Verlag, Berlin Heidelberg New York, pp. 191–246.

El Karoui, N., Quenez, M.C. (1997b) Imperfect markets and backward stochastic differential equations. In: *Numerical Methods in Finance*, L.C.G. Rogers, D. Talay, eds., Cambridge University Press, Cambridge, pp. 181–214.

El Karoui, N., Peng, S., Quenez, M.-C. (1997) Backward stochastic differential equations in finance. *Math. Finance* 7, 1–71.

El Karoui, N. (1999) *Modélisation de l'information.* Lecture notes, CEA-EDF-INRIA.

Elliott, R.J. (1977) Innovation projections of a jump process and local martingale. *Math. Proc. Cambridge Phil. Soc.* 81, 77–90.

Elliott, R.J. (1982) *Stochastic Calculus and Applications.* Springer-Verlag, Berlin Heidelberg New York.

Elliott, R.J., Kopp, P.E. (1999) *Mathematics of Financial Markets.* Springer-Verlag, Berlin Heidelberg New York.

Elliott, R.J., Jeanblanc, M., Yor, M. (2000) On models of default risk. *Math. Finance* 10, 179–195.

Embrechts, P., McNeal, A.J., Straumann, D. (2002) Correlation and dependence in risk management: Properties and pitfalls. In: *Risk Management: Value at Risk and Beyond,* M.A.H. Dempster, ed., Cambridge University Press, Cambridge, pp. 129–144.

Embrechts, P., Lindskog, F., McNeal, A.J. (2003) Modelling dependence with copulas and applications to risk management. In: *Handbook of Heavy Tailed Distributions in Finance,* S. Rachev, ed., Elsevier, pp. 329–384

Eom, Y.H., Helwege, J., Huang, J.-Z. (2003) Structural models of corporate bond pricing: An empirical analysis. Forthcoming in *Rev. Finan. Studies.*

Ericsson, J. (2000) Asset substitution, debt pricing, optimal leverage and maturity. *Finance* 21(2), 39–70.

Ericsson, J., Renault, O. (2000) Liquidity and credit risk. Working paper, McGill University and Université Catholique de Louvain.

Ericsson, J., Reneby, J. (1998) A framework for valuing corporate securities. *Appl. Math. Finance* 5, 143–163.

Ericsson, J., Reneby, J. (1999) A note on contingent claims pricing with non-traded assets. Working paper, Université Catholique de Louvain and Stockholm School of Economics.

Erlenmaier, U., Gersbach, H. (2000) Default probabilities and default correlations. Working paper, University of Heidelberg.

Fabozzi, F.J. (2000) *The Handbook of Fixed Income Securities.* 6th edition. McGraw-Hill, New York.

Fan, H., Sundaresan, S. (1997) Debt valuation, strategic debt service and optimal dividend policy. Working paper, Columbia University.

Finger, C.C. (2000) A comparison of stochastic default rate models. Working paper, RiskMetrics Group.

Finnerty, J.D. (1999) Adjusting the binomial model for default risk. *J. Portfolio Management* 25(2), 93–103.

Fisher, L. (1959) Determinants of risk premium on corporate bonds. *J. Political Econom.* 67, 217–237.

Fons, J.S. (1987) The default premium and corporate bond experience. *J. Finance* 42, 81–97.

Fons, J.S. (1994) Using default rates to model the term structure of credit risk. *Finan. Analysts J.* 50(5), 25–32.

Föllmer, H., Leukert, P. (1999) Quantile hedging. *Finance Stochast.* 3, 251–273.

Föllmer, H., Leukert, P. (2000) Efficient hedging: Cost versus shortfall risk. *Finance Stochast.* 4, 117–146.

Fooladi, I.J., Roberts, G.S., Skinner, F.S. (1997) Duration for bonds with default risk. *J. Bank. Finance* 21, 1–16.

Foss, G.W. (1995) Quantifying risk in the corporate bond markets. *Finan. Analysts J.* 51(2), 29–34.

Francis, J.C., Frost, J.A., Whittaker, J.G. (1999) *Handbook of Credit Derivatives.* Irwin/McGraw-Hill, New York.

François, P. (1996) Bond evaluation with default risk: A review of continuous time approach. Working paper, ESSEC.

Franks, J., Torous, W. (1989) An empirical investigation of U.S. firms in reorganization. *J. Finance* 44, 747–769.

Franks, J., Torous, W. (1994) A comparison of financial recontracting in distressed exchanges and Chapter 11 reorganizations. *J. Finan. Econom.* 35, 349–370.

Frey, R., McNeil, A.J. (2000) Modelling dependent defaults. Working paper, University of Zurich and ETHZ.

Fridson, M.S., Jónsson, J.G. (1995) Spread versus Treasuries and the riskiness of high-yield bonds. *J. Fixed Income* 5(3), 79–88.

Frydman, H., Kallberg, J.G., Kao, D.L. (1985) Testing the adequacy of Markov chains and mover-stayer models as representations of credit behavior. *Operations Res.* 33, 1203–1214.

Galai, D., Masulis, R.W. (1976) The option pricing model and the risk factor of stock. *J. Finan. Econom.* 3, 53–81.

Galai D., Schneller, M.I. (1978) Pricing of warrants and the value of the firm. *J. Finance* 33, 1333-1342.

Gauthier, C., de la Noue, P., Rouzeau, E. (1998) Analyzing corporate credit risk: A quantitative approach. *Quants No.29*, Recherche and Innovation, Crédit Commercial de France.

Geman, H., El Karoui, N., Rochet, J.C. (1995) Changes of numeraire, changes of probability measures and pricing of options. *J. Appl. Probab.* 32, 443–458.

Gersbach, H., Lipponer, A. (1997a) The correlation effect. Working paper, University of Heidelberg.

Gersbach, H., Lipponer, A. (1997b) Default correlations, macroeconomic risk and credit portfolio management. Working paper, University of Heidelberg.

Geske, R. (1977) The valuation of corporate liabilities as compound options. *J. Finan. Quant. Anal.* 12, 541–552.

Geske, R. (1979) The valuation of compound options. *J. Finan. Econom.* 7, 63–81.

Geske, R., Johnson, H.E. (1984) The valuation of corporate liabilities as compound options: A correction. *J. Finan. Quant. Anal.* 19, 231–232.

Gill, R.D. (1999) Applications of product-integration in survival analysis. In: *Proc. Workshop on Product Integrals and Pathwise Integration,* University of Aarhus.

Glasserman, P., Zhao, X. (1999) Fast greeks by simulation in forward LIBOR models. *J. Comput. Finance* 3(1), 5–39.

Glasserman, P., Zhao, X. (2000) Arbitrage-free discretization of lognormal forward Libor and swap rate model. *Finance Stochast.* 4, 35–68.

Gordy, M.B. (2000) A comparative anatomy of credit risk models. *J. Bank. Finance* 24, 119–149.

Greenfield, Y. (2000) *Hedging of Credit Risk Embedded in Derivative Transactions.* Ph.D. dissertation, Carnegie Mellon University.

Grinblatt, M. (2001) An analytical solution for interest rate swap spreads. Forthcoming in *Rev. Intern. Finance.*

Grundke, P. (2001) Pricing defaultable securities in firm value and intensity models: A comparison. Working paper, University of Cologne.

Gupton, G.M., Finger, C.C., Bhatia, M. (1997) *CreditMetrics: Technical Document.* J.P. Morgan & Incorporated, New York [www.riskmetrics.com/research].

Harrison, J.M., Pliska, S.R. (1981) Martingales and stochastic integrals in the theory of continuous trading. *Stochastic Process. Appl.* 11, 215–260.

He, H. (1999) Modeling term structures of swap spreads. Working paper, Yale School of Management.

He, H., Keirstead, W.P., Rebholz, J. (1998) Double lookbacks. *Math. Finance* 8, 201–228.

He, S.W, Wang, J.G, Yan, J.A. (1992) *Semimartingale Theory and Stochastic Calculus.* Science Press and CRC Press Inc.

Heath, D.C., Jarrow, R.A., Morton, A. (1990) Bond pricing and the term structure of interest rates: A discrete time approximation. *J. Finan. Quant. Anal.* 25, 419–440.

Heath, D.C., Jarrow, R.A., Morton, A. (1992) Bond pricing and the term structure of interest rates: A new methodology for contingent claim valuation. *Econometrica* 60, 77–105.

Helwege, J., Turner, C. (1999) The slope of the credit yield curve for speculative-grade issuers. *J. Finance* 54, 1869–1884.

Ho, T.S., Singer, R.F. (1982) Bond indenture provisions and the risk of corporate debt. *J. Finan. Econom.* 10, 375–406.

Ho, T.S., Singer, R.F. (1984) The value of corporate debt with a sinking-fund provision. *J. Business* 57, 315–336.

Hodges, S., Webber, N., Wong, M.C.W. (2000) Pricing of corporate bonds with call and default features. Working paper, Warwick School of Business.

Houweling, P., Hoek, J., Kleibergen, F. (2001) The joint estimation of term structures and credit spreads. *J. Empirical Finance* 8, 297–323.

Huang, J.-Z. (1997) The option to default and optimal debt service. Working paper, Stern School of Business, New York University.

Hübner, G. (2001) The analytic pricing of asymmetric defaultable swaps. *J. Bank. Finance* 25, 295–316.

Huge, B., Lando, D. (1999) Swap pricing with two-sided default risk in a rating-based model. *European Finance Review* 3, 239–268.

Hughston, L.P. (1996) Pricing of credit derivatives. *Financial Derivatives and Risk Management* 5, 11–16.

Hughston, L.P. (1997) *Pricing Models for Credit Derivatives.* Lecture notes, Merrill Lynch, London.

Hughston, L.P., Turnbull, S. (2000) Credit derivatives made simple. *Risk* 13(10), 36–43.

Hull, J.C. (1997) *Options, Futures, and Other Derivatives.* 3$^{rd}$ edition. Prentice-Hall, Englewood Cliffs (New Jersey).

Hull, J.C., White, A. (1995) The impact of default risk on the prices of options and other derivative securities. *J. Bank. Finance* 19, 299–322.

Hull, J.C., White, A. (2000) Valuing credit default swaps I: No counterparty credit risk. *J. Derivatives* 8(1), 29–40.

Hull, J.C., White, A. (2001) Valuing credit default swaps II: Modeling default correlations. *J. Derivatives* 8(3), 12–22.

Hunt, P.J., Kennedy, J.E. (2000) *Financial Derivatives in Theory and Practice.* J. Wiley, Chichester.

Hunt, P.J., Kennedy, J.E., Pelsser, A. (2000) Markov-functional interest rate models. *Finance Stochast.* 4, 391–408.

Ikeda, N., Watanabe, S. (1989) *Stochastic Differential Equations and Diffusion Processes.* 2$^{nd}$ edition. North-Holland, Amsterdam.

Israel, R.B., Rosenthal, J.S., Wei, J.Z. (2001) Finding generators for Markov chains via empirical transition matrices, with applications to credit ratings. *Math. Finance* 11, 245–265.

Itô, K., McKean, H.P. (1965) *Diffusion Processes and Their Sample Paths.* Springer-Verlag, Berlin Heidelberg New York.

Iyengar, S. (1985) Hitting lines with two-dimensional Brownian motion. *SIAM J. Appl. Math.* 45, 983–989.

Jacod, J. (1979) *Calcul stochastique et problèmes de martingales. Lecture Notes in Math. 714.* Springer-Verlag, Berlin Heidelberg New York.

Jacod, J., Shiryaev, A.N. (1987) *Limit Theorems for Stochastic Processes.* Springer-Verlag, Berlin Heidelberg New York.

James, J., Webber, N. (2000) *Interest Rate Modelling.* J. Wiley, Chichester.

Jamshidian, F. (1989) An exact bond option pricing formula. *J. Finance* 44, 205–209.

Jamshidian, F. (1997) LIBOR and swap market models and measures. *Finance Stochast.* 1, 293–330.

Jarrow, R.A., Madan, D.B. (1995) Option pricing using the term structure of interest rates to hedge systematic discontinuities in asset returns. *Math. Finance* 5, 311–336.

Jarrow, R.A., Turnbull, S.M. (1995) Pricing derivatives on financial securities subject to credit risk. *J. Finance* 50, 53–85.

Jarrow, R.A., Turnbull, S.M. (2000a) *Derivative Securities.* 2$^{nd}$ edition. Southwestern Publishers, Cincinnati (Ohio).

Jarrow, R.A., Turnbull, S.M. (2000b) The intersection of market and credit risk. *J. Bank. Finance* 24, 271–299.

Jarrow, R.A., van Deventer, D.R. (1998) Integrating interest rate risk and credit risk in asset and liability management. Working paper, Kamakura Corporation.

Jarrow, R.A., Yu, F. (2001) Counterparty risk and the pricing of defaultable securities. *J. Finance* 56, 1765-1799.

Jarrow, R.A., Lando, D., Turnbull, S.M. (1997) A Markov model for the term structure of credit risk spreads. *Rev. Finan. Stud.* 10, 481–523.

Jarrow, R.A., Lando, D., Yu, F. (2000) Default risk and diversification: Theory and applications. Working paper.

Jeanblanc, M., Rutkowski, M. (2000a) Modelling of default risk: An overview. In: *Mathematical Finance: Theory and Practice,* Higher Education Press, Beijing, pp. 171–269.

Jeanblanc, M., Rutkowski, M. (2000b) Modelling of default risk: Mathematical tools. Working paper, Université d'Évry and Warsaw University of Technology.

Jeanblanc, M., Rutkowski, M. (2002) Default risk and hazard process. In: *Mathematical Finance – Bachelier Congress 2000,* H. Geman, D. Madan, S.R. Pliska, T. Vorst, eds., Springer-Verlag, Berlin Heidelberg New York, pp. 281–312.

Jennen, C., Lerche, H.R. (1981) First exit densities of Brownian motion through one-sided moving boundaries. *Z. Wahrsch. verw. Gebiete* 55, 133–148.

Jeulin, T. (1980) *Semi-martingales et grossissement de filtration. Lecture Notes in Math.* 833, Springer-Verlag, Berlin Heidelberg New York.

Jeulin, T., Yor, M. (1978) Grossissement d'une filtration et semi-martingales: formules explicites. In: *Lecture Notes in Math.* 649, Springer-Verlag, Berlin Heidelberg New York, pp. 78–97.

Johnson, H., Stulz, R. (1987) The pricing of options with default risk. *J. Finance* 42, 267–280.

Johnson, R.E. (1967) Term structure of corporate bond yields as a function of risk of default. *J. Finance* 22, 313–345.

Jonkhart, M.J.L. (1979) On the term structure of interest rates and the risk of default: An analytical approach. *J. Bank. Finance* 3, 253–262.

Jones, E., Mason, S., Rosenfeld, E. (1984) Contingent claim analysis of corporate capital structures: An empirical investigation. *J. Finance* 39, 611–625.

Kao, D.L. (2000) Estimating and pricing credit risk: An overview. *Finan. Analysts J.* 56(4), 50–66.

Karatzas, I., Shreve, S. (1991) *Brownian Motion and Stochastic Calculus.* 2$^{nd}$ edition. Springer-Verlag, Berlin Heidelberg New York.

Karatzas, I., Shreve, S. (1998) *Methods of Mathematical Finance.* Springer-Verlag, Berlin Heidelberg New York.

Kavvathas, D. (2000) Estimating credit rating transition probabilities for corporate bonds. Working paper, University of Chicago.

Kealhofer, S., Kwok, S., Weng, W. (1998) Uses and abuses of bond default rates. Working paper, KMV Corporation.

Kiefer, N.M. (1988) Economic duration data and hazard functions. *J. Econ. Literature* 26, 646–679.

Kiesel, R., Perraudin, W., Taylor, A. (1999) The structure of credit risk. Working paper, Birbeck College.

Kiesel, R., Perraudin, W., Taylor, A. (2002) Credit and interest rate risk. In: *Risk Management: Value at Risk and Beyond*, M.A.H. Dempster, ed., Cambridge University Press, Cambridge, pp. 129–144.

Kijima, M. (1998) Monotonicity in a Markov chain model for valuing coupon bond subject to credit risk. *Math. Finance* 8, 229–247.

Kijima, M. (1999) A Gaussian term structure model of credit risk spreads and valuation of yield-spread options. Working paper, Tokyo Metropolitan University.

Kijima, M. (2000) Valuation of a credit swap of the basket type. *Rev. Derivatives Res.* 4, 81–97.

Kijima, M., Komoribayashi, K. (1998) A Markov chain model for valuing credit risk derivatives. *J. Derivatives* 6, Fall, 97–108.

Kijima, M., Muromachi, Y. (2000) Credit events and the valuation credit derivatives of basket type. *Rev. Derivatives Res.* 4, 55–79.

Kim, I.J., Ramaswamy, K., Sundaresan, S. (1993a) The valuation of corporate fixed income securities. Working paper, Wharton School, University of Pennsylvania.

Kim, I.J., Ramaswamy, K., Sundaresan, S. (1993b) Does default risk in coupons affect the valuation of corporate bonds? *Finan. Management* 22, 117–131.

Klein, P. (1996) Pricing Black-Scholes options with correlated credit risk. *J. Bank. Finance* 20, 1211–1129.

Klein, P., Inglis, M. (2001) Pricing vulnerable European options when the option's payoff can increase the risk of financial distress. *J. Bank. Finance* 25, 993–1012.

Krahnen, J.P., Weber, M. (2001) Generally accepted rating principles: A primer. *J. Bank. Finance* 25, 3–23.

Krylov, N.V. (1995) *Introduction to the Theory of Diffusion Processes.* American Mathematical Society, Providence.

Küchler, U., Sørensen, M. (1997) *Exponential Families of Stochastic Processes.* Springe Verlag, New York.

Kusuoka, S. (1999) A remark on default risk models. *Adv. Math. Econom.* 1, 69–82.

Lamberton, D., Lapeyre, B. (1996) *Introduction to Stochastic Calculus Applied to Finance.* Chapman and Hall, London.

Lando, D. (1993) A continuous-time Markov model of the term structure of credit risk spread. Working paper, Cornell University.

Lando, D. (1994) *Three Essays on Contingent Claim Pricing.* Ph.D. dissertation, Cornell University.

Lando, D. (1997) Modelling bonds and derivatives with credit risk. In: *Mathematics of Derivative Securities,* M. Dempster, S. Pliska, eds., Cambridge University Press, Cambridge, pp. 369–393.

Lando, D. (1998) On Cox processes and credit-risky securities. *Rev. Derivatives Res.* 2, 99–120.

Lando, D. (2000a) Some elements of rating-based credit risk modeling. In: *Advanced Fixed-Income Valuation Tools,* N. Jegadeesh, B. Tuckman, eds., J. Wiley, Chichester, pp. 193–215.

Lando, D. (2000b) On correlated defaults in a rating-based model – common state variables versus simultaneous defaults. Working paper.

Lando, D., Skødeberg, T. (2002) Analyzing rating transitions and rating drift with continuous observations. *J. Bank. Finance* 26, 423–444.

Last, G., Brandt, A. (1995) *Marked Point Processes on the Real Line. The Dynamic Approach.* Springer-Verlag, Berlin Heidelberg New York.

Laurent, J.-P. (2000) Default swap and credit spread options. Working paper, BNP Paribas, London.

Lehrbass, F. (1997) Defaulters get intense. *Risk – Credit Risk Supplement*, July, 56–59.

Leland, H.E. (1994) Corporate debt value, bond covenants, and optimal capital structure. *J. Finance* 49, 1213–1252.

Leland, H.E., Toft, K. (1996) Optimal capital structure, endogenous bankruptcy, and the term structure of credit spreads. *J. Finance* 51, 987–1019.

Leland, H.E. (1998) Agency costs, risk management, and capital structure. *J. Finance* 53, 1213–1244.

Li, H. (1998) Pricing swaps with default risk. *Rev. Derivatives Res.* 2, 231–250.

Li, D.X. (1998) Constructing a credit curve. *Credit Risk: Risk Special Report*, November, 40–44.

Li, D.X. (1999a) The valuation of basket credit derivatives. *CreditMetrics Monitor*, April, 34–50.

Li, D.X. (1999b) The valuation of the *i*th-to-default basket credit derivatives. Working paper, RiskMetrics Group.

Li, D.X. (2000) On default correlation: A copula function approach. *J. Fixed Income* 9(4), 43–54.

Lindskog, F. (2000) *Modelling Dependence with Copulas and Applications to Risk Management.* Master thesis, Swiss Federal Institute of Technology.

Linetsky, V. (1997) Modeling defaultable securities by diffusions with killing. Working paper.

Linetsky, V. (1999) Step options. *Math. Finance* 9, 55–65.

Liptser, R.S., Shiryaev, A.N. (1978) *Statistics of Random Processes, Volume II Applications.* Springer-Verlag, Berlin Heidelberg New York.

Litterman, R., Iben, T. (1991) Corporate bond valuation and the term structure of credit spreads. *J. Portfolio Management* 17(3), 52–64.

Litzenberger, R. (1992) Swaps: Plain and fanciful. *J. Finance* 47, 831–850.

Liu, J., Longstaff, F.A., Mandell, R.E. (2000) The market price of credit risk: An empirical analysis of interest rate swap spreads. Working paper, UCLA.

Longstaff, F.A. (1995) How much can marketability affect security values. *J. Finance* 50, 1767–1774.

Longstaff, F.A. (2001) Optimal portfolio choice and the valuation of illiquid securities. *Rev. Finan. Stud.* 14, 407–431.

Longstaff, F.A., Schwartz, E.S. (1995) A simple approach to valuing risky fixed and floating rate debt. *J. Finance* 50, 789–819.

Lotz, C. (1998) Locally minimizing the credit risk. Working paper, University of Bonn.

Lotz, C. (1999) Optimal shortfall hedging of credit risk. Working paper, University of Bonn.

Lotz, C., Schlögl, L. (2000) Default risk in a market model. *J. Bank. Finance* 24, 301–327.

Lucas, D.J. (1995) Default correlations and credit analysis. *J. Fixed Income* 4(4), 76–87.

Ma, J., Yong, J. (1999) *Forward-Backward Stochastic Differential Equations and Their Applications.* Springer-Verlag, Berlin Heidelberg New York.

Madan, D.B. (2000) Pricing the risks of default: A survey. Working paper, University of Maryland.

Madan, D.B., Unal, H. (1998) Pricing the risk of default. *Rev. Derivatives Res.* 2, 121–160.

Madan, D.B., Unal, H. (2000) A two-factor hazard-rate model for pricing risky debt and the term structure of credit spreads. *J. Finan. Quant. Anal.* 35, 43–65.

Maksymiuk, R., Gątarek, D. (1999) Applying HJM to credit risk. *Risk* 12(5), 67–68.

Martin, M. (1997) *Credit Risk in Derivative Products.* Ph.D. dissertation, London Business School, University of London.

Martellini, L., Priaulet, P. (2001) *Fixed-Income Securities. Dynamic Methods for Interest Rate Risk Pricing and Hedging.* J. Wiley, Chichester.

Mason, S.P., Bhattacharya, S. (1981) Risky debt, jump processes, and safety covenants. *J. Finan. Econom.* 9, 281–307.

Mazziotto, G., Szpirglas, J. (1979) Modèle général de filtrage non linéaire et équations différentielles stochastiques associées. *Ann. Inst. H. Poincaré* 15, 147–173.

McNulty, C., Levin, R. (2000) Modeling credit migration. Working paper, J.P. Morgan Securities, Inc., New York.

Miltersen, K., Sandmann, K., Sondermann, D. (1997) Closed form solutions for term structure derivatives with log-normal interest rates. *J. Finance* 52, 409–430.

Mella-Barral, P. (1999) The dynamics of default and debt reorganization. *Rev. Finan. Stud.* 12, 535–578.

Mella-Barral, P., Perraudin, W. (1997) Strategic debt service. *J. Finance* 52, 531–556.

Mella-Barral, P., Tychon, P. (1999) Default risk in asset pricing. *Finance* 20(1).

Mel'nikov, A.V. (1999) *Financial Markets. Stochastic Analysis and the Pricing of Derivative Securities.* American Mathematical Society, Providence.

Merton, R.C. (1973) Theory of rational option pricing. *Bell J. Econom. Manag. Sci.* 4, 141–183.

Merton, R.C. (1974) On the pricing of corporate debt: The risk structure of interest rates. *J. Finance* 29, 449–470.

Merton, R.C. (1976) Option pricing when underlying stock returns are discontinuous. *J. Finan. Econom.* 3, 125–144.

Merton, R.C. (1990) *Continuous-Time Finance.* Basil Blackwell, Oxford.

Miltersen, K., Sandmann, K., Sondermann, D. (1997) Closed form solutions for term structure derivatives with log-normal interest rates. *J. Finance* 52, 409–430.

Monkkonen, H. (1997) *Modeling Default Risk: Theory and Empirical Evidence.* Ph.D. dissertation, Queen's University.

Monkkonen, H. (2000) Margining the spread. *Risk* 13(10), 109–112.

Moody's Investment Services (1997) *The Binomial Expansion Technique. Technical Document.*

Musiela, M., Rutkowski, M. (1997a) *Martingale Methods in Financial Modelling.* Springer-Verlag, Berlin Heidelberg New York.

Musiela, M., Rutkowski, M. (1997b) Continuous-time term structure models: forward measure approach. *Finance Stochast.* 1, 261–291.

Musiela, M., Sondermann, D. (1993) Different dynamical specifications of the term structure of interest rates and their implications. Working paper, University of Bonn.

Nakazato, D. (1997) Gaussian term structure model with credit rating classes. Working paper, Industrial Bank of Japan.

Neftci, S.N. (1996) *An Introduction to the Mathematics of Financial Derivatives.* Academic Press, New York.

Nelken, I. (1999) *Implementing Credit Derivatives. Strategies and Techniques for Using Credit Derivatives in Risk Management.* Irwin/McGraw-Hill, New York.

Nelsen, R. (1999) *An Introduction to Copulas.* Springer-Verlag, Berlin Heidelberg New York.

Nickel, P., Perraudin, W., Varotto, S. (2000) Stability of rating transitions. *J. Bank. Finance* 24, 203–227

Nielsen, S.S., Ronn, E.I. (1997) The valuation of default risk in corporate bonds and interest rate swaps. *Adv. Futures Options Res.* 9, 175–196.

Nielsen, T.N., Saá-Requejo, J., Santa-Clara, P. (1993) Default risk and interest rate risk: The term structure of default spreads. Working paper, INSEAD.

Nyfeler, M.A. (2000) Modeling dependencies in credit risk management. Diploma thesis, ETHZ.

Øksendal, B. (1998) *Stochastic Differential Equations.* 5$^{th}$ edition. Springer-Verlag, Berlin Heidelberg New York.

Pardoux, E., Peng, S. (1990) Adapted solutions of a backward stochastic differential equations. *Systems Control Lett.* 14, 55–61.

Patel, N. (2001) Credit derivatives: vanilla volumes challenged. *Risk* 14(2), 32–35.

Pelsser, A. (2000) *Efficient Methods for Valuing Interest Rate Derivatives.* Springer-Verlag, Berlin Heidelberg New York.

Peng, S. (1993) Backward stochastic differential equation and its application in optimal control. *Appl. Math. Optim.* 27, 125–144.

Pierides, Y.A. (1997) The pricing of credit risk derivatives. *J. Econom. Dynamics Control* 21, 1579–1611.

Pitts, C., Selby, M. (1983) The pricing of corporate debt: A further note. *J. Finance* 38, 1311–1313.

Pliska, S.R. (1997) *Introduction to Mathematical Finance: Discrete Time Models.* Blackwell Publishers, Oxford.

Protter, P. (1990) *Stochastic Integration and Differential Equations. A New Approach.* Springer-Verlag, Berlin Heidelberg New York.

Pugachevsky, D. (1999) Generalizing with HJM. *Risk* 12(8), 103–105.

Pye, G. (1974) Gauging the default premium. *Finan. Analysts J.* 30(1), 49–52.

Rachev, S., Schwartz, E., Khindanova, I. (2000) Stable modeling of credit risk. Working paper, UCLA.

Ramaswamy, K., Sundaresan, S.M. (1986) The valuation of floating-rate instruments, theory and evidence. *J. Finan. Econom.* 17, 251–272.

Rebholz, J.A. (1994) *Planar Diffusions with Applications to Mathematical Finance.* Ph.D. dissertation, University of California, Berkeley.

Rebonato, R. (1999) *Volatility and Correlation in the Pricing of Equity, FX and Interest-Rate Options.* J. Wiley, Chichester.

Rebonato, R. (2000) On the simultaneous calibration of multifactor lognormal interest rate models to Black volatilities and to the correlation matrix. *J. Comput. Finance* 2(4), 5–27.

Rendleman, R.J. (1992) How risks are shared in interest rate swaps. *J. Finan. Services Res.* 5–34.

Revuz, D., Yor, M. (1999) *Continuous Martingales and Brownian Motion.* 3$^{rd}$ edition. Springer-Verlag, Berlin Heidelberg New York.

Rich, D. (1994) A note on the valuation and hedging of equity swaps. *J. Finan. Engineering* 5, 323–334.

Rich, D., Leipus, R. (1997) An option-based approach to analyzing financial contracts with multiple indenture provisions. *Adv. Futures Options Res.* 9, 1–36.

Rogers, L.C.G. (1999) Modelling credit risk. Working paper, University of Bath.

Rogers, L.C.G., Williams, D. (2000) *Diffusions, Markov Processes and Martingales.* 2$^{nd}$ edition. Cambridge University Press.

Rolski, T., Schmidli, H., Schmidt, V., Teuggels, J. (1998) *Stochastic Processes for Insurance and Finance.* J. Wiley, Chichester.

Rutkowski, M. (1999) Models of forward Libor and swap rates. *Appl. Math. Finance* 6, 29–60.

Rutkowski, M. (2001) Modelling of forward Libor and swap rates. In: *Option Pricing, Interest Rates and Risk Management,* E. Jouini, J. Cvitanić, M. Musiela, eds., Cambridge University Press, Cambridge, pp. 336–395.

Saá-Requejo, J., Santa-Clara, P. (1999) Bond pricing with default risk. Working paper, UCLA.

Sandmann, K., Sondermann, D. (1994) On the stability of lognormal interest rate models and the pricing of Eurodollar futures. Working paper, University of Bonn.

Sarig, O., Warga, A. (1989) Some empirical estimates of the risk structure of interest rates. *J. Finance* 46, 1351–1360.

Sarkar, S. (2001) Probability of call and likelihood of the call feature in a corporate bond. *J. Bank. Finance* 25, 505–533.

Saunders, A. (1999) *Credit Risk Measurements: New Approaches to Value at Risk and Other Paradigms.* J. Wiley, Chichester.

Schlögl, E. (1999) A multicurrency extension of the lognormal interest rate market model. Working paper, University of Technology, Sydney.

Schlögl, L. (1998) An exposition of intensity-based models of securities and derivatives with default risk. Working paper, University of Bonn.

Schmid, B., Zagst, R. (2000) A three-factor defaultable term structure model. *J. Fixed Income* 10(2), 63–79.

Schmidt, W.M. (1998) Modelling correlated defaults. Working paper.

Schönbucher, P.J. (1996) The term structure of defaultable bond prices. Working paper, University of Bonn.

Schönbucher, P.J. (1998a) Term structure modelling of defaultable bonds. *Rev. Derivatives Res.* 2, 161–192.

Schönbucher, P.J. (1998b) Pricing credit risk derivatives. Working paper, University of Bonn.

Schönbucher, P.J. (1999) A tree implementation of a credit spread model for credit derivatives. Working paper, University of Bonn.

Schönbucher, P.J. (2000a) *Credit Risk Modelling and Credit Derivatives.* Ph.D. dissertation, University of Bonn.

Schönbucher, P.J. (2000b) A Libor market model with default risk. Working paper, University of Bonn.

Schoenmakers, J. and Coffey, B. (1999) Libor rates models, related derivatives and model calibration. Working paper, WIAS, Berlin.

Schwartz, T. (1998) Estimating the term structures of corporate debt. *Rev. Derivatives Res.* 2, 193–230.

Schweizer, M. (2001) A guided tour through quadratic hedging approaches. In: *Option Pricing, Interest Rates and Risk Management,* E. Jouini, J. Cvitanić, M. Musiela, eds., Cambridge University Press, Cambridge, pp. 538–574.

Scott, J. (1981) The probability of bankruptcy: A comparison of early predictions and theoretical models. *J. Bank. Finance* 5, 317–344.

Shaked, M., Shanthikumar, J.G. (1987) The multivariate hazard construction. *Stochastic Process. Appl.* 24, 241–258.

Shimko, D., ed. (1999) *Credit Risk: Models and Management.* Risk Books, London.

Shimko, D., Tejima, N., van Deventer, D.R. (1993) The pricing of risky debt when interest rates are stochastic. *J. Fixed Income* 3(2), 58–65.

Shirakawa, H. (1999) Evaluation of yield spread for credit risk. *Adv. Math. Econom.* 1, 83–97.

Shiryaev, A.N. (1999) *Essentials of Stochastic Finance: Facts, Models, Theory.* World Scientific Publ.

Shumway, T. (2001) Forecasting bankruptcy more accurately: A simple hazard model. *J. Business* 74, 101–124.

Sidenius, J. (2000) LIBOR market models in practice. *J. Comput. Finance* 3(3), 5–26.

Skinner, F.S. (1998) Hedging bonds subject to credit risk. *J. Bank. Finance* 22, 321–345.

Steele, M. (2000) *Stochastic Calculus and Financial Applications.* Springer-Verlag, Berlin Heidelberg New York.

Sun, T., Sundaresan. S., Wang, C. (1993) Interest rate swaps: An empirical investigation. *J. Finan. Econom.* 34, 77–99.

Syski, R. (1992) *Passage Times for Markov Chains.* IOS Press, Amsterdam.

Szatzschneider, W. (2000) CIR model in financial markets. Working paper, Anahuac University.

Taurén, M. (1999) A comparison of bond pricing models in the pricing of credit risk. Working paper, Indiana University.

Tavakoli, J.M. (1998) *Credit Derivatives: A Guide to Instruments and Applications.* J. Wiley, Chichester.

Thomas, L.C., Allen, D.E., Morkel-Kingsbury, N. (1998) A hidden Markov chain model for the term structure of bond credit risk spreads. Working paper.

Uhrig, M. (1996) An empirical examination of the Longstaff-Schwartz bond option valuation model. *J. Derivatives* 4, 41–54.

Vasicek, O. (1977) An equilibrium characterisation of the term structure. *J. Finan. Econom.* 5, 177–188.

Vasicek, O. (1984) Credit valuation. Working paper, KMV Corporation.

Wang, D.F. (1999a) Pricing defaultable debt: Some exact results. *Internat. J. Theor. Appl. Finance* 2, 95–99.

Wang, S.S. (1999b) Aggregation of correlated risk portfolios: Models and algorithms. Working paper.

Wei, J.Z. (2000) A multi-factor, Markov chain model for credit migrations and credit spreads. Working paper, University of Toronto.

Wei, D.G., Guo, D. (1997) Pricing risky debt: An empirical comparison of the Longstaff and Schwartz, and Merton models. *J. Fixed Income* 7(2), 8–28.

Weiss, L.A. (1990) Bankruptcy resolution: Direct costs and violation of priority of claims. *J. Finan. Econom.* 27, 285–314.

Williams, D. (1991) *Probability with Martingales.* Cambridge University Press, Cambridge.

Wilson, T. (1997a) Portfolio credit risk. I and II *Risk* 10(9,10), 111–117, 56–61.

Wong, D. (1998) A unifying credit model. Working paper, Research Advisory Services, Capital Markes Group, Scotia Capital Markets.

Yawitz, J. (1977) An analytical model of interest rate differentials and different default recoveries. *J. Finan. Quant. Anal.* 12, 481–490.

Yin, G.G., Zhang, Q. (1997) *Continuous-Time Markov Chains and Applications. A Singular Perturbation Approach.* Springer-Verlag, Berlin Heidelberg New York.

Yor, M. (1997) *Some Aspects of Brownian Motion. Part II: Some Recent Martingale Problems,* Lectures in Mathematics. ETH Zürich. Birkhäuser, Basel.

Yor, M. (2001) *Exponential Functionals of Brownian Motion and Related Processes.* Springer-Verlag, Berlin Heidelberg New York.

Yu, H., Kwok, Y.K. (2002) Contingent claim approach for analyzing the credit risk of defaultable currency swaps. *AMS/IP Stud. in Adv. Math.* 26, pp. 79–92.

Zheng, C.K. (2000) Understanding the default-implied volatility for credit spreads. *J. Derivatives* 7(4), 67–76.

Zhou, C. (1996) A jump-diffusion approach to modeling credit risk and valuing defaultable securities. Working paper, Federal Reserve Board, Washington.

Zhou, C. (2001) An analysis of default correlations and multiple defaults. *Rev. Finan. Stud.* 14, 555–576.

# Basic Notation

# Subject Index

Lightning Source UK Ltd.
Milton Keynes UK
UKHW02f1202160418
321121UK00003B/67/P